Mechanical Engineering Series

Frederic F. Ling
Series Editor

Springer
New York
Berlin
Heidelberg
Hong Kong
London
Milan
Paris
Tokyo

Mechanical Engineering Series

J. Angeles, **Fundamentals of Robotic Mechanical Systems: Theory, Methods, and Algorithms, 2nd ed.**

P. Basu, C. Kefa, and L. Jestin, **Boilers and Burners: Design and Theory**

J.M. Berthelot, **Composite Materials: Mechanical Behavior and Structural Analysis**

I.J. Busch-Vishniac, **Electromechanical Sensors and Actuators**

J. Chakrabarty, **Applied Plasticity**

K.K. Choi and N.H. Kim, **Structural Sensitivity Analysis and Optimization 1: Linear Systems**

K.K. Choi and N.H. Kim, **Structural Sensitivity Analysis and Optimization 2: Nonlinear Systems and Applications**

G. Chryssolouris, **Laser Machining: Theory and Practice**

V.N. Constantinescu, **Laminar Viscous Flow**

G.A. Costello, **Theory of Wire Rope, 2nd ed.**

K. Czolczynski, **Rotordynamics of Gas-Lubricated Journal Bearing Systems**

M.S. Darlow, **Balancing of High-Speed Machinery**

J.F. Doyle, **Nonlinear Analysis of Thin-Walled Structures: Statics, Dynamics, and Stability**

J.F. Doyle, **Wave Propagation in Structures: Spectral Analysis Using Fast Discrete Fourier Transforms, 2nd ed.**

P.A. Engel, **Structural Analysis of Printed Circuit Board Systems**

A.C. Fischer-Cripps, **Introduction to Contact Mechanics**

A.C. Fischer-Cripps, **Nanoindentation, 2nd ed.**

J. García de Jalón and E. Bayo, **Kinematic and Dynamic Simulation of Multibody Systems: The Real-Time Challenge**

W.K. Gawronski, **Advanced Structural Dynamics and Active Control of Structures**

W.K. Gawronski, **Dynamics and Control of Structures: A Modal Approach**

G. Genta, **Dynamics of Rotating Systems**

K.C. Gupta, **Mechanics and Control of Robots**

J. Ida and J.P.A. Bastos, **Electromagnetics and Calculations of Fields**

(continued after index)

Kyung K. Choi
Nam H. Kim

Structural Sensitivity Analysis and Optimization 1

Linear Systems

With 134 Figures

Springer

Kyung K. Choi
Department of Mechanical and
Industrial Engineering
The University of Iowa
Iowa City, IA 5224
USA
kkchoi@ccad.uiowa.edu

Nam Ho Kim
Department of Mechanical and
Aerospace Engineering
The University of Florida
POB 116250
Gainsville, FL 32611-6250
USA
nkim@ufl.edu

Series Editor
Frederick F. Ling
Ernest F. Gloyna Regents Chair in Engineering, Emeritus
Department of Mechanical Engineering
The University of Texas at Austin
Austin, TX 78712-1063, USA
and
William Howard Hart Professor Emeritus
Department of Mechanical Engineering,
Aeronautical Engineering and Mechanics
Renssalaer Polytechnic Institute
Troy, NY 12180-3590, USA

Library of Congress Cataloging-in-Publication Data
Choi, Kyung K.
Structural sensitivity analysis and optimization / Kyung K. Choi, Nam H. Kim.—1st ed.
p. cm. — (Mechanical engineering series)
Includes bibliographical references and index.
ISBN 0-387-23232-X (alk. paper) — ISBN 0-387-23336-9 (alk. paper)
1. Structural analysis (Engineering) I. Kim, Nam H. II. Title. III. Mechanical engineering series (Berlin, Germany)
TA645.C48 2005
624.1′71—dc22 2004062574

ISBN 0-387-23232-X Printed on acid-free paper.

Printed in the United States of America. (EB)

9 8 7 6 5 4 3 2 1 SPIN 10949825

springeronline.com

Dedicated to our wives
Ho-Youn
Jee-Hyun

Preface

Structural design sensitivity analysis concerns the relationship between design variables available to the design engineer and structural responses determined by the laws of mechanics. The dependence of response measures such as displacement, stress, strain, natural frequency, buckling load, acoustic response, frequency response, noise-vibration-harshness (NVH), thermoelastic response, and fatigue life on the material property, sizing, component shape, and configuration design variables is implicitly defined through the governing equations of structural mechanics. In this text, first- and second-order design sensitivity analyses are presented for static and dynamics responses of both linear and nonlinear structural systems, including elastoplastic and frictional contact problems.

Prospective readers or users of the text are seniors and graduate students in mechanical, civil, biomedical, industrial, and engineering mechanics, aerospace, and mechatronics; graduate students in mathematics; researchers in these same fields; and structural design engineers in industry.

A substantial literature exists on the technical aspects of structural design sensitivity analysis. While some studies directly address the topic of design sensitivity, the vast majority of research is imbedded within texts and papers devoted to structural optimization. The premise of this text is that a comprehensive theory of structural design sensitivity analysis for linear and nonlinear structures can be treated in a unified way. The objective is therefore to provide a complete treatment of the theory and practical numerical methods of structural design sensitivity analysis. Design sensitivity supports optimality criteria methods of structural optimization and serves as the foundation for iterative methods of structural optimization. One of the most common structural design methods involves decisions made by the design engineer based on experience and intuition. This conventional mode of structural design can be substantially enhanced if the design engineer is provided with design sensitivity information that explains design change influences, without requiring a trial and error process.

Such advanced, state-of-the-art analysis methods as finite element analysis, boundary element analysis, and meshfree analysis provide reliable tools for the evaluation of the structural design. However, they give the design engineer little help in identifying ways to modify the design to either avoid problems or improve desired qualities. Using design sensitivity information generated by methods that exploit finite element, boundary element, or meshfree formulations, the design engineer can carry out systematic trade-off analysis and improve the design. This text presents design sensitivity analysis (DSA) theory and numerical implementation to create advanced design methodologies for mechanical systems and structural components, which will permit economical designs that are strong, stable, reliable, and have long service life. The design methodologies can be used by design engineers in the academia, industry, and government to obtain optimal structural designs for ground vehicles, aircraft, space systems, ships, heavy equipment, machinery, biomedical devices, etc. Extensive numerical methods for computing design sensitivity are included in this text for practical application and software development. More importantly, the numerical method allows seamless integration of CAD-FEA-DSA software tools, so that design optimization can be carried out using CAD geometric models instead of FEA models. This capability allows integration of CAD-CAE-CAM so that optimized designs can be manufactured effectively.

The book is organized into two volumes, four parts, and fourteen chapters. Parts I and II are in Volume 1: Linear Systems, and Parts III and IV are in Volume 2: Nonlinear Systems and Applications.

Part I introduces structural design concepts that include the CAD-based design model, design parameterization, performance measures, costs, and constraints. Based on the design model, an analysis model is introduced using finite element analysis. A broad overview of design sensitivity analysis methods is provided. By relying on energy principles to develop design sensitivity analysis theory, the design sensitivity method is developed without requiring highly sophisticated mathematics. The energy method is introduced in order to develop the variational equation and its relationship to the finite element method. Chapters 2 and 3 are essentially a review for students who have already learned energy methods in structural mechanics. The finite element method is explained as a technique based on a piecewise polynomial approximation of the displacement field and as an application of the variational method for approximating a solution to the governing boundary-value problem. In Part II, this relationship is successfully used in the development of discrete and continuum design sensitivity analysis methods and their relationships.

Part II treats design sensitivity analysis of linear structural systems. Both discrete and continuum design sensitivity analysis methods are explained. Chapter 4 describes finite-dimensional problems in which the structural response is a finite-dimensional vector of structural displacements, and the design variable is a finite-dimensional vector of design parameters. Governing structural equations are matrix equations. Direct design differentiation and adjoint variable methods of design sensitivity analysis are presented, along with the design derivatives of eigenvalues and eigenvectors. The computational aspects of implementing these methods are treated in some detail in conjunction with finite-element analysis codes. Chapters 5 through 7 treat continuum problems in which response and design variables are functions (displacement field and material distribution) and governing structural equations are the variational equations introduced in Chapters 2 and 3. Sizing, shape, and configuration design variables are treated separately in Chapters 5 through 7, respectively. Both the direct differentiation and adjoint variable method of design sensitivity analysis are developed, and design derivatives of eigenvalues are derived. Analytical solutions to simple examples and numerical solutions to more complex examples are presented. For both shape and configuration design variables, the material derivative concept is taken from continuum mechanics to predict the effect of design changes on the structural response. For a structural component with curvature, a more general configuration design theory is presented in Section 7.5 of Chapter 7. For shape design sensitivity, the adjoint variable method is used to derive expressions for differentials of the structural response, either as boundary integrals (the boundary method) or domain integrals (the domain method). A similar method is used for the shape design sensitivity of eigenvalues.

Part III treats design sensitivity analysis of nonlinear structural systems using continuum design sensitivity analysis methods. As with Chapters 2 and 3, the equilibrium equations for nonlinear structural systems are derived using the principles of virtual work from Chapter 8. Both geometric and material nonlinearities are treated. Nonlinear elasticity, buckling, hyperelasticity, elastoplasticity, nonlinear transient dynamics, and frictional contact problems are included. In nonlinear structural analysis, total and updated Lagrangian approaches have been introduced. The equilibrium equations are then linearized at the previously known configuration to yield incremental formulations for nonlinear analysis. The linearized equilibrium equation plays a key role in design sensitivity analysis in subsequent chapters, since the first-order variation with respect to the design parameter includes linearization of the energy form. The linearized form that appears during design sensitivity analysis is the same as the linearized form for nonlinear analysis. Sizing, shape, and configuration design variables are treated separately in Chapters 9 through 11, respectively. Both adjoint variable and direct differentiation methods are given for the nonlinear elastic problem. However, for nonlinear elastoplastic problems, only the direct differentiation method is used, since the design sensitivity is path-dependent. Analytical derivations of design sensitivity expressions for structural components are presented, along with numerical examples of sensitivity computations.

Part IV is devoted to practical design tools and applications: sizing and shape design parameterization, design velocity field computation, numerical implementation of the sensitivity for general-purpose code development, and various other practical design applications. In Chapter 12, sizing design parameterization for line and surface design components is introduced. For shape design parameterization, a three-step process is developed. One important aspect of shape design parameterization is the connection between the design parameterization and the computation of the design velocity field, as explained in Chapter 13. In Chapter 13, the computational aspects of design sensitivity analysis are considered, using the finite element method to solve the original governing and adjoint equations. The numerical method allows seamless integration of CAD-FEA-DSA, so that design optimization can be carried out using CAD models instead of FEA ones. Chapter 14 includes a number of practical design applications of linear and nonlinear structural systems with additional applications in thermoelastic analysis and fatigue design optimization to aid application-oriented readers.

A final comment on the notation used in this text. The structural design engineer may find that the notation conventionally used in structural mechanics has not always been adhered to. The field of design sensitivity analysis presents a dilemma regarding notation since it draws from fields as diverse as structural mechanics, differential calculus, calculus of variations, control theory, differential operator theory, and functional analysis. Unfortunately, the literature in each of these fields assigns a different meaning to the same symbol. Consequently, it is at times necessary to use symbols that look identical in an equation, but that come from very different notational systems. As a result, some notational compromise is required. The authors have adhered to standard notation except where ambiguity would arise, in which case the notation being used is indicated.

This book has been made possible due to contributions from the authors' former students, namely, Drs. R.-J. Yang, H.G. Lee, H.G. Seong, B. Dopker, T.-M. Yao, J.L.T. Santos, J.-S. Park, J. Lee, S.-L. Twu, K.-H. Chang, M. Godse, S.M. Wang, I. Shim, C.-J. Chen, Y.-H. Park, H.-Y. Hwang, X. Yu, W. Duan, S.-H. Cho, B.S. Choi, I. Grindeanu, J. Tu, B.-D. Youn, and Y. Yuan. Special appreciation is given to Professor K.-H. Chang at University of Oklahoma for his contributions to numerical methods and his examples in shape design sensitivity analysis and optimization. In addition, the authors value the contributions of colleagues Drs. J. Cea, B. Roussellet, J.P. Zolesio, R. Haftka, B.M. Kwak, G.W. Hou, H.L. Lam, and Y.M. Yoo. Finally, special thanks to Mr. R. Watkins for his outstanding work editing the manuscript.

Kyung K. Choi
Nam H. Kim
December 2004

Contents
1: Linear Systems

Contents
2: Nonlinear Systems and Applications

PART 1
Structural Design and Analysis

1
Introduction to Structural Design

1.1 Elements of Structural Design

The design of a structural system has two categories: designing a new structure and improving the existing structure to perform better. The design engineer's experience and creative ideas are required in the development of a new structure, since it is very difficult to quantify a new design using mathematical measures. Recently, limited inroads have been made in the creative work of the structural design using mathematical tools [1]. However, the latter evolutionary process is encountered much more frequently in engineering designs. For example, how many times does an automotive company design a new car using a completely different concept? The majority of a design engineer's work concentrates on improving the existing vehicle so that the new car can be more comfortable, more durable, and safer. In this text, we will focus on a design's evolutionary process by using mathematical models and computational tools.

Structural design is a procedure to improve or enhance the performance of a structure by changing its parameters. A *performance measure* can be quite general in engineering fields, and can include: the weight, stiffness, and compliance of a structure; the fatigue life of a mechanical component; the noise in the passenger compartment; the vibration level; the safety of a vehicle in a crash, etc. However, this text does not address such aesthetic measures as whether a car or a structural design is attractive to customers. All performance measures are presumed to be measurable quantities. System parameters are variables that a design engineer can change during the design process. For example, the thickness of a vehicle body panel can be changed to improve vehicle performance. The cross section of a beam can be changed in designing bridge structures. System parameters that can be changed during the design process are called *design variables*, even including the geometry of the structure.

Recently, the simulation-based design process has emerged as the future tool of the product development and manufacturing process, since it allows one to achieve a higher quality product, through a reduction in development time in introducing new products to the market, a reduction in testing cycles, and a reduction in total development costs. As noted in the scholarly treatment of product performance by Clark and Fujimoto [2], essentially all development activities prior to the operation of a vehicle are simulations. In this sense, simulation can involve either mathematical models, or physical experiments that are created to emulate environments and conditions experienced by the product in its actual use.

Great strides have been made during the past decade in computer-aided design (CAD) and computer-aided engineering (CAE) tools for mechanical system development. Discipline-oriented simulation capabilities in structures, mechanical system dynamics, aerodynamics, control systems, and numerous related fields are being used to support a broad range of mechanical system design applications. Integration of these tools to create a robust simulation-based design capability, however, remains a challenge. Based on their extensive survey of the automotive industry in the mid-1980s, Clark and Fujimoto [2] concluded that simulation tools in support of vehicle development were on the horizon, but not yet ready for pervasive application. The explosion in computer, software, and modeling and simulation technology that has occurred since the mid-1980s suggests that high-fidelity tools for simulation-based design are now at hand. Properly integrated, they can resolve uncertainties and significantly impact mechanical system design.

An example of an integrated concurrent engineering environment for development of large-scale wheeled and tracked vehicle systems is illustrated in Fig. 1.1 [3]. It comprises simulation and modeling tools and an integration infrastructure to: (1) support design analysis, supportability analysis, operation analysis, and development process control; (2) establish connectivity between all application tools, with tool interactions transparent to the user; (3) refine product requirements; and (4) conduct trade-off analyses and make informed decisions to yield a robust optimal design. The integrated test bed shown supports concurrent design, operator-in-the-loop driving simulation, dynamic performance analysis, durability prediction, structural design sensitivity analysis and optimization, maintainability analysis, and design process management. The test bed permits all members of the development team to simulate the performance and effectiveness of product and process designs, at a level of fidelity comparable to that which would be achieved in physical prototyping.

Using the test bed, an integrated simulation-based design process for the fatigue life of vehicle components can be developed [3], as shown in Fig. 1.2. The process includes CAD modeling, dynamic analysis, fatigue analysis, design sensitivity analysis, and design optimization. The CAD-based design model is critically important for multidisciplinary analysis and design optimization. The process allows engineers to create a CAD model of a vehicle system and automatically translate the CAD model into a dynamics model. Dynamic simulations of the vehicle model are then carried out over typical road profiles to obtain load histories at selected components. In the meantime, CAD models of the selected vehicle components are created for design parameterization and translated into finite element analysis (FEA) models. The computation of the fatigue life of a component consists of two parts: dynamic stress computation and fatigue life prediction. The dynamic stress can be obtained either from experiments (mounting sensors or transducers on a physical component) or from simulation. Fatigue analysis is performed using the low cycle fatigue approach. Once the fatigue life of the vehicle components is obtained, design sensitivity analysis with respect to shape design variables defined in the CAD model is performed. With the design sensitivity information, design optimization can be carried out to obtain an optimum design.

As shown in Figs. 1.1 and 1.2, modern developments of structural design are closely related to concurrent engineering environments by which multidisciplinary simulation, design, and manufacturing are possible. Even though concurrent engineering is not the focus of this text, we want to emphasize structural design as a component of concurrent engineering. Figure 1.1 shows an example of concurrent engineering environments used in structural design. An important feature of Fig. 1.1 is database management using the CAD tool. Structural modeling and most interfaces are achieved using the CAD tool. Thus, design parameterization and structural model updates have to be carried out in the CAD model. Through the design parameterization, CAD, CAE, and CAM procedures are interrelated to form a concurrent engineering environment.

The engineering design of the structural system in the simulation-based design process consists of structural modeling, design parameterization, structural analysis, design problem definition, design sensitivity analysis, and design optimization. Figure 1.3 is a flow chart of the structural design process in which computational analysis and mathematical programming play essential roles in the design. The success of the system-level, simulation-based design process shown in Fig. 1.2 strongly depends on a consistent design parameterization, an accurate structural and design sensitivity analysis, and an efficient mathematical programming algorithm.

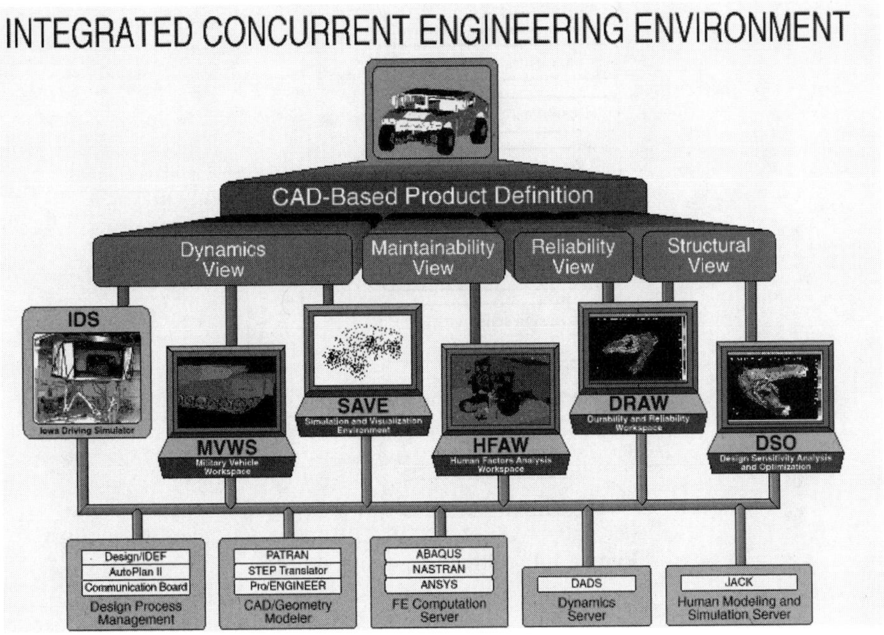

Figure 1.1. Integrated concurrent engineering environment.

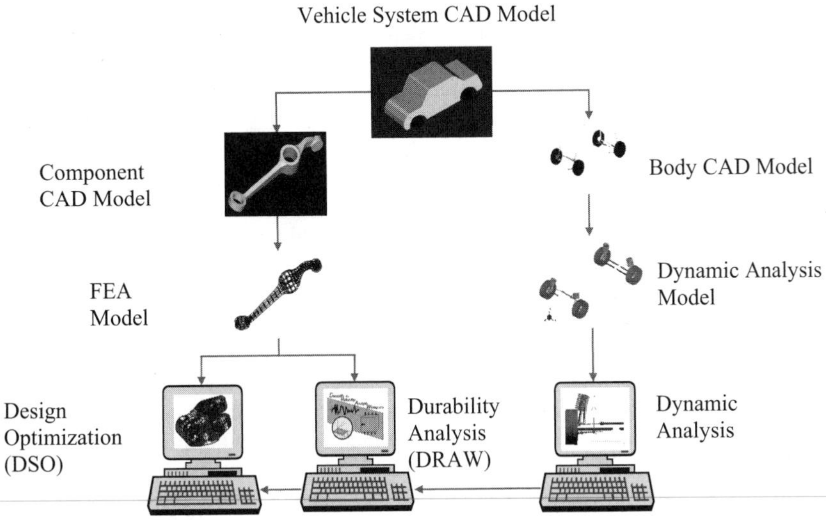

Figure 1.2. Simulation-based design optimization process for fatigue life.

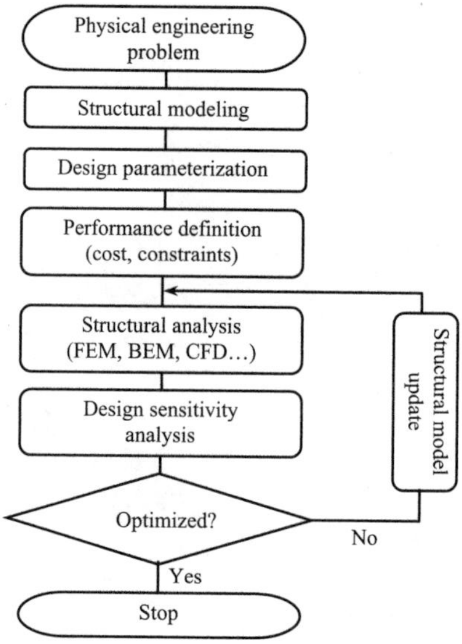

Figure 1.3. Structural design process.

A design engineer simplifies the physical engineering problem into a mathematical model that can represent the physical problem up to the desired level of accuracy. A mathematical model has parameters that are related to the system parameters of the physical problem. A design engineer identifies those design variables to be used during the design process. *Design parameterization*, which allows the design engineer to define the geometric properties for each design component of the structural system being designed, is one of the most important steps in the structural design process. The principal role of design parameterization is to define the geometric parameters that characterize the structural model and to collect a subset of the geometric parameters as design variables. Design parameterization forces engineering teams in design, analysis, and manufacturing to interact at an early design stage, and supports a unified design variable set to be used as the common ground to carry out all analysis, design, and manufacturing processes. Only proper design parameterization will yield a good optimum design, since the optimization algorithm will search within a design space that is defined for the design problem. The design space is defined by the type, number, and range of design variables. Depending on whether it is a concept or detailed design, selected design variables could be non-CAD based parameters. An example of such a design variable is a tire stiffness characteristic in vehicle dynamics during an early vehicle design stage.

Structural analysis can be carried out using experiments in actual or reduced scale, which is a straightforward and still prevalent method for industrial applications. However, the expense and the inefficiency involved in fabricating prototypes make this approach difficult to apply. The analytical method may resolve these difficulties, since it approximates the structural problem as a mathematical model and solves it in a simplified form. In this text, a mathematical model is used to evaluate the performance measures of a structural problem. However, the analytical method has limitations even for very simple structural problems.

With the emergence of various computational capabilities, most analytical approaches to mathematical problems have been converted to numerical approaches, which are able to solve very complicated, real engineering applications. Finite element analysis, boundary element analysis (BEA), and meshfree analysis are a short list of mathematical tools used in structural analysis. The development of FEA is one of the most remarkable successes in structural analysis. The governing differential equation of the structural problem is converted to its integral form and then solved using FEA. Vast amounts of literature are published regarding FEA; for example, refer to [4] and references therein. The complex structural domain is discretized by a set of non-overlapping, simple-shaped finite elements, and an equilibrium condition is imposed on each element. By solving a linear system of matrix equations, the performance measures of a structure are computed in the approximated domain. The accuracy of the approximated solution can be improved by reducing the size of finite elements and/or increasing the order of approximation within an element.

Selection of a design space and an analysis method must be carefully determined since the analysis, both in terms of accuracy and efficiency, must be able to handle all possible designs in the chosen design space. That is, the larger the design space, the more sophisticated the analysis capability must be. For example, if larger shape design changes are expected during design optimization, mesh distortion in FEA could be a serious problem and a finite element model that can handle large shape design changes must be used.

A performance measure in a simulation-based design is the result of structural analysis. Based on the evaluation of analysis results, such engineering concerns as high stress, clearance, natural frequency, or mass can be identified as performance measures for design improvement. Typical examples of performance measures are mass, volume, displacement, stress, compliance, buckling, natural frequency, noise, fatigue life, and crashworthiness. A definition of performance measures permits the design engineer to specify the structural performance from which the sensitivity information can be computed.

Cost and constraints can be defined by combining certain performance measures with appropriate constraint bounds for interactive design optimization. *Cost function*, sometimes called the *objective function*, is minimized (or maximized) during optimization. Selection of a proper cost function is an important decision in the design process. A valid cost function has to be influenced by the design variables of the problem; otherwise, it is not possible to reduce the cost by changing the design. In many situations, an obvious cost function can be identified. In other situations, the cost function is a combination of different structural performance measures. This is called a multiobjective cost function.

Constraint functions are the criteria that the system has to satisfy for each feasible design. Among all design ranges, those that satisfy the constraint functions are candidates for the optimum design. For example, a design engineer may want to design a bridge whose weight is minimized and whose maximum stress is less than the yield stress. In this case, the cost function, or weight, is the most important criterion to be minimized. However, as long as stress, or constraint, is less than the yield stress, the stress level is not important.

Design sensitivity analysis, which is a main focus of this text, is used to compute the sensitivity of performance measures with respect to design variables. This is one of the most expensive and complicated procedures in the structural optimization process. Structural design sensitivity analysis is concerned with the relationship between design variables available to the engineer and the structural response determined by the laws of mechanics. Design sensitivity information provides a quantitative estimate of desirable design change, even if a systematic design optimization method is not used. Based on the design sensitivity results, an engineer can decide on the direction and amount of design change needed to improve the performance measures. In addition, design sensitivity information can provide answers to "what if" questions by predicting performance measure perturbations when the perturbations of design variables are provided.

Substantial literature has emerged in the field of structural design sensitivity analysis [5]. Design sensitivity analysis of structural systems and machine components has emerged as a much needed design tool, not only because of its role in optimization algorithms, but also because design sensitivity information can be used in a CAE environment for early product trade-off in a concurrent design process.

Recently, the advent of powerful graphics-based engineering workstations with increasing computational power has created an ideal environment for making interactive design optimization a viable alternative to more monolithic batch-based design optimization. This environment integrates design processes by letting the design engineer create a geometric model, build a finite element model, parameterize the geometric model, perform FEA, visualize FEA results, characterize performance measures, and carry out design sensitivity analysis and optimization.

Design sensitivity information can be used during a postprocessing of the interactive design process. The principal objective of the postprocessing design stage is to utilize the design sensitivity information to improve the design. Figure 1.4 shows the four-step interactive design process: (1) to visually display design sensitivity information, (2) to carry out what-if studies, (3) to make trade-off determinations, and (4) to execute interactive design optimization. The first three design steps, which are interactive modes of the design process, help the design engineer improve the design by providing structural behavior information at the current design stage. The last design step, which could be either interactive or a batch mode of the design process, launches a mathematical programming algorithm to perform design optimization. Depending on the design problem, the design engineer could use some or all of the four design steps to improve the design at each iterative step. As a result, new designs could be obtained from what-if, trade-off, or interactive optimization design steps.

For the purposes of design optimization, a mathematical programming technique is often used to find an optimum design that can best improve the cost function within a feasible region. Mathematical programming generates a set of design variables that

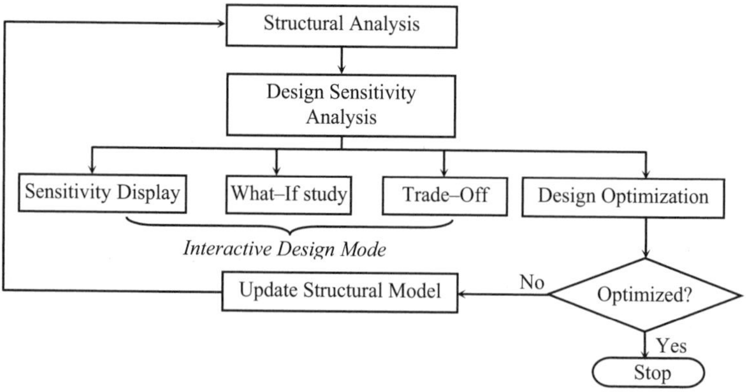

Figure 1.4. Postprocessing design stage.

require performance values from structural analysis and sensitivity information from design sensitivity analysis to find an optimum design. Thus, the structural model has to be updated for a different set of design variables supplied by mathematical programming. If the cost function reaches a minimum with all constraint requirements satisfied, then an optimum design is obtained.

In the following sections, each element of the design process is discussed in detail.

1.2 Structural Modeling and Design Parameterization

The first step in the design process is structural modeling and design parameterization. The physical engineering problem is converted to a mathematical model and the parameters that define the mathematical model have to be identified. Then, the goal of the design process is to find the proper set of design variables to produce the desired performance.

1.2.1 Structural Modeling

When engineers analyze a structural problem, they need to convert the physical problem into a mathematical representation. Many analysis tools can be used to solve this ideal mathematical problem. After arriving at a solution of the mathematical problem, the meaning of the solution has to be correctly interpreted in its physical sense. Thus, if there is an error in the mathematical representation of the physical problem, then it is impossible to properly analyze the physical problem no matter what analysis tools are used. This mathematical representation of the structural problem is called *structural modeling.*

The reliability of the analysis results strongly depends on the assumptions and idealization used in structural modeling. However, a too-complex representation of the physical problem may make it difficult to solve the mathematical problem. For example, when an engineer wants to determine the height and width of a bridge, it may not be important to model every bolt, because the desired results will consist of global flexibility and the bridge's maximum degree of deflection. If each bolt is modeled for the entire bridge structure, then the analysis cost dramatically increases. It may be presumed that sections are constructed continuously without any breaks. However, after determining the size of the bridge, the engineer may want to design the number and size of each section of the bridge. The maximum magnitude of load carried by each bolt would then be of major concern. In this case, the size and distance between bolts would be important and structural modeling would need to include bolt strength. Consequently, different concerns require different structural models. It is the engineer's responsibility to find an appropriate trade-off between accuracy and computational costs of analysis.

1.2.2 Design Parameterization

In structural modeling, the physical problem is represented by mathematical expressions, which contain parameters for defining that problem. For example, the cantilever beam in Fig. 1.5 has parameters including the length, l; the radius of cross section, r; and Young's modulus, E. These parameters define the system and are called *design variables*. If design variables are determined, then the structural problem can be analyzed. Obviously, different design variable values usually yield different analysis results. The aim of the structural design process is to find the values of design variables that satisfy all requirements.

All design variables must satisfy the physical requirements of the problem. For example, length l of the cantilever beam in Fig. 1.5 cannot have a negative value. Physical requirements define the design variable bounds. Valid design variables may have to take into account various manufacturing requirements. For example, the radius of a cantilever beam satisfies its physical requirement if r is a positive number. However, in real applications, the circular cross-sectional beam may not be manufactured if its radius is bigger than r^0. Thus, the range of feasible design can be stated as $0 < r \le r^0$. (Note: In general, the bounds of design variables are denoted as $r^L \le r \le r^U$, where r^L is called the lower bound and r^U is called the upper bound, respectively.) In addition, the design engineer may want to impose certain design constraints on the problem. For example, the maximum stress of the beam may not exceed σ^0 and the maximum tip displacement of the beam must not be greater than z^0. A set of design variables that satisfies the constraints is called a *feasible design*, while a set that does not satisfy constraints is called an *infeasible design*. It is difficult to determine whether a current design is feasible, unless the structural problem is analyzed. For complicated structural problems, it may not be simple to choose the appropriate design constraints so that the feasible region is not empty.

There are two types of design variables: continuous and discrete. Many design optimization algorithms consider design variables to be continuous. In this text, we presume that all design variables are continuous within their lower and upper bound limits. However, discrete design problems often appear in real engineering problems. For example, due to manufacturing limitations, the structural components of many engineering systems are only available in fixed shapes and sizes. Discrete design variables can be thought of as continuous design variables with constraints. As a result, it is more expensive to obtain an optimum design for a problem with discrete design variables. It is possible, however, to solve the problem assuming continuous design variables. After obtaining an optimum solution for the design problem, the nearest discrete values of the optimum design variables can be tested for feasibility. If the nearest discrete design variables are not feasible, then several iterations can be carried out to find the nearest feasible design.

It is convenient to classify design variables according to their characteristics. In the design of structural systems made of truss, beam, membrane, shell, and elastic solid members, there are five kinds of design variables: material property design variables, such as Young's modulus; sizing design variables, such as thickness and cross-sectional area; shape design variables, such as length and geometric shape; configuration design variables, such as orientation and location of structural components; and topological design variables.

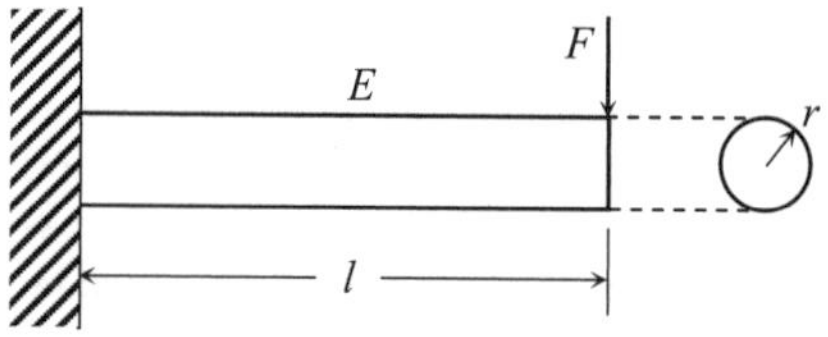

Figure 1.5. Parameters defining circular cross-sectional cantilever beam.

Material Property Design Variable

In structural modeling, the material property is used as a parameter of the structural problem. Young's modulus and Poisson's ratio, for example, are required in the linear elastic problem. If these material properties are subject to change, then they are called *material property design variables*. These kinds of design variables do not appear in regular design problems, since in most cases material properties are presumed to be constant. Analysis using such a constant material property is called the deterministic approach. Another approach uses probability and assumes that material properties are not constant but distributed within certain ranges. This is called the probabilistic approach and is more practical, since a number of experiments will usually yield a number of different test results. In this case, material properties are no longer considered to be constant and can therefore be used as design variables.

Sizing Design Variable

Sizing design variables are related to the geometric parameter of the structure. For example, most automotive and airplane parts are made from plate/shell components. It is natural that a design engineer wants to change the thickness (or gauge) of the plate/shell structure in order to reduce the weight of the vehicle. For a structural model, plate thickness is considered a parameter. However, the global geometry of the structure does not change. Plate thickness can be considered a *sizing design variable*. The sizing design variable is similar to the material property design variable in the sense that both variables change the parameters of the structural problem.

Another important type of sizing design variable is the cross-sectional geometry of the beam and truss. Figure 1.6 provides some examples of the shapes and parameters that define these cross sections. In the structural analysis of truss, for example, the cross-sectional area is required as a parameter of the problem. If a rectangular cross section is used, then the area would be defined as $A = b \times h$. Thus, without any loss of generality, b and h can be considered design variables of the design problem. Detailed discussions of sizing design problems are discussed in Chapter 5 using the distributed parameter approach.

Shape Design Variable

While material property and the sizing design variables are related to the parameters of the structural problem, the shape design variable is related to the structure's geometry. The shape of the structure does not explicitly appear as a parameter in the structural formulation. Although the design variables in Fig. 1.6 determine the cross-sectional shape, they are not shape design variables, since these cross-sectional shapes are considered parameters in the structural problem. However, the length of the truss or beam

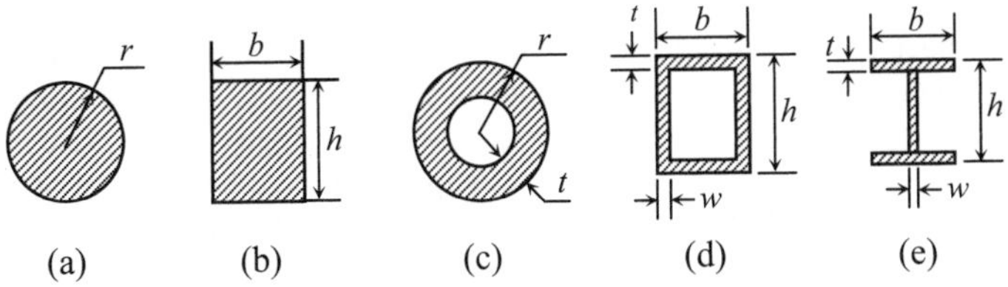

Figure 1.6. Sizing design variables for cross-sectional areas of truss and beam. (a) Solid circular, (b) rectangular, (c) circular tube, (d) rectangular tube, (e) I-section.

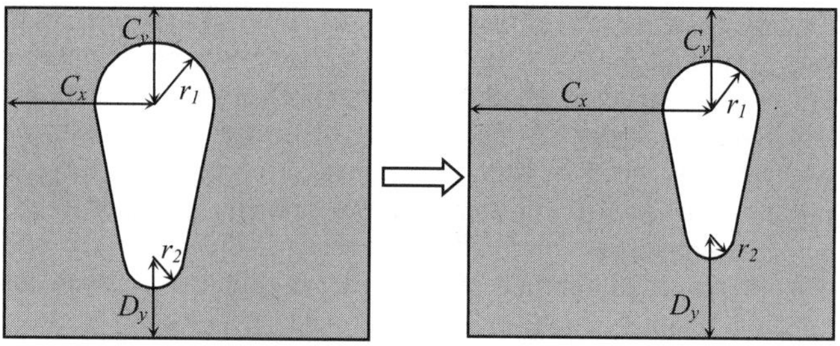

Figure 1.7. Shape design variables.

should be treated as a shape design variable. Usually, the shape design variable defines the domain of integration in structural analysis. Thus, it is not possible to extract shape design variables from a structural model and to use them as sizing design variables.

Consider a rectangular block with a slot, as presented in Fig. 1.7. The location and size of the slot is determined by the geometric values of C_x, C_y, D_y, r_1, and r_2, which are shape design variables. Different values of shape design variables yield different structural shapes. However, these shape design variables do not explicitly appear in the structural problem. If the finite element method is used to perform structural analysis, then integration is carried out over the structural domain (the gray area), which is the shape design variable. Since shape design variables do not explicitly appear in the structural problem, the shape design problem is more difficult to solve than the sizing design problem. Detailed discussions of the shape design problem are presented in Chapter 6 using the material derivative concept of continuum mechanics.

Configuration Design Variable

For those built-up structures made of truss, beam, and shell components, there is another type of design variable in addition to shape design called the configuration design variable, which is related to the structural component's orientation. These components have local coordinate systems fixed on the body of the structure, and state variables of the problem are described in local coordinate systems. If several different components are connected together for the built-up structure, the state variables described in the local coordinate system are transformed to the global coordinate system. If the structural components change their orientation in space, the transformation between the local and global coordinates also changes. Thus, this transformation can be considered the configuration design variable. Since configuration design variables are defined for built-up structures, they are inherently coupled with shape design variables. That is, in order to allow one member of the built-up structure to rotate, another member's shape needs to be changed. The configuration design variable is not applicable to solid components in which all rotations can be expressed in terms of shape changes. A simple configuration design variable will be explained using the example of a three-bar truss in Section 1.2.3. More detailed discussions of the configuration design problem are presented in Chapter 7 using the material derivative concept in continuum mechanics.

Topology Design Variable

If shape and configuration design variables represent changes in structural geometry and orientation, then topology design determines the structure's layout. For example, in Fig.

1.7, shape design can change the size and location of the slot within the block. However, shape design cannot completely remove the slot from the block, or introduce a new slot. Topology design determines whether the slot can be removed or an additional slot is required.

The choice of the topology design variable is nontrivial compared with other design variables. Which parameter is capable of representing the birth or death of the structural layout? Early developments in topology design focused on truss structures. For a given set of points in space, design engineers tried to connect these points using truss structures, in order to find the best layout to support the largest load. Thus, the on-off types of topology design variables are used. These kinds of designs, however, could turn out to be discontinuous and unstable.

Recent developments in topology design are strongly related to finite element analysis. The candidate design domain is modeled using finite elements, and then the material property of each element is controlled. If it is necessary to remove a certain region, then the material property value (e.g., Young's modulus) will approach zero, such that there will be no structural contribution from the removed region. Thus, material property design variables could be used for the purpose of topology design variables. The on-off type of design variable can be approximated by using continuous polynomials in order to remove the difficulties associated with discrete design variables.

In many applications, topology design is used at the concept design stage such that the layout of the structure is determined. After the layout is determined, sizing and shape designs are used to determine the detailed geometry of the structure.

A final comment on design parameterization: it is desirable to have a linearly independent set of design variables. If one does not, then relations between design variables must be imposed as constraints, which may make the design optimization process expensive, as the number of design variables and constraints increase. Furthermore, if design variable constraints are not properly established, meaningless design results will be obtained after an extensive amount of computational effort. As mentioned before, this problem is strongly related to structural modeling, since a well-defined structural model should have an independent set of parameters to define the entire system. Even if defining a good model is not an easy task for a complicated design problem, the design engineer nevertheless has to define a proper and independent set of parameters as much as possible in the structural modeling stage.

1.2.3 Three-Bar Truss Example

In this section, a simple example is introduced to discuss design parameterization, which includes material property, sizing, shape, and configuration design variables. This example will be used repeatedly in subsequent sections to explain the structural analysis and design process. The three-bar truss consists of three truss components, as shown in Fig. 1.8.

For truss components, only one material parameter is involved, which is Young's modulus, E. Thus, the material design parameter is $\boldsymbol{u} = [E]$. On the other hand, the cross-sectional area of each component can be chosen to represent the sizing design variables, stated as $\boldsymbol{u} = [b_1, b_2, b_3]^T$. As explained in Fig. 1.6, the dimensions that determine the cross-sectional shape can be represented as a sizing design. However, for general truss structures, it is purely a matter of convenience whether the cross-sectional dimensions or the cross-sectional area is chosen as the sizing design. As far as analysis is concerned, only cross-sectional area information is required. For example, if the cross section of component 1 is a solid circular shape with radius r, then the relation between the cross-sectional area and the radius would be $b_1 = \pi r^2$.

In Fig. 1.8, for example, if the length of member 1 is changed from l to $l + \delta l$, then the integration domain, which is the shape design, would change. However, since all the members are interconnected, the third member's length and orientation must be modified to satisfy geometric requirements (Fig. 1.9). The change of length involves the shape design, while the rotation of a member affects the configuration design.

As previously pointed out, shape and configuration design variables are closely coupled in the three-bar truss structure. The shape and configuration design variable is

$$\delta \mathbf{x} = [\delta x_1, \delta x_2, \delta x_3, \delta x_4, \delta x_5, \delta x_6]^T. \tag{1.1}$$

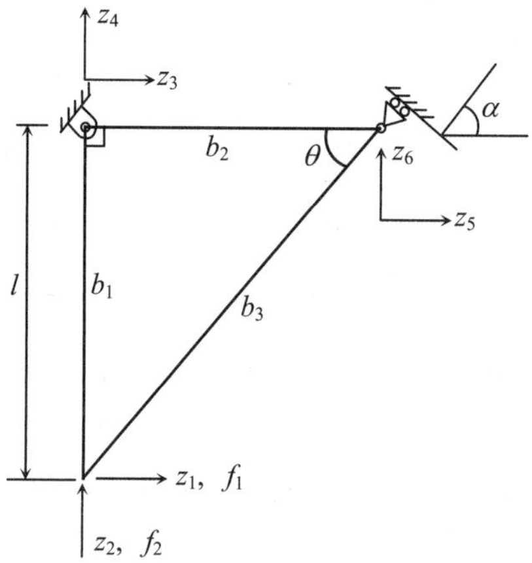

Figure 1.8. Three-bar truss structure.

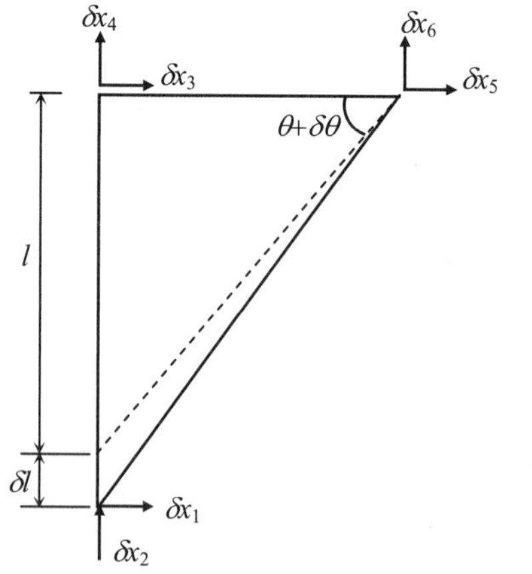

Figure 1.9. Shape and configuration design variable of truss structure.

1.3 Structural Analysis

Structural analysis is solving the mathematical model of the physical problem. In this section, a *variational method* or *energy method* is introduced by using a simple truss structure. The structural equilibrium is viewed as a stationary condition of the total potential energy. For the positive definite quadratic energy, this condition becomes the global minimum condition. Combined with finite element discretization, this method is one of the most popular approaches in structural analysis.

Consider a one-dimensional truss structure under a distributed load $f(x)$ and with point loads F_1 at $x = 0$ and F_2 at $x = l$, as shown in Fig. 1.10. In this text, $z(x)$ denotes a displacement function; the more common notation u used for displacement is reserved as the design variable. For the moment, let the cross-sectional area $u = A(x)$ vary along the truss structural component.

For given x, we assume that stress σ is constant over the cross section $A(x)$. Thus, stress is a function of x alone. The same assumption is given to strain ε such that $\varepsilon(x) = dz/dx$. The stress-strain relation is linear elastic, stated as

$$\sigma(x) = E\varepsilon(x), \tag{1.2}$$

where E is Young's modulus.

As loads are applied to the structure, the structure deforms to resist them. If all loads are removed, then the structure recovers its original shape. Thus, energy is stored in the deformed structure and is called the *strain energy*, defined as

$$U \equiv \frac{1}{2}\int_0^l A\sigma\varepsilon\,dx = \frac{1}{2}\int_0^l EA\left(\frac{dz}{dx}\right)^2 dx. \tag{1.3}$$

If the applied load accompanies the deformation, then work is done to the structure, and we can define it as

$$W \equiv \int_0^l fz\,dx + F_1 z(0) + F_2 z(l). \tag{1.4}$$

We can define the total potential energy of the structure as the difference between U and W, given as

$$\Pi = U - W = \frac{1}{2}\int_0^l EA\left(\frac{dz}{dx}\right)^2 dx - \int_0^l fz\,dx - F_1 z(0) - F_2 z(l). \tag{1.5}$$

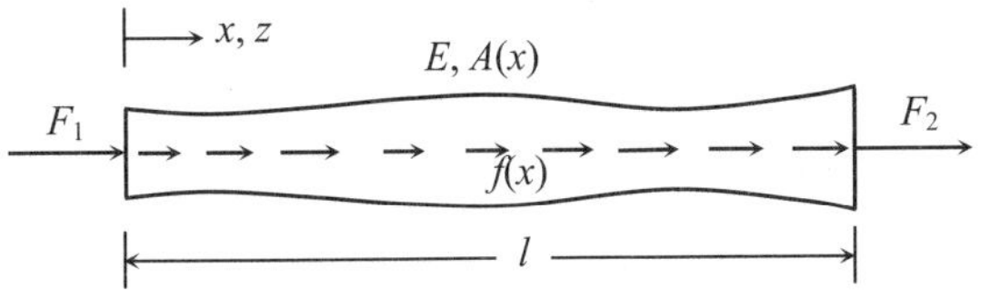

Figure 1.10. Truss structural component.

When the structure is in equilibrium, the forces generated by the structural deformation are the same as the externally applied loads. It is equally true that the total potential energy in (1.5) becomes stationary, which means that the first variation vanishes, so that

$$\delta \Pi = \int_0^l EA\left(\frac{dz}{dx}\right)\left(\frac{d\bar{z}}{dx}\right)dx - \int_0^l f\,\bar{z}\,dx - F_1\bar{z}(0) - F_2\bar{z}(l) = 0, \tag{1.6}$$

where $\bar{z}$ is the first-order variation of displacement z. A detailed explanation of the variation is provided in Chapter 2. For the moment, $\bar{z}$ can be thought of as a small, arbitrary perturbation of z. If z is fixed at a point x, then $\bar{z}$ vanishes at the same point. The structural problem is to solve for z in a way that satisfies (1.6) for all arbitrary $\bar{z}$.

If the kinematic boundary conditions are given at some point for displacement z, then the possible candidates for the solution are limited to those that satisfy the displacement boundary conditions. For example, if the truss structure in Fig. 1.10 is fixed at $x = 0$, then the candidates for the solution belong to the following solution space:

$$Z = \left\{ z \in H^1(0,l) \mid z(0) = 0 \right\}, \tag{1.7}$$

where H^1 is a Sobolev space [6] of order one whose elements are continuous functions, and the first derivative of the function is square integrable in the domain. For a more detailed discussion of basic function spaces, refer to Appendix A.2. For the moment, readers can think of H^1 as a space of smooth functions. In addition, the displacement variation $\bar{z}$ should satisfy (1.7). Thus, the term $F_1\bar{z}(0)$ in (1.6) would vanish. Physically, if the displacement is fixed, then there would be no work done to the structure by the applied load.

1.4 Finite Element Analysis

The analytical solution to (1.6) is nontrivial, even for a simple built-up structure. For a general-shaped structure, an approximation of (1.6) is required by using finite elements. The finite element method approximates the domain of the structure as a simple geometry set, and then establishes the equilibrium conditions for each finite element. By combining all finite elements, a global system of matrix equations is obtained.

Consider a truss finite element with a constant cross-sectional area as shown in Fig. 1.11. For simplicity, the distributed load is removed. Displacement of the truss element is represented by two end-displacements, namely, z_1 and z_2.

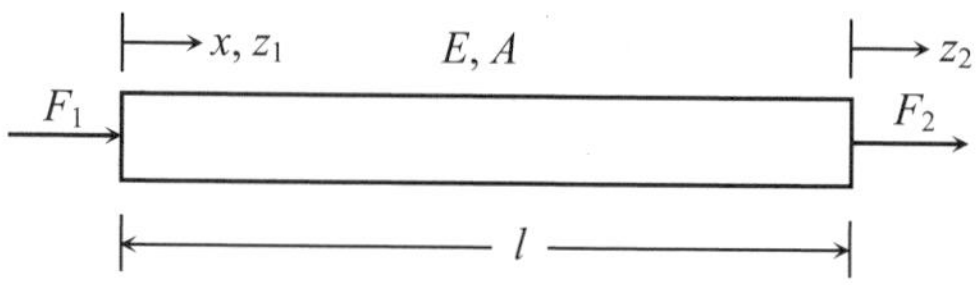

Figure 1.11. Truss finite element.

The displacement at a point x in the domain $(0, l)$ is interpolated by using z_1 and z_2 as

$$z(x) = N_1(x)z_1 + N_2(x)z_2,$$
$$N_1(x) = \frac{l-x}{l}, \quad N_2(x) = \frac{x}{l}, \tag{1.8}$$

where $N_1(x)$ and $N_2(x)$ are called the *shape functions* corresponding to nodes 1 and 2, respectively. Note that $z(0) = z_1$ and $z(l) = z_2$. The derivative of displacement, or strain, can be directly obtained from (1.8) as

$$\frac{dz}{dx} = \frac{1}{l}(z_2 - z_1),$$
$$\frac{d\bar{z}}{dx} = \frac{1}{l}(\bar{z}_2 - \bar{z}_1), \tag{1.9}$$

in which the second equation is the derivative of the displacement variation. By using (1.8) and (1.9), the variation of the total potential energy in (1.6) can be written as

$$\delta\Pi = \frac{EA}{l}(z_2 - z_1)(\bar{z}_2 - \bar{z}_1) - F_1\bar{z}_1 - F_2\bar{z}_2 = 0, \tag{1.10}$$

for all $\bar{z}_1$ and $\bar{z}_2$ in $Z = R^2$. To express (1.10) systematically, it is necessary to define the following vectors

$$\boldsymbol{z} = \begin{bmatrix} z_1 \\ z_2 \end{bmatrix}, \quad \bar{\boldsymbol{z}} = \begin{bmatrix} \bar{z}_1 \\ \bar{z}_2 \end{bmatrix}, \quad \boldsymbol{f} = \begin{bmatrix} F_1 \\ F_2 \end{bmatrix}, \quad \boldsymbol{k} = \frac{EA}{l}\begin{bmatrix} 1 & -1 \\ -1 & 1 \end{bmatrix}, \tag{1.11}$$

where $\boldsymbol{z}$ is the nodal displacement vector and $\bar{\boldsymbol{z}}$ its variation; $\boldsymbol{f}$ is the nodal force vector; and $\boldsymbol{k}$ is the element stiffness matrix. The variational equation (1.10) for the truss element can be written as

$$\bar{\boldsymbol{z}}^T \boldsymbol{k}\boldsymbol{z} = \bar{\boldsymbol{z}}^T \boldsymbol{f}, \tag{1.12}$$

for all $\bar{\boldsymbol{z}}$ in $Z = R^2$. Since (1.12) is satisfied for all $\bar{\boldsymbol{z}}$ in Z, it is equivalent to solving the following matrix equation:

$$\boldsymbol{k}\boldsymbol{z} = \boldsymbol{f}, \tag{1.13}$$

which is called the *local finite element equation.* Equation (1.13) is applicable to one finite element. However, for a built-up structure, many truss elements are connected together to make a complete structure. In this case, the local finite element equation in (1.13) has to be combined to construct the global finite element equation, and this process is called *assembly*.

To see the assembly and solution processes of the truss element, consider the three-bar truss example with a multipoint boundary condition, as shown previously in Fig. 1.8. The displacement and load vectors can be written in the global coordinate system as

$$\boldsymbol{z}_g = [z_1 \quad z_2 \quad z_3 \quad z_4 \quad z_5 \quad z_6]^T, \tag{1.14}$$

$$\boldsymbol{F}_g = [f_1 \quad f_2 \quad 0 \quad 0 \quad 0 \quad 0]^T. \tag{1.15}$$

The element stiffness matrix in the body-fixed local coordinate system must transformed to the global coordinate system. For this purpose, let $\boldsymbol{d}^i$ denote the element local coordinate, and $\boldsymbol{q}^i$ represent the globally oriented element coordinate, as illustrated in Fig. 1.12. In each element, the transformation between the body-fixed and the globally oriented coordinate is

$$
\begin{aligned}
\boldsymbol{d}^1 &= \boldsymbol{q}^1, \\
\boldsymbol{d}^2 &= \boldsymbol{q}^2, \\
\boldsymbol{d}^3 &= \begin{bmatrix} c & s & 0 & 0 \\ -s & c & 0 & 0 \\ 0 & 0 & c & s \\ 0 & 0 & -s & c \end{bmatrix} \boldsymbol{q}^3,
\end{aligned}
$$

where $c = \cos\theta$ and $s = \sin\theta$. In addition, the globally oriented coordinates are transformed into the global coordinate z_g by using Boolean matrices as

$$
\boldsymbol{q}^1 = \begin{bmatrix} 0 & 1 & 0 & 0 & 0 & 0 \\ 1 & 0 & 0 & 0 & 0 & 0 \\ 0 & 0 & 0 & 1 & 0 & 0 \\ 0 & 0 & 1 & 0 & 0 & 0 \end{bmatrix} z_g,
$$

$$
\boldsymbol{q}^2 = \begin{bmatrix} 0 & 0 & 1 & 0 & 0 & 0 \\ 0 & 0 & 0 & 1 & 0 & 0 \\ 0 & 0 & 0 & 0 & 1 & 0 \\ 0 & 0 & 0 & 0 & 0 & 1 \end{bmatrix} z_g,
$$

$$
\boldsymbol{q}^3 = \begin{bmatrix} 1 & 0 & 0 & 0 & 0 & 0 \\ 0 & 1 & 0 & 0 & 0 & 0 \\ 0 & 0 & 0 & 0 & 1 & 0 \\ 0 & 0 & 0 & 0 & 0 & 1 \end{bmatrix} z_g.
$$

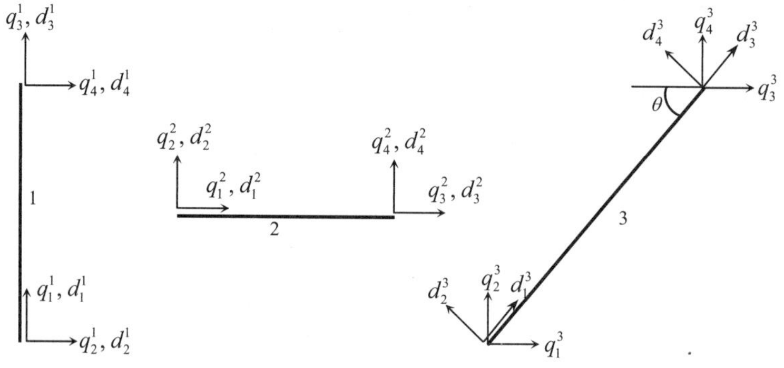

Figure 1.12. Body-fixed and globally oriented element coordinate system.

After the assembly process, the generalized global stiffness matrix is expressed in terms of the global displacement coordinate z_g as

$$\boldsymbol{K}_g(\boldsymbol{u}) = \frac{E}{l}\begin{bmatrix} b_3c^2s & b_3cs^2 & 0 & 0 & -b_3c^2s & -b_3cs^2 \\ b_3cs^2 & b_1+b_3s^3 & 0 & -b_1 & -b_3cs^2 & -b_3s^3 \\ 0 & 0 & b_2s/c & 0 & -b_2s/c & 0 \\ 0 & -b_1 & 0 & b_1 & 0 & 0 \\ -b_3c^2s & -b_3cs^2 & -b_2s/c & 0 & b_2s/c+b_3c^2s & b_3cs^2 \\ -b_3cs^2 & -b_3s^3 & 0 & 0 & b_3cs^2 & b_3s^3 \end{bmatrix}. \tag{1.16}$$

In (1.16), $\boldsymbol{u} = [b_1, b_2, b_3]^T$ denotes the design variable vector, which is the cross-sectional area of each truss element. Note that the global stiffness matrix $\boldsymbol{K}_g$ is singular, since it has a rigid body motion that can be removed by applying boundary conditions. As shown in Fig. 1.8, the displacement variables z_3 and z_4 are fixed. In addition, z_5 and z_6 are dependent on each other. Solution candidates must satisfy these conditions. In this problem, space Z of kinematically admissible displacements is

$$Z_h = \{ \boldsymbol{z}_g \in R^6 : z_3 = z_4 = 0,\ z_5\cos\alpha + z_6\sin\alpha = 0 \}, \tag{1.17}$$

and $\boldsymbol{K}_g(\boldsymbol{u})$ is the positive definite in Z_h, although it is not positive definite in all of R^6. Thus, the global variational equation of three-bar truss example is obtained as

$$\bar{\boldsymbol{z}}_g^T \boldsymbol{K}_g \boldsymbol{z}_g = \bar{\boldsymbol{z}}_g^T \boldsymbol{F}_g, \qquad \forall \bar{\boldsymbol{z}}_g \in Z_h, \tag{1.18}$$

where $\forall \bar{\boldsymbol{z}}_g \in Z_h$ denotes for "all $\bar{\boldsymbol{z}}_g$ in Z_h."

In actual computation, the global variational equation is modified to explicitly eliminate the boundary condition. However, this is not the general case. In many FEA codes, the size of $\boldsymbol{K}_g$ is retained. Instead of explicitly removing rows and columns corresponding to the boundary conditions, equivalent relations are substituted to make $\boldsymbol{K}_g$ positive definite. Since z_3 and z_4 are prescribed, they can be eliminated from the variational equation. In addition, since z_5 and z_6 has a relation, z_6 can be expressed in terms of z_5, as in (1.17).

Consider the case in which $\theta = 45^\circ$ and $\alpha = 30^\circ$, and define the reduced global displacement vector as $\boldsymbol{z} = [z_1, z_2, z_5]^T$. Accordingly, the reduced global load vector is defined as $\boldsymbol{F} = [f_1, f_2, 0]^T$. By removing the third and the fourth rows and columns of $\boldsymbol{K}_g$, and by substituting the relation $z_6 = -\sqrt{3}z_5$, the reduced stiffness matrix in this example would be

$$\boldsymbol{K}(\boldsymbol{u}) = \frac{E}{2\sqrt{2}l}\begin{bmatrix} b_3 & b_3 & (\sqrt{3}-1)b_3 \\ b_3 & 2\sqrt{2}b_1 + b_3 & (\sqrt{3}-1)b_3 \\ (\sqrt{3}-1)b_3 & (\sqrt{3}-1)b_3 & 2\sqrt{2}b_2 + (4-2\sqrt{3})b_3 \end{bmatrix}. \tag{1.19}$$

Thus, the reduced global matrix equation is written as

$$\boldsymbol{K}\boldsymbol{z} = \boldsymbol{F}. \tag{1.20}$$

If $f_1 = f_2 = 1$ and $l = 1$, then the solution of the reduced global matrix equation (1.20) is obtained as

$$\boldsymbol{z} = \begin{bmatrix} \dfrac{4-2\sqrt{3}}{Eb_2} + \dfrac{2\sqrt{2}}{Eb_3} & 0 & \dfrac{1-\sqrt{3}}{Eb_2} \end{bmatrix}^T. \tag{1.21}$$

The solution method in (1.20) is easier than that of (1.18). However, for general, complex kinematic constraints, it may not be easy to explicitly construct the reduced matrix $\boldsymbol{K}$. In addition, many FEA codes do not generate a reduced matrix $\boldsymbol{K}$ during the solution procedure. Thus, the use of (1.18) is clear. In the next section, we will discuss how solution z in (1.21) can be changed as a function of design vector $\boldsymbol{b}$.

1.5 Structural Design Sensitivity Analysis

Design sensitivity analysis is used to compute the rate of performance measure change with respect to design variable changes. Obviously, the performance measure is presumed to be a differentiable function of the design, at least in the neighborhood of the current design point. For complex engineering applications, it is not simple to prove a performance measure's differentiability with respect to the design. Consequently, the question of differentiability will be postponed until Chapters 5 and 6. For most problems in this text, one can assume that the performance measure is differentiable with respect to the design.

In general, a structural performance measure depends on the design. For example, a change in the cross-sectional area of a beam would affect the structural weight. This type of dependence is simple if the expression of weight in terms of the design variables is known. For example, the weight of a straight beam with a circular cross section can be expressed as

$$W(r) = \pi r^2 l,$$

where $u = r$ is the radius and l is the length of the beam. If the radius is a design variable, then the design sensitivity of W with respect to r would be

$$\frac{dW}{dr} = 2\pi r l.$$

This type of function is *explicitly dependent* on the design, since the function can be explicitly written in terms of that design. Consequently, only algebraic manipulation is involved and no finite element analysis is required to obtain the design sensitivity of an explicitly dependent performance measure.

However, in most cases, a structural performance measure does not explicitly depend on the design. For example, when the stress of a beam is considered as a performance measure, there is no simple way to express the design sensitivity of stress explicitly in terms of the design variable r. In the linear elastic problem, the stress of the structure is determined from the displacement, which is a solution to the finite element analysis. Thus, the sensitivity of stress $\sigma(z)$ can be written as

$$\frac{d\sigma}{dr} = \frac{d\sigma}{dz}^T \frac{dz}{dr}, \tag{1.22}$$

where z is the displacement of the beam. Since the expression of stress as a function of displacement is known, $d\sigma/dz$ can be easily obtained. The only difficulty is the computation of dz/dr, which is the state variable (displacement) sensitivity with respect to the design variable r.

When a design engineer wants to compute the design sensitivity of performance measures such as stress $\sigma(z)$ in (1.22), structural analysis (finite element analysis, for example) has presumably already been carried out. Assume that the structural problem is governed by the following linear algebraic equation

$$\boldsymbol{K}(u)z = \boldsymbol{f}(u). \tag{1.23}$$

Equation (1.23) is a matrix equation of finite elements if $\boldsymbol{K}$ and $\boldsymbol{f}$ are understood to be the stiffness matrix and load vector, respectively. Suppose the explicit expressions of $\boldsymbol{K}(u)$ and $\boldsymbol{f}(u)$ are known and differentiable with respect to u. Since the stiffness matrix $\boldsymbol{K}(u)$ and load vector $\boldsymbol{f}(u)$ depend on the design u, solution z also depends on the design u. However, it is important to note that this dependency is implicit, which is why we need to develop a design sensitivity analysis methodology. As shown in (1.22), dz/du must be computed using the governing equation of (1.23). This can be achieved by differentiating (1.23) with respect to u, as

$$\boldsymbol{K}(u)\frac{dz}{du} = \frac{d\boldsymbol{f}}{du} - \frac{d\boldsymbol{K}}{du}z. \tag{1.24}$$

Assuming that the explicit expressions of $\boldsymbol{K}(u)$ and $\boldsymbol{f}(u)$ are known, $d\boldsymbol{K}/du$ and $d\boldsymbol{f}/du$ can be evaluated. Thus, if solution z in (1.23) is known, then dz/du can be computed from (1.24), which can then be substituted into (1.22) to compute $d\sigma/du$. Note that the stress performance measure is *implicitly dependent* on the design through state variable z.

In this text, it is assumed that the general performance measure ψ depends on the design explicitly and implicitly. That is, the performance measure ψ is presumed to be a function of design u, and state variable $z(u)$, as

$$\psi = \psi(z(u), u). \tag{1.25}$$

The sensitivity of ψ can thus be expressed as

$$\frac{d\psi(z(u),u)}{du} = \left.\frac{\partial \psi}{\partial u}\right|_{z=\text{const}} + \left.\frac{\partial \psi}{\partial z}\right|_{u=\text{const}}^{T} \frac{dz}{du}. \tag{1.26}$$

The only unknown term in (1.26) is dz/du. Various computational methods to obtain dz/du are introduced in the following sections.

1.5.1 Methods of Structural Design Sensitivity Analysis

Various methods employed in design sensitivity analysis are listed in Fig. 1.13. Three approaches are used to obtain the design sensitivity: the approximation, discrete, and continuum approaches. In the approximation approach, design sensitivity is obtained by either the *forward finite difference* or the *central finite difference method*. In the discrete method, design sensitivity is obtained by taking design derivatives of the discrete governing equation. For this process, it is necessary to take the design derivative of the stiffness matrix. If this derivative is obtained analytically using the explicit expression of the stiffness matrix with respect to the design variable, it is an *analytical method*, since the analytical expressions of $\boldsymbol{K}(u)$ and $\boldsymbol{f}(u)$ are used. However, if the derivative is obtained using a finite difference method, the method is called a *semianalytical method*. In the continuum approach, the design derivative of the variational equation is taken before it is discretized. If the structural problem and sensitivity equations are solved as a continuum problem, then it is called the *continuum-continuum method*. However, only very simple, classical problems can be solved analytically. Thus, the continuum sensitivity equation is solved by discretization in the same way that structural problems are solved. Since differentiation is taken at the continuum domain and is then followed by discretization, this method is called the *continuum-discrete method*. These methods will be explained in detail in the following sections.

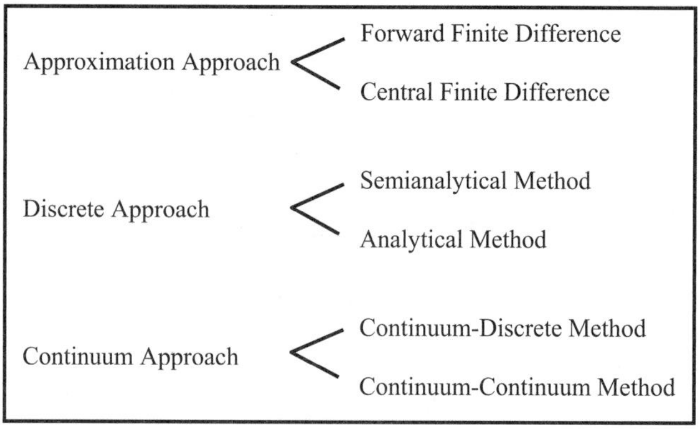

Figure 1.13. Approaches to design sensitivity analysis.

1.5.2 Finite Difference Method

The easiest way to compute sensitivity information of the performance measure is by using the finite difference method. Different designs yield different analysis results and, thus, different performance values. The finite difference method actually computes design sensitivity of performance by evaluating performance measures at different stages in the design process. If u is the current design, then the analysis results provide the value of performance measure $\psi(u)$. In addition, if the design is perturbed to $u + \Delta u$, where Δu represents a small change in the design, then the sensitivity of $\psi(u)$ can be approximated as

$$\frac{d\psi}{du} \approx \frac{\psi(u + \Delta u) - \psi(u)}{\Delta u}. \tag{1.27}$$

Equation (1.27) is called the *forward difference method* since the design is perturbed in the direction of $+\Delta u$. If $-\Delta u$ is substituted in (1.27) for Δu, then the equation is defined as the *backward difference method.* Additionally, if the design is perturbed in both directions, such that the design sensitivity is approximated by

$$\frac{d\psi}{du} \approx \frac{\psi(u + \Delta u) - \psi(u - \Delta u)}{2\Delta u}, \tag{1.28}$$

then the equation is defined as the *central difference method.*

The advantage of the finite difference method is obvious. If structural analysis can be performed and the performance measure can be obtained as a result of structural analysis, then the expressions in (1.27) and (1.28) are virtually independent of the problem types considered. Consequently, this method is still popular in engineering design.

However, sensitivity computation costs become the dominant concern in the design process. If n represents the number of designs, then $n + 1$ analyses have to be carried out for either the forward or backward difference method, and $2n + 1$ analyses are required for the central difference method. For modern, practical engineering applications, the cost of structural analysis is rather expensive. Thus, this method is infeasible for large-scale problems containing many design variables.

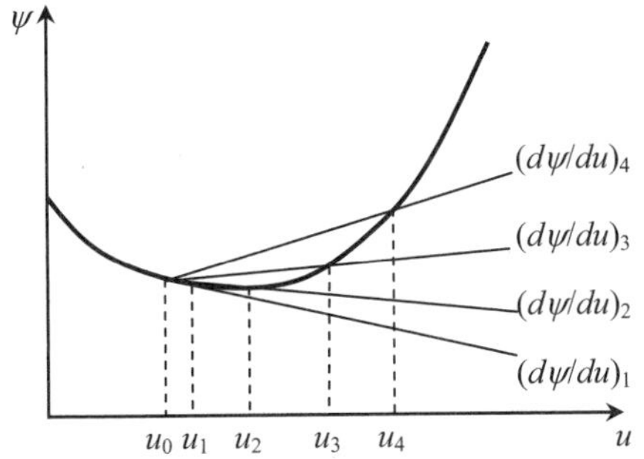

Figure 1.14. Influence of step size in forward finite difference method.

Another major disadvantage of the finite difference method is the accuracy of its sensitivity results. In (1.27), accurate results can be expected when Δu approaches zero. Figure 1.14 shows some sensitivity results using the finite difference method. The tangential slope of the curve at u_0 is the exact sensitivity value. Depending on perturbation size, we can see that sensitivity results are quite different. For a mildly nonlinear performance measure, relatively large perturbation provides a reasonable estimation of sensitivity results. However, for highly nonlinear performances, a large perturbation yields completely inaccurate results. Thus, the determination of perturbation size greatly affects the sensitivity result. And even though it may be necessary to choose a very small perturbation, numerical noise becomes dominant for a too-small perturbation size. That is, with a too-small perturbation, no reliable difference can be found in the analysis results. For example, if up to five digits of significant numbers are valid in a structural analysis, then any design perturbation in the finite difference that is smaller than the first five significant digits cannot provide meaningful results. As a result, it is very difficult to determine design perturbation sizes that work for all problems.

Example 1.1. Three-Bar Truss (Finite Difference Method). Consider the three-bar truss example shown previously in Fig. 1.8. In this example, the finite element matrix equation, (1.20), can be solved analytically with the solution given in (1.21) as a function of the design variable vector $\boldsymbol{u} = [b_1, b_2, b_3]^T$, which is the cross-sectional area of the truss elements. For simplicity, if the current value of the design is $\boldsymbol{u} = [1, 1, 1]^T$, and $E = 1$, then the solution becomes

$$z = \left[4 - 2\sqrt{3} + 2\sqrt{2},\ \ 0,\ \ 1 - \sqrt{3} \right]^T. \tag{1.29}$$

Let us compute the design sensitivity of z_1 by using the finite difference method. Since the dependence of z_1 on the design is explicitly given, z_1 can be straightforwardly computed at different design stages. Table 1.1 shows the sensitivities of z_1 with different perturbation sizes. As the perturbation size decreases, the sensitivity value using finite difference method approaches an exact sensitivity value. In many cases, the central finite difference method is more accurate than the forward/backward finite difference method, as shown in Table 1.1, although for the central finite difference method, two performance measure evaluations are involved. Note that, in the latter case, since (1.21) is the exact

Table 1.1. Sensitivity results of finite difference method.

Design	Forward FDM			Central FDM			Exact
	$\Delta b = 0.5$	$\Delta b = 0.1$	$\Delta b = 0.01$	$\Delta b = 0.5$	$\Delta b = 0.1$	$\Delta b = 0.01$	
b_1	0	0	0	0	0	0	0
b_2	–0.35727	–0.48718	–0.53059	–0.71453	–0.54131	–0.53595	–0.53590
b_3	–1.88561	–2.57130	–2.8004	–3.77124	–2.85700	–2.82871	–2.82843

solution of the matrix equation in (1.20), there is no concern with numerical noise. That is, as the design perturbation size decreases, the finite difference results will converge to the exact design sensitivities.

1.5.3 Discrete Method

A structural problem is often discretized in finite dimensional space in order to solve complex problems, as shown with the finite element method in Section 1.4. The discrete method computes the performance design sensitivity of the discretized problem, where the governing equation is a system of linear equations, as in (1.23). If the explicit form of the stiffness matrix $\boldsymbol{K}(u)$ and the load vector $\boldsymbol{f}(u)$ are known, and if solution $\boldsymbol{z}$ of matrix equation $\boldsymbol{Kz} = \boldsymbol{f}$ is obtained, then the design sensitivity of the displacement vector can also be obtained, as

$$\boldsymbol{K}(u)\frac{d\boldsymbol{z}}{du} = \frac{d\boldsymbol{f}}{du} - \frac{d\boldsymbol{K}}{du}\boldsymbol{z}. \tag{1.30}$$

This is a discrete approach to the *analytical method*, since the explicit expressions of $\boldsymbol{K}(u)$ and $\boldsymbol{f}(u)$ are used to obtain design derivatives of the stiffness matrix and load vector. Even if the expression of (1.30) is in the global system matrix, actual computation of these derivatives can still be carried out on the element level in order to avoid a massive amount of calculation related to global stiffness matrix $\boldsymbol{K}$. An in-depth discussion of this method is presented in Chapter 4.

It is not difficult to compute $d\boldsymbol{f}/du$, since the applied force is usually either independent of the design, or it has a simple expression. However, the computation of $d\boldsymbol{K}/du$ in (1.30) depends on the type of problem. In addition, modern advances in the finite element method use numerical integration in the computation of $\boldsymbol{K}$. In this case, the explicit expression of $\boldsymbol{K}$ in terms of u may not be available. Moreover, in the case of the shape design variable, computation of the analytical derivative of the stiffness matrix is quite costly. Because of this, the semianalytical method is a popular choice for discrete shape design sensitivity analysis approaches. However, Barthelemy and Haftka [7] show that the semianalytical method can have serious accuracy problems for shape design variables in structures modeled by beam, plate, truss, frame, and solid elements. They found that accuracy problems occur even for a simple cantilever beam. Moreover, errors in the early stage of approximation multiply during the matrix equation solution phase. As a remedy, Olhoff et al. [8] proposed an exact numerical differentiation method when the analytical form of the element stiffness matrix is available.

Example 1.2. Three-Bar Truss (Discrete Method). To obtain discrete design sensitivity with respect to sizing design variables, consider the three-bar truss problem. More complicated design variables, namely, shape and configuration, will be considered in Chapters 6 and 7. Let us begin with the reduced global stiffness matrix given in (1.19). Since the explicit form of $\boldsymbol{K}(\boldsymbol{u})$ is given as a function of design variable $\boldsymbol{u}$, its derivative can be obtained as

$$\frac{d\boldsymbol{K}}{db_1} = \begin{bmatrix} 0 & 0 & 0 \\ 0 & 1 & 0 \\ 0 & 0 & 0 \end{bmatrix}, \quad \frac{d\boldsymbol{K}}{db_2} = \begin{bmatrix} 0 & 0 & 0 \\ 0 & 0 & 0 \\ 0 & 0 & 1 \end{bmatrix}, \quad \frac{d\boldsymbol{K}}{db_3} = \frac{1}{2\sqrt{2}} \begin{bmatrix} 1 & 1 & \sqrt{3}-1 \\ 1 & 1 & \sqrt{3}-1 \\ \sqrt{3}-1 & \sqrt{3}-1 & 4-2\sqrt{3} \end{bmatrix}, \tag{1.31}$$

and load vector $\boldsymbol{f}$ is independent of the design, that is, $d\boldsymbol{f}/d\boldsymbol{u} = 0$. The right side of (1.30) can be obtained by multiplying (1.31) with (1.29) for each design variable. If $\boldsymbol{F}^u$ denotes the right side of (1.30), then its explicit expression would be

$$\boldsymbol{F}^{b_1} = \begin{bmatrix} 0 \\ 0 \\ 0 \end{bmatrix}, \quad \boldsymbol{F}^{b_2} = \begin{bmatrix} 0 \\ 0 \\ \sqrt{3}-1 \end{bmatrix}, \quad \boldsymbol{F}^{b_3} = \begin{bmatrix} -1 \\ -1 \\ 1-\sqrt{3} \end{bmatrix}. \tag{1.32}$$

Since the right side, corresponding to the first design variable, vanishes, the sensitivity of displacement with respect to b_1 also vanishes. By solving (1.30) with respect to $d\boldsymbol{z}/d\boldsymbol{u}$, we obtain

$$\frac{d\boldsymbol{z}}{db_1} = \begin{bmatrix} 0 \\ 0 \\ 0 \end{bmatrix}, \quad \frac{d\boldsymbol{z}}{db_2} = \begin{bmatrix} 2\sqrt{3}-4 \\ 0 \\ \sqrt{3}-1 \end{bmatrix}, \quad \frac{d\boldsymbol{z}}{db_3} = \begin{bmatrix} -2\sqrt{2} \\ 0 \\ 0 \end{bmatrix}, \tag{1.33}$$

which is the same result as the direct computation of sensitivity in (1.21).

Example 1.3. Cantilever Beam (Discrete Method). Consider the cantilever beam in Fig. 1.15 with point load p at $x = l$. If one finite element is used to discretize the structure, as shown in Fig. 1.15(b), then displacement $z(x)$ can be approximated as

$$z(x) = \boldsymbol{N}^T \boldsymbol{z}_g = [N_1 \quad N_2 \quad N_3 \quad N_4] \begin{bmatrix} z_1 \\ \theta_1 \\ z_2 \\ \theta_2 \end{bmatrix}, \tag{1.34}$$

where z_1 and z_2 are nodal displacement, θ_1 and θ_2 are nodal rotations, and N_i's are corresponding shape functions of the approximation, defined as

$$\begin{aligned} N_1 &= 1 - \frac{3x^2}{l^2} + \frac{2x^3}{l^3}, & N_2 &= x - \frac{2x^2}{l} + \frac{x^3}{l^2}, \\ N_3 &= \frac{3x^2}{l^2} - \frac{2x^3}{l^3}, & N_4 &= -\frac{x^2}{l} + \frac{x^3}{l^2}. \end{aligned} \tag{1.35}$$

The variational equation of the beam bending problem in a continuum model can be written as

$$\int_0^l EIz_{,xx}\bar{z}_{,xx}\, dx = \int_0^l p\delta(x-l)\bar{z}\, dx, \qquad \forall \bar{z} \in Z, \tag{1.36}$$

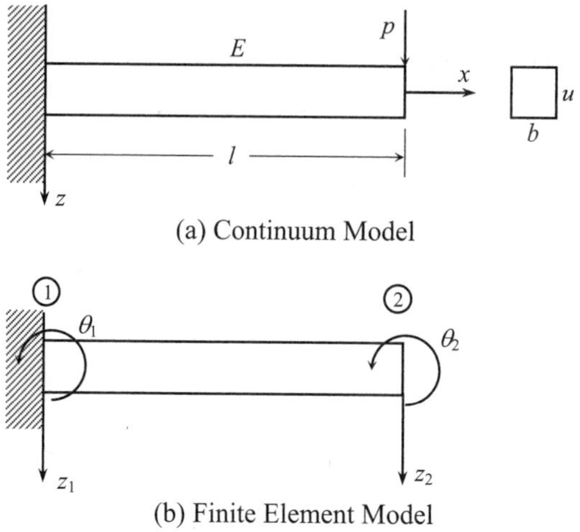

Figure 1.15. Cantilever beam.

where E is Young's modulus, $I = bu^3/12$ is the second moment of inertia, $\delta(x - l)$ is the Dirac delta measure, which has a value of infinity at $x = l$ and zero otherwise, and Z is the space of kinematically admissible displacements. For the moment, Z can be thought of as the space of smooth functions that satisfy the boundary condition $z(0) = \theta(0) = 0$.

To obtain a finite element equation, the approximation in (1.34) is substituted into variational (1.36) to obtain

$$\bar{\boldsymbol{z}}_g^T \boldsymbol{K}_g \boldsymbol{z}_g = \bar{\boldsymbol{z}}_g^T \boldsymbol{F}_g, \qquad \forall \bar{\boldsymbol{z}}_g \in Z_h, \tag{1.37}$$

where $Z_h = \{\boldsymbol{z} \in R^4 \mid z_1(0) = \theta_1(0) = 0\}$ is the discretized version of Z. $\boldsymbol{K}_g$ and $\boldsymbol{F}_g$ are given as

$$\begin{aligned} \boldsymbol{K}_g &= \frac{EI}{\ell^3}\begin{bmatrix} 12 & 6\ell & -12 & 6\ell \\ 6\ell & 4\ell^2 & -6\ell & 2\ell^2 \\ -12 & -6\ell & 12 & -6\ell \\ 6\ell & 2\ell^2 & -6\ell & 4\ell^2 \end{bmatrix}, \\ \boldsymbol{F}_g &= [0 \quad 0 \quad p \quad 0]^T. \end{aligned} \tag{1.38}$$

For the cantilever beam shown in Fig. 1.15, the boundary condition is given such that $z_1(0) = \theta_1(0) = 0$. Thus, matrix $\boldsymbol{K}_g$ and vector $\boldsymbol{F}_g$ can be reduced only for z_2 and θ_2, as

$$\boldsymbol{K}\boldsymbol{z} \equiv \frac{EI}{\ell^3}\begin{bmatrix} 12 & -6\ell \\ -6\ell & 4\ell^2 \end{bmatrix}\begin{bmatrix} z_2 \\ \theta_2 \end{bmatrix} = \begin{bmatrix} p \\ 0 \end{bmatrix} \equiv \boldsymbol{F}. \tag{1.39}$$

The solution $\boldsymbol{z}$, and thus $\boldsymbol{z}_g$, can be obtained by solving (1.39) as

$$\boldsymbol{z}_g = \begin{bmatrix} 0 & 0 & \dfrac{p\ell^3}{3EI} & \dfrac{p\ell^2}{2EI} \end{bmatrix}^T, \tag{1.40}$$

and by using approximation in (1.34), displacement function $z(x)$ can be obtained as

$$z(x) = \boldsymbol{N}^T \boldsymbol{z}_g = \frac{p}{6EI}(-x^3 + 3\ell x^2). \tag{1.41}$$

Note that $z(x)$ in (1.41) is the exact solution in this special example.

The discrete method of design sensitivity can be obtained by differentiating the finite element matrix (1.39) with respect to the design. Consider height u of the cross-sectional dimension as a design variable. The design sensitivity equation can be obtained from (1.39) using a procedure similar to that in (1.30), as

$$\boldsymbol{K}(u)\frac{d\boldsymbol{z}}{du} = \frac{d\boldsymbol{F}}{du} - \frac{d\boldsymbol{K}}{du}\boldsymbol{z} \equiv \boldsymbol{F}^u, \tag{1.42}$$

where $d\boldsymbol{F}/du = \boldsymbol{0}$, since $\boldsymbol{F}$ is independent of the design, and $d\boldsymbol{K}/du$ is calculated from (1.39) as

$$\frac{d\boldsymbol{K}}{du} = \frac{3EI}{\ell^3 u}\begin{bmatrix} 12 & -6\ell \\ -6\ell & 4\ell^2 \end{bmatrix}. \tag{1.43}$$

Thus, the right side of (1.42) can be computed as

$$\boldsymbol{F}^u = \frac{d\boldsymbol{F}}{du} - \frac{d\boldsymbol{K}}{du}\boldsymbol{z} = -\frac{3EI}{\ell^3 u}\begin{bmatrix} 12 & -6\ell \\ -6\ell & 4\ell^2 \end{bmatrix}\begin{bmatrix} \frac{p\ell^3}{3EI} \\ \frac{p\ell^2}{2EI} \end{bmatrix} = \begin{bmatrix} -\frac{3p}{u} \\ 0 \end{bmatrix}, \tag{1.44}$$

and the design sensitivity of the displacement vector can be solved from (1.42) as

$$\frac{d\boldsymbol{z}}{du} = \frac{\ell}{12EI}\begin{bmatrix} 4\ell^2 & 6\ell \\ 6\ell & 12 \end{bmatrix}\begin{bmatrix} -\frac{3p}{u} \\ 0 \end{bmatrix} = \begin{bmatrix} -\frac{p\ell^3}{EIu} \\ -\frac{3p\ell^2}{2EIu} \end{bmatrix}. \tag{1.45}$$

Since the shape function of the finite element approximation is independent of the design, the interpolation in (1.34) is valid for displacement sensitivity. Thus, the displacement function sensitivity can be obtained as

$$\frac{dz(x)}{du} = \boldsymbol{N}^T\frac{d\boldsymbol{z}}{du} = \frac{p}{2EIu}x^2(x - 3\ell). \tag{1.46}$$

Equation (1.46) can be verified by directly differentiating the exact solution in (1.41).

1.5.4 Continuum Method

In the continuum method, the design derivative of the variational equation (the continuum model of the structure) is taken before discretization. Since differentiation is taken before any discretization takes place, this method provides more accurate results than the discrete approach. In addition, profound mathematical proofs are available regarding the existence and uniqueness of the design sensitivity. Most discussions in this text focus on the continuum method, in which analytical expressions of design sensitivity are obtained in the continuum setting.

Sizing design variables are distributed parameters of the continuum equation. For shape design variables, the material derivative concept of continuum mechanics is used to relate variations in structural shape to the structural performance measures [5]. Using the continuum design sensitivity analysis approach, design sensitivity expressions are obtained in the form of integrals, with integrands written in terms of such physical quantities as displacement, stress, strain, and domain shape change. If exact solutions to the continuum equations are used to evaluate these design sensitivity expressions, then this procedure is referred to as the continuum-continuum method. On the other hand, if approximation methods such as the finite element, boundary element, or meshfree method are used to evaluate these terms, then this procedure is called the continuum-discrete method. The continuum-continuum method provides the exact design sensitivity of the exact model, whereas the continuum-discrete method provides an approximate design sensitivity of the exact model. When FEA is used to evaluate the structural response, then the same discretization method as structural analysis has to be used to compute the design sensitivity of performance measures in the continuum-discrete method.

Example 1.4. Cantilever Beam (Continuum Method). Continuum-based design sensitivity analysis is used to differentiate the variational (1.36) for the cantilever beam discussed in Example 1.3. As with the discrete method, the right side of (1.36) is independent of the design. Let us define differentiation or variation as

$$z' = \frac{dz}{du}\delta u, \tag{1.47}$$

where δu is the amount of perturbation. The left side of (1.36) can be differentiated with respect to design u as

$$\frac{d}{du}\left[\int_0^l EIz_{,xx}\bar{z}_{,xx}\,dx\right]\delta u = \int_0^l EIz'_{,xx}\bar{z}_{,xx}\,dx + \int_0^l \frac{3EI}{u} z_{,xx}\bar{z}_{,xx}\delta u\,dx. \tag{1.48}$$

Thus, the continuum-based design sensitivity equation is obtained as, with $\delta u = 1$,

$$\int_0^l EIz'_{,xx}\bar{z}_{,xx}\,dx = -\int_0^l \frac{3EI}{u} z_{,xx}\bar{z}_{,xx}\,dx, \qquad \forall \bar{z} \in Z, \tag{1.49}$$

which yields the solution $z' = dz/du$. The continuum-continuum method solves (1.49) to obtain z' directly, whereas the continuum-discrete method first discretizes (1.49), following the same procedure as the finite element method. If the same approximation is used for z' as displacement function z in (1.34), then the left side of (1.49) becomes equivalent to (1.36) by considering $\mathbf{z}_g$ as $\mathbf{z}'_g$. The discretized design sensitivity equation therefore becomes

$$\bar{\mathbf{z}}_g \mathbf{K}_g \mathbf{z}'_g = \bar{\mathbf{z}}_g \mathbf{F}_g^f, \qquad \forall \bar{\mathbf{z}}_g \in Z_h, \tag{1.50}$$

which can be solved using the same procedure as in Example 1.3.

Note that in the continuum method it is neither necessary to differentiate the stiffness matrix $d\mathbf{K}/du$, nor to use any matrix multiplication procedure to calculate $d\mathbf{K}/du \cdot \mathbf{z}$, which involves a large amount of additional computational cost.

One frequently asked question is: "Are the discrete and continuum-discrete methods equivalent?" To answer this question, four conditions have to be given. First, the same discretization (shape function) used in the FEA method must be used for continuum design sensitivity analysis. Second, an exact integration (instead of a numerical integration) must be used in the generation of the stiffness matrix and in the evaluation of continuum-based design sensitivity expressions. Third, the exact solution (and not a numerical solution) of the finite element matrix equation and the adjoint equation should be used to compare these two methods. Fourth, the movement of discrete grid points must be consistent with the design parameterization method used in the continuum method.

For the sizing design variable, it is shown in [5] that the discrete and continuum-discrete methods are equivalent under the conditions given above, using a beam as the structural component. It has also been argued that the discrete and continuum-discrete methods are equivalent in shape design problems under the conditions given above [9]. One point to note is that these four conditions are not easy to satisfy; in many cases, numerical integration is used and exact solutions of the FE matrix equation cannot be obtained.

1.5.5 Summary of Design Sensitivity Analysis Approaches

As explained in previous sections, the design sensitivity analysis method has been developed along two fundamentally different paths, as shown in Fig. 1.13. In the discrete method, design derivatives of a discretized structural FEA equation are taken to obtain design sensitivity information. In the continuum method, design derivatives of the variational governing equation are taken to obtain explicit design sensitivity expressions in integral form with integrands written in terms of the following variations: material property, sizing, shape, and configuration design variables, and such natural physical quantities as displacement, stress, and strain [5]. The explicit design sensitivity expressions are then numerically evaluated using the analysis results of FEA codes. Unlike the finite difference method, the continuum method provides accurate design sensitivity information without recourse to the uncertainties of perturbation size. In addition, the continuum method does not require the derivatives of stiffness, mass, and damping matrices, as with the discrete method shown in Fig. 1.16. Another advantage of the continuum method is that it provides unified, structural design sensitivity analysis capability, so that it is possible to develop one design sensitivity analysis software system that works with a number of well-established analysis methods, such as FEA, the boundary element method, the p-method of FEA, and the meshfree method.

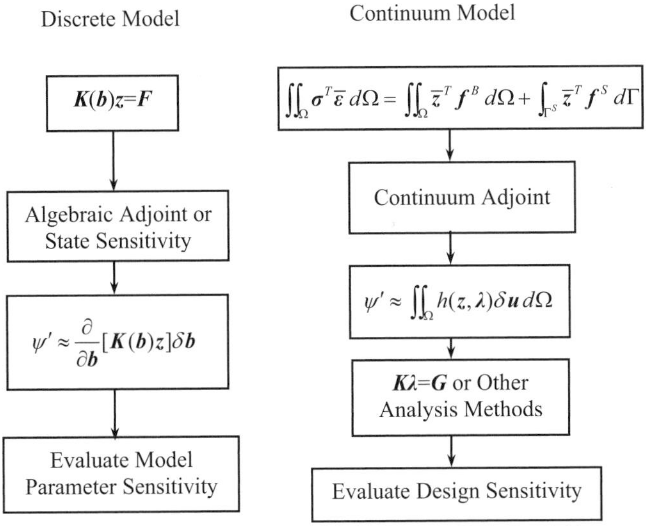

Figure 1.16. Design sensitivity analysis methods.

One important advantage of the continuum-based design sensitivity analysis method is that it is possible to compute the results of design sensitivity analysis of the established FEA/BEA/meshfree codes with respect to the geometric design variables employed by CAD tools. For example, connections to CAD tools can be made by providing the design sensitivity of performance measures with respect to those design variables defined on the CAD tool. Using the same CAD-based design parameters in manufacturing tools lays the foundation for concurrent engineering. And once models are based on the same CAD tool, an integrated CAD-FEA-DSA system can be used to develop a design tool for a concurrent engineering method, such that design and manufacturing engineers can perform trade-off analysis in the early stages of the design process. A connection can also be made to multibody dynamic simulation, computational fluid dynamics, and other CAE tools, if these tools use the same CAD modeler as explained in Figs. 1.1 and 1.2.

1.6 Second-Order Design Sensitivity Analysis

First-order design sensitivity analysis, which was introduced in the previous section, is a linear approximation of the performance measure in terms of the design. However, if a higher-order approximation is used, then the accuracy of the approximation obviously increases. Second-order design sensitivity analysis uses a quadratic formula in order to approximate the performance change. Let us consider a Taylor series expansion of $\psi(u)$ up to the quadratic terms, as

$$\psi(u + \Delta u) \approx \psi(u) + \left(\frac{d\psi}{du}\right)^T \Delta u + \frac{1}{2}\Delta u^T \frac{d^2\psi}{du^2}\Delta u. \tag{1.51}$$

This approximation is exact if ψ is a quadratic function of u. For general nonlinear performance, the quadratic approximation in (1.51) is much more accurate than linear approximation. In (1.51) the term $d^2\psi/du^2$ is called the *second-order design sensitivity* of ψ. For a general n dimensional design vector $\boldsymbol{u}$, the second-order design sensitivity becomes a $n \times n$ symmetric Hessian matrix, which involves $n(n+1)/2$ calculations.

Example 1.5. Three-bar Truss. Consider the three-bar truss example given in the previous section. For second-order design sensitivity, the design derivative is taken from the first-order sensitivity equation in (1.24), to obtain

$$\boldsymbol{K}\frac{d^2z}{du^2} = \frac{d^2\boldsymbol{f}}{du^2} - \frac{d^2\boldsymbol{K}}{du^2}z - 2\frac{d\boldsymbol{K}}{du}\frac{dz}{du}. \tag{1.52}$$

The derivatives of stiffness matrix with respect to design in (1.31) are constants. Thus, the second-order derivative of stiffness matrix $d^2\boldsymbol{K}/du^2$ vanishes, along with $d^2\boldsymbol{f}/du^2$, and (1.52) is simplified to

$$\boldsymbol{K}\frac{d^2z}{du^2} = -2\frac{d\boldsymbol{K}}{du}\frac{dz}{du}. \tag{1.53}$$

By using (1.31) and (1.33), (1.53) can be solved for d^2z/du^2, yielding

$$\frac{d^2z}{db_2^2} = \begin{bmatrix} 4(2-\sqrt{3}) \\ 0 \\ 2(1-\sqrt{3}) \end{bmatrix}, \quad \frac{d^2z}{db_3^2} = \begin{bmatrix} 4\sqrt{2} \\ 0 \\ 0 \end{bmatrix}, \tag{1.54}$$

which can be verified by differentiating the explicit expression of displacement in (1.21) twice with respect to the design. All other terms are zero.

Second-order sensitivity information is very useful for an optimization algorithm, since quadratic convergence can be achieved near the solution points if the Hessian information is available. However, computing second-order sensitivity results in quite large computational costs. Thus, the design engineer has to decide between computational cost and the optimization algorithm's efficiency.

1.7 Design Optimization

The purpose of many structural design problems is to find the best design among many possible candidates. For this reason, the design engineer has to specify the best possible design as well as the best possible candidates. As discussed in Section 1.2, a possible candidate must exist within a feasible design region to satisfy problem constraints. Every design in the feasible region is an acceptable design, even if it is not the best one. The best design is usually the one that minimizes (or maximizes) the cost function of the design problem. Thus, the goal of the design optimization problem is to find a design that minimizes the cost function among all feasible designs. Although design sensitivity analysis is the main focus of this text, because many optimized designs will be presented, design optimization algorithms are briefly introduced in this section. However, this brief discussion is by no means a complete treatment of optimization methods. For a more detailed treatment, refer to [10] through [12].

Most gradient-based optimization algorithms are based on the mathematical programming method, which requires the function values and sensitivity information at given design variables. For a given design variable that defines the structural model, structural analysis provides the values of the cost and constraint functions for the algorithm. Design sensitivities of the cost and constraint functions must also be supplied to the optimization algorithm. Then, the optimization algorithms, discussed in this section, calculate the best possible design of the problem. Each algorithm has its own advantages and disadvantages. The performance of an optimization algorithm critically depends on the characteristics of the design problem and the types of cost and constraint functions.

1.7.1 Linear Programming Method

The linear programming method can be used when cost and constraints are linear functions of the design variables [13]. Most structural design problems, however, are nonlinear with respect to their design variables. Thus, the linear programming method is not of much use for structural problems. However, a nonlinear problem can be solved by approximating a sequence of linear problems, which will be discussed in Section 1.7.3. The standard form of a linear programming problem is

$$\begin{aligned} \text{minimize} \quad & f = \boldsymbol{c}^T \boldsymbol{u} \\ \text{subject to} \quad & \boldsymbol{A}\boldsymbol{u} = \boldsymbol{b} \\ & \boldsymbol{u} \geq \boldsymbol{0}, \end{aligned} \tag{1.55}$$

where $\boldsymbol{c} = [c_1, c_2, \ldots, c_n]^T$ is the coefficient of the cost function, $\boldsymbol{A}$ is the $m \times n$ matrix, and $\boldsymbol{b}$ is the $m \times 1$ vector. Inequality constraints can be treated as equality constraints by introducing slack variables. Since all functions are linear, the feasible regions defined by linear equalities are convex, along with the cost function. Thus, if any optimum solution of (1.55) exists, then it is a global minimum solution of the problem. The reason for introducing the linear problem here is that a very efficient method exists for solving linear programming problems, namely *the simplex method.* A positive feature of a linear programming problem is that the solution always lies on the boundary of the feasible region. Thus, the simplex method finds a solution by moving each corner point of the convex boundary.

1.7.2 Unconstrained Optimization Problems

When cost and/or constraints are nonlinear functions of the design, the design problem is called a *nonlinear programming method*, as contrasted to the linear programming method discussed in the previous section. Most engineering problems fall into the former category. Because the properties of nonlinear programming are nonlinear, this method is frequently solved using the numerical, rather than the analytical, method.

When there are no constraints on the design problem, it is referred to as an *unconstrained optimization problem.* Even if most engineering problems have constraints, these problems can be transformed into unconstrained ones by using the penalty method, or the Lagrange multiplier method. The unconstrained optimization problem sometimes contains the lower and upper limits of a design variable, since this type of constraint can be treated in simple way. The standard form of an unconstrained optimization problem can be written as

$$\begin{aligned} &\text{minimize} \quad f(\boldsymbol{u}), \\ &\text{subject to} \quad u_k^L \le u_k \le u_k^U, \qquad k = 1, \ldots, n. \end{aligned} \tag{1.56}$$

In the following subsections, numerical methods for solving (1.56) are discussed.

Steepest Descent Method

The numerical procedure for solving (1.56) is an iterative update of design $\boldsymbol{u}$. If $\boldsymbol{u}^k$ is the value of the design at the kth iteration, then the new design at the $(k+1)$th iteration can be obtained by

$$\boldsymbol{u}^{k+1} = \boldsymbol{u}^k + \alpha \boldsymbol{d}^{k+1}, \tag{1.57}$$

where $\boldsymbol{d}^{k+1}$ is called the descent direction and α is a step size, used to determine the amount of movement in the direction of $\boldsymbol{d}^{k+1}$. If the descent direction is given, then parameter α is determined by using the line search procedure to find the minimum value of a cost function in the descent direction. The steepest descent method uses the gradient of the cost function as the descent direction, such that

$$\boldsymbol{d}^{k+1} = -\frac{\partial f(\boldsymbol{u}^k)}{\partial \boldsymbol{u}^k} = -\nabla f(\boldsymbol{u}^k), \tag{1.58}$$

which is the design sensitivity of the cost function. This method suffers from a slow convergence near the optimum design, since it does not use any information from the previous design, and only first-order information of the function is used. Note that $\boldsymbol{d}^k$ and $\boldsymbol{d}^{k+1}$ are always orthogonal, such that a zigzagging pattern appears in the optimization process.

Conjugate Gradient Method

The conjugate gradient method developed by Fletcher and Reeves [14] improves the rate of slow convergence in the steepest descent method by using gradient information from the previous iteration. The difference in this method is the computation of $\boldsymbol{d}^{k+1}$ in (1.57). The new descent direction is computed by

$$\boldsymbol{d}^{k+1} = -\nabla f(\boldsymbol{u}^k) + \beta_k^2 \boldsymbol{d}^{k-1}, \tag{1.59}$$

where

$$\beta_k = \frac{\left\|\nabla f(\boldsymbol{u}^k)\right\|}{\left\|\nabla f(\boldsymbol{u}^{k-1})\right\|}, \tag{1.60}$$

and where the first iteration is the same as (1.57). This method tends to select the descent direction as a diagonal of two orthogonal steepest descent directions, such that a zigzagging pattern can be eliminated. This method always has better convergence than the steepest descent method.

Newton Method

The previous methods we have examined use first-order information (first-order design sensitivity) of the cost function to find the optimum design, which is called linear approximation. The Newton method uses second-order information (second-order design sensitivity) to approximate the cost function as a quadratic function of the design. The major concern is how to compute the second-order design sensitivity (or Hessian) matrix. Let us define the Hessian matrix as second-order design sensitivity, defined in (1.51) as

$$\boldsymbol{H}(\boldsymbol{u}^k) \equiv \left[\frac{\partial f(\boldsymbol{u}^k)}{\partial u_i^k \partial u_j^k}\right], \quad i, j = 1, \ldots, n. \tag{1.61}$$

The new design can then be determined, as

$$\boldsymbol{u}^{k+1} = \boldsymbol{u}^k + \Delta \boldsymbol{u}^{k+1}, \tag{1.62}$$

where

$$\Delta \boldsymbol{u}^{k+1} = -\boldsymbol{H}^{-1}(\boldsymbol{u}^k) \nabla f(\boldsymbol{u}^k). \tag{1.63}$$

If the current estimated design $\boldsymbol{u}^k$ is sufficiently close to the optimum design, then Newton's method will show a quadratic convergence. However, the greater the number of design variables, the greater the cost of computing $\boldsymbol{H}(\boldsymbol{u}^k)$ in (1.61). In addition, Newton's method does not guarantee a convergence. Thus, several modifications are available. For example, the design update algorithm in (1.62) can be modified to include a step size by using a line search, as in (1.57).

Quasi-Newton Method

Although Newton's method has a quadratic convergence, the cost of computing the Hessian matrix and the lack of a guaranteed convergence, are drawbacks to this method. The quasi-Newton method has an advantage over the steepest descent method and Newton's method: it only requires first-order sensitivity information, and it approximates the Hessian matrix to speed up the convergence.

The DFP (Davidon-Fletcher-Powell [14]) method approximates the inverse of the Hessian matrix using first-order sensitivity information. By initially assuming that the inverse of the Hessian is the identity matrix, this method updates the inverse of the Hessian matrix during design iteration. A nice feature of this method is that the positive definiteness of the Hessian matrix is preserved.

The BFGS (Broydon-Fletcher-Goldfarb-Shanno [15]) method updates the Hessian matrix directly, rather than updating its inverse as with the DFP method. Starting from the identity matrix, the Hessian matrix remains positive definite if an exact line search is used.

1.7.3 Constrained Optimization Problems

Most engineering problems have constraints that must be satisfied during the design optimization process. These two types of constraints are handled separately: equality and inequality constraints. The standard form of the design optimization problem in constrained optimization can be written as

$$\begin{aligned} &\text{minimize} \quad f(\boldsymbol{u}) \\ &\text{subject to} \quad h_i(\boldsymbol{u}) = 0, \qquad i = 1, \dots, p \\ &\qquad\qquad g_j(\boldsymbol{u}) \le 0, \qquad j = 1, \dots, m \\ &\qquad\qquad u_l^L \le u_l \le u_l^U, \qquad l = 1, \dots, n. \end{aligned} \tag{1.64}$$

The computational method to find a solution to (1.64) has two phases: first, to find a direction $\boldsymbol{d}$ that can reduce the cost $f(\boldsymbol{u})$, while correcting for any constraint violations that are violated; and second, to determine the step size of movement α in the direction of $\boldsymbol{d}$.

Sequential Linear Programming (SLP)

The SLP method approximates the nonlinear problem as a sequence of linear programming problems such that the simplex method in Section 1.7.1 may be used to find the solution to each iteration. By using function values and sensitivity information, the nonlinear problem in (1.64) is linearized in a similar way as Taylor's expansion method in the first order, as

$$\begin{aligned} &\text{minimize} \quad f(\boldsymbol{u}^k) + \nabla f^T \Delta \boldsymbol{u}^k \\ &\text{subject to} \quad h_i(\boldsymbol{u}^k) + \nabla h_i^T \Delta \boldsymbol{u}^k = 0, \qquad i = 1, \dots, p \\ &\qquad\qquad g_i(\boldsymbol{u}^k) + \nabla g_i^T \Delta \boldsymbol{u}^k \le 0, \qquad j = 1, \dots, m \\ &\qquad\qquad u_l^L \le u_l \le u_l^U, \qquad l = 1, \dots, n. \end{aligned} \tag{1.65}$$

Since all functions and their sensitivities at $\boldsymbol{u}^k$ are known, the linear programming problem in (1.65) can be solved using the simplex method for $\Delta\boldsymbol{u}^k$. Even if the sensitivity information is not used to solve a linear programming problem, design sensitivity information is required in order to approximate the nonlinear problem as a linear one with SLP. In solving (1.65) for $\Delta\boldsymbol{u}^{\mathrm{k}}$, the move limit $\Delta \boldsymbol{u}_L^k \le \Delta \boldsymbol{u}^k \le \Delta \boldsymbol{u}_U^k$ is critically important for convergence.

Sequential Quadratic Programming (SQP)

Compared with previous methods, which use first-order sensitivity information to determine the search direction $\boldsymbol{d}$, SQP solves a quadratic subproblem to find that search direction, which has both quadratic cost and linear constraints:

$$\begin{aligned} &\text{minimize} \quad f(\boldsymbol{u}^k) + \nabla f^T \boldsymbol{d} + \frac{1}{2}\boldsymbol{d}^T \boldsymbol{H} \boldsymbol{d} \\ &\text{subject to} \quad h_i(\boldsymbol{u}^k) + \nabla h_i^T \boldsymbol{d} = 0, \qquad i = 1, \dots, p \\ &\qquad\qquad g_j(\boldsymbol{u}^k) + \nabla g_j^T \boldsymbol{d} \le 0, \qquad j = 1, \dots, m. \end{aligned} \tag{1.66}$$

This special form of the quadratic problem can be effectively solved, for example, by using the Kuhn-Tucker condition and the simplex method. Starting from the identity matrix, the Hessian matrix $\boldsymbol{H}$ is updated at each iteration by using the aforementioned methods in unconstrained optimization algorithms. The advantage of solving (1.66) in this way is that for positive definite $\boldsymbol{H}$ the problem is convex and the solution is unique. Moreover, this method does not require the move limit as in SLP.

Constrained Steepest Descent Method

In the unconstrained optimization process detailed in Section 1.7.2, the descent direction $\boldsymbol{d}$ is obtained from the cost function sensitivity. When constraints exist, this descent direction has to be modified in order to include their effect. If constraints are violated, then these constraints are added to the cost function using a penalty method. Design sensitivity of the penalized cost function combines the effects of the original cost function and the violated constraint functions.

Constrained Quasi-Newton Method

If the linear approximation of constraints in SQP is substituted for a quadratic approximation, then the convergence rate of (1.66) will be improved. However, solving the optimization problem for quadratic cost and constraints is not an easy process. The constrained quasi-Newton method combines the Hessian information of constraints with the cost function by using the Lagrange multiplier method. Nevertheless, it is still necessary to compute the constraint function Hessian. The main purpose of the constrained quasi-Newton method is to approximate the Hessian matrix by using first-order sensitivity information. The extended cost function is

$$L(\boldsymbol{u},\boldsymbol{v},\boldsymbol{w}) = f(\boldsymbol{u}) + \sum_{i=1}^{p} v_i h_i(\boldsymbol{u}) + \sum_{j=1}^{m} w_i g_i(\boldsymbol{u}), \tag{1.67}$$

where both $\boldsymbol{v} = [v_1, v_2, \ldots, v_p]^T$ and $\boldsymbol{w} = [w_1, w_2, \ldots, w_m]^T$ are the Lagrange multipliers for equality and inequality constraints, respectively. Note that $\boldsymbol{w} > \boldsymbol{0}$. Let the second-order design sensitivity of L be $\nabla^2 L$. The extended quadratic programming problem of (1.66) thus becomes

$$\begin{aligned} &\text{minimize} \quad f(\boldsymbol{u}^k) + \nabla f^T \boldsymbol{d} + \frac{1}{2}\boldsymbol{d}^T \nabla^2 L \boldsymbol{d} \\ &\text{subject to} \ h_i(\boldsymbol{u}^k) + \nabla h_i^T \boldsymbol{d} = 0, \qquad i = 1, \ldots, p \\ &\qquad\qquad g_j(\boldsymbol{u}^k) + \nabla g_j^T \boldsymbol{d} \le 0, \qquad j = 1, \ldots, m \\ &\qquad\qquad w_l \ge 0, \qquad\qquad\qquad\qquad l = 1, \ldots, m. \end{aligned} \tag{1.68}$$

Feasible Direction Method

The feasible direction method is designed to allow design movement within the feasible region in each iteration. Based on the previous design, the updated design reduces the cost function and remains in the feasible region. Since all designs are feasible, a design at any iteration can be used, even if it is not an optimum design. Since this method uses the linearization of functions as in SLP, it is difficult to maintain nonlinear equality constraints. Thus, this approach is used exclusively for inequality constraints. Search direction $\boldsymbol{d}$ can be found by solving the following linear subproblem:

$$
\begin{aligned}
&\text{minimize} \quad \beta \\
&\text{subject to} \quad \nabla f^T \boldsymbol{d} \le \beta \\
&\qquad \nabla g_i^T \boldsymbol{d} \le \beta, \quad i = 1, \ldots, m_{active} \\
&\qquad -1 \le d_j \le 1, \quad j = 1, \ldots, n,
\end{aligned}
\tag{1.69}
$$

where m_{active} is the number of active inequality constraints. After finding a direction $\boldsymbol{d}$ that can reduce cost function and maintain feasibility, a line search is used to determine step size α.

Gradient Projection Method

The feasible direction method solves the linear programming problem to find the direction of the design change. The gradient projection method, however, uses a simpler method for computing this direction. The direction obtained by the steepest descent method is projected on the constraint boundary, such that the new design can move along the constraint boundary. Thus, the direction of the design change reduces the cost function while maintaining the constraint along its boundary. For a general nonlinear constraint, however, a small movement along the tangent line of the boundary will violate this constraint. Thus, in actual implementation, a correction algorithm has to be followed. The gradient projection method behaves well when the constraint boundary is moderately nonlinear.

2
Variational Methods of Structural Systems

In design sensitivity analysis, design derivatives of the governing equation are taken to obtain sensitivities of structural responses with respect to the design variables. Thus, it is very important to understand the characteristics of governing equations. In this and subsequent chapters, an introduction to linear structural problems is presented by using the variational formulation. The purpose of this chapter is not a rigorous development of the structural analysis method, but rather a brief introduction to structural analysis from a design sensitivity analysis viewpoint. Static, eigenvalue, thermal, and dynamic analyses equations are derived from the energy principle. Energy bilinear and load linear forms are introduced in the continuum model. Specific expressions of those forms that correspond to each structural component will be introduced in Chapter 3.

In order to take advantage of the variational method in design sensitivity analysis, it is essential to work with *the space of kinematically admissible displacement fields*. Readers who are primarily interested in applications can restrict their attention to smooth function spaces without being concerned with more general function spaces. Nevertheless, rigorous derivations are introduced in this chapter in order to extend smooth function space to a more general function space. Examples that are treated later in the text are first presented and analyzed in this chapter.

2.1 Introduction

Classically, a structural problem is formulated using a differential equation that is satisfied at every point in the domain. Force equilibrium is imposed on an arbitrary infinitesimal element of the structure in order to obtain the boundary-value problem. The smoothness of the solution in the boundary-value problem depends on the order of the differential equation. For example, truss and continuum problems require continuous second-order derivatives of the solution, while beam and plate bending problems require continuous fourth-order derivatives. However, this chapter will show that these orders of differentiability are not necessary in order to represent many types of mechanical behaviors. In contrast, the variational approach reduces the solution's smoothness requirements, and provides a general interpretation of the solution. Even though the classical differential equation may fail to yield a solution, the variational problem will provide a generalized solution, which is in fact the *natural solution* to the structural problem. In addition, for the purpose of design sensitivity analysis, a variational formulation is more natural than a differential equation in representing structural deformation. Furthermore, a variational formulation that has been mathematically obtained can be rigorously related to a virtual work or energy principle in mechanics.

A complete mathematical theory related to the existence and uniqueness of the solution, see [16] through [18], was developed using the Sobolev space and the properties of a bounded, elliptic, linear operator. However, mathematical comprehension of this functional analysis requires a good deal of effort, with some physical insights. In contrast, a relatively simple theory is available that can formulate the structural problem using the energy principle. If the structural system is conservative, then it has a potential energy. Structural equilibrium is considered to be a stationary configuration of the total potential energy (principle of minimum total potential energy). Since the potential energy of many structural problems is the positive definite quadratic function of a state variable (that is, displacement), the stationary condition yields a unique global minimum solution. In Section 2.2, a variational method is developed from the differential operator equation for a conservative structural system. An important result is then shown, namely, that if the solution for a structural differential equation exists, then that solution is the minimizing solution of the total potential energy. In addition, even if the structural differential problem does not have a solution, the solution that minimizes the total potential energy may exist and would provide a natural solution to the structural problem. The energy principles presented here will be restricted to small strains and displacements so that strain-displacement relationships can be expressed in linear equations; such displacements and corresponding strains obviously have additive properties. A nonlinear elastic stress-strain relationship will be discussed in Part III of this text.

The energy-based formulation of the structural problem in Section 2.2 is generalized to the principle of virtual work in Section 2.3, which can handle arbitrary constitutive relations. The principle of virtual work is the equilibrium of the work done by both internal and external forces with the small, arbitrary virtual displacements that satisfy kinematic constraints. For a conservative system, the same results are obtained as with the principle of minimum total potential energy in Section 2.2. The unified approach to various structural problems is made possible by introducing energy bilinear and load linear forms. As long as they share the same properties, all structural problems in this text can be treated in the same manner, even those with different differential operators. The existence and uniqueness of a solution can be shown through rigorous mathematical proofs. The concept of Sobolev space and the bounded property of an energy bilinear form are required in the proof. In this text, however, such rigorous mathematical proofs are avoided and corresponding references are instead cited.

The variational formulation can be generalized for problems in which the time-dependent force is applied to the structure. The inertia effect is considered as a kinetic energy. Structural equilibrium is the stationary condition of the difference between the total potential energy and the kinetic energy. The time-integrated form of the variational equation is obtained as Hamilton's principle in Section 2.4. However, if the inertia property is considered as a body force acting in the negative direction of acceleration, and the principle of virtual work is used in the structural domain, then a second-order ordinary differential equation with respect to time is obtained, which corresponds to the instantaneous expression of Hamilton's principle. This formulation is used frequently for computational purposes by integrating the time domain as an initial-boundary-value problem.

The eigenvalue problem represents the natural vibration and column buckling of the structure. Although the characteristics of these two types of problems are quite different, the mathematical representations of them are similar. The variational formulation of the eigenvalue problem is derived in Section 2.5 from the homogeneous dynamic equation. The solution space of the eigenvalue problem is the same as the static problem, since the same differential operator is involved. Since the set of eigenfunctions is complete in the solution space, it is possible to represent the solution of the static problem as a linear combination of eigenfunctions.

The dynamic frequency response of a mechanical or structural system is of interest in design problems that are subjected to harmonically varying external loads caused by a reciprocating power train or other rotating machinery, such as a motor, fan, compressor, or forging hammer [19]. For example, airplane body and wing structures are subjected to a harmonic load transmitted from the propulsion system. Similarly, ship vibration resulting from propeller and engine excitation can cause noise problems, cracks, fatigue failure of the tailshaft, and discomfort for the crew. When a machine, or any structure, oscillates in some form of periodic or random motion, alternating pressure waves are generated that propagate from the moving surface at the speed of sound. For example, the interior sound pressure oscillation in an automobile compartment can occur when the input forces transmitted from the road and the power train excite the boundary panels of the vehicle compartment. Such motion with frequencies between 20 Hz and 20 kHz stimulates the hearing mechanism of a human [20]. In Section 2.6.1, a variational formulation of the frequency response problem using complex variables is developed in continuum elasticity. The frequency response formulation is coupled with the acoustic problem in Section 2.6.2 to solve the structure-induced noise problem where the boundary structure vibration stimulates the interior air of the vehicle and causes interior noise.

The thermoelastic problem is particularly important in analysis of the automotive engine part and the turbine blade where temperature-induced deformation is significant. When the structural problem undergoes a temperature change, the thermal effect changes the state response. The material property and the constitutive relation also depend on temperature change. In addition to the displacement, temperature is added to the state response variable. Variational formulations of the steady-state heat conduction equation and the elasticity equilibrium equation with thermal load are developed in Section 2.7. Although the effects of temperature and displacement are coupled, by treating thermal effect as an external load to the structure, decoupled variational equations can be relatively easy to obtain, which can then be solved sequentially.

2.2 Energy Method

Mathematical models of many structural problems are formulated as differential equations that are satisfied at every point in the domain. These differential equations are usually obtained from the three fundamental laws of mechanics: conservation of mass, conservation of linear momentum, and conservation of angular momentum. Conservation of mass can be easily satisfied for a Lagrangian description of the problem, and the conservation of an angular momentum concludes the symmetry of the stress tensor. Thus, the conservation of linear momentum, which is a differential equation used to satisfy the force equilibrium, is the major consideration in the structural problem.

Minimum Principles for Operator Equations

The differential equation of the structural problem can be represented as a differential operator equation. If the linear problem is considered, then the differential operator is also linear. Let A be a positive definite linear operator that represents the structural problem under consideration. For a continuum problem, A is a second-order differential operator, while for beam and plate problems, A is a fourth-order operator. The structural problem is to find the solution $z \in D_A$ that also satisfies

$$Az = f, \tag{2.1}$$

where f is the applied force and D_A is the solution space, which has a component that satisfies the required smoothness of the solution and one that satisfies the boundary conditions of the problem. We want to emphasize here that the operator A is closely tied to the domain D_A. That is, even if another operator B has exactly the same differential formula in its definition as the operator A, if the domain D_B is different from D_A, then B is a different operator. The smoothness of the solution is determined by the order of operator A. For example, C^2 functions are required for the second-order differential operator and C^4 functions for the fourth-order operator. (Note: $C^m(\Omega)$ is the set of functions whose values and first m derivatives are continuous on Ω. For an introductory treatment of such function spaces, refer to Appendix A.2.) D_A can be considered a collection of solution candidates.

The *positive definite* property of operator A means that A is symmetric and, for all $z \neq 0$ in D_A, the following condition holds:

$$(Az, z) \equiv \iint_\Omega z^T Az \, d\Omega > 0, \tag{2.2}$$

where $(\bullet,\bullet)$ is the $L_2(\Omega)$-scalar product of functions, and Ω is the structural domain under consideration. (Note: $L_2(\Omega)$ is the space of square integrable functions such that $L_2(\Omega) = \{f : \iint_\Omega |f(x)|^2 \, d\Omega < \infty\}$.) In fact, (2.2) satisfies the requirement of the scalar product and the norm. Thus, $(Az,z)^{1/2} \equiv \|z\|_A$ is called the energy norm, and $(Az,\boldsymbol{w}) \equiv [z,\boldsymbol{w}]_A$ is called the energy scalar product. Also, A is said to be a *positive bounded below linear operator* if a constant $c > 0$ exists, such that

$$(Az, z) \geq c(z, z) = c\|z\|^2, \tag{2.3}$$

for all $z \neq 0$ in D_A. In many mechanics problems, z is the deflection and Az is the force, so (Az,z) is proportional to the energy that is required to produce the deflection z. The property $(Az,z) > 0$, for all admissible $z \neq 0$, states that a positive amount of energy is required to produce a nonzero deflection. The property $(Az,z) \geq c\|z\|^2$, for $c > 0$, states that a lower bound exists for the amount of energy that must be expended in order to achieve a nonzero displacement, which implies system stability. In other words, large deflections can be produced only by large expenditures of energy. If A is positive definite, but not positive bounded below, then a large deflection can be produced with a very small expenditure of energy. Such behavior appears in unstable structures.

Example 2.1. String Problem. To further illustrate the aforementioned discussion, consider a string of length l with a distributed load of $f(x)$, as in Fig. 2.1. In this case, displacement and distributed load are scalar functions. The governing differential equation is

$$Az \equiv -\frac{\partial^2 z}{\partial x^2} = f(x), \qquad x \in (0,l),$$
$$z(0) = 0,$$
$$z(l) = 0,$$

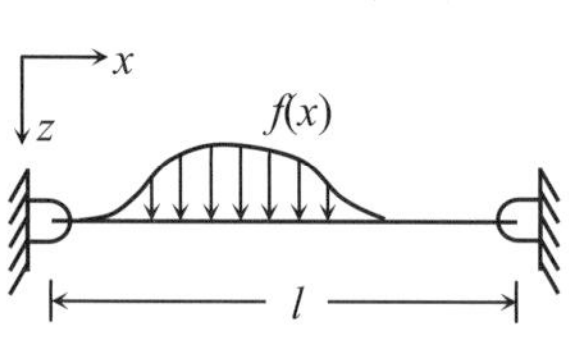

Figure 2.1. String under distributed load.

where A is the second-order differential operator. The state variable z is scalar for this problem. The solution space of this problem is

$$D_A = \{ z \in C^2(0,l) \mid z(0) = z(l) = 0 \}.$$

It is easy to show that A is linear, since $A(\alpha z + \beta w) = \alpha Az + \beta Aw$ for any z and w in D_A and for any scalar α and β. Also, from an integration by parts it can be shown that the linear operator A is symmetric, for any z and w in D_A, as

$$\begin{aligned}(Az,w) &= \int_0^l \left(-\frac{d^2 z}{dx^2}\right) w\,dx \\ &= \int_0^l \frac{dz}{dx}\frac{dw}{dx}dx - \left[\frac{dz}{dx}w\right]_0^l \\ &= \int_0^l \left(-\frac{d^2 w}{dx^2}\right) z\,dx + \left[\frac{dw}{dx}z\right]_0^l \\ &= (z, Aw),\end{aligned}$$

where the boundary terms vanish from the homogeneous boundary conditions. Note that the symmetric property of operator A is strongly dependent on the boundary conditions, since the operator A is closely tied to the domain D_A. Thus, if different boundary conditions are given, A may not be symmetric. In fact, it should be a different operator. The positive definiteness of operator A can be shown easily from

$$(Az,z) = \int_0^l \left(\frac{dz}{dx}\right)^2 dx \geq 0. \tag{2.4}$$

Also, $(Az,z) = 0$ implies that $dz/dx = 0$, or that $z =$ constant. Then, due to the homogeneous boundary conditions, $z(x) = 0$. Thus, A is positive definite for the given solution space. A more detailed discussion shows that A is positive bounded below [21].

The differential equation in (2.1) requires that the solution be in D_A, where the second-order derivatives are continuous and boundary conditions are satisfied. These requirements are too restrictive to represent solutions for many types of applied loads. For example, when the applied load is a point load in the string problem, then there is no solution that satisfies (2.1). However, physically a solution exists for the point load and the solution is a C^0-continuous function. Thus, the notion of a solution space has to be extended to represent a variety of loading conditions. The energy approach presented below provides a more general method than that of operator theory to represent a solution.

Let A be a positive definite linear operator with domain D_A. Then, consider the following energy functional, defined as

$$\Pi(\mathbf{z}) = (A\mathbf{z},\mathbf{z}) - 2(\mathbf{z},\mathbf{f}) = \iint_\Omega (\mathbf{z}^T A\mathbf{z} - 2\mathbf{z}^T \mathbf{f})\, d\Omega. \tag{2.5}$$

It will be shown that the solution to (2.1) is unique and minimizes $\Pi(\mathbf{z})$ for all $\mathbf{z}$ in D_A, so that the operator equation is converted into an optimization problem. Let us presume that (2.1) has a solution $\mathbf{z}$. In order to show that this solution is unique, assume that there are two solutions in D_A, $\mathbf{z}$ and $\mathbf{w}$. Then, $\mathbf{z} - \mathbf{w}$ satisfies $A(\mathbf{z} - \mathbf{w}) = 0$. Thus,

$$(A(\mathbf{z}-\mathbf{w}), (\mathbf{z}-\mathbf{w})) = 0.$$

Since A is positive definite, $\mathbf{z} - \mathbf{w} = 0$. Thus, if a solution to (2.1) exists, it is unique. Note that the uniqueness of the operator equation comes from the positive definiteness of the operator A. This property is similar to the matrix theory in so far as for the positive definite matrix $\mathbf{A}_{[n\times n]}$, the linear system of equations $\mathbf{Az} = \mathbf{f}$ has a unique solution $\mathbf{z}$.

Next, let z_0 in D_A be the solution to (2.1), so that $Az_0 = f$. Then, z_0 minimizes the energy functional in (2.5). To show this, making this substitution for f into (2.5) yields

$$\begin{aligned}\Pi(z) &= (Az, z) - 2(Az_0, z) \\ &= [z, z]_A - 2[z_0, z]_A \\ &= [z - z_0, z - z_0]_A - [z_0, z_0]_A \\ &= \| z - z_0 \|_A^2 - \| z_0 \|_A^2 .\end{aligned}$$

It is clear that $\Pi(z)$ has its minimum value if and only if $z = z_0$. Thus, the solution to (2.1) minimizes the energy functional $\Pi(z)$.

Now, let us show that the inverse of the previous statement is also true. Suppose that there exists a function z_0 in D_A that minimizes the functional $\Pi(z)$ of (2.5). Let $\mathbf{v}(\mathbf{x})$ be an arbitrary function from D_A and let α be a real number. Then, $\Pi(z_0 + \alpha\mathbf{v}) - \Pi(z_0) \geq 0$. Using the symmetry of the operator A, the function

$$\Pi(z_0 + \alpha \mathbf{v}) - \Pi(z_0) = 2\alpha(Az_0 - f, \mathbf{v}) + \alpha^2 (A\mathbf{v}, \mathbf{v}) \geq 0$$

of α takes on its minimum value of zero at $\alpha = 0$. Thus, its derivative, with respect to α at $\alpha = 0$, must be zero from the stationary condition of the optimization theory, that is,

$$2(Az_0 - f, \mathbf{v}) = 0,$$

for all $\mathbf{v}$ in D_A. Since $\mathbf{v}$ is an arbitrary function in D_A, it follows that $Az_0 - f = 0$. Thus, z_0 is the solution of the operator (2.1). Thus, the solution that minimizes $\Pi(z)$, if it is in D_A, is a unique solution of the operator (2.1). The method to minimize the energy functional in (2.5) is called the *energy approach*.

However, we have to be careful of the situation in which a z_0 that minimizes $\Pi(z)$ may not belong to D_A. For the string problem in Example 2.1 with unit point load $f = 1$ at $x = \xi$, the solution to the problem is

$$z_0(x) = \begin{cases} (\xi - 1)x, & 0 \leq x \leq \xi \\ \xi x - \xi, & \xi \leq x \leq 1, \end{cases}$$

which minimizes the functional $\Pi(z)$ in (2.5), and clearly $z_0 \in C^0$ does not belong to D_A. More specifically, the solution space D_A of the string problem needs to be extended to include all continuous functions with square integrable first derivatives, so that the energy norm in (2.4) is well defined. On this extended solution space, the operator A cannot be properly defined, and only the solution that minimizes the energy functional in (2.5) exists. Consequently, the energy approach provides a broader range of solutions than the operator approach. For that reason, the solution to (2.1) is called the *classical solution* (or *strong solution*) and the solution that minimizes (2.5) is called the *generalized solution* (or *weak solution*). However, we want to emphasize that if a solution exists to (2.1), then it is always the minimizing solution to (2.5).

As will become evident in later examples, the functional $\Pi(z)$ is proportional to the total potential energy of the system under consideration. The previous discussion in this chapter provides rigorous proof of the principle of minimum total potential energy. Under specific boundary conditions, this principle allows the problem of integrating a differential equation to be replaced with the problem of seeking a function that minimizes the functional of (2.5). Methods of solving continuum mechanics problems that involve the minimization of the functional of (2.5) are referred to as *energy methods*, or *variational methods*. This allows numerical approaches such as the finite element method to be developed.

Principle of Minimum Total Potential Energy

Consider the linear elastic structure in Fig. 2.2 under the applied surface-traction load $\boldsymbol{f}^s$ on the boundary Γ^s, and under the body force $\boldsymbol{f}^b$ in the domain. The whole boundary Γ is decomposed into $\Gamma = \Gamma^h \cup \Gamma^s$ and $\Gamma^h \cap \Gamma^s = \varnothing$. The motion of the structure is fixed (or prescribed) on the essential boundary Γ^h. Due to the applied load, the elastic structure experiences deformation (or displacement) $\boldsymbol{z}(\boldsymbol{x}) = [z_1, z_2, z_3]^T$ for $\boldsymbol{x} \in \Omega$, although the nominal configuration of the structure resists any deformation by generating internal forces. It is assumed that this internal force is proportional to the amount of deformation. For a given applied load, if the internal force is smaller than the applied force, then the structure continues to deform in order to equilibrate the two forces. Many structural problems consist of computing the displacement due to force equilibrium conditions between the applied load and internal forces.

If the concept of structural force equilibrium is extended to energy formulation, then a good deal of physical insight can be obtained. Let the displacements be used as state variables of the problem considered. The internal force, generated during deformation, can be thought of as energy that is stored in the structure. As the structure deforms, not only does the internal force increase, but the energy of the structure also increases. This stored energy is called the *strain energy* of the structure, defined as

$$U(\boldsymbol{z}) \equiv \frac{1}{2} \iint_{\Omega} \sigma_{ij}(\boldsymbol{z}) \varepsilon_{ij}(\boldsymbol{z})\, d\Omega, \tag{2.6}$$

where i, j=1, …, N; N is the dimension of space (1, 2, or 3), and summation is used for the repeated indices i and j. In (2.6),

$$\begin{aligned} \varepsilon_{ij}(\boldsymbol{z}) &= \frac{1}{2}\left(\frac{\partial z_i}{\partial x_j} + \frac{\partial z_j}{\partial x_i}\right) \\ &= \frac{1}{2}(z_{i,j} + z_{j,i}) \end{aligned} \tag{2.7}$$

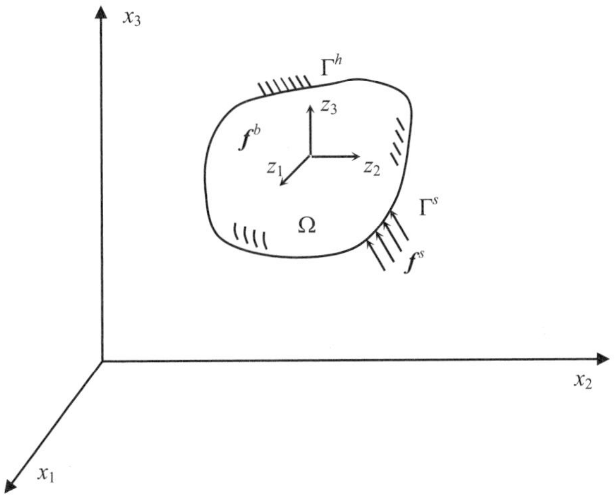

Figure 2.2. Structure under equilibrium.

and

$$\sigma_{ij}(z) = C_{ijkl}\varepsilon_{kl}(z) \tag{2.8}$$

are the strain and stress tensors, respectively. In (2.7), the subscribed comma represents the derivative with respect to the spatial coordinate; i.e., $z_{i,j} = \partial z_i/\partial x_j$. In (2.8), C_{ijkl} is a fourth-order constitutive tensor that defines the relation between strain and stress tensors. For linear problems, C_{ijkl} is constant. The strain energy $U(z)$ is the energy required to produce the displacement z. For elastic problems, since $U(z)$ does not depend on the path chosen for deformation, it is a function of the configuration z.

If force is applied to the structure and the structure deforms in the direction of the applied force, then work is done by the applied force. The work done by the applied load can be defined as

$$W(z) = \iint_{\Omega} z^T \boldsymbol{f}^b \, d\Omega + \int_{\Gamma^s} z^T \boldsymbol{f}^s \, d\Gamma. \tag{2.9}$$

The first integral in (2.9) represents the work done by the body force $\boldsymbol{f}^b$, while the second integral is the work done by the surface traction load $\boldsymbol{f}^s$. The integrals are evaluated over the whole domain Ω and over the traction boundary Γ^s. If any concentrated force $\boldsymbol{f}$ is applied externally, then (2.9) may include Dirac delta measure as in Example 1.3 in Chapter 1. Note that $U(z)$ is a quadratic function of z, while $W(z)$ is a linear function of z.

Since the strain energy $U(z)$ is independent of the deformation path, it is potential energy that is stored in the structure. If the applied force in (2.9) is conservative, then (2.9) defines the negative value of potential energy generated by the applied loads. The applied load is considered conservative if it is independent of deformation, so that the work done by a system of applied forces in traversing any closed path in displacement space has to be zero. The total potential energy of the structure is the difference between the strain energy and the work done by the applied loads, written as

$$\begin{aligned} \Pi(z) &= U(z) - W(z) \\ &= \frac{1}{2}\iint_{\Omega} \sigma_{ij}(z)\varepsilon_{ij}(z)\, d\Omega \\ &\quad - \iint_{\Omega} z^T \boldsymbol{f}^b \, d\Omega - \int_{\Gamma^s} z^T \boldsymbol{f}^s \, d\Gamma. \end{aligned} \tag{2.10}$$

Note that the definition of $\Pi(z)$ in (2.10) is actually half of the energy functional defined in (2.5). However, the absolute value of the total potential energy does not have any meaning here; only the relative quantity is important.

The principle of minimum total potential energy is as follows: for all displacements that satisfy the boundary conditions, known as kinematically admissible displacements, those which satisfy the operator equation of linear elasticity, if they exist,

$$\begin{aligned} Az \equiv -div\boldsymbol{\sigma}(z) &= \boldsymbol{f}^b, \quad & \boldsymbol{x} \in \Omega \\ z &= 0, \quad & \boldsymbol{x} \in \Gamma^h \\ \boldsymbol{\sigma n} &= \boldsymbol{f}^s, \quad & \boldsymbol{x} \in \Gamma^s, \end{aligned} \tag{2.11}$$

make the total potential energy in (2.10) stationary on

$$D_A = \left\{ z \in [C^2(\Omega)]^3 \mid z = 0 \text{ on } \boldsymbol{x} \in \Gamma^h, \ \boldsymbol{\sigma n} = \boldsymbol{f}^s \text{ on } \boldsymbol{x} \in \Gamma^s \right\}. \tag{2.12}$$

In (2.11), $div\boldsymbol{\sigma} = \sigma_{ij,j}$ is the divergence of the stress tensor and $\boldsymbol{n}$ is an outward unit normal vector to the surface Γ^s. Due to the conservation of angular momentum, the stress tensor $\boldsymbol{\sigma}$ is symmetric ($\sigma_{ij} = \sigma_{ji}$).

For the generalized solution that minimizes the total potential energy in (2.10), the solution space D_A of (2.12) has to be extended so that the potential energy in (2.10) can be well defined. This space is called the *space of the finite energy* or the *space of kinematically admissible displacements*, defined as

$$Z = \left\{ \boldsymbol{z} \in [H^1(\Omega)]^3 \mid \boldsymbol{z} = 0 \text{ on } \boldsymbol{x} \in \Gamma^h \right\}, \tag{2.13}$$

where $H^m(\Omega)$ is the Sobolev space of order m. (Note: $H^m(\Omega)$ is the Sobolev space of the order m, whose functions are continuously differentiable up to $m - 1$, and mth partial derivatives belong to $L_2(\Omega)$. For an introductory treatment of such function spaces, refer to Appendix A.2.) It is important to point out that the traction boundary condition $\boldsymbol{\sigma n} = \boldsymbol{f}^s$ is not required to define the space of kinematically admissible displacements. The condition $\boldsymbol{z} = 0$ is called the *essential boundary condition*, while $\boldsymbol{\sigma n} = \boldsymbol{f}^s$ is referred to as the *natural boundary condition*. The natural boundary condition is included in the work done by the applied load in (2.10). Thus, it is easier to construct the space of kinematically admissible displacements than it is to construct D_A. Generally, if we let the order of differential operator A equal $2m$, then the boundary conditions that contain $(m-1)$th order derivatives are called the essential boundary conditions and derivatives of a higher order than $(m-1)$ are called the natural boundary conditions.

The principle of minimum total potential energy provides a generalized solution to the differential equation. This generalized solution is in Z. In addition, the uniqueness of the solution is proved by using the positive definite property of operator A in (2.2) on D_A and the fact that $D_A(\Omega) \subset Z \subset L_2(\Omega)$ and $D_A(\Omega)$ is dense in $L_2(\Omega)$ [Appendix A.2]. (Note: A subset W of V is said to be dense in V, if for a given arbitrary small ε and any $f \in V$, there exists a $g \in W$ such that $\|f - g\| < \varepsilon$.) For stable structures, the differential operator is positive bounded below and thus, a unique solution can be expected.

2.3 Variational Formulation and the Principle of Virtual Work

The principle of minimum total potential energy discussed in the previous section requires obtaining a stationary condition for the total potential energy. This principle is closely related to the variational formulation presented in this section.

Let the state variable $\boldsymbol{z} \in Z$ be the solution to the structural problem that uniquely minimizes $\Pi(\boldsymbol{z})$. The *virtual displacement*, or the variation of $\boldsymbol{z}$, is a small, arbitrary perturbation of $\boldsymbol{z}$ in Z. Let the virtual displacement be $\boldsymbol{\eta}(\boldsymbol{x})$ with a small scalar τ so that the perturbed state is $\boldsymbol{z} + \tau\boldsymbol{\eta}$. Since the perturbed state has to be in Z, the perturbation $\boldsymbol{\eta}(\boldsymbol{x})$ has to vanish at its essential boundary; i.e., $\boldsymbol{\eta}(\boldsymbol{x})$ satisfies all homogeneous essential boundary conditions. Consider a functional Π that is defined on Z. For a sufficiently small τ, if the limit

$$\begin{aligned} \delta\Pi(\boldsymbol{z};\boldsymbol{\eta}) &= \lim_{\tau \to 0} \frac{1}{\tau}[\Pi(\boldsymbol{z} + \tau\boldsymbol{\eta}) - \Pi(\boldsymbol{z})] \\ &= \left. \frac{d}{d\tau} \Pi(\boldsymbol{z} + \tau\boldsymbol{\eta}) \right|_{\tau=0} \end{aligned} \tag{2.14}$$

exists, then it is called the *first variation* of Π at $\boldsymbol{z}$ in the direction of $\boldsymbol{\eta}$. If this limit exists for every $\boldsymbol{\eta}$ in Z, Π is said to be differentiable (i.e., Fréchet differentiable) at $\boldsymbol{z}$. The variation of a functional is also very important in design sensitivity analysis in which the first variation with respect to design is considered. This procedure will be discussed in depth in Part II of this text.

If a functional has a first variation, then quantitative criteria can be defined for its minimization. The focus here is on the necessary conditions for extrema. Presume that $\mathbf{z}$ in Z is such that

$$\Pi(\mathbf{z}) \le \Pi(\mathbf{w}), \tag{2.15}$$

for all $\mathbf{w}$ in Z. Then, $\mathbf{z}$ is said to minimize Π over Z. If (2.15) holds for all $\mathbf{w}$ in Z that satisfy $\|\mathbf{w} - \mathbf{z}\| \le d$, for some $d > 0$, Π is said to have a relative minimum value at $\mathbf{z}$.

From (2.15), for any $\boldsymbol{\eta}(\mathbf{x})$ in Z and for any sufficiently small τ, if Π has a relative minimum at $\mathbf{z}$, then

$$\Pi(\mathbf{z}) = \min_{\tau} \Pi(\mathbf{z} + \tau\boldsymbol{\eta}) = \Pi(\mathbf{z} + \tau\boldsymbol{\eta})\big|_{\tau=0},$$

that is, for fixed $\mathbf{z}$ and $\boldsymbol{\eta}$, the real value function $\Pi(\mathbf{z} + \tau\boldsymbol{\eta})$ of the real parameter τ is a minimum at $\tau = 0$. If the functional has a first variation, then $\Pi(\mathbf{z} + \tau\boldsymbol{\eta})$ is a differentiable function of τ, and a necessary condition for a minimum of Π at $\mathbf{z}$ is

$$\delta\Pi(\mathbf{z};\boldsymbol{\eta}) = \frac{d}{d\tau}\Pi(\mathbf{z} + \tau\boldsymbol{\eta})\bigg|_{\tau=0} = 0, \tag{2.16}$$

for all $\boldsymbol{\eta}$ in Z. The notation $\delta\Pi(\mathbf{z};\boldsymbol{\eta})$ represents a variation of Π at $\mathbf{z}$ in the direction of $\boldsymbol{\eta}$. Thus, the principle of minimum total potential energy is equivalent to the condition of (2.16) for all kinematically admissible $\boldsymbol{\eta}$.

The function $\boldsymbol{\eta}(\mathbf{x})$ can be thought of as a variation of the displacement $\mathbf{z}$ placed in the same context as in (2.14) since

$$\delta\mathbf{z} = \frac{d}{d\tau}(\mathbf{z} + \tau\boldsymbol{\eta})\bigg|_{\tau=0} = \boldsymbol{\eta} \equiv \bar{\mathbf{z}}. \tag{2.17}$$

Since the variation $\boldsymbol{\eta}$ is related to the displacement $\mathbf{z}$, the notation $\bar{\mathbf{z}}$ is used instead of $\delta\mathbf{z}$ to denote the variation of displacement $\mathbf{z}$ in this text. This notational system is preferred in order to avoid an excessive usage of δ, which typically denotes the Dirac delta measure.

The total potential energy in (2.10) is composed of the strain energy and the work done by the applied load. Thus, a variational formulation of the structural problem can be written, using the first variation of $\Pi(\mathbf{z})$, as

$$\delta\Pi(\mathbf{z};\bar{\mathbf{z}}) = \delta U(\mathbf{z};\bar{\mathbf{z}}) - \delta W(\mathbf{z};\bar{\mathbf{z}}) = 0, \tag{2.18}$$

for all $\bar{\mathbf{z}}$ in Z. Equation (2.18) is called the *variational equation* of the structural problem under consideration. The first term in (2.18) is obtained from the definition of $U(\mathbf{z})$ in (2.6) and the stress-strain relation in (2.8) as

$$\begin{aligned}\delta U(\mathbf{z};\bar{\mathbf{z}}) &= \iint_{\Omega} \varepsilon_{ij}(\bar{\mathbf{z}}) C_{ijkl} \varepsilon_{kl}(\mathbf{z})\, d\Omega \\ &\equiv a(\mathbf{z},\bar{\mathbf{z}}),\end{aligned} \tag{2.19}$$

where $a(\mathbf{z},\bar{\mathbf{z}})$ is called the *energy bilinear form* since it is bilinear with respect to its two arguments $\mathbf{z}$ and $\bar{\mathbf{z}}$. $\varepsilon_{ij}(\bar{\mathbf{z}})$ is the same as $\varepsilon_{ij}(\mathbf{z})$ in (2.7) by substituting $\bar{\mathbf{z}}$ into $\mathbf{z}$. Thus, the energy bilinear form is symmetric with respect to its arguments. The variation of the work done by the applied load can be written as

$$\begin{aligned}\delta W(z;\bar{z}) &= \iint_{\Omega} \bar{z}^T f^b \, d\Omega + \int_{\Gamma^s} \bar{z}^T f^s \, d\Gamma \\ &\equiv \ell(\bar{z}),\end{aligned} \tag{2.20}$$

where $\ell(\bar{z})$ is called the *load linear form*. Initially, only conservative loads are considered such that $\ell(\bar{z})$ is independent of displacement. Thus, the variational formulation of the structural problem in (2.18) can be written as

$$a(z,\bar{z}) = \ell(\bar{z}), \qquad \forall \bar{z} \in Z, \tag{2.21}$$

where $\forall \bar{z} \in Z$ represents "for all $\bar{z}$ in Z." Since the variational formulation in (2.21) is equivalent to the principle of minimum potential energy, the existence and uniqueness of the solution remains, as discussed in the previous section. Thus, if the load linear form on the right side of (2.21) is continuous in the space Z and if the energy bilinear form on the left side of (2.21) is positive definite on Z, then (2.21) has a unique solution $z \in Z$.

The advantage of using notation in (2.21) is that even for different structural problems, the same symbolic notation can be used as long as the problems share the same properties. As shown in Chapter 3, the expressions of $a(z,\bar{z})$ and $\ell(\bar{z})$ are different, depending on the structural component (such as truss, beam, plate, etc). However, the variational (2.21) can represent all structural problems in this text. Thus, by using (2.21), a unified consideration of the structural problem is possible.

The existence and uniqueness of the solution to (2.21) is well established using the *Sobolev space*. Of particular importance in the mathematical analysis of linear elastic problems using the variational method is the *Sobolev imbedding theorem*. The existence theory for variational equations in the form of (2.21) guarantees that there will be a solution $z \in [H^1(\Omega)]^3$, where $H^1(\Omega)$ is the Sobolev space of order one. The Sobolev imbedding theorem asserts that $z \in C^0(\Omega)$. This fact explains why kinematic boundary conditions in (2.13), which are given in terms of the function value, are preserved in convergence of sequences of functions in $H^1(\Omega)$. For a concise introduction to Sobolev space and its application to structural mechanics, the reader is referred to the outstanding article of Fichera [17]. For a comprehensive treatment of the subject, see Adams [22]. The engineer who is primarily interested in applications need not be concerned with these functional analysis generalizations. However, the following results need to be emphasized. If a solution exists to differential equation (2.1), then it is also the solution to variational equation (2.21). But, a solution to (2.1) does not exist if the distributed function f is a Dirac delta measure, which means that f is the applied point load. Nevertheless, the variational equation (2.21) still has a solution in this case, which is called a generalized solution.

Principle of Virtual Work

The variational formulation provided by (2.21), obtained from the principle of minimum total potential energy, is limited in solving linear elastic problems. In the principle of virtual work, the constitutive relations, including the elastoplasticity, can be quite general since we are not assuming that potential energy exists. Let the differential problem in (2.11) be satisfied and let the integration by parts be justified. Consider a virtual displacement $\bar{z}$ that satisfies the essential boundary condition, i.e., $\bar{z} = 0$ on Γ^h. Note that the displacement variation $\bar{z}$ in (2.17) is related to displacement z. Even if the same notation $\bar{z}$ is used here, virtual displacement is considered a small, arbitrary continuous field that satisfies the problem's kinematic constraints, while the applied load is kept constant. Since the differential equation (2.11) is satisfied in the domain Ω, by multiplying $\bar{z}$ on both sides of the differential equation and integrating it, we have

$$\iint_{\Omega} \bar{z}_i(\sigma_{ij,j} + f_i^b)\,d\Omega = 0, \tag{2.22}$$

for any $\bar{z}$ in Z. (Note: In strict mathematical terms, this statement is true when the space Z is dense in $L_2(\Omega)$, which is the case.) In (2.22), equilibrium of the structural problem is sought in the sense of integration. The pointwise requirement of differential equation has no meaning in the variational approach. Since the differential equation (2.11) is obtained from the force equilibrium relation, the term $\sigma_{ij,j} + f_i^b$ represents unbalanced force, while (2.22) represents the virtual work done by the system during virtual displacement. Thus, structural equilibrium is considered a vanishing condition of the virtual work. After integrating by parts, the principle of virtual work is obtained by using the symmetric property of the stress tensor $\boldsymbol{\sigma}$, the boundary conditions of (2.11), and the constitutive relation in (2.8) as

$$\begin{aligned}
&\iint_{\Omega} \bar{z}_i\left(\sigma_{ij,j} + f_i^b\right)d\Omega \\
&\quad = -\iint_{\Omega} \bar{z}_{i,j}\sigma_{ij}\,d\Omega + \iint_{\Omega} \bar{z}_i f_i^b\,d\Omega + \int_{\Gamma^h \cup \Gamma^s} \bar{z}_i\sigma_{ij}n_j\,d\Gamma \\
&\quad = -\iint_{\Omega} \bar{z}_{i,j}\sigma_{ij}\,d\Omega + \iint_{\Omega} \bar{z}_i f_i^b\,d\Omega + \int_{\Gamma^s} \bar{z}_i f_i^s\,d\Gamma \\
&\quad = -\iint_{\Omega} \varepsilon_{ij}(\bar{z})C_{ijkl}\varepsilon_{kl}(z)\,d\Omega + \iint_{\Omega} \bar{z}_i f_i^b\,d\Omega + \int_{\Gamma^s} \bar{z}_i f_i^s\,d\Gamma \\
&\quad = 0.
\end{aligned}$$

We use the fact that $\sigma_{ij}n_j = f_i^s$ on Γ^s. By using definitions of the energy bilinear form and the load linear form, the principle of virtual work can be stated as

$$a(z,\bar{z}) = \ell(\bar{z}), \qquad \forall \bar{z} \in Z. \tag{2.23}$$

Equation (2.23) is the same as the variational formulation in (2.21). In the principle of virtual work, the left side of (2.23) is interpreted as virtual work done by internal force, while the right side is seen as virtual work done by external applied force. Thus, (2.23) states that the structure is in equilibrium when internal and external virtual works are equal during all virtual displacements.

In the derivation of the principle of virtual work in (2.22) it is assumed that the differential equation is satisfied at every point within the structure, which is an unnecessary requirement. Further, consider a virtual work

$$\delta W = \iint_{\Omega} \bar{z}_i f_i^b\,d\Omega + \int_{\Gamma} \bar{z}_i f_i^s\,d\Gamma. \tag{2.24}$$

Since $\bar{z} = \boldsymbol{0}$ on Γ^h, the whole boundary $\Gamma = \Gamma^h \cup \Gamma^s$ is used instead of Γ^s. Using the relation of $f_i^s = \sigma_{ij}n_j$ and Gauss' theorem, the virtual work in (2.24) can be extended to

$$\iint_{\Omega} \bar{z}_i f_i^b\,d\Omega + \int_{\Gamma} \bar{z}_i f_i^s\,d\Gamma = \iint_{\Omega} \bar{z}_i(f_i^b + \sigma_{ij,j})\,d\Omega + \iint_{\Omega} \sigma_{ij}\varepsilon_{ij}(\bar{z})\,d\Omega. \tag{2.25}$$

Again, the symmetric property of the stress tensor is used in the above derivation. The first integral on the right side of the above equation is the same as in (2.22), which vanishes. Thus, the same principle of virtual work as in (2.23) is obtained. A subtle difference in this approach is that it is unnecessary to assume pointwise satisfaction of the differential equation. As long as the first integral on the right side vanishes, the principle of virtual work is well defined.

The difference between the variational formulation and the principle of virtual work cannot be clearly seen from the conservative system or from the linear elastic structural problem. However, in developing the variational formulation, we assumed that potential energy exists in the structure. Thus, the variational formulation is limited to linear problems. For most of the problems discussed in Part III, the potential energy of the structural problem does not exist. For those problems, the principle of virtual work has to be used. However, proving the existence and uniqueness of a solution in (2.23) is a difficult procedure that goes beyond the scope of this text. For a proof of existence and uniqueness, the reader is referred to the articles of Aubin [16] and Fichera [17].

2.4 Hamilton's Principle

A large number of problems in mechanics can be described using the variational principle known as Hamilton's principle. By removing the time-dependent variables, the variational formulation discussed in the previous section is a special case of Hamilton's principle. In this section, Hamilton's principle is developed for linear problems by introducing inertia properties that will enable us to describe the dynamic characteristics of the structure.

If a time-dependent load is applied to the structure, then the structure's response depends on time. As each particle moves, the velocity of the structure generates a *kinetic energy* that is different from the potential energy, defined as

$$T(\mathbf{z}_{,t}) \equiv \frac{1}{2}\iint_{\Omega} \rho\, \mathbf{z}_{,t}^T \mathbf{z}_{,t}\, d\Omega, \tag{2.26}$$

where $\rho(\mathbf{x})$ is the mass density of the structure and the subscribed comma denotes the derivative, such that $\mathbf{z}_{,t}$ is the derivative of the displacement with respect to time; i.e., the velocity. It is presumed that the kinetic energy in (2.26) is well defined for all kinematically admissible displacements. To connect this equation to the variational formulation discussed in Section 2.2, the first variation of $T(\mathbf{z})$ can be obtained for a small, arbitrary virtual displacement $\bar{\mathbf{z}}$ as

$$\delta T(\mathbf{z}_{,t};\bar{\mathbf{z}}_{,t}) \equiv \iint_{\Omega} \rho\, \bar{\mathbf{z}}_{,t}^T \mathbf{z}_{,t}\, d\Omega. \tag{2.27}$$

In evaluating (2.27), it is presumed that the time derivative is independent of the variation so that they can be exchanged. Equation (2.27) contains the velocity of $\bar{\mathbf{z}}(\mathbf{x},t)$, i.e., the virtual velocity. However, this variation is inappropriate since the variational formulation is given in terms of the virtual displacement. To convert (2.27) into its virtual displacement form, let us consider a virtual displacement that satisfies the kinematically admissible conditions in Z, and the following additional conditions:

$$\bar{\mathbf{z}}(\mathbf{x},0) = \bar{\mathbf{z}}(\mathbf{x},t_T) = 0, \tag{2.28}$$

where t_T is the terminal time of the problem. By integrating (2.27) over the time interval and using integration by parts in time, we can obtain a suitable form for the variational purpose as

$$\begin{aligned}
\int_0^{t_T} \delta T(\mathbf{z}_{,t};\bar{\mathbf{z}}_{,t})\,dt &= \int_0^{t_T} \Big(\iint_{\Omega} \rho\, \bar{\mathbf{z}}_{,t}^T \mathbf{z}_{,t}\, d\Omega\Big)\,dt \\
&= \iint_{\Omega} \rho(\bar{\mathbf{z}}^T \mathbf{z}_{,t})\Big|_0^{t_T} d\Omega - \int_0^{t_T} \Big(\iint_{\Omega} \rho\, \bar{\mathbf{z}}^T \mathbf{z}_{,tt}\, d\Omega\Big)\,dt \\
&= -\int_0^{t_T} \Big(\iint_{\Omega} \rho\, \bar{\mathbf{z}}^T \mathbf{z}_{,tt}\, d\Omega\Big)\,dt \\
&\equiv -\int_0^{t_T} d(\mathbf{z}_{,tt},\bar{\mathbf{z}})\,dt,
\end{aligned} \tag{2.29}$$

where $z_{,tt}$ is the second-order derivative of the displacement; i.e., acceleration. In (2.29), the integrand $d(z_{,tt},\bar{z})$ is called the *kinetic energy bilinear form*.

With the notations defined above, a general form of Hamilton's principle is obtained, which is suitable for design sensitivity analysis. Only a conservative system is considered between time 0 and t_T. For an elastic system subject to conservative dynamic loading, the Hamilton's principle states that the integral $\int_0^{t_T}[\Pi(z) - T(z_{,t})]dt$ becomes stationary. Following classical approaches [23] through [25], the variational form of Hamilton's principle requires that

$$\delta\int_0^{t_T}[\Pi(z) - T(z_{,t})]dt = 0, \tag{2.30}$$

for all times between 0 and t_T and for all kinematically admissible virtual displacements $\bar{z}$ that satisfy the additional conditions in (2.28).

In terms of the load linear form in (2.20) and the strain energy and kinetic energy bilinear forms of (2.19) and (2.29), respectively, (2.30) can be written as

$$\int_0^{t_T}\{d(z_{,tt},\bar{z}) + a(z,\bar{z})\}\,dt = \int_0^{t_T}\ell(\bar{z})\,dt, \tag{2.31}$$

for all kinematically admissible virtual displacements $\bar{z}$ that satisfy (2.21) and (2.28). Note that the variational formulation in Section 2.3 is a special case of (2.31), in which $d(z_{,tt},\bar{z}) = 0$ and time dependencies are removed.

This general formulation of Hamilton's principle provides the variational equations of structural dynamics, which can be used to extend the theory presented in Section 5.4 for the transient dynamic design sensitivity analysis of structures.

Hamilton's principle can also be obtained from the principle of virtual work. In this approach, the inertia force is considered as a body force that acts in the negative direction of acceleration (d'Alembert's principle). Thus, the structural differential (2.11) can be written, considering the inertia force, as

$$\mathrm{div}\boldsymbol{\sigma}(z) + \boldsymbol{f}^b = \rho z_{,tt}, \qquad \boldsymbol{x} \in \Omega, \tag{2.32}$$

with the boundary conditions from (2.11) and the initial conditions

$$\begin{aligned} z(\boldsymbol{x},0) &= z^0(\boldsymbol{x}), && \boldsymbol{x} \in \Omega \\ z_{,t}(\boldsymbol{x},0) &= z_{,t}^0(\boldsymbol{x}), && \boldsymbol{x} \in \Omega. \end{aligned} \tag{2.33}$$

Equation (2.32) with boundary conditions from (2.11) and initial conditions from (2.33) defines the initial-boundary-value problem (IBVP) in structural dynamics.

As in the case of the principle of virtual work, the differential equation (2.32) is multiplied by a virtual displacement $\bar{z}$ and then integrated over the domain. After integrating by parts and applying boundary conditions, we have

$$d(z_{,tt},\bar{z}) + a(z,\bar{z}) = \ell(\bar{z}), \qquad \forall \bar{z} \in Z. \tag{2.34}$$

If $d(z_{,tt},\bar{z})$ is moved to the right side of (2.34), then it can be considered the virtual work of the inertia force. When the time effect vanishes, then the principle of virtual work of the static problem is easily recovered as in (2.23).

Previous development of IBVP in structural dynamics can be generalized to include a damping effect, such as viscous damping. The differential equation of structural dynamics can be written in the form

$$\rho z_{,tt} + C z_{,t} + A z = f(x,t), \tag{2.35}$$

where ρ and C represent mass and damping effects in the structure, respectively, and $f(x,t)$ is the applied load. The operator A is the differential operator encountered in the static and eigenvalue problems of elastic systems. The state variable $z(x,t)$ is a function of both space and time. Boundary conditions for the problem are left unspecified at the present time, since they depend on the characteristics of a specific structural system. Initial conditions are given in (2.33).

The variational equation can be derived from (2.35) by multiplying an arbitrary virtual displacement $\bar{z}$ and integrating over both space and time, to obtain

$$\begin{aligned} &\int_0^T \iint_\Omega [\rho \bar{z}^T z_{,tt} + C \bar{z}^T z_{,t}] \, d\Omega \, dt \\ &\quad + \int_0^T \iint_\Omega \bar{z}^T A z \, d\Omega \, dt = \int_0^T \iint_\Omega \bar{z}^T f \, d\Omega \, dt. \end{aligned} \tag{2.36}$$

Since the integral involving the operator A on the left side of (2.36) is defined as the bilinear form of the structure, (2.36) may be written in the form

$$\int_0^T \left\{ \iint_\Omega [\rho \bar{z}^T z_{,tt} + C \bar{z}^T z_{,t}] \, d\Omega + a(z, \bar{z}) \right\} dt = \int_0^T \iint_\Omega \bar{z}^T f \, d\Omega \, dt, \tag{2.37}$$

which must hold for all $\bar{z} \in Z$, the space of kinematically admissible displacements for the structure. A rigorous mathematical theory regarding such variational equations may be found in the pioneering text of Lions and Magenes [26]. Roughly speaking, the variational form of (2.37), with the initial conditions from (2.33), is equivalent to the initial-boundary-value problem.

2.5 Eigenvalue Problem

The eigenvalue problem frequently appears in structural analysis when the vibration and buckling of structural components are taken into consideration. Natural frequencies and mode shapes of free vibration are determined by the eigenvalue problem. The buckling load and buckling shape are also determined by this problem. The conventional differential operator version of the eigenvalue problem is presented in this section and is then extended to a more flexible and rigorous variational formulation. Technical justification of the variational formulation then follows in the same format used to treat static problems in Section 2.3.

Vibration of a Structure

Natural vibration is a way to describe the behavior of the structure when no external force is applied. Usually, after applying an impact force or an initial velocity, the structure vibrates based on the characteristics of its material property and geometric shape. Using (2.35) without any body force and damping, the governing equation of motion can be written as

$$\rho z_{,tt}(x,t) + A z(x,t) = 0, \tag{2.38}$$

where A is the differential operator encountered in the static response problem, which is independent of time, and the structural response is a function of the spatial coordinates and time. Equation (2.38) is a second-order, homogeneous differential equation with respect to time. By using the separation of variables method, let us assume that the solution $z(x,t)$ is composed of

$$z(x,t) = y(x)e^{j\omega t}, \tag{2.39}$$

where $j = \sqrt{-1}$ is the complex variable and ω^2 is a natural frequency that depends on the problem considered. By substituting (2.39) into (2.38) and factoring out all $e^{j\omega t}$ terms, we can obtain a time-independent equation of natural vibration as

$$Ay = \rho\omega^2 y. \tag{2.40}$$

Equation (2.40) is a homogeneous equation of $y(x)$ that yields a trivial solution $y = 0$. However, we are interested in the nontrivial solution $y \neq 0$.

In a general vibrating structure, the formal operator equation of the eigenvalue problem is

$$Ay = \zeta By, \quad y \neq 0, \tag{2.41}$$

where the operator B is a simpler continuous operator than structural part A, except for buckling problems (often B is a scalar function as in (2.40)). The vector function $y(x)$ is used to distinguish the eigenfunction from the static response z, and $\zeta = \omega^2$ is the associated eigenvalue. Since (2.41) is homogeneous with respect to y, for any scalar $\alpha \neq 0$, αy is also a solution to (2.41). Thus, the following normalization condition can be used to remove this arbitrariness:

$$(By, y) = 1. \tag{2.42}$$

Equation (2.41) is called the generalized eigenvalue problem, whereas it is called the standard eigenvalue problem if B is the identity operator. In structural vibration problems, operator B does not contain any differentiation. Thus, the smoothness of the solution to (2.41) depends on the order of differentiation involved in A, which has the same property as the static problems in Section 2.3. The eigenfunction y in (2.42) has to satisfy the boundary conditions. Thus, the solution space of (2.41) is the same as that of the static problem. Like the static problem, the unnatural requirement of smoothness of the eigenfunction y can be reduced by introducing the variational formulation of the eigenvalue problem.

Even if the phenomenon of vibration involves time, the formulation in (2.41) is independent of time. Only the static characteristics of the structure are used. Thus, the same assumptions and procedures used in the principle of virtual work can be used to obtain the variational formulation of the eigenvalue problem. The L_2-scalar product on both sides of (2.41) with a smooth function $\bar{y}$ that satisfies the same boundary conditions as y may be formed to obtain the variational equation of the eigenvalue problem as

$$a(y,\bar{y}) \equiv (Ay,\bar{y}) = \zeta(By,\bar{y}) \equiv \zeta d(y,\bar{y}). \tag{2.43}$$

Conversely, if (2.43) holds for all $\bar{y}$ in a smooth class of functions, and if y is sufficiently regular, then y and ζ constitute the solution of the eigenvalue problem in (2.41), see [16], [17], and [27].

As in Section 2.2, for the generalized solution, the space Z of kinematically admissible displacements requires the smoothness of the solution so that the strain energy U and kinetic energy T are well defined or finite. Usually the requirements of finite kinetic energy T are not significant compared with that of strain energy U, since the operator B is simpler than A. The generalized solution $y \in Z$ ($y \neq 0$) is then characterized by the variational equation

$$a(y,\bar{y}) = \zeta d(y,\bar{y}), \quad \forall \bar{y} \in Z. \tag{2.44}$$

The boundary-value problem presented here is known as a Sturm-Liouville problem [21]. Such problems are important because the sets of orthogonal eigenfunctions generated by them are complete in $L_2(\Omega)$ [23]. Further, a positive statement can be made about pointwise convergence for the series representation of a sufficiently well-behaved function $f(\boldsymbol{x})$ in terms of the eigenfunctions.

Buckling of a Horizontal Rod

The buckling of a structure is a representative example of structural instability. Small perturbation from equilibrium configuration can result in global collapse. The governing solution of a buckling problem has similar behavior to a vibration problem, although the physical interpretation is completely different.

Consider that the horizontal rod in Fig. 2.3 is subject to an axial compressive load F. From a static analysis viewpoint, the rod can reach an equilibrium state by generating an internal force that corresponds to F. Thus, axial normal stress is the only component of internal force (stress). However, if a small perturbation δ exists, then the bending moment $F\delta$ is generated. If axial force F is small enough that the bending moment $F\delta$ is smaller than the restoring moment of the rod, then the perturbation returns to its equilibrium state. On the other hand, if F is larger than the critical value (buckling load) such that the perturbation δ grows and, as a result, the bending moment increases, then there will be a total collapse of the structure. The main purpose of buckling analysis is to predict the critical value of the axial force F and the buckling shape.

Let the length of the rod be l, the moment of inertia be I, and Young's modulus of the structure be E. Based on the linear properties of the problem, compressive and bending parts can be superposed. Since the compressive part does not contribute to buckling, only the bending part is considered. At a point $x \in (0, l)$, the bending moment is $M = F(\delta - y(x))$ and the deflection equation is

$$EIy_{,xx} + Fy = F\delta,$$

which is a second-order nonhomogeneous linear differential equation. From the classical literature on differential equations, the particular solution y_P to above equation can be obtained from the characteristics of the applied moment $F\delta$. The homogeneous solution y_H is the solution to the fourth-order differential equation by removing the right side of above equation and taking derivative twice,

$$(EIy_{,xx})_{,xx} = -Fy_{,xx}, \tag{2.45}$$

which has a similar structure to (2.41) by identifying $\zeta = F$ and B is a second-order differential operator $By = -y_{,xx}$.

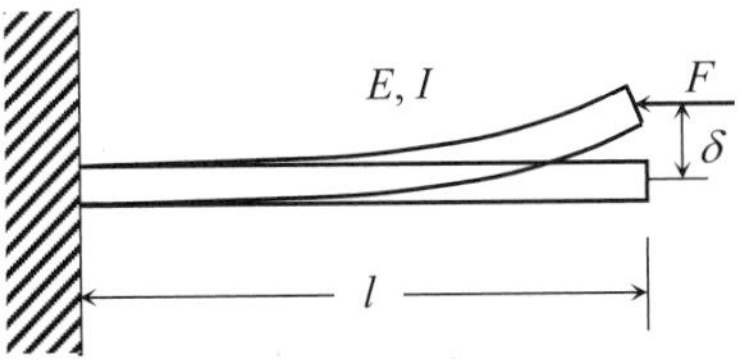

Figure 2.3. Buckling of a load by axial compressive load.

Thus, in the case of general buckling problems, the differential operator equation can be written as

$$Ay = \zeta By, \quad y \neq 0, \tag{2.46}$$

where A is the same as the structural differential operator in (2.1). As seen in the problem above, operator B in the buckling problem is more complex than operator B in vibration. In most cases, however, B is simpler than A, because a lower order of differentiation appears in B. The vector function $\boldsymbol{y}$ denotes an eigenfunction, which represents the shape of the buckling mode, and ζ is the associated eigenvalue, which is the critical load. Since (2.46) is homogeneous with respect to $\boldsymbol{y}$, the following normalization condition is used to remove any arbitrariness in the solution:

$$(By, y) = 1. \tag{2.47}$$

The variational formulation of the buckling problem follows the same procedure as in the vibration problem. Consequently, (2.44) will yield the variational formulation of the buckling problem provided the meaning of the eigenvalue and eigenfunction are appropriately interpreted.

2.6 Frequency Response Problem

In this and subsequent sections, variational formulations for specific engineering applications are developed. These applications, combined with design sensitivity analysis developed in later chapters, will provide examples of practical design applications.

A structural dynamic problem differs from the corresponding static problem due to the time-varying nature of the excitation. Over a time interval, a dynamic load varies in magnitude, direction, and/or applied location, and the resulting time-varying displacement and stress yield dynamic responses. In most structural dynamic problems, damping is present and the primary effect of the damping is the removal of energy from the structural system. There are many different mechanisms that can cause damping, such as internal friction, sliding friction, or viscous flow. Structural damping effects due to internal friction are often taken into account in solving structural problems.

2.6.1 Structural Response

Consider the transient dynamics formulation presented in Section 2.4 with a damping effect that is proportional to the velocity of the structure. The equation of motion for structural dynamics with viscous damping has the form

$$\rho z_{,tt}(\boldsymbol{x},t) + Cz_{,t}(\boldsymbol{x},t) + Az(\boldsymbol{x},t) = \boldsymbol{F}(\boldsymbol{x},t), \qquad \boldsymbol{x} \in \Omega,\ t > 0, \tag{2.48}$$

where Ω is the domain of the structure, $z(\boldsymbol{x},t)$ is the displacement, A is a linear partial differential operator presented in Section 2.3, $\rho(\boldsymbol{x})$ and $C(\boldsymbol{x})$ are the structural mass density and viscous damping effects, respectively, and $\boldsymbol{F}(\boldsymbol{x},t)$ is the applied harmonic load. The initial conditions of the dynamic problem are the same as in (2.33). The viscous damping force can be considered proportional to the velocity of the particle point.

The solution to (2.48) consists of the sum of two parts: a forced motion, which is directly related to $\boldsymbol{F}(\boldsymbol{x},t)$, and a natural motion, which is necessary to satisfy the initial conditions. Mathematically, the forced motion is a particular solution, whereas the natural motion is a homogeneous solution. In the case of harmonic excitation, the forced motion is referred to as the steady-state response. Since many single- and multi-degree of freedom structures are subjected to harmonic excitation, the steady-state response is very important. The displacement and phase angle that result from a steady-state response is called the *dynamic frequency response* of the system [28] and [29].

For the steady-state response, it is necessary to remove the time-dependent terms from (2.48). Since the harmonic load is considered, $\boldsymbol{F}(\boldsymbol{x},t)$ can be expressed as

$$\boldsymbol{F}(\boldsymbol{x},t) = \boldsymbol{f}(\boldsymbol{x})e^{j\omega t}, \tag{2.49}$$

where $\boldsymbol{f}(\boldsymbol{x})$ is the vector function of magnitude of the harmonic load and ω is the load frequency, which is considered a constant. The steady-state response has the same frequency as the applied load but may have a different phase angle. Using the complex variable method, the displacement $\boldsymbol{z}(\boldsymbol{x},t)$ is expressed as

$$\begin{aligned} \boldsymbol{z}(\boldsymbol{x},t) &= \boldsymbol{y}(\boldsymbol{x})e^{-j\alpha(\boldsymbol{x})}e^{j\omega t} \\ &= [\boldsymbol{y}^1(\boldsymbol{x}) - j\boldsymbol{y}^2(\boldsymbol{x})]e^{j\omega t} \\ &= \boldsymbol{z}(\boldsymbol{x})e^{j\omega t}, \end{aligned} \tag{2.50}$$

where $\boldsymbol{y}(\boldsymbol{x})$ is the dynamic frequency response displacement magnitude, $\boldsymbol{z}(\boldsymbol{x})$ is the complex displacement, and $\alpha(\boldsymbol{x})$ is the phase angle. Thus, the structure oscillates with shape $\boldsymbol{z}(\boldsymbol{x})$ and frequency ω. Although we use the same symbol $\boldsymbol{z}(\boldsymbol{x})$ as in the static problem, the unknown $\boldsymbol{z}(\boldsymbol{x})$ of the frequency response problem is a complex variable.

Time dependency can be eliminated by substituting (2.49) and (2.50) into (2.48) to obtain the spatial state operator equation

$$-\omega^2 \rho \boldsymbol{z}(\boldsymbol{x}) + j\omega C\boldsymbol{z}(\boldsymbol{x}) + A\boldsymbol{z}(\boldsymbol{x}) = \boldsymbol{f}(\boldsymbol{x}), \qquad \boldsymbol{x} \in \Omega, \tag{2.51}$$

with its appropriate boundary conditions. Note that, in (2.51), even though $\boldsymbol{f}(\boldsymbol{x})$ is real, the state variable $\boldsymbol{z}(\boldsymbol{x})$ is complex.

Since complex variables appear in the frequency response problem, a new definition of the linear and bilinear forms must be defined. In the case of a complex variable, semilinear and sesquilinear forms are used instead of linear and bilinear forms, respectively [30]. Let $\mathbb{C}$ be a field of complex numbers, and let $\mathbb{Q}$ be a vector space over the field $\mathbb{C}$. A map f from $\mathbb{Q}$ to $\mathbb{C}$ is called a *semilinear form* if

$$f(\alpha \boldsymbol{x} + \beta \boldsymbol{y}) = \alpha^* f(\boldsymbol{x}) + \beta^* f(\boldsymbol{y}), \tag{2.52}$$

for all $\alpha, \beta \in \mathbb{C}$ and $\boldsymbol{x}, \boldsymbol{y} \in \mathbb{Q}$. In (2.52), α^* and β^* represent the complex conjugates of α and β, respectively. If $\mathbb{C}$ is the field of real numbers, then the semilinear form is equivalent to a linear form. Let $\mathbb{Q}$ and $\mathbb{N}$ be two vector spaces over the same field $\mathbb{C}$. A map $(\boldsymbol{x}, \boldsymbol{y}) \mapsto a(\boldsymbol{x}, \boldsymbol{y})$ from $\mathbb{Q} \times \mathbb{N}$ into $\mathbb{C}$ is called a *sesquilinear form* if for every $\boldsymbol{y} \in \mathbb{N}$ the map $\boldsymbol{x} \mapsto a(\boldsymbol{x}, \boldsymbol{y})$ is a linear form on $\mathbb{Q}$, and for every $\boldsymbol{x} \in \mathbb{Q}$ the map $\boldsymbol{y} \mapsto a(\boldsymbol{x}, \boldsymbol{y})$ is a semilinear form on $\mathbb{N}$. If $\boldsymbol{y} \mapsto a(\boldsymbol{x}, \boldsymbol{y})$ is a linear form, then the sesquilinear and bilinear forms are the same. Note that the sesquilinear form is not symmetric in its arguments.

As has been mentioned, the variational formulation of (2.51) is similar to the static problem presented in Section 2.3. However, since the complex variable $\boldsymbol{z}(\boldsymbol{x})$ is used for the state variable, the complex conjugate $\bar{\boldsymbol{z}}^*$ is used for the displacement variation. By multiplying both sides of (2.51) with $\bar{\boldsymbol{z}}^*$ and integrating it over the domain Ω, after integration by parts for differential operator A, the corresponding variational equation with viscous damping can be derived as

$$\begin{aligned}&\iint_\Omega [-\omega^2 \rho z^T + j\omega C z^T]\bar{z}^* \, d\Omega + a(z,\bar{z}) \\ &\quad = \iint_\Omega f^{bT}\bar{z}^* \, d\Omega + \int_{\Gamma^s} f^{sT}\bar{z}^* \, d\Gamma \equiv \ell(\bar{z}), \qquad \forall \bar{z}^* \in \mathbb{Z},\end{aligned} \tag{2.53}$$

where $\bar{z}^*$ is the complex conjugate of the kinematically admissible virtual displacement $\bar{z}$, $\mathbb{Z}$ is the complex space of kinematically admissible virtual displacements, $a(z,\bar{z})$ is the energy sesquilinear form, and $\ell(\bar{z})$ is the load semilinear form. Equation (2.53) provides the variational equation of the dynamic frequency response under an oscillating excitation with frequency ω. The first integrand on the left side of (2.53) has a similar structure to the kinetic energy bilinear form in (2.34) if acceleration $z_{,tt}$ is substituted into displacement z to yield $-\omega^2 d(z,\bar{z})$. The first integrand on the left side of (2.53) yields the *mass sesquilinear form* as

$$d(z,\bar{z}) = \iint_\Omega \rho(\boldsymbol{x};\boldsymbol{u}) z^T \bar{z}^* \, d\Omega. \tag{2.54}$$

The second integrand on the left side of (2.53) can be denoted by using the structural *damping sesquilinear form*, as

$$c(z,\bar{z}) = \iint_\Omega C(\boldsymbol{x}) z^T \bar{z}^* \, d\Omega. \tag{2.55}$$

Note that the complex conjugate $\bar{z}^*$ of the second argument $\bar{z}$ of these sesquilinear forms $d(z,\bar{z})$, $c(z,\bar{z})$, and $a(z,\bar{z})$ are used in integrations that define them. Likewise, the complex conjugate of the argument of the semilinear form $\ell(\bar{z})$ is used in integration that defines it, as shown in (2.53). This rule will be consistently used for frequency response analysis, including design sensitivity analysis, as will be shown in Sections 5.4 and 7.4 of Chapters 5 and 7, respectively. By using (2.53) and (2.55), the variational equation of the dynamics response problem can then be obtained as

$$-\omega^2 d(z,\bar{z}) + j\omega c(z,\bar{z}) + a(z,\bar{z}) = \ell(\bar{z}), \qquad \forall \bar{z}^* \in \mathbb{Z}. \tag{2.56}$$

Structural damping, a variant of viscous damping, is caused by internal material friction or by the connections between structural components. It has been experimentally observed that, per cycle of vibration, the dissipated energy of the material is proportional to displacement [31]. When the damping coefficient is small, as in the case of structures, damping is primarily effective at those frequencies that are close to the resonance. The variational equation with structural damping effect is

$$-\omega^2 d(z,\bar{z}) + (1 + j\varphi) a(z,\bar{z}) = \ell(\bar{z}), \qquad \forall \bar{z}^* \in \mathbb{Z}, \tag{2.57}$$

where φ is the structural damping coefficient, such that $\omega c(z,\bar{z}) = \varphi a(z,\bar{z})$. That is, the structural damping is proportional to the displacement but in phase with the velocity [32].

2.6.2 Acoustic Response

Interior noise and the structural vibration of such motorized vehicles as automobiles, aircraft, and marine vehicles, are of increasing concern due to their lightweight design. Vibration of a structural component is undesirable either because excessive vibration causes fatigue problems, or because the vibration produces sound waves in adjacent fluid regions. For example, noise in an automobile interior occurs because forces transmitted from the suspension and power train excite the boundary panels of the vehicle compartment. In this section, a variational formulation of the structural-acoustic system is developed.

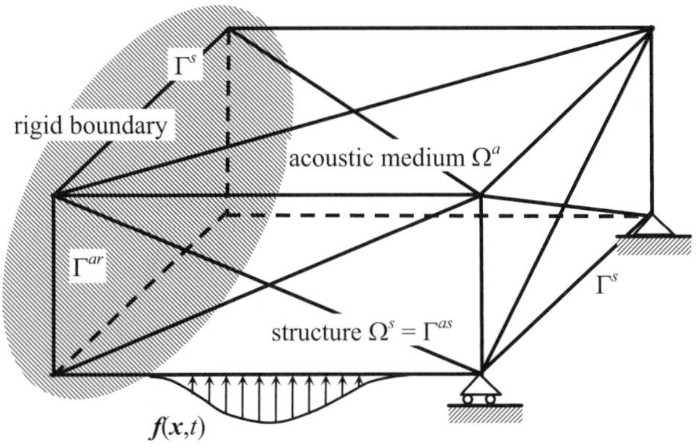

Figure 2.4. Structural-acoustic system.

A structural-acoustic system with fully enclosed volume is shown in Fig. 2.4. All members of the structure are assumed to be plates and/or beams in three-dimensional space. The structure encloses a three-dimensional acoustic medium whose dynamic response is coupled to that of the structure.

Let Ω^a and Ω^s be the domain of the acoustic medium and the structure, respectively. The acoustic domain has a boundary $\Gamma = \Gamma^{ar} \cup \Gamma^{as}$, where Γ^{ar} is the rigid boundary, and Γ^{as} is the structural boundary as the acoustic medium interfaces the structure. Thus, $\Gamma^{as} = \Omega^s$. The coupled dynamic motion of the structure and the acoustic medium can be described using the following system of differential equations [33]:

Structure:

$$\begin{aligned}\rho z_{,tt}(\boldsymbol{x},t) + Cz_{,t}(\boldsymbol{x},t) + Az(\boldsymbol{x},t) \\ = \boldsymbol{f}(\boldsymbol{x},t) + \boldsymbol{f}^p(\boldsymbol{x},t), \quad \boldsymbol{x} \in \Omega^s, t \geq 0,\end{aligned} \tag{2.58}$$

with boundary condition

$$Gz = 0, \quad \boldsymbol{x} \in \Gamma^s, \tag{2.59}$$

and initial condition

$$\begin{aligned}z(\boldsymbol{x},0) = 0, \quad \boldsymbol{x} \in \Omega^s \\ z_{,t}(\boldsymbol{x},0) = 0, \quad \boldsymbol{x} \in \Omega^s.\end{aligned} \tag{2.60}$$

Acoustic Medium:

$$\frac{1}{\beta} p_{,tt}(\boldsymbol{x},t) - \frac{1}{\rho_0} \nabla^2 p(\boldsymbol{x},t) = 0, \quad \boldsymbol{x} \in \Omega^a, t \geq 0, \tag{2.61}$$

with boundary condition

$$\nabla p \boldsymbol{n} = 0, \quad x \in \Gamma^{ar}, \tag{2.62}$$

and initial condition

$$\begin{aligned} p(\boldsymbol{x},0) &= 0, \qquad \boldsymbol{x} \in \Omega^a \\ p_{,t}(\boldsymbol{x},0) &= 0, \qquad \boldsymbol{x} \in \Omega^a. \end{aligned} \tag{2.63}$$

Interface Conditions:

$$\boldsymbol{f}^p(\boldsymbol{x},t) = p(\boldsymbol{x},t)\boldsymbol{n}, \qquad \boldsymbol{x} \in \Gamma^{as} \equiv \Omega^s \tag{2.64}$$

$$\nabla p \boldsymbol{n} = -\rho_0 \boldsymbol{z}_{,tt}^T \boldsymbol{n}, \qquad \boldsymbol{x} \in \Gamma^{as} \equiv \Omega^s. \tag{2.65}$$

Equation (2.58) describes a structural vibration where Ω^s is the domain of the structure, $\rho(\boldsymbol{x})$ is the mass density of the structure, $C(\boldsymbol{x})$ is the linear differential operator that corresponds to the damping of the structure, A is the fourth-order symmetric partial differential operator of the structure, $\boldsymbol{f}(\boldsymbol{x},t)$ is the time-dependent applied load, $\boldsymbol{f}^p$ is the acoustic pressure applied to the structure at the structure-acoustic medium interface, and $\boldsymbol{n}$ is the outward unit normal vector at the boundary of the acoustic medium. The dynamic response $\boldsymbol{z}(\boldsymbol{x},t) = [z_1, z_2, z_3]^T$ is the displacement field of the structure. The boundary condition in (2.59) is imposed on the structural boundary Γ^s using the trace operator G [5]. The structural component is governed by the transient dynamic response developed in (2.48) with the acoustic pressure considered as an applied load.

Equation (2.61) describes the propagation of linear acoustic waves in the acoustic medium Ω^a, where $\beta = \rho_0 c_0^2$ is the adiabatic bulk modulus, ρ_0 is the equilibrium density of the medium, and c_0 is the acoustic velocity. The acoustic wave (2.61) will be modified in order to make it similar to structural equations [34]. The dynamic response $p(\boldsymbol{x},t)$ is the acoustic pressure. The normal gradient of the pressure vanishes at the rigid wall Γ^{ar}, as shown in (2.62).

Structure-acoustic medium interaction can be seen in (2.64) and (2.65). In (2.64), the structural load $\boldsymbol{f}^p$ is imposed by acoustic pressure. Equation (2.65) is the interface condition in which the normal components of the pressure gradient and the structural acceleration vector are proportional to each other. As can be seen in Fig. 2.4, the structure-acoustic medium interface Γ^{as} is the same as the domain Ω^s of the structure.

When the harmonic force $\boldsymbol{f}(\boldsymbol{x},t)$ with frequency ω is applied to the structure of the coupled system, the corresponding dynamic responses $\boldsymbol{z}(\boldsymbol{x},t)$ and $p(\boldsymbol{x},t)$ are also harmonic functions with the same frequency ω. These can be represented using complex harmonic functions, as

$$\begin{aligned} \boldsymbol{f}(\boldsymbol{x},t) &= \boldsymbol{f}(\boldsymbol{x})e^{j\omega t}, \\ \boldsymbol{z}(\boldsymbol{x},t) &= \boldsymbol{z}(\boldsymbol{x})e^{j\omega t}, \\ p(\boldsymbol{x},t) &= p(\boldsymbol{x})e^{j\omega t}, \end{aligned} \tag{2.66}$$

where $\boldsymbol{f}(\boldsymbol{x})$ is the magnitude vector of the harmonic load, and $\boldsymbol{z}(\boldsymbol{x})$ and $p(\boldsymbol{x})$ are complex variables and independent of time. To obtain the differential equation without time variable, substitute (2.66) into (2.58) through (2.65) and then evaluate the time differentiation. After factoring out $e^{j\omega t}$ terms, the following time-independent, steady-state system of equations will be obtained:

Structure:

$$-\omega^2 \rho \boldsymbol{z} + j\omega C \boldsymbol{z} + A\boldsymbol{z} = \boldsymbol{f}(\boldsymbol{x}) + \boldsymbol{f}^p(\boldsymbol{x}), \qquad \boldsymbol{x} \in \Omega^s, \tag{2.67}$$

with boundary condition

$$Gz = 0, \quad \boldsymbol{x} \in \Gamma^s. \tag{2.68}$$

Acoustic Medium:

$$Bp = -\frac{\omega^2}{\beta} p - \frac{1}{\rho_0} \nabla^2 p = 0, \qquad \boldsymbol{x} \in \Omega^a, \tag{2.69}$$

with boundary condition

$$\nabla p \boldsymbol{n} = 0, \qquad x \in \Gamma^{ar}. \tag{2.70}$$

Interface Conditions:

$$\boldsymbol{f}^p = p\boldsymbol{n}, \qquad \boldsymbol{x} \in \Gamma^{as} \equiv \Omega^s \tag{2.71}$$

$$\nabla p \boldsymbol{n} = \omega^2 \rho_0 z^T \boldsymbol{n}, \qquad \boldsymbol{x} \in \Gamma^{as} \equiv \Omega^s. \tag{2.72}$$

Equations (2.67) through (2.72) constitute differential equations for the structural-acoustic system. The differential equation (2.67) of the structural part and the differential equation (2.69) of the acoustic part are coupled through the interface conditions given in (2.71) and (2.72).

To develop a variational formulation of the structural-acoustic system, it is necessary to define $\overline{z}$ and $\overline{p}$ as the kinematically admissible virtual states of the displacement z and pressure p, respectively. The variational formulations of (2.67) and (2.69) can be obtained by multiplying both sides of (2.67) and (2.69) with the complex conjugates $\overline{z}^*$ and $\overline{p}^*$ of $\overline{z} \in \mathbb{Z}$ and $\overline{p} \in P$, respectively, integrating by parts over each physical domain, adding them, and using the boundary and interface conditions [28] and [29],

$$q(z,\overline{z}) + b(p,\overline{p}) - \chi(p,\overline{z}) - \phi(z,\overline{p}) = \ell(\overline{z}), \tag{2.73}$$

which must hold for all kinematically admissible virtual states $\{\overline{z},\overline{p}\} \in Q$ where Q is a complex vector space that satisfies the boundary and interface conditions as

$$Q = \left\{ (z,p) \in \mathbb{Z} \times P \mid \boldsymbol{f}^p = p\boldsymbol{n} \text{ and } \nabla p^T \boldsymbol{n} = \omega^2 \rho_0 z^T \boldsymbol{n},\ \boldsymbol{x} \in \Gamma^{as} \equiv \Omega^s \right\} \tag{2.74}$$

and

$$\begin{aligned} \mathbb{Z} &= \left\{ z \mid \mathrm{Re}(z) \in [H^2(\Omega^s)]^3, \mathrm{Im}(z) \in [H^2(\Omega^s)]^3, Gz = 0,\ \boldsymbol{x} \in \Gamma^s \right\} \\ P &= \left\{ p \mid \mathrm{Re}(p) \in H^1(\Omega^a), \mathrm{Im}(p) \in H^1(\Omega^a), \nabla p\boldsymbol{n} = 0,\ \boldsymbol{x} \in \Gamma^{ar} \right\}. \end{aligned} \tag{2.75}$$

In (2.75), Re(z), Re(p), Im(z), and Im(p) are the real and imaginary parts of the complex variables z and p, respectively.

In (2.73), the sesquilinear forms $q(\bullet,\bullet)$, $b(\bullet,\bullet)$, $\phi(\bullet,\bullet)$, and $\chi(\bullet,\bullet)$ are defined as

$$q(z,\overline{z}) \equiv -\omega^2 d(z,\overline{z}) + j\omega c(z,\overline{z}) + a(z,\overline{z}) \tag{2.76}$$

$$b(p,\overline{p}) = \iiint_{\Omega^a} \left(-\frac{\omega^2}{\beta} p\overline{p}^* + \frac{1}{\rho_0} \nabla p^T \nabla \overline{p}^*\right) d\Omega^a \tag{2.77}$$

$$\phi(z,\bar{p}) = \omega^2 \iint_{\Omega^s} \bar{p}^* \boldsymbol{n}^T z \, d\Omega^s \tag{2.78}$$

$$\chi(p,\bar{z}) = \iint_{\Omega^s} p\boldsymbol{n}^T \bar{z}^* \, d\Omega^s . \tag{2.79}$$

As explained in Section 2.6.1, the complex conjugates $\bar{z}^*$ and $\bar{p}^*$ of the second arguments $\bar{z}$ and $\bar{p}$ of these sesquilinear forms are used in integrations that define them. If there is no acoustic medium, then the variational equation (2.73) can be simplified by dropping all terms corresponding to the acoustic medium, including the interface conditions, and the result will be the same as (2.56).

2.7 Thermoelastic Problem

In this section, variational formulations of the steady- state heat conduction equation and the elasticity equilibrium equation with thermal load are developed. This type of problem is particularly important in the analysis of automotive engine parts and turbine blades where temperature-induced deformations are significant. Although the effects of temperature and structural deformation are coupled, by treating thermal effect as an external load to the structure, a relatively simple decoupled variational equation can be obtained, which can then be solved sequentially.

2.7.1 Thermal Analysis

Consider a three-dimensional thermoelastic, isotropic and homogeneous solid, as shown in Fig. 2.5.

The steady-state heat conduction equation and the boundary conditions are given as

$$\begin{aligned} -k\theta_{,ii} &= g && \text{in } \Omega \\ \theta &= \theta^0 && \text{on } \Gamma_\theta^0 \\ k\theta_{,i}n_i &= q && \text{on } \Gamma_\theta^1 \\ k\theta_{,i}n_i + h(\theta - \theta^\infty) &= 0 && \text{on } \Gamma_\theta^2 , \end{aligned} \tag{2.80}$$

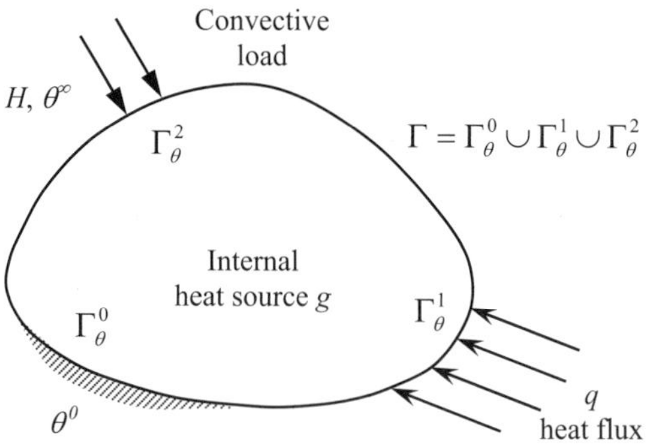

Figure 2.5. Thermal analysis model.

where $\theta = T - T^0$ and the following conditions are given: T is the absolute temperature, T^0 is the reference temperature of the stress-free state of the solid body, θ^0 is the prescribed temperature, θ^∞ is the ambient temperature, n_i is ith component of the unit normal vector on the boundary, k is the heat conductivity of the body, h is the convective heat transfer coefficient, q is the heat flux vector, g is the internal heat source, Γ_θ^0 is the boundary where the temperature is prescribed, Γ_θ^1 is the boundary where the heat flux is prescribed, and, finally, Γ_θ^2 is the boundary where the heat convection is prescribed.

By multiplying both sides of the heat conduction equation with a virtual temperature field $\bar{\theta}$, integrating over the physical domain Ω, using integration by parts, and using boundary conditions, the following equation

$$\begin{aligned} &\iiint_\Omega k\theta_{,i}\bar{\theta}_{,i}\,d\Omega + \iint_{\Gamma_\theta^2} h\theta\bar{\theta}\,d\Gamma \\ &= \iiint_\Omega g\bar{\theta}\,d\Omega + \iint_{\Gamma_\theta^1} q\bar{\theta}\,d\Gamma + \iint_{\Gamma_\theta^2} h\theta_\infty\bar{\theta}\,d\Gamma, \qquad \forall\bar{\theta}\in\Theta \end{aligned} \tag{2.81}$$

is obtained, where Θ is the space of kinematically admissible temperatures

$$\Theta = \left\{\theta \in H^1(\Omega)\,\middle|\,\theta = 0, x \in \Gamma_\theta^0\right\}, \tag{2.82}$$

and $H^1(\Omega)$ is the Sobolev space of the first order (see Appendix A.2). The bilinear and the linear forms of the heat problem are defined as

$$\begin{aligned} a_\theta(\theta,\bar{\theta}) &\equiv \iiint_\Omega k\theta_{,i}\bar{\theta}_{,i}\,d\Omega + \iint_{\Gamma_\theta^2} h\theta\bar{\theta}\,d\Gamma \\ \ell_\theta(\bar{\theta}) &\equiv \iiint_\Omega g\bar{\theta}\,d\Omega + \iint_{\Gamma_\theta^1} q\bar{\theta}\,d\Gamma + \iint_{\Gamma_\theta^2} h\theta^\infty\bar{\theta}\,d\Gamma. \end{aligned} \tag{2.83}$$

The variational form of the heat equation is then obtained as

$$a_\theta(\theta,\bar{\theta}) = \ell_\theta(\bar{\theta}), \qquad \forall\bar{\theta}\in\Theta. \tag{2.84}$$

2.7.2 Elastic Analysis

The structural part of the problem is the same as the linear elasticity described in Section 2.3, except that the material property and response state are dependent on the temperature field. The equilibrium equation and the boundary conditions for a general three-dimensional elasticity can be written as

$$\begin{aligned} div\boldsymbol{\sigma}(\boldsymbol{z}) + \boldsymbol{f}^b &= 0, && \boldsymbol{x}\in\Omega \\ \boldsymbol{z} &= 0, && \boldsymbol{x}\in\Gamma^h \\ \boldsymbol{\sigma n} &= \boldsymbol{f}^s, && \boldsymbol{x}\in\Gamma^s, \end{aligned} \tag{2.85}$$

where $div\boldsymbol{\sigma} = \sigma_{ij,j}$, $\boldsymbol{\sigma} = [\sigma_{ij}]$ is the stress tensor, $\boldsymbol{f}^b$ is the body force, $\boldsymbol{f}^s$ is the surface traction, and $\boldsymbol{n}$ is the outward unit normal vector to surface Γ^s. Due to the conservation of angular momentum, the stress tensor $\boldsymbol{\sigma}$ is symmetric ($\sigma_{ij} = \sigma_{ji}$).

As the temperature of the structure increases, the stress induced from structural strain is reduced. Thus, the change of temperature can be considered to be a reduction of strain. Since no directional effect exists for a temperature field, the strain induced by thermal effect is always volumetric. Besides the two constants of linear elasticity, only one additional constant α, the *coefficient of linear expansion*, is required. In general, however, the linear elasticity coefficients are considered functions of temperature, which is typical for a problem with a large temperature change. In (2.85), the stress tensor is defined as

$$\begin{aligned}\boldsymbol{\sigma}(\boldsymbol{z},\theta) &= \boldsymbol{C}(T):[\boldsymbol{\varepsilon}(\boldsymbol{z}) - \alpha\theta\boldsymbol{1}] \\ &= \boldsymbol{C}(T):\boldsymbol{\varepsilon}(\boldsymbol{z}) - \beta(T)\theta\boldsymbol{1},\end{aligned} \tag{2.86}$$

where $\theta = T - T_0$, $\boldsymbol{\varepsilon} = [\varepsilon_{ij}] = ½(z_{i,j} + z_{j,i})$ is the strain tensor, and $\beta(T) = \alpha E(T)/(1-2\nu(T))$ is the thermal modulus. The elasticity tensor in (2.86) is given in (2.8). Note that Young's modulus $E(T)$ and Poisson's ratio $\nu(T)$ depend on absolute temperature T.

The weak form of the elasticity equation is

$$\iiint_\Omega \boldsymbol{\sigma}(\boldsymbol{z},\theta):\boldsymbol{\varepsilon}(\bar{\boldsymbol{z}})\, d\Omega = \iiint_\Omega \bar{\boldsymbol{z}}^T \boldsymbol{f}^B\, d\Omega + \iint_{\Gamma^S} \bar{\boldsymbol{z}}^T \boldsymbol{f}^S\, d\Gamma, \quad \forall \bar{\boldsymbol{z}} \in Z, \tag{2.87}$$

where Z is the space of kinematically admissible virtual displacements:

$$Z = \{\boldsymbol{z} \in [H^1(\Omega)]^3 : \boldsymbol{z} = 0,\ \ \boldsymbol{x} \in \Gamma^g\}. \tag{2.88}$$

From the constitutive relation in (2.86) and the fact that material properties are functions of temperature, (2.87) is a coupled problem of displacement and temperature. For the sake of simplicity, small displacement and small temperature change can be assumed for the coupled theromoelastic problem in (2.84) and (2.87). Using the stress-strain relationship from (2.86), the variational form of (2.87) becomes

$$\begin{aligned}&\iiint_\Omega \boldsymbol{\varepsilon}(\bar{\boldsymbol{z}}):\boldsymbol{C}(T):\boldsymbol{\varepsilon}(\boldsymbol{z})\, d\Omega \\ &= \iiint_\Omega (\bar{\boldsymbol{z}}^T\boldsymbol{f}^B + \beta(T)\theta div(\bar{\boldsymbol{z}}))\, d\Omega + \iint_{\Gamma^S} \bar{\boldsymbol{z}}^T \boldsymbol{f}^S\, d\Gamma, \quad \forall \bar{\boldsymbol{z}} \in Z.\end{aligned} \tag{2.89}$$

The temperature field computed from (2.84) is considered an external load for the purpose of structural analysis. The thermal problem and the elasticity problem are decoupled so that the elasticity problem is still linear, even though Young's modulus and Poisson's ratio are assumed to be dependent on temperature. By defining an energy bilinear form as

$$a(\boldsymbol{z},\bar{\boldsymbol{z}}) \equiv \iiint_\Omega \boldsymbol{\varepsilon}(\bar{\boldsymbol{z}}):\boldsymbol{C}(T):\boldsymbol{\varepsilon}(\boldsymbol{z})\, d\Omega, \tag{2.90}$$

and a load linear form as

$$\ell(\bar{\boldsymbol{z}}) \equiv \iiint_\Omega (\bar{\boldsymbol{z}}^T\boldsymbol{f}^B + \beta(T)\theta div(\bar{\boldsymbol{z}}))\, d\Omega + \iint_{\Gamma^S} \bar{\boldsymbol{z}}^T \boldsymbol{f}^S\, d\Gamma, \tag{2.91}$$

Equation (2.89) becomes

$$a(\boldsymbol{z},\bar{\boldsymbol{z}}) = \ell(\bar{\boldsymbol{z}}), \quad \forall \bar{\boldsymbol{z}} \in Z. \tag{2.92}$$

Thus, the analysis procedure is made rather simple by decoupling the displacement and the temperature fields. The thermal problem in (2.84) is first solved for $\theta = T - T^0$. Absolute temperature T is then obtained from given T^0 and θ. The material properties of elasticity are obtained for given absolute temperature T, and linear elasticity is then solved using (2.92).

3
Variational Equations and Finite Element Methods

The mathematical theory of the boundary-value problem that describes deformation, buckling, and harmonic vibration of elastic structures has been turned into a powerful variational approach [16] and [35]. As shown in Chapter 2, this theory begins with the classical boundary-value problem and reduces to a variational, or energy-related formulation. The result is a rigorous existence and uniqueness theory, providing a foundation for the finite element method [36] and [37]. In retrospect, the variational formulation obtained may be viewed as the principle of virtual work, or the Galerkin method for solving boundary-value problems [38] and [39]. This chapter introduces the variational equation for various structural components, such as truss, beam, plate, and solid. Since the purpose of introducing structural components is to use them for design sensitivity analysis, special emphasis is given to design parameters that can be chosen from structural components. Energy bilinear forms of specific structural components are derived in Section 3.1.

Development of finite element methods using structural analysis was preceded by a more physically based matrix approach to structural analysis pioneered in the 1960s by Pipes [40], Langhaar [41], and a group of engineers concerned with applications. A formal distinction that can be drawn between finite element theory [4], [36], [37], and [42] and the matrix theory of structural analysis is the perspective taken in modeling the structure. In the case of a matrix structural analysis, the structure is dissected into bite-size pieces, each of which is characterized by a set of nodal displacements and an associated force-displacement relationship. In contrast, if a continuum viewpoint is adopted, then the displacement field associated with the structure is characterized by a set of differential equations of equilibrium and applied load. The finite element technique is then based on a piecewise polynomial approximation of the displacement field and on an application of variational methods for approximating a solution to the governing boundary-value problem. The finite element approach is employed throughout this text.

This chapter concentrates on a class of structures that can be readily described by finite-element matrix equations. The fundamental concepts of the finite-element structural analysis method are presented in Section 3.3, which includes a discussion of the variational principles upon which the derivation of structural equations is based. These concepts are used in the subsequent chapters to carry out design sensitivity analysis of static response, eigenvalues, and the dynamic response of a structure. Element-based matrix equations are assembled in Section 3.4 to construct a global system of matrix equations. The variational principles of continuum setting in Chapter 2 will be revisited in the discrete system of matrix equations.

3.1 Energy Bilinear and Load Linear Forms of Static Problems

3.1.1 Truss Component

A truss structural component supports the axial load. Consider the truss component in Fig. 3.1 with distributed axial load $f(x)$ per unit length. The coordinate system is established such that the x-axis is parallel to the axial direction of the truss. Let E be Young's modulus of the elastic material and let $A(x)$ be the cross-sectional area, which is a function of coordinate x. From a design point of view, E and $A(x)$ can be design variables for material property and sizing problems, respectively. If the length l is changed, the domain of integration is also changed, which is a shape design variable. If the local coordinate x is rotated with respect to the fixed global coordinate system, the rotation can be considered a configuration design variable. As discussed in Section 1.2 of Chapter 1, a sizing design variable can be the cross-sectional $A(x)$ itself and/or section dimensions that construct a cross section of the truss component. Although details of each design parameter are discussed in the subsequent chapters, consider a sizing design vector, denoted as

$$\boldsymbol{u} = [E, A(x)]^T \in U \equiv R \times C^0(0,l). \tag{3.1}$$

This notation simply means that E is real (in R) and $A(x)$ is in $C^0(0,l)$.

For the truss component, the deformation over the cross section is presumed constant. Deformation is a function of x alone and all other components are zero. Let z be the deformation along the x-axis. From the force equilibrium of an infinitesimal element, the governing differential equation of the truss component can be written as

$$\begin{aligned} -(EA(x)z_{,1})_{,1} &= f(x), \quad x \in (0,l) \\ z(0) &= 0 \\ z_{,1}(l) &= 0, \end{aligned} \tag{3.2}$$

where the subscribed comma denotes differentiation with respect to the spatial coordinate, i.e., $z_{,1} = \partial z/\partial x$. The second-order differential equation, (3.2), is well defined when the cross-sectional area $A(x)$ is a C^1 function, distributed load $f(x)$ is a C^0 function, and the displacement function $z(x)$ is a C^2 function. In addition, the solution $z(x)$ has to satisfy two boundary conditions: one for the essential boundary condition ($z(0) = 0$) and the other for the natural boundary condition ($z_{,1}(l) = 0$).

For many engineering applications, there are limitations in solving differential equation (3.2) directly. For example, this equation does not have any meaning if the point load is applied to the structure, which frequently happens during an approximation of the

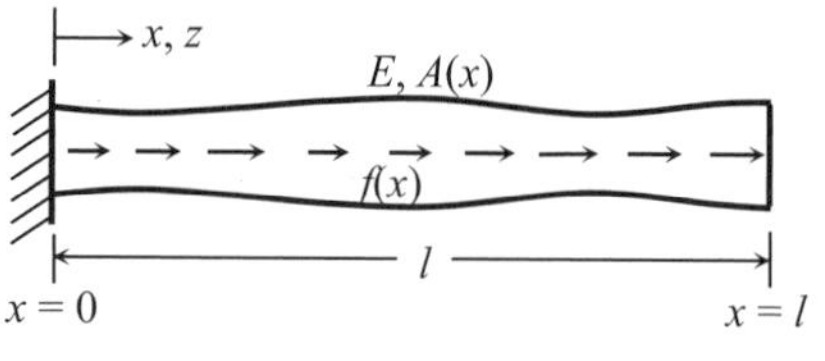

Figure 3.1. Truss structural component.

structural modeling. In addition, a discontinuous $A(x)$ may occur when different cross sections are welded together in a large structure. Thus, both the problem and the solution have to be extended to represent a variety of engineering applications by using the principle of virtual work.

The principle of virtual work, as presented in Section 2.3 of Chapter 2, is obtained by multiplying (3.2) with an arbitrary function $\bar{z}$ (called virtual displacement) and then integrating over the domain as

$$\int_0^l z_{,1}\bar{z}_{,1}\,dx - \int_0^l f\,\bar{z}\,dx = [EAz_{,1}\bar{z}]\big|_0^l, \tag{3.3}$$

where integration by parts is used once. Equation (3.3) is called the *variational identity*, which will be used for shape design sensitivity analysis in Chapter 6. Among arbitrary $\bar{z}$, let us choose those that satisfy the homogeneous essential boundary condition, that is, $\bar{z}(0) = 0$. Thus, the space of kinematically admissible displacements is defined as

$$Z = \left\{ z \in H^1(0,l) \mid z(0) = 0 \right\}, \tag{3.4}$$

where H^1 is the Sobolev space of the first order. Note that Z contains the homogeneous essential boundary condition but not the natural boundary condition. Since the derivative of the solution vanishes at $x = l$, the following variational equation is obtained from (3.3)

$$\int_0^l EAz_{,1}\bar{z}_{,1}\,dx = \int_0^l f\,\bar{z}\,dx, \tag{3.5}$$

for all $\bar{z}$ in Z. Note that the above variational problem is well defined for the integrable cross-sectional area $A(x)$ as well as for the continuous displacement function $z(x)$ whose first derivative is in $L_2(\Omega)$. Therefore, smoothness requirements for this variational problem are much less than for the classical differential equation.

For the homogeneous boundary condition, the solution space is the same as Z. From (3.5), the structural energy bilinear and load linear forms are defined as

$$a_u(z,\bar{z}) = \int_0^l EAz_{,1}\bar{z}_{,1}\,dx \tag{3.6}$$

and

$$\ell_u(\bar{z}) = \int_0^l f\bar{z}\,dx, \tag{3.7}$$

where subscript $\boldsymbol{u}$ denotes the dependence of these two forms on the design. The variational equation (3.5) of the truss component can be represented using the energy bilinear and load linear forms as

$$a_u(z,\bar{z}) = \ell_u(\bar{z}), \qquad \forall \bar{z} \in Z. \tag{3.8}$$

Note that $a_u(\bullet,\bullet)$ is symmetrical with respect to its arguments. Important properties of $a_u(\bullet,\bullet)$ for proving the existence and uniqueness of the solution will be discussed in Section 3.1.7. The expressions of $a_u(z,\bar{z})$ and $\ell_u(\bar{z})$ are different for different structural problems. However, using the variational equation (3.8), a unified approach to static analysis is possible.

Solutions to (3.2) and (3.8) need to be explained. If a solution to the differential equation (3.2) exists, then that is also the solution to the variational equation (3.8). However, the solution to (3.2) does not exist if the distributed function f is a Dirac delta measure, which is the applied point load. In this case, the variational equation (3.8) has a solution called a *generalized solution*. As a result, the variational formulation is more appropriate from the point of view of mechanics than the second-order differential equation (3.2).

3.1.2 Beam Component

A beam component supports transverse loads through its bending deflection. Two beam formulations are introduced in this section: the Bernoulli-Euler and the Timoshenko beam theories. The former assumes that bending stress alone is the dominant contribution to the structure, whereas the latter also takes into account the transverse shear effect. The advantages and disadvantages of each formulation with respect to design sensitivity analysis are discussed.

Bernoulli-Euler Beam Theory (C^1 Approach)

The Bernoulli-Euler beam theory (frequently called the technical beam theory) was developed for a beam structure in which the cross-sectional dimension is small compared with the span so that transverse shear deformation is ignored. Consider the beam component in Fig. 3.2 with an axial coordinate x, clamped supports on both sides, and a variable moment of inertia $I(x)$. Distribution of the moment of inertia, $I(x) > 0$, may be taken as a smooth function belonging to $C^0(0,l)$. When different cross-sectional sizes are connected through welding, piecewise continuous $I(x)$ can also be used and still be mathematically meaningful. In terms of design variables, E and $I(x)$ are the material property and the sizing design variables, respectively. If length l is changed, then the domain of integration, which is a shape design variable, is changed. If the local coordinate x is rotated with respect to the fixed global coordinate system, then the rotation can be considered a configuration design variable. As discussed in Section 1.2 of Chapter 1, the sizing design variable can be the moment of inertia $I(x)$ itself or cross-sectional dimensions that construct the moment of inertia of the beam component.

Denoting $z(x)$ as a vertical displacement function, the governing differential equation of the beam bending problem of Fig. 3.2 can be formally written as

$$\begin{gathered} (EI(x)z_{,11})_{,11} = f(x), \quad x \in (0,l) \\ z(0) = z(l) = 0 \\ z_{,1}(0) = z_{,1}(l) = 0, \end{gathered} \tag{3.9}$$

where $f(x)$ is the distributed load per unit length and the subscribed comma indicates derivatives with respect to x, i.e., $z_{,11} = \partial^2 z/\partial x^2$. The fourth-order differential equation, (3.9), is well defined when $I(x) \in C^2(0,l)$, $f(x) \in C^0(0,l)$, and $z(x) \in C^4(0,l)$. As mentioned

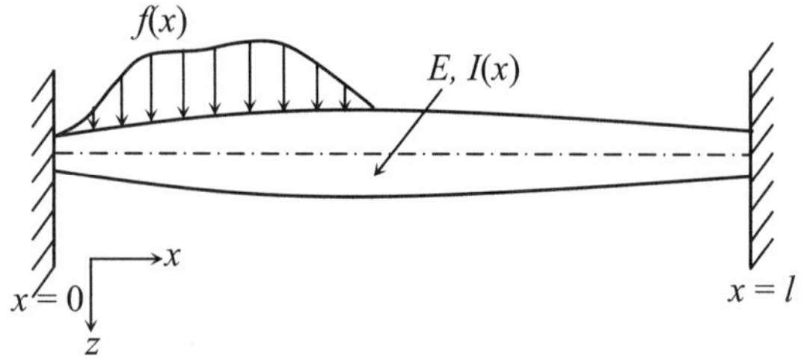

Figure 3.2. Clamped beam of variable cross-sectional area.

in the previous section, these requirements are unnecessary and inappropriate for general engineering applications.

The material constant E and the second moment of inertia $I(x)$ may be viewed as sizing design variables, since they specify the structure and may be selected by the design engineer. To simplify the notation, they are denoted as a design vector:

$$\boldsymbol{u} = [E, I(x)]^T \in U \equiv R \times C^0(0,l). \tag{3.10}$$

It is clear that the solution to (3.9) depends on the design $\boldsymbol{u}$. This dependence may be denoted by $z(x;\boldsymbol{u})$, that is, a displacement function defined in $0 \le x \le l$ that depends on $\boldsymbol{u}$.

Considered in its classical form, in which all functions are assumed to be sufficiently smooth, both sides of (3.9) can be multiplied by an arbitrary function $\bar{z}$ and integrated over the structural domain to obtain

$$\int_0^l [(EI(x)z_{,11})_{,11} - f]\bar{z}\, dx = 0, \tag{3.11}$$

which must hold for any integrable function $\bar{z}$. Conversely, if (3.11) holds for all twice continuously differentiable functions $\bar{z}$ that satisfy the boundary conditions of (3.9), and if $z \in C^4(0,l)$, then the differential equation (3.9) is satisfied. This is true since the space of kinematically admissible displacements,

$$Z = \{ z \in H^2(0,l) \,|\, z(0) = z(l) = z_{,1}(0) = z_{,1}(l) = 0 \}, \tag{3.12}$$

is dense in $L_2(0,l)$. Two integrations by parts can now be carried out in the first term in (3.11) to obtain

$$\int_0^l EIz_{,11}\bar{z}_{,11}\, dx - \int_0^l f\bar{z}\, dx = \left[EIz_{,11}\bar{z}_{,1} - (EIz_{,11})_{,1}\bar{z} \right]\Big|_0^l, \tag{3.13}$$

which is called the *variational identity*, and is derived without imposing boundary conditions and will be used for shape design sensitivity analysis in Chapter 6. Note that any assumptions regarding kinematically admissible displacements are not yet used in the variational identity. The right side of (3.13) vanishes because $\bar{z} \in Z$ is required to satisfy the boundary conditions in (3.9). Energy bilinear and load linear forms are now defined as

$$a_{\boldsymbol{u}}(z,\bar{z}) = \int_0^l EIz_{,11}\bar{z}_{,11}\, dx \tag{3.14}$$

and

$$\ell_{\boldsymbol{u}}(\bar{z}) = \int_0^l f\bar{z}\, dx. \tag{3.15}$$

After employing the boundary condition in (3.9), (3.13) therefore becomes

$$a_{\boldsymbol{u}}(z,\bar{z}) = \ell_{\boldsymbol{u}}(\bar{z}), \quad \forall \bar{z} \in Z. \tag{3.16}$$

As discussed in Chapter 2, it is important to note that the restriction of $I(x)$ to $C^2(0,l)$ and z to $C^4(0,l)$ is unnecessary. The energy bilinear form is well defined for $I(x) \in C^0(0,l)$ or $L^\infty(0,l)$ and for z and $\bar{z}$ that are in $H^2(0,l)$. Thus, the variational equation in (3.16) may be satisfied by a function z that has only one continuous derivative, with a possible second derivative that is only required in $L^2(0,l)$ and that satisfies the boundary conditions of (3.9). Such a function is called the variational or *generalized solution* to the boundary-value problem.

An alternative to the variational formulation of the beam equation may be obtained from the minimum total potential energy principle of beam bending, that is, the displacement $z \in Z$ minimizes the total potential energy of the beam structure:

$$\Pi = \int_0^l \left[\frac{1}{2} EI(z_{,11})^2 - fz \right] dx. \tag{3.17}$$

It is clear that the potential energy is well defined as long as $z_{,11} \in L^2(0,l)$ and it does not require z to be $C^4(0,l)$, as in (3.9). Equating the first variation of Π to zero, in which the variation $\bar{z}(x)$ has the second-order derivative, $\bar{z}_{,11} \in L^2(0,l)$, and assuming that $\bar{z}$ satisfies the boundary conditions of (3.9), we obtain

$$\begin{aligned} \delta\Pi &= \frac{d}{d\tau} \int_0^l \left[\frac{1}{2} EI(z_{,11} + \tau\bar{z}_{,11})^2 - f(z + \tau\bar{z}) \right] dx \bigg|_{\tau=0} \\ &= \int_0^l \left[EIz_{,11}\bar{z}_{,11} - f\bar{z} \right] dx = 0. \end{aligned} \tag{3.18}$$

But, this is the same as (3.16). Virtually no knowledge of the complicated Sobolev space theory is required if the problem is looked at as a stationary condition in optimization theory. As was proved in Section 2.2 of Chapter 2, the positive definite and positive bounded below properties of the energy bilinear form can be used to prove the existence and uniqueness of the solution to (3.18).

Recovery of the differential equation (3.9) is only possible if integration by parts can be justified, requiring either restrictive and physically unjustifiable assumptions on differentiability of z and I, or the introduction of distributional derivatives [16], [17], and [35], which in reality make the boundary-value problem of (3.9) equivalent to the variational equation of (3.16). Thus, the variational formulation is more natural from the point of view of mechanics than the fourth-order differential equation of (3.9).

While the foregoing analysis has been carried out with the clamped-clamped beam of Fig. 3.2 and with the boundary conditions of (3.9), the same results are valid for many other boundary conditions, including the following support conditions and associated boundary conditions:

1. Simply supported

$$z(0) = z_{,11}(0) = z(l) = z_{,11}(l) = 0, \tag{3.19}$$

2. Cantilevered

$$z(0) = z_{,1}(0) = z_{,11}(l) = [EI(l)z_{,11}(l)]_{,1} = 0, \tag{3.20}$$

3. Clamped–simply supported

$$z(0) = z_{,1}(0) = z(l) = z_{,11}(l) = 0. \tag{3.21}$$

Note that since the boundary terms in (3.13) vanish if z and $\bar{z}$ satisfy these boundary conditions, the bilinear form $a_u(z, \bar{z})$ of (3.14) is applicable to all boundary conditions in (3.19) through (3.21). It can be shown that the variational characterization of the solution in (3.16) is valid only if z and $\bar{z}$ satisfy the *kinematic boundary conditions* of (3.19), (3.20), or (3.21), which involve the first-order derivative. In other words, boundary conditions involving second-order or third-order derivatives are *natural boundary conditions* and need not be satisfied by z and $\bar{z}$ in (3.16).

Timoshenko Beam Theory (C^0 Approach)

The C^0 beam component has many attractive features in comparison with the technical beam theory. The major advantage is that it permits constant shear deformation across the cross section. Thus, this theory is well matched to a relatively thick beam where shear deformation is not negligible compared with its bending deformation. Although the principle of virtual work was used in the previous section to derive the variational equation for the technical beam theory, the principle of minimum total potential energy will instead be used in this section. Figure 3.3 shows the deformation kinematics of a Timoshenko beam in R^2 with a clamped boundary and distributed load. Only two strain components are nonzero, which are caused by bending and shear deformation, and are written as

$$\begin{aligned} \varepsilon_{11} &= -x_3\theta_{,1} \\ \gamma_{13} &= 2\varepsilon_{13} = (z_{,1} - \theta), \end{aligned} \tag{3.22}$$

where z is a vertical displacement of the neutral axis, θ is a fiber rotation, and x_3 is the thickness direction coordinate. Note that the additional variable θ is added to represent the strain in (3.22), while the order of differentiation is reduced to one, which is comparable to the technical beam. From the small deformation and linear elastic assumption, the stress-strain relation can be obtained as

$$\begin{aligned} \sigma_{11} &= E\varepsilon_{11} \\ \sigma_{13} &= k\mu\gamma_{13}, \end{aligned} \tag{3.23}$$

where E is Young's modulus, μ is the shear modulus, and k is the shear correction factor, needed to compensate for the constant shear strain assumption across the cross section.

From the general definition of stress and strain, the total potential energy of the beam structure can be obtained by applying the relation in (3.22) and (3.23) to (2.10) in Chapter 2 as

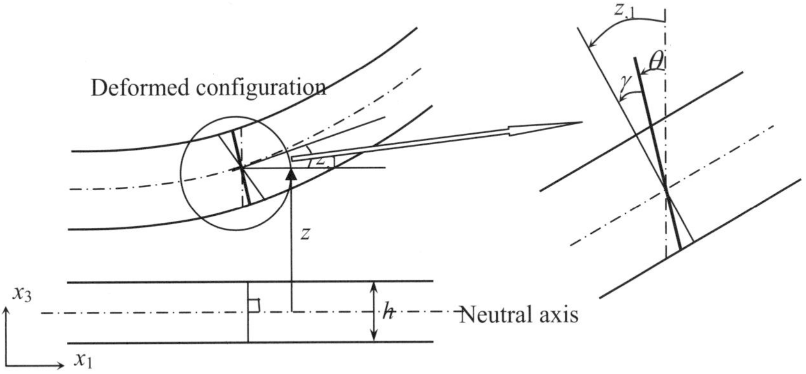

Figure 3.3. Deformation kinematics of timoshenko beam.

$$\begin{aligned}\Pi &= \frac{1}{2}\iiint_{\Omega}\sigma^T\varepsilon\, d\Omega - \iiint_{\Omega} z^T f\, d\Omega\\ &= \frac{1}{2}\int_0^l\int_A [Ex_3^2(\theta_{,1})^2 + k\mu(z_{,1}-\theta)^2]\, dA dx - \int_0^l z^T f\, dx \\ &= \frac{1}{2}\int_0^l [EI(\theta_{,1})^2 + k\mu A(z_{,1}-\theta)^2]\, dx - \int_0^l z^T f\, dx,\end{aligned} \tag{3.24}$$

where $\mathbf{z} = [z, \theta]^T$, $I(x) = \int_A x_3^2\, dA$ is the second moment of inertia, and $\boldsymbol{f} = [f, m]^T$ is the distributed load and moment in the x direction per unit length. If (3.24) is compared with the technical beam theory in (3.17), then $EI(\theta_{,1})^2$ corresponds to the bending effect in (3.14), and $k\mu A(z_{,1} - \theta)^2$ represents the contribution of shear strain energy. Note that if $z_{,1} = \theta$, then (3.24) is equivalent to (3.17). The contribution of shear strain energy should vanish as $h \to 0$ in order to match physical observation. Thus $k\mu A(z_{,1} - \theta)^2$ plays the role of a penalty function for the thin cross section of the beam. Note that $I(x)$ is proportional to h^3, whereas $A(x)$ is proportional to h. Thus, the bending strain energy term decreases faster than the shear strain energy term as $h \to 0$, unless $(z_{,1} - \theta)$ approaches zero faster than $O(h^2)$. However, if the same interpolation function is used for z and θ, then, in general, $(z_{,1} - \theta)$ cannot be zero. This situation is called shear locking, since $z_{,1}$ contains the derivative of a function and θ has a function value. Many numerical studies have been done in an attempt to remove the shear locking phenomena [4] and [43]. Since θ and z are independent variables in (3.24), the first variation of Π includes that of θ and z, written as

$$\delta\Pi = \int_0^l [\bar{\theta}_{,1}EI\theta_{,1} + (\bar{z}_{,1} - \bar{\theta})k\mu A(z_{,1} - \theta)]\, dx - \int_0^l \bar{\mathbf{z}}^T f\, dx = 0. \tag{3.25}$$

Let us define state variable $\mathbf{z} = [z, \theta]^T$ to include all unknowns of the problem, and let us define the energy bilinear and load linear forms as

$$a(\mathbf{z},\bar{\mathbf{z}}) = \int_0^l [\bar{\theta}_{,1}EI\theta_{,1} + (\bar{z}_{,1} - \bar{\theta})k\mu A(z_{,1} - \theta)]\, dx \tag{3.26}$$

$$\ell(\bar{\mathbf{z}}) = \int_0^l \bar{\mathbf{z}}^T f\, dx. \tag{3.27}$$

Note that only the first derivative is involved in the energy bilinear form in (3.26), which makes the problem simpler than the C^1 approach. The variational equation of the beam problem then becomes

$$a(\mathbf{z},\bar{\mathbf{z}}) = \ell(\bar{\mathbf{z}}), \quad \forall \bar{\mathbf{z}} \in Z, \tag{3.28}$$

where $Z \subset [H^1(0,l)]^2$ is the space of kinematically admissible displacements that satisfy all homogeneous boundary conditions:

1. Simply supported

$$Z = \left\{ \mathbf{z} = [z,\theta]^T \mid \mathbf{z} \in [H^1(0,l)]^2,\ z(0) = z(l) = 0 \right\}, \tag{3.29}$$

2. Cantilevered

$$Z = \left\{ \mathbf{z} = [z,\theta]^T \mid \mathbf{z} \in [H^1(0,l)]^2,\ z(0) = \theta(0) = 0 \right\}, \tag{3.30}$$

3. Clamped-simply supported

$$Z = \left\{ \mathbf{z} = [z,\theta]^T \mid \mathbf{z} \in [H^1(0,l)]^2,\ \mathbf{z}(0) = z(l) = 0 \right\}. \tag{3.31}$$

3.1.3 Plate Component

In this section, a brief review of plate formulation is presented for design sensitivity analysis in subsequent chapters. The plate component is an extension of the beam component into R^2. Two formulations are introduced: the thin plate theory, where transverse bending effect is ignored, and the thick plate theory based on the Mindlin/Reissner formulation, where constant shear deformation along the thickness is considered.

Thin Plate Component (C^1 Approach)

In the thin plate theory, a plane stress condition is assumed such that stress components in the thickness direction are ignored, since the thickness is very small compared with the span of the plate. Therefore, bending stress is the only contribution to the structure. Consider a plate component of thickness $h \in C^0(\Omega)$ or $L^\infty(\Omega)$ ($h(x) \geq h_0 > 0$), as shown in Fig. 3.4, where the x_1-x_2 coordinate plane corresponds to the midsurface. On the top face of the plate, a distributed load $f(x_1, x_2)$ is applied. The formal boundary-value problem for vertical displacement $z \in C^4(\Omega)$ of the midsurface is written as

$$[D(\boldsymbol{u})(z_{,11} + \nu z_{,22})]_{,11} + [D(\boldsymbol{u})(z_{,22} + \nu z_{,11})]_{,22} + 2(1-\nu)[D(\boldsymbol{u})z_{,12}]_{,12} = f, \tag{3.32}$$

in domain Ω, where $D(\boldsymbol{u})$ is the flexural rigidity defined as

$$D(\boldsymbol{u}) = \frac{Eh^3}{12(1-\nu^2)}, \tag{3.33}$$

where $E \geq E_0 > 0$ is Young's modulus, ν is Poisson's ratio, and $\boldsymbol{u} = [E, h(\boldsymbol{x})]^T$ is the design vector. At the boundary Γ, one can define a new coordinate system whose axis is normal and tangential to the boundary. Let n and s denote the components of those directions. The clamped boundary condition can be described as

$$\left.\begin{aligned} z &= 0 \\ \frac{\partial z}{\partial n} &= 0 \end{aligned}\right\} \quad \text{on } \Gamma. \tag{3.34}$$

Note that since $z = 0$ along boundary Γ, $\partial z/\partial s = 0$.

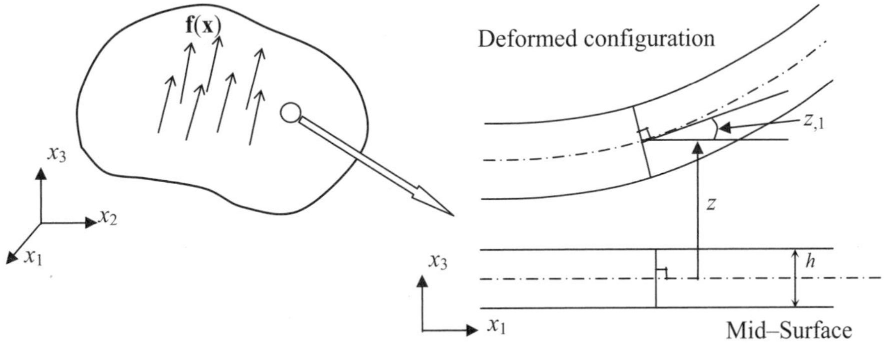

Figure 3.4. Thin plate component.

The variational formulation is obtained by multiplying both sides of differential equation (3.32) by an arbitrary function $\bar{z}$ and integrating over Ω, as

$$\iint_\Omega \bar{z}\{[D(\boldsymbol{u})(z_{,11}+\nu z_{,22})]_{,11}+[D(\boldsymbol{u})(z_{,22}+\nu z_{,11})]_{,22}+2(1-\nu)[D(\boldsymbol{u})z_{,12}]_{,12}-f\}\,d\Omega=0. \quad (3.35)$$

For convenience, let us define the following differential operators [16] through [18] as:

$$Mz = D\frac{\partial^2 z}{\partial n^2}+D\nu\left(\frac{1}{r}\frac{\partial z}{\partial n}+\frac{\partial^2 z}{\partial s^2}\right) \quad (3.36)$$

and

$$\begin{aligned}-Nz &= [D(z_{,11}+\nu z_{,22})]_{,1}n_1+[D(z_{,22}+\nu z_{,11})]_{,2}n_2\\ &\quad +(1-\nu)[(Dz_{,12})_{,2}n_1+(Dz_{,12})_{,1}n_2]\\ &\quad +(1-\nu)\frac{\partial}{\partial s}\left(D\frac{\partial^2 z}{\partial n\partial s}\right),\end{aligned} \quad (3.37)$$

where Mz represents the bending moment over the cross section, and Nz represents the transverse shear force. In (3.36), r is the radius of curvature of the boundary. In (3.37), $\boldsymbol{n} = [n_1, n_2]^T$ is the outward unit normal vector of boundary Γ. After integration by parts in (3.35), we have

$$\iint_\Omega \boldsymbol{\kappa}(\bar{z})^T \boldsymbol{C}^b \boldsymbol{\kappa}(z)\,d\Omega - \iint_\Omega \bar{z}f\,d\Omega = \int_\Gamma \frac{\partial\bar{z}}{\partial n}Mz\,d\Gamma + \int_\Gamma \bar{z}Nz\,d\Gamma, \quad (3.38)$$

which is a *variational identity* for the plate bending problem. In (3.38), the curvature vector $\boldsymbol{\kappa}$ and bending stiffness matrix $\boldsymbol{C}^b$ are defined as

$$\boldsymbol{\kappa}(z) = \begin{bmatrix} z_{,11}\\ z_{,22}\\ 2z_{,12}\end{bmatrix} \quad (3.39)$$

and

$$\boldsymbol{C}^b = \frac{h^3}{12}\boldsymbol{C} = \frac{Eh^3}{12(1-\nu^2)}\begin{bmatrix}1 & \nu & 0\\ \nu & 1 & 0\\ 0 & 0 & \frac{1}{2}(1-\nu)\end{bmatrix}, \quad (3.40)$$

where $\boldsymbol{C}$ is the stiffness matrix of the plane stress problem in linear elasticity. If the virtual displacement $\bar{z}$ satisfies the boundary conditions in (3.34), then the boundary integrals on the right side of (3.38) vanish. If we define the energy bilinear and load linear forms as

$$a_{\boldsymbol{u}}(z,\bar{z}) \equiv \iint_\Omega \boldsymbol{\kappa}(\bar{z})^T \boldsymbol{C}^b \boldsymbol{\kappa}(z)\,d\Omega \quad (3.41)$$

and

$$\ell_{\boldsymbol{u}}(\bar{z}) = \iint_\Omega \bar{z}f\,d\Omega, \quad (3.42)$$

then the variational equation of the plate bending problem becomes

$$a_{\boldsymbol{u}}(z,\bar{z}) = \ell_{\boldsymbol{u}}(\bar{z}), \quad \forall \bar{z}\in Z, \quad (3.43)$$

where Z is the space of kinematically admissible displacements that satisfies the essential boundary conditions in (3.34), written as

$$Z = \{ z \in H^2(\Omega) \mid z = \partial z / \partial n = 0 \text{ on } \Gamma \}. \tag{3.44}$$

The variational equation (3.43) is valid for a plate with $h \in L^\infty(\Omega)$. As in the case of the beam, it is unnatural and unnecessary to restrict variational equation solutions to $C^4(\Omega)$.

The variational equation (3.43) is still valid for other types of boundary conditions so that the right side of the variational identity in (3.38) vanishes. The following boundary conditions can be considered:

1. Clamped

$$z = \frac{\partial z}{\partial n} = 0, \quad \text{on } \Gamma, \tag{3.45}$$

2. Simply supported

$$z = Mz = 0, \quad \text{on } \Gamma, \tag{3.46}$$

3. Free edge

$$Mz = Nz = 0, \quad \text{on } \Gamma. \tag{3.47}$$

Note that the variational equation (3.43) contains the second-order derivative of displacement in curvature vector $\boldsymbol{\kappa}$. If the finite element method is used to approximate (3.43), continuity across the element boundary has to be preserved for the displacement and its derivative (called the C^1 approach). Major difficulties can be expected in constructing a C^1-continuous design velocity field for shape and configuration design sensitivity analyses (Chapter 7), whereas only C^0-continuity of the design velocity field is required for the following thick plate formulation.

Thick Plate Component (C^0 Approach)

The thick plate theory (the Mindlin/Reissner plate) allows for transverse shear deformation effects, and thus offers an attractive alternative to the classical Kirchhoff thin plate theory (see Fig. 3.5). The main assumptions of the plate component are essentially the same as those of the thick beam:

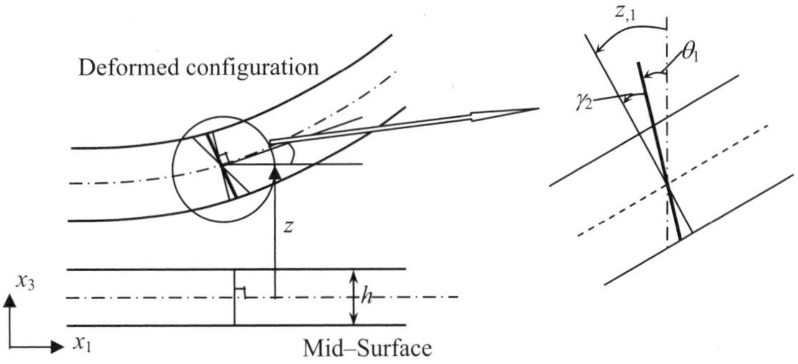

Figure 3.5. Thick plate component.

1. The normal stress on the midsurface of the plates is negligible (plane stress),
2. Fiber cross sections remain on the plane after deformation (constant shear deformation), and finally,
3. The fiber cross section is not necessarily normal to the midsurface after deformation.

Let the coordinate system be given such that the x_3-axis is normal to the midsurface. The first assumption provides that $\sigma_{33} = 0$, i.e., the plane stress condition. Only five components of the stress tensor are considered in the following derivation, with a vector notation for convenience.

Note that the rotations represented by θ_1 and θ_2 are independent, although they are expressed using displacement derivatives from the thin plate theory. Strain is divided into bending and transverse shear parts, and is written

$$\boldsymbol{\varepsilon} = \begin{bmatrix} \varepsilon_{11} \\ \varepsilon_{22} \\ 2\varepsilon_{12} \end{bmatrix} = -x_3 \begin{bmatrix} \theta_{1,1} \\ \theta_{2,2} \\ \theta_{1,2} + \theta_{2,1} \end{bmatrix} = -x_3 \boldsymbol{\kappa} \tag{3.48}$$

$$\boldsymbol{\gamma} = \begin{bmatrix} 2\varepsilon_{23} \\ 2\varepsilon_{13} \end{bmatrix} = \begin{bmatrix} z_{,2} - \theta_2 \\ z_{,1} - \theta_1 \end{bmatrix}. \tag{3.49}$$

Note that the strain resultants given in (3.48), namely, $\theta_{1,1}$, $\theta_{2,2}$, and $(\theta_{1,2} + \theta_{2,1})$, are direct curvatures in the x_1 and x_2 directions and the twisting curvature, respectively. In (3.49), $z_{,2} - \theta_2$ and $z_{,1} - \theta_1$ are the shear rotations in the 2-3 and the 1-3 planes, respectively. In (3.48), the same notation $\boldsymbol{\kappa}$ is used as in (3.39), since they both represent the plate curvature. However, different definitions should be used in this formulation. The stress-strain relation can be written for the linear elastic material as

$$\boldsymbol{\sigma} = \begin{bmatrix} \sigma_{11} \\ \sigma_{22} \\ \sigma_{12} \end{bmatrix} = \boldsymbol{C\varepsilon} = -x_3 \boldsymbol{C\kappa} \tag{3.50}$$

$$\boldsymbol{\tau} = \begin{bmatrix} \sigma_{23} \\ \sigma_{13} \end{bmatrix} = k\boldsymbol{D}\boldsymbol{\gamma}, \tag{3.51}$$

where $\boldsymbol{C}$ has the same form as in the plane stress problem in (3.40); k (= 5/6) is the shear correction factor, compensating for the assumed constant shear strain along the cross section; and $\boldsymbol{D}$ is the elastic modulus, which corresponds to the transverse shear, defined as

$$\boldsymbol{D} = \begin{bmatrix} \mu & 0 \\ 0 & \mu \end{bmatrix}, \quad \boldsymbol{C}^s = hk\boldsymbol{D}, \tag{3.52}$$

where μ is the shear modulus.

The total potential energy for a typical Mindlin/Reissner plate is then given as

$$\begin{aligned} \Pi &= \frac{1}{2}\iint_\Omega \int_{-\frac{h}{2}}^{\frac{h}{2}} (\boldsymbol{\sigma}^T\boldsymbol{\varepsilon} + \boldsymbol{\tau}^T\boldsymbol{\gamma})\, dx_3 d\Omega - \iint_\Omega \boldsymbol{z}^T \boldsymbol{f}\, d\Omega \\ &= \frac{1}{2}\iint_\Omega \boldsymbol{\kappa}^T \boldsymbol{C}^b \boldsymbol{\kappa}\, d\Omega + \frac{1}{2}\iint_\Omega \boldsymbol{\gamma}^T \boldsymbol{C}^s \boldsymbol{\gamma}\, d\Omega - \iint_\Omega \boldsymbol{z}^T \boldsymbol{f}\, d\Omega, \end{aligned} \tag{3.53}$$

where $z = [z, \theta_1, \theta_2]^T$ and $f = [f, m_1, m_2]^T$. The first integral on the right side of (3.53) denotes bending strain energy, while the second indicates the transverse shear strain energy. Shear strain energy can be rewritten as

$$\frac{1}{2}\iint_\Omega hk\mu[(z_{,2}-\theta_2)^2+(z_{,1}-\theta_1)^2]\,d\Omega. \tag{3.54}$$

This formulation can be regarded as a penalty term for the thin plate theory in order to reduce shear strain as the plate thickness–to–span ratio is reduced. The structural variational equation is obtained from the principle of minimum total potential energy as

$$\begin{aligned} a_u(z,\bar{z}) &\equiv \iint_\Omega \boldsymbol{\kappa}(\bar{\boldsymbol{\theta}})^T \boldsymbol{C}^b \boldsymbol{\kappa}(\boldsymbol{\theta})\,d\Omega \\ &+\iint_\Omega \boldsymbol{\gamma}(\bar{z})^T \boldsymbol{C}^s \boldsymbol{\gamma}(z)\,d\Omega \\ &= \iint_\Omega \bar{z}^T \boldsymbol{f}\,d\Omega \equiv \ell_u(\bar{z}), \end{aligned} \tag{3.55}$$

for all $\bar{z} = [\bar{z}, \bar{\theta}_1, \bar{\theta}_2]^T \in [H^1(\Omega)]^3$ in kinematically admissible displacements that satisfy the homogeneous boundary conditions.

Variational equation (3.55) contains the first-order derivative of displacement and rotation. Thus, continuity of all state variables is required across the finite element boundary, which is called the C^0 approach. From a design sensitivity point of view, there are several benefits obtained from the C^0 approach as compared with the C^1 approach. However, there are some computational difficulties in solving (3.55) numerically that are related to shear locking and membrane locking. For a more detailed discussion, refer to Hughes' text on this topic [43].

3.1.4 Elastic Solid

Consider the three-dimensional linear elasticity problem of an arbitrarily shaped body, as shown in Fig. 3.6. Except for the material property, the solid component has no sizing design parameters. Since the material point coincides with the configuration, the solid component is frequently used in the shape design problem. Note that no configuration design need be defined since the rotational effect of a solid can be represented by its material point movement, which is the shape design problem. Three components of displacement $z = [z_1, z_2, z_3]^T$ characterize the displacement at each point $\boldsymbol{x} = [x_1, x_2, x_3]^T$ in the elastic body. Let the domain of the body be denoted as Ω, and let boundary Γ consist of traction boundary Γ^s and displacement boundary Γ^h, such that $\Gamma^s \cup \Gamma^h = \Gamma$ and $\Gamma^s \cap \Gamma^h = \varnothing$. The displacement is prescribed on Γ^h and the traction force is prescribed on Γ^s. The governing differential equations of equilibrium for the elastic body are

$$\begin{aligned} -div\boldsymbol{\sigma}(z) &= \boldsymbol{f}^b,\ \ \boldsymbol{x}\in\Omega, \\ z &= 0,\quad \boldsymbol{x}\in\Gamma^h, \\ \boldsymbol{\sigma n} &= \boldsymbol{f}^s,\ \ \boldsymbol{x}\in\Gamma^s, \end{aligned} \tag{3.56}$$

where $div\boldsymbol{\sigma} = \sigma_{ij,j}$, $\boldsymbol{\sigma} = [\sigma_{ij}]$ is the stress tensor, $\boldsymbol{f}^b$ is a body force, $\boldsymbol{f}^s$ is a surface traction, and $\boldsymbol{n}$ is an outward unit normal vector to the traction surface Γ^s. Due to the conservation of angular momentum, the stress tensor $\boldsymbol{\sigma}$ is symmetrical ($\sigma_{ij} = \sigma_{ji}$).

Let the strain tensor be $\boldsymbol{\varepsilon} = [\varepsilon_{ij}] = ½(z_{i,j} + z_{j,i})$. As with previous beam and plate components, the formal differential equation (3.56) can be reduced to a variational problem by multiplying both sides by an arbitrary virtual displacement vector $\bar{z} = [\bar{z}_1, \bar{z}_2, \bar{z}_3]^T \in [H^1(\Omega)]^3$ and integrating by parts, to obtain

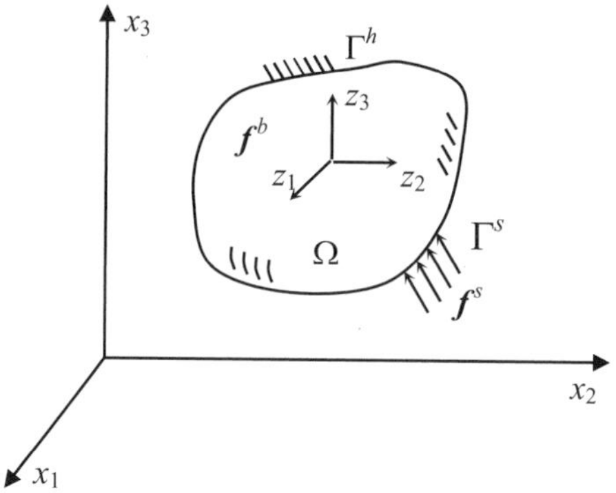

Figure 3.6. Three-dimensional elastic solid.

$$\iiint_{\Omega} \sigma(z) : \varepsilon(\bar{z})\, d\Omega - \iiint_{\Omega} \bar{z}^T f^b\, d\Omega = \iint_{\Gamma} \bar{z}^T \sigma(z) n\, d\Gamma, \qquad \forall \bar{z} \in [H^1(\Omega)]^3, \tag{3.57}$$

where ":" is the standard contraction operator of tensors, i.e., $\boldsymbol{a} : \boldsymbol{b} = a_{ij} b_{ij}$, and $H^1(\Omega)$ is Sobolev space of order one as defined in Section 2.2 of Chapter 2. Note that the boundary condition in (3.56) has not yet been applied. Equation (3.57) is a *variational identity* and will be used in the shape design sensitivity analysis in Chapter 6. The variational equation can be obtained by choosing $\bar{z}$ such that the homogeneous kinematic boundary condition of (3.56) is satisfied. On Γ^h, the right side of (3.57) vanishes from the second relation of (3.56) and the traction force $\boldsymbol{\sigma n}$ is given as $\boldsymbol{f}^s$ on Γ^s. Let us define the energy bilinear and load linear forms as

$$a_u(z, \bar{z}) = \iiint_{\Omega} \sigma(z) : \varepsilon(\bar{z})\, d\Omega \tag{3.58}$$

and

$$\ell_u(\bar{z}) = \iiint_{\Omega} \bar{z}^T f^b\, d\Omega + \iint_{\Gamma^s} \bar{z}^T f^s\, d\Gamma. \tag{3.59}$$

The variational equation corresponding to the differential equaqtion (3.56) then becomes

$$a_u(z, \bar{z}) = \ell_u(\bar{z}), \qquad \forall \bar{z} \in Z, \tag{3.60}$$

where Z is the space of kinematically admissible displacements that satisfy the kinematic boundary condition,, that is,

$$Z = \left\{ z \in [H^1(\Omega)]^3 \,\middle|\, z(x) = \boldsymbol{0},\ x \in \Gamma^h \right\}. \tag{3.61}$$

Equation (3.60) is a generalization of the boundary-value problem in (3.56), in the sense that if a solution to the boundary-value problem exists, then it satisfies (3.60) for all displacement fields $\bar{z} \in Z$. Conversely, solution z in (3.60), which exists for all displacement fields $\bar{z}$ satisfying (3.61), also solves the boundary-value problem, if a solution to that problem exists. However, in certain situations a solution to the boundary-value problem in (3.56) does not exist. For example, when $\boldsymbol{f}^b$ is a concentrated force, such that the Dirac delta measure is used in (3.56), then no solution exists in the classical sense. In that case, a solution to variational equation (3.60) still exists as a *generalized solution* within the Sobolev space $[H^1(\Omega)]^3$.

The energy bilinear form $a_{\boldsymbol{u}}(\bullet,\bullet)$ is symmetrical and linear with respect to its arguments for a linear elastic constitutive relation where the stress-strain relation is given as

$$\begin{aligned} \sigma_{ij} &= \lambda \varepsilon_{kk} \delta_{ij} + 2\mu \varepsilon_{ij} \\ &= C_{ijkl} \varepsilon_{kl} \\ C_{ijkl} &= \lambda \delta_{ij} \delta_{kl} + \mu(\delta_{ik}\delta_{jl} + \delta_{il}\delta_{jk}), \end{aligned} \tag{3.62}$$

where λ and μ are Lame's constants of the isotropic material and δ_{ij} is the Kronecker delta symbol, i.e., having a value of one when $i = j$ and otherwise remaining at zero. Only two independent constants are used to describe the constitutive relation of linear homogeneous materials. Lame's constants can be replaced by other engineering-oriented constants for convenience, as illustrated in the following example.

The tensor notation used in the previous derivations could be somewhat complicated to follow since a fourth-order constitutive relation is used in (3.59). For readers who are familiar with vector notation, the strain vector can be defined in the following form:

$$\boldsymbol{\varepsilon} = \begin{bmatrix} \varepsilon_{11} \\ \varepsilon_{22} \\ \varepsilon_{33} \\ 2\varepsilon_{12} \\ 2\varepsilon_{23} \\ 2\varepsilon_{13} \end{bmatrix} = \begin{bmatrix} z_{1,1} \\ z_{2,2} \\ z_{3,3} \\ z_{1,2} + z_{2,1} \\ z_{2,3} + z_{3,2} \\ z_{1,3} + z_{3,1} \end{bmatrix}, \tag{3.63}$$

where the subscribed comma denotes the derivative with respect to x_i. The stress-strain relation in (3.62) (generalized Hooke's law) is given with a vector notation of the stress tensor as

$$\boldsymbol{\sigma} = \begin{bmatrix} \sigma_{11} \\ \sigma_{22} \\ \sigma_{33} \\ \sigma_{12} \\ \sigma_{23} \\ \sigma_{13} \end{bmatrix} = \boldsymbol{C\varepsilon}, \tag{3.64}$$

where $\boldsymbol{C}$ is the 6×6 symmetric elastic modulus matrix,

$$\boldsymbol{C} = \frac{E}{(1+\nu)(1-2\nu)} \begin{bmatrix} 1-\nu & \nu & \nu & 0 & 0 & 0 \\ \nu & 1-\nu & \nu & 0 & 0 & 0 \\ \nu & \nu & 1-\nu & 0 & 0 & 0 \\ 0 & 0 & 0 & \dfrac{1-2\nu}{2} & 0 & 0 \\ 0 & 0 & 0 & 0 & \dfrac{1-2\nu}{2} & 0 \\ 0 & 0 & 0 & 0 & 0 & \dfrac{1-2\nu}{2} \end{bmatrix}, \tag{3.65}$$

and E and ν are Young's modulus and Poisson's ratio in linear elasticity, respectively. The relations between E, ν, and λ, μ are:

$$\nu = \frac{\lambda}{2(\lambda+\mu)}, \qquad E = \frac{\mu(3\lambda+2\mu)}{\lambda+\mu}$$
$$\lambda = \frac{E\nu}{(1+\nu)(1-2\nu)}, \qquad \mu = \frac{E}{2(1+\nu)}, \tag{3.66}$$

and in order for $\lambda > 0$, it is clear that the condition $0 < \nu < ½$ is required. Using (3.63), (3.64), and (3.65), the energy bilinear form in vector notation therefore becomes

$$a_u(z,\bar{z}) \equiv \iiint_\Omega \varepsilon(\bar{z})^T \boldsymbol{C} \varepsilon(z)\, d\Omega, \tag{3.67}$$

where the linear property of $a_u(\bullet,\bullet)$ can be clearly identified. In this text, vector notation is preferred unless it is necessary to use tensor notation.

An alternative view of variational equation (3.60) can be obtained from the total potential energy of the three-dimensional solid structure, defined as

$$\Pi = \frac{1}{2}\iiint_\Omega \varepsilon(z)^T \boldsymbol{C} \varepsilon(z)\, d\Omega - \iiint_\Omega z^T \boldsymbol{f}^b\, d\Omega - \iint_{\Gamma^s} z^T \boldsymbol{f}^s\, d\Gamma. \tag{3.68}$$

The first term corresponds to the strain energy stored in the system due to deformation and the other two terms correspond to the work done by external force. The principle of minimum total potential energy in Chapter 2 says that the equilibrium of the structural problem corresponds to the stationary condition in (3.68), which is the vanishing condition of the first variation. By taking the first variation of (3.68), we have

$$\begin{aligned} a_u(z,\bar{z}) &= \iiint_\Omega \varepsilon(\bar{z})^T \boldsymbol{C} \varepsilon(z)\, d\Omega \\ &= \iiint_\Omega \bar{z}^T \boldsymbol{f}^b\, d\Omega + \iint_{\Gamma^s} \bar{z}^T \boldsymbol{f}^s\, d\Gamma \equiv \ell_u(\bar{z}), \end{aligned} \tag{3.69}$$

for all $\bar{z} \in Z$, which is same as (3.60), obtained from the principle of virtual work.

Plane Stress Elastic Solid

In certain situations, the three-dimensional elasticity problem specializes in lower-dimensional problems. For example, the *plane strain* problem can be inferred when the thickness of the x_3-coordinate is dominant or when deformations in the x_3-coordinate are fixed. In plane strain problems, the variational formulation in (3.60) is still valid by restricting the index of the coordinate to the first and second dimensions. Even if the σ_{33} component is nonzero, its effect vanishes, since ε_{33} is zero. In contrast, with thin elastic solids, stress components normal to the plane where solids lie are often essentially zero (*plane stress*). The plane stress problem has a different coefficient from the three-dimensional elasticity or plane strain problem even if only two spatial coordinates are used.

Consider a plane stress problem, in which all of the components of stress in the x_3-direction are zero. From (3.62), this yields

$$\begin{aligned} \sigma_{13} &= 2\mu\varepsilon_{13} = 0 \\ \sigma_{23} &= 2\mu\varepsilon_{23} = 0 \\ \sigma_{33} &= \lambda(\varepsilon_{11}+\varepsilon_{22}+\varepsilon_{33}) + 2\mu\varepsilon_{33} = 0. \end{aligned} \tag{3.70}$$

It is straightforward to say that $\varepsilon_{23} = \varepsilon_{32} = \varepsilon_{13} = \varepsilon_{31} = 0$. From the last relation in (3.70), ε_{33} can be expressed in terms of ε_{11} and ε_{22}, as

$$\varepsilon_{33} = -\frac{\lambda}{\lambda + 2\mu}(\varepsilon_{11} + \varepsilon_{22}). \tag{3.71}$$

By substituting these relations into the general stress-strain relation of (3.64), plane stress-strain relations are produced as

$$\boldsymbol{\sigma} = \begin{bmatrix} \sigma_{11} \\ \sigma_{22} \\ \sigma_{12} \end{bmatrix} = \boldsymbol{C\varepsilon}, \tag{3.72}$$

where

$$\boldsymbol{\varepsilon} = \begin{bmatrix} \varepsilon_{11} \\ \varepsilon_{22} \\ 2\varepsilon_{12} \end{bmatrix} = \begin{bmatrix} z_{1,1} \\ z_{2,2} \\ z_{1,2} + z_{2,1} \end{bmatrix} \tag{3.73}$$

is the strain vector and $\boldsymbol{C}$ is the 3×3 symmetrical elastic modulus matrix for the plane-stress problem, as given in (3.40). As a result of this notation, (3.60) remains valid as a variational equation of elasticity, with the only indices limit being from 1 to 2. Note that even though no dependence on x_3 appears in this problem, the stress-strain relation in (3.72) is not obtained by simply suppressing the third index in (3.62).

Consider the variable thickness $h(\boldsymbol{x})$, a thin elastic slab with in-plane loading and fixed edges, as shown in Fig. 3.7, where Ω is a subset of R^2 and Γ is its boundary. The thickness of the slab is bounded such that $h(\boldsymbol{x}) \geq h_0 > 0$. Defining $\boldsymbol{f} = [f_1, f_2]^T$ as the body force per unit area, one can integrate over the x_3-coordinate in (3.60), using (3.72), to obtain the following variational equation:

$$a_{\boldsymbol{u}}(\boldsymbol{z},\bar{\boldsymbol{z}}) = \ell_{\boldsymbol{u}}(\bar{\boldsymbol{z}}), \quad \forall \bar{\boldsymbol{z}} \in Z, \tag{3.74}$$

where the energy bilinear form is defined as

$$a_{\boldsymbol{u}}(\boldsymbol{z},\bar{\boldsymbol{z}}) \equiv \iint_{\Omega} h(\boldsymbol{x})\boldsymbol{\varepsilon}(\bar{\boldsymbol{z}})^T \boldsymbol{C}\boldsymbol{\varepsilon}(\boldsymbol{z})\, d\Omega, \tag{3.75}$$

and Z is the space of kinematically admissible displacements, which satisfy the essential boundary condition, that is,

$$Z = \left\{ \boldsymbol{z} \in [H^1(\Omega)]^2 \;\middle|\; \boldsymbol{z}(\boldsymbol{x}) = 0,\ \boldsymbol{x} \in \Gamma \right\}, \tag{3.76}$$

and the design variable $\boldsymbol{u} = [h(\boldsymbol{x})]$ is the variable thickness of the slab.

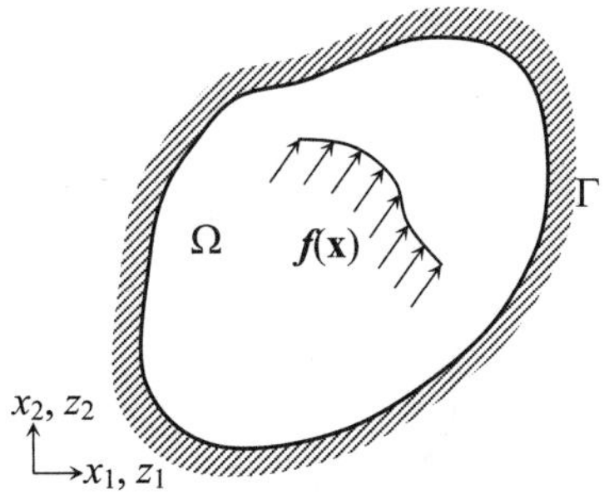

Figure 3.7. Clamped elastic solid of variable thickness $h(\boldsymbol{x})$ in R^2.

The examples in this section have been selected to illustrate the way in which design dependence arises in a consistent way with a certain class of distributed-parameter structural components. In each case, Dirichlet boundary conditions are treated in detail. The selection of these boundary conditions is a convenience rather than a requirement. If the *trace boundary operator* theory were used in its full generality [16] and [22], the Neumann and mixed boundary conditions that naturally arise in applications could also be treated, with the only penalty being the analytic and algebraic complexity of the problem.

3.1.5 Deflection of a Membrane

Consider the membrane shown in Fig. 3.8, with uniform tension T, mass density $\rho(\boldsymbol{x}) \in C^1(\bar{\Omega})$ per unit area, and lateral load $f \in C^1(\bar{\Omega})$, where $\bar{\Omega}$ is the closure of Ω. The formal differential equation for membrane deflection is

$$-T\nabla^2 z = f(\boldsymbol{x}), \qquad \boldsymbol{x} \in \Omega, \tag{3.77}$$

where ∇^2 is the Laplace operator, such that $\nabla^2 g = \partial^2 g/\partial x_1^2 + \partial^2 g/\partial x_2^2$. The boundary condition is $z = 0$ on Γ. By multiplying arbitrary function $\bar{z}$ and integrating by parts, the following variational identity is obtained for the membrane deflection problem:

$$T\iint_\Omega \nabla z^T \nabla \bar{z}\, d\Omega - \iint_\Omega f\,\bar{z}\, d\Omega = T\int_\Gamma \frac{\partial z}{\partial n}\bar{z}\, d\Gamma, \qquad \forall \bar{z} \in H^1(\Omega), \tag{3.78}$$

where $\partial z/\partial n$ is the normal component of ∇z on the boundary. If $\bar{z}$ satisfies the kinematic boundary condition in the above equation, such that $\bar{z} \in Z = H_0^1(\Omega)$ (Note: $H_0^m(\Omega)$ is a subspace of functions from $H^m(\Omega)$ that vanish along with their derivatives up to order $m-1$ on the boundary of Ω. See Appendix A.2.6.), and if the energy bilinear and load linear forms are defined as

$$a_\Omega(z,\bar{z}) = T\iint_\Omega \nabla z^T \nabla \bar{z}\, d\Omega \tag{3.79}$$

and

$$\ell_\Omega(\bar{z}) = \iint_\Omega f\bar{z}\, d\Omega, \tag{3.80}$$

then the structural variational equation for the membrane deflection problem can be obtained as

$$a_u(z,\bar{z}) = \ell_u(\bar{z}), \qquad \forall \bar{z} \in Z. \tag{3.81}$$

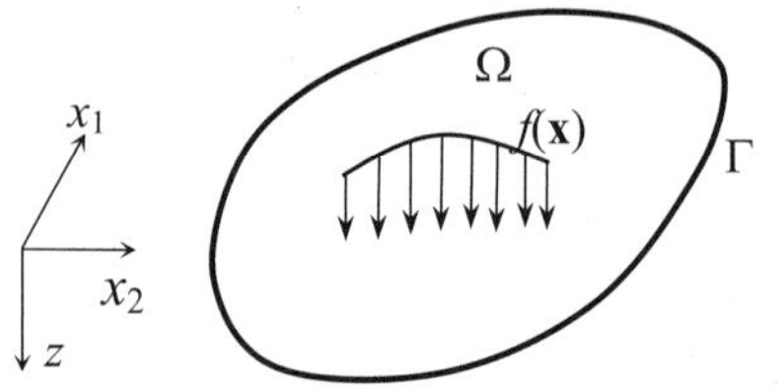

Figure 3.8. Deflection of membrane.

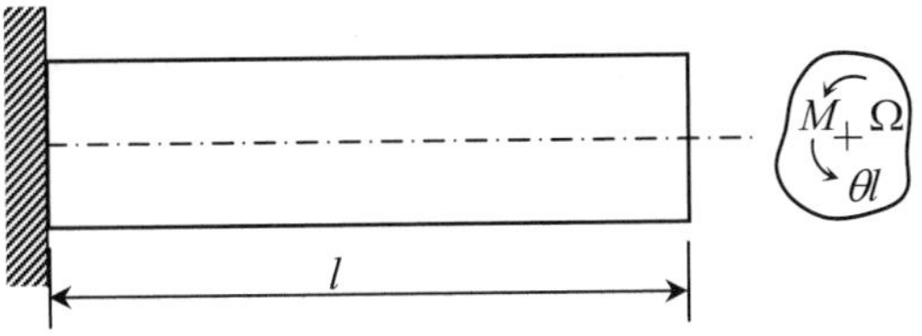

Figure 3.9. Torsion of elastic shaft.

3.1.6 Torsion of an Elastic Shaft

Consider the shaft torsion problem, as illustrated in Fig. 3.9. Torque M is applied to the shaft at its free end, resulting in a unit twist angle θ. According to the St. Vernant theory of torsion [44], the shaft's elastic deformation is governed by the following formal boundary-value problem:

$$\begin{aligned} Az \equiv -\nabla^2 z = 2, \quad & \boldsymbol{x} \in \Omega \\ z = 0, \quad & \boldsymbol{x} \in \Gamma, \end{aligned} \tag{3.82}$$

where z is the Prandtl stress function. The torque-angular deflection relation is given by $M = \mu J \theta$, where μ is the shear modulus of the shaft material, and J is the torsional rigidity, given by

$$J = 2 \iint_\Omega z \, d\Omega. \tag{3.83}$$

By comparing (3.77) and (3.82), it is easy to note that they are exactly the same if $f/T = 2$, which is the basis for the membrane analogy approach [44]. Hence, the variational identity for the shaft is

$$\iint_\Omega \nabla z^T \nabla \bar{z} \, d\Omega - 2 \iint_\Omega \bar{z} \, d\Omega = \int_\Gamma \frac{\partial z}{\partial n} \bar{z} \, d\Gamma, \qquad \forall \bar{z} \in H^1(\Omega). \tag{3.84}$$

If the kinematic boundary condition is imposed in (3.84), the variational equation then becomes

$$a_\Omega(z, \bar{z}) \equiv \iint_\Omega \nabla z^T \nabla \bar{z} \, d\Omega = 2 \iint_\Omega \bar{z} \, d\Omega \equiv \ell_\Omega(\bar{z}), \qquad \forall \bar{z} \in Z = H_0^1(\Omega). \tag{3.85}$$

3.1.7 General Form of Static Variational Equations

In all previous examples, the boundary-value problem for deformation due to applied load was written as it appears in the literature on mechanics. For classical solutions to make sense, a high degree of smoothness of design and state (displacement) functions must be assumed. In each example, however, both sides of the differential equation could be multiplied by an arbitrary *virtual displacement* $\bar{z}$ that satisfies the *kinematic boundary conditions*, integrated over the domain of the component, and integrated by parts to reduce the order of the derivatives of z that appear, so that z and $\bar{z}$ are differentiated to the same order. The result is a *variational equation* of the form

$$a(z,\bar{z}) = \ell(\bar{z}), \tag{3.86}$$

which must hold for all *kinematically admissible* smooth virtual displacements $\bar{z} \in Z$. As noted in each example, (3.86) can be viewed as the principle of virtual work and can be derived directly from the variational principle of mechanics.

Specific forms of (3.86) are given for a truss in (3.8), for a beam in (3.16) and (3.28), for a plate in (3.43) and (3.55), and for a linear elastic solid in (3.60). While the details of specific formulas differ, they are all in the form of (3.86). This form is adequate for an engineering design sensitivity analysis using the adjoint variable method, which will be presented in Section 5.2 of Chapter 5. From a mathematical point of view, however, it should be noted that for each specific case, the state z does not need to be restricted to a classical smooth space of displacement, but can extend to a subspace Z of an appropriate Sobolev space of functions that satisfy *kinematic boundary conditions*. Likewise, the design space can be extended to a nonsmooth design space U. The ability to extend design space is important and theoretically valuable if one wishes to admit nonsmooth designs.

In each of the examples studied, it can be observed that positive constants K and γ exist, such that

$$a_u(z,\bar{z}) \le K\|z\|_Z \|\bar{z}\|_Z, \qquad \forall z,\bar{z} \in Z \tag{3.87}$$

and

$$a_u(z,z) \ge \gamma\|z\|_Z^2, \qquad \forall z \in Z, \tag{3.88}$$

where $K < \infty$, $\gamma > 0$, and $\boldsymbol{u} \in U$ is restricted to be uniformly nonzero. Here, $\|\bullet\|_Z$ denotes the appropriate Sobolev norm [Appendix A.2]. Equation (3.87) is an upper bound on the bilinear form, while (3.88) is a lower bound on the strain energy, which is called the *strong ellipticity* or *Z-ellipticity* property (i.e., positive bounded below). To observe the physical significance of these inequalities, place $\bar{z} = z$ into (3.87), and use (3.88) to obtain

$$\gamma\|z\|_Z^2 \le a_u(z,z) \le K\|z\|_Z^2, \qquad \forall z \in Z. \tag{3.89}$$

Since the value of $a_u(z,z)$ is twice the strain energy in each example, (3.89) shows that strain energy defines an energy norm that is equivalent to the Sobolev norm. This important fact has been used to advantage by Mikhlin [38] and [39] and others to develop powerful variational methods. Any stronger or weaker norm would destroy the bounds of either (3.87) or (3.88), therefore spoiling the equivalence between energy and function space norms.

Furthermore, the inequalities in (3.87) and (3.88) as well as the Lax-Milgram theorem [16] of functional analysis guarantee the existence of a unique solution $z(\boldsymbol{x};\boldsymbol{u})$ to (3.86). Again, a stronger or weaker norm in Z would spoil either the existence or the uniqueness of the solution. Thus, the Sobolev space setting is "just right," from both a physical and mathematical point of view.

For the purpose of design sensitivity analysis, the variational formulation in (3.86) and the inequalities in (3.87) and (3.88) are the foundation for a proof with the following results: solution $z(\boldsymbol{x};\boldsymbol{u})$ is differentiable with respect to the design; (3.86) can be differentiated with respect to the design; and the result can be used to write variations of cost and constraint functions explicitly. An adjoint variable method for implementing this technique is presented and illustrated in Section 5.2 of Chapter 5. While its theoretical foundations require the use of the Sobolev space setting, this method is implemented and its calculations are carried out without a rigorous use of functional analysis.

3.2 Vibration and Buckling of Elastic Systems

The eigenvalue problem discussed in Section 2.5 of Chapter 2 is extended in this section for various structural components. The variational equation of the eigenvalue problem is

$$a_u(\boldsymbol{y},\bar{\boldsymbol{y}}) = \zeta d_u(\boldsymbol{y},\bar{\boldsymbol{y}}), \qquad \forall \bar{\boldsymbol{y}} \in Z, \tag{3.90}$$

where ζ is the eigenvalue, $\boldsymbol{y}$ is the eigenfunction, and Z is the space of kinematically admissible displacements. In this section, the expressions of $a_u(\boldsymbol{y},\bar{\boldsymbol{y}})$ and $d_u(\boldsymbol{y},\bar{\boldsymbol{y}})$ are developed for each structural component.

Vibration of a String

A perfectly flexible string of variable mass density per unit length, $\rho \in C^0(0,l)$ or $L^\infty(0,l)$ ($\rho(x) \geq \rho_0 > 0$) and tension $T \geq T_0 > 0$, is shown in Fig. 3.10. The design vector is $\boldsymbol{u} = [\rho(x), T]^T$. The differential equation of the eigenvalue problem is

$$-Ty_{,11} = \zeta \rho y, \tag{3.91}$$

where $\zeta = \omega^2$, with ω as the natural frequency of the vibration and the eigenfunction $y(x)$ determining the vibrating shape of the string. The boundary conditions are

$$y(0) = y(l) = 0. \tag{3.92}$$

As with the static problem, multiply the differential equation (3.91) with an arbitrary function $\bar{y}$ and integrate over the domain $(0,l)$. After integrating by parts, the following variational identity can be obtained:

$$T\int_0^l y_{,1}\bar{y}_{,1}\,dx - \zeta\int_a^b \rho y\bar{y}\,dx = [-Ty_{,1}\bar{y}]\Big|_0^l. \tag{3.93}$$

An identical form of variational equation (3.90) can be obtained by defining the bilinear forms, as

$$a_u(y,\bar{y}) = T\int_0^l y_{,1}\bar{y}_{,1}\,dx \tag{3.94}$$

$$d_u(y,\bar{y}) = \int_0^l \rho y\bar{y}\,dx, \tag{3.95}$$

and by applying the boundary condition in (3.92). Unlike the static problems discussed in Section 3.1, the notation $y(x;\boldsymbol{u})$ will be used for the eigenfunction. Since only first-order

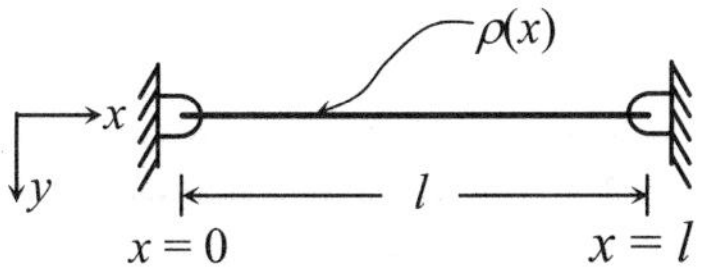

Figure 3.10. Vibrating string with linear mass density $\rho(x)$.

derivatives appear in the formula for $a_u(y,\bar{y})$, it is logical to select $Z \subset H^1(0,l)$. The boundary conditions of (3.92) are satisfied in a generalized sense [16] and [22] if space Z of kinematically admissible displacements is restricted to $Z = H_0^1(0,l)$. It is readily verifiable [16] and [17] that the form $a_u(y,y)$ is Z-elliptic, so the theory discussed in Section 3.1 also holds true for this problem.

Vibration of a Beam

For a technical beam with variable cross-sectional area $A(x)$, let $A \in C^0(0,l)$ or $L^\infty(0,l)$ $(A(x) \geq A_0 > 0)$, such that the moment of inertia is $I(x) = \alpha A(x)^2$. Young's modulus $E \geq E_0 > 0$ and the mass density $\rho \geq \rho_0 > 0$ also serve as design variables. Here, the design vector is $\boldsymbol{u} = [A(x), E, \rho]^T$. A beam with clamped-clamped supports is shown in Fig. 3.11. The formal differential equation of the eigenvalue problem is

$$(E\alpha A^2 y_{,11})_{,11} = \zeta \rho A y, \tag{3.96}$$

where $\zeta = \omega^2$, with ω as the natural frequency. Boundary conditions for the clamped-clamped beam are

$$y(0) = y_{,1}(0) = y(l) = y_{,1}(l) = 0. \tag{3.97}$$

As with the string problem, the variational identity is obtained by multiplying an arbitrary function $\bar{y}$ with (3.96) and integrating by parts, as

$$\int_0^l E\alpha A^2 y_{,11}\bar{y}_{,11}\,dx - \zeta \int_0^l \rho A y \bar{y}\,dx = \left[E\alpha A^2 y_{,11}\bar{y}_{,1} - (E\alpha A^2 y_{,11})_{,1}\bar{y}\right]\Big|_0^l. \tag{3.98}$$

After applying the boundary condition in (3.97) and defining the following bilinear forms

$$a_u(y,\bar{y}) = \int_0^l E\alpha A^2 y_{,11}\bar{y}_{,11}\,dx \tag{3.99}$$

$$d_u(y,\bar{y}) = \rho \int_0^l A y \bar{y}\,dx, \tag{3.100}$$

the variational eigenvalue equation, (3.90), is obtained for the beam vibration problem. Since only second derivatives arise in $a_u(y,\bar{y})$, it is logical to select $Z = H_0^2(0,l)$. The boundary conditions in (3.96) are satisfied in a generalized sense [16] and [17] if space Z of kinematically admissible displacements is defined as $Z = H_0^2(0,l)$. All the properties of $a_u(y,\bar{y})$ of interest are shown in Section 3.1. Note that the smoothness requirement of $d_u(y,\bar{y})$ is less than that of $a_u(y,\bar{y})$. As noted in Section 3.1, the bilinear forms of (3.99) and (3.100) are equally valid for other boundary conditions given in (3.19) through (3.21).

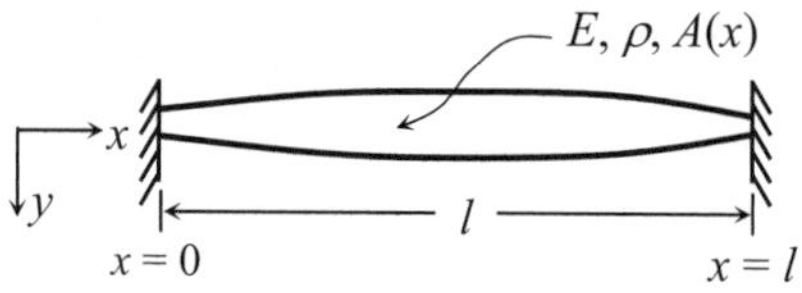

Figure 3.11. Clamped-clamped vibrating beam with variable moment of inertia $I(x)$.

Buckling of a Column

If a column is subjected to an axial load P, as shown in Fig. 3.12, then buckling can occur if P is larger than the critical load ζ. Using the same design variables as the beam vibration, the formal differential equation of the eigenvalue problem is obtained in Section 2.5 of Chapter 2 as

$$(E\alpha A^2 y_{,11})_{,11} = -\zeta y_{,11}, \tag{3.101}$$

with the same boundary conditions as (3.97). Since mass density does not arise in column buckling, the design vector is $\boldsymbol{u} = [A(x),\ E]^T$. By multiplying $\bar{y} \in Z$ and performing an integration by parts for differential equation (3.101), the variational identity and the bilinear forms in (3.90) can be obtained for the column buckling problem as

$$\int_0^l E\alpha A^2 y_{,11}\bar{y}_{,11}\,dx - \zeta \int_0^l y_{,1}\bar{y}_{,1}\,dx = \left[E\alpha A^2 y_{,11}\bar{y}_{,1} - (E\alpha A^2 y_{,11})_{,1}\bar{y} - \zeta y_{,1}\bar{y} \right]\Big| \tag{3.102}$$

$$a_u(y,\bar{y}) = \int_0^l E\alpha A^2 y_{,11}\bar{y}_{,11}\,dx \tag{3.103}$$

$$d_u(y,\bar{y}) = \int_0^l y_{,1}\bar{y}_{,1}\,dx, \tag{3.104}$$

where y and $\bar{y}$ satisfy the boundary conditions in (3.97). Since $a_u(y,\bar{y})$ and $d_u(y,\bar{y})$ involve derivatives of y and $\bar{y}$ that are no higher than the second order, and since the boundary conditions are the same as the vibrating beam, space Z of kinematically admissible displacements may again be defined as $Z = H_0^2(0,l)$.

Vibration of a Membrane

Consider a vibrating membrane with variable mass density $\rho(\mathbf{x})$ per unit area, $\rho \in C^0(\Omega)$ or $L^\infty(\Omega)$ ($\rho(\mathbf{x}) \ge \rho_0 > 0$) and with membrane tension T, as shown in Fig. 3.13. The formal differential equation of the eigenvalue problem would be

$$-T\nabla^2 y = \zeta \rho y, \tag{3.105}$$

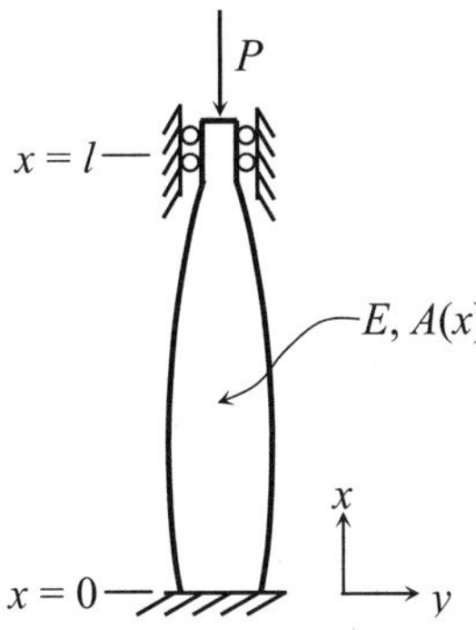

Figure 3.12. Clamped-clamped column with variable moment of inertia $I(x)$.

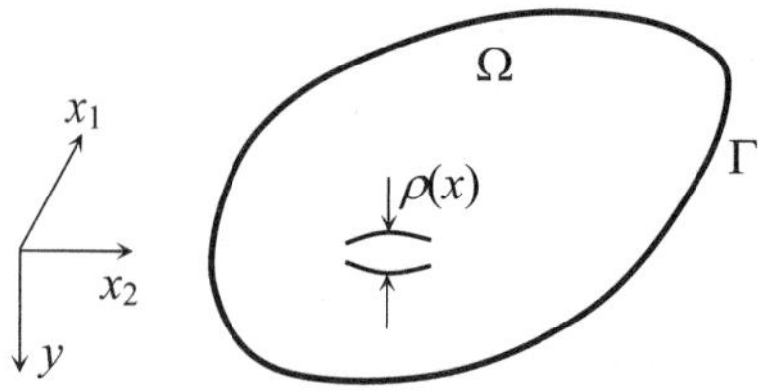

Figure 3.13. Membrane of variable mass density $\rho(\boldsymbol{x})$.

where ∇^2 is the Laplace operator, such that $\nabla^2 g = \partial^2 g/\partial x_1^2 + \partial^2 g/\partial x_2^2$, and $\zeta = \omega^2$, with ω being the natural frequency, and with the following boundary condition:

$$y = 0, \quad \text{on } \Gamma. \tag{3.106}$$

Here $\boldsymbol{u} = [\rho(\boldsymbol{x}),\ T]^T$ is the design variable, and the variational identity and the bilinear forms of (3.90) are

$$T \iint_\Omega \nabla y^T \nabla \bar{y}\, d\Omega - \zeta \iint_\Omega \rho\, y\, \bar{y}\, d\Omega = T \int_\Gamma \frac{\partial y}{\partial n} \bar{y}\, d\Gamma \tag{3.107}$$

$$a_{\boldsymbol{u}}(y, \bar{y}) = T \iint_\Omega \nabla y^T \nabla \bar{y}\, d\Omega \tag{3.108}$$

$$d_{\boldsymbol{u}}(y, \bar{y}) = \iint_\Omega \rho\, y\, \bar{y}\, d\Omega, \tag{3.109}$$

where y and $\bar{y}$ satisfy the boundary condition of (3.106). As with a vibrating string, $Z = H_0^1(\Omega)$, and the bilinear form $a_{\boldsymbol{u}}(y, \bar{y})$ is Z-elliptic [16] and [17].

Vibration of a Plate

Consider a clamped vibrating plate of variable thickness, $h \in C^0(\Omega)$ or $L^\infty(\Omega)$ ($h(\boldsymbol{x}) \geq h_0 > 0$), with Young's modulus $E \geq E_0 > 0$, and a mass density $\rho \geq \rho_0 > 0$, as shown in Fig. 3.14. The formal differential equation of the eigenvalue problem is

$$[D(\boldsymbol{u})(y_{,11} + \nu y_{,22})]_{,11} + [D(\boldsymbol{u})(y_{,22} + \nu y_{,11})]_{,22} + 2(1-\nu)[D(\boldsymbol{u}) y_{,12}]_{,12} = \zeta \rho h y, \tag{3.110}$$

where $D(\boldsymbol{u})$ is the flexural rigidity given in (3.33), $\zeta = \omega^2$ with ω being the natural frequency, $0 < \nu < 0.5$ is Poisson's ratio, and the boundary condition for a clamped plate is

$$y = \frac{\partial y}{\partial n} = 0, \quad \text{on } \Gamma, \tag{3.111}$$

where $\partial y/\partial n$ is the normal derivative of y on Γ. Here the design vector is $\boldsymbol{u} = [h(\boldsymbol{x}), E, \rho]^T$.

By multiplying (3.110) by $\bar{y}$, integrating over Ω, and integrating by parts, the following variational identity and bilinear forms of (3.90) are obtained:

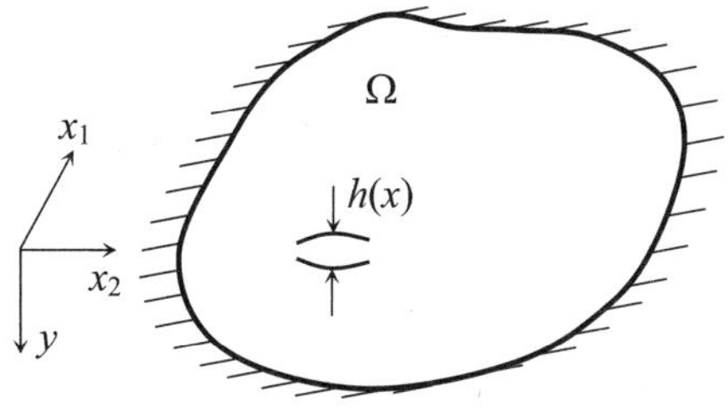

Figure 3.14. Clamped plate of variable thickness $h(\boldsymbol{x})$.

$$\iint_{\Omega} \boldsymbol{\kappa}(\bar{z})^T \boldsymbol{C}^b \boldsymbol{\kappa}(z)\, d\Omega - \zeta\rho \iint_{\Omega} hy\bar{y}\, d\Omega = \int_{\Gamma} \bar{y} My\, d\Gamma + \int_{\Gamma} \frac{\partial \bar{y}}{\partial n} Ny\, d\Gamma \tag{3.112}$$

$$a_{\boldsymbol{u}}(y,\bar{y}) = \iint_{\Omega} \boldsymbol{\kappa}(\bar{y})^T \boldsymbol{C}^b \boldsymbol{\kappa}(y)\, d\Omega \tag{3.113}$$

$$d_{\boldsymbol{u}}(y,\bar{y}) = \rho \iint_{\Omega} hy\bar{y}\, d\Omega, \tag{3.114}$$

where y and $\bar{y}$ satisfy the boundary conditions of (3.111). As with a vibrating beam, the natural domain of the energy bilinear form $a_{\boldsymbol{u}}(y,\bar{y})$ is $Z = H_0^2(\Omega)$.

Given these bilinear forms, the variational formula presented at the beginning of this section characterizes the eigenvalue behavior of each of the five problems discussed. They all have the same basic variational structure, and all bilinear forms share the same degree of regularity in terms of their design dependence. In each problem studied in this section, eigenvalue ζ depends on design $\boldsymbol{u}$, since the differential equations and variational equations depend on $\boldsymbol{u}$ (i.e., $\zeta = \zeta(\boldsymbol{u})$). The objective is to determine how ζ depends on $\boldsymbol{u}$. An analysis of the sensitivity of ζ with respect to changes in $\boldsymbol{u}$ is somewhat more complicated than the static displacement problem in Section 3.1, since eigenfunction y also depends on $\boldsymbol{u}$ (i.e., $y = y(\boldsymbol{x};\boldsymbol{u})$) and its sensitivity must also be taken into consideration.

General Form of Eigenvalue Variational Equations

A unified variational form for each eigenvalue problem is obtained in the form of (3.90), much as the variational form for static response was obtained in Section 3.1. While detailed expressions of the bilinear forms are different for each example, the same general properties hold true. The most general function space setting is given in each example, but an engineer primarily interested in applications may assume that design and state variables are as smooth as desired. For readers who want rigorous derivations, detailed Sobolev space settings are used in Haug et al. [5] to prove the differentiability of eigenvalues and to derive formulas to calculate eigenvalue derivatives with respect to design.

3.3 Finite Element Structural Equations

A structural analysis based on the finite element approach is introduced in this section, using beam, truss, plate, and solid elements as models. Apart from more intricate algebra that is required for more complex elements, the basic approach for deriving element equations is identical to the process illustrated in this section. Finite element methods of structural analysis require knowledge of the behavior of each element in the structure. Once each element is described, the governing equations of the entire structure may then be derived.

3.3.1 Truss Element

Consider a finite element discretization of the truss component using a linear polynomial function. Figure 3.15 shows a truss finite element with the linear cross-sectional area $A(x)$. For manufacturing purposes it is useful in the structural design of a truss component if $A(x)$ remains constant within an element. Even if several elements are connected, each of which has a different cross-sectional area, the structural energy form in (3.6) will be well defined, since it is represented by the integration of discontinuous functions.

An important component of the finite element method is transforming the general-shaped element into a reference element from which a unified formulation can be obtained. The relation between finite and reference elements can be obtained from the mapping relation. Let a position of x in the finite element be mapped onto $\xi \in [-1,1]$ in the reference element. The mapping relation between x and ξ would be

$$
\begin{gathered}
x = N_1(\xi)x_1 + N_2(\xi)x_2 \\
N_1(\xi) = \frac{1-\xi}{2}, \quad N_2(\xi) = \frac{1+\xi}{2},
\end{gathered}
\tag{3.115}
$$

where x_1 and x_2 are the locations of nodes 1 and 2, and N_1 and N_2 are corresponding shape functions. If the finite element is isoparametric, the displacement is approximated in the same way as in (3.115). For convenience, let us define the following vector notations:

$$
\boldsymbol{x} = \begin{bmatrix} x_1 \\ x_2 \end{bmatrix}, \quad \boldsymbol{d} = \begin{bmatrix} z_1 \\ z_2 \end{bmatrix}, \quad \boldsymbol{N}(\xi) = \begin{bmatrix} N_1 \\ N_2 \end{bmatrix}. \tag{3.116}
$$

Displacement z, which corresponds to the reference coordinate ξ in the element, can be approximated as

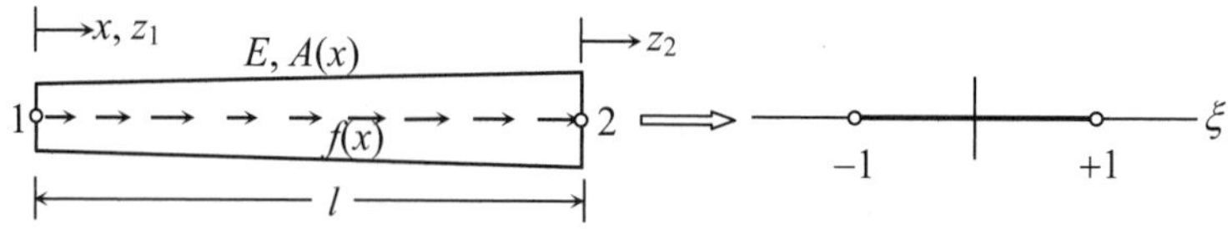

(a) Finite Element (b) Reference

Figure 3.15. Linear truss finite element.

$$z(x) = \boldsymbol{N}(\xi)^T \boldsymbol{d}. \tag{3.117}$$

In the Galerkin approximation of a structural problem, the same discretization is used for the displacement variation, such that

$$\bar{z}(x) = \boldsymbol{N}(\xi)^T \bar{\boldsymbol{d}}, \tag{3.118}$$

where $\bar{\boldsymbol{d}} = [\bar{z}_1, \bar{z}_2]^T$ is the nodal displacement variation.

To approximate the variational equation (3.8), we have to obtain the derivative of the displacement in (3.117), with respect to the spatial coordinate x. Since the reference coordinate is a function of ξ, the following is obtained using the chain rule of differentiation:

$$\frac{dz}{dx} = \boldsymbol{d}^T \frac{d\boldsymbol{N}}{d\xi}\frac{d\xi}{dx}, \tag{3.119}$$

where $d\xi/dx$ can be obtained from the inverse of the mapping relation in (3.115), as

$$\begin{aligned} \frac{dx}{d\xi} &= \boldsymbol{x}^T \frac{d\boldsymbol{N}}{d\xi} = \frac{1}{2}(x_2 - x_1) = \frac{l}{2} \\ \frac{d\xi}{dx} &= \frac{2}{l}. \end{aligned} \tag{3.120}$$

Thus,

$$\begin{aligned} \frac{dz}{dx} &= \frac{1}{l}[-1 \quad 1]\boldsymbol{d} \\ &= \boldsymbol{B}\boldsymbol{d}, \end{aligned} \tag{3.121}$$

where $\boldsymbol{B}$ is the discrete *displacement-strain matrix*, even though in this special example, it is a row vector. By using (3.121), the structural energy bilinear form of (3.6) can be approximated by

$$\begin{aligned} a_u(z,\bar{z}) &= \bar{\boldsymbol{d}}^T \int_0^l EA(x)\boldsymbol{B}^T\boldsymbol{B}\,dx\,\boldsymbol{d} \\ &= \bar{\boldsymbol{d}}^T \boldsymbol{k}\,\boldsymbol{d}, \end{aligned} \tag{3.122}$$

where $\boldsymbol{k}$ is the 2×2 *element stiffness matrix*. In the simplest case in which the constant cross-sectional area A is used, the integrand of (3.122) is constant. Thus, explicit computation of the $\boldsymbol{k}$ matrix can be obtained as

$$\boldsymbol{k} = \frac{EA}{l}\begin{bmatrix} 1 & -1 \\ -1 & 1 \end{bmatrix}. \tag{3.123}$$

The load linear form in (3.7) can easily be approximated for the linear and constant distribution of load $f(x)$. In the case of constant load distribution, (3.7) is approximated as

$$\begin{aligned} \ell_u(\bar{z}) &= \int_0^l \bar{z} f\,dx \\ &\approx \bar{\boldsymbol{d}}^T \int_0^l \boldsymbol{N}(\xi) f\,dx, \end{aligned} \tag{3.124}$$

where a transformation between x and ξ is required. Since a one-to-one mapping relation exists between x and ξ, we have the relation $\int dx = \int J d\xi$, where J is the determinant of the Jacobian of the mapping relation in (3.120), and is defined as

$$J = \left| \frac{dx}{d\xi} \right| = \frac{l}{2}. \tag{3.125}$$

Integration of the load linear form is therefore transformed to the reference coordinate as

$$\begin{aligned} \ell_u(\bar{z}) &\approx \bar{\boldsymbol{d}}^T \int_{-1}^{1} \begin{bmatrix} N_1 \\ N_2 \end{bmatrix} J \, d\xi \, f \\ &= \bar{\boldsymbol{d}}^T \begin{bmatrix} 1 \\ 1 \end{bmatrix} \frac{fl}{2} \\ &\equiv \bar{\boldsymbol{d}}^T \boldsymbol{f}, \end{aligned} \tag{3.126}$$

where $\boldsymbol{f}$ is the *element force vector*. In contrast, if the concentrated nodal force is applied to the element, then $\boldsymbol{f}$ would simply be $\boldsymbol{f} = [f_1, f_2]^T$. From (3.122) and (3.126), the discrete variational equation of the truss component would therefore be

$$\bar{\boldsymbol{d}}^T \boldsymbol{k} \boldsymbol{d} = \bar{\boldsymbol{d}}^T \boldsymbol{f}, \tag{3.127}$$

for all $\bar{\boldsymbol{d}}$ in Z_h where $Z_h \subset R^2$ is the discrete space of kinematically admissible displacements.

3.3.2 Beam Element

Bernoulli-Euler Beam (C^1 Approach)

Consider the linear beam element shown in Fig. 3.16, with Young's modulus E, shear modulus μ, second moment of inertia I, cross-sectional area A, element length $l = x_2 - x_1$, and linearly distributed force $f(x)$ per unit length. In the Bernoulli-Euler beam theory, the transverse displacement and rotation are nodal degrees-of-freedom, which are related by $\theta_1 = z_{1,1}$ and $\theta_2 = z_{2,1}$. In the reference domain, a different mapping scheme used for the truss element in (3.115) is employed for computational convenience, such that reference coordinate $\xi \in [0,1]$. Thus, the Jacobian of (3.125) can be obtained from

$$\xi = \frac{1}{l}(x - x_1) \quad \Rightarrow \quad J = \left| \frac{dx}{d\xi} \right| = l. \tag{3.128}$$

Since each element has four independent variables, the transverse displacement z and the slope θ can be approximated using a cubic polynomial, as

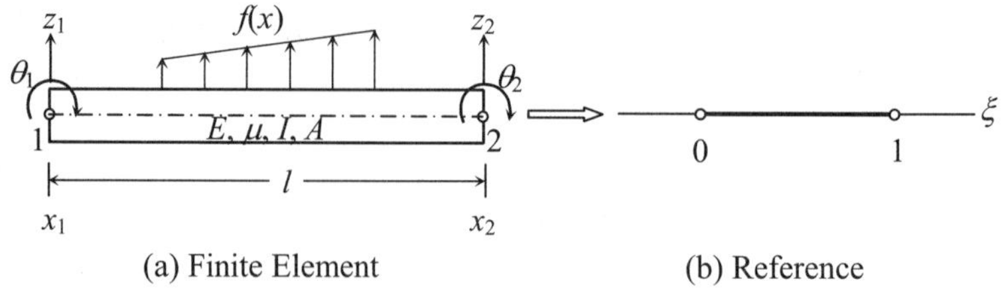

Figure 3.16. Technical beam finite element.

$$
\begin{aligned}
z(\xi) &= a_0 + a_1\xi + a_2\xi^2 + a_3\xi^3 \\
\theta(\xi) &= z_{,1}(\xi) = \frac{dz}{d\xi}\frac{d\xi}{dx} = \frac{1}{l}(a_1 + 2a_2\xi + 3a_3\xi^2),
\end{aligned}
\tag{3.129}
$$

and, by imposing the following conditions,

$$
\begin{aligned}
z(0) &= z_1 \\
\theta(0) &= \theta_1 \\
z(1) &= z_2 \\
\theta(1) &= \theta_2,
\end{aligned}
\tag{3.130}
$$

the unknown coefficient a_i can be found. After algebraic manipulation, the following interpolation relation can be obtained:

$$
z(\xi) = [N_1 \quad N_2 \quad N_3 \quad N_4]\begin{bmatrix} z_1 \\ \theta_1 \\ z_2 \\ \theta_2 \end{bmatrix} \equiv \boldsymbol{N}^T\boldsymbol{d},
\tag{3.131}
$$

where $\boldsymbol{d}$ contains all unknowns of the element and N_i are Hermite shape functions, defined as

$$
\begin{aligned}
N_1(\xi) &= 1 - 3\xi^2 + 2\xi^3 \\
N_2(\xi) &= l(\xi - 2\xi^2 + \xi^3) \\
N_3(\xi) &= 3\xi^2 - 2\xi^3 \\
N_4(\xi) &= l(-\xi^2 + \xi^3).
\end{aligned}
\tag{3.132}
$$

Figure 3.17 plots the Hermite shape functions along the neutral axis in the reference element.

To construct the stiffness matrix of a beam element, the second derivative of transverse displacement z has to be computed from the approximation in (3.131), as

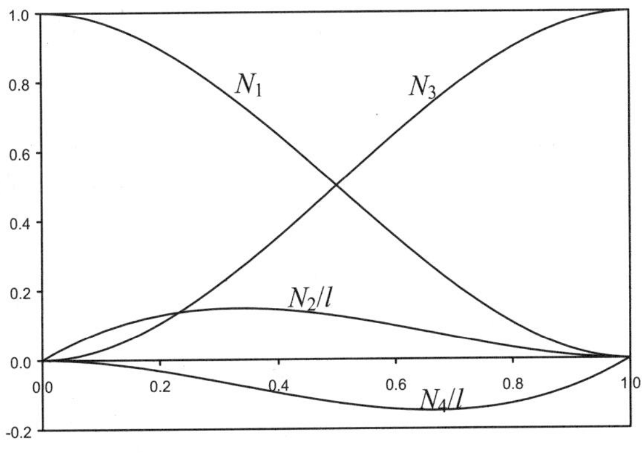

Figure 3.17. Hermite shape functions of Bernoulli-Euler beam element.

$$z_{,11} = \frac{1}{l^2}[-6+12\xi, l(-4+6\xi), 6-12\xi, l(-2+6\xi)]\boldsymbol{d} \equiv \boldsymbol{B}\boldsymbol{d}, \tag{3.133}$$

where $\boldsymbol{B}$ is the discrete *displacement-strain matrix*, although in this special example, it is a row vector. By using (3.133), the structural energy bilinear form in (3.14) can be approximated by

$$\begin{aligned} a_u(z,\bar{z}) &= \bar{\boldsymbol{d}}^T \int_0^1 EI\,\boldsymbol{B}^T\boldsymbol{B}\,J\,d\xi\,\boldsymbol{d} \\ &= \bar{\boldsymbol{d}}^T\boldsymbol{k}\,\boldsymbol{d}, \end{aligned} \tag{3.134}$$

where $\boldsymbol{k}$ is the *element stiffness matrix* of the technical beam element. For the simplest case in which the constant, second moment of inertia I is used, the element stiffness matrix in (3.134) can be explicitly obtained as

$$\boldsymbol{k} = \frac{EI}{l^3}\begin{bmatrix} 12 & 6l & -12 & 6l \\ & 4l^2 & -6l & 2l^2 \\ & & 12 & -6l \\ & sym. & & 4l^2 \end{bmatrix}. \tag{3.135}$$

For the load linear form in (3.15), let us consider a uniformly distributed load f per unit length. The load linear form in (3.15) is approximated by

$$\begin{aligned} \ell_u(\bar{z}) &= \int_{x_1}^{x_2} \bar{z}f\,dx \\ &\approx \bar{\boldsymbol{d}}^T \int_0^1 \boldsymbol{N}J\,d\xi\,f \\ &= \bar{\boldsymbol{d}}^T \begin{bmatrix} fl/2 \\ fl^2/12 \\ fl/2 \\ -fl^2/12 \end{bmatrix} \\ &\equiv \bar{\boldsymbol{d}}^T\boldsymbol{f}, \end{aligned} \tag{3.136}$$

where $\boldsymbol{f}$ is the element force vector. If the concentrated nodal force and moment are applied to the nodes in (3.136), then $\boldsymbol{f} = [f_1, m_1, f_2, m_2]^T$. From (3.134) and (3.136), the discrete variational equation of the technical beam element can be obtained as

$$\bar{\boldsymbol{d}}^T\boldsymbol{k}\boldsymbol{d} = \bar{\boldsymbol{d}}^T\boldsymbol{f}, \tag{3.137}$$

for all $\bar{\boldsymbol{d}}$ in $Z_h \subset R^4$, and Z_h satisfies all essential boundary conditions.

For transient dynamics or eigenvalue problems, the kinetic energy form in (3.100) is required, whose discretization becomes

$$\begin{aligned} d_u(z,\bar{z}) &\approx \bar{\boldsymbol{d}}^T \int_0^1 \rho A \boldsymbol{N}\boldsymbol{N}^T J\,d\xi\,\boldsymbol{d} \\ &= \frac{\rho A l}{420}\bar{\boldsymbol{d}}^T \begin{bmatrix} 156 & 22l & 54 & -13l \\ 22l & 4l^2 & 13l & -3l^2 \\ 54 & 13l & 156 & -22l \\ -13l & -3l^2 & -22l & 4l^2 \end{bmatrix} \boldsymbol{d} \\ &\equiv \bar{\boldsymbol{d}}^T\boldsymbol{m}\,\boldsymbol{d}, \end{aligned} \tag{3.138}$$

where $\boldsymbol{m}$ is the consistent *element mass matrix*. Since the kinetic energy of the beam element is positive for any nonzero deformation, it is expected that $\boldsymbol{m}$ will be positive definite; hence, nonsingular. These properties can be analytically verified.

For the column buckling problem, the bilinear form $d_u(z,\bar{z})$ differs from (3.138). In (3.104), first-order derivatives are involved whose discretization yields a geometric stiffness matrix, as

$$\begin{aligned} d_u(z,\bar{z}) &\approx \bar{\boldsymbol{d}}^T \int_0^1 \boldsymbol{N}_{,1}\boldsymbol{N}_{,1}^T J\, d\xi\, \boldsymbol{d} \\ &= \bar{\boldsymbol{d}}^T \frac{1}{60l}\begin{bmatrix} 36 & 3l & -36 & 3l \\ 3l & 4l^2 & -3l & -l^2 \\ -36 & -3l & 36 & -3l \\ 3l & -l^2 & -3l & 4l^2 \end{bmatrix}\boldsymbol{d} \\ &\equiv \bar{\boldsymbol{d}}^T \boldsymbol{b}\boldsymbol{d}. \end{aligned} \tag{3.139}$$

It is important to note that the beam-element stiffness matrix in (3.135) depends on length l of the beam element, cross-sectional area A, and moment of inertia I of the cross-sectional area. If the cross-sectional dimensions of the beam element are taken as design variables, as is the case when the structural element sizes are regarded as design variables, then the element stiffness matrix also depends on design variables. If the geometry of the structure is changed, then element length l depends on the design variables and is also involved in a nonlinear way with the element stiffness matrix.

Somewhat more complex computations will show that the element stiffness matrix in (3.135) is positive semidefinite and of rank two. Matrix rank is associated with the physical observable fact that the element shown in Fig. 3.16 has two rigid-body degrees of freedom. That is, it is possible to move the element in the plane with two kinematic degrees of freedom without any deformation or strain energy. Such an element is said to have a free energy motion. However, if the left end of the beam element shown in Fig. 3.16 were fixed (i.e., $z_1 = \theta_1 = 0$), then the energy bilinear form calculated by (3.14) would be positive definite in variables z_2 and θ_2. As illustrated throughout this text, positive definiteness of the system strain energy plays a crucial role in the mathematical theory of design sensitivity analysis.

Timoshenko Beam (C^0 Approach)

Consider the linear beam element in Fig. 3.18 with Young's modulus E, shear modulus μ, second moment of inertia I, cross-sectional area A, element length $l = x_2 - x_1$, and linearly distributed force $f(x)$. The C^0 beam element as shown in Fig. 3.18 has transverse displacement and rotation as independent nodal degrees-of-freedom. Furthermore, since only first-order derivatives are involved, linear trial solutions can be used. Similar to the linear truss element in the previous section, the Lagrange interpolation formula can be used for writing shape functions. Thus, the mapping relation between x and ξ is the same as in (3.115).

If the finite element is isoparametric, then the transverse displacement and rotation are approximated in the same way as in (3.117). For convenience, the following vector notations are defined:

$$\boldsymbol{d} = \begin{bmatrix} z_1 \\ \theta_1 \\ z_2 \\ \theta_2 \end{bmatrix}, \quad \boldsymbol{N}_z(\xi) = \begin{bmatrix} N_1 \\ 0 \\ N_2 \\ 0 \end{bmatrix}, \quad \boldsymbol{N}_\theta(\xi) = \begin{bmatrix} 0 \\ N_1 \\ 0 \\ N_2 \end{bmatrix}, \tag{3.140}$$

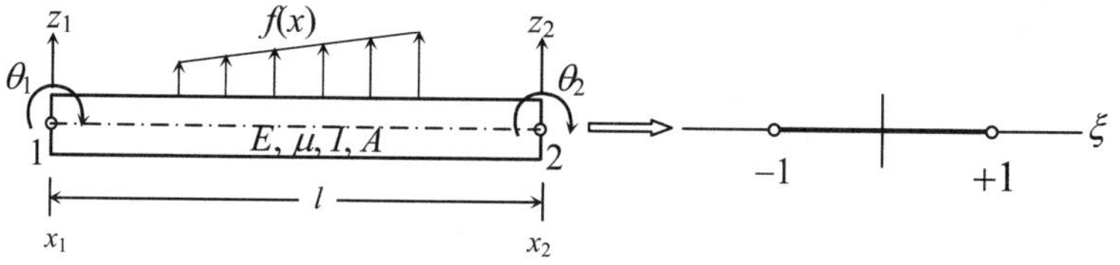

Figure 3.18. Linear beam finite element.

where $\boldsymbol{d}$ contains all unknowns of the element, $\boldsymbol{N}_z(\xi)$ is the shape function of transverse displacement, and $\boldsymbol{N}_\theta(\xi)$ is the shape function of the rotation. Displacement z and rotation θ correspond to reference coordinate ξ in the element, and can be approximated as

$$z(x) = \boldsymbol{N}_z(\xi)^T \boldsymbol{d} \tag{3.141}$$

and

$$\theta(x) = \boldsymbol{N}_\theta(\xi)^T \boldsymbol{d}. \tag{3.142}$$

To approximate the variational equation (3.28), the derivatives of displacement and rotation in (3.141) and (3.142) must be computed with respect to spatial coordinate x. Since the reference coordinate is a function of ξ, the chain rule of differentiation can be used to obtain

$$\frac{dz}{dx} = \boldsymbol{d}^T \frac{d\boldsymbol{N}_z}{d\xi}\frac{d\xi}{dx} \tag{3.143}$$

and

$$\frac{d\theta}{dx} = \boldsymbol{d}^T \frac{d\boldsymbol{N}_\theta}{d\xi}\frac{d\xi}{dx}, \tag{3.144}$$

where $d\xi/dx$ can be obtained from (3.120). Thus, the curvature and shear deformation term in (3.26) can be interpolated as

$$\frac{d\theta}{dx} = \frac{1}{l}[0, -1, 0, 1]\boldsymbol{d} \equiv \boldsymbol{B}_d\boldsymbol{d} \tag{3.145}$$

and

$$\frac{dz}{dx} - \theta = \left[-\frac{1}{l}, -\frac{1-\xi}{2}, \frac{1}{l}, -\frac{1+\xi}{2}\right]\boldsymbol{d} \equiv \boldsymbol{B}_s\boldsymbol{d}, \tag{3.146}$$

where $\boldsymbol{B}_d$ and $\boldsymbol{B}_s$ are the discrete *displacement-strain matrices* of bending and shear deformation, respectively. By using (3.145) and (3.146), the structural energy bilinear form in (3.26) can be approximated as

$$
\begin{aligned}
a(z,\bar{z}) &\approx \bar{\boldsymbol{d}}^T \int_0^l EI(x)\,\boldsymbol{B}_b^T \boldsymbol{B}_b\, dx\, \boldsymbol{d} \\
&+ \bar{\boldsymbol{d}}^T \int_0^l k\mu A(x)\,\boldsymbol{B}_s^T \boldsymbol{B}_s\, dx\, \boldsymbol{d} \\
&\equiv \bar{\boldsymbol{d}}^T \boldsymbol{k}_b\, \boldsymbol{d} + \bar{\boldsymbol{d}}^T \boldsymbol{k}_b\, \boldsymbol{d} \\
&\equiv \bar{\boldsymbol{d}}^T \boldsymbol{k}\, \boldsymbol{d},
\end{aligned} \tag{3.147}
$$

where $\boldsymbol{z} = [z, \theta]^{\mathrm{T}}$ and $\boldsymbol{k} = \boldsymbol{k}_b + \boldsymbol{k}_s$ is the *stiffness matrix* of the Timoshenko beam element. For the simplest case in which a constant cross-sectional area A is used, the explicit computation of the $\boldsymbol{k}$ matrix can be obtained as

$$
\boldsymbol{k} = \boldsymbol{k}_b + \boldsymbol{k}_s = \begin{bmatrix}
\frac{k\mu A}{l} & \frac{k\mu A}{2} & -\frac{k\mu A}{l} & \frac{k\mu A}{2} \\
\frac{k\mu A}{2} & \frac{EI}{l} + \frac{k\mu Al}{3} & -\frac{k\mu A}{2} & -\frac{EI}{l} + \frac{k\mu Al}{6} \\
-\frac{k\mu A}{l} & -\frac{k\mu A}{2} & \frac{k\mu A}{l} & -\frac{k\mu A}{2} \\
\frac{k\mu A}{2} & -\frac{EI}{l} + \frac{k\mu Al}{6} & -\frac{k\mu A}{2} & \frac{EI}{l} + \frac{k\mu Al}{3}
\end{bmatrix}, \tag{3.148}
$$

where $\boldsymbol{k}_b$ and $\boldsymbol{k}_s$ are the bending and transverse shear stiffness matrices, respectively, defined by

$$
\boldsymbol{k}_b = \frac{EI}{l} \begin{bmatrix}
0 & 0 & 0 & 0 \\
0 & 1 & 0 & -1 \\
0 & 0 & 0 & 0 \\
0 & -1 & 0 & 1
\end{bmatrix} \tag{3.149}
$$

and

$$
\boldsymbol{k}_s = \frac{k\mu A}{l} \begin{bmatrix}
1 & \frac{l}{2} & -1 & \frac{l}{2} \\
\frac{l}{2} & \frac{l^2}{3} & -\frac{l}{2} & \frac{l^2}{6} \\
-1 & -\frac{l}{2} & 1 & -\frac{l}{2} \\
\frac{l}{2} & \frac{l^2}{6} & -\frac{l}{2} & \frac{l^2}{3}
\end{bmatrix}. \tag{3.150}
$$

The load linear form in (3.27) can easily be approximated for linear and constant load distribution $f(x)$. In the case of a constant distribution, (3.27) is approximated as

$$
\begin{aligned}
\ell(\bar{z}) &= \int_0^l \bar{z} f\, dx \\
&\approx \bar{\boldsymbol{d}}^T \int_{-1}^{1} \boldsymbol{N}_z J\, d\xi\, f \\
&= \bar{\boldsymbol{d}}^T \begin{bmatrix} 1 \\ 0 \\ 1 \\ 0 \end{bmatrix} \frac{fl}{2} \equiv \bar{\boldsymbol{d}}^T \boldsymbol{f},
\end{aligned} \tag{3.151}
$$

where $\boldsymbol{f}$ is the *element force vector*. However, if the concentrated nodal force and moment are applied to the element, then we can simply say $\boldsymbol{f} = [f_1, m_1, f_2, m_2]^T$. From (3.147) and

(3.151), the discrete variational equation of the truss component is

$$\bar{\boldsymbol{d}}^T \boldsymbol{k}\,\boldsymbol{d} = \bar{\boldsymbol{d}}^T \boldsymbol{f}, \tag{3.152}$$

for all $\bar{\boldsymbol{d}}$ in $Z_h \subset R^4$, where Z_h is the discrete space of kinematically admissible displacements.

3.3.3 Plate Element

A finite element based on thin plate theory requires C^1-continuity across the element boundary, which is extremely difficult for a general-shaped element. Only the rectangular geometry of a finite element that is still nonconforming is considered. In contrast, a C^0-continuous plate element can be easily developed using various interpolation schemes.

Thin Plate Element

Consider a rectangular flat plate element, as shown in Fig. 3.19. The midsurface of the plate is taken to be on the x_1-x_2 plane. For each corner of the element, vertical displacement z and two rotations (θ_x and θ_y) are considered as unknown nodal variables. Since z, θ_x, and θ_y are related in the thin plate theory, only an approximation of the vertical displacement is considered. Since the plate element contains 12 unknowns, the following bases are chosen for interpolation:

$$[1, \xi, \eta, \xi^2, \xi\eta, \eta^2, \xi^3, \xi^2\eta, \xi\eta^2, \eta^3, \xi^3\eta, \xi\eta^3], \tag{3.153}$$

where $\boldsymbol{\xi} = [\xi, \eta]^T$ is the vector of the reference coordinate. Note that for computational convenience, the reference coordinates are defined as $\xi = x/a$ and $\eta = y/b$. After imposing an interpolation condition at each node similar to the beam element in (3.130), the following approximation can be obtained:

$$z(\boldsymbol{\xi}) = \boldsymbol{N}^T(\boldsymbol{\xi})\boldsymbol{d}, \tag{3.154}$$

where $\boldsymbol{d} = [z_1, \theta_{x1}, \theta_{y1}, \ldots, z_4, \theta_{x4}, \theta_{y4}]^T$ represents the element's unknown variables and

$$\boldsymbol{N}(\boldsymbol{\xi}) = \begin{bmatrix} 1 - \xi\eta - (3-2\xi)\xi^2(1-\eta) - (1-\xi)(3-2\eta)\eta^2 \\ (1-\xi)\eta(1-\eta)^2 b \\ -\xi(1-\xi)^2(1-\eta)a \\ (1-\xi)(3-2\eta)\eta^2 + \xi(1-\xi)(1-2\xi)\eta \\ -(1-\xi)(1-\eta)\eta^2 b \\ -\xi(1-\xi)^2\eta a \\ (3-2\xi)\xi^2\eta - \xi\eta(1-\eta)(1-2\eta) \\ -\xi(1-\eta)\eta^2 b \\ (1-\xi)\xi^2\eta a \\ (3-2\xi)\xi^2(1-\eta) + \xi\eta(1-\eta)(1-2\eta) \\ \xi\eta(1-\eta)^2 b \\ (1-\xi)\xi^2(1-\eta)a \end{bmatrix} \tag{3.155}$$

is the shape function of approximation. Equation (3.155) consists of Hermite shape functions. Figure 3.20 plots the Hermite shape function in the element domain, which is conceptually the same as the Hermite function of the beam element (see Fig. 3.17).

The displacement function represented by (3.154) ensures that boundary displacements and tangential slope along the boundary between adjacent plate elements are compatible. However, the normal slope along the boundary is not compatible. Consequently, discontinuities in the first-order derivatives of displacement exist across element boundaries.

Integration of the structural energy form over the plate element in (3.41) results in the following plate element stiffness matrix [45]:

$$\boldsymbol{k} = \frac{Eh^3}{12(1-\nu^2)ab}\begin{bmatrix} \boldsymbol{k}_{11} & sym \\ \boldsymbol{k}_{21} & \boldsymbol{k}_{22} \end{bmatrix}_{12\times 12}, \tag{3.156}$$

where E is Young's modulus; ν is Poisson's ratio; and 6×6 submatrices $\boldsymbol{k}_{11}$, $\boldsymbol{k}_{21}$, and $\boldsymbol{k}_{22}$ are defined in (3.157), (3.158), and (3.159), respectively. In (3.157) through (3.159), $\gamma = b/a$, the ratio between horizontal and vertical plate dimensions. Note that the stiffness matrix depends on material properties and plate thickness, both of which may be taken as design parameters.

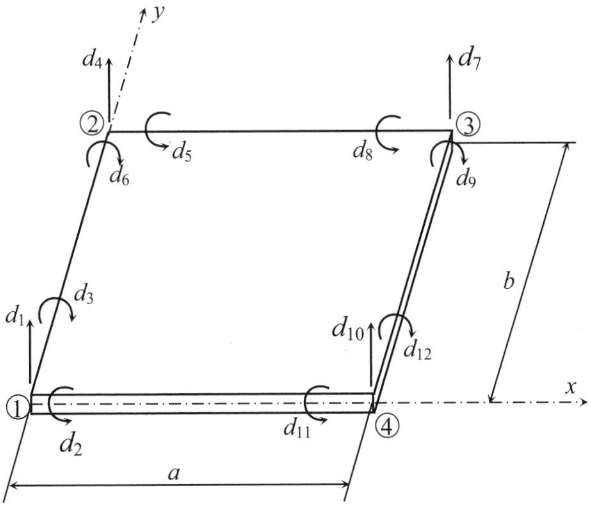

Figure 3.19. Rectangular plate element.

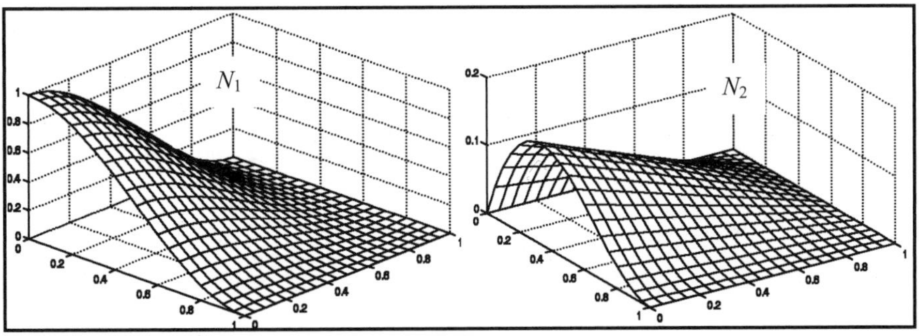

Figure 3.20. Three-dimensional surface plots of shape functions.

$$
\boldsymbol{k}_{11} = \begin{bmatrix}
4(\gamma^2+\gamma^{-2})+\frac{1}{5}(14-4\nu) & & & & & \\
[2\gamma^{-2}+\frac{1}{5}(1+4\nu)]b & [\frac{4}{3}\gamma^{-2}+\frac{4}{15}(1-\nu)]b^2 & & & \textit{symmetric} & \\
-[2\gamma^{2}+\frac{1}{5}(1+4\nu)]a & -\nu ab & [\frac{4}{3}\gamma^{2}+\frac{4}{15}(1-\nu)]a^2 & & & \\
2(\gamma^2-2\gamma^{-2})-\frac{1}{5}(14-4\nu) & -[2\gamma^{-2}+\frac{1}{5}(1-\nu)]b & [-\gamma^{2}+\frac{1}{5}(1+4\nu)]a & 4(\gamma^2+\gamma^{-2})+\frac{1}{5}(14-4\nu) & & \\
[2\gamma^{-2}+\frac{1}{5}(1-\nu)]b & [\frac{2}{3}\gamma^{-2}-\frac{1}{15}(1-\nu)]b^2 & 0 & -[2\gamma^{-2}+\frac{1}{5}(1+4\nu)]b & [\frac{4}{3}\gamma^{-2}+\frac{4}{15}(1-\nu)]b^2 & \\
[-\gamma^{2}+\frac{1}{5}(1+4\nu)]a & 0 & [\frac{2}{3}\gamma^{2}-\frac{4}{15}(1-\nu)]a^2 & -[2\gamma^{2}+\frac{1}{5}(1+4\nu)]a & \nu ab & [\frac{4}{3}\gamma^{2}+\frac{4}{15}(1-\nu)]a^2
\end{bmatrix} \tag{3.157}
$$

$$
\boldsymbol{k}_{21} = \begin{bmatrix}
-2(\gamma^2+\gamma^{-2})+\frac{1}{5}(14-4\nu) & [-\gamma^{-2}+\frac{1}{5}(1-\nu)]b & [\gamma^{2}-\frac{1}{5}(1-\nu)]a & -2(2\gamma^2-\gamma^{-2})-\frac{1}{5}(14-4\nu) & [-\gamma^{-2}+\frac{1}{5}(1+4\nu)]b & [2\gamma^{2}+\frac{1}{5}(1-\nu)]a \\
[\gamma^{-2}+\frac{1}{5}(1-\nu)]b & [\frac{1}{3}\gamma^{-2}+\frac{1}{15}(1-\nu)]b^2 & 0 & [-\gamma^{-2}+\frac{1}{5}(1+4\nu)]b & [\frac{2}{3}\gamma^{-2}-\frac{4}{15}(1-\nu)]b^2 & 0 \\
[-\gamma^{2}+\frac{1}{5}(1-\nu)]b & 0 & [\frac{1}{3}\gamma^{2}+\frac{1}{15}(1-\nu)]a^2 & -[2\gamma^{2}+\frac{1}{5}(1-\nu)]a & 0 & [\frac{2}{3}\gamma^{2}-\frac{1}{15}(1-\nu)]a^2 \\
-2(2\gamma^2-\gamma^{-2})-\frac{1}{5}(14-4\nu) & [\gamma^{-2}-\frac{1}{5}(1+4\nu)]b & [2\gamma^{2}+\frac{1}{5}(1-\nu)]a & -2(\gamma^2+\gamma^{-2})+\frac{1}{5}(14-4\nu) & [\gamma^{-2}-\frac{1}{5}(1-\nu)]b & [\gamma^{-2}-\frac{1}{5}(1-\nu)]a \\
[\gamma^{-2}-\frac{1}{5}(1+4\nu)]b & [\frac{2}{3}\gamma^{-2}-\frac{4}{15}(1-\nu)]b^2 & 0 & [-\gamma^{-2}+\frac{1}{5}(1-\nu)]b & [\frac{1}{3}\gamma^{-2}+\frac{1}{15}(1-\nu)]b^2 & 0 \\
-[2\gamma^{2}+\frac{1}{5}(1-\nu)]a & 0 & [\frac{2}{3}\gamma^{2}-\frac{1}{15}(1-\nu)]a^2 & [-\gamma^{2}+\frac{1}{5}(1-\nu)]a & 0 & [\frac{1}{3}\gamma^{2}+\frac{1}{15}(1-\nu)]a^2
\end{bmatrix} \tag{3.158}
$$

$$
\boldsymbol{k}_{22} = \begin{bmatrix}
4(\gamma^2+\gamma^{-2})+\frac{1}{5}(14-4\nu) & & & & & \\
-[2\gamma^{-2}+\frac{1}{5}(1+4\nu)]b & [\frac{4}{3}\gamma^{-2}+\frac{4}{15}(1-\nu)]b^2 & & & \textit{symmetric} & \\
[2\gamma^{2}+\frac{1}{5}(1+4\nu)]a & -\nu ab & [\frac{4}{3}\gamma^{2}+\frac{4}{15}(1-\nu)]a^2 & & & \\
2(\gamma^2-2\gamma^{-2})-\frac{1}{5}(14-4\nu) & [2\gamma^{-2}+\frac{1}{5}(1-\nu)]b & [\gamma^{2}-\frac{1}{5}(1+4\nu)]a & 4(\gamma^2+\gamma^{-2})+\frac{1}{5}(14-4\nu) & & \\
-[2\gamma^{-2}+\frac{1}{5}(1-\nu)]b & [\frac{2}{3}\gamma^{-2}-\frac{1}{15}(1-\nu)]b^2 & 0 & [2\gamma^{-2}+\frac{1}{5}(1+4\nu)]b & [\frac{4}{3}\gamma^{-2}+\frac{4}{15}(1-\nu)]b^2 & \\
[\gamma^{2}-\frac{1}{5}(1+4\nu)]a & 0 & [\frac{2}{3}\gamma^{2}-\frac{4}{15}(1-\nu)]a^2 & [2\gamma^{2}+\frac{1}{5}(1+4\nu)]a & \nu ab & [\frac{4}{3}\gamma^{2}+\frac{4}{15}(1-\nu)]a^2
\end{bmatrix} \tag{3.159}
$$

Mindlin/Reissner Plate Element

Discretization of the Mindlin/Reissner plate is discussed in this section. Since cross-sectional rotation and transverse displacement are independent of each other, three separate interpolations for z, θ_1, and θ_2 must be assumed. Furthermore, since only first-order derivatives are involved in the structural energy form in (3.55), a linear polynomial can be used in interpolation. As a result, it is easier to choose shape functions using this theory than the C^1 approach employed in the previous section. Lagrangian interpolation formulas and isoparametric mapping can now be used, whereas Hermite shape functions are used for a C^1 plate element. A four-node Mindlin/Reissner plate element, together with its reference element for writing shape functions, is shown in Fig. 3.21.

Let (ξ_I, η_I) be the value of the reference coordinate corresponding to node I whose value is ± 1, as shown in Fig. 3.21. The Lagrangian shape functions of the reference element can be written as

$$N_I(\boldsymbol{\xi}) = \frac{1}{4}(1 + \xi\xi_I)(1 + \eta\eta_I), \tag{3.160}$$

where $\boldsymbol{\xi} = [\xi, \eta]^T$ is the reference coordinate vector and $I = 1, \ldots, 4$ is the number of nodes in the element. If $\boldsymbol{z} = [z, \theta_1, \theta_2]^T$ is the vector of vertical displacement and cross-section rotation, then $\boldsymbol{z}$ is interpolated using nodal values $\boldsymbol{d} = [z_1, \theta_{11}, \theta_{12}, z_2, \theta_{21}, \theta_{22}, \ldots, \theta_{44}]^T$, as

$$\boldsymbol{z} = \begin{bmatrix} z \\ \theta_1 \\ \theta_2 \end{bmatrix} = \begin{bmatrix} N_1 & 0 & 0 & N_2 & 0 & 0 & N_3 & 0 & \\ 0 & N_1 & 0 & 0 & N_2 & 0 & 0 & N_3 & \cdots \\ 0 & 0 & N_1 & 0 & 0 & N_2 & 0 & 0 & \end{bmatrix} \boldsymbol{d}. \tag{3.161}$$

For the isoparametric mapping element, the displacement and the spatial coordinate are interpolated in the same method, such that

$$\begin{aligned} x(\boldsymbol{\xi}) &= \sum_{I=1}^{4} N_I(\boldsymbol{\xi}) x_I, \\ y(\boldsymbol{\xi}) &= \sum_{I=1}^{4} N_I(\boldsymbol{\xi}) y_I, \end{aligned} \tag{3.162}$$

where x_I and y_I are the nodal coordinates of node I. From (3.162), the Jacobian matrix of the transformation can be obtained as

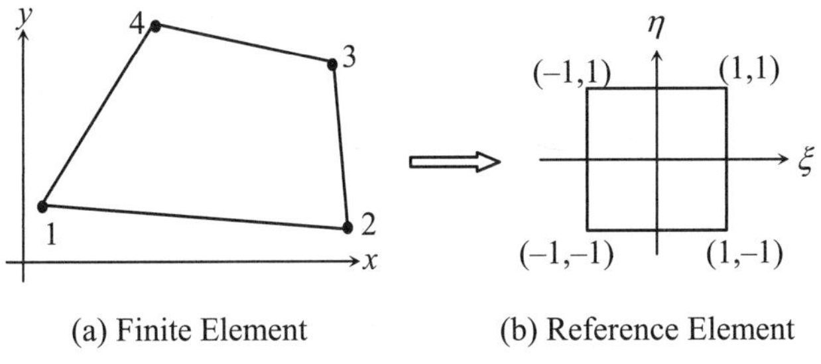

Figure 3.21. Quadrilateral Mindlin/Reissner element.

$$\boldsymbol{J} = \begin{bmatrix} \dfrac{\partial x}{\partial \xi} & \dfrac{\partial y}{\partial \xi} \\ \dfrac{\partial x}{\partial \eta} & \dfrac{\partial y}{\partial \eta} \end{bmatrix}. \tag{3.163}$$

Using the inverse relation of (3.163), spatial derivatives can be expressed in terms of the derivatives in the reference coordinates, as

$$\begin{bmatrix} \dfrac{\partial}{\partial x} \\ \dfrac{\partial}{\partial y} \end{bmatrix} = \frac{1}{|\boldsymbol{J}|} \begin{bmatrix} \dfrac{\partial y}{\partial \eta} & -\dfrac{\partial y}{\partial \xi} \\ -\dfrac{\partial x}{\partial \eta} & \dfrac{\partial x}{\partial \xi} \end{bmatrix} \begin{bmatrix} \dfrac{\partial}{\partial \xi} \\ \dfrac{\partial}{\partial \eta} \end{bmatrix}, \tag{3.164}$$

where $\partial/\partial\xi$ and $\partial/\partial\eta$ can be easily obtained by taking the derivative of the shape function in (3.160).

Since the energy bilinear form in (3.55) depends on curvature and shear strain vectors, the calculations for vectors $\boldsymbol{\kappa}$ and $\boldsymbol{\gamma}$ in (3.48) and (3.49) can be organized as

$$\boldsymbol{\kappa}(\xi) = \boldsymbol{B}_b(\xi)\boldsymbol{d} \tag{3.165}$$

and

$$\boldsymbol{\gamma}(\xi) = \boldsymbol{B}_s(\xi)\boldsymbol{d}, \tag{3.166}$$

where $\boldsymbol{B}_b$ and $\boldsymbol{B}_s$ are displacement-strain matrices of bending and shear deformation, respectively, and whose expressions are calculated as

$$\boldsymbol{B}_b = \begin{bmatrix} 0 & N_{1,1} & 0 & 0 & N_{2,1} & 0 & \\ 0 & 0 & N_{1,2} & 0 & 0 & N_{2,2} & \cdots \\ 0 & N_{1,2} & N_{1,1} & 0 & N_{2,2} & N_{2,1} & \end{bmatrix} \tag{3.167}$$

and

$$\boldsymbol{B}_s = \begin{bmatrix} N_{1,2} & 0 & -N_1 & N_{2,2} & 0 & -N_2 & \cdots \\ N_{1,1} & -N_1 & 0 & N_{2,1} & -N_2 & 0 & \end{bmatrix}. \tag{3.168}$$

Thus, from the definition of the structural energy form in (3.55), we obtain

$$\begin{aligned} a_u(z,\bar{z}) &\approx \bar{\boldsymbol{d}}^T \iint_\Omega \boldsymbol{B}_b^T \boldsymbol{C}^b \boldsymbol{B}_b \, d\Omega \boldsymbol{d} \\ &\quad + \bar{\boldsymbol{d}}^T \iint_\Omega \boldsymbol{B}_s^T \boldsymbol{C}^s \boldsymbol{B}_s \, d\Omega \boldsymbol{d} \\ &= \bar{\boldsymbol{d}}^T \boldsymbol{k}_b \boldsymbol{d} + \bar{\boldsymbol{d}}^T \boldsymbol{k}_s \boldsymbol{d} \\ &= \bar{\boldsymbol{d}}^T \boldsymbol{k} \, \boldsymbol{d}, \end{aligned} \tag{3.169}$$

where $\boldsymbol{k}_b$ is the 12×12 element bending matrix, and $\boldsymbol{k}_s$ is the 12×12 transverse shear stiffness matrix.

If only constant vertical force f is considered for the load linear form in (3.55) without any moments, then the load linear form in (3.55) can be approximated as

$$\begin{aligned} \ell_u(\bar{z}) &\approx \bar{\boldsymbol{d}}_I^T \iint_\Omega N_I \, d\Omega f \\ &= \bar{\boldsymbol{d}}^T \boldsymbol{f}. \end{aligned} \tag{3.170}$$

From (3.169) and (3.170), the discrete variational equation of the plate element is

$$\bar{\boldsymbol{d}}^T \boldsymbol{k} \boldsymbol{d} = \bar{\boldsymbol{d}}^T \boldsymbol{f}, \qquad \forall \bar{\boldsymbol{d}} \in Z_h, \tag{3.171}$$

where $Z_h \subset R^{12}$ is the discrete space of kinematically admissible displacements.

3.3.4 Three-Dimensional Elastic Solid

There are many ways to interpolate a solid component using a finite element. Here, only an eight-node isoparametric quadrilateral element is taken as an example. For a more detailed discussion of solid elements, refer to the additional literature on this topic [4], [43], and [46]. Figure 3.22 depicts an eight-node, three-dimensional, isoparametric solid element.

Let $\boldsymbol{z}_I = [z_{I1}, z_{I2}, z_{I3}]^T$ be the displacement vector at node $I = 1, \ldots, 8$ and let $\xi_I = [\xi_I, \eta_I, \zeta_I]^T$ be the corresponding reference coordinate. For the isoparametric element, the coordinate and the displacement of the element can be expressed by

$$\boldsymbol{x} = \sum_{I=1}^{8} N_I(\xi) \boldsymbol{x}_I \tag{3.172}$$

and

$$\boldsymbol{z} = \sum_{I=1}^{8} N_I(\xi) \boldsymbol{z}_I, \tag{3.173}$$

where $\boldsymbol{x}_I$ is the nodal coordinate and $N_I(\xi)$ is the isoparametric shape function, defined as

$$N_I(\xi) = \frac{1}{8}(1 + \xi\xi_I)(1 + \eta\eta_I)(1 + \zeta\zeta_I), \tag{3.174}$$

where (ξ_I, η_I, ζ_I) are the values of the reference coordinate corresponding to node I, and where their values are ± 1, as shown in Fig. 3.22.

The transformation from physical to reference elements can be defined using a mapping relation. The Jacobian matrix of the transformation can be obtained by taking the derivative of (3.172) as

$$\boldsymbol{J}_{3\times 3} = \frac{d\boldsymbol{x}}{d\xi} = \sum_{I=1}^{8} \boldsymbol{x}_I \frac{dN_I(\xi)}{d\xi}. \tag{3.175}$$

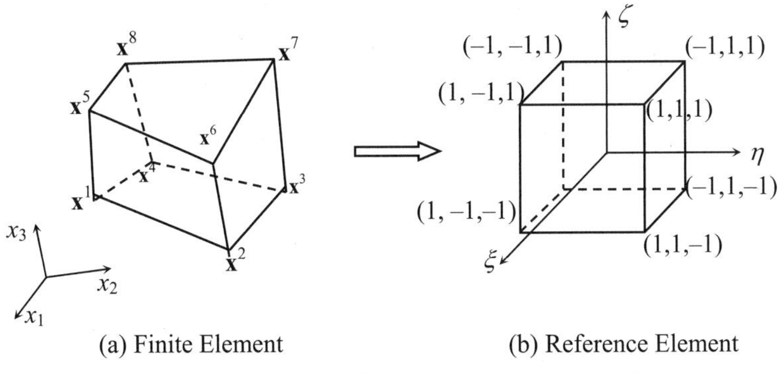

Figure 3.22. Eight-node three-dimensional isoparametric solid element.

Note that $dN_I/d\xi$ is a (1×3) row vector. The initial geometry of the element has to be well shaped, such that $|\boldsymbol{J}| > 0$. To compute the strain vector, differentiation of the approximation in (3.173) has to be taken. By using the inverse relation of (3.175), the spatial derivative of the shape function can be obtained as

$$\frac{dN_I}{d\boldsymbol{x}} = \frac{dN_I}{d\xi}\boldsymbol{J}^{-T}. \tag{3.176}$$

The strain vector can thus be obtained in the form

$$\varepsilon(\boldsymbol{z}) = \sum_{I=1}^{8} \boldsymbol{B}_I \boldsymbol{z}_I, \tag{3.177}$$

where

$$\boldsymbol{B}_I = \begin{bmatrix} N_{I,1} & 0 & 0 \\ 0 & N_{I,2} & 0 \\ 0 & 0 & N_{I,3} \\ N_{I,2} & N_{I,1} & 0 \\ 0 & N_{I,3} & N_{I,2} \\ N_{I,3} & 0 & N_{I,1} \end{bmatrix} \tag{3.178}$$

is the discrete displacement-strain matrix of a solid element. The approximation of $\varepsilon(\bar{\boldsymbol{z}})$ can be obtained in a similar way.

Note that all variables in the physical element are transformed into the reference element. Thus, it will be helpful if the integration over element domain Ω can be transformed into an integration over the reference element, which can be achieved using the following relation:

$$\iiint_{\Omega} d\Omega = \int_{-1}^{1}\int_{-1}^{1}\int_{-1}^{1} |\boldsymbol{J}|\, d\xi d\eta d\zeta. \tag{3.179}$$

Thus, the energy bilinear form of the element in (3.67) can be approximated as

$$\begin{aligned} a_u(\boldsymbol{z}, \bar{\boldsymbol{z}}) &\approx \sum_{I=1}^{8}\sum_{J=1}^{8} \bar{\boldsymbol{z}}_I^T \int_{-1}^{1}\int_{-1}^{1}\int_{-1}^{1} \boldsymbol{B}_I^T \boldsymbol{C} \boldsymbol{B}_J |\boldsymbol{J}|\, d\xi d\eta d\zeta\, \boldsymbol{z}_J \\ &\equiv \bar{\boldsymbol{d}}^T \boldsymbol{k}\, \boldsymbol{d}, \end{aligned} \tag{3.180}$$

where $\boldsymbol{d} = [z_{11}, z_{12}, z_{13}, z_{21}, z_{22}, z_{23}, \ldots, z_{81}, z_{82}, z_{83}]^T$ is the unknown nodal displacement vector, and $\boldsymbol{k}$ is the 24×24 element stiffness matrix.

For simplicity, the load linear form will be discretized with no traction force by

$$\begin{aligned} \ell(\bar{\boldsymbol{z}}) &\approx \sum_{I=1}^{8} \bar{\boldsymbol{z}}_I^T \int_{-1}^{1}\int_{-1}^{1}\int_{-1}^{1} N_I(\xi) \boldsymbol{f}^b |\boldsymbol{J}|\, d\xi d\eta d\zeta \\ &\equiv \bar{\boldsymbol{d}}^T \boldsymbol{f}. \end{aligned} \tag{3.181}$$

Thus, the discrete variational equation of a solid component becomes

$$\bar{\boldsymbol{d}}^T \boldsymbol{k}\boldsymbol{d} = \bar{\boldsymbol{d}}^T \boldsymbol{f}, \quad \forall \bar{\boldsymbol{d}} \in Z_h, \tag{3.182}$$

where $Z_h \subset R^{24}$ is the discrete space of kinematically admissible displacements.

For the eigenvalue problem, the kinetic energy bilinear form can be approximated by

$$\begin{aligned} d(z,\bar{z}) &\approx \sum_{I=1}^{8}\sum_{J=1}^{8} \bar{z}_I^T \int_{-1}^{1}\int_{-1}^{1}\int_{-1}^{1} \rho N_I N_J |\boldsymbol{J}| d\xi d\eta d\zeta \, z_J \\ &\equiv \bar{\boldsymbol{d}}^T \boldsymbol{m}\, \boldsymbol{d}, \end{aligned} \tag{3.183}$$

where $\boldsymbol{m}$ is the *consistent mass matrix*. The discrete eigenvalue problem of solid components can be obtained as

$$\bar{\boldsymbol{d}}^T \boldsymbol{k}\, \boldsymbol{d} = \zeta \bar{\boldsymbol{d}}^T \boldsymbol{m}\, \boldsymbol{d}, \qquad \forall \bar{\boldsymbol{d}} \in Z_h, \tag{3.184}$$

where ζ is the eigenvalue.

3.4 Global Matrix Equations for the Finite Element Method

3.4.1 Construction of Global Matrices

In the previous section, a matrix equation was developed using the finite element method at each element level. General, complex geometry of engineering applications is approximated using a set of finite elements. Finite elements are connected to each other with adjacent elements through common nodes. To construct a global system of matrix equations in which all elements are connected, an assembly procedure has to be followed. The global stiffness matrix that is generated by such a procedure is usually positive semidefinite, since a rigid body motion exists in the structure, which can be removed by imposing boundary conditions. The direct removal of fixed degrees-of-freedom, or the substitution of an equivalent relation, can impose essential boundary conditions.

Global Stiffness and Mass Matrices

The total strain and kinetic energy of a structure may be obtained by adding together the strain and kinetic energy of all elements that make up the structure. Before a meaningful expression can be written for the total system strain and kinetic energy, it is first necessary to define a system of global displacements for all nodes in the structure, relative to the global coordinate system. Let $z_g \in R^n$ denote this *global displacement vector*.

Transformation from Local to Global Coordinates

Since the individual elements of the structure have their own inherent displacement vectors in the body-fixed coordinate system, as illustrated in Figs. 3.15 and 3.16, displacement must first be transformed from the element's body-fixed coordinate to a coordinate that parallels the global coordinates. Let $\boldsymbol{d}^i$ denote the nodal displacement coordinate vector of the ith element in its body-fixed system. A *rotation matrix* $\boldsymbol{S}^i$ may be used to define these local displacement coordinates in terms of global coordinates, which are denoted as $\boldsymbol{q}^i$, that is,

$$\boldsymbol{d}^i = \boldsymbol{S}^i \boldsymbol{q}^i. \tag{3.185}$$

The transformed element displacement now coincides with components of the global displacement vector z_g. Therefore, it is possible to define a *Boolean transformation matrix* $\boldsymbol{\beta}^i$ that consists of only zeros and ones, and gives the relation

$$\boldsymbol{q}^i = \boldsymbol{\beta}^i z_g. \tag{3.186}$$

Note that if $\boldsymbol{q}^i$ is an r-vector and $\boldsymbol{z}_g$ is an n-vector ($n > r$), then $\boldsymbol{\beta}^i$ is an $r \times n$ matrix that only consists of r unit components, with zeros as the remaining entries.

Example 3.1. Coordinate Transformation of Three-bar Truss Structure. In order to explain coordinate transformation between local and global coordinates, consider element 3 of the three-bar truss example given in Section 1.2.3 of Chapter 1. Since the direction of elements 1 and 2 are already aligned with the global coordinate system, no rotation is required; only Boolean transformation is required for these elements. However, for element 3 the body-fixed coordinate is rotated at angle θ from the global coordinates. In this case, the rotational matrix and the Boolean transformation matrix are defined as

$$\boldsymbol{S}^3 = \begin{bmatrix} \cos\theta & \sin\theta & 0 & 0 \\ 0 & 0 & \cos\theta & \sin\theta \end{bmatrix}$$

and

$$\boldsymbol{\beta}^3 = \begin{bmatrix} 1 & 0 & 0 & 0 & 0 & 0 \\ 0 & 1 & 0 & 0 & 0 & 0 \\ 0 & 0 & 0 & 0 & 1 & 0 \\ 0 & 0 & 0 & 0 & 0 & 1 \end{bmatrix}.$$

Generalized Global Stiffness Matrix

By denoting the ith element stiffness matrix as $\boldsymbol{k}^i$, the strain energy in the ith element may be written as

$$U^i = \frac{1}{2}\boldsymbol{d}^{iT}\boldsymbol{k}^i\boldsymbol{d}^i. \tag{3.187}$$

Substituting from (3.185) and (3.186), this formula becomes

$$U^i = \frac{1}{2}\boldsymbol{q}^{iT}\boldsymbol{S}^{iT}\boldsymbol{k}^i\boldsymbol{S}^i\boldsymbol{q}^i = \frac{1}{2}\boldsymbol{z}_g^T\boldsymbol{\beta}^{iT}\boldsymbol{S}^{iT}\boldsymbol{k}^i\boldsymbol{S}^i\boldsymbol{\beta}^i\boldsymbol{z}_g. \tag{3.188}$$

The strain energy of the entire structure is now obtained by adding the strain energy from all NE elements in the structure, to obtain

$$\begin{aligned} U &= \sum_{i=1}^{NE} U^i \\ &= \frac{1}{2}\boldsymbol{z}_g^T\left[\sum_{i=1}^{NE}\boldsymbol{\beta}^{iT}\boldsymbol{S}^{iT}\boldsymbol{k}^i\boldsymbol{S}^i\boldsymbol{\beta}^i\right]\boldsymbol{z}_g \\ &\equiv \frac{1}{2}\boldsymbol{z}_g^T\boldsymbol{K}_g\boldsymbol{z}_g, \end{aligned} \tag{3.189}$$

where $\boldsymbol{K}_g$ is the *generalized global stiffness matrix*,

$$\boldsymbol{K}_g = \sum_{i=1}^{NE}\boldsymbol{\beta}^{iT}\boldsymbol{S}^{iT}\boldsymbol{k}^i\boldsymbol{S}^i\boldsymbol{\beta}^i. \tag{3.190}$$

Reduced Global Stiffness Matrix

If all boundary conditions associated with the structure have been imposed so that no rigid-body degree-of-freedom exists, then generalized global stiffness matrix $\boldsymbol{K}_g$ is positive definite, denoted simply by $\boldsymbol{K}$, and is called the *reduced global stiffness matrix*. However, if the generalized global stiffness matrix is assembled without any consideration of boundary conditions, it will generally not be positive definite. As will be seen later, it is important to note this distinction since many formulations and computer codes use matrix methods that employ the generalized global stiffness matrix and impose constraints during the solution process. Such processes do not explicitly eliminate dependent displacement coordinates, so the positive definite reduced global stiffness matrix $\boldsymbol{K}$ is not constructed, and thus, not available for design sensitivity calculations.

Generalized Global Mass Matrix

As in the case of strain energy, the kinetic energy of the ith element may be written in terms of generalized velocities. Since matrices $\boldsymbol{S}^i$ and $\boldsymbol{\beta}^i$ do not depend on generalized coordinates, the following holds true:

$$\begin{aligned} \boldsymbol{d}^i_{,t} &= \boldsymbol{S}^i \boldsymbol{q}^i_{,t} \\ \boldsymbol{q}^i_{,t} &= \boldsymbol{\beta}^i \boldsymbol{z}_{g,t}, \end{aligned} \tag{3.191}$$

where the subscribed comma denotes a time differentiation, i.e., $\boldsymbol{d}_{,t} = \partial \boldsymbol{d}/\partial t$. Using these relations, the kinetic energy of the ith element may be written as

$$T^i = \frac{1}{2}\boldsymbol{d}^{iT}_{,t}\boldsymbol{m}^i\boldsymbol{d}^i_{,t} = \frac{1}{2}\boldsymbol{q}^{iT}_{,t}\boldsymbol{S}^{iT}\boldsymbol{m}^i\boldsymbol{S}^i\boldsymbol{q}^i_{,t} = \frac{1}{2}\boldsymbol{z}^T_{g,t}\,\boldsymbol{\beta}^{iT}\boldsymbol{S}^{iT}\boldsymbol{m}^i\boldsymbol{S}^i\,\boldsymbol{\beta}^i\boldsymbol{z}_{g,t}. \tag{3.192}$$

Summing up the kinetic energy for all elements, the total kinetic energy for the system can be written as

$$\begin{aligned} T &= \sum_{i=1}^{NE} T^i \\ &= \frac{1}{2}\boldsymbol{z}^T_{g,t}\left[\sum_{i=1}^{NE}\boldsymbol{\beta}^{iT}\boldsymbol{S}^{iT}\boldsymbol{m}^i\boldsymbol{S}^i\,\boldsymbol{\beta}^i\right]\boldsymbol{z}_{g,t} \\ &= \frac{1}{2}\boldsymbol{z}^T_{g,t}\boldsymbol{M}_g\boldsymbol{z}_{g,t}, \end{aligned} \tag{3.193}$$

where $\boldsymbol{M}_g$ is the *generalized global mass matrix*,

$$\boldsymbol{M}_g = \sum_{i=1}^{NE}\boldsymbol{\beta}^{iT}\boldsymbol{S}^{iT}\boldsymbol{m}^i\boldsymbol{S}^i\,\boldsymbol{\beta}^i. \tag{3.194}$$

Presuming that all structural elements have mass, it is impossible to obtain a nonzero velocity without investing a finite amount of kinetic energy. Therefore, a global mass matrix will always be positive definite.

Reduced Global Mass Matrix

If boundary conditions have been taken into account before the global displacement vector is defined, then the *reduced global mass matrix* will be denoted as $\boldsymbol{M}$, as in the case of the corresponding reduced global stiffness matrix $\boldsymbol{K}$. It is important to note that the global stiffness matrix and mass matrix both depend on design variables that appear in element stiffness mass matrices, as in the case of member-size design variables, and on geometric design variables that appear in the rotation matrices $\boldsymbol{S}^i$. It is clear that the dependence on geometric variables for global stiffness and mass matrices is much more complex than the dependence on member-size design variables.

Example 3.2. Two-bar Truss. As a simple illustration of the previously described transformations, consider the two-bar truss, shown in Fig. 3.23. Since rotation at the ends of the truss elements does not occur either in strain or in kinetic energy expressions, they are simply suppressed. The transformation from body-fixed to globally oriented element displacement coordinates can be described as $\boldsymbol{d}^1$ and $\boldsymbol{d}^2$:

$$\boldsymbol{d}^1 = \begin{bmatrix} \sin\theta & \cos\theta & 0 & 0 \\ -\cos\theta & \sin\theta & 0 & 0 \\ 0 & 0 & \sin\theta & \cos\theta \\ 0 & 0 & -\cos\theta & \sin\theta \end{bmatrix} \boldsymbol{q}^1 \equiv \boldsymbol{S}^1\boldsymbol{q}^1$$

$$\boldsymbol{d}^2 = \begin{bmatrix} \cos\theta & -\sin\theta & 0 & 0 \\ -\sin\theta & -\cos\theta & 0 & 0 \\ 0 & 0 & \cos\theta & -\sin\theta \\ 0 & 0 & -\sin\theta & -\cos\theta \end{bmatrix} \boldsymbol{q}^2 \equiv \boldsymbol{S}^2\boldsymbol{q}^2.$$

The mappings from globally oriented element coordinates to global coordinates are

$$\boldsymbol{q}^1 = \begin{bmatrix} 0 & 0 & 0 & 1 & 0 & 0 \\ 0 & 0 & 1 & 0 & 0 & 0 \\ 0 & 1 & 0 & 0 & 0 & 0 \\ 1 & 0 & 0 & 0 & 0 & 0 \end{bmatrix} \boldsymbol{z}_g \equiv \boldsymbol{\beta}^1\boldsymbol{z}_g$$

$$\boldsymbol{q}^2 = \begin{bmatrix} 1 & 0 & 0 & 0 & 0 & 0 \\ 0 & 1 & 0 & 0 & 0 & 0 \\ 0 & 0 & 0 & 0 & 1 & 0 \\ 0 & 0 & 0 & 0 & 0 & 1 \end{bmatrix} \boldsymbol{z}_g \equiv \boldsymbol{\beta}^2\boldsymbol{z}_g.$$

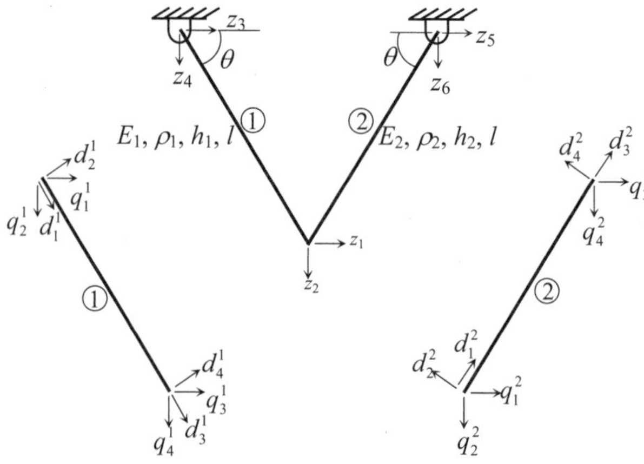

Figure 3.23. Two-bar truss.

Using these transformations and the expressions of (3.190) and (3.194) for the generalized global stiffness and mass matrices, the global matrices can be obtained as

$$
\boldsymbol{K}_g = \frac{1}{l}\begin{bmatrix} (E_1h_1+E_2h_2)c^2 & (E_1h_1-E_2h_2)sc & -E_1h_1c^2 & -E_1h_1sc & -E_2h_2c^2 & E_2h_2sc \\ & (E_1h_1+E_2h_2)s^2 & -E_1h_1sc & -E_1h_1s^2 & -E_2h_2sc & -E_2h_2s^2 \\ & & E_1h_1c^2 & E_1h_1sc & 0 & 0 \\ & & & E_1h_1s^2 & 0 & 0 \\ & symmetric & & & E_2h_2c^2 & -E_2h_2sc \\ & & & & & E_2h_2s^2 \end{bmatrix} \tag{3.195}
$$

and

$$
\boldsymbol{M}_g = \frac{l}{6}\begin{bmatrix} 2(\rho_1h_1+\rho_2h_2) & 0 & \rho_1h_1 & 0 & \rho_2h_2 & 0 \\ & 2(\rho_1h_1+\rho_2h_2) & 0 & \rho_1h_1 & 0 & \rho_2h_2 \\ & & 2\rho_1h_1 & 0 & 0 & 0 \\ & & & 2\rho_2h_2 & 0 & 0 \\ & symmetric & & & 2\rho_2h_2 & 0 \\ & & & & & 2\rho_2h_2 \end{bmatrix}, \tag{3.196}
$$

where $c = \cos\theta$ and $s = \sin\theta$. If the pin joints at the top of the truss are fixed, then the boundary conditions for this structure are $z_3 = z_4 = z_5 = z_6 = 0$. After imposing these boundary conditions, the strain energy and the kinetic energy are obtained only in terms of two displacement coordinates z_1 and z_2. Thus, rows and columns that correspond to fixed displacement coordinates are deleted in the generalized global stiffness and mass matrices in (3.195) and (3.196). As a result, the reduced stiffness and mass matrices are obtained as

$$
\boldsymbol{K} = \frac{1}{l}\begin{bmatrix} (E_1h_1+E_2h_2)c^2 & (E_1h_1-E_2h_2)sc \\ (E_1h_1-E_2h_2)sc & (E_1h_1+E_2h_2)s^2 \end{bmatrix} \tag{3.197}
$$

and

$$
\boldsymbol{M} = \frac{l}{3}\begin{bmatrix} \rho_1h_1+\rho_2h_2 & 0 \\ 0 & \rho_1h_1+\rho_2h_2 \end{bmatrix}. \tag{3.198}
$$

Note that while the generalized global stiffness matrix $\boldsymbol{K}_g$ in (3.195) is singular and, in fact, has a rank deficiency of 4, reduced stiffness matrix $\boldsymbol{K}$ in (3.197) is positive definite.

While the two-bar truss example is simple, it describes a systematic procedure for assembling global stiffness and mass matrices. Since this assembly procedure is systematic, numerous computer codes have been developed to automate the process of constructing $\boldsymbol{K}_g$ and $\boldsymbol{M}_g$. Depending on the nature of the boundary conditions, it is possible to systematically collapse the generalized global stiffness and mass matrices to the reduced stiffness and mass matrices $\boldsymbol{K}$ and $\boldsymbol{M}$, as was done in this example. In many applications, however, more complex constraints among generalized coordinates arise, such as multipoint constraints, making the reduction process more complicated. Numerical techniques, including systematic reduction and the application of Lagrange multipliers, are used to solve such problems [36], [42], and [45].

In the previous section, the design dependence of the element stiffness matrix is obvious. However, due to boundary conditions and/or constraints, the design dependence of the global stiffness matrix may not be easy to obtain. For design sensitivity purposes it is recommended that a matrix equation be used before boundary conditions are imposed, as shown in Chapter 4.

3.4.2 Variational Principles for Discrete Structural Systems

In the previous chapter, the variational principle of continuum systems was considered. In this section, the same variational principle is discussed for discrete systems.

Potential Energy

Structural systems considered in this chapter are *conservative* in nature, that is, the work done by a system of applied force in traveling through any closed path in displacement space must be equal to zero. Denoting $\boldsymbol{F}_g$ as a vector of force components that are consistent with the global displacement vector z_g, the conservative condition can be written as

$$\int_C \boldsymbol{F}_g^T d\boldsymbol{z}_g = 0, \tag{3.199}$$

where C is any closed path in the space of displacement-generalized coordinates. As is well known [41], for a force field $\boldsymbol{F}_g(z_g) = [F_{g1}, F_{g2}, \cdots, F_{gn}]^T$ to be conservative, an analytical condition exists as

$$\frac{\partial F_{gi}}{\partial z_{gj}} = \frac{\partial F_{gj}}{\partial z_{gi}}, \qquad i, j = 1, \ldots, n. \tag{3.200}$$

Assuming that the force field $\boldsymbol{F}_g(z_g)$ is conservative, a *potential energy* function $W(z_g)$ exists, such that

$$\boldsymbol{F}_g(\boldsymbol{z}_g) = -\frac{\partial W(\boldsymbol{z}_g)}{\partial \boldsymbol{z}_g}, \qquad i = 1, \ldots, n. \tag{3.201}$$

For constant applied force $\boldsymbol{F}_g$, the condition of (3.200) is easily satisfied, and the potential energy can be written as

$$W(\boldsymbol{z}_g) = -\boldsymbol{F}_g^T \boldsymbol{z}_g. \tag{3.202}$$

In this case, (3.201) can be easily verified as true.

In the case of a structural buckling problem, displacement at the point of an applied load P is given as a quadratic form in displacement z_g. Based on the similar procedure for a beam element in (3.139), the geometric stiffness term can be written as

$$\Delta = \frac{1}{2} \boldsymbol{z}_g^T \boldsymbol{D}_g \boldsymbol{z}_g. \tag{3.203}$$

It is presumed that the *global geometric stiffness matrix* $\boldsymbol{D}_g$ has been transformed into its symmetric form, which is always possible for a quadratic form. The potential energy of the load P is thus

$$Q = -P\Delta = -\frac{P}{2} \boldsymbol{z}_g^T \boldsymbol{D}_g \boldsymbol{z}_g, \tag{3.204}$$

where P and Δ are given in the same positive direction.

For a conservative mechanical system, it is possible to obtain a potential energy function for all applied loads. The *total potential energy* of a structural system is defined as the sum of the strain energy of the structure and the potential energy of all applied loads, that is,

$$\Pi = U + W + Q. \tag{3.205}$$

For a linear structural system, the strain energy is given by the quadratic form in (3.189), and the potential energy of applied loads is the sum of terms arising from (3.202) and (3.204). The total potential energy can thus be written as

$$\Pi = \frac{1}{2} z_g^T K_g z_g - F_g^T z_g - \frac{P}{2} z_g^T D_g z_g. \tag{3.206}$$

Theorem of Minimum Total Potential Energy

By denoting Z as the vector space of all kinematically admissible displacements for the structural system, and presuming homogeneous boundary and interface conditions, the following theorem of minimum total potential energy is true [41] and [45].

Theorem 3.1. Minimum Total Potential Energy. The displacement $z_g \in Z$ that occurs due to an externally applied conservative load acting on an elastic structure minimizes the total potential energy of the structural system over all kinematically admissible displacements.

It is important to note that this statement of minimum total potential energy does not require that the displacement coordinates z_{gi} (i=1, …, n) be independent, although it is presumed that they are related by homogeneous linear equations. While this limitation is not essential in the theory of structural mechanics, it is adequate for the purposes of this text.

Lagrange's Equation of Motion

The second major variational principle of structural mechanics employed here provides a variational form for the dynamic equations of motion. Presuming that the applied forces F_g depend only on time, that is, that $F_g = F_g(t)$, the *Lagrangian* of a dynamic system can be defined as

$$L = T(z_{g,t}, z_{g,t}) - \Pi(z_g), \tag{3.207}$$

where $T(z_{g,t}, z_{g,t})$ is the kinetic energy of the system, which is a quadratic form of $z_{g,t}$. Neglecting the effect of the last term in (3.206), the Lagrangian for a linear structural system can be written as

$$L = \frac{1}{2} z_{g,t}^T M_g z_{g,t} - \frac{1}{2} z_g^T K_g z_g + F_g^T z_g. \tag{3.208}$$

In terms of the Lagrangian, the motion of a conservative structural system with a subspace Z of kinematically admissible displacements may be characterized by the following theorem [47].

Theorem 3.2. Variational Form of Lagrange's Equation. The equation of motion in a conservative system, for $z_g(t)$ in the space Z of kinematically admissible displacements, may be written in the form

$$\bar{z}_g(t)^T \left[\frac{d}{dt}\left(\frac{\partial L}{\partial z_{g,t}} \right) - \frac{\partial L}{\partial z_g} \right] = 0, \tag{3.209}$$

which is valid for all *virtual displacement* $\bar{z}_g(t)$ that is consistent with its constraint, i.e., $\bar{z}_g(t) \in Z$.

The variational form of Lagrange's equation of motion is valid even for dependent state variables. For the case in which kinematic admissible conditions are employed in order to algebraically reduce the global displacement vector z_g to an independent form (of dimension m), (3.209) may be written in the following reduced form:

$$\frac{d}{dt}\left(\frac{\partial L}{\partial z_{,t}}\right) - \frac{\partial L}{\partial z} = \mathbf{0}. \tag{3.210}$$

Before (3.210) is used, it is critical to verify that displacement coordinate z is independent, since this form of Lagrange's equation of motion is invalid if the displacement coordinates are dependent.

3.4.3 Reduced Matrix Equation of Structural Mechanics

Displacement Due to Static Load

Consider a linear structural system described by the reduced stiffness matrix $\boldsymbol{K}$, the mass matrix $\boldsymbol{M}$, and the applied load $\boldsymbol{F}$. For such a system, kinematic constraints have been used to eliminate dependent displacement coordinates, thus yielding an independent *reduced displacement vector* z. In this case, the theorem of minimum total potential energy requires that the gradient of the total potential energy must be equal to zero at equilibrium. Using (3.206) with $P = 0$,

$$\boldsymbol{K} z = \boldsymbol{F}. \tag{3.211}$$

Further, with boundary conditions and interface conditions explicitly eliminated, the reduced stiffness matrix $\boldsymbol{K}$ is positive definite, and (3.211) is both necessary and a sufficient condition for stable equilibrium.

Buckling

In structural buckling, a potential energy term in the form of (3.204) arises, and no other externally applied force is considered. In such a situation, the theorem of minimum total potential energy for stable equilibrium yields the condition

$$\boldsymbol{K} z - P\boldsymbol{D} z = \mathbf{0}. \tag{3.212}$$

If $P = 0$ for a positive definite reduced stiffness matrix $\boldsymbol{K}$, then the only possible solution to (3.212) is $z = 0$, that is, the only stable equilibrium state of the system with no externally applied load is zero displacement. As P increases, particularly since $\boldsymbol{D}$ is generally positive semidefinite, a point will be reached at which the matrix $\boldsymbol{K} - P\boldsymbol{D}$ ceases to be positive definite; hence, it becomes singular. The smallest load P for which this occurs is called the fundamental *buckling load* of the structure.

Since the coefficient matrix of z in (3.212) becomes singular, a nontrivial solution exists, but not one that is unique. Therefore, the solution is an eigenvector that corresponds to the eigenvalue P. In order to distinguish the eigenvector associated with buckling from the static displacement state, the eigenvector is denoted as $\boldsymbol{y}$ (called a *buckling mode*), rather than z, and yields the following *generalized eigenvalue problem*:

$$\boldsymbol{K}\,\boldsymbol{y} = P\boldsymbol{D}\,\boldsymbol{y}. \tag{3.213}$$

The matrix $\boldsymbol{K}$ is taken to be positive definite, and $\boldsymbol{D}$ is positive semidefinite. Thus, all eigenvalues in (3.213) are strictly positive.

Dynamic Response

Consider a dynamic response with no boundary or interface conditions, that is, a structure with independent generalized coordinates. Lagrange's equation in (3.210) applies here and may be written in matrix form, using (3.208) with $\boldsymbol{F} = \boldsymbol{F}_g$, $\boldsymbol{M} = \boldsymbol{M}_g$, and $\boldsymbol{K} = \boldsymbol{K}_g$, as

$$\boldsymbol{M}\, z_{,tt} + \boldsymbol{K}\, z - \boldsymbol{F} = \boldsymbol{0}. \tag{3.214}$$

The *initial conditions* of motion for such a system consist of specifying the position and velocity of the system at some initial time, say $t = 0$, that is,

$$\begin{aligned} z(0) &= z^0 \\ z_{,t}(0) &= z_{,t}^0. \end{aligned} \tag{3.215}$$

Natural Vibration

The natural vibration of a structure is defined as the harmonic motion of the structural system with no applied load. A *natural frequency* ω is sought such that the solution $z(t)$ to (3.214) with $\boldsymbol{F} = 0$ is harmonic, that is,

$$z(t) = \boldsymbol{y} \sin(\omega t + \alpha), \tag{3.216}$$

where $\boldsymbol{y}$ is a constant vector defining a *mode shape* of vibration. Substituting $z(t)$ from (3.216) into (3.214) with $\boldsymbol{F} = 0$, we obtain

$$[-\omega^2 \boldsymbol{M}\, \boldsymbol{y} + \boldsymbol{K}\, \boldsymbol{y}] \sin(\omega t + \alpha) = \boldsymbol{0}, \tag{3.217}$$

which must hold for all time t. Therefore, the generalized eigenvalue problem is

$$\boldsymbol{K}\, \boldsymbol{y} = \zeta\, \boldsymbol{M}\, \boldsymbol{y}, \tag{3.218}$$

where $\zeta = \omega^2$. Equation (3.218) is an eigenvalue problem for natural frequency ω and associated mode shape $\boldsymbol{y}$, just as (3.213) was an eigenvalue problem for buckling load P and mode shape $\boldsymbol{y}$. In both cases, the reduced stiffness matrix $\boldsymbol{K}$ is positive definite, and both $\boldsymbol{D}$ and $\boldsymbol{M}$ are at least positive semidefinite. These mathematical properties of matrices that arise in structural equations play an essential role for both the theoretical properties of solutions and the computational methods for constructing solutions.

3.4.4 Variational Equations for Discrete Structural Systems

Variational Equilibrium Equation

It is not necessary to eliminate the dependent displacement coordinates in order to obtain the governing equations of a structural system. For example, let Z be the vector space of kinematically admissible displacements. First, consider a structure with externally applied load $\boldsymbol{F}_g$ and potential energy provided by (3.205). The theorem of minimum total potential energy is still valid for displacement in vector space Z. Let z_g be the equilibrium position that minimizes Π in (3.205) over vector space Z. Next, consider an arbitrary *virtual displacement* $\bar{z}_g \in Z$, and evaluate the total potential energy at an arbitrary point neighboring z_g, that is, for small ε and fixed $\bar{z}_g$,

$$\Pi(z_g + \varepsilon \bar{z}_g) \equiv H(\varepsilon). \tag{3.219}$$

Since the total potential energy has a minimum at z_g, the function $H(\varepsilon)$ defined by (3.219) has a minimum at $\varepsilon = 0$ for any $\bar{z}_g \in Z$. It is therefore necessary that the derivative of H with respect to ε be zero at $\varepsilon = 0$. Using the total potential energy formula from (3.206), but with the last term deleted, the following is obtained:

$$\bar{z}_g^T \boldsymbol{K}_g z_g - \bar{z}_g^T \boldsymbol{F}_g = 0, \qquad \forall \bar{z}_g \in Z, \tag{3.220}$$

which is the *variational equation of equilibrium*.

In order to take advantage of the mathematical form of this problem, it is necessary to define the *energy bilinear form*:

$$a(z_g, \bar{z}_g) = \bar{z}_g^T \boldsymbol{K}_g z_g \tag{3.221}$$

and the *load linear form*, which is defined by the externally applied load $\boldsymbol{F}_g$ as

$$\ell(\bar{z}_g) = \bar{z}_g^T \boldsymbol{F}_g. \tag{3.222}$$

Using this notation, the variational equation (3.220) can be written as

$$a(z_g, \bar{z}_g) = \ell(\bar{z}_g), \qquad \forall \bar{z}_g \in Z. \tag{3.223}$$

Under the hypothesis that the strain energy quadratic form is positive definite on vector space Z of kinematically admissible displacements, the following theorem is true:

Theorem 3.3. Theorem of Virtual Work. Assume that

$$a(z_g, z_g) > 0, \qquad \forall z_g \in Z, \quad z_g \neq 0. \tag{3.224}$$

Then, (3.223) has the unique solution $z_g \in Z$.

Proof. The proof directly follows the Lax-Milgram theorem of functional analysis [16] and the positive definite property of $a(z_g, z_g)$. An alternative version of the proof uses the fact that $a(z_g, z_g)$ is convex on Z, and the fact that (3.223) is the necessary and sufficient condition for z_g to be the minimum point. These results can be found in the optimization theory [48].

■

The unique solution to (3.223) guaranteed by Theorem 3.3 is the same obtained by first eliminating the dependent displacement coordinates, constructing the reduced global stiffness matrix, and finally solving (3.211). The final step is executed numerically in finite element computer code. The variational form of the structural equations in (3.223) has a substantial theoretical advantage in design sensitivity analysis.

Reduction of Variational Equilibrium Equation to Matrix Form

Equation (3.223) can be used to generate a matrix equation in order to construct a numerical solution. Let $\boldsymbol{\phi}^i \in Z \subset R^n$ (i=1, ···, m; $m < n$) be a *basis* of vector space Z, that is, a *linearly independent* set of vectors that *span* Z. The solution to (3.223) may then be written

$$z_g = \sum_{i=1}^{m} c_i \boldsymbol{\phi}^i = \boldsymbol{\Phi c}, \tag{3.225}$$

where $\boldsymbol{\Phi} = [\boldsymbol{\phi}^1, \ldots, \boldsymbol{\phi}^m]$ and the coefficients of c_i are uniquely determined. Substituting this representation for z_g into (3.223) and evaluating (3.223) as $\bar{z}_g = \boldsymbol{\phi}^j \, (j = 1, \ldots, m)$, the following relation is produced:

$$\sum_{i=1}^{m} a(\boldsymbol{\phi}^i, \boldsymbol{\phi}^j) c_i = \ell(\boldsymbol{\phi}^j), \quad j = 1, \ldots, m, , \tag{3.226}$$

with the definitions

$$\begin{aligned} \hat{\boldsymbol{K}} &\equiv [a(\boldsymbol{\phi}^i, \boldsymbol{\phi}^j)]_{m \times m} = [\boldsymbol{\phi}^{iT} \boldsymbol{K}_g \boldsymbol{\phi}^j]_{m \times m} = \boldsymbol{\Phi}^T \boldsymbol{K}_g \boldsymbol{\Phi} \\ \hat{\boldsymbol{F}} &= [\ell(\boldsymbol{\phi}^i)]_{m \times 1} = [\boldsymbol{\phi}^{iT} \boldsymbol{F}_g]_{m \times 1} = \boldsymbol{\Phi}^T \boldsymbol{F}_g \\ \boldsymbol{c} &= [c_i]_{m \times 1}. \end{aligned} \tag{3.227}$$

Equation (3.226) may be written in matrix form as

$$\hat{\boldsymbol{K}} \boldsymbol{c} = \hat{\boldsymbol{F}}. \tag{3.228}$$

This equation has a unique solution $\boldsymbol{c}$, since $\hat{\boldsymbol{K}}$ is positive definite (due to the assumption of positive definiteness of the energy bilinear form on Z). It is also clear that matrices $\hat{\boldsymbol{K}}$ and $\hat{\boldsymbol{F}}$ depend on the basis of space Z that is chosen. Different basis choices yield different matrices, although the resulting solution is nevertheless unique. The foregoing argument can be reversed to construct a proof of Theorem 3.3.

Variational Equation of Buckling

Consider a structural buckling problem, in which the potential energy of the load is given by (3.204). As with the preceding discussion, the total potential energy must be minimized over space Z. By using the total potential energy expression in (3.206) with $\boldsymbol{F}_g = 0$, and by using (3.219), the derivative with respect to ε must equal zero, yielding the *variation equation of buckling*:

$$\bar{\boldsymbol{y}}_g^T \boldsymbol{K}_g \boldsymbol{y}_g = P \bar{\boldsymbol{y}}_g^T \boldsymbol{D}_g \boldsymbol{y}_g, \quad \forall \bar{\boldsymbol{y}}_g \in Z, \tag{3.229}$$

where the solution is denoted by the vector $\boldsymbol{y}_g$. Defining the bilinear form $d(\bullet,\bullet)$ as

$$d(\boldsymbol{y}_g, \bar{\boldsymbol{y}}_g) = \bar{\boldsymbol{y}}_g^T \boldsymbol{D}_g \boldsymbol{y}_g, \tag{3.230}$$

(3.229) may be written in the more compact form as

$$a(\boldsymbol{y}_g, \bar{\boldsymbol{y}}_g) = P d(\boldsymbol{y}_g, \bar{\boldsymbol{y}}_g), \quad \forall \bar{\boldsymbol{y}}_g \in Z. \tag{3.231}$$

This is the variational form of the eigenvalue problem for structural buckling.

Reduction of the Variational Equation of Buckling to Matrix Form

As in the case of structural equilibrium, the variational equation (3.231) can be reduced to a matrix equation, using a basis for space Z. This yields the generalized eigenvalue problem

$$\hat{\boldsymbol{K}} \boldsymbol{c} = P \hat{\boldsymbol{D}} \boldsymbol{c}, \tag{3.232}$$

where components of vector $\boldsymbol{c}$ are coefficients of (3.225). Expanding the eigenvector $\boldsymbol{y}_g$ in terms of the basis $\boldsymbol{\phi}^i$ and the matrix $\boldsymbol{D}_g$, we have

$$\hat{\boldsymbol{D}} \equiv [d(\boldsymbol{\phi}^i, \boldsymbol{\phi}^j)]_{m \times m} = [\boldsymbol{\phi}^{iT} \boldsymbol{D}_g \boldsymbol{\phi}^j]_{m \times m} = \boldsymbol{\Phi}^T \boldsymbol{D}_g \boldsymbol{\Phi}. \tag{3.233}$$

As will normally be the case, the matrix $\boldsymbol{D}_g$ is positive definite on vector space Z, so the matrices $\hat{\boldsymbol{D}}$ and $\hat{\boldsymbol{K}}$ are positive definite, resulting in important theoretical and computational properties.

Variational Equation of Vibration

Consider the variational form of Lagrange's equation of motion in (3.209), with the Lagrangian defined by (3.208). In vector form, (3.209) becomes

$$\bar{z}_g(t)^T[\boldsymbol{M}_g z_{g,tt}(t) + \boldsymbol{K}_g z_g(t) - \boldsymbol{F}_g(t)] = 0, \qquad \forall \bar{z}_g(t) \in Z, \tag{3.234}$$

and must hold true for all values of time t.

In the case of harmonic motion with $\boldsymbol{F}_g = \boldsymbol{0}$, a solution to (3.234) in the form of (3.216) is sought. Substituting (3.216) into (3.234) (with $\boldsymbol{y}_g$ replacing z_g) gives

$$[-\omega^2 \bar{\boldsymbol{y}}_g^T \boldsymbol{M}_g \boldsymbol{y}_g + \bar{\boldsymbol{y}}_g^T \boldsymbol{K}_g \boldsymbol{y}_g]\sin(\omega t + \alpha) = 0, \qquad \forall \bar{\boldsymbol{y}}_g \in Z, \tag{3.235}$$

which must hold for all time t. Thus, it is necessary that

$$a(\boldsymbol{y}_g, \bar{\boldsymbol{y}}_g) = \zeta d(\boldsymbol{y}_g, \bar{\boldsymbol{y}}_g), \ \forall \bar{\boldsymbol{y}}_g \in Z, \tag{3.236}$$

where $\zeta = \omega^2$. The bilinear form $a(\bullet,\bullet)$ is as given in (3.221), and the bilinear form $d(\bullet,\bullet)$ is defined as

$$d(\boldsymbol{y}_g, \bar{\boldsymbol{y}}_g) = \bar{\boldsymbol{y}}_g^T \boldsymbol{M}_g \boldsymbol{y}_g. \tag{3.237}$$

Since the generalized mass matrix $\boldsymbol{M}_g$ is positive definite and the strain energy bilinear form $a(\bullet,\bullet)$ is normally positive definite on Z, desirable mathematical properties are associated with the variational equation given in (3.236).

Reduction of the Variational Equation of Vibration to Matrix Form

As with the foregoing analysis of the buckling eigenvalue problem, a matrix equation in the form of (3.232) may be obtained for the vibration problem. Thus, vibration and buckling problems have a similar form and share many mathematical properties.

While it is clear that the finite-dimensional structural analysis problem can be reduced to its matrix equation form, it will be shown in Section 4.1.4 of Chapter 4 that the variational form as developed in this chapter is better suited to the needs of structural design sensitivity analysis.

3.4.5 Numerical Integration

The finite element formulation requires integration over the domain or over the boundary during the construction of the stiffness matrix and force vector. Analytical integration, as used for the examples of Chapter 1, is limited to simple one-dimensional problems. Most integrals cannot be evaluated explicitly, and it is often faster to integrate them numerically rather than evaluating them exactly. Among many numerical integration methods that have been proposed, a Gauss integration rule is commonly used in the finite element formulation, due to its simplicity and accuracy. In this section, a brief introduction to the Gauss integration rule is provided. A rigorous study of numerical integration, including error estimates, can be found in Chapter 5 of Atkinson [49].

Consider one-dimensional integration of a function $f(\xi)$ over the interval $[-1, 1]$. Although the integration interval can be arbitrary, the interval $[-1, 1]$ is used without loss of generality because it is convenient to apply the reference element in the finite element formulation. A general form of Gauss integration can be written as

$$\int_{-1}^{1} f(\xi)d\xi \approx \sum_{i=1}^{NG} \omega_i f(\xi_i), \tag{3.238}$$

where NG is the number of integration points, ξ_i is the integration point, and ω_i is the nonnegative integration weight. The integration points and weights are chosen such that the right side of (3.238) equals the left side for polynomials $f(\xi)$ as much as possible. In general, an NG-point Gauss integration method integrates $(2NG - 1)$-order polynomials exactly. Such method is extremely accurate in most cases, and is the one most frequently used in modern finite element formulations. Table 3.1 summarizes the integration points and weights for Gauss integration.

A multidimensional integration can be constructed by employing the one-dimensional integration rule on each coordinate separately. In two- and three-dimensional domains, the Gauss integration rule can be written as, respectively,

$$\int_{-1}^{1}\int_{-1}^{1} f(\xi,\eta)\,d\xi d\eta = \sum_{i=1}^{NG}\sum_{j=1}^{NG} \omega_i \omega_j f(\xi_i,\eta_j) \tag{3.239}$$

and

$$\int_{-1}^{1}\int_{-1}^{1}\int_{-1}^{1} f(\xi,\eta,\zeta)\,d\xi d\eta d\zeta = \sum_{i=1}^{NG}\sum_{j=1}^{NG}\sum_{k=1}^{NG} \omega_i \omega_j \omega_k f(\xi_i,\eta_j,\zeta_k). \tag{3.240}$$

Figure 3.24 illustrates the integration points in two-dimensional reference elements. The computational cost of Gauss integration is proportional to $(NG)^2$ for two-dimensional problems and $(NG)^3$ for three-dimensional problems.

Table 3.1. Gauss integration points and weights.

NG	Integration Points (ξ_i)	Weights (ω_i)
1	0.0	2.0
2	±.5773502692	1.0
3	±.7745966692	.5555555556
	0.0	.8888888889
4	±.8611363116	.3478546451
	±.3399810436	.6521451549
5	±.9061798459	.2369268851
	±.5384693101	.4786286705
	0.0	.5688888889

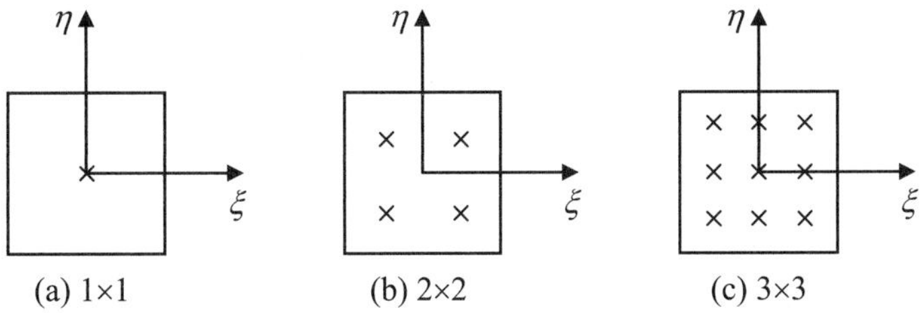

Figure 3.24. Gauss integration points in two-dimensional reference elements.

In conjunction with design sensitivity analysis, the numerical integration method can result in difficulties in the analytical differentiation of the stiffness matrix with respect to the design variable, because the explicit expression of the design variable does not appear in the numerically integrated stiffness matrix. In such a case, the continuum expression corresponding to the matrix is first differentiated, and then the resulting expression is numerically integrated using the same method as the initial matrix. Thus, the discrete formulation of design sensitivity analysis in Chapter 4 encounters a possible difficulty in which differentiation of the global stiffness matrix is required. As a remedy, the discrete expression of the element stiffness matrix may need to be differentiated before numerical integration. Afterwards, the formula is assembled to construct the differentiated global stiffness matrix. This method, however, can be considered a hybrid, since the domain concept from the continuum approach is used during differentiation.

PART II
Design Sensitivity Analysis of Linear Structural Systems

4
Discrete Design Sensitivity Analysis

In this chapter, design sensitivity analysis of a discrete matrix equation is introduced for structural problems. The design represents a structural parameter that can affect the results of the analysis. For example, when the cross-sectional area of a truss component changes, the displacement results change for the fixed applied force because the stiffness matrix changes. In such cases, the cross-sectional area of the truss component can be a design. There are many types of designs, including the thickness of a plate, the length of a beam, and the area of a two-dimensional solid. As a given design value, structural analysis provides a unique performance value, in such forms as displacement, stress, and frequency. However, different designs usually produce different performance measures. If two plates have different thickness values, namely h_1 and h_2, then the stress values at a point on each plate will be different. We can therefore say that the performance measure (i.e., stress) depends on the design (i.e., plate thickness). Design sensitivity analysis computes the performance's dependence on the design. In other words, design sensitivity is the differentiation of the performance measure with respect to the design.

There are two types of design dependence: *explicit* and *implicit*. When the expression of the performance measure contains the design, one can easily compute the performance sensitivity with respect to the design by applying a differentiation rule. For example, when the structural volume is the performance measure and the plate thickness is the design, then the performance value will be the plate area multiplied by its thickness. In this case, design sensitivity can be obtained simply by differentiating the volume with respect to the thickness in order to obtain the plate area as a sensitivity value. This is called *explicit dependence*. However, when beam stress is a performance measure, the derivative of stress with respect to the design is not easily obtained. If we let the displacement vector be a state variable and let stress be computed from the displacement, then we can say that stress is a function of displacement. The derivative of stress with respect to the design thus includes the derivative of displacement with respect to the design, which has to be computed another way. Thus, the dependence of stress on the design is *implicit*. The essential procedure of structural design sensitivity analysis is to obtain this implicit dependence on the design by differentiating the structural equation.

As shown in Chapter 3, in the static response analysis the linear system of matrix equations is solved in order to compute state variable z. Since the stiffness matrix and the force vector depend on the design, the state variable implicitly depends on the design. Many structural performance measures, such as stress, displacement, and frequency, depend on the state variable. Thus, the design sensitivity of performance functions clearly depends implicitly on the design through the state variable. This chapter presents a design sensitivity analysis in order to obtain this implicit dependence on the design.

The design sensitivity information of a general performance measure can be computed either with the *direct differentiation method* or with the *adjoint variable method*. The former directly solves for the design dependency of a state variable, and then computes performance sensitivity using the chain rule of differentiation. This method clearly shows the implicit dependence on the design, and a very simple sensitivity expression can be obtained. The latter method, however, constructs an adjoint problem that solves for the adjoint variable, which contains all implicitly dependent terms.

Through the assembly procedure in finite element analysis, the generalized global stiffness matrix is obtained with dimensions ($n \times n$). Although this generalized stiffness matrix contains information of all elements, it does not contain any information of the boundary and interface conditions. Thus, a rigid body motion exists in the structure. Mathematically, the generalized stiffness matrix is singular. However, the dependence of the generalized stiffness matrix on the design can be easily computed by following the element stiffness formulation and the assembly procedure in the global stiffness matrix. For structural analysis, after applying such kinematic constraints as displacement boundary and interface conditions, the rigid body motion of the structure is removed. Technically, the dimensions of the global stiffness matrix are reduced to ($m \times m$), where $m < n$ and $n - m$ is the number of independent boundary and interface conditions. The reduced stiffness matrix is now positive definite. Positive definiteness is a very attractive property of a linear system because it yields a unique solution. However, the reduced stiffness matrix's dependence on the design is not easily obtained if the boundary and interface conditions are not simple, or if the analysis code does not explicitly generate a reduced stiffness matrix. Different analysis software may use different methods for imposing the boundary conditions; consequently, the design sensitivity analysis code depends on the specific analysis code. Thus, a generalized stiffness matrix is preferable for design sensitivity analysis, although additional theories must be considered.

In order to enhance the convergence rate during design iteration, second-order design sensitivity analysis provides important information to the optimization algorithm in conjunction with first-order sensitivity results. However, the cost of evaluating second-order sensitivity information needs to be considered, since a significant amount of computational costs is involved in repeatedly solving a linear system of equations. A very efficient sensitivity expression can be obtained using a hybrid method, which combines the direct differentiation and the adjoint variable methods.

The design sensitivity analysis of an eigenvalue problem for a simple, unrepeated eigenvalue can be simply expressed without solving an adjoint equation. Only the stiffness matrix, mass matrix, and eigenvector are required to calculate the eigenvalue derivative with respect to the design. However, eigenvector design sensitivity requires a significant amount of knowledge in mathematical theory. The direct solution of a linear system, as well as the eigenvector expansion method using a Ritz vector will be discussed later in this chapter. For the repeated eigenvalue problem, it is shown that the repeated eigenvalue is only directionally differentiable, and the repeated eigenvector is not differentiable with respect to the design.

Transient dynamic analysis can be formulated using either direct time integration or eigenvector expansion. For a linear problem, the eigenvector expansion method is predominantly used, since it is more efficient and provides a physical interpretation of the problem. Design sensitivity analysis of transient dynamics is presented in Section 4.3 using a system of matrix equations. The adjoint variable method yields a terminal-value problem, compared with the initial-value problem of response analysis. For computational efficiency, the direct differentiation method using a linear combination of eigenvectors and/or Ritz vectors can be used to approximate the design sensitivity of a transient response.

4.1 Static Response Design Sensitivity

4.1.1 Statement of the Problem

Design sensitivity analysis begins by defining the design parameters. Let $\boldsymbol{b} = [b_1, \ldots, b_k]^T$ denote the $(k \times 1)$ vector of member-size design variables and material properties. Even though material properties are not sizing design variables, in this text they will be categorized as sizing design variables, since the sensitivity analysis procedure is the same as sizing design variables. The design includes the cross-sectional area of a truss, the plate thickness, the moment of inertia of a beam, and Young's modulus, etc. As explained in Section 3.4 of Chapter 3, when the member size and the geometric variables are taken as design variables, the generalized stiffness matrix and the load vector are functions of the design variables, that is,

$$\begin{aligned} \boldsymbol{K}_g &= \boldsymbol{K}_g(\boldsymbol{b}) \\ \boldsymbol{F}_g &= \boldsymbol{F}_g(\boldsymbol{b}), \end{aligned} \tag{4.1}$$

where $\boldsymbol{K}_g(\boldsymbol{b})$ is the $(n \times n)$ matrix, and $\boldsymbol{F}_g(\boldsymbol{b})$ is the $(n \times 1)$ vector. For example, a change in plate thickness produces a change in the stiffness matrix $\boldsymbol{K}_g$. In the gravity field, the cross-sectional area of a truss changes the structure's body force. Equation (4.1) expresses all dependence of the structure on design $\boldsymbol{b}$. For the moment, the stiffness matrix and the force vector are assumed to be continuously differentiable with respect to design $\boldsymbol{b}$. It is presumed that kinematic admissibility conditions (boundary conditions and interface conditions) are not explicit functions of the design. Such is the case when member size variables are chosen as design variables. Fixed displacement conditions at each end of the beam will not change as the cross-sectional area changes. This type of design is called the *sizing design*. However, when the location of the nodes at the end of the beam is changed, then the boundary conditions change. The case in which kinematic admissibility conditions are functions of the design is included in the shape design sensitivity formulation of Chapter 6.

Since the generalized stiffness matrix and the load vector are dependent on the design, the energy bilinear form in (3.221) and the load linear form in (3.222) also depend on the design. They are denoted here in the following variational equation:

$$\begin{aligned} a_{\boldsymbol{b}}(\boldsymbol{z}_g, \bar{\boldsymbol{z}}_g) &\equiv \bar{\boldsymbol{z}}_g^T \boldsymbol{K}_g(\boldsymbol{b}) \boldsymbol{z}_g \\ \ell_{\boldsymbol{b}}(\bar{\boldsymbol{z}}_g) &\equiv \bar{\boldsymbol{z}}_g^T \boldsymbol{F}_g(\boldsymbol{b}) \\ a_{\boldsymbol{b}}(\boldsymbol{z}_g, \bar{\boldsymbol{z}}_g) &= \ell_{\boldsymbol{b}}(\bar{\boldsymbol{z}}_g), \qquad \forall \bar{\boldsymbol{z}}_g \in Z, \end{aligned} \tag{4.2}$$

where Z is the space of kinematically admissible displacements that satisfy homogeneous boundary conditions. In (4.2), $\boldsymbol{K}_g(\boldsymbol{b})$ is an $(n \times n)$ matrix while all other vectors have $(n \times 1)$ dimensions. Recall that a unique solution $\boldsymbol{z}_g$ exists in (4.2). For a given design $\boldsymbol{b}$, $\boldsymbol{z}_g$ is uniquely determined. However, a different design will yield a different solution to the variational equation. Since these equations explicitly depend on the design, it is clear that the solution $\boldsymbol{z}_g$ is dependent on the design, that is,

$$\boldsymbol{z}_g = \boldsymbol{z}_g(\boldsymbol{b}). \tag{4.3}$$

Note that the dependence of $\boldsymbol{z}_g$ on the design is implicit through the variational equation (4.2).

In most structural design problems, a cost function is minimized or maximized, subject to constraints on stresses, displacements, and eigenvalues. Most gradient-based optimization algorithms, which find the optimum value of the cost function while satisfying all constraints, require the sensitivity information of the cost and constraints at each design iteration. In this text, we refer to the cost and constraints as *performance measures*. Performance measures can include the weight of a structure, the displacement at a point, and stress in a certain region. These measures depend on the design in their expressions and/or through the state variable z_g, which also depends on the design, as denoted in (4.3). The main purpose of design sensitivity analysis is to find the gradient information of the performance measures with respect to design variables. Consider a general function that may represent any of these performance measures, written in the form

$$\psi = \psi(\boldsymbol{b}, \boldsymbol{z}_g(\boldsymbol{b})). \tag{4.4}$$

This function depends on the design in two ways: first, through explicit design dependence, and, second, through implicit dependence, which comes from solution z_g to the state equation. The objective of design sensitivity analysis is to determine the total – explicit and implicit – dependence of such functions on the design, i.e., to compute $d\psi/d\boldsymbol{b}$. To this end, two fundamental questions need to be answered: Given that the function ψ is differentiable in its arguments, is the total dependence of ψ on the design differentiable? In addition, if the solution z_g is differentiable with respect to the design, then how can the derivative of ψ be calculated with respect to the design?

4.1.2 Design Sensitivity Analysis with Reduced Stiffness Matrix

For sensitivity analysis purposes, it is simpler to use a reduced stiffness matrix. However, since $\boldsymbol{K}(\boldsymbol{b})$ is constructed from $\boldsymbol{K}_g(\boldsymbol{b})$ by imposing boundary and interface conditions, it is occasionally difficult to find the dependence of $\boldsymbol{K}(\boldsymbol{b})$ on the design. In this section, we assume that the expression of $\boldsymbol{K}(\boldsymbol{b})$ and $\boldsymbol{F}(\boldsymbol{b})$ are known with respect to design $\boldsymbol{b}$. Consider a structural formulation in which dependent variables have been directly eliminated using boundary conditions, and a set of structural equations can be obtained that resemble (3.211), in the form

$$\boldsymbol{K}(\boldsymbol{b})\boldsymbol{z} = \boldsymbol{F}(\boldsymbol{b}), \tag{4.5}$$

where $\boldsymbol{K}(\boldsymbol{b})$ is the reduced stiffness matrix($m \times m$), $\boldsymbol{F}(\boldsymbol{b})$ is the reduced load vector($m \times 1$), and m is the number of independent degrees-of-freedom. In this section, (4.5) will be referred to as a structural problem or response problem. Recall that the reduced stiffness matrix $\boldsymbol{K}(\boldsymbol{b})$ is positive definite, hence nonsingular. It is assumed that all entries in $\boldsymbol{K}(\boldsymbol{b})$ and $\boldsymbol{F}(\boldsymbol{b})$ are "s" times continuously differentiable with respect to the design. The implicit function theorem [50] thus guarantees that solution $\boldsymbol{z} = \boldsymbol{z}(\boldsymbol{b})$ to (4.5) is also s times continuously differentiable. Most design sensitivity analyses in this text are developed under the premise that $s = 1$, except for second-order design sensitivity, in which $s = 2$. Thus, the preceding question concerning differentiability of $\boldsymbol{z}$ with respect to the design has been answered. But, the problem of computing the total derivative of ψ in (4.4) with respect to the design remains to be solved.

Direct Differentiation Method

The direct differentiation method evaluates the implicit dependence of $\boldsymbol{z}$ on design $\boldsymbol{b}$ by using the derivative of structural equation (4.5). Let k be the dimension of design vector $\boldsymbol{b}$. Using the chain rule of differentiation and matrix calculus notations, the total derivative of $\psi(\boldsymbol{b},\boldsymbol{z}(\boldsymbol{b}))$ with respect to $\boldsymbol{b}$ can be calculated as

$$\frac{d\psi}{d\boldsymbol{b}} = \frac{\partial\psi}{\partial\boldsymbol{b}} + \frac{\partial\psi}{\partial z}\frac{dz}{d\boldsymbol{b}}, \tag{4.6}$$

where $\partial\psi/\partial\boldsymbol{b}$ ($1 \times k$ row vector) represents explicit dependence on design $\boldsymbol{b}$, and the second term shows the implicit dependence through displacement z. The derivative of a scalar function with respect to the vector gives a row vector with the same dimension as the vector. For a more detailed explanation of vector and matrix algebra, refer to Appendix A.1. For a given expression of ψ, only the $dz/d\boldsymbol{b}$ term ($m \times k$) is unknown, which can be computed from the differentiation of the state equation (4.5). By differentiating both sides of (4.5) with respect to design $\boldsymbol{b}$, the following is obtained:

$$\boldsymbol{K}(\boldsymbol{b})\frac{dz}{d\boldsymbol{b}} = \frac{\partial\boldsymbol{F}(\boldsymbol{b})}{\partial\boldsymbol{b}} - \frac{\partial}{\partial\boldsymbol{b}}[\boldsymbol{K}(\boldsymbol{b})\tilde{z}], \tag{4.7}$$

where the superposed tilde (~) indicates a variable held constant during the partial differentiation process. In the last term of (4.7), the dependence of z on $\boldsymbol{b}$ is suppressed in order to evaluate partial derivatives; thus, only the dependence of $\boldsymbol{K}(\boldsymbol{b})$ needs to be considered. Since the reduced stiffness matrix $\boldsymbol{K}(\boldsymbol{b})$ is nonsingular, (4.7) may be solved for $dz/d\boldsymbol{b}$ as

$$\frac{dz}{d\boldsymbol{b}} = \boldsymbol{K}^{-1}(\boldsymbol{b})\left[\frac{\partial\boldsymbol{F}(\boldsymbol{b})}{\partial\boldsymbol{b}} - \frac{\partial}{\partial\boldsymbol{b}}[\boldsymbol{K}(\boldsymbol{b})\tilde{z}]\right]. \tag{4.8}$$

The result may now be substituted into (4.6) to obtain the total derivative of $\psi(\boldsymbol{b},z(\boldsymbol{b}))$ as

$$\frac{d\psi}{d\boldsymbol{b}} = \frac{\partial\psi}{\partial\boldsymbol{b}} + \frac{\partial\psi}{\partial z}\boldsymbol{K}^{-1}(\boldsymbol{b})\frac{\partial}{\partial\boldsymbol{b}}[\boldsymbol{F}(\boldsymbol{b}) - \boldsymbol{K}(\boldsymbol{b})\tilde{z}]. \tag{4.9}$$

Since the expression of ψ with respect to $\boldsymbol{b}$ and z is known from its definition, $\partial\psi/\partial\boldsymbol{b}$ and $\partial\psi/\partial z$ can be readily obtained. Also, assuming the expressions of $\boldsymbol{K}(\boldsymbol{b})$ and $\boldsymbol{F}(\boldsymbol{b})$ are known, the partial derivative in the last part of (4.9) can be calculated. From the fact that $\boldsymbol{K}(\boldsymbol{b})$ is a positive definite matrix, the inverse of $\boldsymbol{K}(\boldsymbol{b})$ can be calculated such that the total derivative of ψ can be obtained with respect to design $\boldsymbol{b}$. However, it is questionable how useful (4.9) is, since the direct computation of $\boldsymbol{K}^{-1}(\boldsymbol{b})$ is impractical for real applications. Alternatively, (4.7) can be numerically solved for $dz/d\boldsymbol{b}$ and substituted into (4.6) to obtain the desired result. For each component b_i of $\boldsymbol{b}$, the right side of (4.7) is a column vector that serves as a fictitious force corresponding to dz/db_i. This procedure is known as the *direct differentiation method*, which has been extensively used in structural optimization due to its straightforward derivations. Computational aspects of this approach will be discussed in Section 4.1.4.

Adjoint Variable Method

In order to avoid the calculation of $dz/d\boldsymbol{b}$ in (4.7), the adjoint variable method will be developed. In (4.9), all terms can be easily calculated from their definition, except for $(\partial\psi/\partial z)\boldsymbol{K}^{-1}(\boldsymbol{b})$, which is a ($1 \times m$) row vector. Also, $(\partial\psi/\partial z)\boldsymbol{K}^{-1}(\boldsymbol{b})$ is not related to the design derivative, that is, $(\partial\psi/\partial z)\boldsymbol{K}^{-1}(\boldsymbol{b})$ is constant and only needs to be computed once for all $\partial\psi/\partial b_i$, $i = 1, \ldots, k$. The main idea is to directly compute this term by defining it as the *adjoint variable* $\boldsymbol{\lambda}$:

$$\boldsymbol{\lambda} \equiv \left[\frac{\partial\psi}{\partial z}\boldsymbol{K}^{-1}(\boldsymbol{b})\right]^T = \boldsymbol{K}^{-1}(\boldsymbol{b})\frac{\partial\psi}{\partial z}^T, \tag{4.10}$$

where the symmetric property of matrix $\boldsymbol{K}(\boldsymbol{b})$ has been used. Symmetry of $\boldsymbol{K}(\boldsymbol{b})$ is an important property in the adjoint variable method, associated with the response problem in (4.5). Rather than directly evaluating $\boldsymbol{\lambda}$ from (4.10), which involves computing $\boldsymbol{K}^{-1}(\boldsymbol{b})$, both sides of (4.10) can be multiplied by matrix $\boldsymbol{K}(\boldsymbol{b})$ to obtain the following *adjoint equation in* $\boldsymbol{\lambda}$:

$$\boldsymbol{K}(\boldsymbol{b})\boldsymbol{\lambda} = \frac{\partial \psi}{\partial \boldsymbol{z}}^T. \tag{4.11}$$

Note that the same stiffness matrix is used in the adjoint and response problem, because of the symmetric property of $\boldsymbol{K}(\boldsymbol{b})$. The adjoint problem is the same as the response problem, except that the former has an explicit dependence of ψ on $\boldsymbol{z}$ with $(\partial\psi/\partial\boldsymbol{z})$ as a load vector. The right side of (4.11) is sometimes called an *adjoint load*. Equation (4.11) may be solved for $\boldsymbol{\lambda}$ and the result substituted into (4.9) using (4.10), to obtain

$$\frac{d\psi}{d\boldsymbol{b}} = \frac{\partial\psi}{\partial\boldsymbol{b}} + \boldsymbol{\lambda}^T \frac{\partial}{\partial\boldsymbol{b}}[\boldsymbol{F}(\boldsymbol{b}) - \boldsymbol{K}(\boldsymbol{b})\tilde{\boldsymbol{z}}]. \tag{4.12}$$

Note that the computation of $\boldsymbol{\lambda}$ in (4.11) is independent of the design parameter b_i. Since $\boldsymbol{K}(\boldsymbol{b})$ and $(\partial\psi/\partial\boldsymbol{z})$ are known, (4.11) can be solved for $\boldsymbol{\lambda}$, and this solution can be used for all design parameters b_i. A somewhat more convenient form for derivative calculation purposes can be written as

$$\frac{d\psi}{d\boldsymbol{b}} = \frac{\partial\psi}{\partial\boldsymbol{b}} + \frac{\partial}{\partial\boldsymbol{b}}\left[\tilde{\boldsymbol{\lambda}}^T\boldsymbol{F}(\boldsymbol{b}) - \tilde{\boldsymbol{\lambda}}^T\boldsymbol{K}(\boldsymbol{b})\tilde{\boldsymbol{z}}\right]. \tag{4.13}$$

This approach is called the *adjoint variable method* for design sensitivity analysis. The direct differentiation method in (4.9) is more closely related to the design vector, whereas the adjoint variable method in (4.13) is more closely related to the performance measure. Different performance measures have different adjoint load expressions. The computational aspects of this approach will be discussed in Section 4.1.4.

4.1.3 Design Sensitivity Analysis with Generalized Stiffness Matrix

If the reduced stiffness matrix $\boldsymbol{K}(\boldsymbol{b})$, and the reduced force vector $\boldsymbol{F}(\boldsymbol{b})$, are readily available, either of the two methods yields a complete solution to the design sensitivity problem. However, for complicated kinematic admissibility conditions (or boundary conditions), and in particular for those multipoint constraints that involve linear combinations of several state variables, $\boldsymbol{K}(\boldsymbol{b})$ and $\boldsymbol{F}(\boldsymbol{b})$ are not explicitly generated. Since direct reduction of the stiffness matrix requires a significant amount of computer memory, many structural analysis codes do not reduce the dimension of matrix $\boldsymbol{K}_g(\boldsymbol{b})$ in order to impose boundary conditions. Instead, they usually substitute boundary conditions directly into $\boldsymbol{K}_g(\boldsymbol{b})$ to make it positive definite. Thus, the computation of partial derivatives on the right-hand side of (4.9), or in (4.13), is difficult. It is therefore preferable to develop a design sensitivity formulation that works directly with a singular, generalized stiffness matrix. As will be shown in Chapter 5, this approach of design sensitivity analysis is closely related to the continuum design sensitivity analysis method.

Differentiability of Global Displacement

Let $(n - m)$ be the number of independent homogeneous boundary conditions, and let an explicit form of vector space Z be given for all kinematically admissible displacements, as

$$Z = \{\boldsymbol{z}_g \in R^n \mid \boldsymbol{G}\boldsymbol{z}_g = \boldsymbol{0}\}, \tag{4.14}$$

where $\boldsymbol{G}$ is an $(n-m) \times n$ matrix that defines the boundary conditions and does not depend on the design. With a basis vector $\boldsymbol{\phi}^i$ (i=1, …, m) of Z that is independent of the design, a solution z_g to variational equation (3.220) may be represented in the form of (3.225), where coefficient vector $\boldsymbol{c}$ is determined by (3.228), and is written in the form of

$$\hat{\boldsymbol{K}}(\boldsymbol{b})\boldsymbol{c} = \hat{\boldsymbol{F}}(\boldsymbol{b}). \tag{4.15}$$

Note that the dependence of $\hat{\boldsymbol{K}}$ and $\hat{\boldsymbol{F}}$ on design $\boldsymbol{b}$ is explicitly defined in terms of $\boldsymbol{K}_g(\boldsymbol{b})$ and $\boldsymbol{F}_g(\boldsymbol{b})$ in (3.227). Therefore, $\hat{\boldsymbol{K}}(\boldsymbol{b})$ and $\hat{\boldsymbol{F}}(\boldsymbol{b})$ are differentiable with respect to the design, and $\hat{\boldsymbol{K}}(\boldsymbol{b})$ is nonsingular in nominal design $\boldsymbol{b}$ and its neighboring designs. The derivative of $\boldsymbol{c}$ with respect to the design can then be obtained by either of the previously described methods. Once $d\boldsymbol{c}/d\boldsymbol{b}$ is determined, (3.225) may be used to obtain

$$\frac{dz_g}{d\boldsymbol{b}} = \boldsymbol{\Phi}\frac{d\boldsymbol{c}}{d\boldsymbol{b}}, \tag{4.16}$$

since $\boldsymbol{\Phi}$ does not depend on $\boldsymbol{b}$. Thus, the question of the differentiability of z_g is resolved. Computation of the required derivative $dz_g/d\boldsymbol{b}$ may be carried out by using the variational formulation in (3.220), written using the notation of (4.2) as

$$a_{\boldsymbol{b}}(z_g, \bar{z}_g) = \ell_{\boldsymbol{b}}(\bar{z}_g), \qquad \forall \bar{z}_g \in Z. \tag{4.17}$$

Directional Derivatives

In order to take advantage of variational equation (4.17), it is helpful to introduce a directional derivative notation that will be used throughout the remainder of this text. Consider a nominal design $\boldsymbol{b}$, and those neighboring designs that are described by arbitrary design variation $\delta\boldsymbol{b}$ and small parameter $\tau > 0$, as

$$\boldsymbol{b}_\tau = \boldsymbol{b} + \tau\delta\boldsymbol{b}. \tag{4.18}$$

Thus, for given design variation $\delta\boldsymbol{b}$, the design perturbation is controlled by one parameter τ. Similar to the first variation in the calculus of variations, the following directional derivative notations are employed:

$$\begin{aligned} z_g' &= z_g'(\boldsymbol{b}, \delta\boldsymbol{b}) \equiv \frac{d}{d\tau} z_g(\boldsymbol{b} + \tau\delta\boldsymbol{b})\bigg|_{\tau=0} = \frac{dz_g}{d\boldsymbol{b}}\delta\boldsymbol{b} \\ a'_{\delta\boldsymbol{b}}(z_g, \bar{z}_g) &\equiv \frac{d}{d\tau} a_{\boldsymbol{b}+\tau\delta\boldsymbol{b}}(z_g(\boldsymbol{b}), \bar{z}_g)\bigg|_{\tau=0} \\ &= \frac{\partial}{\partial \boldsymbol{b}}(\bar{z}_g^T \boldsymbol{K}_g(\boldsymbol{b})\tilde{z}_g)\delta\boldsymbol{b} \\ \ell'_{\delta\boldsymbol{b}}(\bar{z}_g) &\equiv \frac{d}{d\tau} \ell_{\boldsymbol{b}+\tau\delta\boldsymbol{b}}(\bar{z}_g)\bigg|_{\tau=0} \\ &= \frac{\partial}{\partial \boldsymbol{b}}(\bar{z}_g^T \boldsymbol{F}_g(\boldsymbol{b}))\delta\boldsymbol{b}, \end{aligned} \tag{4.19}$$

where the prime (′) denotes the *differential* (or *variation*) of a function with respect to $\boldsymbol{b}$ in the direction of $\delta\boldsymbol{b}$. If the result is linear in $\delta\boldsymbol{b}$, then the function whose differential has been taken is *differentiable*. Otherwise, it is only *directionally differentiable*, in the sense of the Gateaux differential. With this notation, the prime may be explicitly employed, including argument $\delta\boldsymbol{b}$, in order to emphasize dependence on the design variation. Note that the design dependence of displacement z_g caused by design perturbation τ is eliminated in $a'_{\delta b}(z_g, \bar{z}_g)$. Thus, the total derivative of $a_b(z_g, \bar{z}_g)$ will contain the generalized displacement variation, in addition to $a'_{\delta b}(z_g, \bar{z}_g)$, as shown below.

Since matrix $\boldsymbol{G}$ in (4.14), which defines vector space Z of kinematically admissible displacements, does not depend on the design, arbitrary vector $\bar{z}_g \in Z$ in (4.17) also need not depend on the design. By taking the total variation of both sides of (4.17), and by using the chain rule of differentiation, the following design sensitivity equation can be obtained:

$$a_b(z'_g, \bar{z}_g) = \ell'_{\delta b}(\bar{z}_g) - a'_{\delta b}(z_g, \bar{z}_g), \qquad \forall \bar{z}_g \in Z. \tag{4.20}$$

In (4.20), the bilinear property of energy form $a_b(\bullet,\bullet)$ is used such that

$$\begin{aligned}
&\frac{d}{d\tau} a_{b+\tau\delta b}(z_g(\boldsymbol{b}+\tau\delta\boldsymbol{b}), \bar{z}_g)\Big|_{\tau=0} \\
&= \frac{\partial}{\partial \boldsymbol{b}}(\bar{z}_g^T \boldsymbol{K}_g(\boldsymbol{b})\tilde{z}_g)\delta\boldsymbol{b} + \bar{z}_g^T \boldsymbol{K}_g(\boldsymbol{b})\frac{dz_g}{d\boldsymbol{b}}\delta\boldsymbol{b} \\
&= a'_{\delta b}(z_g, \bar{z}_g) + a_b(z'_g, \bar{z}_g).
\end{aligned}$$

The right side of (4.20) can be computed if solution z_g has been obtained from (4.17), and if the explicitly dependent terms are obtained from the definition in (4.19). The only remaining question is whether the solution to (4.20) belongs to the space of kinematically admissible displacements. Note that for $z_g(\boldsymbol{b}) \in Z$, $\boldsymbol{G}z_g(\boldsymbol{b}) = 0$. Taking the variation of both sides of this equation, we obtain

$$\boldsymbol{G}z'_g(\boldsymbol{b}, \delta\boldsymbol{b}) = \boldsymbol{0}. \tag{4.21}$$

Thus, z'_g belongs to the space Z for any design variation $\delta\boldsymbol{b}$. Equation (4.20) thus has a unique solution, z'_g, just as the response problem in (4.17) has a unique solution. It is interesting to note that z_g and its variation z'_g are in the same kinematically admissible displacement space. Readers should not confuse the derivative in the structural domain with the derivative in the design domain. In general, the regularity of a function is reduced after differentiation. However, since the design space is independent of the structural space, z'_g can have the same regularity as z_g, even if the definition of z'_g contains a derivative as in (4.19). Actually, z'_g should be understood as a variation rather than a partial derivative.

Direct Differentiation Method

By taking $\delta\boldsymbol{b}$ as a unit vector in the ith design coordinate direction, (4.20) may be solved for dz_g/db_i. Repeating this process with $i = 1, 2, \ldots, k$ will yield all partial derivatives of z_g with respect to $\boldsymbol{b}$. Specifically, (4.20) may be written in terms of the ith component of $\boldsymbol{b}$, as

$$\bar{z}_g^T \boldsymbol{K}_g(\boldsymbol{b})\frac{\partial z_g}{\partial b_i} = \frac{\partial}{\partial b_i}(\bar{z}_g^T \boldsymbol{F}_g(\boldsymbol{b})) - \frac{\partial}{\partial b_i}(\bar{z}_g^T \boldsymbol{K}_g(\boldsymbol{b})\tilde{z}_g), \qquad i = 1, \ldots, k. \tag{4.22}$$

This may be interpreted as solving the original structural equation with an artificially applied load that is the coefficient of $\bar{z}_g^T$ on the right side of (4.22), which is the fictitious load.

Adjoint Variable Method

Consider the last term in (4.6) with z_g, namely, $(\partial\psi/\partial z_g)(dz_g/d\boldsymbol{b})$, which is to be obtained without evaluating matrix $dz_g/d\boldsymbol{b}$. The process of the adjoint variable method is to regard the coefficient $\partial\psi/\partial z_g$ of $dz_g/d\boldsymbol{b}$ as a load vector, which is the adjoint load $(\partial\psi/\partial z_g)^T$. The adjoint variable $\boldsymbol{\lambda}_g \in Z$ associated with this adjoint load needs to be determined, that is,

$$a_{\boldsymbol{b}}(\boldsymbol{\lambda}_g,\bar{\boldsymbol{\lambda}}_g) = \frac{\partial\psi}{\partial z_g}\bar{\boldsymbol{\lambda}}_g, \qquad \forall\bar{\boldsymbol{\lambda}}_g \in Z. \tag{4.23}$$

Note that this equation is nothing more than the structural equation of displacement $\boldsymbol{\lambda}_g$ due to the applied load vector $(\partial\psi/\partial z_g)^T$. Therefore, it may be readily solved.

By evaluating (4.23) at $\bar{\boldsymbol{\lambda}}_g = z_g'$ (recall that $z_g' \in Z$), and by using the notation introduced in the first line of (4.19), we obtain

$$\frac{\partial\psi}{\partial z_g}z_g' = \frac{\partial\psi}{\partial z_g}\frac{dz_g}{d\boldsymbol{b}}\delta\boldsymbol{b} = a_{\boldsymbol{b}}(\boldsymbol{\lambda}_g, z_g'). \tag{4.24}$$

Similarly, evaluating (4.20) at $\bar{z}_g = \boldsymbol{\lambda}_g$, we also obtain

$$a_{\boldsymbol{b}}(z_g',\boldsymbol{\lambda}_g) = \ell'_{\delta\boldsymbol{b}}(\boldsymbol{\lambda}_g) - a'_{\delta\boldsymbol{b}}(z_g,\boldsymbol{\lambda}_g). \tag{4.25}$$

Noting that the energy bilinear form $a_{\boldsymbol{b}}(\bullet,\bullet)$ is symmetric, (4.24) and (4.25) yield the following important relation:

$$\frac{\partial\psi}{\partial z_g}\frac{dz_g}{d\boldsymbol{b}}\delta\boldsymbol{b} = \ell'_{\delta\boldsymbol{b}}(\boldsymbol{\lambda}_g) - a'_{\delta\boldsymbol{b}}(z_g,\boldsymbol{\lambda}_g). \tag{4.26}$$

The left side of (4.26) represents the differential of ψ through the implicit dependence. Thus, this implicitly dependent component is expressed in terms of adjoint variable $\boldsymbol{\lambda}_g$ in (4.26). From (4.4), the total differential of function ψ can be denoted as

$$\frac{d\psi}{d\boldsymbol{b}}\delta\boldsymbol{b} = \left[\frac{\partial\psi}{\partial\boldsymbol{b}} + \frac{\partial\psi}{\partial z_g}\frac{\partial z_g}{\partial\boldsymbol{b}}\right]\delta\boldsymbol{b}. \tag{4.27}$$

Substituting (4.26) into the second term on the right of (4.27), and employing the second and third lines of (4.19), the total differential of ψ can be written as an explicit function of design variation $\delta\boldsymbol{b}$:

$$\begin{aligned}\frac{d\psi}{d\boldsymbol{b}}\delta\boldsymbol{b} &= \left[\frac{\partial\psi}{\partial\boldsymbol{b}}\delta\boldsymbol{b} + \ell'_{\delta\boldsymbol{b}}(\boldsymbol{\lambda}_g) - a'_{\delta\boldsymbol{b}}(z_g,\boldsymbol{\lambda}_g)\right]\\ &= \left[\frac{\partial\psi}{\partial\boldsymbol{b}} + \frac{\partial}{\partial\boldsymbol{b}}\left(\tilde{\boldsymbol{\lambda}}_g^T\boldsymbol{F}_g(\boldsymbol{b})\right) - \frac{\partial}{\partial\boldsymbol{b}}\left(\tilde{\boldsymbol{\lambda}}_g^T\boldsymbol{K}_g(\boldsymbol{b})\tilde{z}_g\right)\right]\delta\boldsymbol{b}.\end{aligned} \tag{4.28}$$

Since (4.28) holds for all design variations $\delta\boldsymbol{b}$, we can use the following equivalent relation:

$$\frac{d\psi}{d\boldsymbol{b}} = \frac{\partial\psi}{\partial\boldsymbol{b}} - \frac{\partial}{\partial\boldsymbol{b}}\left[\tilde{\boldsymbol{\lambda}}_g^T\boldsymbol{K}_g(\boldsymbol{b})\tilde{z}_g - \tilde{\boldsymbol{\lambda}}_g^T\boldsymbol{F}_g(\boldsymbol{b})\right]. \tag{4.29}$$

It is interesting to note that even though the generalized stiffness matrix $\boldsymbol{K}_g$ is singular, the load vector used in the direct differentiation approach in (4.22) is in the same form as the one that appears in computing the reduced stiffness matrix in (4.7). Similarly, in the adjoint variable method, the single load vector that is employed for adjoint computation in (4.23) is in exactly the same form as the load vector in the matrix adjoint equation in (4.11), which is used in the reduced stiffness matrix. Computational considerations associated with these observations will now be discussed.

4.1.4 Computational Considerations

In most structural design problems, numerous load conditions must be accounted for in the design process. For example, the design of a bridge may need to take into account such load conditions as the self-weight load, the distributed load due to vehicles traveling on the bridge, and the wind load. Therefore, instead of a single load discussed in preceding sections, many kinds of loads appear, denoted as $\boldsymbol{F}_g^j$ ($j = 1, \ldots, NL$), where NL is the number of applied loads. The same stiffness matrix is applicable for all load conditions, but structural equations yield different displacement vectors z_g^j ($j = 1, \ldots, NL$) associated with different applied force vectors.

Further, in realistic design problems numerous performance constraints must be taken into account in the design process. For example, the design of a bridge may require the maximum value of allowed stress and displacement. Even though there may be a multitude of constraints under consideration, the design engineer normally evaluates constraints in a trial design and wants to obtain design sensitivity information only for those constraints that are active or violated. For comparative purposes, designate the active design constraints under consideration as ψ_i ($i = 1, \ldots, NC$), where NC is the number of active design constraints. In addition, let some constraints be active for each load condition. Load conditions that have no influence on any active constraint may be eliminated for design sensitivity analysis purposes. Design sensitivity analysis computations that are required for the direct differentiation and adjoint variable approach may now be summarized for both the matrix and the variational analysis methods discussed in Sections 4.1.2 and 4.1.3.

Direct Differentiation Method

To calculate the total derivative of each constraint ψ_i using the direct differentiation approach, (4.7) must be solved for each load condition, yielding the following set of equations:

$$\boldsymbol{K}(\boldsymbol{b})\frac{dz^j}{d\boldsymbol{b}} = \frac{\partial \boldsymbol{F}^j(\boldsymbol{b})}{\partial \boldsymbol{b}} - \frac{\partial}{\partial \boldsymbol{b}}[\boldsymbol{K}(\boldsymbol{b})\tilde{z}^j], \quad j = 1, \ldots, NL. \tag{4.30}$$

Since each of the equations in (4.30) represents k number of equations for dz^j/db_i ($i = 1, \ldots, k$), there are $k \times NL$ equations to be solved. These solutions are quite efficiently obtained, since the reduced stiffness matrix $\boldsymbol{K}$ has been previously factored into the structural analysis process. With all dz^j/db_i from (4.30), design sensitivity may now be directly calculated from (4.6).

Adjoint Variable Method

Consider the adjoint variable method in which (4.11) must be solved for each constraint under consideration, that is,

$$\boldsymbol{K}(\boldsymbol{b})\boldsymbol{\lambda}^i = \frac{\partial \psi_i^T}{\partial z^i}, \quad i = 1, \ldots, NC, \tag{4.31}$$

where it is presumed that each constraint $\psi_i(z^i)$ involves only displacement z^i corresponding to the ith load. Thus, there are exactly NC number of equations to be solved for vectors $\boldsymbol{\lambda}^i$ ($i = 1, \ldots, NC$). Once this computation is complete, the design sensitivities of each constraint are calculated directly from (4.13), requiring only a moderate amount of computation. Note that the coefficient matrix in (4.31) is the reduced stiffness matrix, which was factored during structural analysis. Therefore, the amount of computational effort required to solve (4.31) is also moderate.

Comparison of the Direct Differentiation and Adjoint Variable Methods

In determining which approach will be employed, only the number of equations to be solved and the number of vectors to be stored and operated on during design sensitivity analysis need be compared. If $k \times NL < NC$, then the direct differentiation method in (4.30) is preferable. On the other hand, if $k \times NL > NC$, then it is preferable to use the adjoint variable method in (4.31). In structural optimization, since the number of active NC constraints must be no greater than the number of k design variables, the adjoint variable approach will be the most efficient method, even for a single loading condition. With a multiple loading condition, NC is normally much smaller than $k \times NL$; therefore, in most structural applications the adjoint variable method is more efficient. However, there may be applications in a preliminary design stage in which the design engineer considers a small number of design variables with a large number of constraints. In that case, the direct differentiation approach is preferable.

Precisely the same counting process is applicable to the variational analysis approach in (4.17). In this approach, a $k \times NL$ number of equations are solved in (4.22) for the derivatives of the state variables with respect to the design. Similarly, using the adjoint variable technique, an NC number of adjoint equations in (4.23) are solved for adjoint variables associated with each active constraint. Thus, precisely the same criteria are involved in determining which approach is best suited to the design problem under consideration.

Computation of Design Derivatives

A comparison between the reduced matrix approach (Section 4.1.2) and the variational approach (Section 4.1.3) is also possible. Since most structural analysis computer codes either numerically construct or completely avoid the reduced stiffness matrix $\boldsymbol{K}(\boldsymbol{b})$, an explicit form of $\boldsymbol{K}(\boldsymbol{b})$ is not generally available. Therefore, computation of the derivatives of $\boldsymbol{K}(\boldsymbol{b})$ with respect to the design, required in (4.9) for the direct differentiation approach, and in (4.13) for the adjoint variable approach, encounters some difficulty. While transformations may be written that reduce the generalized stiffness matrix $\boldsymbol{K}_g$ to reduced stiffness matrix $\boldsymbol{K}$, the transformations differ from one computer code to another. Therefore, implementation of design sensitivity analysis using the reduced stiffness matrix becomes code dependent, and may be numerically inefficient.

If the variational approach is employed, then the derivatives of $\boldsymbol{K}_g(\boldsymbol{b})$ with respect to the design can be calculated without difficulty. Such derivatives are required for the direct differentiation approach in (4.22), and for the adjoint variable approach in (4.29). In fact, using (3.190), the derivative required in (4.29) may be written as the sum of all element matrix derivatives, as

$$\begin{aligned}\frac{\partial}{\partial \boldsymbol{b}}\left(\tilde{\boldsymbol{\lambda}}_g^T \boldsymbol{K}_g(\boldsymbol{b})\tilde{\boldsymbol{z}}_g \right) &= \frac{\partial}{\partial \boldsymbol{b}}\left[\sum_{i=1}^{NE} \tilde{\boldsymbol{\lambda}}_g^T \boldsymbol{\beta}^{iT} \boldsymbol{S}^{iT} \boldsymbol{k}^i(\boldsymbol{b}) \boldsymbol{S}^i \boldsymbol{\beta}^i \tilde{\boldsymbol{z}}_g \right] \\ &= \frac{\partial}{\partial \boldsymbol{b}}\left[\sum_{i=1}^{NE} \tilde{\boldsymbol{\lambda}}^{iT} \boldsymbol{S}^{iT} \boldsymbol{k}^i(\boldsymbol{b}) \boldsymbol{S}^i \tilde{\boldsymbol{z}}^i \right],\end{aligned} \tag{4.32}$$

where $\boldsymbol{\lambda}^i$ and $\boldsymbol{z}^i$ are components of the adjoint and displacement vectors associated with the ith element. The practicality of this computation is clear from the following two observations. First, for each element, the element stiffness matrix $\boldsymbol{k}^i(\boldsymbol{b})$ and the geometric matrix $\boldsymbol{S}^i(\mathbf{b})$ will only depend on a small number of design variables associated with the given element and its nodes. Thus, in the sum of (4.32), only a few terms will be nonzero. Second, an evaluation of the design derivatives of the element bilinear forms in (4.32) only requires a moderate amount of calculation for those nonzero terms.

A similar argument may be made when computing the design derivatives in (4.22) using the direct differentiation approach, but with the following caveat: all components of the $dz_g/d\boldsymbol{b}$ matrix ($n \times k$) are now required in order to perform a complete design sensitivity analysis. However, the summation form in (4.32), in which $\bar{\lambda}$ is replaced by $\bar{z}$, can be employed in order to reduce the computational burden when evaluating the right side of (4.22). This is an important practical consideration when implementing a large-scale analysis code for derivative computations. The directness with which computations can be performed using the variational approach makes it a favorable choice in terms of generality and numerical effectiveness.

Another practical consideration involves calculating the design derivatives of element matrices that are implicitly generated [4] and [42]. Rather than using closed form expressions with respect to design variables, such as those presented in Section 3.4.5, many modern finite element formulations perform a numerical integration in order to evaluate the element stiffness and mass matrices. For implicitly generated element matrices, design differentiation can be performed throughout the sequence of calculations used to generate element matrices, thus leading to implicit design derivative routines.

A simple alternative is to perturb one design variable at a time, and to use finite differences in order to approximate the element matrix derivatives, for example,

$$\frac{\partial \boldsymbol{k}^i}{\partial b_j} \approx \frac{\boldsymbol{k}^i(\boldsymbol{b}+\tau \boldsymbol{e}^j) - \boldsymbol{k}^i(\boldsymbol{b})}{\tau},$$

where $\boldsymbol{e}^j$ has a one in the jth position and zeros elsewhere, and τ is a small perturbation in b_j. This is the semianalytical method described in Section 1.5.3 of Chapter 1, which is very convenient from a numerical implementation point of view, since the element stiffness matrix has already been constructed in response analysis. Simple computation of one more element stiffness matrix is required to obtain the derivative of the element stiffness matrix. However, as explained in Section 1.5.2 of Chapter 1, there is no universal criterion by which the size of perturbation τ can be determined. If a too-small perturbation is used, such that the difference in the numerator is the same magnitude as the computer's numerical accuracy, then a numerical error becomes the dominant factor in determining the element matrix derivatives. In contrast, if a too-large perturbation is used, then an approximation error becomes the dominant factor. Thus, the derivative's accuracy strongly depends on the problem type and previous experience.

4.1.5 Second-Order Design Sensitivity Analysis

As shown in Sections 4.1.2 and 4.1.3, if the applied load vector and system stiffness matrix has s continuous derivatives with respect to the design, then state variable z also has s continuous derivatives with respect to the design. Assuming that function ψ also has s continuous derivatives, it is possible to calculate up to s-order of partial derivatives of ψ, with respect to the design variables. Consider the specific case $s = 2$, which is a second-order design sensitivity. Such an analysis is especially important in modern design optimization algorithms where many nonlinear performance measures are approximated with second-order algebraic functions. The second-order sensitivity information is sometimes called a Hessian, or a curvature of the performance measure. If a performance measure is a second-order function of the design, then an exact minimum value of the performance measure can be found in the first iteration by supplying the

Hessian information. Thus, the second-order sensitivity information can enhance the speed of convergence in a nonlinear optimization algorithm.

Direct Differentiation Method with Reduced Stiffness Matrix

Consider the case of a structural formulation with a reduced stiffness matrix, that is, the kinematic constraints have been explicitly eliminated, and all components of the displacement vector are independent. The chain rule of differentiation may then be used to obtain the second-order derivative of ψ with respect to the components of design variables b_i and b_j, as

$$\begin{aligned} \frac{d^2\psi}{db_i db_j} &= \frac{d}{db_j}\left[\frac{\partial\psi}{\partial b_i} + \frac{\partial\psi}{\partial z}\frac{dz}{db_i}\right] \\ &= \frac{\partial^2\psi}{\partial b_i \partial b_j} + \frac{\partial^2\psi}{\partial b_i \partial z}\frac{dz}{db_j} + \frac{\partial^2\psi}{\partial z \partial b_j}\frac{dz}{db_i} \\ &+ \frac{dz}{db_j}^T \frac{\partial^2\psi}{\partial z \partial z}\frac{dz}{db_i} + \frac{\partial\psi}{\partial z}\frac{d^2 z}{db_i db_j}. \end{aligned} \tag{4.33}$$

This notation needs some explanation. The derivative of ψ with respect to b_i in the bracket is in reality a partial derivative of ψ with respect to b_i, accounting for the direct dependence of ψ on $\boldsymbol{b}$ and of ψ on $z(\boldsymbol{b})$. The terms on the right side include the partial derivatives of ψ with respect to its explicit dependence on b_i. Note that $\partial^2\psi/\partial z\partial z$ in (4.33) is a $(m \times m)$ matrix. From the expression of ψ with respect to its arguments, the unknown terms in (4.33) are $dz/d\boldsymbol{b}$ and d^2z/db_idb_j, which can be computed from the derivative of the response problem.

Since both first- and second-order derivatives of z with respect to the design appear in (4.33), consider calculating them by using the structural equation, as it is presented in Section 4.1.2:

$$\boldsymbol{K}(\boldsymbol{b})z = \boldsymbol{F}(\boldsymbol{b}). \tag{4.34}$$

As in Section 4.1.2, the same property is presumed for $\boldsymbol{K}(\boldsymbol{b})$ and $\boldsymbol{F}(\boldsymbol{b})$. Using the direct design differentiation approach, differentiate (4.34) with respect to b_i to obtain

$$\boldsymbol{K}(\boldsymbol{b})\frac{dz}{db_i} = \frac{\partial \boldsymbol{F}(\boldsymbol{b})}{\partial b_i} - \frac{\partial}{\partial b_i}[\boldsymbol{K}(\boldsymbol{b})\tilde{z}]. \tag{4.35}$$

Since matrix $\boldsymbol{K}(\boldsymbol{b})$ is nonsingular, or positive definite, (4.35) may be numerically solved to obtain the first derivative of z with respect to the design. This is the same direct differentiation method used in first-order design sensitivity analysis. To obtain the second-order sensitivity expression of the state variable, differentiate (4.35) with respect to b_j, as

$$\begin{aligned} \boldsymbol{K}(\boldsymbol{b})\frac{d^2 z}{db_i db_j} &= \frac{\partial^2 \boldsymbol{F}(\boldsymbol{b})}{\partial b_i \partial b_j} - \frac{\partial^2}{\partial b_i \partial b_j}[\boldsymbol{K}(\boldsymbol{b})\tilde{z}] \\ &- \frac{\partial}{\partial b_i}\left[\boldsymbol{K}(\boldsymbol{b})\frac{d\tilde{z}}{db_j}\right] - \frac{\partial}{\partial b_j}\left[\boldsymbol{K}(\boldsymbol{b})\frac{d\tilde{z}}{db_i}\right]. \end{aligned} \tag{4.36}$$

Again, note that the coefficient matrix of the second derivatives of the state variable with respect to the design is nonsingular, so that the second derivatives may be computed numerically if the response result z and the first-order sensitivity result $dz/d\boldsymbol{b}$ are available, along with the explicit expression of $\boldsymbol{K}(\boldsymbol{b})$ and $\boldsymbol{F}(\boldsymbol{b})$ in terms of design $\boldsymbol{b}$. Next, solve (4.35) and (4.36) for both the first- and second-order derivatives of z, and substitute the results into (4.33) to obtain the second-order derivatives of ψ.

While the above approach is conceptually simple, it requires a large amount of computation. If k is the dimension of the design variable, then one original finite element analysis and $(3k + k^2)/2$ finite element reanalyses, which include a k number of finite element reanalyses for the first-order design sensitivity, and $k(k + 1)/2$ finite element reanalyses [see (4.36)] for the second-order design and mixed design sensitivities. For example, when $k = 10$, which is very common for an optimization problem, evaluating (4.35) and (4.36) will require a solution procedure that is repeated 66 times. Even if the stiffness matrix is factorized in the response analysis, a large amount of computation is still necessary. As will be seen in the following paragraph, considerably better results are achieved using the adjoint variable approach for second-order design sensitivity analysis.

Adjoint Variable Method with Reduced Stiffness Matrix

From (4.13), the derivative of ψ with respect to b_i may be written as

$$\frac{d\psi}{db_i} = \frac{\partial \psi}{\partial b_i} - \boldsymbol{\lambda}^T \frac{\partial \boldsymbol{K}(\boldsymbol{b})}{\partial b_i}\boldsymbol{z} + \boldsymbol{\lambda}^T \frac{\partial \boldsymbol{F}(\boldsymbol{b})}{\partial b_i}. \tag{4.37}$$

It is important to note that in order for (4.37) to be valid $\boldsymbol{z}$ must be the solution to (4.34), and $\boldsymbol{\lambda}$ must be the solution to

$$\boldsymbol{K}(\boldsymbol{b})\boldsymbol{\lambda} = \frac{\partial \psi}{\partial \boldsymbol{z}}^T, \tag{4.38}$$

which follows from (4.11). Thus, both $\boldsymbol{z}$ and $\boldsymbol{\lambda}$ in (4.37) are dependent on the design and in the calculation of the second-order derivatives of ψ, design dependence must be accounted for in $\boldsymbol{z}$ and $\boldsymbol{\lambda}$. By using the chain rule of calculation, the second-order derivative of ψ with respect to the design is

$$\begin{aligned}\frac{d^2\psi}{db_i db_j} &= \frac{\partial^2 \psi}{\partial b_i \partial b_j} - \frac{\partial}{\partial b_j}\left[\tilde{\boldsymbol{\lambda}}^T \frac{\partial \boldsymbol{K}}{\partial b_i}\tilde{\boldsymbol{z}}\right] + \frac{\partial}{\partial b_j}\left[\tilde{\boldsymbol{\lambda}}^T \frac{\partial \boldsymbol{F}}{\partial b_i}\right] \\ &+ \left[\frac{\partial^2 \psi}{\partial b_i \partial \boldsymbol{z}} - \boldsymbol{\lambda}^T \frac{\partial \boldsymbol{K}}{\partial b_i}\right]\frac{d\boldsymbol{z}}{db_j} + \left[\frac{\partial \boldsymbol{F}}{\partial b_i}^T - \boldsymbol{z}^T \frac{\partial \boldsymbol{K}}{\partial b_i}\right]\frac{d\boldsymbol{\lambda}}{db_j}.\end{aligned} \tag{4.39}$$

In order to evaluate the second-order derivatives in (4.39), $d\boldsymbol{z}/d\boldsymbol{b}$ and $d\boldsymbol{\lambda}/d\boldsymbol{b}$ must be accounted for. Differentiating both sides of (4.38) with respect to the design and pre-multiplying by $\boldsymbol{K}^{-1}(\boldsymbol{b})$, the following result is obtained:

$$\frac{d\boldsymbol{\lambda}}{db_j} = \boldsymbol{K}^{-1}(\boldsymbol{b})\left[\frac{\partial^2 \psi^T}{\partial \boldsymbol{z}\partial b_j} - \frac{\partial}{\partial b_j}\left(\boldsymbol{K}(\boldsymbol{b})\tilde{\boldsymbol{\lambda}}\right) + \frac{\partial^2 \psi}{\partial \boldsymbol{z}\partial \boldsymbol{z}}\frac{d\boldsymbol{z}}{db_j}\right]. \tag{4.40}$$

This result may be inserted into (4.39), followed by the definition of the adjoint variable $\boldsymbol{\gamma}^i$ for terms containing $\boldsymbol{K}^{-1}(\boldsymbol{b})$, to remove any computational costs involved in the matrix inverse:

$$\begin{aligned}\frac{d^2\psi}{db_i db_j} &= \frac{\partial^2 \psi}{\partial b_i \partial b_j} - \frac{\partial}{\partial b_j}\left[\tilde{\boldsymbol{\lambda}}^T \frac{\partial \boldsymbol{K}}{\partial b_i}\tilde{\boldsymbol{z}}\right] + \frac{\partial}{\partial b_j}\left[\tilde{\boldsymbol{\lambda}}^T \frac{\partial \boldsymbol{F}}{\partial b_i}\right] + \left[\frac{\partial^2 \psi}{\partial b_i \partial \boldsymbol{z}} - \boldsymbol{\lambda}^T \frac{\partial \boldsymbol{K}}{\partial b_i}\right]\frac{d\boldsymbol{z}}{db_j} \\ &+ \underbrace{\left[\frac{\partial \boldsymbol{F}}{\partial b_i}^T - \boldsymbol{z}^T \frac{\partial \boldsymbol{K}}{\partial b_i}\right]\boldsymbol{K}^{-1}(\boldsymbol{b})}_{\boldsymbol{\gamma}^{iT}}\left[\frac{\partial^2 \psi}{\partial \boldsymbol{z}\partial b_j} - \frac{\partial}{\partial b_j}\left(\boldsymbol{K}(\boldsymbol{b})\tilde{\boldsymbol{\lambda}}\right) + \frac{\partial^2 \psi}{\partial \boldsymbol{z}\partial \boldsymbol{z}}\frac{d\boldsymbol{z}}{db_j}\right].\end{aligned}$$

Thus, the adjoint equation related to $d\boldsymbol{\lambda}/d\boldsymbol{b}$ can be defined with the solution $\boldsymbol{\gamma}^i$, as

$$\boldsymbol{K}(\boldsymbol{b})\boldsymbol{\gamma}^i = \frac{\partial \boldsymbol{F}}{\partial b_i} - \frac{\partial \boldsymbol{K}(\boldsymbol{b})}{\partial b_i} \boldsymbol{z}. \tag{4.41}$$

Substituting the result of (4.41) into (4.39), and using (4.40), which is the same procedure used in obtaining (4.12), we obtain

$$\begin{aligned}\frac{d^2\psi}{db_i db_j} &= \frac{\partial^2 \psi}{\partial b_i \partial b_j} - \frac{\partial}{\partial b_j}\left[\tilde{\boldsymbol{\lambda}}^T \frac{\partial \boldsymbol{K}}{\partial b_i} \tilde{\boldsymbol{z}}\right] + \frac{\partial}{\partial b_j}\left[\tilde{\boldsymbol{\lambda}}^T \frac{\partial \boldsymbol{F}}{\partial b_i}\right] \\ &+ \boldsymbol{\gamma}^{iT}\left[\frac{\partial^2 \psi^T}{\partial \boldsymbol{z} \partial b_j} - \frac{\partial}{\partial b_j}(\boldsymbol{K}(\boldsymbol{b})\tilde{\boldsymbol{\lambda}})\right] + \left[\frac{\partial^2 \psi^T}{\partial b_i \partial \boldsymbol{z}} - \boldsymbol{\lambda}^T \frac{\partial \boldsymbol{K}}{\partial b_i} + \boldsymbol{\gamma}^{iT} \frac{\partial^2 \psi}{\partial \boldsymbol{z} \partial \boldsymbol{z}}\right] \frac{d\boldsymbol{z}}{db_j}.\end{aligned} \tag{4.42}$$

The only unknown term in (4.42) is $d\boldsymbol{z}/d\boldsymbol{b}$, which can be computed by defining another adjoint equation. Using the direct differentiation method in (4.8), the expression of $d\boldsymbol{z}/d\boldsymbol{b}$ is substituted into the last term of (4.42), to yield

$$\underbrace{\left[\frac{\partial^2 \psi}{\partial b_i \partial \boldsymbol{z}} - \boldsymbol{\lambda}^T \frac{\partial \boldsymbol{K}}{\partial b_i} + \boldsymbol{\gamma}^{iT} \frac{\partial^2 \psi}{\partial \boldsymbol{z} \partial \boldsymbol{z}}\right] \boldsymbol{K}^{-1}(\boldsymbol{b})}_{\boldsymbol{\eta}^{iT}} \left[\frac{\partial \boldsymbol{F}(\boldsymbol{b})}{\partial b_j} - \frac{\partial}{\partial b_j}[\boldsymbol{K}(\boldsymbol{b})\tilde{\boldsymbol{z}}]\right].$$

The preceding sequence of computations may now be repeated, defining a new adjoint variable $\boldsymbol{\eta}^i$ as the solution to

$$\boldsymbol{K}(\boldsymbol{b})\boldsymbol{\eta}^i = \frac{\partial^2 \psi^T}{\partial \boldsymbol{z} \partial b_i} - \frac{\partial \boldsymbol{K}(\boldsymbol{b})}{\partial b_i} \boldsymbol{\lambda} + \frac{\partial^2 \psi}{\partial \boldsymbol{z} \partial \boldsymbol{z}} \boldsymbol{\gamma}^i. \tag{4.43}$$

The last term in (4.42) can be replaced by a directly computable expression from (4.35) and (4.43). After substitution, the desired result is obtained as

$$\begin{aligned}\frac{d^2\psi}{db_i db_j} &= \frac{\partial^2 \psi}{\partial b_i \partial b_j} - \frac{\partial^2}{\partial b_i \partial b_j}[\tilde{\boldsymbol{\lambda}}^T \boldsymbol{K}(\boldsymbol{b})\tilde{\boldsymbol{z}}] + \frac{\partial^2}{\partial b_i \partial b_j}(\tilde{\boldsymbol{\lambda}}^T \boldsymbol{F}) \\ &+ \boldsymbol{\gamma}^{iT} \frac{\partial^2 \psi^T}{\partial \boldsymbol{z} \partial b_j} - \frac{\partial}{\partial b_j}[\tilde{\boldsymbol{\gamma}}^{iT} \boldsymbol{K}(\boldsymbol{b})\tilde{\boldsymbol{\lambda}}] + \boldsymbol{\eta}^{iT} \boldsymbol{K}(\boldsymbol{b})\tilde{\boldsymbol{\gamma}}^j,\end{aligned} \tag{4.44}$$

where only the jth component of the second derivative of ψ is included.

Assuming that the cost of evaluating a partial derivative of explicitly dependent terms is negligible compared with solving a linear system of equations, the computational cost of (4.44) results from computing $\boldsymbol{z}$, $\boldsymbol{\lambda}$, $\boldsymbol{\gamma}^i$, and $\boldsymbol{\eta}^i$ ($i = 1, \ldots, k$). Equation (4.44) provides an explicit formula for all second derivatives of ψ with respect to the design, requiring a solution from a total $1 + NC + k + (NC \times k)$ number of equations, which under normal circumstances is considerably less than the $1 + 3k/2 + k^2/2$ number of solutions required using the direct differentiation approach from (4.33).

A Hybrid Direct Differentiation − Adjoint Variable Method

By combining the direct differentiation and adjoint method, Haftka [51] introduced a refinement that has a computational advantage by a factor of two. From (4.36), the second-order derivative of $\boldsymbol{z}$ can be represented by

$$\frac{d^2 \boldsymbol{z}}{db_i db_j} = \boldsymbol{K}^{-1}\left[\frac{\partial^2 \boldsymbol{F}}{\partial b_i \partial b_j} - \frac{\partial^2}{\partial b_i \partial b_j}(\boldsymbol{K}(\boldsymbol{b})\tilde{\boldsymbol{z}}) - \frac{\partial}{\partial b_i}\left(\boldsymbol{K}(\boldsymbol{b})\frac{d\tilde{\boldsymbol{z}}}{db_j}\right) - \frac{\partial}{\partial b_j}\left(\boldsymbol{K}(\boldsymbol{b})\frac{d\tilde{\boldsymbol{z}}}{db_i}\right)\right]. \tag{4.45}$$

From (4.38), recall that

$$\lambda^T = \frac{\partial \psi}{\partial z} K^{-1}. \tag{4.46}$$

Now, substitute the second derivatives of z from (4.45) and (4.46) into (4.33) to obtain

$$\begin{aligned}\frac{d^2\psi}{db_i db_j} &= \frac{\partial^2 \psi}{\partial b_i \partial b_j} + \frac{\partial^2 \psi}{\partial b_i \partial z}\frac{dz}{db_j} + \frac{\partial^2 \psi}{\partial z \partial b_j}\frac{dz}{db_i} + \frac{dz}{db_j}^T \frac{\partial^2 \psi}{\partial z \partial z}\frac{dz}{db_i} \\ &+ \lambda^T \left[\frac{\partial^2 F}{\partial b_i \partial b_j} - \frac{\partial^2}{\partial b_i \partial b_j}(K(b)\tilde{z}) - \frac{\partial}{\partial b_i}\left(K(b)\frac{d\tilde{z}}{db_j}\right) - \frac{\partial}{\partial b_j}\left(K(b)\frac{d\tilde{z}}{db_i}\right)\right].\end{aligned} \tag{4.47}$$

If the direct differentiation method is employed to solve (4.35) for dz/db_i ($i = 1, \ldots, k$), all terms on the right side of (4.47) can be evaluated. Note that z, k vectors of dz/db_i, and λ are needed, for a total $1 + NC + k$ number of solutions, much less than the $1 + NC + k + (NC \times k)$ solutions in the pure adjoint variable method.

Computational Considerations

The practicality involved in using (4.44) and (4.47) should be evaluated based on the number of computations. Therefore, let us only consider member-size design variables (fixed geometry). Of course, the first term on the right side of (4.44) must be calculated directly. The second term may be calculated using the summation form of the reduced stiffness matrix of (4.5), as

$$\frac{\partial^2}{\partial b_i \partial b_j}[\tilde{\lambda}^T K(b)\tilde{z}] = \sum_{i=1}^{NE} \lambda^T \beta^{iT} S^{iT} \frac{\partial k^i(b)}{\partial b_i \partial b_j} S^i \beta^i z. \tag{4.48}$$

Note that most terms in the summation of element stiffness matrix derivatives are equal to zero. The third term on the right side of (4.44) involves second derivatives of the load vector with respect to the design. If the load vector is constant, these derivatives are all zero. If the load vector depends on the design, then the expressions of second derivatives of the load vector must be calculated. Similar observations follow for the evaluation of the terms in (4.47).

Analysis with Generalized Stiffness Matrix

The foregoing analysis requires the explicit computation of the reduced stiffness matrix and its first and second derivatives with respect to the design. However, in applications involving complicated kinematic admissibility conditions, difficulties arise with such computations.

The second derivatives of ψ with respect to the design may be written in the same way as (4.33)

$$\begin{aligned}\frac{d^2\psi}{db_i db_j} &= \frac{\partial \psi}{\partial z_g}\frac{d^2 z_g}{db_i db_j} + \frac{\partial^2 \psi}{\partial z_g \partial b_j}\frac{dz_g}{db_i} \\ &+ \frac{dz_g}{db_j}^T \frac{\partial^2 \psi}{\partial z_g \partial z_g}\frac{dz_g}{db_i} + \frac{\partial^2 \psi}{\partial b_i \partial z_g}\frac{dz_g}{db_j} + \frac{\partial^2 \psi}{\partial b_i \partial b_j},\end{aligned} \tag{4.49}$$

where the total derivative notation on the left emphasizes the inclusion of the design dependence of z_g, which appears in the performance measure. In order to treat the first term on the right side of (4.49), consider the ith component of (4.22) and differentiate both sides with respect to b_j in order to obtain the identity

$$
\begin{aligned}
\bar{z}_g^T \boldsymbol{K}_g(\boldsymbol{b}) \frac{d^2 z_g}{db_i db_j} &= -\frac{\partial^2}{\partial b_i \partial b_j}(\bar{z}_g^T \boldsymbol{K}_g(\boldsymbol{b}) \tilde{z}_g) - \frac{\partial}{\partial b_i}\left(\bar{z}_g^T \boldsymbol{K}_g(\boldsymbol{b}) \frac{d\tilde{z}_g}{db_j} \right) \\
&\quad - \frac{\partial}{\partial b_j}\left(\bar{z}_g^T \boldsymbol{K}_g(\boldsymbol{b}) \frac{d\tilde{z}_g}{db_i} \right) + \bar{z}_g^T \frac{\partial^2 \boldsymbol{F}_g}{\partial b_i \partial b_j}, \qquad \forall \bar{z}_g \in Z.
\end{aligned} \tag{4.50}
$$

Observe that (4.23) may be evaluated at $\bar{\lambda}_g = d^2 z_g / db_i db_j$, and (4.50) may be evaluated at $\bar{z}_g = \lambda_g$ to obtain an expression for the first term on the right side of (4.49). Substituting into (4.49) gives

$$
\begin{aligned}
\frac{d^2 \psi}{db_i db_j} &= -\frac{\partial^2}{\partial b_i \partial b_j}(\tilde{\lambda}_g^T \boldsymbol{K}_g(\boldsymbol{b}) \tilde{z}_g) - \frac{\partial}{\partial b_i}\left(\tilde{\lambda}_g^T \boldsymbol{K}_g(\boldsymbol{b}) \frac{d\tilde{z}_g}{db_j} \right) \\
&\quad - \frac{\partial}{\partial b_j}\left(\tilde{\lambda}_g^T \boldsymbol{K}_g(\boldsymbol{b}) \frac{d\tilde{z}_g}{db_i} \right) + \lambda_g^T \frac{\partial^2 \boldsymbol{F}_g}{\partial b_i \partial b_j} + \frac{\partial^2 \psi}{\partial z_g \partial b_j} \frac{dz_g}{db_i} \\
&\quad + \frac{dz_g}{db_j}^T \frac{\partial^2 \psi}{\partial z_g \partial z_g} \frac{dz_g}{db_i} + \frac{\partial^2 \psi}{\partial b_i \partial z_g} \frac{dz_g}{db_j} + \frac{\partial^2 \psi}{\partial b_i \partial b_j}.
\end{aligned} \tag{4.51}
$$

Note that the second derivative forms calculated in (4.51) and (4.47) are identical. It is important to note, however, that (4.51) is valid even for the singular stiffness matrix $\boldsymbol{K}_g(\boldsymbol{b})$, whereas the derivation in (4.47) relies heavily on the inverse of the reduced stiffness matrix $\boldsymbol{K}(\boldsymbol{b})$. These computations, needed to construct the terms in (4.51), are identical to those in (4.47). However, (4.51) has a desirable property: it is only necessary to compute the design derivatives of the generalized stiffness matrix, and not those of the reduced stiffness matrix.

4.1.6 Examples

Beam

Consider a clamped-clamped beam of unit length that is subject to an applied load $f(x)$ and a self-weight $\gamma h(x)$, where γ is the weight density of the beam. For simplicity, let us presume that the cross-sectional dimensions are similar in all directions, such that the moment of inertia can be expressed as $I(x) = \alpha h^2(x)$ around the beam's neutral axis, and α is a positive constant. For example, when the solid, circular cross section in Table 12.1 of Chapter 12 is considered, $h = \pi r^2$ and $I = \pi r^4/4$, yielding $\alpha = 1/(4\pi)$ [52]. If a stepped beam is considered, as shown in Fig. 4.1, then

$$
h(x) = b_i, \qquad \frac{i-1}{n} < x < \frac{i}{n}, \tag{4.52}
$$

where the beam has been subdivided into n sections, each with a constant cross-sectional area b_i. The b_i $(i = 1, \ldots, n)$ areas, and Young's modulus $E = b_{n+1}$ may be viewed as design variables.

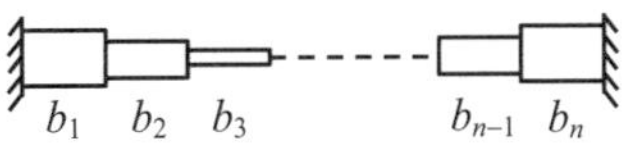

Figure 4.1. Stepped beam.

Consider compliance as a performance measure, with constant *f*, given as

$$\begin{aligned}\psi &= \int_0^1 (f + \gamma h) w(x)\, dx \\ &= \sum_{i=1}^{n} \int_{(i-1)/n}^{i/n} (f + \gamma b_i) w(x)\, dx.\end{aligned} \tag{4.53}$$

Using the shape function in (3.131) for the *i*th element and using the coordinate transformation in (3.191), the following can be written:

$$w(x) = \boldsymbol{N}\boldsymbol{d}^i = \boldsymbol{N}\boldsymbol{S}^i \boldsymbol{\beta}^i \boldsymbol{z}_g, \tag{4.54}$$

where $\boldsymbol{N}$ is the shape function, $\boldsymbol{S}^i$ is the rotation matrix, and $\boldsymbol{\beta}^i$ is the Boolean matrix. For the one-dimensional beam problem, $\boldsymbol{S}^i$ can be the identity matrix, and (4.53) becomes

$$\begin{aligned}\psi &= \left(\sum_{i=1}^{n} \int_{(i-1)/n}^{i/n} (f + \gamma b_i) N \beta^i \, dx \right) \boldsymbol{z}_g \\ &= \boldsymbol{F}_g(\boldsymbol{b})^T \boldsymbol{z}_g.\end{aligned} \tag{4.55}$$

For the structural equation,

$$\bar{\boldsymbol{z}}_g^T \boldsymbol{K}_g \boldsymbol{z}_g = \bar{\boldsymbol{z}}_g^T \boldsymbol{F}_g, \qquad \forall \bar{\boldsymbol{z}}_g \in Z, \tag{4.56}$$

where elements of Z satisfy the clamped boundary conditions. Using the adjoint variable method from (4.23) in Section 4.1.3, the following adjoint equation is obtained:

$$\bar{\boldsymbol{\lambda}}_g^T \boldsymbol{K}_g \boldsymbol{\lambda}_g = \bar{\boldsymbol{\lambda}}_g^T \boldsymbol{F}_g, \qquad \forall \bar{\boldsymbol{\lambda}}_g \in Z. \tag{4.57}$$

Since the adjoint load on the right side of (4.57) is the same as the load for the beam problem in (4.56), adjoint solution $\boldsymbol{\lambda}$ is identical to the initial solution $\boldsymbol{z}$, and thus, $\boldsymbol{\lambda}_g = \boldsymbol{z}_g$.

From (4.29), the sensitivity formula is given as

$$\begin{aligned}\frac{d\psi}{db_i} &= \frac{\partial \psi}{\partial b_i} + \frac{\partial}{\partial b_i}[\tilde{\boldsymbol{z}}_g^T \boldsymbol{F}_g(\boldsymbol{b}) - \tilde{\boldsymbol{z}}_g^T \boldsymbol{K}_g(\boldsymbol{b})\tilde{\boldsymbol{z}}_g] \\ &= \int_{(i-1)/n}^{i/n} 2\gamma w\, dx - \frac{\partial}{\partial b_i}\left[\sum_{i=1}^{n} \tilde{\boldsymbol{z}}_g^T \boldsymbol{\beta}^{i^T} \boldsymbol{k}^i \boldsymbol{\beta}^i \tilde{\boldsymbol{z}}_g \right] \\ &= \int_{(i-1)/n}^{i/n} 2\gamma w\, dx - \int_{i-1/n}^{i/n} 2E\alpha b_i \boldsymbol{q}_i^T \boldsymbol{N}_{xx}^T \boldsymbol{N}_{xx} \boldsymbol{q}_i\, dx \\ &= \int_{(i-1)/n}^{i/n} [2\gamma w - 2E\alpha b_i (w_{xx})^2]\, dx,\end{aligned} \tag{4.58}$$

for $i = 1, 2, \ldots, n$, where, from (3.135)

$$\boldsymbol{k}^i = \int_{(i-1)/n}^{i/n} \boldsymbol{N}_{xx}^T E\alpha b_i^2 \boldsymbol{N}_{xx}\, dx. \tag{4.59}$$

Also,

$$\begin{aligned}\frac{d\psi}{db_{n+1}} &= -\frac{\partial}{\partial b_{n+1}}[\tilde{\boldsymbol{z}}_g^T \boldsymbol{K}_g(\boldsymbol{b})\tilde{\boldsymbol{z}}_g] \\ &= -\sum_{i=1}^{n} \int_{(i-1)/n}^{i/n} \alpha b_i^2 \boldsymbol{q}_i^T \boldsymbol{N}_{xx}^T \boldsymbol{N}_{xx} \boldsymbol{q}_i\, dx \\ &= -\sum_{i=1}^{n} \int_{(i-1)/n}^{i/n} \alpha b_i^2 (w_{xx})^2\, dx.\end{aligned} \tag{4.60}$$

Hence,

$$\begin{aligned}\psi' &= \sum_{i=1}^{n}\left(\int_{(i-1)/n}^{i/n}[2\gamma w - 2E\alpha b_i (w_{xx})^2]\,dx\right)\delta b_i \\ &\quad -\left(\sum_{i=1}^{n}\int_{(i-1)/n}^{i/n}\alpha b_i^2 (w_{xx})^2\,dx\right)\delta E.\end{aligned} \tag{4.61}$$

Three-Bar Truss

Consider a simple three-bar truss with multipoint boundary conditions, as shown in Fig. 4.2. The design variables for this structure are cross-sectional areas b_i of the truss members. The generalized stiffness matrix is obtained in Section 1.4 of Chapter 1, as

$$\boldsymbol{K}_g(\boldsymbol{b}) = \frac{E}{l}\begin{bmatrix} b_3c^2s & b_3cs^2 & 0 & 0 & -b_3c^2s & -b_3cs^2 \\ b_3cs^2 & b_1+b_3s^2 & 0 & -b_1 & -b_3cs^2 & -b_3s^3 \\ 0 & 0 & b_2s/c & 0 & -b_2s/c & 0 \\ 0 & -b_1 & 0 & b_1 & 0 & 0 \\ -b_3c^2s & -b_3cs^2 & -b_2s/c & 0 & b_2s/c+b_3c^2s & b_3cs^2 \\ -b_3cs^2 & -b_3s^3 & 0 & 0 & b_3cs^2 & b_3s^3 \end{bmatrix}, \tag{4.62}$$

where $c = \cos\theta$ and $s = \sin\theta$. In this problem, space Z of kinematically admissible displacements is given as

$$Z = \left\{ z_g \in R^6 : z_3 = z_4 = 0,\ z_5\cos\alpha + z_6\sin\alpha = 0 \right\}, \tag{4.63}$$

and $\boldsymbol{K}_g(\boldsymbol{b})$ is positive definite on Z, even though it is not positive definite on R^6.

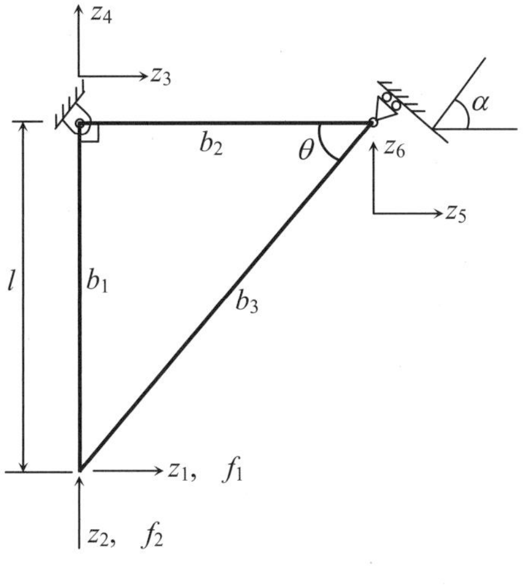

Figure 4.2. Three-bar truss.

If $\theta = 45^\circ$ and $\alpha = 30^\circ$, then with $\mathbf{z} = [z_1, z_2, z_5]^T$ the reduced stiffness matrix in this elementary example becomes

$$\boldsymbol{K}(\boldsymbol{b}) = \frac{E}{2\sqrt{2}l}\begin{bmatrix} b_3 & b_3 & (\sqrt{3}-1)b_3 \\ b_3 & 2\sqrt{2}b_1 + b_3 & (\sqrt{3}-1)b_3 \\ (\sqrt{3}-1)b_3 & (\sqrt{3}-1)b_3 & 2\sqrt{2}b_2 + (4-2\sqrt{3})b_3 \end{bmatrix}. \tag{4.64}$$

If $f_1 = f_2 = 1$ and $l = 1$, then the solution to the reduced matrix formulation in (4.5) is

$$\mathbf{z} = \begin{bmatrix} \frac{4-2\sqrt{3}}{Eb_2} + \frac{2\sqrt{2}}{Eb_3} & 0 & \frac{1-\sqrt{3}}{Eb_2} \end{bmatrix}^T. \tag{4.65}$$

If $\psi = z_1$, then the adjoint equation of (4.11) is

$$\boldsymbol{K}(\boldsymbol{b})\boldsymbol{\lambda} = \partial\psi^T/\partial\mathbf{z} = [1 \quad 0 \quad 0]^T, \tag{4.66}$$

with a solution of

$$\boldsymbol{\lambda} = \begin{bmatrix} \frac{1}{Eb_1} + \frac{4-2\sqrt{3}}{Eb_2} + \frac{2\sqrt{2}}{Eb_3} & -\frac{1}{Eb_1} & \frac{1-\sqrt{3}}{Eb_2} \end{bmatrix}^T. \tag{4.67}$$

Using $\mathbf{z}$ and $\boldsymbol{\lambda}$ from (4.65) and (4.67), the reduced matrix design sensitivity formula in (4.13) produces

$$\frac{d\psi}{d\boldsymbol{b}} = -\frac{\partial}{\partial \boldsymbol{b}}(\tilde{\boldsymbol{\lambda}}^T \boldsymbol{K}(\boldsymbol{b})\tilde{\mathbf{z}}) = \begin{bmatrix} 0 & \frac{2\sqrt{3}-4}{Eb_2^2} & -\frac{2\sqrt{2}}{Eb_3^2} \end{bmatrix}. \tag{4.68}$$

This can be verified by taking a derivative of z_1 in (4.65) with respect to the design parameter $\boldsymbol{b}$.

If a generalized matrix formulation is employed, then the solution $\mathbf{z}_g$ to (4.17) must be found, which is

$$\mathbf{z}_g = \begin{bmatrix} \frac{4-2\sqrt{3}}{Eb_2} + \frac{2\sqrt{2}}{Eb_3} & 0 & 0 & 0 & \frac{1-\sqrt{3}}{Eb_2} & \frac{3-\sqrt{3}}{Eb_2} \end{bmatrix}^T. \tag{4.69}$$

For $\psi = z_1$, the adjoint equation of (4.23) is

$$\boldsymbol{\lambda}_g^T \boldsymbol{K}_g(\boldsymbol{b})\bar{\boldsymbol{\lambda}}_g = [1 \quad 0 \quad 0 \quad 0 \quad 0 \quad 0]\bar{\boldsymbol{\lambda}}_g, \quad \forall \bar{\boldsymbol{\lambda}}_g \in Z, \tag{4.70}$$

with the solution

$$\boldsymbol{\lambda}_g = \begin{bmatrix} \frac{1}{Eb_1} + \frac{4-2\sqrt{3}}{Eb_2} + \frac{2\sqrt{2}}{Eb_3} & -\frac{1}{Eb_1} & 0 & 0 & \frac{1-\sqrt{3}}{Eb_2} & \frac{3-\sqrt{3}}{Eb_2} \end{bmatrix}^T. \tag{4.71}$$

Then, the design sensitivity formula in (4.29) produces

$$\frac{d\psi}{d\boldsymbol{b}} = -\frac{\partial}{\partial \boldsymbol{b}}(\tilde{\boldsymbol{\lambda}}_g^T \boldsymbol{K}_g(\boldsymbol{b})\tilde{\mathbf{z}}_g) = \begin{bmatrix} 0 & \frac{2\sqrt{3}-4}{Eb_2^2} & -\frac{2\sqrt{2}}{Eb_3^2} \end{bmatrix}, \tag{4.72}$$

which is identical to the result obtained in (4.68).

For second-order design sensitivity, solving (4.22) for $d\mathbf{z}_g/db_i$ $(i = 1, 2, 3)$ gives

$$\frac{dz_g}{db_1} = [0 \quad 0 \quad 0 \quad 0 \quad 0 \quad 0]^T$$
$$\frac{dz_g}{db_2} = \left[\frac{2\sqrt{3}-4}{Eb_2^2} \quad 0 \quad 0 \quad 0 \quad \frac{\sqrt{3}-1}{Eb_2^2} \quad \frac{\sqrt{3}-3}{Eb_2^2}\right]^T \tag{4.73}$$
$$\frac{dz_g}{db_3} = \left[-\frac{2\sqrt{2}}{Eb_3^2} \quad 0 \quad 0 \quad 0 \quad 0 \quad 0\right]^T.$$

For $\psi = z_1$, from (4.51),

$$\frac{\partial^2 \psi}{\partial b_2^2} = -2\frac{\partial}{\partial b_2}(\tilde{\lambda}_g^T \boldsymbol{K}_g(\boldsymbol{b})\frac{d\tilde{z}_g}{db_2}) = \frac{8-4\sqrt{3}}{Eb_2^3} \tag{4.74}$$

and

$$\frac{\partial^2 \psi}{\partial b_3^2} = -2\frac{\partial}{\partial b_3}(\tilde{\lambda}_g^T \boldsymbol{K}_g(\boldsymbol{b})\frac{d\tilde{z}_g}{db_3}) = \frac{4\sqrt{2}}{Eb_3^3}. \tag{4.75}$$

The remaining second derivatives are zero. Hence, the Hessian of ψ is a diagonal matrix. These results can be verified by taking the derivative of $d\psi/d\boldsymbol{b}$ in (4.72) with respect to the design $\boldsymbol{b}$.

Ten-Member Cantilever Truss

To illustrate the foregoing method, a 10-member cantilever truss, as shown in Fig. 4.3, will be used. Young's modulus of elasticity for the truss is $E = 1.0 \times 10^7$ psi, and the weight density is $\gamma = 0.1$ lb/in^3.

This problem has been examined in the literature [48] in order to compare various optimal design techniques. The problem is to choose a cross-sectional area for each truss member that minimizes its weight, subject to stress, displacement, and member-size constraints. The cost function is a linear function of the design variables, written as

$$\psi_0 = \sum_{i=1}^{m} \gamma_i l_i b_i, \tag{4.76}$$

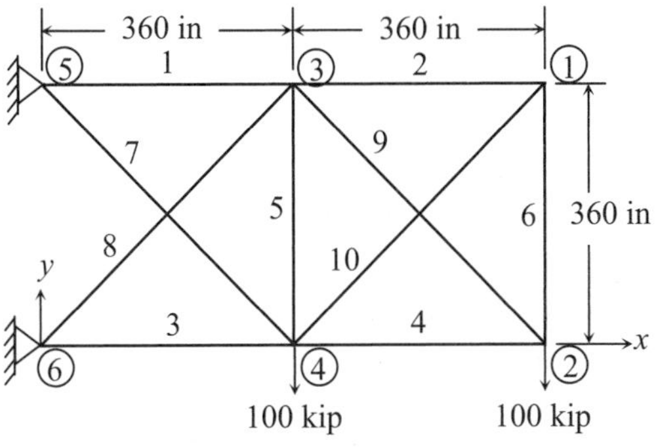

Figure 4.3. Ten-member cantilever truss.

where γ_i is the weight density, l_i is the length, and b_i is the cross-sectional area of the ith member. Stress and displacement constraints for the problem are expressed as

$$\psi_i = \frac{|\sigma_i|}{\sigma_i^a} - 1.0 \le 0, \quad i = 1, 2, \ldots, m \tag{4.77}$$

$$\psi_{j+m} = \frac{|z_j|}{z_j^a} - 1.0 \le 0, \quad j = 1, 2, \ldots, n, \tag{4.78}$$

where σ_i and σ_i^a are the calculated and allowable stresses for the ith member, and z_j and z_j^a are the calculated and allowable jth nodal displacements. Allowable stresses are given as $\sigma_i^a = 2.5 \times 10^4$ psi, while displacements are given as $z_j^a = 2.0$ in.

For the cost function, direct calculation of the design derivatives yields

$$\frac{d\psi_0}{db_i} = \gamma_i l_i, \tag{4.79}$$

and no adjoint problem needs to be defined. For stress constraints,

$$\sigma_i = \frac{E\Delta l_i}{l_i}, \quad i = 1, 2, \ldots, m, \tag{4.80}$$

where Δl_i is the change in l_i, which must be expressed in terms of nodal displacement $\boldsymbol{z}$. The adjoint equation of (4.11) is then

$$\boldsymbol{K\lambda} = \frac{\partial \psi_i^T}{\partial \boldsymbol{z}} = \frac{E}{l_i \sigma_i^a} \frac{\partial |\Delta l_i|^T}{\partial \boldsymbol{z}}, \quad i = 1, 2, \ldots, m, \tag{4.81}$$

which is nothing more than the structural equation for the displacement $\boldsymbol{\lambda}$ due to load vector $\partial \psi_i^T / \partial \boldsymbol{z}$. Therefore, solution $\boldsymbol{\lambda}^{(i)}$ can be found, where superscript (i) denotes the association of $\boldsymbol{\lambda}$ with constraint ψ_i. The reduced stiffness matrix design sensitivity formula of (4.13) gives

$$\frac{d\psi_i}{d\boldsymbol{b}} = -\frac{\partial}{\partial \boldsymbol{b}}(\tilde{\boldsymbol{\lambda}}^{(i)T} \boldsymbol{K}(\boldsymbol{b})\tilde{\boldsymbol{z}}). \tag{4.82}$$

For displacement constraints ψ_{j+m}, the adjoint equation is

$$\boldsymbol{K\lambda} = \frac{\partial \psi_{j+m}^T}{\partial \boldsymbol{z}} = \operatorname{sgn}(z_j)\left[0 \ \cdots \ 0 \ \ \frac{1}{z_j^a} \ \ 0 \ \cdots \ 0\right]^T, \tag{4.83}$$

where

$$\operatorname{sgn}(z_j) = \begin{cases} +1, & \text{if } z_j > 0 \\ -1, & \text{if } z_j < 0. \end{cases}$$

Note that the adjoint load in (4.83) is a point load of magnitude $\pm 1/z_j^a$ in the jth nodal displacement direction. As before, the solution $\boldsymbol{\lambda}^{(j+m)}$ to (4.83) may be found. Then, the sensitivity formula in (4.13) gives

$$\frac{\partial \psi_{j+m}}{\partial \boldsymbol{b}} = -\frac{\partial}{\partial \boldsymbol{b}}\left(\tilde{\boldsymbol{\lambda}}^{(j+m)T} \boldsymbol{K}(\boldsymbol{b})\tilde{\boldsymbol{z}}\right). \tag{4.84}$$

Table 4.1. Design derivatives of constraints for 10-member cantilever truss.

Number	Design	$d\psi_1^T/d\mathbf{b}$	$d\psi_2^T/d\mathbf{b}$	$d\psi_3^T/d\mathbf{b}$
1	28.6	0.0082	–0.0009	–0.0093
2	0.2	–0.0696	–0.0284	0.0109
3	23.6	–0.0104	0.0012	–0.0062
4	15.4	–0.0006	–0.0003	–0.0076
5	0.2	–2.3520	–0.9601	0.1402
6	0.2	–0.0696	–0.0284	0.0109
7	3.0	–0.8369	–0.4398	–0.0177
8	21.0	0.0231	–0.0026	–0.0128
9	21.8	–0.0009	–0.0004	–0.0108
10	0.2	–0.1968	–0.0803	0.0308

Table 4.2. Comparison of sensitivity calculation.

Constraint	ψ_i^1	ψ_i^2	$\Delta\psi_i = \psi_i^1 - \psi_i^2$	ψ_i'	$\psi_i' / \Delta\psi_i \times 100\%$
ψ_1	1.6038	1.4798	–0.1240	–0.1302	105.0
ψ_2	0.6094	0.5325	–0.0769	–0.0807	105.0
ψ_3	0.0472	–0.0027	–0.0499	–0.0524	105.0

Using these sensitivity formulas, the design derivatives of some constraints are calculated at the initial design, as given in the second column of Table 4.1. The vectors $d\psi_1^T/d\boldsymbol{b}$ and $d\psi_2^T/d\boldsymbol{b}$ are design derivatives of normalized stresses in members 5 and 7, respectively, and $d\psi_3^T/d\boldsymbol{b}$ is the derivative of the normalized displacement in the y direction at node 2.

Define ψ_i^1 and ψ_i^2 as the constraint function values for initial design $\boldsymbol{b}$ and modified design $\boldsymbol{b} + \Delta\boldsymbol{b}$, respectively. Let $\Delta\psi_i$ be the difference between ψ_i^1 and ψ_i^2, and let $\psi_i' = (d\psi_i/d\boldsymbol{b})\Delta b_i$ be the difference predicted by design sensitivity calculations. The ratio of ψ_i' and $\Delta\psi_i$ multiplied by 100 is used as a measure of sensitivity accuracy. For example, 100% means that the predicted change is exactly the same as the actual change. Note that this accuracy measure will not give correct information when $\Delta\psi_i$ is much smaller than ψ_i^1 and ψ_i^2, because $\Delta\psi_i$ may be beyond the significant digits of ψ_i^1 and ψ_i^2. Numerical results with a 5% design change, $\Delta\boldsymbol{b} = 0.05\boldsymbol{b}$, are given in Table 4.2.

As a second numerical example, consider the same 10-member cantilever truss, but with multipoint boundary conditions, as shown in Fig. 4.4. In this problem, the space Z of kinematically admissible displacements is

$$Z = \left\{ \boldsymbol{z}_g \in R^{12} : z_9 = z_{10} = 0,\ z_{11}\cos\alpha + z_{12}\sin\alpha = 0 \right\}, \tag{4.85}$$

where $\alpha = 30°$. The same constraints given in (4.77) and (4.78) are considered in this problem. For the stress constraints of (4.77), the adjoint equation of (4.23) is

$$\boldsymbol{\lambda}_g^T \boldsymbol{K}\bar{\boldsymbol{\lambda}}_g = \frac{\partial \psi_i}{\partial \boldsymbol{z}_g}\bar{\boldsymbol{\lambda}}_g = \frac{E}{l_i \sigma_i^a}\frac{\partial |\Delta l_i|}{\partial \boldsymbol{z}}\bar{\boldsymbol{\lambda}}_g, \quad \forall \bar{\boldsymbol{\lambda}}_g \in Z, \tag{4.86}$$

with solution $\boldsymbol{\lambda}_g^{(i)}$ ($i = 1, 2, \ldots, m$). The design sensitivity formula of (4.29) becomes

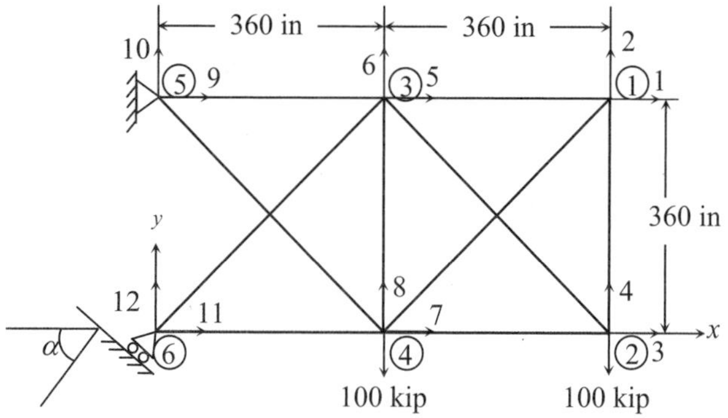

Figure 4.4. Ten-Member cantilever truss with multipoint boundary condition.

Table 4.3. Comparison of sensitivity calculation (multipoint boundary condition).

Constraint	ψ_i^1	ψ_i^2	$\Delta\psi_i = \psi_i^1 - \psi_i^2$	ψ_i'	$\psi_i' / \Delta\psi_i \times 100\%$
ψ_1	11.1476	10.6963	–0.5848	–0.6140	105.0
ψ_2	–0.4176	–0.4488	–0.0312	–0.0330	105.6
ψ_3	1.1134	2.0127	–0.1006	–0.1057	105.0

$$\frac{d\psi_i}{d\boldsymbol{b}} = -\frac{\partial}{\partial \boldsymbol{b}}\left(\tilde{\boldsymbol{\lambda}}_g^{(i)T} \boldsymbol{K}(\boldsymbol{b})\tilde{\boldsymbol{z}}_g \right). \tag{4.87}$$

For displacement constraints ψ_{j+m}, the adjoint equation is

$$\begin{aligned} \boldsymbol{\lambda}_g^T \boldsymbol{K}\bar{\boldsymbol{\lambda}}_g &= \frac{\partial \psi_{j+m}}{\partial \boldsymbol{z}_g}\bar{\boldsymbol{\lambda}}_g \\ &= \operatorname{sgn}(z_j)[0 \ \cdots \ 0 \ \ 1/z_j^a \ \ 0 \ \cdots \ 0]\bar{\boldsymbol{\lambda}}_g, \quad \forall \bar{\boldsymbol{\lambda}}_g \in Z, \end{aligned} \tag{4.88}$$

with the solution $\boldsymbol{\lambda}_g^{(j+m)}$ ($j = 1, 2, \cdots, n$). The design sensitivity formula of (4.29) becomes

$$\frac{d\psi_{j+m}}{d\boldsymbol{b}} = -\frac{\partial}{\partial \boldsymbol{b}}\left(\tilde{\boldsymbol{\lambda}}_g^{(j+m)T} \boldsymbol{K}(\boldsymbol{b})\tilde{\boldsymbol{z}}_g \right). \tag{4.89}$$

A comparison of the difference between actual and predicted design sensitivity changes derived from the sensitivity formulas of constraint values, and with a 5% overall change in design variables, is presented in Table 4.3.

4.2 Design Sensitivity of the Eigenvalue Problem

As shown in Section 3.2 of Chapter 3, the natural frequency of vibration and buckling load are eigenvalues of a generalized eigenvalue problem; hence, they depend on the design. The purpose of this section is to obtain design derivatives of such eigenvalues, and to explore the important exceptional case in which repeated eigenvalues appear as a solution to an optimal design problem. It will be shown that eigenvalue sensitivity with respect to design may be calculated without solving an adjoint equation. It will also be shown that the simple eigenvector is differentiable and can be obtained by solving a linear system of equations in the orthogonal subspace of the eigenvector. The efficiency and accuracy of eigenvector sensitivity results can be improved by using the Ritz vector approach. Due to the singularity of the characteristic matrix associated with an eigenvalue, some technical complexities arise in eigenvalue and eigenvector design sensitivity analysis that do not appear with a static problem.

4.2.1 Eigenvalue Design Sensitivity Analysis

In this section, it is assumed that all eigenvalues and eigenvectors are simple; they are not repeated. A separate section will be devoted to repeated eigenvalues.

Reduced Stiffness and Mass Matrices

Consider the formulation for natural frequency or buckling (for buckling problems, $\boldsymbol{M}(\boldsymbol{b})$ is the geometric stiffness matrix) described by the following eigenvalue problem:

$$\boldsymbol{K}(\boldsymbol{b})\,\boldsymbol{y} = \zeta \boldsymbol{M}(\boldsymbol{b})\,\boldsymbol{y}, \tag{4.90}$$

where eigenvector $\boldsymbol{y}$ is normalized by the condition

$$\boldsymbol{y}^T \boldsymbol{M}(\boldsymbol{b})\,\boldsymbol{y} = 1. \tag{4.91}$$

As with the static problem, the reduced stiffness matrix $\boldsymbol{K}(\boldsymbol{b})$ is a function of design $\boldsymbol{b}$. In addition, the mass matrix $\boldsymbol{M}(\boldsymbol{b})$ is a function of the design. With plate thickness design, the structural mass depends on the plate thickness. It is presumed here that the reduced stiffness and mass (or geometric stiffness) matrices are positive definite and differentiable with respect to the design. Under these conditions, the following theorem holds true.

Theorem 4.1. If symmetric, positive definite matrices $\boldsymbol{K}(\boldsymbol{b})$ and $\boldsymbol{M}(\boldsymbol{b})$ in (4.90) are continuously differentiable with respect to the design, and if eigenvalue ζ is simple (not repeated), then the eigenvalue and its associated eigenvector in (4.90) and (4.91) are continuously differentiable with respect to the design.

Proof. For a direct proof, see section II.6 of Kato [53]. A more general theorem, which specifically addresses the results stated here, is proven in Section 4.2.5 of this text. ■

By premultiplying (4.90) with the transpose of an arbitrary vector $\bar{\boldsymbol{y}}$, the following identity can be obtained:

$$\bar{\boldsymbol{y}}^T \boldsymbol{K}(\boldsymbol{b})\,\boldsymbol{y} = \zeta\, \bar{\boldsymbol{y}}^T \boldsymbol{M}(\boldsymbol{b})\,\boldsymbol{y}, \qquad \forall \bar{\boldsymbol{y}} \in R^m. \tag{4.92}$$

To develop a design sensitivity analysis, perturb design $\boldsymbol{b}$ in the direction of $\delta\boldsymbol{b}$, to define the perturbed design $\boldsymbol{b}_\tau$ as a function of a scalar parameter τ

$$\boldsymbol{b}_\tau = \boldsymbol{b} + \tau\,\delta\boldsymbol{b}. \tag{4.93}$$

After substituting $\boldsymbol{b}_\tau$ into (4.92) and differentiating both sides with respect to τ, we obtain the following relation:

$$
\begin{aligned}
&\frac{\partial}{\partial \boldsymbol{b}}[\bar{\boldsymbol{y}}^T \boldsymbol{K}(\boldsymbol{b})\,\tilde{\boldsymbol{y}}]\delta \boldsymbol{b} + \bar{\boldsymbol{y}}^T \boldsymbol{K}(\boldsymbol{b})\,\boldsymbol{y}' \\
&= \zeta' \bar{\boldsymbol{y}}^T \boldsymbol{M}(\boldsymbol{b})\,\boldsymbol{y} + \zeta \frac{\partial}{\partial \boldsymbol{b}}[\bar{\boldsymbol{y}}^T \boldsymbol{M}(\boldsymbol{b})\,\tilde{\boldsymbol{y}}]\delta \boldsymbol{b} + \zeta \bar{\boldsymbol{y}}^T \boldsymbol{M}(\boldsymbol{b})\,\boldsymbol{y}', \qquad \forall \bar{\boldsymbol{y}} \in R^m,
\end{aligned} \tag{4.94}
$$

where $\bar{\boldsymbol{y}}$ is independent of member-size design $\boldsymbol{b}$ and, as in (4.19),

$$
\begin{aligned}
\boldsymbol{y}' &= \boldsymbol{y}'(\boldsymbol{b},\delta\boldsymbol{b}) \equiv \frac{d}{d\tau}\boldsymbol{y}(\boldsymbol{b}+\tau\delta\boldsymbol{b})\Big|_{\tau=0} = \frac{d\boldsymbol{y}}{d\boldsymbol{b}}\delta\boldsymbol{b} \\
\zeta' &= \zeta'(\boldsymbol{b},\delta\boldsymbol{b}) \equiv \frac{d}{d\tau}\zeta(\boldsymbol{b}+\tau\delta\boldsymbol{b})\Big|_{\tau=0} = \frac{d\zeta}{d\boldsymbol{b}}\delta\boldsymbol{b}
\end{aligned} \tag{4.95}
$$

are *differentials* (or *variations*) of the eigenvector and eigenvalue at given design $\boldsymbol{b}$ in the direction of $\delta\boldsymbol{b}$. Since (4.94) must hold for arbitrary vector $\bar{\boldsymbol{y}}$, substitute $\bar{\boldsymbol{y}} = \boldsymbol{y}$ into (4.94), using (4.91), to obtain

$$
\zeta' = \left\{ \frac{\partial}{\partial \boldsymbol{b}}[\tilde{\boldsymbol{y}}^T \boldsymbol{K}(\boldsymbol{b})\,\tilde{\boldsymbol{y}}] - \zeta \frac{\partial}{\partial \boldsymbol{b}}[\tilde{\boldsymbol{y}}^T \boldsymbol{M}(\boldsymbol{b})\,\tilde{\boldsymbol{y}}] \right\} \delta\boldsymbol{b} + \boldsymbol{y}'^T[\boldsymbol{K}(\boldsymbol{b})\boldsymbol{y} - \zeta \boldsymbol{M}(\boldsymbol{b})\boldsymbol{y}]. \tag{4.96}
$$

Note that the last term in (4.96) is zero since $\boldsymbol{y}$ is an eigenvector of (4.90). Thus, (4.96) is reduced to the desired result—the differentiation of the eigenvalue with respect to the design—as

$$
\frac{\partial \zeta}{\partial \boldsymbol{b}} = \frac{\partial}{\partial \boldsymbol{b}}[\tilde{\boldsymbol{y}}^T \boldsymbol{K}(\boldsymbol{b})\,\tilde{\boldsymbol{y}}] - \zeta \frac{\partial}{\partial \boldsymbol{b}}[\tilde{\boldsymbol{y}}^T \boldsymbol{M}(\boldsymbol{b})\,\tilde{\boldsymbol{y}}]. \tag{4.97}
$$

It is interesting to note that this eigenvalue derivative with respect to the design may be calculated without solving an adjoint equation or obtaining an eigenvector derivative. In addition to the result from the eigenvalue problem, the explicit dependence of $\boldsymbol{K}(\boldsymbol{b})$ and $\boldsymbol{M}(\boldsymbol{b})$ is required to evaluate (4.97). Thus, once the eigenvalue problem has been solved for a simple (nonrepeated) eigenvalue, the eigenvalue derivative can be directly calculated using (4.97). In this sense, the differentiation of eigenvalues is simpler than the differentiation of those structural performance functions that involve the static response, unless multiple (repeated) eigenvalues are encountered.

Generalized Stiffness and Mass Matrices

Consider the variational formulation of the eigenvalue problem as presented in (3.231) and (3.236), written in the form

$$
a_b(\boldsymbol{y}_g, \bar{\boldsymbol{y}}_g) \equiv \bar{\boldsymbol{y}}_g^T \boldsymbol{K}_g(\boldsymbol{b})\boldsymbol{y}_g = \zeta \bar{\boldsymbol{y}}_g^T \boldsymbol{M}_g(\boldsymbol{b})\boldsymbol{y}_g \equiv \zeta d_b(\boldsymbol{y}_g, \bar{\boldsymbol{y}}_g), \ \forall \bar{\boldsymbol{y}}_g \in Z. \tag{4.98}
$$

Recall that bilinear forms, $a_b(\bullet,\bullet)$ and $d_b(\bullet,\bullet)$, are positive definite in space $Z \subset R^n$ of kinematically admissible displacements, that is,

$$
\begin{aligned}
a_b(\boldsymbol{y}_g, \boldsymbol{y}_g) &> 0, \\
d_b(\boldsymbol{y}_g, \boldsymbol{y}_g) &> 0,
\end{aligned} \qquad \forall \boldsymbol{y}_g \in Z, \qquad \boldsymbol{y}_g \neq 0. \tag{4.99}
$$

In order to obtain the simplest possible derivation of eigenvalue design sensitivity in this setting, basis vector $\boldsymbol{\phi}^i$ (i = 1, …, m) of Z may be introduced. It is presumed here that kinematic constraints do not explicitly depend on the design, so that vector $\boldsymbol{\phi}^i$ is independent of the design. Also, recall that the dimension of space $Z \subset R^n$ is $m < n$. Any vector $\boldsymbol{y}_g \in Z$ may be approximated as a linear combination of $\boldsymbol{\phi}^i$, that is,

$$y_g = \sum_{i=1}^{m} c_i \boldsymbol{\phi}^i = \boldsymbol{\Phi} \boldsymbol{c}, \tag{4.100}$$

where $\boldsymbol{\Phi} = [\boldsymbol{\phi}^1, \boldsymbol{\phi}^2, \cdots, \boldsymbol{\phi}^m]$ is the characteristic matrix, and the coefficients c_i are still to be determined. Substituting the expression of $\boldsymbol{y}_g$ into (4.98), and evaluating (4.98) with $\bar{\boldsymbol{y}}_g = \boldsymbol{\phi}^j$ ($j = 1, \ldots, m$), the following system of equations is produced for the coefficients of c_i:

$$\sum_{i=1}^{m} a_b(\boldsymbol{\phi}^i, \boldsymbol{\phi}^j) c_i = \zeta \sum_{i=1}^{m} d_b(\boldsymbol{\phi}^i, \boldsymbol{\phi}^j) c_i, \qquad j = 1, \ldots, m \tag{4.101}$$

In matrix form, these equations may be written as

$$\begin{aligned} \hat{\boldsymbol{K}}(\boldsymbol{b})\boldsymbol{c} &= \zeta \hat{\boldsymbol{M}}(\boldsymbol{b})\boldsymbol{c} \\ \boldsymbol{c}^T \hat{\boldsymbol{M}}(\boldsymbol{b})\boldsymbol{c} &= 1, \end{aligned} \tag{4.102}$$

which is another eigenvalue problem with eigenvalue ζ and eigenvector $\boldsymbol{c}$ in reduced space R^m. In (4.102), the following definitions are used:

$$\begin{aligned} \hat{\boldsymbol{K}}(\boldsymbol{b}) &= \boldsymbol{\Phi}^T \boldsymbol{K}_g(\boldsymbol{b}) \boldsymbol{\Phi} \\ \hat{\boldsymbol{M}}(\boldsymbol{b}) &= \boldsymbol{\Phi}^T \boldsymbol{M}_g(\boldsymbol{b}) \boldsymbol{\Phi}. \end{aligned} \tag{4.103}$$

Note that since matrix $\boldsymbol{\Phi}$ is independent of the design, if $\boldsymbol{K}_g(\boldsymbol{b})$ and $\boldsymbol{M}_g(\boldsymbol{b})$ are differentiable with respect to the design, then the matrices $\hat{\boldsymbol{K}}(\boldsymbol{b})$ and $\hat{\boldsymbol{M}}(\boldsymbol{b})$ are also differentiable.

Using the conditions given by (4.99), the matrices $\hat{\boldsymbol{K}}(\boldsymbol{b})$ and $\hat{\boldsymbol{M}}(\boldsymbol{b})$ may be shown to be positive definite. Thus, the result from (4.97) can be applied in order to obtain the eigenvalue derivative with respect to the design. Since ζ is the solution to the eigenvalue problem in (4.102) with eigenvector $\boldsymbol{c}$, the derivative of eigenvalue ζ with respect to the design is

$$\frac{\partial \zeta}{\partial \boldsymbol{b}} = \frac{\partial}{\partial \boldsymbol{b}}[\tilde{\boldsymbol{c}}^T \hat{\boldsymbol{K}}(\boldsymbol{b})\tilde{\boldsymbol{c}}] - \zeta \frac{\partial}{\partial \boldsymbol{b}}[\tilde{\boldsymbol{c}}^T \hat{\boldsymbol{M}}(\boldsymbol{b})\tilde{\boldsymbol{c}}]. \tag{4.104}$$

The use of (4.104) is questionable since the matrices $\hat{\boldsymbol{K}}(\boldsymbol{b})$ and $\hat{\boldsymbol{M}}(\boldsymbol{b})$ are not generated during the structural analysis process. Only the explicit dependence of $\boldsymbol{K}_g(\boldsymbol{b})$ and $\boldsymbol{M}_g(\boldsymbol{b})$ on the design is known. In order to use the result from (4.104), it is necessary to note that (4.100) and the second equation in (4.102) yield

$$1 = \boldsymbol{c}^T \hat{\boldsymbol{M}}(\boldsymbol{b})\boldsymbol{c} = \boldsymbol{c}^T \boldsymbol{\Phi}^T \boldsymbol{M}_g(\boldsymbol{b}) \boldsymbol{\Phi} \boldsymbol{c} = \boldsymbol{y}_g^T \boldsymbol{M}_g(\boldsymbol{b}) \boldsymbol{y}_g = d_b(\boldsymbol{y}_g, \boldsymbol{y}_g). \tag{4.105}$$

Furthermore, substituting the matrices $\hat{\boldsymbol{K}}(\boldsymbol{b})$ and $\hat{\boldsymbol{M}}(\boldsymbol{b})$ from (4.103) into (4.104) gives

$$\frac{\partial \zeta}{\partial \boldsymbol{b}} = \frac{\partial}{\partial \boldsymbol{b}}[\tilde{\boldsymbol{c}}^T \boldsymbol{\Phi}^T \boldsymbol{K}_g(\boldsymbol{b}) \boldsymbol{\Phi} \tilde{\boldsymbol{c}}] - \zeta \frac{\partial}{\partial \boldsymbol{b}}[\tilde{\boldsymbol{c}}^T \boldsymbol{\Phi}^T \boldsymbol{M}_g(\boldsymbol{b}) \boldsymbol{\Phi} \tilde{\boldsymbol{c}}].$$

With (4.100), the desired result can be obtained as

$$\frac{\partial \zeta}{\partial \boldsymbol{b}} = \frac{\partial}{\partial \boldsymbol{b}}[\tilde{\boldsymbol{y}}_g^T \boldsymbol{K}_g(\boldsymbol{b}) \tilde{\boldsymbol{y}}_g] - \zeta \frac{\partial}{\partial \boldsymbol{b}}[\tilde{\boldsymbol{y}}_g^T \boldsymbol{M}_g(\boldsymbol{b}) \tilde{\boldsymbol{y}}_g]. \tag{4.106}$$

Given the expression of $\boldsymbol{K}_g(\boldsymbol{b})$ and $\boldsymbol{M}_g(\boldsymbol{b})$ on the design and given the solution to the eigenvalue problem in (4.98), the derivative of ζ with respect to design $\boldsymbol{b}$ can readily be obtained from (4.106). Note that the form of (4.106) is identical to that obtained with the reduced stiffness matrix in (4.97). However, the computational advantages associated with (4.106) are considerable. Generalized stiffness and mass matrices can be used to calculate the design sensitivity of a simple eigenvalue without resorting to matrix manipulations that transform generalized matrices into reduced matrices.

4.2.2 Design Sensitivity Analysis of Eigenvectors

Design sensitivity analysis of eigenvectors has been an active research area since the earlier work by Fox and Kapoor [54] because of the increasing importance of the eigenvector sensitivity in the development of structural design optimization, dynamic system identification, and dynamic control. There are several methods to compute the derivatives of eigenvectors: the overall finite difference method, the continuum method [5], [55], and [56], the modal method [54], the modified modal method [57], Nelson's direct method [58], Ritz vector method [59], and the iterative method [60]. In this text, a design sensitivity formulation of eigenvectors is derived in the orthogonal subspace, and Ritz vector method is introduced to approximate eigenvector sensitivity.

Generalized Stiffness and Mass Matrices

As with the static response problem, since $\boldsymbol{\Phi}$ in (4.100) does not depend on $\boldsymbol{b}$, $\boldsymbol{y}_g' = (d\boldsymbol{y}_g/d\boldsymbol{b})\delta\boldsymbol{b} = \boldsymbol{\Phi}(d\boldsymbol{c}/d\boldsymbol{b})\delta\boldsymbol{b}$. Thus, an eigenvector $\boldsymbol{y}_g$ that corresponds to a simple eigenvalue is differentiable with respect to the design. In order to obtain directional derivative $\boldsymbol{y}_g'$ of eigenvector $\boldsymbol{y}_g$, which corresponds to the smallest simple eigenvalue of (4.98), take the total variation of both sides of (4.98) and use the chain rule of differentiation to obtain

$$\begin{aligned} a_b(\boldsymbol{y}_g', \bar{\boldsymbol{y}}_g) - \zeta d_b(\boldsymbol{y}_g', \bar{\boldsymbol{y}}_g) &= \zeta d_{\delta b}'(\boldsymbol{y}_g, \bar{\boldsymbol{y}}_g) - a_{\delta b}'(\boldsymbol{y}_g, \bar{\boldsymbol{y}}_g) + \zeta' d_b(\boldsymbol{y}_g, \bar{\boldsymbol{y}}_g) \\ &= \zeta d_{\delta b}'(\boldsymbol{y}_g, \bar{\boldsymbol{y}}_g) - a_{\delta b}'(\boldsymbol{y}_g, \bar{\boldsymbol{y}}_g) \\ &\quad + [a_{\delta b}'(\boldsymbol{y}_g, \boldsymbol{y}_g) - \zeta d_{\delta b}'(\boldsymbol{y}_g, \boldsymbol{y}_g)] d_b(\boldsymbol{y}_g, \bar{\boldsymbol{y}}_g), \quad \forall \bar{\boldsymbol{y}}_g \in Z, \end{aligned} \tag{4.107}$$

where (4.106) is used in the last equality.

The bilinear form of the left side of (4.107) no longer needs to be positive definite on Z, since it is the difference of two positive definite forms. It is therefore not clear that a unique solution exists for (4.107). However, note that (4.107) is easily satisfied for $\bar{\boldsymbol{y}}_g = \boldsymbol{y}_g$. A subspace W of Z that is d_b-orthogonal to $\boldsymbol{y}_g$ may be defined, and Z may be written as the direct sum of W and $\boldsymbol{y}_g$, that is,

$$Z = W \oplus \{\boldsymbol{y}_g\},$$

where $\{\boldsymbol{y}_g\}$ is the one-dimensional subspace of Z spanned by $\boldsymbol{y}_g$ and

$$W = \{\boldsymbol{v} \in Z \mid d_b(\boldsymbol{v}, \boldsymbol{y}_g) = 0\}. \tag{4.108}$$

The notation $\oplus$ means that since $d_b(\bullet,\bullet)$ is positive definite on Z, every vector $\boldsymbol{w} \in Z$ can be uniquely written in the form

$$\boldsymbol{w} = \boldsymbol{v} + \alpha \boldsymbol{y}_g, \quad \boldsymbol{v} \in W, \ \alpha \in R^1.$$

Thus, W is the subspace of Z that is d_b-orthogonal to $\{\boldsymbol{y}_g\}$.

Since (4.107) is valid for all $\bar{\boldsymbol{y}}_g \in Z$, every element of Z can be uniquely written as the sum of elements from W and $\{\boldsymbol{y}_g\}$, and since (4.107) is easily satisfied for $\bar{\boldsymbol{y}}_g = \boldsymbol{y}_g$, (4.107) is reduced to

$$a_b(y'_g, \bar{y}_g) - \zeta d_b(y'_g, \bar{y}_g) = \zeta d'_{\delta b}(y_g, \bar{y}_g) - a'_{\delta b}(y_g, \bar{y}_g), \qquad \bar{y}_g \in W. \tag{4.109}$$

It now remains to show that the bilinear form on the left side of (4.109) is positive definite in the subspace W.

Using the Rayleigh quotient representation of eigenvalues in (4.98), it is well known that the second eigenvalue minimizes the Rayleigh quotient over all vectors of $\boldsymbol{v} \in W$ [38]. Since the second eigenvalue is larger than the smallest simple eigenvalue ζ,

$$\zeta < \frac{a_b(\boldsymbol{v},\boldsymbol{v})}{d_b(\boldsymbol{v},\boldsymbol{v})}, \qquad \forall \boldsymbol{v} \in W, \quad \boldsymbol{v} \neq 0$$

or,

$$a_b(\boldsymbol{v},\boldsymbol{v}) - \zeta d_b(\boldsymbol{v},\boldsymbol{v}) > 0, \ \forall \boldsymbol{v} \in W, \quad \boldsymbol{v} \neq 0. \tag{4.110}$$

This shows that the bilinear form on the left side of (4.109) is positive definite on W. Thus, (4.109) has the unique solution $\boldsymbol{y}'_g \in W$ for the directional derivative of eigenvector $\boldsymbol{y}_g$, which corresponds to the smallest simple eigenvalue. This argument can be extended to any simple eigenvalue, replacing W by the subspace of Z that is d_b-orthogonal to all eigenvectors corresponding to eigenvalues smaller than ζ.

By letting $\delta\boldsymbol{b}$ be a vector that contains a one in the jth position and zeros elsewhere, $\boldsymbol{y}'_g$ becomes $d\boldsymbol{y}_g/db_j$, and (4.109) becomes

$$\bar{\boldsymbol{y}}_g^T[\boldsymbol{K}_g(\boldsymbol{b}) - \zeta \boldsymbol{M}_g]\frac{d\boldsymbol{y}_g}{db_j} = \zeta \frac{\partial}{\partial b_j}[\bar{\boldsymbol{y}}_g^T \boldsymbol{M}_g(\boldsymbol{b})\tilde{\boldsymbol{y}}_g] - \frac{\partial}{\partial b_j}[\bar{\boldsymbol{y}}_g^T \boldsymbol{K}_g(\boldsymbol{b})\tilde{\boldsymbol{y}}_g], \qquad \forall \bar{\boldsymbol{y}}_g \in W. \tag{4.111}$$

Note that the design sensitivity of the eigenvector does not require the eigenvalue design sensitivity ζ'. Several numerical techniques exist for solving either (4.109) for $\boldsymbol{y}'_g$, or (4.111) for $d\boldsymbol{y}_g/db_j$. Nelson [58] presents a direct computational technique that uses the reduced stiffness matrix, and is effective for computations in which reduced system matrices are known. The potential exists for directly applying such numerical techniques as subspace iteration [46] in order to construct a solution to (4.109) related to the basic eigenvalue problem. Wang and Choi proposed the following Ritz vector method [59].

Ritz Vector Method for Reduced Stiffness and Mass Matrices

It is difficult to construct an orthogonal subspace W for a general eigenvalue problem by using the solution method in (4.111), even for reduced stiffness and mass matrices. In this section, the design sensitivity of a simple (not repeated) eigenvector is discussed in the reduced matrix equations. Two methods are usually used: first, solve a linear system of equations, and second, expand using eigenvectors. In the first method, matrix multiplication is a lengthy operation and destroys any banded structure associated with the original eigensystem. The second method is undesirable for large eigensystems, since eigenvector calculation for large structures is very expensive. In this section, the Ritz vector expansion method [59] is introduced to improve the accuracy of sensitivity results with a small number of eigenvectors.

Assume that $\boldsymbol{y}^i$ is the ith eigenvector ($i = 1, \ldots, q$) and $\boldsymbol{\psi}^j$ is the jth Ritz vector ($j = 1, \ldots, r$) [59]. We can construct the Ritz vectors that are $\boldsymbol{M}(\boldsymbol{b})$-orthonormal to the eigenvectors. $\boldsymbol{\phi}^k$ can then be defined as

$$\begin{aligned} \boldsymbol{\phi}^i &\equiv \boldsymbol{y}^i, & i &= 1,\ldots,q \\ \boldsymbol{\phi}^{q+k} &\equiv \boldsymbol{\psi}^k, & k &= 1,\ldots,r. \end{aligned} \tag{4.112}$$

The eigenvalue problem in (4.90) is rewritten as

$$\boldsymbol{K}(\boldsymbol{b})\boldsymbol{y}^i = \zeta_i \boldsymbol{M}(\boldsymbol{b})\boldsymbol{y}^i, \quad i = 1,\ldots,q, \tag{4.113}$$

and the normalizing condition that is employed is written as

$$\boldsymbol{y}^{i^T}\boldsymbol{M}(\boldsymbol{b})\boldsymbol{y}^j = \boldsymbol{\psi}^{i^T}\boldsymbol{M}(\boldsymbol{b})\boldsymbol{\psi}^j = \boldsymbol{\phi}^{i^T}\boldsymbol{M}(\boldsymbol{b})\boldsymbol{\phi}^j = \delta_{ij}, \tag{4.114}$$

for all i and j. From (4.111), the sensitivity equation is obtained as

$$(\boldsymbol{K} - \zeta_i \boldsymbol{M})\boldsymbol{y}^{i'} = \zeta_i \frac{\partial}{\partial \boldsymbol{b}}(\boldsymbol{M}(\boldsymbol{b})\tilde{\boldsymbol{y}}^i)\delta\boldsymbol{b} - \frac{\partial}{\partial \boldsymbol{b}}(\boldsymbol{K}(\boldsymbol{b})\tilde{\boldsymbol{y}}^i)\delta\boldsymbol{b} \equiv \boldsymbol{F}_f. \tag{4.115}$$

Assuming that the eigenvalue problem is solved for ζ_i and $\boldsymbol{y}^i$, $\boldsymbol{F}_f$ on the right side of (4.115) can be evaluated. As discussed before, the matrix $(\boldsymbol{K} - \zeta_i\boldsymbol{M})$ on the left side of (4.115) is positive definite on W and the solution $\boldsymbol{y}^{i'} \in W$ is $\boldsymbol{M}(\boldsymbol{b})$-orthogonal to $\boldsymbol{y}^i$.

The Ritz vector method represents the design sensitivity of the eigenvector $\boldsymbol{y}^i$ as a linear combination of eigenvectors and Ritz vectors, that is, if $\boldsymbol{\Phi} = [\boldsymbol{\phi}^k] = [\boldsymbol{y}^1, \ldots, \boldsymbol{y}^q, \boldsymbol{\psi}^1, \ldots, \boldsymbol{\psi}^r]$, then the design sensitivity of the eigenvector is expressed with coefficient $\boldsymbol{c}$ as

$$\boldsymbol{y}^{i'} = \boldsymbol{\Phi}\boldsymbol{c}. \tag{4.116}$$

The advantage of using Ritz vectors in (4.116) is that only a small number of eigenvectors can be used in (4.116), such that for a large eigensystem, $q + r \ll m$. Substituting (4.116) into (4.115), and premultiplying $\boldsymbol{\Phi}^T$, the following is obtained:

$$\boldsymbol{\Phi}^T(\boldsymbol{K}(\boldsymbol{b}) - \zeta_i\boldsymbol{M}(\boldsymbol{b}))\boldsymbol{\Phi}\boldsymbol{c} = \boldsymbol{\Phi}^T\boldsymbol{F}_f. \tag{4.117}$$

Equation (4.117) can be written in partitioned form as

$$\begin{bmatrix} \boldsymbol{A}^{11}_{q\times q} & \boldsymbol{0}_{q\times r} \\ \boldsymbol{0}_{r\times q} & \boldsymbol{A}^{22}_{r\times r} \end{bmatrix} \begin{bmatrix} \boldsymbol{c}^1 \\ \boldsymbol{c}^2 \end{bmatrix} = \begin{bmatrix} \boldsymbol{F}^1 \\ \boldsymbol{F}^2 \end{bmatrix}, \tag{4.118}$$

where, using the property in (4.114) and the formula $\boldsymbol{y}^{k^T}\boldsymbol{K}\boldsymbol{y}^j = \zeta_j\delta_{jk}$, the following terms are obtained:

$\boldsymbol{A}^{11}$ = a $q \times q$ diagonal matrix with diagonal terms $\zeta_j - \zeta_i$, $j = 1, \ldots, q$
$\boldsymbol{A}^{22}$ = a $r \times r$ nonsingular full matrix $\boldsymbol{\psi}^{k^T}\boldsymbol{K}(\boldsymbol{b})\boldsymbol{\psi}^j - \zeta_i\delta_{kj}$
$\boldsymbol{c}^1 = [c_1^1, \ldots, c_q^1]^T, \quad \boldsymbol{c}^2 = [c_1^2, \ldots, c_r^2]^T$
$\boldsymbol{F}^1 = [\boldsymbol{y}^{1^T}\boldsymbol{F}_f, \ldots, \boldsymbol{y}^{q^T}\boldsymbol{F}_f]^T, \quad \boldsymbol{F}^2 = [\boldsymbol{\psi}^{1^T}\boldsymbol{F}_f, \ldots, \boldsymbol{\psi}^{r^T}\boldsymbol{F}_f]^T.$

Thus, the coefficients of the expansion in (4.116) are obtained as

$$\begin{aligned} c_j^1 &= \frac{\boldsymbol{y}^{j^T}\boldsymbol{F}_f}{\zeta_j - \zeta_i}, \qquad j \neq i,\ j = 1,\ldots,q \\ c_i^1 &= -\frac{1}{2}\frac{\partial}{\partial \boldsymbol{b}}(\tilde{\boldsymbol{y}}^{iT}\boldsymbol{M}(\boldsymbol{b})\tilde{\boldsymbol{y}}^i) \\ \boldsymbol{c}^2 &= (\boldsymbol{A}^{22})^{-1}\boldsymbol{F}^2. \end{aligned} \tag{4.119}$$

Note that if $r = 0$, then A^{22} does not exist and $c^2 = \mathbf{0}$, which becomes an eigenvector expansion method. When $j = i$, $y^{iT}M(b)y^i = 1$ is differentiated with respect to design $\boldsymbol{b}$, and (4.116) is then substituted in to compute c_i. Wang and Choi [59] show that two Ritz vectors ($r = 2$) are enough for most applications that go through numerical tests. Consequently, only a 2×2 matrix inversion is involved in order to solve $\boldsymbol{c}^2$.

4.2.3 Second-Order Design Sensitivity of a Simple Eigenvalue

From (4.106), the smallest eigenvalue sensitivity with respect to design b_i may be written as

$$\frac{\partial \zeta}{\partial b_i} = \frac{\partial}{\partial b_i}[\tilde{y}_g^T K_g(b)\tilde{y}_g] - \zeta \frac{\partial}{\partial b_i}[\tilde{y}_g^T M_g(b)\tilde{y}_g]. \tag{4.120}$$

Differentiating with respect to b_j gives

$$\begin{aligned}\frac{\partial^2 \zeta}{\partial b_i \partial b_j} &= \frac{\partial^2}{\partial b_i \partial b_j}[\tilde{y}_g^T K_g(b)\tilde{y}_g] - \zeta \frac{\partial^2}{\partial b_i \partial b_j}[\tilde{y}_g^T M_g(b)\tilde{y}_g] \\ &\quad - \left\{ \frac{\partial}{\partial b_j}[\tilde{y}_g^T K_g(b)\tilde{y}_g] - \zeta \frac{\partial}{\partial b_j}[\tilde{y}_g^T M_g(b)\tilde{y}_g] \right\} \frac{\partial}{\partial b_i}[\tilde{y}_g^T M_g(b)\tilde{y}_g] \\ &\quad + 2\frac{\partial}{\partial b_i}[\tilde{y}_g^T K_g(b)]\frac{dy_g}{db_j} - 2\zeta \frac{\partial}{\partial b_i}[\tilde{y}_g^T M_g(b)]\frac{dy_g}{db_j}.\end{aligned} \tag{4.121}$$

In order to evaluate the second derivative of ζ in (4.121), $d\boldsymbol{y}_g/db_i$ and $d\boldsymbol{y}_g/db_j$ must be calculated, which can be done by solving (4.111). Once (4.111) is solved, the result may then be substituted into (4.121) to obtain the second design derivative of ζ with respect to design components b_i and b_j.

Note that the computation of all second design derivatives of ζ requires a solution to (4.120) for $i = 1, \ldots, k$. These results may be substituted into (4.121) and the partial derivatives with respect to b_j ($j = 1, \ldots, k$) may be calculated. Thus, all $k^2/2 + k/2$ distinct derivatives of ζ are obtained with respect to the design. In obtaining the derivatives, k sets of equations in (4.111) must be dealt with, and numerical computation is performed in order to evaluate the right side of (4.121). While this presents a substantial amount of computation, the fact that the second design derivative of the eigenvalue is available with respect to the design can be of value in iterative design optimization.

4.2.4 Systematic Occurrence of Repeated Eigenvalues in Structural Optimization

In carrying out vibration and buckling analysis, it is well known that computational difficulties can arise if *repeated eigenvalues* (natural frequencies or buckling loads) arise. However, the possibility that a precisely repeated eigenvalue will occur has often been dismissed on practical grounds.

While repeated eigenvalues may be unlikely to occur in randomly specified structures, they are far more likely to happen in optimized structures. Thompson and Hunt [61] have devoted considerable attention to designs that are constructed with simultaneous buckling failure modes, i.e., repeated eigenvalues. More recently, Olhoff and Rasmussen [62] have shown that a repeated buckling load may occur in an optimized clamped-clamped column. Their results corrected an erroneous solution published much earlier [63]. Subsequent to the Olhoff-Rasmussen finding, Masur and Mroz [64] provided an elegant treatment of optimality criteria for structures in which repeated eigenvalues occur. They showed that a singular (nondifferentiable) optimization problem could arise. In addition, Prager and Prager [65] demonstrated that singular behavior associated with repeated eigenvalues may arise even for a very simple finite-dimensional column model, as with the distributed parameter column of Olhoff and Rasmussen [62]. Simple vibration and buckling problems are introduced in this section in order to show how repeated eigenvalues can arise in structural optimization.

Vibration Example

Consider a spring-mass system, as shown in Fig. 4.5. The eigenvalue equation for small-amplitude vibration of the rigid body is simply derived as

$$\boldsymbol{K}(\boldsymbol{b})\boldsymbol{y} \equiv \begin{bmatrix} 4b_1 + b_2 & b_2 \\ b_2 & 4b_1 + b_2 \end{bmatrix} \begin{bmatrix} y_1 \\ y_2 \end{bmatrix} = \zeta \begin{bmatrix} 2 & 1 \\ 1 & 2 \end{bmatrix} \begin{bmatrix} y_1 \\ y_2 \end{bmatrix} \equiv \zeta \boldsymbol{My}, \tag{4.122}$$

where $\zeta = 2\omega^2 m/3$, m is the mass of the bar, and $I = ml^2/12$ is the moment of inertia. In this example, horizontal bar motion is ignored, and spring constants are regarded as design variables.

The optimal design objective is to find design parameters b_1 and b_2 in order to minimize the spring weight, which is presumed to be in the form

$$\psi_0 = c_1 b_1 + c_2 b_2, \tag{4.123}$$

where c_1 and c_2 are known constants. Minimization is carried out subject to the constraint that eigenvalues are not lower than $\zeta_0 > 0$, and that spring constants are nonnegative. These constraints are given in the inequality constraint form as

$$\begin{aligned} \psi_1 &= \zeta_0 - \zeta_1 \le 0 \\ \psi_2 &= \zeta_0 - \zeta_2 \le 0 \\ \psi_3 &= -b_1 \le 0 \\ \psi_4 &= -b_2 \le 0. \end{aligned}$$

Since the eigenvalues of (4.122) are $\zeta_1 = (4b_1+2b_2)/3$ and $\zeta_2 = 4b_1$, these constraints become

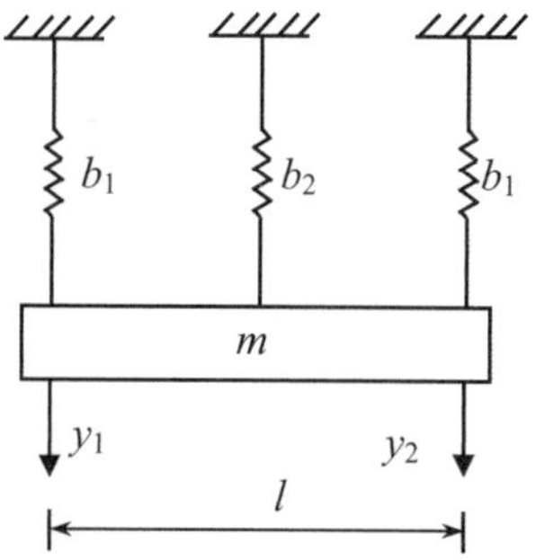

Figure 4.5. Spring-mass system with two degrees-of-freedom.

$$\begin{aligned}\psi_1 &= \zeta_0 - (4b_1 + 2b_2)/3 \le 0 \\ \psi_2 &= \zeta_0 - 4b_1 \le 0 \\ \psi_3 &= -b_1 \le 0 \\ \psi_4 &= -b_2 \le 0.\end{aligned} \tag{4.124}$$

Equations (4.123) and (4.124) define a linear programming problem. The feasible set is shown graphically in Fig. 4.6. Note that the slope of the line connecting points A and B is –2. The level lines of the cost function are straight, with a slope equal to $-c_1/c_2$. The cost function decreases as the level lines of cost move to the lower left. Thus, it is clear that point A (the repeated eigenvalue) is optimum if $c_1/c_2 > 2$, and point B (the simple eigenvalue) is optimum if $c_1/c_2 < 2$.

Column Buckling Example

Consider a column with elastically clamped ends, as shown in Fig. 4.7. The column has five rigid segments of length l and six elastic hinges at the ends of the segments with a bending stiffness of b_0^2. Rotation of the column's end sections by the angle θ_0 is opposed by the clamping moment $M_0 = b_0^2\theta_0$, where b_0 is given as a constant. When $b_0 = 0$ the ends are pin-supported, and when $b_0 = \infty$ they are rigidly clamped. Because the boundary conditions at both ends are identical, bending stiffness at the optimum design will be symmetric with respect to the center of the column, and the buckling modes will be either symmetric or antisymmetric with respect to the center. A column design is specified by the bending stiffness of b_1^2 for hinges 1 and 4, and the bending stiffness of b_2^2 for hinges 2 and 3, as shown in Fig. 4.7. A buckling mode that is known to be symmetric or antisymmetric is specified by deflection y_1 of nodes 1 and 4, and deflection y_2 of nodes 2 and 3. Upward deflections are regarded as positive.

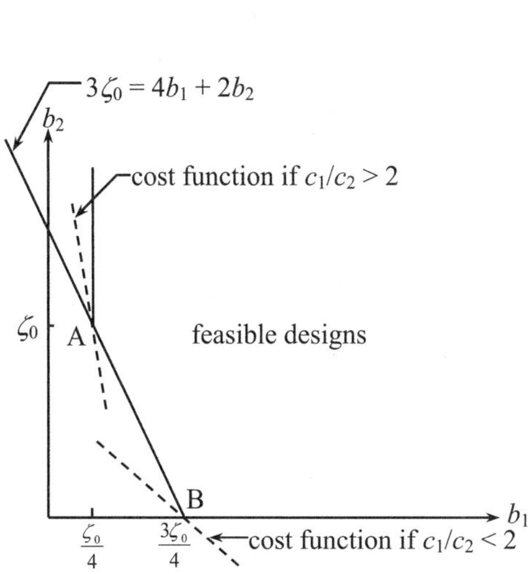

Figure 4.6. Feasible region in design space for systems with two degrees-of-freedom.

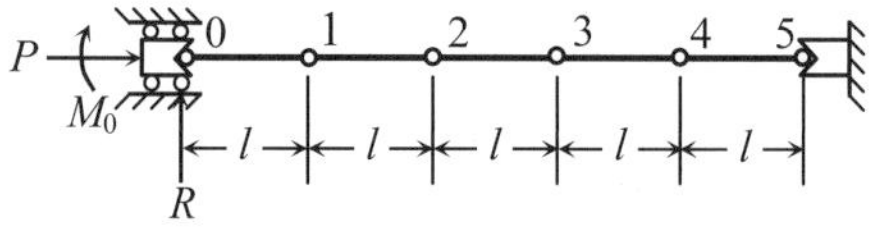

Figure 4.7. Elastically supported column.

At the left end, the column is subject to axial load P, reaction force R, and clamping moment M_0. The bending moment at the ith hinge is

$$M_i = M_0 - ilR - Py_i, \quad i = 0, \ldots, 4, \tag{4.125}$$

where $y_0 = 0$. If θ_i denotes the relative rotation of the segments, which meet at the ith hinge, then they will be considered positive if a counterclockwise rotation of the segment to the right of the ith hinge exceeds the rotation of the segment to the left, as follows:

$$M_i = b_i^2 \theta_i = b_i^2 (y_{i+1} - 2y_i + y_{i-1})/l, \quad i = 0, \ldots, 4, \tag{4.126}$$

where $y_{-1} = y_0 = 0$. At this point, it is convenient to introduce reference stiffness b^{*2} and to define the dimensionless variables as

$$\hat{P} = \frac{Pl}{b^{*2}}, \quad \hat{R} = \frac{Rl}{b^{*2}}, \quad \hat{M}_i = \frac{M_i}{b^{*2}}, \quad \hat{y}_i = \frac{y_i}{l}, \quad \hat{b}_i = \frac{b_i}{b^*}. \tag{4.127}$$

Using these dimensionless variables, and after deleting ^ for the sake of notational simplicity, (4.125) and (4.126) will yield

$$M_0 - iR - Py_i = b_i^2 (y_{i+1} - 2y_i + y_{i-1}), \quad i = 0, 1, 2. \tag{4.128}$$

For a symmetric buckling mode, $y_3 = y_2$ and $R = 0$. For $i = 0, 1, 2$, (4.128) yields

$$\begin{aligned} M_0 &= b_0^2 y_1 \\ M_0 - P_s y_1 &= b_1^2 (y_2 - 2y_1) \\ M_0 - P_s y_2 &= b_2^2 (-y_2 + y_1), \end{aligned} \tag{4.129}$$

where P_s is the buckling load of the symmetric mode. When the value of M_0 is taken from the first of these equations and is substituted into the other two, linear homogeneous equations are obtained for y_1 and y_2 that will only admit a nontrivial solution if

$$P_s^2 - (b_0^2 + 2b_1^2 + b_2^2)P_s + b_0^2(b_1^2 + b_2^2) + b_1^2 b_2^2 = 0. \tag{4.130}$$

The smaller root of this equation is the symmetric buckling load.

In order to find a design with the highest buckling load, let the cost of the design $[b_1, b_2]^T$ be fixed by the relation

$$b_1 + b_2 = 1 \tag{4.131}$$

Taking (4.131) into account, (4.130) can be reduced to

$$P_s^2 - (1 + b_0^2 + 3b_1^2 - 2b_1)P_s + b_0^2(2b_1^2 - 2b_1 + 1) + b_1^2(b_1^2 - 2b_1 + 1) = 0. \tag{4.132}$$

For an antisymmetric buckling mode, $y_3 = -y_2$, and $R = 2M_0/(5l)$, because both the bending moment and the deflection vanish at the center of the column. Proceeding in the same manner as above, a quadratic equation can be obtained for the antisymmetric buckling load P_a, in the form

$$\begin{aligned} &P_a^2 - (3 + 0.6b_0^2 + 5b_1^2 - 6b_1)P_a \\ &\quad + b_0^2(2b_1^2 - 3.6b_1 + 1.8) + 5b_1^2(b_1^2 - 2b_1 + 1) = 0. \end{aligned} \tag{4.133}$$

The smaller root of this equation is the antisymmetric buckling load.

As shown in Fig. 4.8, for fixed values of b_0 the smaller of the buckling loads P_s and P_a is a function of b. To illustrate the important features of this relation, the case in which $b_0 = 0.3$ is presented, for which buckling load variation is shown by the line ABCD. The arcs AB and CD correspond to antisymmetric buckling, while the arc BC corresponds to symmetric buckling. At point B both symmetric and antisymmetric buckling are possible, and the buckling load is given in the form of a repeated eigenvalue. A similar observation is warranted for point C. The arc BC, however, has its highest coordinate at point H, so the optimum design for $b_0 = 0.3$ corresponds to the value of b_1 as 0.39, which causes buckling in a symmetric mode, with the buckling load being a simple eigenvalue. However, the buckling load is a repeated eigenvalue if b_0 is larger than the value $b_0 =$ 0.57, which corresponds to point E, and the optimum design may buckle in a symmetric mode, an antisymmetric mode, or in any linear combination of the two.

Note that a second local maximum value occurs as a repeated eigenvalue for $b_0 \geq$ 1.31, which corresponds to point F. Another local maximum occurs as a simple eigenvalue for $b_0 \geq 0.94$, corresponding to point G. Also, note that the curve in Fig. 4.8 is not concave, but that several relative maximum values occur, and when a repeated eigenvalue occurs at an optimum design, the eigenvalue is not differentiable with respect to the design. Thus, if $dP/db_1 = 0$ was used as an optimality criteria, serious errors would result.

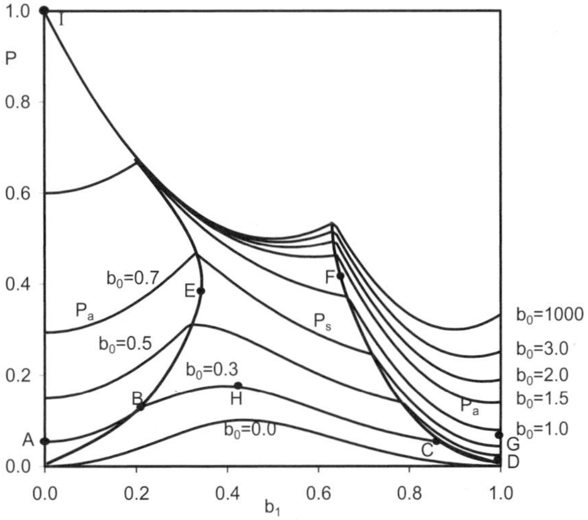

Figure 4.8. Buckling loads for optimum columns.

4.2.5 Directional Derivatives of Repeated Eigenvalues

Analysis with Reduced Global Stiffness and Mass Matrices
First, consider the eigenvalue problems that arise with structural vibration or buckling. Using the reduced stiffness and mass (or geometric stiffness) matrices, the eigenvalue problem can be written as

$$\boldsymbol{K}(\boldsymbol{b})\,\boldsymbol{y} = \zeta \boldsymbol{M}(\boldsymbol{b})\,\boldsymbol{y}, \tag{4.134}$$

where $\boldsymbol{y} \in R^m$. In this problem, $\boldsymbol{K}(\boldsymbol{b})$ and $\boldsymbol{M}(\boldsymbol{b})$ are symmetric, positive definite matrices.

The derivation of design sensitivity in Section 4.2.1 is valid only under the assumption that the eigenvalues and eigenvectors are differentiable with respect to the design, which is true if the eigenvalue is simple. However, even a repeated eigenvalue is *directionally differentiable* (see Appendix A.3), which will be proven in the following theorem.

Let eigenvalue $\zeta(\boldsymbol{b})$ in (4.134) have a multiplicity of $s \geq 1$ at $\boldsymbol{b}$ and let the $s \times s$ matrix $\mathcal{M}$ be defined by the following elements:

$$\mathcal{M}_{ij} = \frac{\partial}{\partial \boldsymbol{b}}[\tilde{\boldsymbol{y}}^{iT} \boldsymbol{K}(\boldsymbol{b})\tilde{\boldsymbol{y}}^{j}]\delta\boldsymbol{b} - \zeta(\boldsymbol{b})\frac{\partial}{\partial \boldsymbol{b}}[\tilde{\boldsymbol{y}}^{iT} \boldsymbol{M}(\boldsymbol{b})\tilde{\boldsymbol{y}}^{j}]\delta\boldsymbol{b}, \quad i, j = 1, \ldots, s, \tag{4.135}$$

where $\{\boldsymbol{y}^i\}$ (i = 1, 2, …, s) is any $\boldsymbol{M}(\boldsymbol{b})$-orthonormal basis of the eigenspace associated with $\zeta(\boldsymbol{b})$. Note that $\mathcal{M}$ depends on the design change direction $\delta\boldsymbol{b}$, i.e., $\mathcal{M} = \mathcal{M}(\boldsymbol{b}, \delta\boldsymbol{b})$. The following theorem characterizes the directional derivatives of repeated eigenvalues.

Theorem 4.2. If the matrices $\boldsymbol{K}(\boldsymbol{b})$ and $\boldsymbol{M}(\boldsymbol{b})$ are symmetric, positive definite, and differentiable, then the directional derivatives $\zeta_i'(\boldsymbol{b},\delta\boldsymbol{b})$ (i = 1, …, s) of a repeated eigenvalue $\zeta(\boldsymbol{b})$ in the direction $\delta\boldsymbol{b}$ exist, and are equal to the eigenvalues of matrix $\mathcal{M}$.

Proof. Since matrices $\boldsymbol{K}(\boldsymbol{b})$ and $\boldsymbol{M}(\boldsymbol{b})$ are positive definite, hence nonsingular, (4.134) may be rewritten as

$$\boldsymbol{C}(\boldsymbol{b})\boldsymbol{y}^i \equiv [\boldsymbol{K}^{-1}(\boldsymbol{b})\boldsymbol{M}(\boldsymbol{b})]\boldsymbol{y}^i = \frac{1}{\zeta}\boldsymbol{y}^i, \quad i = 1, \ldots, s, \tag{4.136}$$

where $(\boldsymbol{y}^i, \boldsymbol{M}\boldsymbol{y}^j) = \delta_{ij}$, and δ_{ij} is the Kronecker delta, which has a value of one if $i = j$, and a value of zero otherwise. Since $\boldsymbol{K}(\boldsymbol{b})$ and $\boldsymbol{M}(\boldsymbol{b})$ are differentiable with respect to $\boldsymbol{b}$, $\boldsymbol{C}(\boldsymbol{b})$ is also differentiable with respect to $\boldsymbol{b}$. In particular,

$$\boldsymbol{C}(\boldsymbol{b} + \tau\delta\boldsymbol{b}) = \boldsymbol{C}(\boldsymbol{b}) + \tau\left[\sum_l \frac{\partial \boldsymbol{C}}{\partial b_l}\delta b_l\right] + o(\tau), \tag{4.137}$$

where $o(\tau)$ denotes a quantity such that

$$\lim_{\tau \to 0} o(\tau)/\tau = 0.$$

By drawing on Theorem 5.11 in Chapter 2 of Kato [53], the following can be written:

$$\zeta_i(\boldsymbol{b} + \tau\delta\boldsymbol{b}) = \zeta_i + \tau\zeta_i'(\boldsymbol{b},\delta\boldsymbol{b}) + o(\tau), \quad i = 1, \ldots, s, \tag{4.138}$$

where $\zeta_i'(\boldsymbol{b},\delta\boldsymbol{b}) = -[\zeta(\boldsymbol{b})]^2\alpha_i'(\boldsymbol{b},\delta\boldsymbol{b})$ and $\alpha_i'(\boldsymbol{b},\delta\boldsymbol{b})$ are eigenvalues of the operator, where argument $\boldsymbol{b}$ has been suppressed for the sake of notational simplification. Now, define

$$N = P\left[\sum_l \frac{\partial C}{\partial b_l}\delta b_l\right]P, \tag{4.139}$$

where P is the M-orthogonal projection matrix that maps R^m onto the eigenspace

$$Y = \left\{ y \in R^m : y = \sum_{i=1}^{s} a_i y^i,\ a_i \text{ real} \right\}.$$

That is, for any $y \in R^m$

$$Py = \sum_{i=1}^{s} (My, y^i) y^i, \tag{4.140}$$

and the scalar product $(\bullet,\bullet)$ is defined as $(v,y) \equiv v^T y = \sum v_i y_i$.

The eigenvalues of the operator N must now be found. Each eigenvector of N can be expressed as

$$\bar{y} = \sum_{j=1}^{s} a_j y^j, \tag{4.141}$$

where not all the values of a_j are zero. Hence,

$$P\left[\sum_l \frac{\partial C}{\partial b_l}\delta b_l\right]P\bar{y} = \alpha'(b,\delta b)\bar{y},$$

or,

$$\sum_{j=1}^{s} a_j P\left[\sum_l \frac{\partial C}{\partial b_l}\delta b_l\right]Py^j = \alpha' \sum_{j=1}^{s} a_j y^j. \tag{4.142}$$

Taking the scalar product of (4.142) with My^i gives

$$\begin{aligned}\sum_{j=1}^{s} a_j \left(P[\sum_l \frac{\partial C}{\partial b_l}\delta b_l]Py^j, My^i \right) &= \alpha' \sum_{j=1}^{s} a_j \left(y^j, My^i \right) \\ &= \alpha' a_i, \quad i = 1, \ldots, s.\end{aligned} \tag{4.143}$$

To have a nontrivial solution a_j, α' must be an eigenvalue of the matrix

$$\hat{N}_{ij} = [(Ny^j, My^i)]_{s\times s}. \tag{4.144}$$

According to the definition of $C = K^{-1}M$ in (4.136),

$$\frac{\partial C}{\partial b_l} = K^{-1}\frac{\partial M}{\partial b_l} - K^{-1}\frac{\partial K}{\partial b_l}K^{-1}M. \tag{4.145}$$

Thus,

$$\hat{N} = \left[\left(My^i, PK^{-1}\sum_l \left(\frac{\partial M}{\partial b_l} - \frac{\partial K}{\partial b_l}C \right)\delta b_l Py^j \right) \right]_{s\times s}. \tag{4.146}$$

Since $Py^j = y^j$ and $Cy^j = (1/\zeta)y^j$,

$$\hat{N} = \left[\left(My^i, PK^{-1}\sum_l \left(\frac{\partial M}{\partial b_l}y^j - \frac{1}{\zeta}\frac{\partial K}{\partial b_l}y^j \right)\delta b_l \right) \right]_{s\times s}. \tag{4.147}$$

Note that for any vector $\boldsymbol{v} \in R^m$,

$$\begin{aligned}(\boldsymbol{My}^i, \boldsymbol{Pv}) &= \sum_{j=1}^{s} \left(\boldsymbol{My}^i, (\boldsymbol{Mv}, \boldsymbol{y}^j)\boldsymbol{y}^j \right) \\ &= (\boldsymbol{Mv}, \boldsymbol{y}^i) = (\boldsymbol{v}, \boldsymbol{My}^i) = \frac{1}{\zeta}(\boldsymbol{v}, \boldsymbol{Ky}^i).\end{aligned} \tag{4.148}$$

Applying this result to (4.147) gives

$$\begin{aligned}\hat{N} &= \left[\left(\frac{1}{\zeta}\boldsymbol{Ky}^i, \boldsymbol{K}^{-1} \sum_l \left(\frac{\partial \boldsymbol{M}}{\partial b_l}\boldsymbol{y}^j - \frac{1}{\zeta}\frac{\partial \boldsymbol{K}}{\partial b_l}\boldsymbol{y}^j \right) \delta b_l \right) \right]_{s\times s} \\ &= \left[\frac{1}{\zeta}\frac{\partial}{\partial \boldsymbol{b}}(\tilde{\boldsymbol{y}}^i, \boldsymbol{M}(\boldsymbol{b})\tilde{\boldsymbol{y}}^j)\delta\boldsymbol{b} - \frac{1}{\zeta^2}\frac{\partial}{\partial \boldsymbol{b}}(\tilde{\boldsymbol{y}}^i, \boldsymbol{K}(\boldsymbol{b})\tilde{\boldsymbol{y}}^j)\delta\boldsymbol{b} \right]_{s\times s}.\end{aligned} \tag{4.149}$$

Noting that $\alpha_i'(\boldsymbol{b},\delta\boldsymbol{b})$ are the eigenvalues of $\hat{N}$ and that $\zeta_i'(\boldsymbol{b},\delta\boldsymbol{b}) = -\zeta^2(\boldsymbol{b})\alpha_i'(\boldsymbol{b},\delta\boldsymbol{b})$, it can be concluded that $\zeta_i'(\boldsymbol{b},\delta\boldsymbol{b})$ are the eigenvalues of $\mathcal{M} = -\zeta^2(\boldsymbol{b})\hat{N}$, which gives

$$\mathcal{M} = \left[\frac{\partial}{\partial \boldsymbol{b}}(\tilde{\boldsymbol{y}}^i, \boldsymbol{K}(\boldsymbol{b})\tilde{\boldsymbol{y}}^j)\delta\boldsymbol{b} - \zeta\frac{\partial}{\partial \boldsymbol{b}}(\tilde{\boldsymbol{y}}^i, \boldsymbol{M}(\boldsymbol{b})\tilde{\boldsymbol{y}}^j)\delta\boldsymbol{b} \right]_{s\times s}. \tag{4.150}$$

Since this is the same matrix defined in (4.135), the proof of the theorem is complete. ■

The notation $\zeta_i'(\boldsymbol{b},\delta\boldsymbol{b})$ is selected from Theorem 4.2 to emphasize the directional derivative's dependence on $\delta\boldsymbol{b}$. It is not surprising that in the vicinity of a design in which the eigenvalue is repeated s times there may be s distinct eigenvalues. A remarkable fact implied by the preceding result is that the eigenvalues of matrix $\mathcal{M}$ do not depend on the $\boldsymbol{M}(\mathbf{b})$-orthonormal basis selected for the eigenspace. Moreover, if eigenvalues $\zeta_i(\boldsymbol{b}+\delta\boldsymbol{b})$ are ordered by increasing magnitude, then their directional derivatives are the eigenvalues of $\mathcal{M}$ in that same order.

In order to illustrate that the eigenvalue's directional derivatives are not generally linear in $\delta\boldsymbol{b}$, consider a double eigenvalue. For the case in which $s = 2$, the characteristic equation for determining the eigenvalues of $\mathcal{M}$ may be written as

$$\begin{vmatrix} \mathcal{M}_{11} - \zeta' & \mathcal{M}_{12} \\ \mathcal{M}_{21} & \mathcal{M}_{22} - \zeta' \end{vmatrix} = \mathcal{M}_{11}\mathcal{M}_{22} - \mathcal{M}_{12}^2 - (\mathcal{M}_{11} + \mathcal{M}_{22})\zeta' + (\zeta')^2 = 0, \tag{4.151}$$

where the fact that $\mathcal{M}_{12} = \mathcal{M}_{21}$ has been used. Solving this characteristic equation for ζ' provides a pair of roots that give the directional derivatives of the eigenvalue as

$$\zeta_i'(\boldsymbol{b}+\delta\boldsymbol{b}) = \frac{1}{2}\{(\mathcal{M}_{11} + \mathcal{M}_{22}) \pm [(\mathcal{M}_{11} + \mathcal{M}_{22})^2 - 4(\mathcal{M}_{11}\mathcal{M}_{22} - \mathcal{M}_{12}^2)]^{1/2}\},\quad i = 1, 2, \tag{4.152}$$

where $i = 1$ corresponds to the "–" sign, and $i = 2$ corresponds to the "+" sign.

This equation orders the directional derivatives of the repeated eigenvalue according to magnitude. Even though $\mathcal{M}_{ij}$ are linear in $\delta\boldsymbol{b}$, it is clear with this ordering that the resulting formula for $\zeta_i'(\boldsymbol{b},\delta\boldsymbol{b})$ is not linear in $\delta\boldsymbol{b}$; consequently, it is not a Fréchet derivative (see Appendix A.3). The fact that $\zeta_i'(\boldsymbol{b},\delta\boldsymbol{b})$ is not a Fréchet derivative may be because the eigenvalues have been ordered according to their magnitude. As indicated by the schematic diagram in Fig. 4.9, even if a smooth ordering of the eigenvalue exists with respect to the design, ordering the eigenvalue by magnitude leads to derivative discontinuity at the point of a repeated eigenvalue.

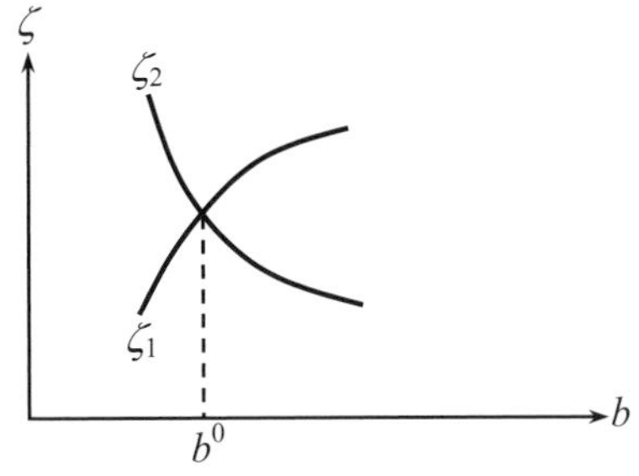

Figure 4.9. Schematic of eigenvalue crossing.

Computation of Directional Derivatives of a Repeated Eigenvalue

An ordering of $\zeta_i(\boldsymbol{b} + \tau\delta\boldsymbol{b})$ exists such that the mapping $\tau \rightarrow \zeta_i(\boldsymbol{b} + \tau\delta\boldsymbol{b})$ is differentiable at $\tau = 0$. In general, however, this ordering depends on $\delta\boldsymbol{b}$. In order to see the dependence, let $s = 2$. A method for determining directional derivatives was introduced by Masur and Mroz [66]. They used an orthogonal transformation of eigenvectors, beginning with a given $\mathcal{M}(\boldsymbol{b})$-orthonormal set $\boldsymbol{y}^1$ and $\boldsymbol{y}^2$ and defined a "rotated" set as follows:

$$\begin{aligned} \breve{\boldsymbol{y}}^1 &= \boldsymbol{y}^1 \cos\phi + \boldsymbol{y}^2 \sin\phi \\ \breve{\boldsymbol{y}}^2 &= -\boldsymbol{y}^1 \sin\phi + \boldsymbol{y}^2 \cos\phi \end{aligned} \tag{4.153}$$

where ϕ is a rotation parameter. A simple calculation shows that if $\boldsymbol{y}^1$ and $\boldsymbol{y}^2$ are $\boldsymbol{M}(\boldsymbol{b})$-orthonormal, then so are $\breve{\boldsymbol{y}}^1$ and $\breve{\boldsymbol{y}}^2$. The transformed eigenvectors may thus be used to evaluate matrix $\mathcal{M}$ in (4.135), denoted as $\breve{\mathcal{M}}$. Since the eigenvalues of $\breve{\mathcal{M}}$ are the same as those for $\mathcal{M}$, a rotation parameter ϕ may be chosen to make matrix $\mathcal{M}$ diagonal. If ϕ can be found, then the diagonal elements of $\breve{\mathcal{M}}$ will be the eigenvalues of the original matrix, and therefore, the directional derivatives of the repeated eigenvalue. Thus, it is a requirement that

$$\begin{aligned} 0 = \breve{\mathcal{M}}_{12} &= \frac{\partial}{\partial \boldsymbol{b}}[\breve{\boldsymbol{y}}^{1^T} \boldsymbol{K}(\boldsymbol{b})\breve{\boldsymbol{y}}^2]\delta\boldsymbol{b} - \zeta(\boldsymbol{b})\frac{\partial}{\partial \boldsymbol{b}}[\breve{\boldsymbol{y}}^{1^T} \boldsymbol{M}(\boldsymbol{b})\breve{\boldsymbol{y}}^2]\delta\boldsymbol{b} \\ &= -\cos\phi\sin\phi\mathcal{M}_{11} + (\cos^2\phi - \sin^2\phi)\mathcal{M}_{12} + \sin\phi\cos\phi\mathcal{M}_{22} \\ &= \tfrac{1}{2}\sin 2\phi(\mathcal{M}_{22} - \mathcal{M}_{11}) + \cos 2\phi\mathcal{M}_{12}. \end{aligned} \tag{4.154}$$

Equation (4.154) may be solved for

$$\phi = \phi(\delta\boldsymbol{b}) = \frac{1}{2}\text{Arctan}\left[\frac{2\mathcal{M}_{12}(\boldsymbol{y}^1, \boldsymbol{y}^2, \delta\boldsymbol{b})}{\mathcal{M}_{11}(\boldsymbol{y}^1, \boldsymbol{y}^1, \delta\boldsymbol{b}) - \mathcal{M}_{22}(\boldsymbol{y}^2, \boldsymbol{y}^2, \delta\boldsymbol{b})}\right], \tag{4.155}$$

where the notation emphasizes that ϕ depends on the direction of design change $\delta\boldsymbol{b}$. Even though $\mathcal{M}_{ij}$ linearly depend on $\delta\boldsymbol{b}$, their ratio on the right side of (4.155) is not linear in $\delta\boldsymbol{b}$. Furthermore, the arctangent function is nonlinear.

Angle ϕ may be used to evaluate $\breve{\mathcal{M}}_{11}$ and $\breve{\mathcal{M}}_{22}$ in order to obtain the directional derivatives of the repeated eigenvalue, that is,

$$\zeta_1'(\boldsymbol{b},\delta\boldsymbol{b}) = \breve{\mathcal{M}}_{11} = \cos^2\phi(\delta\boldsymbol{b})\mathcal{M}_{11}(\delta\boldsymbol{b}) + \sin 2\phi(\delta\boldsymbol{b})\mathcal{M}_{12}(\delta\boldsymbol{b}) + \sin^2\phi(\delta\boldsymbol{b})\mathcal{M}_{22}(\delta\boldsymbol{b}) \tag{4.156}$$

$$\zeta_2'(\boldsymbol{b},\delta\boldsymbol{b}) = \breve{\mathcal{M}}_{22} = \sin^2\phi(\delta\boldsymbol{b})\mathcal{M}_{11}(\delta\boldsymbol{b}) - \sin 2\phi(\delta\boldsymbol{b})\mathcal{M}_{12}(\delta\boldsymbol{b}) + \cos^2\phi(\delta\boldsymbol{b})\mathcal{M}_{22}(\delta\boldsymbol{b}), \tag{4.157}$$

where the notation $\mathcal{M}_{ij}(\delta\boldsymbol{b})$ is used to emphasize dependence on the design change. Note that even though $\mathcal{M}_{ij}(\delta\boldsymbol{b})$ is linear in $\delta\boldsymbol{b}$, since the trigonometric multipliers depend on $\delta\boldsymbol{b}$, the directional derivatives appearing in (4.156) and (4.157) are in general nonlinear in $\delta\boldsymbol{b}$. Thus, the directional derivatives of a repeated eigenvalue are not linear in $\delta\boldsymbol{b}$. Hence, ζ is nondifferentiable. Only if $\mathcal{M}_{12}(\delta\boldsymbol{b})$ is equal to zero for all $\delta\boldsymbol{b}$ with some pair of $\boldsymbol{M}(\boldsymbol{b})$-orthonormal eigenvectors can the repeated eigenvalues be ordered in such a way that they are Fréchet differentiable.

It may be noted in (4.155) that for $\tau \neq 0$, $\phi(\boldsymbol{b},\tau\delta\boldsymbol{b}) = \phi(\boldsymbol{b},\delta\boldsymbol{b})$, that is, $\phi(\boldsymbol{b},\delta\boldsymbol{b})$ is homogeneous of degree zero in $\delta\boldsymbol{b}$. Thus, since $\mathcal{M}_{ij}(\delta\boldsymbol{b})$ are linear in $\delta\boldsymbol{b}$,

$$\zeta_i'(\boldsymbol{b},\tau\delta\boldsymbol{b}) = \tau\zeta_i'(\boldsymbol{b},\delta\boldsymbol{b}), \tag{4.158}$$

that is, the directional derivatives of a repeated eigenvalue are homogeneous of degree one in $\delta\boldsymbol{b}$. This implies that once $\delta\boldsymbol{b}$ is fixed, the eigenvalues can be ordered in such a way that the repeated eigenvalue is differentiable with respect to τ.

While the foregoing approach could also be used to treat a triple eigenvalue, such an analysis would be much more complex. For example, the matrix would be 3 × 3, and a cubic characteristic equation would have to be solved. An alternative is to use a three-parameter family of $\boldsymbol{M}(\boldsymbol{b})$-orthonormal eigenfunctions and to choose three rotation parameters that would cause the off-diagonal terms of $\breve{\mathcal{M}}$ to be zero. This is a complicated task, since three trigonometric equations in three unknowns must be solved. While an analytical solution to the directional derivatives of eigenvalues with a multiplicity greater than two may be difficult, the same basic idea may be employed for numerical calculation.

Analysis Using a Generalized Global Stiffness and Mass Matrix

Consider the reduced formulation of the generalized eigenvalue problem in (4.98), given by (4.102), for a repeated eigenvalue problem with the nonsingular reduced stiffness and mass matrices $\hat{\boldsymbol{K}}(\boldsymbol{b})$ and $\hat{\boldsymbol{M}}(\boldsymbol{b})$, given in (4.103). Let c^i (i = 1, …, s) represent $\hat{\boldsymbol{M}}(\boldsymbol{b})$-orthonormal eigenvectors, with

$$\hat{\boldsymbol{K}}(\boldsymbol{b})\boldsymbol{c}^i = \zeta\hat{\boldsymbol{M}}(\boldsymbol{b})\boldsymbol{c}^i, \quad i = 1,\ldots,s. \tag{4.159}$$

Thus, the vectors $\boldsymbol{y}_g^i = \boldsymbol{\Phi}\boldsymbol{c}^i$ satisfy the relation

$$\delta_{ij} = \boldsymbol{c}^{i^T}\hat{\boldsymbol{M}}(\boldsymbol{b})\boldsymbol{c}^j = \boldsymbol{c}^{i^T}\boldsymbol{\Phi}^T\boldsymbol{M}_g(\boldsymbol{b})\boldsymbol{\Phi}\boldsymbol{c}^j = \boldsymbol{y}_g^{i^T}\boldsymbol{M}_g(\boldsymbol{b})\boldsymbol{y}_g^j, \tag{4.160}$$

where δ_{ij} is the Kronecker delta, so that $\boldsymbol{y}_g^i$ are $\boldsymbol{M}_g(\boldsymbol{b})$-orthonormal.

For the reduced eigenvalue equation of (4.159), use (4.135) to define

$$\hat{\mathcal{M}}_{ij} = \frac{\partial}{\partial\boldsymbol{b}}[\tilde{\boldsymbol{c}}^{i^T}\hat{\boldsymbol{K}}_g(\boldsymbol{b})\tilde{\boldsymbol{c}}^j]\delta\boldsymbol{b} - \zeta(\boldsymbol{b})\frac{\partial}{\partial\boldsymbol{b}}[\tilde{\boldsymbol{c}}^{i^T}\hat{\boldsymbol{M}}_g(\boldsymbol{b})\tilde{\boldsymbol{c}}^j]\delta\boldsymbol{b}. \tag{4.161}$$

Based on Theorem 4.2.2, the directional derivatives $\zeta_i'(\boldsymbol{b},\delta\boldsymbol{b})$ ($i = 1, \ldots, s$) of the repeated eigenvalue ζ from either (4.98), or (4.159), are the eigenvalues of $\hat{\mathcal{M}}$. Using $\boldsymbol{y}_g^i = \boldsymbol{\Phi}\boldsymbol{c}^i$ and (4.103),

$$\begin{aligned}\hat{\mathcal{M}}_{ij} &= \frac{\partial}{\partial \boldsymbol{b}}[\tilde{\boldsymbol{c}}^{i^T}\boldsymbol{\Phi}^T\boldsymbol{K}_g(\boldsymbol{b})\boldsymbol{\Phi}\tilde{\boldsymbol{c}}^j]\delta\boldsymbol{b} - \zeta(\boldsymbol{b})\frac{\partial}{\partial \boldsymbol{b}}[\tilde{\boldsymbol{c}}^{i^T}\boldsymbol{\Phi}^T\boldsymbol{M}_g(\boldsymbol{b})\boldsymbol{\Phi}\tilde{\boldsymbol{c}}^j]\delta\boldsymbol{b} \\ &= \frac{\partial}{\partial \boldsymbol{b}}[\boldsymbol{y}_g^{i^T}\boldsymbol{K}_g(\boldsymbol{b})\boldsymbol{y}_g^j]\delta\boldsymbol{b} - \zeta(\boldsymbol{b})\frac{\partial}{\partial \boldsymbol{b}}[\boldsymbol{y}_g^{i^T}\boldsymbol{M}_g(\boldsymbol{b})\boldsymbol{y}_g^j]\delta\boldsymbol{b}, \quad i, j = 1, \ldots, s.\end{aligned} \tag{4.162}$$

Thus, (4.155) through (4.157) are valid for the directional derivatives of a repeated eigenvalue, with $\mathcal{M}_{ij}$ being replaced by $\hat{\mathcal{M}}$; that is, $\mathcal{M}_{ij}$ is written in terms of the generalized global stiffness and mass matrices in (4.162).

4.2.6 Examples

Three-Bar Truss

To illustrate the results from previous sections, consider the three-bar truss in Section 4.1.6. In order to simplify the calculation, first consider the generalized global lumped mass matrix, given as

$$\boldsymbol{M}_g(\boldsymbol{b}) = \frac{\rho l}{2}\mathrm{diag}[b_1 + \sqrt{2}b_3, b_1 + \sqrt{2}b_3, b_1 + b_2, b_1 + b_2, b_2 + \sqrt{2}b_3, b_2 + \sqrt{2}b_3], \tag{4.163}$$

where ρ is mass density. The space of kinematically admissible displacements is

$$Z = \{\boldsymbol{z}_g \in R^6 : y_3 = y_4 = 0,\ y_5\cos\alpha + y_6\sin\alpha = 0\}, \tag{4.164}$$

and $\boldsymbol{K}_g(\boldsymbol{b})$ from (4.62) is positive definite on Z. If $\theta = 45^\circ$ and $\alpha = 30^\circ$, then with $\boldsymbol{y} = [y_1, y_2, y_5]^T$ the reduced mass matrix is

$$\boldsymbol{M}(\boldsymbol{b}) = \frac{\rho l}{2}\mathrm{diag}[b_1 + \sqrt{2}b_3, b_1 + \sqrt{2}b_3, 4(b_2 + \sqrt{2}b_3)]. \tag{4.165}$$

For the eigenvalue problem, assume that $E = 1$, $\rho = 1$, $b_1 = b_2 = 1$, and $b_3 = 2\sqrt{2}$. Consequently, the fundamental eigenvalue is $\zeta = 0.08038$ and the $\boldsymbol{M}(\boldsymbol{b})$-normalized eigenvector is

$$\boldsymbol{y} = [y_1,\ y_2,\ y_5]^T = [-0.3496,\ 0.08451,\ 0.2601]^T. \tag{4.166}$$

The reduced eigenvalue design sensitivity may now be evaluated from (4.97) as

$$\begin{aligned}\frac{\partial\zeta}{\partial \boldsymbol{b}} &= \frac{\partial}{\partial \boldsymbol{b}}[\tilde{\boldsymbol{y}}^T\boldsymbol{K}(\boldsymbol{b})\tilde{\boldsymbol{y}}] - \zeta\frac{\partial}{\partial \boldsymbol{b}}[\tilde{\boldsymbol{y}}^T\boldsymbol{M}(\boldsymbol{b})\tilde{\boldsymbol{y}}] \\ &= [0.001944,\ 0.05678,\ -0.02076].\end{aligned} \tag{4.167}$$

Even if a generalized global formulation is employed, the same eigenvalue is computed as in the reduced formulation. The $\boldsymbol{M}_g(\boldsymbol{b})$-normalized eigenvector is

$$\boldsymbol{y} = [-0.3496,\ 0.08451,\ 0,\ 0,\ 0.2601,\ -0.45051]^T. \tag{4.168}$$

The eigenvalue design sensitivity formula of (4.106), along with the variational formulation, gives

$$\begin{aligned}\frac{\partial \zeta}{\partial \boldsymbol{b}} &= \frac{\partial}{\partial \boldsymbol{b}}[\tilde{\boldsymbol{y}}_g^T \boldsymbol{K}_g(\boldsymbol{b})\tilde{\boldsymbol{y}}_g] - \zeta \frac{\partial}{\partial \boldsymbol{b}}[\tilde{\boldsymbol{y}}_g^T \boldsymbol{M}_g(\boldsymbol{b})\tilde{\boldsymbol{y}}_g] \\ &= [0.001944, \quad 0.05678, \quad -0.02076],\end{aligned} \tag{4.169}$$

which is the same as (4.167).

Since this example offers no evidence of a design that leads to a repeated eigenvalue, repeated eigenvalue sensitivity formulas have not been written.

Portal Frame

The portal frame shown in Fig. 4.10 is an example of a repeated eigenvalue that occurs at a given design. The structure is modeled using beam elements of length l_i and a uniform cross-sectional area b_i. No axial deformation is considered. The design problem is to find $\boldsymbol{b} \in R^n$ that minimizes the weight

$$\psi_0(\boldsymbol{b}) = \gamma \sum_{i=1}^{n} l_i b_i, \tag{4.170}$$

subject to natural frequency constraints

$$\psi_i = \zeta_0 - \zeta_i \le 0, \qquad i = 1, 2, \tag{4.171}$$

and constraints on the cross-sectional area

$$\psi_{j+2} = c_j - b_j \le 0, \qquad j = 1, 2, \ldots, n,$$

where γ is the weight density of the material, and $\zeta_i = \omega_i^2$.

Numerical results are based on the following data:

1. The length of each member of the portal frame is 10 inches,
2. The moment of inertia of the cross-sectional area is $I_i = \alpha b_i^2$,
3. The geometry of a cross section is circular ($\alpha = 0.08$),
4. Young's modulus of elasticity is $E = 10.3 \times 10^6$ psi,
5. The mass density of the material is $\gamma = 0.26163 \times 10^{-3}$ lb-sec^2/in^4.

The 18-element finite-element model presented in Fig. 4.10 is used in the computation, and the current design, which yields repeated eigenvalues of $\zeta_1 = 3.360591 \times 10^7$ and $\zeta_2 = 3.364971 \times 10^7$, is given in column (a) of Table 4.4.

Perturbation direction $\delta\boldsymbol{b}$, which is used in the calculation of directional derivatives $\zeta_i'(\boldsymbol{b},\delta\boldsymbol{b})$ in (4.156) and (4.157), is given in column (b) in the same table. A comparison of the design sensitivity between actual and predicted changes using the sensitivity formulas from (4.156) and (4.157) is presented in Table 4.5. Since $\zeta_i'(\boldsymbol{b},\delta\boldsymbol{b})$ ($i = 1, 2$) are nonlinear in $\delta\boldsymbol{b}$ for the current design, $d\psi_i/d\boldsymbol{b}$ cannot be found to calculate $\psi_i' = (d\psi_i / d\boldsymbol{b})\delta\boldsymbol{b}$.

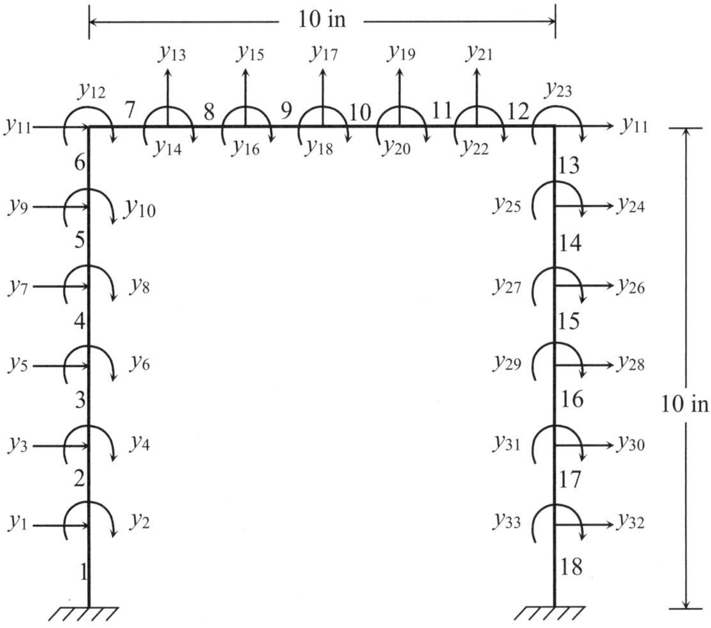

Figure 4.10. Eighteen-element model of portal frame.

Table 4.4. Current design and perturbation.

(a) Current Design		(b) Perturbation	
i	b_i	i	$\delta\mathbf{b}(i)$
1	0.6614E+01	1	0.6906E–01
2	0.4626E+01	2	0.4933E–01
3	0.2747E+01	3	0.2921E–01
4	0.1602E+01	4	0.7251E–02
5	0.9134E+00	5	0.4467E–02
6	0.3709E+00	6	0.1841E–02
7	0.3500E+00	7	0.0000E+00
8	0.3500E+00	8	0.0000E+00
9	0.3500E+00	9	0.0000E+00
10	0.3500E+00	10	0.0000E+00
11	0.3500E+00	11	0.0000E+00
12	0.3500E+00	12	0.0000E+00
13	0.3709E+00	13	–0.2025E–02
14	0.9134E+00	14	–0.5360E–02
15	0.1602E+01	15	–0.9426E–01
16	0.2747E+01	16	–0.4090E–01
17	0.4626E+01	17	–0.7400E–01
18	0.6614E+01	18	–0.1114E+00

Table 4.5. Comparison of sensitivity.

Constraint	$\Delta\psi$	ψ'	$\psi' / \Delta\psi \times 100\%$
ψ_1	–0.1875E+06	–0.2016E+06	107.5
ψ_2	0.8397E+05	0.9968E+05	118.7

4.3 Transient Dynamic Response Design Sensitivity

Thus far, in this chapter, static response and eigenvalues that represent steady-state motion and structural buckling have been treated. Under time-varying loads or nonzero initial conditions, transient dynamic response must also be considered. Design sensitivity analysis for transient dynamic response is presented in this section, first discussing the damping effect, and then concentrating on the results of an undamped structure, yielding substantial computational simplification. In terms of practical computation, the transient response can be approximated by using a linear combination of eigenvectors and/or Ritz vectors. Design sensitivity for this approximation provides an efficient and uncoupled matrix equation.

4.3.1 Design Sensitivity Analysis of Damped Elastic Structures

Consider a structure in which the generalized stiffness and mass matrices have been reduced by accounting for boundary conditions. Let the damping force be represented in the form of $\boldsymbol{C}(\boldsymbol{b})\boldsymbol{z}_{,t}$, where $\boldsymbol{z}_{,t} = d\boldsymbol{z}/dt$ denotes the velocity vector. Under these conditions, *Lagrange's equation of motion* becomes the second-order differential equation, as

$$\boldsymbol{M}(\boldsymbol{b})\boldsymbol{z}_{,tt} + \boldsymbol{C}(\boldsymbol{b})\boldsymbol{z}_{,t} + \boldsymbol{K}(\boldsymbol{b})\boldsymbol{z} = \boldsymbol{F}(t,\boldsymbol{b}), \tag{4.172}$$

with the initial conditions

$$\begin{aligned} \boldsymbol{z}(0) &= \boldsymbol{z}^0 \\ \boldsymbol{z}_{,t}(0) &= \boldsymbol{z}^0_{,t}. \end{aligned} \tag{4.173}$$

For the dynamic structure, the following form of a general performance measure will be considered:

$$\psi = g(\boldsymbol{z}(T),\boldsymbol{b}) + \int_0^T G(\boldsymbol{z},\boldsymbol{b})\, dt, \tag{4.174}$$

where the final time T is determined by a condition in the form

$$\Omega(\boldsymbol{z}(T), \boldsymbol{z}_{,t}(T), \boldsymbol{b}) = 0. \tag{4.175}$$

That is, given a specific design $\boldsymbol{b}$, the equation of motion in (4.172) and (4.173) can be integrated in order to monitor the value of $\Omega(\boldsymbol{z}(t), \boldsymbol{z}_{,t}(t),\boldsymbol{b})$. The time it takes for this quantity to reach zero is defined as final time T. The performance measure in (4.174) can then be evaluated. It is presumed that (4.175) uniquely determines T, at least locally. This requires that the time derivative of Ω is nonzero at T, as

$$\Omega_{,t} = \frac{\partial\Omega}{\partial\boldsymbol{z}}\boldsymbol{z}_{,t}(T) + \frac{\partial\Omega}{\partial\boldsymbol{z}_{,t}}\boldsymbol{z}_{,tt}(T) \neq 0. \tag{4.176}$$

When final time T is prescribed before the response analysis, the relation in (4.175) need not be considered.

It is clear from (4.172) that solution $z = z(t;\boldsymbol{b})$ of the initial-value problem in (4.172) and (4.173) depends on design variable $\boldsymbol{b}$. The nature of this dependence is characterized by the following well-known theorem from ordinary differential equation theory [67].

Theorem 4.3. If matrices $\boldsymbol{M}(\boldsymbol{b})$, $\boldsymbol{C}(\boldsymbol{b})$, and $\boldsymbol{K}(\boldsymbol{b})$ and vector $\boldsymbol{F}(t,\boldsymbol{b})$ are s times continuously differentiable with respect to design $\boldsymbol{b}$, and if matrix $\boldsymbol{M}(\boldsymbol{b})$ is nonsingular, then solution $z = z(t,\boldsymbol{b})$ is s times continuously differentiable with respect to $\boldsymbol{b}$.

Theorem 4.3 guarantees that the dynamic response of a structural system is essentially as smooth as the dependence on $\boldsymbol{b}$ in the equation of motion.

To obtain the design sensitivity of ψ, define a design variation in the form

$$\boldsymbol{b}_\tau = \boldsymbol{b} + \tau\delta\boldsymbol{b}. \tag{4.177}$$

Design $\boldsymbol{b}$ is perturbed in the direction of $\delta\boldsymbol{b}$ with the parameter τ. Substituting $\boldsymbol{b}_\tau$ into (4.174), the derivative of (4.174) can be evaluated with respect to τ at $\tau = 0$. Leibnitz's rule of differentiation of an integral [68] may be used to obtain the following expression:

$$\begin{aligned}\psi' &= \frac{\partial g}{\partial \boldsymbol{b}}\delta\boldsymbol{b} + \frac{\partial g}{\partial z}[z'(T) + z_{,t}(T)T'] \\ &\quad + G(z(T),\boldsymbol{b})T' + \int_0^T \left[\frac{\partial G}{\partial z}z' + \frac{\partial G}{\partial \boldsymbol{b}}\delta\boldsymbol{b}\right] dt,\end{aligned} \tag{4.178}$$

where

$$z' = z'(\boldsymbol{b},\delta\boldsymbol{b}) \equiv \frac{d}{d\tau} z(t,\boldsymbol{b} + \tau\delta\boldsymbol{b})\bigg|_{\tau=0} = \frac{d}{d\boldsymbol{b}}[z(t,\boldsymbol{b})]\delta\boldsymbol{b}$$

$$T' = T'(\boldsymbol{b},\delta\boldsymbol{b}) \equiv \frac{d}{d\tau} T(\boldsymbol{b} + \tau\delta\boldsymbol{b})\bigg|_{\tau=0} = \frac{dT}{d\boldsymbol{b}}\delta\boldsymbol{b}.$$

Note that since the expression in (4.175) that determines T depends on the design, T will also depend on the design. Thus, terms arise in (4.178) that involve the derivative of T with respect to the design. In order to eliminate these terms, differentiate (4.175) with respect to τ and evaluate it at $\tau = 0$ in order to obtain

$$\frac{\partial \Omega}{\partial z}[z'(T) + z_{,t}(T)T'] + \frac{\partial \Omega}{\partial z_{,t}}[z'_{,t}(T) + z_{,tt}(T)T'] + \frac{\partial \Omega}{\partial \boldsymbol{b}}\delta\boldsymbol{b} = 0. \tag{4.179}$$

This equation may also be written as

$$\Omega_{,t}T' = \left[\frac{\partial \Omega}{\partial z}z_{,t}(T) + \frac{\partial \Omega}{\partial z_{,t}}z_{,tt}(T)\right]T' = -\left(\frac{\partial \Omega}{\partial z}z'(T) + \frac{\partial \Omega}{\partial z_{,t}}z'_{,t}(T) + \frac{\partial \Omega}{\partial \boldsymbol{b}}\delta\boldsymbol{b}\right). \tag{4.180}$$

Since it is presumed by (4.176) that $\Omega_{,t} \neq 0$, then

$$T' = -\frac{1}{\Omega_{,t}}\left(\frac{\partial \Omega}{\partial z}z'(T) + \frac{\partial \Omega}{\partial z_{,t}}z'_{,t}(T) + \frac{\partial \Omega}{\partial \boldsymbol{b}}\delta\boldsymbol{b}\right). \tag{4.181}$$

Substituting the result of (4.181) into (4.178), the following is obtained:

$$
\begin{aligned}
\psi' = &\left[\frac{\partial g}{\partial z} - \left(\frac{\partial g}{\partial z} z_{,t}(T) + G(z(T),\boldsymbol{b}) \right) \frac{1}{\Omega_{,t}} \frac{\partial \Omega}{\partial z} \right] z'(T) \\
&- \left[\frac{\partial g}{\partial z} z_{,t}(T) + G(z(T),\boldsymbol{b}) \right] \frac{1}{\Omega_{,t}} \frac{\partial \Omega}{\partial z_{,t}} z'_{,t}(T) \\
&+ \int_0^T \left[\frac{\partial G}{\partial z} z' + \frac{\partial G}{\partial \boldsymbol{b}} \delta \boldsymbol{b} \right] dt \\
&+ \frac{\partial g}{\partial \boldsymbol{b}} \delta \boldsymbol{b} - \left[\frac{\partial g}{\partial z} z_{,t}(T) + G(z(T),\boldsymbol{b}) \right] \frac{1}{\Omega_{,t}} \frac{\partial \Omega}{\partial \boldsymbol{b}} \delta \boldsymbol{b}.
\end{aligned} \tag{4.182}
$$

Note that ψ' depends on z' and $z'_{,t}$ at T, as well as on z' within the integration.

In order to write ψ' in (4.182) explicitly in terms of a design variation, the adjoint variable technique employed in Sections 4.1.2 and 4.1.3 can be used. In the case of a dynamic system, all terms in (4.172) can be multiplied by $\boldsymbol{\lambda}^T(t)$ and integrated over the interval $[0,T]$, to obtain the following identity in $\boldsymbol{\lambda}$:

$$
\int_0^T \boldsymbol{\lambda}^T [\boldsymbol{M}(\boldsymbol{b}) z_{,tt} + \boldsymbol{C}(\boldsymbol{b}) z_{,t} + \boldsymbol{K}(\boldsymbol{b}) z - \boldsymbol{F}(t,\boldsymbol{b})]\, dt = 0. \tag{4.183}
$$

Since this equation must hold for arbitrary $\boldsymbol{\lambda}$, which is now taken to be independent of the design, substitute $\boldsymbol{b}_\tau$ into (4.183) and differentiate it with respect to τ in order to obtain the following relationship:

$$
\int_0^T [\boldsymbol{\lambda}^T \boldsymbol{M}(\boldsymbol{b}) z'_{,tt} + \boldsymbol{\lambda}^T \boldsymbol{C}(\boldsymbol{b}) z'_{,t} + \boldsymbol{\lambda}^T \boldsymbol{K}(\boldsymbol{b}) z' - \frac{\partial R}{\partial \boldsymbol{b}} \delta \boldsymbol{b}]\, dt = 0, \tag{4.184}
$$

where

$$
R = \tilde{\boldsymbol{\lambda}}^T \boldsymbol{F}(t,\boldsymbol{b}) - \tilde{\boldsymbol{\lambda}}^T \boldsymbol{M}(\boldsymbol{b}) \tilde{z}_{,tt} - \tilde{\boldsymbol{\lambda}}^T \boldsymbol{C}(\boldsymbol{b}) \tilde{z}_{,t} - \tilde{\boldsymbol{\lambda}}^T \boldsymbol{K}(\boldsymbol{b}) \tilde{z}, \tag{4.185}
$$

with the superposed tilde (~) denoting variables that are held constant during the differentiation with respect to the design in (4.184).

Since (4.184) contains the time derivatives of z', integrate the first two integrands by parts in order to move the time derivatives to $\boldsymbol{\lambda}$, as

$$
\begin{aligned}
&\boldsymbol{\lambda}^T(T) \boldsymbol{M}(\boldsymbol{b}) z'_{,t}(T) - \boldsymbol{\lambda}^T_{,t}(T) \boldsymbol{M}(\boldsymbol{b}) z'(T) + \boldsymbol{\lambda}^T \boldsymbol{C}(\boldsymbol{b}) z'(T) \\
&\quad + \int_0^T \left\{ [\boldsymbol{\lambda}^T_{,tt} \boldsymbol{M}(\boldsymbol{b}) - \boldsymbol{\lambda}^T_{,t} \boldsymbol{C}(\boldsymbol{b}) + \boldsymbol{\lambda}^T \boldsymbol{K}(\boldsymbol{b})] z' - \frac{\partial R}{\partial \boldsymbol{b}} \delta \boldsymbol{b} \right\} dt = 0.
\end{aligned} \tag{4.186}
$$

The adjoint variable method expresses the unknown terms in (4.182) in terms of the adjoint variable. Since (4.186) must hold for arbitrary functions $\boldsymbol{\lambda}(t)$, $\boldsymbol{\lambda}$ may be chosen so that the coefficients of terms involving $z'(T)$, $z'_{,t}(T)$, and z' in (4.182) and (4.186) are equal. If such a function $\boldsymbol{\lambda}(t)$ can be found, then the unwanted terms in (4.182) involving $z'(T)$, $z'_{,t}(T)$, and z' can be replaced by terms that explicitly depend on $\delta \boldsymbol{b}$ in (4.186). To be more specific, choose a $\boldsymbol{\lambda}(t)$ that satisfies the following:

$$
\boldsymbol{M}(\boldsymbol{b}) \boldsymbol{\lambda}(T) = - \left[\frac{\partial g}{\partial z} z_{,t}(T) + G(z(T),\boldsymbol{b}) \right] \frac{1}{\Omega_{,t}} \frac{\partial \Omega^T}{\partial z_{,t}} \tag{4.187}
$$

$$
\boldsymbol{M}(\boldsymbol{b}) \boldsymbol{\lambda}_{,t}(T) = \boldsymbol{C}^T(\boldsymbol{b}) \boldsymbol{\lambda}(T) - \frac{\partial g^T}{\partial z} + \left[\frac{\partial g}{\partial z} z_{,t}(T) + G(z(T),\boldsymbol{b}) \right] \frac{1}{\Omega_{,t}} \frac{\partial \Omega^T}{\partial z} \tag{4.188}
$$

$$M(b)\lambda_{,tt} - C^T(b)\lambda_{,t} + K(b)\lambda = \frac{\partial G^T}{\partial z}, \qquad 0 \le t \le T. \tag{4.189}$$

Note that once the dynamic equation of (4.172) and (4.173) is solved and (4.175) is used to determine T, then $z(T)$, $z_{,t}(T)$, $\partial\Omega/\partial z$, $\partial\Omega/\partial z_{,t}$, and $\Omega_{,t}$ may be evaluated. Equation (4.187) can then be solved for $\lambda(T)$ since the mass matrix $\boldsymbol{M(b)}$ is nonsingular. Having determined $\lambda(T)$, all terms on the right of (4.188) can be evaluated, and the equation can be solved for $\lambda_{,t}(T)$. Thus, a set of terminal conditions on λ has been determined. Since $\boldsymbol{M(b)}$ is nonsingular, (4.189) may then be integrated from T to 0, yielding the unique solution $\lambda(t)$. Taken as a whole, (4.187) through (4.189) may be thought of as a *terminal-value problem*.

Since the terms involving a variation in the state variable in (4.182) and (4.186) are identical, substitute (4.186) into (4.182) to obtain

$$\begin{aligned} \psi' &= \frac{\partial g}{\partial b}\delta b + \int_0^T \left[\frac{\partial G}{\partial b} + \frac{\partial R}{\partial b} \right] dt\, \delta b \\ &\quad - \left[\frac{\partial g}{\partial z} z_{,t}(T) + G(z(T), b) \right] \frac{1}{\Omega_{,t}} \frac{\partial \Omega}{\partial b} \delta b \\ &\equiv \frac{\partial \psi}{\partial b} \delta b. \end{aligned} \tag{4.190}$$

Every term in this equation can now be calculated. The terms $\partial g/\partial \boldsymbol{b}$, $\partial G/\partial \boldsymbol{b}$, and $\partial\Omega/\partial \boldsymbol{b}$ represent explicit partial derivatives with respect to the design. The term $\partial R/\partial \boldsymbol{b}$, however, must be evaluated from (4.185), thus requiring $\lambda(t)$. Note also that since design variation $\delta \boldsymbol{b}$ does not depend on time, it is taken outside the integral in (4.190).

Since (4.190) must hold for all $\delta \boldsymbol{b}$, the design derivative vector of ψ is

$$\begin{aligned} \frac{d\psi}{db} &= \frac{\partial g}{\partial b}(z(T), b) + \int_0^T \left[\frac{\partial G}{\partial b}(z, b) + \frac{\partial R}{\partial b}(\lambda(t), z(t), z_{,t}(t), z_{,tt}(t), b) \right] dt \\ &\quad - \frac{1}{\Omega_{,t}} \left[\frac{\partial g}{\partial z} z_{,t}(T) + G(z(T), b) \right] \frac{\partial \Omega}{\partial b}. \end{aligned} \tag{4.191}$$

The computational algorithm that leads to the determination of $d\psi/d\boldsymbol{b}$ requires that the initial-value problem in (4.172) and (4.173) be integrated forward in time from 0 to T. Then, the adjoint terminal-value problem presented by (4.187), (4.188), and (4.189) must be integrated backward in time from T to 0. Both sets of calculations can be done with a well-known numerical integration algorithm [49]. Once these initial- and terminal-value problems have been solved, the design derivative of ψ in (4.191) can then be evaluated using a numerical integration formula [49]. Although substantial numerical computation is required, it is clear that the design derivatives of the dynamic response can be computed.

4.3.2 Design Sensitivity Analysis of Undamped Structures

Consider the special case in which structural damping can be neglected, and the initial conditions are homogeneous. In such a case, the initial-value problem is reduced to

$$\begin{aligned} M(b) z_{,tt} + K(b) z &= F(t, b) \\ z(0) &= \mathbf{0} \\ z_{,t}(0) &= \mathbf{0}. \end{aligned} \tag{4.192}$$

While the theoretical considerations are identical to those in Section 4.3.1, an essential computational advantage exists in this formulation.

Consider the generalized eigenvalue problem associated with the initial-value problem in (4.192), written as

$$\boldsymbol{K}(\boldsymbol{b})\boldsymbol{\phi}^i = \zeta_i \boldsymbol{M}(\boldsymbol{b})\boldsymbol{\phi}^i, \qquad i = 1, \ldots, q \le m. \tag{4.193}$$

In general, a q number of calculated eigenvectors is substantially less than an m number of independent degrees-of-freedom in this equation. Furthermore, it is presumed that eigenvectors $\boldsymbol{\phi}^i$ are normalized by the condition

$$\boldsymbol{\phi}^{jT}\boldsymbol{M}(\boldsymbol{b})\boldsymbol{\phi}^i = \delta_{ij}, \qquad i, j = 1, \ldots, q. \tag{4.194}$$

Using these eigenvectors, it is possible to approximate the solution $\boldsymbol{z}(t)$ to (4.192) by using the eigenvector expansion method, as

$$\boldsymbol{z}(t) \approx \sum_{i=1}^{q} c_i(t)\boldsymbol{\phi}^i = \boldsymbol{\Phi}\boldsymbol{c}(t), \tag{4.195}$$

where $\boldsymbol{\Phi} = [\boldsymbol{\phi}^1, \ldots, \boldsymbol{\phi}^q]$ and $\boldsymbol{c} = [c_1, \ldots, c_q]^T$. Note that if $q = m$, then the solution $\boldsymbol{z}(t)$ can be precisely represented by (4.195). In contrast, it is conventional practice in structural dynamics to select q ($< m$) eigenvectors to efficiently approximate the solution. For a discussion of how to determine the number of eigenvectors to retain in a given problem, the reader is referred to Bathe [46].

Substituting (4.195) into differential equation (4.192), and premultiplying by $\boldsymbol{\Phi}^T$, the following system of differential equations for $\boldsymbol{c}(t)$ is obtained:

$$\boldsymbol{\Phi}^T\boldsymbol{M}(\boldsymbol{b})\boldsymbol{\Phi}\boldsymbol{c}_{,tt} + \boldsymbol{\Phi}^T\boldsymbol{K}(\boldsymbol{b})\boldsymbol{\Phi}\boldsymbol{c} = \boldsymbol{\Phi}^T\boldsymbol{F}(t,\boldsymbol{b}) \equiv \hat{\boldsymbol{F}}(t,\boldsymbol{b}), \tag{4.196}$$

where, since the eigenvectors are independent, the initial conditions in (4.192) become

$$\begin{aligned} \boldsymbol{c}(0) &= 0 \\ \boldsymbol{c}_{,t}(0) &= 0. \end{aligned} \tag{4.197}$$

Using the normalizing condition in (4.194), and the eigenvalue relation in (4.193), (4.196) can be reduced to

$$\boldsymbol{c}_{,tt} + \boldsymbol{\Lambda}\boldsymbol{c} = \hat{\boldsymbol{F}}(t,\boldsymbol{b}), \tag{4.198}$$

where

$$\boldsymbol{\Lambda} = \mathrm{diag}[\zeta_1, \ldots, \zeta_q]. \tag{4.199}$$

Since $\boldsymbol{\Lambda}$ is a diagonal matrix, (4.198) is uncoupled, and may be written in scalar form with the initial conditions provided by (4.197). This uncoupled system is given by

$$\begin{aligned} c_{i,tt} + \zeta_i c_i &= \hat{F}_i(t,\boldsymbol{b}) \\ c_i(0) &= 0 \qquad\qquad i = 1, \ldots, q. \\ c_{i,t}(0) &= 0 \end{aligned} \tag{4.200}$$

An explicit solution to each of these uncoupled initial-value problems may be written as

$$c_i(t) = \frac{1}{\sqrt{\zeta_i}} \int_t^T \sin[\sqrt{\zeta_i}(t-\tau)]\hat{F}_i(\tau,\boldsymbol{b})\,d\tau, \quad i = 1, \ldots, q. \tag{4.201}$$

This solution can be verified by differentiation and substitution into (4.200). Thus, in the case of an undamped structure with homogeneous initial conditions, an explicit solution to the dynamic problem may be obtained by evaluating $c_i(t)$ from (4.201) and substituting their values into (4.195).

The homogeneous initial conditions in (4.192) are not restrictive, since nonhomogeneous initial conditions $z(0) = z^0$ and $z_{,t}(0) = z^0_{,t}$ can be treated by defining the particular solution $z_p = z^0 + tz^0_{,t}$. If this particular solution is substituted into the initial-value problem in (4.172), then the same equation as (4.192) will be obtained, with the one additional term $-\boldsymbol{K}(\boldsymbol{b})z_p$ appearing on the right side of the differential equation.

For design sensitivity analysis, consider a special form of the performance measure in (4.174), with $g = 0$ and an explicitly given terminal time T. In this special case, the right sides of (4.187) and (4.188) vanish, and the adjoint terminal-value problem becomes

$$\begin{aligned} \boldsymbol{M}(\boldsymbol{b})\boldsymbol{\lambda}_{,tt} + \boldsymbol{K}(\boldsymbol{b})\boldsymbol{\lambda} &= \frac{\partial G^T}{\partial z}(t, z(t), \boldsymbol{b}), \\ \boldsymbol{\lambda}(T) &= 0, \\ \boldsymbol{\lambda}_{,t}(T) &= 0. \end{aligned} \tag{4.202}$$

Such assumptions are not restrictive, since in general nonhomogeneous terminal conditions $\boldsymbol{\lambda}(T) = \boldsymbol{\lambda}^0$ and $\boldsymbol{\lambda}_{,t}(T) = \boldsymbol{\lambda}^0_{,t}$ can be obtained from (4.187) and (4.188). In addition, the variables can be changed using the particular solution $\boldsymbol{\lambda}_p = \boldsymbol{\lambda}^0 + (t-T)\boldsymbol{\lambda}^0_{,t}$ in order to obtain homogeneous terminal conditions in (4.202), with an additional term $-\boldsymbol{K}(\boldsymbol{b})\boldsymbol{\lambda}_p$ on the right side of the differential equation. This special case avoids the algebra associated with this transformation.

Note that the left side of differential equation (4.202) is identical to the left side of differential equation (4.192). Thus, the eigenvector expansion technique may be employed, which uses precisely the same set of eigenvectors determined from (4.193) and (4.194). The adjoint variable is then approximated as

$$\boldsymbol{\lambda}(t) \approx \sum_{i=1}^{q} e_i(t)\boldsymbol{\phi}^i = \boldsymbol{\Phi}\boldsymbol{e}(t), \tag{4.203}$$

where $\boldsymbol{e} = [e_1, \ldots, e_q]^T$. Substituting this formula into (4.202), and premultiplying by $\boldsymbol{\Phi}^T$, the uncoupled terminal-valued problems are obtained as

$$\left.\begin{aligned} e_{i,tt} + \zeta_i e_i &= \frac{\partial G}{\partial z}(t, z(t), \boldsymbol{b})\boldsymbol{\phi}^i \\ e_i(T) &= 0 \\ e_{i,t}(T) &= 0 \end{aligned}\right\} \quad i = 1, \ldots, q. \tag{4.204}$$

By following the same procedure in (4.201), these equations may be solved in closed form to obtain

$$e_i(t) = \frac{1}{\sqrt{\zeta_i}} \int_t^T \sin[\sqrt{\zeta_i}(t-\tau)] \frac{\partial G}{\partial z}(\tau, z(\tau), \boldsymbol{b})\boldsymbol{\phi}^i \, d\tau, \qquad i = 1, \ldots, q. \tag{4.205}$$

The adjoint variable $\boldsymbol{\lambda}(t)$ may now be constructed from (4.203), and the design derivatives may be evaluated from (4.191). These results are of substantial practical importance, since structural damping may be neglected in many elastic structures, yielding a computationally efficient design sensitivity algorithm. Since structural damping effects are often approximated, such that the damping matrix $\boldsymbol{C}(\boldsymbol{b})$ is proportional to either the stiffness or the mass matrix [4] and [42], results are further generalized. Using such an approximation method, the foregoing sensitivity computation can be extended to treat structures with this special form of damping.

4.3.3 Modal Reduction Method Using Ritz Vectors

Even if the adjoint variable method developed in the previous two sections has computational advantages for a design problem with a large number of design variables, it suffers from several drawbacks when used for a dynamic problem. First, during the forward integration of dynamic equations, information has to be saved that can be later retrieved at an appropriate time step during the backward integration of adjoint equations, which may use different time steps for integration. Another drawback of this method is the difficulty of positive error control during the numerical integration of state and adjoint equations. Since backward integration can be started after forward integration is completed, it is hard to estimate how an error in forward integration will influence the solution during the backward integration process. Thus, from a computational point of view, the direct differentiation method, which will be developed in this section, is much more appropriate for a dynamic problem.

The accuracy and efficiency of dynamic analysis and design sensitivity analysis can be significantly improved by introducing the Ritz vector into the approximation of (4.195) and (4.203). In addition, the computational difficulties associated with the terminal-value problem of the adjoint equation can be lessened by using the direct differentiation method, since the same solution procedure used for a dynamic analysis can be used for design sensitivity purposes.

The eigenvector expansion method used in Section 4.3.2 is extended here, although in this example, structural damping exists. It is assumed that the terminal condition in (4.175) is prescribed, that is, terminal time T is fixed in advance. Solution $z(t)$ of the Lagrange equation of motion from (4.172) is approximated using a q number of eigenvectors and an r number of Ritz vectors, as

$$z(t) = \sum_{i=1}^{m} \boldsymbol{\phi}^i c_i = \boldsymbol{\Phi} \boldsymbol{c}, \tag{4.206}$$

where $\boldsymbol{\Phi} = [\boldsymbol{\phi}^k] = [\boldsymbol{y}^1, \ldots, \boldsymbol{y}^q, \boldsymbol{\psi}^1, \ldots, \boldsymbol{\psi}^r]$ is the basis matrix consisting of eigenvectors $\boldsymbol{y}^i$ and Ritz vectors $\boldsymbol{\psi}^j$. $\boldsymbol{y}^i$ and $\boldsymbol{\psi}^j$ satisfy the orthonormal condition in (4.120). Substituting (4.206) into (4.172), and premultiplying by $\boldsymbol{\Phi}^T$, we obtain the following coupled differential equation:

$$\boldsymbol{c}_{,tt}(t) + \boldsymbol{\Phi}^T \boldsymbol{C} \boldsymbol{\Phi} \boldsymbol{c}_{,t}(t) + \boldsymbol{\Phi}^T \boldsymbol{K} \boldsymbol{\Phi} \boldsymbol{c}(t) = \boldsymbol{\Phi}^T \boldsymbol{F}(t), \tag{4.207}$$

with initial conditions

$$\begin{aligned} \boldsymbol{c}(0) &= \boldsymbol{\Phi}^T \boldsymbol{M} z(0) \\ \boldsymbol{c}_{,t}(0) &= \boldsymbol{\Phi}^T \boldsymbol{M} z_{,t}(0). \end{aligned} \tag{4.208}$$

Note that (4.207) is an ordinary differential equation. For the solution procedure to (4.207), the reader is referred to the literature [67]. Once solution $\boldsymbol{c}(t)$ to (4.207) is obtained, transient response $z(t)$ can be approximated from (4.206).

Unlike the adjoint variable method, the direct differentiation method differentiates the dynamic (4.172) with respect to design $\boldsymbol{b}$, as

$$\boldsymbol{M}(\boldsymbol{b}) z'_{,tt} + \boldsymbol{C}(\boldsymbol{b}) z'_{,t} + \boldsymbol{K}(\boldsymbol{b}) z' = \boldsymbol{F}_f(t), \tag{4.209}$$

where

$$\boldsymbol{F}_f(t) = \frac{\partial \boldsymbol{F}(t)}{\partial \boldsymbol{b}}\delta\boldsymbol{b} - \frac{\partial}{\partial \boldsymbol{b}}(\boldsymbol{M}(\boldsymbol{b})\tilde{\boldsymbol{z}}_{,tt})\delta\boldsymbol{b} - \frac{\partial}{\partial \boldsymbol{b}}(\boldsymbol{C}(\boldsymbol{b})\tilde{\boldsymbol{z}}_{,t})\delta\boldsymbol{b} - \frac{\partial}{\partial \boldsymbol{b}}(\boldsymbol{K}(\boldsymbol{b})\tilde{\boldsymbol{z}})\delta\boldsymbol{b} \tag{4.210}$$

is the term explicitly dependent on the design. Equation (4.209) is the initial-value problem for $\boldsymbol{z}'(t)$, whose initial condition is given as

$$\begin{aligned} \boldsymbol{z}'(0) &= \boldsymbol{0} \\ \boldsymbol{z}'_{,t}(0) &= \boldsymbol{0}. \end{aligned} \tag{4.211}$$

Since sizing design variables are considered, the boundary condition is independent of the design.

To ensure efficient computation of $\boldsymbol{z}'(t)$, the same superposition method used in (4.206) can be employed, that is, let $\boldsymbol{z}'(t)$ be approximated by

$$\boldsymbol{z}'(t) = \sum_{i=1}^{k} \boldsymbol{\phi}^i v_i(t) = \boldsymbol{\Phi}\,\boldsymbol{v}(t), \qquad k < n, \tag{4.212}$$

where k is the number of basis vectors used for the sensitivity analysis. Note that m basis vectors are used for the response analysis.

Substituting (4.212) into (4.209), and premultiplying the resulting matrix equation by $\boldsymbol{\Phi}^T$, the following is obtained:

$$\boldsymbol{v}_{,tt}(t) + \boldsymbol{\Phi}^T\boldsymbol{C}\boldsymbol{\Phi}\,\boldsymbol{v}_{,t}(t) + \boldsymbol{\Phi}^T\boldsymbol{K}\boldsymbol{\Phi}\,\boldsymbol{v}(t) = \boldsymbol{\Phi}^T\boldsymbol{F}_f(t), \tag{4.213}$$

with initial conditions

$$\begin{aligned} \boldsymbol{v}(0) &= \boldsymbol{0} \\ \boldsymbol{v}_{,t}(0) &= \boldsymbol{0}. \end{aligned} \tag{4.214}$$

Note that dimension k in sensitivity (4.213) is not necessarily the same as dimension m in response (4.207). The coupled (4.213) can be solved using the direct integration method to obtain $\boldsymbol{v}(t)$. Once $\boldsymbol{v}(t)$ is obtained, approximated design sensitivity $\boldsymbol{z}'(t)$ can be obtained from (4.212). In addition, the design sensitivity of performance measures can be obtained using the chain rule of differentiation.

In the case of proportional damping, $\boldsymbol{v}(t)$ can be obtained from the uncoupled equation as

$$v_{i,tt} + 2\zeta_i\omega_i v_{i,t} + \omega_i^2 v_i = \boldsymbol{\phi}^{iT}\boldsymbol{F}_f(t) \tag{4.215}$$

instead of the coupled (4.213). For Rayleigh damping, the term $2\zeta_i\omega_i$ in (4.215) is replaced by $\alpha + \beta\omega_i^2$, where α and β are damping coefficients which define the damping matrix $\boldsymbol{C} = \alpha\boldsymbol{M} + \beta\boldsymbol{K}$.

4.3.4 Functionals in a Structural Dynamic Design

The general form of the cost or constraint functional in (4.174) can be used to approximate most quantities that measure structural response. Consider a response and design constraint that must hold for all times, that is,

$$\eta(\boldsymbol{z}(t), \boldsymbol{b}) \le 0, \qquad 0 \le t \le T. \tag{4.216}$$

Such constraints may be approximated in several ways. For example,

$$\psi_1 = \int_0^T [\eta(z,\boldsymbol{b}) + |\eta(z,\boldsymbol{b})|]\, dt = 0. \tag{4.217}$$

Equivalence between (4.216) and (4.217) for continuous functions is easily demonstrated. Use of the functional in (4.217) enables the constraint error in (4.216) to be reduced to near zero [63]. However, as the error approaches zero, the domain over which the integrand in (4.217) is defined is reduced to zero length, and a singular functional occurs. Such behavior limits the precision with which convergence can be obtained in structural optimization calculations.

An alternative treatment of the constraint in (4.216) is to define time t_1 when the left side of (4.216) reaches maximum value. It must satisfy the following condition:

$$\Omega(z(t_1), z_{,t}(t_1), \boldsymbol{b}) \equiv \frac{\partial \eta}{\partial z}(z(t_1), \boldsymbol{b}) z_{,t}(t_1) = 0. \tag{4.218}$$

The constraint of (4.216) may now be replaced by the equivalent constraint

$$\psi_2 = \eta(z(t_1), \boldsymbol{b}) \le 0. \tag{4.219}$$

This function is in the same form as (4.174), with terminal time t_1 determined by (4.218). Thus, the algorithm from the preceding section can be directly applied.

Finally, an averaging multiplier technique may be used in which a characteristic function $m(t,t_1)$ is defined as symmetric around point $t_1 < T$, and is shown to have a unit integral. Function m is defined on a small subdomain of the interval from 0 to T such that as the length of the subdomain approaches 0, m approaches the Dirac delta measure. The value of $\eta(z(t_1),\boldsymbol{b})$ may thus be approximated as

$$\psi_3 = \int_0^T m(t,t_1)\eta(z(t), \boldsymbol{b})\, dt \le 0, \tag{4.220}$$

where

$$\int_0^T m(t,t_1)\, dt = 1. \tag{4.221}$$

While some error is involved in the approximation of (4.220), good numerical results can be obtained by using a function m defined on a finite subdomain around time t_1, at which point the maximum displacement occurs for the nominal design. This formulation has the advantage that sensitivity of time t_1 with respect to the design need not be considered in approximate computations. Thus, only an integral constraint is involved in actual iterative calculations.

5
Continuum Sizing Design Sensitivity Analysis

In contrast to the matrix equation development of design sensitivity presented in Chapter 4, a distributed-parameter (continuum) approach is presented in this chapter. In the continuum method, the member-size parameters (thickness, cross-sectional shape, and moment of inertia) are distributed throughout the domain as functions. A design sensitivity theory for performance measures with respect to these continuous parameters will be developed. The principal distinction between the two approaches lies in the fact that the continuum method uses a displacement field that satisfies the boundary-value problem to characterize the structural deformation, while the matrix equation method relies on nodal displacement for such information.

While the finite-dimension and distributed-parameter approaches are related (the former is an approximation of the latter), both approaches have advantages and disadvantages. From an engineering viewpoint, the principal disadvantage of the distributed-parameter approach is that it requires a higher level of mathematical sophistication, which is associated with the infinite-dimensional function space of displacement and design. However, as will be seen in this and the subsequent two chapters, symmetry and positive definiteness of energy forms associated with elastic structures yield a complete theory that parallels the matrix theory in Chapter 4. The only real penalty associated with the distributed-parameter formulation is that a high level of complexity is required for the technical proofs.

There are several primary advantages of the distributed-parameter approach to structural design sensitivity analysis:

1. A rigorous mathematical theory is obtained, without the uncertainty that is associated with finite-dimensional approximation error, and

2. Explicit relations for design sensitivity are obtained in terms of physical quantities, rather than in terms of sums of derivatives of element matrices.

The former feature is of importance in the development of structural optimization theory, which has provided the principal motivation for theory development. The latter feature is yet to be fully explored in this chapter. The use of the results of this chapter for numerical calculation is discussed in Sections 5.1.4 and 5.2.3.

A final note is in order on the variational (virtual work) viewpoint that will be adopted in this and subsequent chapters. In Chapter 4, both matrix and variational approaches were seen as viable in treating finite-dimensional systems. The matrix approach was used for a reduced system of equations, while the variational approach was used for a generalized system of equations. However, in the distributed-parameter setting, only a variational approach is possible. The use of linear operator theory may be considered to parallel matrix theory, but the operator theory required is based on reducing each problem to its variational form, see [16], [17], [23], [37], [38], [44], and [53]. In fact, as design sensitivity theory develops, the elegance and practicality of the variational approach become increasingly apparent.

In Section 5.1, sizing design sensitivity analysis is developed for static problems. Various design components are considered in the continuum setting. Eigenvalue design sensitivity formulation is presented in Section 5.2, without solving the adjoint system. The lack of differentiability in the case of repeated eigenvalues is also discussed. In Section 5.3, design sensitivity analysis for transient dynamic response is developed without rigorous theoretical discussion. Transient response can be formulated in the frequency domain, as in Section 5.4, when harmonic excitation is applied to the structure. The idea behind Section 5.4 is further extended to structural-acoustic problems in Section 5.5. Since design sensitivity theory critically depends on the design choice, design parameterization for both line and surface design components is introduced in Section 12.1 of Chapter 12. The way design parameters are limited and linked is also discussed.

5.1 Design Sensitivity Analysis of Static Response

The design sensitivities of such structural performance measures as weight, displacement, compliance, and stress are developed with respect to the design variables defined in Section 5.1 using the continuum approach. As noted in Section 3.1 of Chapter 3, the solution to the static problem depends on the design. The differentiability of the solution with respect to the design, which is proved in [5], is employed in this section to derive a direct differentiation method and an adjoint variable method for the design sensitivity analysis of general functionals. An adjoint problem closely related to the original structural problem is obtained, and explicit formulas for structural response design sensitivity are likewise obtained. Using the finite element method, numerical methods for efficiently calculating design sensitivity coefficients are explained. The applications-oriented reader will note that virtually no knowledge of Sobolev space theory is required to implement this method.

5.1.1 Differentiability of Energy Bilinear Forms and Static Response

Design differentiability results for energy bilinear forms and the solution to the static problems are proved in [5]. These differentiability results are cited here to assist in the development of useful design sensitivity formulas. The rationale for not providing the proofs here is that those technical aspects that prove the existence of design derivatives do not provide any additional insight into the applicability of the adjoint variable technique. It is important to realize, however, that the sensitive question of the existence of design derivatives should not be ignored. Formal calculations with directional derivatives that may not exist are sure to lead to erroneous results. The occurrence of repeated eigenvalues and their lack of differentiability, discussed in the context of finite-dimensions in Chapter 4, as well as in Section 5.2, provide a graphic illustration of a very real structural problem in which the structural response is indeed not differentiable. Thus, the reader is cautioned to verify the regularity properties of solutions before using the results of formal calculations.

Let us begin with the definition of a variation that will be frequently used in the following derivations. As discussed in Section 12.1, $\boldsymbol{u}$ denotes a design vector function. Let ψ be a function that depends on current design $\boldsymbol{u}$ and assume that $\psi(\boldsymbol{u})$ is continuous with respect to design $\boldsymbol{u}$. If the current design is perturbed in the direction of $\delta\boldsymbol{u}$ (arbitrary), and τ is a parameter that controls the perturbation size, then the variation of $\psi(\boldsymbol{u})$ in the direction of $\delta\boldsymbol{u}$ is defined as

$$\psi'_{\delta u} \equiv \frac{d}{d\tau}\psi(\boldsymbol{u}+\tau\delta\boldsymbol{u})\bigg|_{\tau=0} = \frac{\partial\psi}{\partial\boldsymbol{u}}\delta\boldsymbol{u}. \tag{5.1}$$

Throughout this text, the prime symbol " ′ " plays precisely the same role as in Chapter 4 and is, in fact, the first variation in the calculus of variations [69]. For convenience, subscribed $\delta\boldsymbol{u}$ will often be ignored. The term "derivative" or "differentiation" will often be used to denote the variation in (5.1). If the variation of a function is continuous and linear with respect to $\delta\boldsymbol{u}$, then the function is differentiable (even more precisely, it is Fréchet differentiable). For complex structural problems, it is difficult to prove the differentiability of a general function with respect to the design. Readers are referred to [5]. However, for those readers who are application oriented, only the results of differentiability are described in this and subsequent sections.

As proved by Theorem 2.4.1 in [5], each of the energy bilinear forms encountered in Section 3.1 is differentiable with respect to the design, that is,

$$a'_{\delta u}(z,\bar{z}) \equiv \frac{d}{d\tau}a_{u+\tau\delta u}(\tilde{z},\bar{z})\bigg|_{\tau=0} \tag{5.2}$$

exists, where $\tilde{z}$ denotes the state variable z, with the dependence on τ being suppressed, and $\bar{z}$ is independent of τ. $a'_{\delta u}(z,\bar{z})$ is the first variation of the energy bilinear form $a_{\boldsymbol{u}}$ in the direction of $\delta\boldsymbol{u}$. This first variation is continuous and linear in $\delta\boldsymbol{u}$; hence, it is the Fréchet derivative (Appendix A.3) of $a_{\boldsymbol{u}}$ with respect to the design, and is evaluated in the direction of $\delta\boldsymbol{u}$. For proof of this result, readers are referred to [5].

The load linear form of the problems presented in Section 3.1 is also differentiable with respect to the design. More specifically,

$$\ell'_{\delta u}(\bar{z}) \equiv \frac{d}{d\tau}\ell_{u+\tau\delta u}(\bar{z})\bigg|_{\tau=0} \tag{5.3}$$

exists. As in the case of the energy bilinear form, the variation of the load linear form is linear in $\delta\boldsymbol{u}$. As in Chapter 4, the prime will be employed to denote the variation of the energy bilinear and load linear forms in (5.2) and (5.3), with explicit inclusion of argument $\delta\boldsymbol{u}$ to emphasize dependence on design variation.

A substantially more powerful result, derived from Theorem 2.4.3 in [5], is that the solution z to the state equations in Section 3.1, given here in the form

$$a_{\boldsymbol{u}}(z,\bar{z}) = \ell_{\boldsymbol{u}}(\bar{z}), \qquad \forall \bar{z} \in Z, \tag{5.4}$$

is differentiable with respect to the design in which Z is the space of kinematically admissible displacements, that is, the variation

$$z' = z'(\boldsymbol{x};\boldsymbol{u},\delta\boldsymbol{u}) \equiv \frac{d}{d\tau}z(\boldsymbol{x};\boldsymbol{u}+\tau\delta\boldsymbol{u})\bigg|_{\tau=0} \tag{5.5}$$

exists, and is the first variation of the solution to (5.4) at design $\boldsymbol{u}$ and in direction $\delta\boldsymbol{u}$ of the design change. Note that z' is a function of independent variable $\boldsymbol{x}$, and that it depends on design $\boldsymbol{u}$ and direction $\delta\boldsymbol{u}$. As shown in Theorem 2.4.3 in [5], z' is linear in $\delta\boldsymbol{u}$ and is in fact the Fréchet derivative of state variable z with respect to the design, evaluated in the direction of $\delta\boldsymbol{u}$. Proof of the validity of this result is not easily obtained, although it might seem intuitive that the state variable of a system should be smoothly dependent on the design.

An important property of the variation of the state variable, as defined in (5.5), is that the sequence in which variation and partial differentiation are taken is not important since they are interchangeable. In the case that the state variable belongs either to $H^1(\Omega)$ and $H^2(\Omega)$, or to the space of smoother functions, this statement implies that

$$\begin{aligned} (\nabla z)' &= \nabla z', & z &\in H^1(\Omega) \\ (\nabla(\nabla z))' &= \nabla(\nabla z'), & z &\in H^2(\Omega). \end{aligned} \tag{5.6}$$

This property is a direct extension of the well-known calculus of variations property, which states that variation and partial differentiation can be interchanged.

It is presumed throughout this chapter that boundary conditions are homogeneous and independent of the design, that is, they are in the form $Gz = 0$, where G is a differential operator that is independent of the design. Using (5.6), we can easily obtain $(Gz)' = Gz' = 0$. Thus, for the solution $z(\boldsymbol{x};\boldsymbol{u}) \in Z$ in (5.4), its variation $z' \in Z$ belongs to the same function space. This important fact will be frequently referred to in the following derivations.

Note that the energy bilinear form $a_{\boldsymbol{u}}(z,\bar{z})$ is linear in z and contains either a first- or a second-order derivative of z, depending on whether the Sobolev space is $H^1(\Omega)$ or $H^2(\Omega)$. Using these properties, one can use the chain rule of differentiation as well as the definition in (5.2) and (5.5), to obtain the following important formula:

$$\frac{d}{d\tau} a_{\boldsymbol{u}+\tau\delta\boldsymbol{u}}\left(z(\boldsymbol{x};\boldsymbol{u}+\tau\delta\boldsymbol{u}),\bar{z}\right)\Big|_{\tau=0} = a'_{\delta\boldsymbol{u}}(z,\bar{z}) + a_{\boldsymbol{u}}(z',\bar{z}). \tag{5.7}$$

The first term on the right side represents the explicit dependence of $a_{\boldsymbol{u}}$ on the design, whereas the second term represents the implicit dependence through the variation of the state variable.

For the first application of these forgoing definitions, one could apply the variation on both sides of (5.4) and use (5.7) for any fixed virtual displacement $\bar{z} \in Z$ to obtain

$$a_{\boldsymbol{u}}(z',\bar{z}) = \ell'_{\delta\boldsymbol{u}}(\bar{z}) - a'_{\delta\boldsymbol{u}}(z,\bar{z}), \qquad \forall \bar{z} \in Z. \tag{5.8}$$

Presuming that state variable z is the solution to (5.4), (5.8) is a variational equation with the same energy bilinear form for its first variation z'. Since (5.8) solves directly for z', it is called the *direct differentiation method*, as contrasted to the adjoint variable method to be presented in the next section. Noting that the right side of (5.8) is linear in $\bar{z}$, and that the energy bilinear form on the left is Z-elliptic, (5.8) has the unique solution z'. The fact that there is a unique solution to (5.8) agrees with the aforementioned statement that the design derivative of the state variable exists. Furthermore, if one selects the design change direction $\delta\boldsymbol{u}$, then (5.8) can be solved using the finite element method in order to numerically construct z', just as the basic state equation in (5.4) could be solved using the finite element method. However, numerical construction of the solution depends on the design change direction $\delta\boldsymbol{u}$, since $\delta\boldsymbol{u}$ appears on the right side of (5.8). Such calculations are unnecessary if one seeks the explicit forms of the design derivatives as a function of $\delta\boldsymbol{u}$.

5.1.2 Adjoint Variable Design Sensitivity Analysis

Next, consider a structural performance measure that can be written in integral form, as

$$\psi = \iint_{\Omega} g(z,\nabla z,\boldsymbol{u})\, d\Omega, \tag{5.9}$$

where for the present $z \in H^1(\Omega)$, $\nabla z = [z_{i,j}]$, and function g is continuously differentiable with respect to its arguments. Function g can be extended to function $z \in H^2(\Omega)$, in which case the second derivative of z will appear in the integrand. Such a situation will be treated as it appears in specific applications. Functionals in the form of (5.9) represent a wide variety of structural performance measures. For example, the volume of a structural component can be written with a g that depends explicitly on $\boldsymbol{u}$; averaged stress over a subset on a plane elastic solid can be written in terms of $\boldsymbol{u}$ and ∇z; and displacement at a point in a beam or plate can be written formally by using the Dirac delta measure multiplied by displacement. These and other examples will be treated in more detail in Section 5.1.3.

To develop the design sensitivity formula, let us take the variation of the functional in (5.9), as

$$\begin{aligned}\psi' &= \frac{d}{d\tau}\left[\iint_\Omega g(z(\boldsymbol{x};\boldsymbol{u}+\tau\delta\boldsymbol{u}),\nabla z(\boldsymbol{x};\boldsymbol{u}+\tau\delta\boldsymbol{u}),\boldsymbol{u}+\tau\delta\boldsymbol{u})\,d\Omega\right]\Bigg|_{\tau=0} \\ &= \iint_\Omega (g_{,z}z' + g_{,\nabla z} : \nabla z' + g_{,\boldsymbol{u}}\delta\boldsymbol{u})\,d\Omega,\end{aligned} \tag{5.10}$$

where the matrix calculus notation from Appendix A.1 is used, specifically

$$g_{,z} = \left[\frac{\partial g}{\partial z_1}\ \ \frac{\partial g}{\partial z_2}\ \ \frac{\partial g}{\partial z_3}\right],$$

$$g_{,\nabla z} = \left[\frac{\partial g}{\partial z_{i,j}}\right],$$

and ":" is a contraction operator such that $\boldsymbol{a} : \boldsymbol{b} = a_{ij}b_{ij}$. Leibnitz's rule allows the derivative with respect to τ to be taken inside the integral, and the chain rule of differentiation, along with (5.6) has been used to calculate the integrand of (5.10). From the definition of function g, the expressions of $g_{,z}$, $g_{,\nabla z}$, and $g_{,\boldsymbol{u}}$ are assumed to be available. Thus, z' and $\nabla z'$ need to be calculated in order to evaluate ψ'. Recall that z' and $\nabla z'$ depend on the design change direction $\delta\boldsymbol{u}$. The objective here is to obtain an explicit expression of ψ' in terms of $\delta\boldsymbol{u}$, which requires rewriting the first two terms on the right of (5.10) explicitly in terms of $\delta\boldsymbol{u}$.

Paralleling the method used for finite-dimensional structures in Section 4.1.3 of Chapter 4, an adjoint equation is introduced by replacing z' in (5.10) with a virtual displacement $\bar{\lambda}$ and by equating the terms involving $\bar{\lambda}$ in (5.10) to the energy bilinear form $a_{\boldsymbol{u}}(\lambda,\bar{\lambda})$, thus yielding the *adjoint equation* for the *adjoint variable* λ:

$$a_{\boldsymbol{u}}(\lambda,\bar{\lambda}) = \iint_\Omega (g_{,z}\bar{\lambda} + g_{,\nabla z} : \nabla\bar{\lambda})\,d\Omega, \qquad \forall\bar{\lambda} \in Z, \tag{5.11}$$

where the solution λ is desired. A simple application of the Schwartz inequality to the right side of (5.11) shows that it is a bounded linear functional of $\bar{\lambda}$ in the $H^1(\Omega)$ norm. Thus, according to the Lax-Milgram theorem [16], a unique solution to (5.11) exists, defined as the adjoint variable associated with the performance measure in (5.9).

The objective is to express the first two terms on the right of (5.10) in terms of the adjoint variable λ that was obtained from (5.11). Since (5.11) satisfies for all $\bar{\lambda} \in Z$, (5.11) may be evaluated at a specific $\bar{\lambda} = z'$, since $z' \in Z$. After substitution, we obtain

$$a_{\boldsymbol{u}}(\lambda,z') = \iint_\Omega (g_{,z}z' + g_{,\nabla z} : \nabla z')\,d\Omega, \tag{5.12}$$

which is nothing other than the terms in (5.10), which it is now desirable to write explicitly in terms of $\delta \boldsymbol{u}$. Similarly, the identity in (5.8) may be evaluated at a specific $\bar{z} = \lambda$, since both variables belong to Z, to obtain

$$a_u(z', \lambda) = \ell'_{\delta u}(\lambda) - a'_{\delta u}(z, \lambda). \tag{5.13}$$

Recalling that $a_u(\bullet,\bullet)$ is symmetric in its arguments, the left sides of (5.12) and (5.13) are equal, thus yielding the following desired relation:

$$\iint_\Omega (g_{,z} z' + g_{,\nabla z} : \nabla z')\, d\Omega = \ell'_{\delta u}(\lambda) - a'_{\delta u}(z, \lambda), \tag{5.14}$$

where the right side is linear in $\delta \boldsymbol{u}$ and can be evaluated once the state variable z and the adjoint variable λ are determined to be the solutions to (5.4) and (5.11), respectively. Substituting the result of (5.14) into (5.10), the explicit design sensitivity of ψ is obtained as

$$\psi' = \ell'_{\delta u}(\lambda) - a'_{\delta u}(z, \lambda) + \iint_\Omega g_{,u} \delta \boldsymbol{u}\, d\Omega, \tag{5.15}$$

where the first two terms on the right depend on the specific problem under investigation. This formula is applicable to any of the examples in Section 3.1.

Equation (5.15) will serve as the principal tool throughout the remainder of this section as well as in future applications for the design sensitivity analysis of functionals that represent structural response under a static load. This formula forms the basis for both analytical expressions of functional derivatives and numerical methods for calculating design sensitivity coefficients using the finite element method.

In design sensitivity analysis, a number of one-to-one correspondences can be found between the continuum formulation presented in this section and the discrete formulation presented in Section 4.1.3 of Chapter 4, as is illustrated in Table 5.1. In the continuum formulation, design is represented by vector function $\boldsymbol{u}$, whereas the discrete formulation

Table 5.1. Comparison of continuum and discrete formulations.

	Continuum Formulation	Discrete Formulation
Response analysis	$a_u(z, \bar{z}) = \ell_u(\bar{z})$	$\bar{z}_g^T \boldsymbol{K}_g z_g = \bar{z}_g^T \boldsymbol{F}_g$
Structural fictitious load	$a'_{\delta u}(z, \bar{z}) \equiv \dfrac{d}{d\tau} a_{u+\tau\delta u}(\tilde{z}, \bar{z})\Big\vert_{\tau=0}$	$\dfrac{\partial}{\partial \boldsymbol{b}}(\bar{z}_g^T \boldsymbol{K}_g \tilde{z}_g)\delta \boldsymbol{b}$
External fictitious load	$\ell'_{\delta u}(\bar{z}) \equiv \dfrac{d}{d\tau} \ell_{u+\tau\delta u}(\bar{z})\Big\vert_{\tau=0}$	$\dfrac{\partial}{\partial \boldsymbol{b}}(\bar{z}_g^T \boldsymbol{F}_g)\delta \boldsymbol{b}$
Design sensitivity equation	$a_u(z', \bar{z}) = \ell'_{\delta u}(\bar{z}) - a'_{\delta u}(z, \bar{z})$	$\bar{z}_g^T \boldsymbol{K}_g \dfrac{dz_g}{d\boldsymbol{b}} = \dfrac{\partial}{\partial \boldsymbol{b}}(\bar{z}_g^T \boldsymbol{F}_g) - \dfrac{\partial}{\partial \boldsymbol{b}}(\bar{z}_g^T \boldsymbol{K}_g \tilde{z}_g)^*$
Adjoint equation	$a_u(\lambda, \bar{\lambda}) = \iint_\Omega (g_{,z}\bar{\lambda} + g_{,\nabla z} : \nabla \bar{\lambda})\, d\Omega$	$\bar{\lambda}_g^T \boldsymbol{K}_g \lambda_g = \dfrac{\partial \psi}{\partial z_g} \bar{\lambda}_g$
Sensitivity of performance	$\psi' = \ell'_{\delta u}(\lambda) - a'_{\delta u}(z, \lambda) + \iint_\Omega g_{,u} \delta \boldsymbol{u}\, d\Omega$	$\dfrac{d\psi}{d\boldsymbol{b}} = \dfrac{\partial \psi}{\partial \boldsymbol{b}} - \dfrac{\partial}{\partial \boldsymbol{b}}[\tilde{\lambda}_g^T \boldsymbol{K}_g \tilde{z}_g - \tilde{\lambda}_g^T \boldsymbol{F}_g]^*$

* In these equations, $\delta \boldsymbol{b}$ is taken as a unit vector in the ith design coordinate direction as in (4.22) and (4.29).

uses discrete design vector $\boldsymbol{b}$. State variable z and adjoint variable λ are continuous functions in the domain; z_g and λ_g are vectors in the discretized domain. Even if two formulations appear similar, fundamental differences still exist on the level of theoretical completeness, accuracy, and efficiency in the sensitivity results.

5.1.3 Analytical Examples of Static Design Sensitivity

The beam, plate, and plane elasticity problems in Section 3.1 are used here as examples with which to derive design sensitivity formulas using the adjoint variable method. Computational considerations will be taken into account in subsequent sections.

Bending of a Beam

Consider the clamped beam in Fig. 3.2 with the cross-sectional area $A(x)$ and length l. For simplicity, let us presume that the cross-sectional dimensions are similar in all directions, such that the moment of inertia can be expressed as $I(x) = \alpha A^2(x)$ around the beam's neutral axis, and α is a positive constant. For example, when the solid, circular cross section in Table 12.1 of Chapter 12 is considered, $A = \pi r^2$ and $I = \pi r^4/4$, yielding $\alpha = 1/(4\pi)$. This simplification is necessary since the cross-sectional area and moment of inertia are dependent on each other. Thus, the independent design vector is obtained as $\boldsymbol{u} = [E, A(x)]^T$. In this formulation, the applied load on the beam in (3.9) is taken to reflect both the externally applied load $F(x)$ and the self-weight $\gamma A(x)$ per unit length, where γ is the weight density of the beam material. For these load components, the applied load can be written as

$$f(x) = F(x) + \gamma A(x). \tag{5.16}$$

From (3.14) and (3.15), the energy bilinear form and load linear form are defined as

$$a_{\boldsymbol{u}}(z,\bar{z}) = \int_0^l E\alpha A^2 z_{,11}\bar{z}_{,11}\,dx \tag{5.17}$$

and

$$\ell_{\boldsymbol{u}}(\bar{z}) = \int_0^l [F + \gamma A]\bar{z}\,dx. \tag{5.18}$$

Let the design be perturbed in the direction $\delta\boldsymbol{u} = [\delta E, \delta A]^T$ with parameter τ. Variations of the energy bilinear and load linear forms from (5.2) and (5.3) can be calculated as

$$\begin{aligned} a'_{\delta\boldsymbol{u}}(z,\bar{z}) &= \frac{d}{d\tau}\left[\int_0^l (E+\tau\delta E)\alpha(A+\tau\delta A)^2 z_{,11}\bar{z}_{,11}\,dx\right]\Bigg|_{\tau=0} \\ &= \int_0^l [\delta E\alpha A^2 + 2E\alpha A\delta A]z_{,11}\bar{z}_{,11}\,dx \end{aligned} \tag{5.19}$$

and

$$\begin{aligned} \ell'_{\delta\boldsymbol{u}}(\bar{z}) &= \frac{d}{d\tau}\left[\int_0^l [F + \gamma(A+\tau\delta A)]\bar{z}\,dx\right]\Bigg|_{\tau=0} \\ &= \int_0^l \gamma\delta A\bar{z}\,dx. \end{aligned} \tag{5.20}$$

Several alternative forms may now be considered as structural response functionals. First, consider the beam's weight, given as

$$\psi_1 = \int_0^l \gamma A \, dx. \tag{5.21}$$

A direct calculation of this variation yields

$$\psi_1' = \int_0^l \gamma \, \delta A \, dx. \tag{5.22}$$

Note that the direct variational calculation gives the explicit form of the structural weight variation in terms of the design variation. Consequently, no adjoint problem needs to be defined for this functional.

Now, let us consider a second functional that represents structural compliance, and can be defined as

$$\psi_2 = \int_0^l fz \, dx = \int_0^l [F + \gamma A] z \, dx. \tag{5.23}$$

Using the definition from (5.10), we can then take the variation of ψ_2 as

$$\psi_2' = \int_0^l [(F + \gamma A) z' + \gamma z \, \delta A] \, dx. \tag{5.24}$$

Note that the first term in the integrand implicitly depends on the design, while the second term is explicit. The adjoint equation may be defined from (5.11) with an implicitly dependent term, which in this case is

$$a_u(\lambda, \bar{\lambda}) = \int_0^l (F + \gamma A) \bar{\lambda} \, dx, \qquad \forall \bar{\lambda} \in Z. \tag{5.25}$$

Note that by interpreting $\bar{\lambda}$ and $\bar{z}$ as arbitrary, the load functional on the right side of (5.25) is in precisely the same form as the load functional for the original beam problem in (5.18). Since the original bilinear form $a_u(\bullet,\bullet)$ is Z-elliptic, (5.25) and the basic beam (3.16) have identical solutions. In this special case of a compliance functional, λ is the displacement of an adjoint beam that is not only identical to the original beam, but is in fact subjected to an identical load, such that $\lambda = z$. Thus, this problem is self-adjoint; there is no need to solve an additional adjoint problem. The explicit design sensitivity result of (5.15), using (5.19) and (5.20) with $z = \lambda$, thus becomes

$$\psi_2' = \int_0^l [2\gamma z - 2E\alpha A (z_{,11})^2] \delta A \, dx - \delta E \int_0^l \alpha A^2 (z_{,11})^2 \, dx. \tag{5.26}$$

The effect of variations can therefore be accounted for in the cross-sectional area and in Young's modulus of the system. It is interesting to note that the variation in Young's modulus may be taken outside the integral in (5.26).

As an example that can be calculated analytically, consider a uniform clamped-clamped beam with $l = 1$ m, $A = A_0 = 0.005$ m^2, $E = E_0 = 2 \times 10^5$ MPa, $\alpha = 1/6$, $F = 49.61$ kN/m, and $\gamma = 77{,}126$ N/m^3. Displacement under the given load is $z(x) = 2.5 \times 10^{-3}[x^2(1 - x)^2]$. Compliance sensitivity in (5.26) may thus be evaluated as

$$\psi_2' = \int_0^1 [385.6x^2(1-x)^2 - \tfrac{25000}{3}(6x^2 - 6x + 1)^2] \delta A \, dx - 2.08 \times 10^{-11} \delta E.$$

A graph of the coefficient of δA in the integral (Fig. 5.1) illustrates how the addition or deletion of material affects compliance. To decrease compliance most effectively, material should be removed in the vicinity of points $x = 0.2$ and 0.8 where sensitivity is small, and added instead to each end of the beam where sensitivity is large negative.

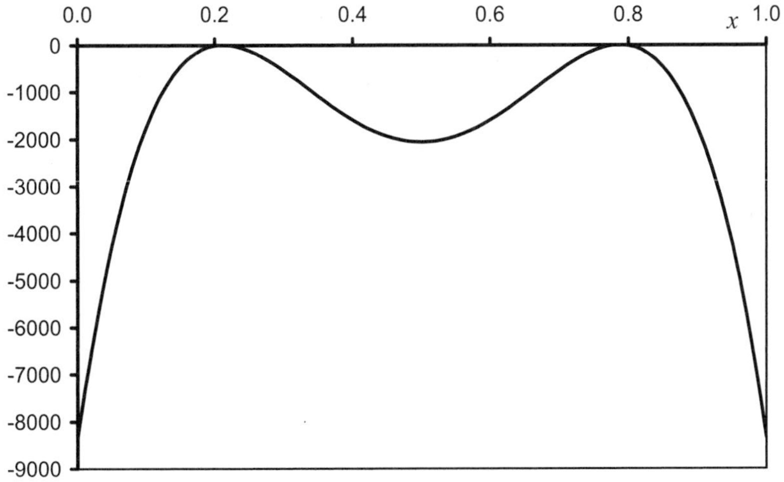

Figure 5.1. Compliance sensitivity $\Lambda^2 = 2\gamma z - 2E\alpha A(z_{,11})^2$.

The generalized result in (5.26) is applicable to arbitrary variation $\delta A(x)$ of the cross-sectional area along the beam. If, however, a parameterized distribution of material is considered along the beam, such as the stepped beam shown in Fig. 4.1, then the cross-sectional area function may be written in the same form as in Section 4.1.6, namely as

$$A(x) = b_i, \quad (i-1)/n < x < i/n, \tag{5.27}$$

where the beam, with l = 1, is subdivided into n sections, each with a constant cross-sectional area. The variation of the design function may thus be written directly as the variation of the design variable b_i, as

$$\delta A(x) = \delta b_i, \quad (i-1)/n < x < i/n. \tag{5.28}$$

This result may now be substituted directly into (5.26) to obtain the explicit design sensitivities that are associated with individual design variables, as

$$\psi_2' = \sum_{i=1}^{n}\left(\int_{(i-1)/n}^{i/n}[2\gamma z - 2E\alpha b_i(z_{,11})^2]dx\right)\delta b_i - \left(\sum_{i=1}^{n}\int_{(i-1)/n}^{i/n}\alpha b_i^2(z_{,11})^2\,dx\right)\delta E. \tag{5.29}$$

The design sensitivity coefficients are then evaluated by numerically calculating those integrals that depend exclusively on the solution to the state equation.

Note that the sensitivity result in (5.29) is the same as the result in (4.61), obtained from the finite-dimension design sensitivity method. That is, the sensitivity result in (4.61) is an approximation of the sensitivity result in (5.26).

Another important functional in the beam design is associated with the strength constraint, usually stated in terms of the allowable stress. Since there may be no continuous second-order derivatives of displacement with an arbitrary load distribution, a pointwise stress constraint may not be meaningful. Therefore, constraints on average stress over small subintervals of the beam are often imposed. Borrowing from elementary beam theory [70], the formula for bending stress is given as

$$\sigma(x) = \frac{h(x)}{2} E z_{,11}(x). \tag{5.30}$$

Since the maximum stress appears either at the top or bottom surface of the beam, the half-depth $h/2$ is used. However, h clearly depends on cross-sectional area A. Thus, for convenience, presume $h/2 = \beta A^{1/2}$, with a constant value for β. For example, if a circular cross section is considered with diameter h, then $\beta = \pi^{-1/2}$. However, such an assumption is restrictive, since for different cross sections, a different relation between cross-sectional area A and half-depth $h/2$ must be used. The averaged stress value over a small, open subinterval $(x_a, x_b) \subset (0, l)$ becomes

$$\psi_3 = \int_0^l \beta A^{1/2}(x) E z_{,11}(x) m_p(x)\, dx, \tag{5.31}$$

where m_p is a characteristic function that is independent of the design, and is only nonzero on the small subinterval (x_a, x_b), defined as

$$m_p = \begin{cases} \dfrac{1}{\int_{x_a}^{x_b} dx} & x \in (x_a, x_b) \\ 0 & x \notin (x_a, x_b). \end{cases}$$

Note that if stress is smooth and if the interval (x_a, x_b) approaches zero length, then m_p becomes the Dirac delta measure, and ψ_3 is the stress value evaluated at a given point. Note also that in the stress constraint formulation, the integrand includes a second-order derivative of the state variable that was not covered in the general derivation method proposed in Section 5.1.2. To illustrate the ease with which the adjoint method can be extended to second-order derivatives, it is first necessary to repeat the calculations that lead to (5.10), as

$$\psi_3' = \int_0^l \beta [A^{1/2} E z_{,11}' + A^{1/2} z_{,11} \delta E + \tfrac{1}{2} A^{-1/2} E z_{,11} \delta A] m_p(x)\, dx. \tag{5.32}$$

Using the same argument that led to a definition of the adjoint equation in (5.11), replace the state variation term z' on the right of (5.32) by virtual displacement $\bar{\lambda}$ to obtain the following adjoint problem:

$$a_u(\lambda, \bar{\lambda}) = \int_0^l \beta A^{1/2} E \bar{\lambda}_{,11} m_p\, dx, \quad \forall \bar{\lambda} \in Z. \tag{5.33}$$

In Sobolev space $H^2(0,l)$, the functional on the right side of (5.33) is a bounded linear functional. According to the Lax-Milgram theorem [16], (5.33) has a unique solution, denoted here as $\lambda^{(3)}$ where superscript (3) represents the association of λ with functional ψ_3. A direct repetition of the argument associated with (5.12) through (5.15) yields

$$\begin{aligned} \psi_3' = &\int_0^l [\tfrac{1}{2} \beta A^{-1/2} E z_{,11} m_p + \gamma \lambda^{(3)} - 2E\alpha A z_{,11} \lambda_{,11}^{(3)}] \delta A\, dx \\ &+ \delta E \int_0^l [\beta A^{1/2} z_{,11} - \alpha A^2 z_{,11} \lambda_{,11}^{(3)}] dx. \end{aligned} \tag{5.34}$$

To physically explain the adjoint problem, it may be helpful to rewrite the adjoint equation (5.33) more explicitly, using (5.17) for $a_u(\bullet,\bullet)$, as

$$\int_0^l E\alpha A^2 \{\lambda_{,11} - [\beta / \alpha A^{3/2}] m_p\} \bar{\lambda}_{,11}\, dx = 0, \quad \forall \bar{\lambda} \in Z.$$

This is exactly the same as the equation of virtual work for deflection λ of an adjoint beam with an initial curvature $[\beta/\alpha A^{3/2}]m_p$ and with no externally applied load. Such an interpretation of the adjoint equation (5.33) as an adjoint structure may be helpful in understanding the significance of λ from a physical point of view. As will be seen in Section 5.1.4, the solution to (5.33) can be efficiently carried out using the finite element method, but without using the idea of an adjoint structure. The concept of an adjoint structure has been introduced by Dems and Mroz [71] for a variety of structural optimization problems.

Note that (5.34) provides a linear first variation of the locally averaged stress functional as a variation of the cross-sectional distribution function A and of Young's modulus E. A parameterization of the cross-sectional area variation $A(x)$, such as the one shown in Fig. 4.1, could now be introduced in the sensitivity formula in (5.34), which would then be reduced exclusively to parameter variations.

Next, consider a special functional that defines the displacement value at an isolated point $\hat{x}$, that is,

$$\psi_4 \equiv z(\hat{x}) = \int_0^l \delta(x-\hat{x})z(x)\,dx, \tag{5.35}$$

where $\delta(x)$ is the Dirac delta measure at zero. According to the Sobolev imbedding theorem [22], it is apparent that this functional is continuous, and that the preceding analysis may be directly applied to interpret $\delta(x)$ as a function. The variation of this functional is thus written as

$$\psi_4' = \int_0^l \delta(x-\hat{x})z'(x)\,dx. \tag{5.36}$$

In this case, the adjoint equation can be obtained from (5.11) as

$$a_u(\lambda,\bar{\lambda}) = \int_0^l \delta(x-\hat{x})\bar{\lambda}\,dx, \qquad \forall\bar{\lambda}\in Z. \tag{5.37}$$

Since the right side of this equation defines a bounded linear functional in $H^2(0,l)$, a unique solution exists for the equation, denoted here as $\lambda^{(4)}$. Interpreting the Dirac delta measure as a unit load applied at point $\hat{x}$, a physical interpretation of $\lambda^{(4)}$ can be immediately obtained as the beam displacement due to a positive unit load at $\hat{x}$. Thus, in this case the adjoint beam is the same original beam with a different load.

The direct evaluation of the design sensitivity of ψ_4, using (5.15), (5.19), and (5.20), yields

$$\psi_4' = \int_0^l [\gamma\lambda^{(4)} - 2E\alpha A z_{,11}\lambda^{(4)}_{,11}]\delta A\,dx - \delta E\int_0^l [\alpha A^2 z_{,11}\lambda^{(4)}_{,11}]\,dx. \tag{5.38}$$

To illustrate how this result could be employed, consider the clamped-clamped beam examined earlier in this section. The solution to the state equation is $z = 2.5\times 10^{-3}[x^2(1-x)^2]$. If the design sensitivity of displacement at the center of the beam is desired, then $\hat{x} = ½$. Thus, the adjoint load in (5.37) is nothing but a unit point load at the center of the beam. The adjoint variable is thus obtained by solving the beam equation with this load, to obtain

$$\lambda^{(4)} = 2.5\times 10^{-8}[8 < x-\tfrac{1}{2} >^3 - 4x^3 + 3x^2],$$

where

$$< x-\tfrac{1}{2} > = \begin{cases} 0 & \text{for } 0\le x\le \tfrac{1}{2} \\ x-\tfrac{1}{2} & \text{for } \tfrac{1}{2}\le x\le 1. \end{cases}$$

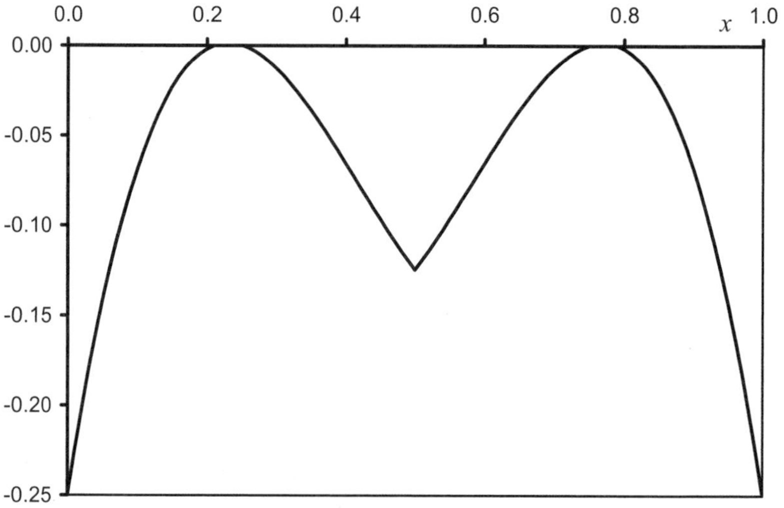

Figure 5.2. Displacement sensitivity $\Lambda^4 = \gamma\lambda^{(4)} - 2E\alpha Az_{,11}\lambda^{(4)}{}_{,11}$.

These expressions may be substituted into (5.38) to obtain the displacement sensitivity as

$$\begin{aligned}\psi_4' = \int_0^l [&1.93\times10^{-3}(8 < x-\tfrac{1}{2} >^3 -4x^3+3x^2)\\ &-2.5\times10^{-1}(6x^2-6x+1)(8<x-\tfrac{1}{2}>-4x+1)]\delta A\,dx\\ &-7.81\times10^{-16}\delta E.\end{aligned}$$

To determine how the material that is added to or deleted from the beam influences displacement at the center, the coefficient of δA can be graphed (Fig. 5.2). In order to decrease $z(\frac{1}{2})$ most effectively, material should be removed near $x = 0.22$ and 0.78 where the sensitivity is small, and added near $x = 0$ and 1 where the sensitivity is large negative.

As a final example of the beam problem, consider the slope of the beam at an isolated point $\hat{x}$ defined by the functional

$$\begin{aligned}\psi_5 \equiv z_{,1}(\hat{x}) &= \int_0^l \delta(x-\hat{x})z_{,1}(x)\,dx\\ &= -\int_0^l \delta_{,1}(x-\hat{x})z(x)\,dx.\end{aligned} \tag{5.39}$$

According to the Sobolev imbedding theorem [22], since ψ_5 is a continuous linear functional in $H^2(0,l)$, the preceding analysis may be applied. The last equality in (5.39) represents an integration by parts that defines the derivative of the Dirac delta measure. In beam theory, it is well known that the derivative of the Dirac delta measure is a point moment applied at point $\hat{x}$. The preceding analysis may now be directly repeated, replacing δ with $-\delta_{,1}$ to define the following adjoint equation:

$$a_u(\lambda,\bar{\lambda}) = -\int_0^l \delta_{,1}(x-\hat{x})\bar{\lambda}\,dx, \quad \forall\bar{\lambda}\in Z, \tag{5.40}$$

where the unique solution is denoted as $\lambda^{(5)}$. Physically, $\lambda^{(5)}$ is the displacement in an adjoint beam that is the original beam with a negative unit moment applied at point $\hat{x}$. As with the preceding process, the next step is evaluating (5.15), as

$$\psi_5' = \int_0^l [\gamma\lambda^{(5)} - 2E\alpha Az_{,11}\lambda_{,11}^{(5)}]\delta A\,dx - \delta E\int_0^l [\alpha A^2 z_{,11}\lambda_{,11}^{(5)}]dx. \tag{5.41}$$

It is interesting to note that for other boundary conditions in (3.19) through (3.21), the sensitivity formulas for ψ_1 through ψ_5 are still valid because, as mentioned in Section 3.1.2, the variational (5.4) is valid for all other boundary conditions.

To illustrate the use of (5.41), consider the same clamped-clamped beam previously discussed. If the design sensitivity of the slope at the beam center is desired, then the adjoint load in (5.40) is nothing but a negative unit moment at the beam center. Thus, the adjoint variable is obtained as

$$\lambda^{(5)} = 1.5\times10^{-7}[-4 < x - \tfrac{1}{2} >^2 + 2x^3 - x^2].$$

Equation (5.41) may now be evaluated to obtain

$$\begin{aligned}\psi_5' = \int_0^1 [&1.16\times10^{-2}(-4 < x - \tfrac{1}{2} >^2 + 2x^3 - x^2)\\ &-0.5(6x^2 - 6x + 1)(-4 < x - \tfrac{1}{2} >^0 + 6x - 1)]\delta A\,dx,\end{aligned}$$

where $<x - ½>^0 = 0$ if $x < ½$, and $<x - ½>^0 = 1$ if $x > ½$.

One interesting aspect of the sensitivity result is that the slope at the beam center, which has a present uniform design $A = 0.005$ m^2, is independent of variation δE. To see how material added to or deleted from the beam influences the slope at the center, the coefficient of δA may be graphed (Fig. 5.3). Figure 5.3 indicates that if material is added or removed symmetrically with respect to $\hat{x}$ = ½, then the slope remains at zero, which is obvious. Adding more material to the left of $\hat{x}$ = ½ increases the slope (note downward direction is the positive z direction as shown in Fig. 3.2), while adding more material to the right decreases it.

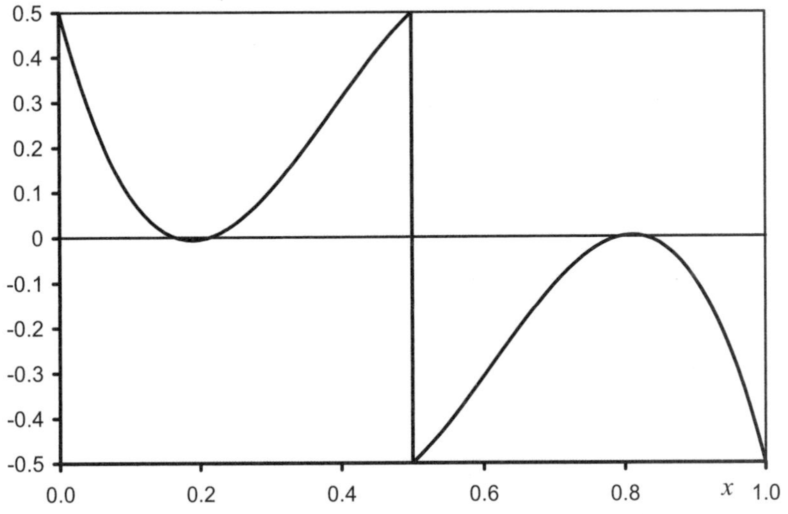

Figure 5.3. Slope sensitivity $\Lambda^5 = \gamma\lambda^{(5)} - 2E\alpha Az_{,11}\lambda^{(5)}{}_{,11}$.

Bending of a Plate

Consider the clamped plate in R^2 with variable thickness $h(\boldsymbol{x})$ and variable Young's modulus E, as was shown in Fig. 3.4. This plate has a distributed load, which consists of externally applied pressure $F(\boldsymbol{x})$ and self-weight, given by

$$f(\boldsymbol{x}) = F(\boldsymbol{x}) + \gamma h(\boldsymbol{x}), \tag{5.42}$$

where γ is the weight density of the plate. For this design-dependent load, the energy bilinear and the load linear forms, given in (3.41) and (3.42) respectively, are

$$a_{\boldsymbol{u}}(z,\bar{z}) = \iint_{\Omega} D(\boldsymbol{u})\boldsymbol{\kappa}(\bar{z})^T \boldsymbol{C}_S \boldsymbol{\kappa}(z)\, d\Omega \tag{5.43}$$

and

$$\ell_{\boldsymbol{u}}(\bar{z}) = \iint_{\Omega} [F + \gamma h]\bar{z}\, d\Omega, \tag{5.44}$$

where $\boldsymbol{u} = [E, h(\boldsymbol{x})]^T$, and where

$$\boldsymbol{\kappa}(z) = \begin{bmatrix} z_{,11} \\ z_{,22} \\ 2z_{,12} \end{bmatrix}, \quad D(\boldsymbol{u}) = \frac{Eh^3}{12(1-\nu^2)}, \quad \boldsymbol{C}_S = \begin{bmatrix} 1 & \nu & 0 \\ \nu & 1 & 0 \\ 0 & 0 & \frac{1}{2}(1-\nu) \end{bmatrix}. \tag{5.45}$$

Thus, design dependence is determined through the flexural rigidity $D(\boldsymbol{u})$. Note that the bending stiffness matrix $\boldsymbol{C}^b$ in (3.40) is represented by $\boldsymbol{C}^b = D(\boldsymbol{u})\boldsymbol{C}_S$ to separate design-dependent $D(\boldsymbol{u})$ from the stiffness matrix.

First, consider the functional defining weight of the plate, as

$$\psi_1 = \iint_{\Omega} \gamma h\, d\Omega. \tag{5.46}$$

Taking a direct variation of this weight yields

$$\psi_1' = \iint_{\Omega} \gamma\, \delta h\, d\Omega. \tag{5.47}$$

Since no variation of the state variable appears in this expression, no adjoint problem needs to be defined, and the explicit design derivative of the weight is obtained.

Next, consider the compliance functional of the plate, as

$$\psi_2 = \iint_{\Omega} [F + \gamma h] z\, d\Omega. \tag{5.48}$$

If the first variation is taken, then the equation yields

$$\psi_2' = \iint_{\Omega} [(F + \gamma h)z' + \gamma z\, \delta h]\, d\Omega. \tag{5.49}$$

Note that the first term in the integrand implicitly depends on the design, while the second term is explicit. Following the procedure in (5.11), one can then define the adjoint equation as

$$a_{\boldsymbol{u}}(\lambda,\bar{\lambda}) = \iint_{\Omega} (F + \gamma h)\bar{\lambda}\, d\Omega, \qquad \forall \bar{\lambda} \in Z. \tag{5.50}$$

Note that (5.50) is identical to plate equation (3.43) for displacement. Therefore, the adjoint plate and load are identical to the original problem $\lambda = z$, and (5.50) does not need to be solved separately.

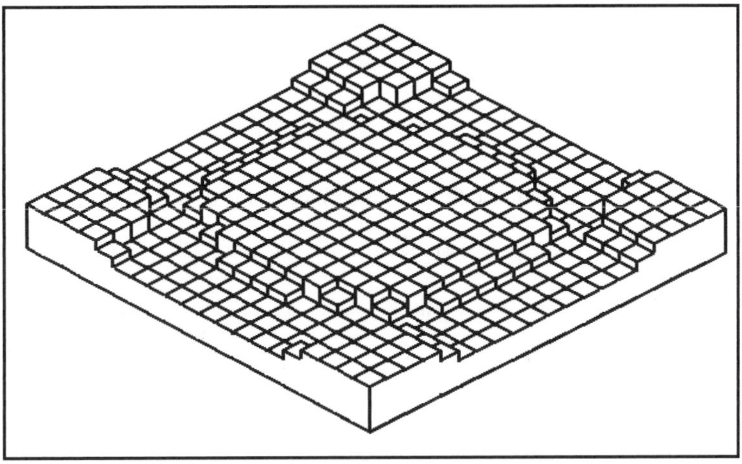

Figure 5.4. Piecewise uniform plate.

As a step in preparing for the evaluation of design sensitivity, the definitions in (5.2) and (5.3) for the plate problem are used to obtain the variations of the energy bilinear form and the load linear form, as

$$\begin{aligned} a'_{\delta u}(z,\bar{z}) &= \iint_{\Omega} \frac{Eh^2\,\delta h}{4(1-\nu^2)} \boldsymbol{\kappa}(\bar{z})^T \boldsymbol{C}_S \boldsymbol{\kappa}(z)\, d\Omega \\ &+ \delta E \iint_{\Omega} \frac{h^3}{12(1-\nu^2)} \boldsymbol{\kappa}(\bar{z})^T \boldsymbol{C}_S \boldsymbol{\kappa}(z)\, d\Omega \end{aligned} \tag{5.51}$$

and

$$\ell'_{\delta u}(\bar{z}) = \iint_{\Omega} \gamma \bar{z}\, \delta h\, d\Omega. \tag{5.52}$$

Direct application of (5.15) to compute the sensitivity of ψ_2 with $\lambda = z$ yields

$$\begin{aligned} \psi'_2 &= \iint_{\Omega} \left\{ 2\gamma z - \frac{Eh^2}{4(1-\nu^2)} \boldsymbol{\kappa}(z)^T \boldsymbol{C}_S \boldsymbol{\kappa}(z) \right\} \delta h\, d\Omega \\ &- \left\{ \iint_{\Omega} \frac{h^3}{12(1-\nu^2)} \boldsymbol{\kappa}(z)^T \boldsymbol{C}_S \boldsymbol{\kappa}(z)\, d\Omega \right\} \delta E. \end{aligned} \tag{5.53}$$

As in the case of beam compliance, it should be observed that this sensitivity result consists of both a first term, which accounts for the effect of variation $\delta h(x)$ on the plate shape function $h(x)$, and a second term, which is a scalar multiplied by the variation δE.

Consider an application of (5.53) to a plate of piecewise constant thickness (Fig. 5.4), where b_i is the constant thickness of the ith rectangular element. The thickness function is thus parameterized as

$$h(\boldsymbol{x}) = b_i, \qquad \boldsymbol{x} \in \Omega_i, \tag{5.54}$$

where Ω_i is the ith rectangular element in Fig. 5.4. The thickness variation δh is thus written as $\delta h(x) = \delta b_i$, $(x \in \Omega_i)$. Consequently, (5.53) becomes

$$\psi_2' = \sum_{i=1}^{n}\left[\iint_{\Omega}\left\{2\gamma z - \frac{Eb_i^2}{4(1-\nu^2)}\boldsymbol{\kappa}(z)^T \boldsymbol{C}_S \boldsymbol{\kappa}(z)\right\} d\Omega\right]\delta b_i$$
$$-\left[\sum_{i=1}^{n}\iint_{\Omega}\frac{b_i^3}{12(1-\nu^2)}\boldsymbol{\kappa}(z)^T \boldsymbol{C}_S \boldsymbol{\kappa}(z)\, d\Omega\right]\delta E. \tag{5.55}$$

Now, consider a plate response functional defined as the displacement at a point $\hat{\boldsymbol{x}}$,

$$\psi_3 \equiv z(\hat{\boldsymbol{x}}) = \iint_{\Omega}\delta(\boldsymbol{x}-\hat{\boldsymbol{x}})z(\boldsymbol{x})\,d\Omega, \tag{5.56}$$

where $\delta(\boldsymbol{x})$ is the Dirac delta measure in the plane, acting at the origin. According to the Sobolev imbedding theorem [22], we see that this functional is continuous, and the foregoing analysis applies. By taking the first variation of (5.56), the following result is produced:

$$\psi_3' = \iint_{\Omega}\delta(\boldsymbol{x}-\hat{\boldsymbol{x}})z'(\boldsymbol{x})\,d\Omega. \tag{5.57}$$

Following the general adjoint formulation from (5.11), the adjoint equation is defined as

$$a_{\boldsymbol{u}}(\lambda,\bar{\lambda}) = \iint_{\Omega}\delta(\boldsymbol{x}-\hat{\boldsymbol{x}})\bar{\lambda}\,d\Omega, \qquad \forall\bar{\lambda}\in Z. \tag{5.58}$$

This equation has a unique solution, denoted as $\lambda^{(3)}$. Since the load functional on the right side of (5.58) is physically interpreted as a unit point load that acts at point $\hat{\boldsymbol{x}}$, the solution $\lambda^{(3)}$ to the adjoint plate problem is simply the original plate displacement due to this load.

Since $\lambda^{(3)}$ has been determined, the general result of (5.15) may now be applied, using the variations in bilinear and linear forms defined in (5.51) and (5.52), to obtain

$$\psi_3' = \iint_{\Omega}\left\{\gamma\lambda^{(3)} - \frac{Eh^2}{4(1-\nu^2)}\boldsymbol{\kappa}(\lambda^{(3)})^T \boldsymbol{C}_S \boldsymbol{\kappa}(z)\right\}\delta h\, d\Omega$$
$$-\left\{\iint_{\Omega}\frac{h^3}{12(1-\nu^2)}\boldsymbol{\kappa}(\lambda^{(3)})^T \boldsymbol{C}_S \boldsymbol{\kappa}(z)\,d\Omega\right\}\delta E. \tag{5.59}$$

The stress performance measure is frequently used as a design optimization criterion. The maximum stress for a plate occurs on the plate surface and is given in the following form [23]:

$$\begin{aligned}\sigma_{11} &= -\frac{Eh}{2(1-\nu^2)}(z_{,11}+\nu z_{,22})\\ \sigma_{22} &= -\frac{Eh}{2(1-\nu^2)}(z_{,22}+\nu z_{,11})\\ \sigma_{12} &= -\frac{Eh}{2(1+\nu)}z_{,12}.\end{aligned} \tag{5.60}$$

The von Mises stress [23] is written as

$$g(\boldsymbol{\sigma}) = \frac{1}{2}[(\sigma_{11}+\sigma_{22})^2 + 3(\sigma_{11}-\sigma_{22})^2 + 12(\sigma_{12})^2]^{1/2}. \tag{5.61}$$

For simplicity, assume that stress σ_{11} in (5.60) is taken as a strength constraint instead of the von Mises stress. Once the design sensitivity expression has been obtained in this simplified form, the idea can be extended to the von Mises stress without any theoretical difficulty. As with the beam problem, it is necessary to define a characteristic function $m_p(\boldsymbol{x})$ as an averaging multiplier, which is nonzero only on the open, small region Ω_p of Ω, and which has an integral of one. Then, the average value of σ_{11} over this small region is

$$\begin{aligned}\psi_4 &= \iint_\Omega \sigma_{11} m_p \, d\Omega \\ &= -\frac{1}{2(1-\nu^2)} \iint_\Omega Eh(z_{,11} + \nu z_{,22}) m_p \, d\Omega.\end{aligned} \tag{5.62}$$

As in the case of a beam stress functional, take the variation of functional ψ_4 to obtain

$$\psi_4' = -\frac{1}{2(1-\nu^2)} \iint_\Omega [Eh(z'_{,11} + \nu z'_{,22}) + h(z_{,11} + \nu z_{,22})\delta E + E(z_{,11} + \nu z_{,22})\delta h] m_p \, d\Omega. \tag{5.63}$$

Replace the state variation term z' on the right side of (5.63) by virtual displacement to obtain the following adjoint equation:

$$a_u(\lambda, \bar{\lambda}) = -\frac{1}{2(1-\nu^2)} \iint_\Omega Eh(\bar{\lambda}_{,11} + \nu \bar{\lambda}_{,22}) m_p \, d\Omega, \qquad \forall \bar{\lambda} \in Z. \tag{5.64}$$

It can be shown that the right side of (5.64) is a bounded linear functional. Hence, according to the Lax-Milgram theorem [16], (5.64) has the unique solution $\lambda^{(4)}$. By using the same procedure that was outlined in (5.12) through (5.15), the following formula can be derived:

$$\begin{aligned}\psi_2' = &\iint_\Omega \left\{ -\frac{E}{2(1-\nu^2)}(z_{,11} + \nu z_{,22}) m_p + \gamma \lambda^{(4)} \right. \\ &\left. - \frac{Eh^2}{4(1-\nu^2)} \boldsymbol{\kappa}(\lambda^{(4)})^T \boldsymbol{C}_S \boldsymbol{\kappa}(z)] \right\} \delta h \, d\Omega \\ - &\iint_\Omega \left\{ \frac{h}{2(1-\nu^2)}(z_{,11} + \nu z_{,22}) m_p \right. \\ &\left. + \frac{h^3}{12(1-\nu^2)} \boldsymbol{\kappa}(\lambda^{(4)})^T \boldsymbol{C}_S \boldsymbol{\kappa}(z) \right\} d\Omega \, \delta E.\end{aligned} \tag{5.65}$$

Plane Elasticity

At this point, let us turn to the plane elastic slab, which was first treated in Section 3.1.4. From (3.75) and (3.59), the energy bilinear form and a load linear form without surface traction are given as follows:

$$a_u(z, \bar{z}) = \iint_\Omega h(\boldsymbol{x}) \varepsilon(\bar{z})^T \boldsymbol{C} \varepsilon(z) \, d\Omega \tag{5.66}$$

and

$$\ell_u(\bar{z}) = \iint_\Omega h(\boldsymbol{x}) \bar{z}^T \boldsymbol{f} \, d\Omega, \tag{5.67}$$

where $h(\boldsymbol{x})$ is the thickness of the plane elastic slab, $z = [z_1, z_2]^T$ is the displacement vector, $\boldsymbol{f} = [f_1, f_2]^T$ is the in-plane body force, $\boldsymbol{C}$ is the plane stress stiffness matrix given in (3.40), and $\varepsilon(z)$ is the strain vector, defined as

$$\varepsilon(z) = \begin{bmatrix} z_{1,1} \\ z_{2,2} \\ z_{1,2} + z_{2,1} \end{bmatrix}. \tag{5.68}$$

The stress vector can be easily obtained from the linear elastic relation as

$$\sigma(z) = \begin{bmatrix} \sigma_{11} \\ \sigma_{22} \\ \sigma_{12} \end{bmatrix} = C\varepsilon(z). \tag{5.69}$$

The design variable is taken here only as the variable thickness $h(\boldsymbol{x})$ of the elastic slab.

To develop design sensitivity of the performance measures, begin with the functional defining weight of the slab

$$\psi_1 = \iint_\Omega \gamma h \, d\Omega. \tag{5.70}$$

Since this functional does not involve z, its variation can be simply calculated as

$$\psi_1' = \iint_\Omega \gamma \, \delta h \, d\Omega \tag{5.71}$$

and requires no adjoint solution.

Now, let us discuss a locally averaged stress functional, which might involve principal stresses, von Mises stress, or some other material failure criteria. By defining a characteristic averaging function $m_p(\boldsymbol{x})$ that is nonzero and constant over a small, open subset $\Omega_p \subset \Omega$, is zero outside of Ω_p, and has an integral of one, the averaged stress functional can be written in the general form as

$$\psi_2 = \iint_\Omega g(\sigma(z)) m_p \, d\Omega. \tag{5.72}$$

While this expression could be written explicitly in terms of the gradients of z, it will be evident in the following discussion that it is more effective to continue with the present notation. Since the components of the stress vector given by (5.69) are linear in z, and since the order in which variation and partial derivatives are taken can be reversed, as was shown in (5.6), the variation of the functional in (5.72) can be written in the following form:

$$\psi_2' = \iint_\Omega g_{,\sigma}(z) \sigma(z') m_p \, d\Omega. \tag{5.73}$$

As with the general derivation of the adjoint (5.11), the variation z' can be replaced by a virtual displacement $\bar{\lambda}$ on the right side of (5.73). As a result, a load functional can be defined for the adjoint equation much as in (5.11) to obtain

$$a_u(\lambda^{(2)}, \bar{\lambda}) = \iint_\Omega g_{,\sigma}(z) \sigma(\bar{\lambda}) m_p \, d\Omega, \qquad \forall \bar{\lambda} \in Z. \tag{5.74}$$

It may be directly shown that the linear form of $\bar{\lambda}$ on the right side of (5.74) is bounded in $H^1(\Omega)$, so that (5.74) has a unique solution $\lambda^{(2)}$ for a displacement field, with the right side of (5.74) defining the load functional. By integrating the right side of (5.74) by parts, one can obtain a formula that could be interpreted as a distributed load acting on the elastic solid. Such a calculation, however, would cause considerable practical and theoretical difficulty, since $g_{,\sigma(z)}$ depends on stress, and since the derivatives of stress do not generally exist in $L^2(\Omega)$. Thus, the linear form on the right side of (5.74) is left as defined.

To provide a physical interpretation of the adjoint problem, (5.74) may be written in the following form, using the symmetry of $a_u(\bullet,\bullet)$:

$$\iint_\Omega h[\varepsilon(\boldsymbol{\lambda}^{(2)})^T - \frac{m_p}{h} g_{,\sigma}(\boldsymbol{z})]\sigma(\bar{\boldsymbol{\lambda}})\,d\Omega = 0, \qquad \forall \bar{\boldsymbol{\lambda}} \in Z,$$

which is nothing other than the elasticity equation for the displacement $\boldsymbol{\lambda}^{(2)}$ of a slab with an initial strain field $(g_{,\sigma(z)} m_p)/h$, and no externally applied load. Thus, this is a physical interpretation of the adjoint plate problem, which may assist in interpreting the significance and properties of adjoint displacement $\boldsymbol{\lambda}^{(2)}$.

In order to eliminate $\boldsymbol{z}'$ from (5.73), it is necessary to define the variation of the energy bilinear form in (5.66) and the load linear form in (5.67) as

$$a'_{\delta u}(\boldsymbol{z},\boldsymbol{\lambda}) = \iint_\Omega \sigma(\boldsymbol{z})^T \varepsilon(\boldsymbol{\lambda}^{(2)})\delta h\, d\Omega \tag{5.75}$$

and

$$\ell'_{\delta u}(\boldsymbol{\lambda}) = \iint_\Omega \boldsymbol{f}^T \boldsymbol{\lambda}^{(2)} \delta h\, d\Omega, \tag{5.76}$$

respectively. By using these two forms along with the symmetric property of the energy bilinear form, and by repeating the sequence of calculations in (5.12) through (5.15), (5.73) can be rewritten as

$$\psi'_2 = \iint_\Omega \left[\boldsymbol{f}^T \boldsymbol{\lambda}^{(2)} - \sigma(\boldsymbol{z})^T \varepsilon(\boldsymbol{\lambda}^{(2)})\right]\delta h\, d\Omega. \tag{5.77}$$

This expression yields the desired explicit sensitivity of the stress functional in (5.72) in terms of the structural solution $\boldsymbol{z}$, and in terms of the adjoint solution $\boldsymbol{\lambda}^{(2)}$.

The analytical examples considered in this section show that for each static problem studied in Section 3.1, direct calculation leads to explicit formulas for the design sensitivity of the functionals under consideration. In most cases, this calculation requires the solution of an adjoint problem that can be interpreted as the original elasticity problem, which contains an artificially defined applied load or an initial strain field. This interpretation can be valuable in taking advantage of existing finite element structural analysis codes, as will be discussed in the next section, as well as for visualizing the properties of adjoint displacement.

5.1.4 Numerical Considerations

Before proceeding from analytical derivations to numerical examples, it is helpful to consider the numerical aspects of computing design sensitivity expressions. Since for digital computation functions must be approximated in the finite-dimensional subspaces of an associated function space, it is important to define the parameterization that will be used in design sensitivity analysis, as explained in Section 12.1. Second, in carrying out actual computations, the finite element method of structural analysis is the most commonly employed computational tool. Therefore, the relationships between the design sensitivity procedure and the finite element method for solving boundary-value problems should be established.

Parameterization of the Design

The piecewise uniform beam and plate, shown in Figs. 4.1 and 5.4 respectively, represent the simplest examples of design parameterization. For a more complicated example, consider a beam with appropriate boundary conditions, in which the family of designs being considered is characterized by the finite-dimensional variable vector $\boldsymbol{b} = [b_1, \ldots, b_m]^T$. As functions of these variables, the moment of inertia and the cross-sectional area may be written in the form

$$\begin{aligned} I &= I(x;\boldsymbol{b}) \\ A &= A(x;\boldsymbol{b}). \end{aligned} \tag{5.78}$$

The energy bilinear form and load linear form for the beam are then expressed as

$$\begin{aligned} a_{\boldsymbol{b}}(z,\bar{z}) &= \int_0^l EI(x;\boldsymbol{b}) z_{,11}\bar{z}_{,11}\,dx \\ \ell_{\boldsymbol{b}}(\bar{z}) &= \int_0^l [F(x) + \gamma A(x;\boldsymbol{b})]\bar{z}\,dx. \end{aligned} \tag{5.79}$$

The notation used here illustrates the fact that the energy forms are functions of design variable $\boldsymbol{b}$ rather than design function $\boldsymbol{u}$. Using the definition of variation of these forms from (5.2) and (5.3), we can obtain their variations with respect to $\boldsymbol{b}$ as

$$\begin{aligned} a'_{\delta\boldsymbol{b}}(z,\bar{z}) &\equiv \frac{d}{d\tau} a_{\boldsymbol{b}+\tau\delta\boldsymbol{b}}(\tilde{z},\bar{z})\bigg|_{\tau=0} = \left[\int_0^l EI_{,\boldsymbol{b}} z_{,11}\bar{z}_{,11}\,dx\right]\delta\boldsymbol{b} \\ \ell'_{\delta\boldsymbol{b}}(\bar{z}) &\equiv \frac{d}{d\tau} \ell_{\boldsymbol{b}+\tau\delta\boldsymbol{b}}(\bar{z})\bigg|_{\tau=0} = \left[\int_0^l [\gamma A_{,\boldsymbol{b}}\bar{z}\,dx\right]\delta\boldsymbol{b}, \end{aligned} \tag{5.80}$$

where design variation $\delta\boldsymbol{b}$ can be taken outside the integrals since it is constant.

Now, consider a general response functional in the form

$$\psi = \int_0^l g(z, z_{,1}, z_{,11}, \boldsymbol{b})\,dx. \tag{5.81}$$

A variation of this functional may be taken to obtain

$$\psi' = \int_0^l [g_{,z}z' + g_{,z_{,1}}z'_{,1} + g_{,z_{,11}}z'_{,11}]dx + \left[\int_0^l g_{,\boldsymbol{b}}\,dx\right]\delta\boldsymbol{b}. \tag{5.82}$$

Let us define an adjoint variable as the solution to the following adjoint variational equation:

$$a_{\boldsymbol{b}}(\lambda,\bar{\lambda}) = \int_0^l [g_{,z}\bar{\lambda} + g_{,z_{,1}}\bar{\lambda}_{,1} + g_{,z_{,11}}\bar{\lambda}_{,11}]dx, \qquad \forall \bar{\lambda} \in Z. \tag{5.83}$$

Since the right side of (5.83) is continuous in $H^2(0,l)$, there is a unique solution for λ. Repeating the sequence of calculations carried out in (5.12) through (5.15), the result obtained is

$$\psi' = \left\{\int_0^l [g_{,\boldsymbol{b}} + \gamma A_{,\boldsymbol{b}}\lambda - EI_{,\boldsymbol{b}} z_{,11}\lambda_{,11}]dx\right\}\delta\boldsymbol{b}. \tag{5.84}$$

This expression gives the sensitivity coefficients of ψ associated with any variations in the design. It is interesting to note that once state and adjoint variables have been determined, the evaluation of this design sensitivity result only requires numerical calculation of the integral. Furthermore, the form of dependence that the beam cross-sectional area and the moment of inertia have on the design can be selected by the design engineer, and only partial derivatives $A_{,\boldsymbol{b}}$ (from the cross-sectional area) and $I_{,\boldsymbol{b}}$ (from the moment of inertia) need to be calculated.

For example, consider the stepped beam in Fig. 4.1, where each uniform segment of the beam makes up an *I*-beam with the section properties shown in Fig. 5.5. Here, the

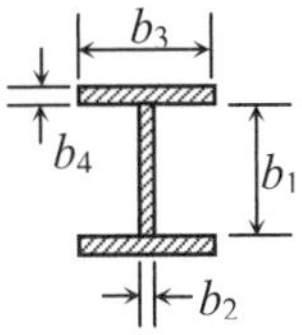

Figure 5.5. I-section beam element.

superscript i ($i = 1, \ldots, n$) denotes the numbering of uniform segments of the beam, and the subscript denotes the four design variables of each segment, for a total of $4n$ design variables. For the ith segment, the following conditions are given:

$$
\begin{aligned}
I^i(\boldsymbol{b}^i) &= \tfrac{1}{12}[b_3^i(8b_4^{i3} + 6b_1^{i2}b_4^i + 12b_1^i b_4^{i2}) + b_1^{i3}b_2^i] \\
A^i(\boldsymbol{b}^i) &= 2b_3^i b_4^i + b_1^i b_2^i.
\end{aligned}
\tag{5.85}
$$

The integral in (5.84) may also be written as a sum of integrals across each segment, yielding

$$
\psi_2' = -\iint_\Omega \frac{Eh^2}{4(1-\nu^2)} \boldsymbol{\kappa}(\lambda^{(i)})^T \boldsymbol{C}\boldsymbol{\kappa}(z)\delta h \, d\Omega, \tag{5.86}
$$

where $\delta \boldsymbol{b}^i = [\delta b_1^i, \delta b_2^i, \delta b_3^i, \delta b_4^i]^T$. This simple formula, evaluated with the aid of numerical integration, produces the design sensitivity of a general functional with respect to all section properties associated with the beam.

Equations (5.84) and (5.86) illustrate a method for automating design sensitivity computations in terms of a design shape function. Equation (5.27) shows the simplest possible form of a design shape function, namely, a piecewise constant design shape. A piecewise linear or a piecewise polynomial function could be considered as a design shape function to describe the material distribution in terms of the design variables as described in Section 5.1. Using the general design sensitivity results from Section 5.1.3 and parameterization of the type introduced in this section, results in the form of (5.86) are expected. Given the expression of (5.86), an algorithm can be written to evaluate the integrals across a typical segment, yielding a form for total design sensitivity. The same process is applicable for different design parameterizations described in Section 5.1. This systematic approach to design sensitivity analysis, which uses distributed parameter sensitivity results and design shape functions, appears as a very promising avenue of inquiry, particularly in the way it can be integrated with the finite element method. If design sensitivity is calculated using this approach, then the need for calculating and storing the design derivative of the system stiffness matrix (described in Chapter 4) is eliminated.

Coupling Design Sensitivity with Finite Element Structural Analysis

From a mathematical point of view, the structural finite element method may be viewed as an application of the Galerkin method [36] and [37] for a solution to boundary-value problems, with the coordinate functions defined as piecewise polynomials within segments (elements). For finite element analysis, the structural domain is partitioned into subdomains called elements. Coordinate functions, defined as polynomials within elements, associated with the nodal values of the state variable, and vanished from elements not adjacent to a given node, are defined based on element shapes, polynomial orders, and smoothness characteristics. For thorough expositions of this approach to the finite element method, readers are referred to texts by Strang and Fix [36], and Ciarlet [37]. For a more engineer-oriented introduction, see the text by Mitchell and Wait [72].

To explain the finite element method using a coordinate function, consider a structural problem whose solution z is a scalar function. If we let $\phi^i(x) \in Z$ ($i = 1, \ldots, n$) be linearly independent coordinate functions, a solution to the structural problem can be approximated as

$$z(x) = \sum_{j=1}^{n} c_j \phi^j(x). \tag{5.87}$$

Recall that solution z must satisfy a variational equation of the form

$$a_b(z, \bar{z}) = \ell_b(\bar{z}), \qquad \forall \bar{z} \in Z. \tag{5.88}$$

We can then substitute the approximation from (5.87) into this variational equation to obtain

$$a_u(\lambda, \bar{\lambda}) = \iint_\Omega E \bar{\lambda}_{,11} m_i \, d\Omega, \qquad \forall \bar{\lambda} \in Z. \tag{5.89}$$

Since the actual solution to (5.88) cannot be exactly written using the finite number of coordinate functions provided by (5.87), (5.89) cannot be satisfied for all $\bar{z} \in Z$. Therefore, it is necessary to find coefficient c_i in the approximate solution, such that (5.89) holds true for every $\bar{z}$ equal to each coordinate function. To put this idea into mathematical form, it is required that

$$\sum_{j=1}^{n} a_b(\phi^i, \phi^j) c_j = \ell_b(\phi^i), \qquad i = 1, \ldots, n. \tag{5.90}$$

If a matrix associated with the left side of (5.90) is defined as

$$\boldsymbol{A} = [a_b(\phi^i, \phi^j)]_{n \times n}, \tag{5.91}$$

and a column vector associated with the right side of (5.90) is defined as

$$\boldsymbol{B} = [\ell_b(\phi^i)]_{n \times 1}, \tag{5.92}$$

then, with coefficient vector $\boldsymbol{c} = [c_1, \ldots, c_n]^T$, (5.90) may be written in matrix form as

$$\boldsymbol{A}\boldsymbol{c} = \boldsymbol{B}. \tag{5.93}$$

Matrix $\boldsymbol{A}$ is precisely the stiffness matrix from Chapter 3, and column vector $\boldsymbol{B}$ represents a load vector applied to the structural system. Without going into great detail, it should be remembered that the entries in the stiffness matrix only require integration over elements adjacent to those nodes in which both ϕ^i and ϕ^j are nonzero. This fact eliminates integration over all but a small subset of the structural domain in order to evaluate the terms that contribute to the stiffness matrix. Furthermore, because the energy bilinear form is Z-elliptic, if the coordinate functions are linearly independent, matrix $\boldsymbol{A}$ is positive definite, and hence, nonsingular.

The idea of using design shape functions to evaluate (5.87) now becomes a possibility. Let each uniform segment of the beam be understood as a finite element. Coordinate functions ϕ^i are used to represent both the state variable z, as in (5.87), and the adjoint variable λ, as

$$\lambda(x) = \sum_{j=1}^{n} d_j \phi^j(x). \tag{5.94}$$

Substituting (5.87) for the state variable and (5.94) for the adjoint variable into (5.86), we can represent the performance sensitivity in terms of coordinate functions as

$$\begin{aligned}\psi' = \sum_{i=1}^{n} \Bigg[& \int_{(i-1)/n}^{i/n} g^i_{,b^i} dx + \gamma \sum_{j=1}^{n} d_j \int_{(i-1)/n}^{i/n} A^i_{,b^i} \phi^j dx \\ & - E \sum_{j=1}^{n} \sum_{k=1}^{n} c_k d_j \int_{(i-1)/n}^{i/n} I^i_{,b^i} \phi^k_{,11} \phi^j_{,11} dx \Bigg] \delta b^i . \end{aligned} \tag{5.95}$$

Although many integrals appear in evaluating the coefficients of (5.95), the reader familiar with finite element methods will note that these calculations are routinely performed in finite element analysis. Incorporating a standard design shape function, represented by the functions A^i and I^i, and a set of piecewise polynomial shape functions ϕ^i, the integration required in (5.95) may be efficiently carried out. In many cases, piecewise linear polynomials will be adequate, and the order of the polynomials appearing in the integration will be very low, allowing for a closed-form evaluation of the integrals and a tabulation of terms in (5.95) as design sensitivity finite elements. As discussed in Section 3.3.2, in the case of a beam, Hermite cubic polynomials are commonly used as displacement shape functions. In this situation, if the linear variation of cross-sectional area and quadratic variation of moment of inertia are incorporated in the design shape function, then the degree of polynomials arising in (5.95) cannot be higher than four. A closed-form integration designed to obtain and tabulate design sensitivity finite elements appears to be a practical objective. In more complex structures, such as plates and plane elastic solids, higher-order polynomials in more than one variable may be required, leading to the need for numerical generation of design sensitivity finite elements. However, these calculations are not intrinsically more tedious than calculations currently carried out for any finite element code. The potential thus seems to exist for a systematic finite element design sensitivity analysis formulation, employing both design and displacement shape functions.

One essential advantage of an integrated design finite element formulation is the ability to identify the effect of numerical error associated with finite element mesh. It has been observed in calculations that the use of distributed-parameter design sensitivity formulas and the finite element method leads to numerical errors in sensitivity coefficients, which may be identified during the process of iterative redesign and reanalysis. The effect of a design change can be predicted using design sensitivity analysis. When reanalysis is carried out, the predicted performance measure change can be compared with the actual change. If there is disagreement, then error has crept into the finite element approximation. While error might appear to be a problem, in fact it can be a blessing in disguise. If the approach outlined in Chapter 4 is followed, in which the structure is discretized and the design variables are imbedded in the global stiffness matrix, then any error inherent in the finite element model will be consistently parameterized without being reported to the user. Precise design sensitivity coefficients of the matrix model are obtained without any realization that substantial inherent error may exist in the original model. In fact, as optimization proceeds, the optimization algorithm may systematically exploit this error, leading to erroneous designs. However, in their current formulation, design sensitivity formulas derived from the distributed-parameter theory and the finite element model can be used to warn the user that an approximation error may be creeping into the calculation.

5.1.5 Numerical Examples

Beam

Consider a simply supported beam with a rectangular cross section and a point load of $f(x) = 100\ \delta(x-\hat{x})$ lb (Fig. 5.6). The material properties are given as $E = 30 \times 10^6$ psi and $\nu = 0.25$. The weight density of the material is ignored. The rectangular beam has unit width, and the depth b_i of element i is taken as a design variable.

Consider a stress performance measure on the top surface of the element i, defined as

$$\psi_i = -\int_0^l \frac{b_i}{2} E z_{,11}(x) m_i \, dx, \tag{5.96}$$

where $b_i/2$ is the half-depth of element i and m_i is the characteristic function applied to element i. Referring back to (5.83), the adjoint equation can be defined as

$$a_b(\lambda,\bar{\lambda}) = -\int_0^l \frac{b_i}{2} E\bar{\lambda}_{,11}(x) m_i \, dx, \quad \forall \bar{\lambda} \in Z, \tag{5.97}$$

and can denote a solution as $\lambda^{(i)}$. Consequently, from (5.84) the first variation of the functional ψ_i can be written as

$$\psi_i' = -\left(\int_0^l \frac{E}{2} z_{,11} m_i \, dx\right)\delta b_i - \sum_{k=1}^{n}\left[\int_{(k-1)l/n}^{kl/n} \frac{b_k^2}{4} E z_{,11}\lambda_{,11}^{(i)} \, dx\right]\delta b_k. \tag{5.98}$$

Note that in the above three equations a constant thickness over a single element is assumed.

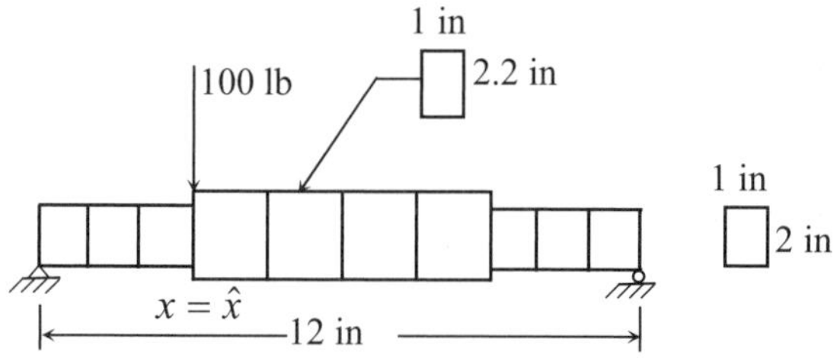

Figure 5.6. Simply supported beam.

Table 5.2. Comparison of sensitivity for beam.

Element Number	ψ_i^1	ψ_i^2	$\Delta\psi_i$	ψ_i'	$\psi_i'/\Delta\psi_i \times 100$
1	6.3000E+01	5.7143E+01	–5.7571E+00	–6.3000E+00	107.6
2	1.8900E+02	1.7143E+02	–1.7571E+01	–1.8900E+01	107.6
3	3.1500E+02	2.8571E+02	–2.9286E+01	–3.1500E+01	107.6
4	2.9008E+02	2.6311E+02	–2.6969E+01	–2.9008E+01	107.6
5	2.4545E+02	2.2263E+02	–2.2820E+01	–2.4545E+01	107.6
6	2.0083E+02	1.8216E+02	–1.8671E+01	–2.0083E+01	107.6
7	1.5620E+02	1.4168E+02	–1.4522E+01	–1.5620E+01	107.6
8	1.3500E+02	1.2245E+02	–1.2551E+01	–1.3500E+01	107.6
9	8.1000E±01	7.3469E+01	–7.5306E+01	–8.1000E+00	107.6
10	2.7000E+01	2.4490E+01	–2.5102E+00	–2.7000E+00	107.6

As shown in Fig. 5.6, a 10-member finite element model of the beam with a cubic shape function is employed to arrive at design sensitivity calculations. Uniform and accurate design sensitivity estimates are obtained, as shown in Table 5.2, for the averaged stress on each element with a 5% overall change in design variable magnitudes. In Table 5.2, the second and third columns represent averaged stress values at the initial design b_i, and at the perturbed design $1.05 \times b_i$, respectively. The fourth column is the finite difference of these two values. The fifth column is the estimated averaged stress value by using the formula in (5.98). The finite difference value and the estimated value are compared in the last column.

Plate

To account for the effect of variation in plate thickness on compliance and displacement functions, consider an application of (5.53) and (5.59) at the discrete point $\hat{x}$. As an example, consider a clamped square plate with a dimension of 1m and a uniform thickness of $h = 0.05$ m, with $E = 200$ GPa, $\nu = 0.3$, $F = 2.22$ MPa, and $\gamma = 7.7 \times 10^4$ N/m^3. If a piecewise constant thickness is assumed with b_i, the constant thickness of the ith element, instead of with (5.53), then (5.55) can be used for the compliance functional.

For numerical calculations, a nonconforming rectangular plate element with 12 degrees-of-freedom is used [45]. A graph of the coefficient of δb_i (Fig. 5.7) shows how the addition or deletion of material affects compliance. The maximum value of the coefficient of δb_i is $\Lambda^2_{\max} = -1.305$ at the corner elements. The minimum value occurs at the middle point between the edge elements, with a value of $\Lambda^2_{\min} = -5.625 \times 10^2$. Thus, to decrease compliance most effectively, material should be moved from the vicinity of the four corners and added close to the middle of the four edges.

If the design sensitivity of displacement is desired at the plate center point $\hat{x} = [\frac{1}{2}, \frac{1}{2}]^T$, then the adjoint load from (5.58) is just a unit load at the plate center point. To see how material added to or deleted from the plate influences displacement at the center, the coefficient of δb_i may be graphed using (5.59) to obtain the result shown in Fig. 5.8. The maximum value of the coefficient of δb_i is $\Lambda^2_{\max} = -3.678 \times 10^{-8}$ at the corner elements, while the minimum value occurs at the center elements with a value of $\Lambda^2_{\min} = -1.167 \times 10^3$. To decrease $z(\hat{x})$ most effectively, material should be removed near the four corners and added to the center of the plate.

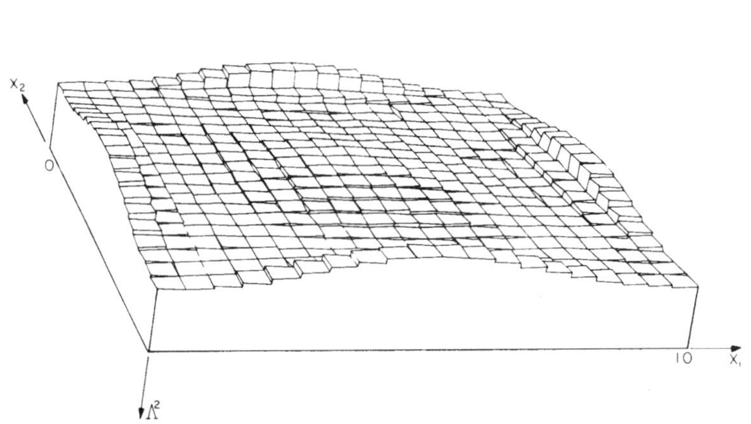

Figure 5.7. Compliance sensitivity for plate.

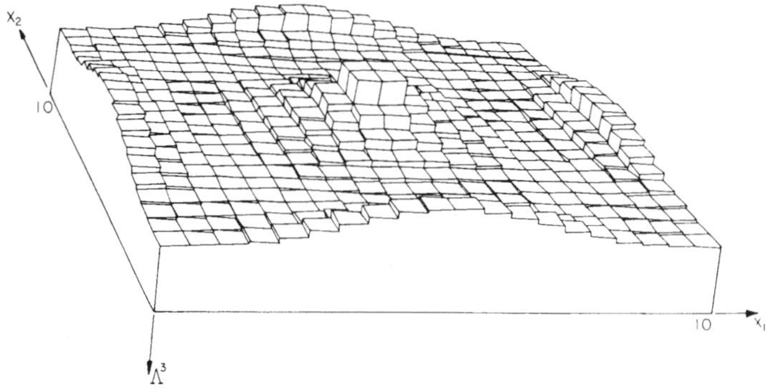

Figure 5.8. Displacement sensitivity for plate.

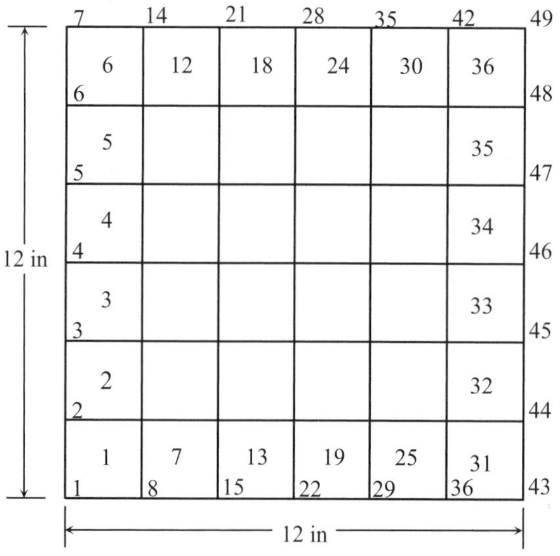

Figure 5.9. Finite element of simply supported plate.

To calculate the plate's stress sensitivity, consider the simply supported square plate shown in Fig. 5.9, with $E = 30 \times 10^6$ psi and $\nu = 0.25$. Let plate thickness $h(x)$ be a design variable and assume that γ can be ignored so that the load linear form is independent of the design.

Consider a stress functional in the form

$$\psi_i = \iint_\Omega Ez_{,11} m_i \, d\Omega, \tag{5.99}$$

where m_i is the characteristic function applied on finite element i. Referring to (5.64), the adjoint equation can be defined as

$$a_u(\lambda,\bar{\lambda}) = \iint_\Omega E\bar{\lambda}_{,11} m_i \, d\Omega, \qquad \forall \bar{\lambda} \in Z, \tag{5.100}$$

with a solution of $\lambda^{(i)}$. From (5.65), the first variation of (5.99) is obtained as

$$\psi_2' = -\iint_\Omega \frac{Eh^2}{4(1-\nu^2)} \boldsymbol{\kappa}(\lambda^{(i)})^T \boldsymbol{C\kappa}(z)\delta h \, d\Omega. \tag{5.101}$$

If piecewise constant thickness is assumed for each finite element, (5.101) can be rewritten as

$$\psi_2' = -\sum_{k=1}^{n} \frac{Eb_k^2}{4(1-\nu^2)} \iint_{\Omega_k} \boldsymbol{\kappa}(\lambda^{(i)})^T \boldsymbol{C\kappa}(z) \, d\Omega \, \delta b_k. \tag{5.102}$$

As before, a nonconforming rectangular plate element with 12 degrees of freedom [45] is employed for numerical calculation. The geometrical configuration and finite element grid are shown in Fig. 5.9. The length of each side of the square plate is 12 in, and the thickness is a uniform 0.1 in. The model has 36 elements, 49 nodal points, and 95 degrees of freedom. Applied loads consist of a point load of 100 lb at the center, and a uniformly distributed load of 100 psi. Results given in Table 5.3 demonstrate that the design sensitivity for each element is excellent, with a 0.1% overall change in design variables. Note that because of symmetry, only those sensitivity results for one quarter of the plate are given in Table 5.3.

Torque Arm

For an example involving a plane elastic component, consider the automotive rear suspension torque arm, as shown in Fig. 5.10. For simplicity, a single, nonsymmetric, static traction load has been proposed. Fixed displacement constraints are applied around the larger hole on the right in order to simulate attachment to a solid rear axle. Torque arm thickness has been chosen as a design variable. The variational equation of the torque arm is

$$\begin{aligned} a_u(z,\bar{z}) &\equiv \iint_\Omega h(x)\boldsymbol{\varepsilon}(\bar{z})^T \boldsymbol{C\varepsilon}(z) \, d\Omega \\ &= \int_{\Gamma^2} \bar{z}^T \boldsymbol{f}^s \, d\Gamma \equiv \ell_u(\bar{z}), \qquad \forall \bar{z} \in Z, \end{aligned} \tag{5.103}$$

Table 5.3. Comparison of stress sensitivity for plate.

Element Number	ψ_i^1	ψ_i^2	$\Delta\psi_i$	ψ_i'	$\psi_i'/\Delta\psi_i\times 100$
1	−7.7010E+05	−7.6779E+05	2.3057E+03	2.3030E+03	99.9
2	−1.7690E+06	−1.7637E+06	5.2965E+03	5.3094E+03	100.2
3	−2.2571E+06	−2.2503E+06	6.7576E+03	6.7702E+03	100.2
7	−1.3671E+06	−1.3630E+06	4.0930E+03	4.0869E+03	99.9
8	−3.6338E+06	−3.6229E+06	1.0880E+04	1.0906E+04	100.2
9	−4.8362E+06	−4.8217E+06	1.4480E+04	1.4508E+04	100.2
13	−1.5622E+06	−1.5575E+06	4.6772E+03	4.6859E+03	100.2
14	−4.2347E+06	−4.2220E+04	1.2679E+04	1.2706E+04	100.2
15	−5.7639E+06	−5.7466E+06	1.7257E+04	1.7293E+04	100.2

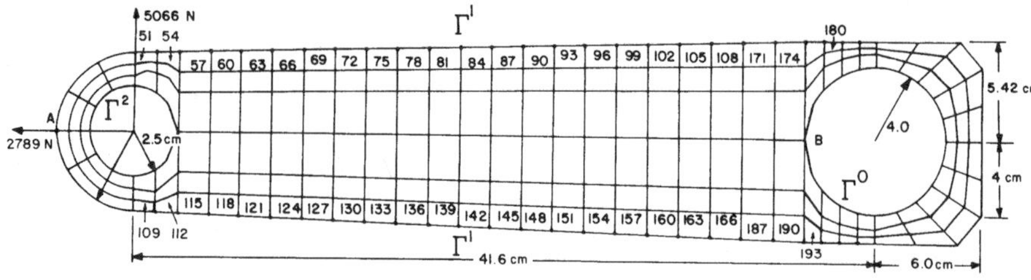

Figure 5.10. Geometry and finite element of torque arm.

Table 5.4. Comparison of von Mises stress sensitivity for torque arm.

Element Number	ψ_i^1	ψ_i^2	$\Delta\psi_i$	ψ_i'	$\psi_i'/\Delta\psi_i\times 100$
54	–9.7690E–01	–9.7693E–01	–2.3075E–05	–2.3098E–05	100.1
66	–9.6734E–01	–9.6737E–01	–3.2632E–05	–3.2665E–05	100.1
75	–9.5025E–01	–9.5030E–01	–4.9699E–05	–4.9748E–05	100.1
87	–9.3080E–01	–9.3087E–01	–6.9130E–05	–6.9199E–05	100.1
96	–9.1860E–01	–9.1868E–01	–8.1317E–05	–8.1398E–05	100.1
105	–9.0812E–01	–9.0821E–01	–9.1786E–05	–9.1878E–05	100.1
115	–9.7021E–01	–9.7024E–01	–2.9756E–05	–2.9786E–05	100.1
127	–9.5415E–01	–9.5420E–01	–4.5805E–05	–4.5850E–05	100.1
145	–9.2374E–01	–9.2382E–01	–7.6183E–05	–7.6259E–05	100.1
160	–9.0483E–01	–9.0493E–01	–9.5073E–05	–9.5169E–05	100.1
171	–9.0491E–01	–9.0500E–01	–9.4997E–05	–9.4997E–05	100.1
180	–9.2579E–01	–9.2587E–01	–7.4134E–05	–7.4208E–05	100.1
187	–8.9958E–01	–8.9968E–01	–1.0032E–04	–1.0042E–04	100.1
193	–9.1117E–01	–9.1126E–01	–8.8743E–05	–8.8831E–05	100.1

where

$$Z = \left\{ z = [z_1, z_2]^T \in [H^1(\Omega)]^2 : \; z = 0 \text{ on } \Gamma^h \right\}. \tag{5.104}$$

Consider a von Mises stress functional of the form

$$\psi_k = \iint_\Omega \frac{(\sigma_Y - \sigma_A)}{\sigma_A} m_k \, d\Omega = \iint_\Omega g m_k \, d\Omega, \tag{5.105}$$

where $g = (\sigma_Y - \sigma_A)/\sigma_A$, σ_A is the allowable stress, m_k is the characteristic function defined on finite element k, and σ_Y is the von Mises yield stress, defined as

$$\sigma_Y = [(\sigma_{11})^2 + (\sigma_{22})^2 + 3(\sigma_{12})^2 - \sigma_{11}\sigma_{22}]^{1/2}. \tag{5.106}$$

For this stress functional, the adjoint equation from (5.74) is

$$a_u(\lambda^{(k)}, \bar{\lambda}) = \iint_\Omega g_{,\sigma} \sigma(\bar{\lambda}) m_k \, d\Omega, \qquad \forall \bar{\lambda} \in Z, \tag{5.107}$$

with solution $\lambda^{(k)}$. The first variation of the functional ψ_k from (5.77) becomes

$$\psi_k' = \iint_{\Omega} \boldsymbol{\varepsilon}(\boldsymbol{\lambda}^{(k)})^T \boldsymbol{C}\boldsymbol{\varepsilon}(\boldsymbol{z})\delta h \, d\Omega. \tag{5.108}$$

If piecewise constant thickness b_l is assumed for finite element l, then (5.108) can be rewritten as

$$\psi_k' = -\sum_{l=1}^{n} \left[\iint_{\Omega_l} \boldsymbol{\varepsilon}(\boldsymbol{\lambda}^{(k)})^T \boldsymbol{C}\boldsymbol{\varepsilon}(\boldsymbol{z}) \, d\Omega \right] \delta b_l. \tag{5.109}$$

The finite element model shown in Fig 5.10, which includes 204 elements, 707 nodal points, 1332 degrees of freedom, and an eight-node isoparametric element, is used for numerical calculation. Applied loads and dimensions are also shown. Young's modulus is 207.4 GPa, Poisson's ratio is 0.25, and allowable stress is 81 MPa. A uniform thickness of 1 cm is used for the initial design. Numerical results for stresses in selected boundary elements are shown in Table 5.4. With a 0.1% uniform change of design variables, excellent sensitivity results are obtained.

5.2 Eigenvalue Design Sensitivity

The examples presented in Section 3.2 clearly show that eigenvalues, which represent the natural frequencies and buckling loads of structures, depend on the design. The objective in this section is to obtain the design sensitivity of eigenvalues. For conservative systems, it happens that no adjoint equation is necessary, and eigenvalue sensitivities are expressed in terms of the eigenvalues, the eigenvectors, and variations in the bilinear forms. Theorems that establish the differentiability of simple eigenvalues and the directional differentiability of repeated eigenvalues are first stated, and their significance is then discussed. Using the differentiability results, explicit formulas for the design variations of eigenvalues, both simple and repeated, are obtained. Using examples from Section 3.2, analytical calculations are carried out to illustrate how the method is used. Numerical considerations associated with the computation of eigenvalue design sensitivity are discussed, and numerical examples are provided.

5.2.1 Differentiability of Energy Bilinear Forms and Eigenvalues

Basic results concerning the differentiability of eigenvalues are developed in detail in Section 2.5 in [5]. The purpose of this section is to summarize those key results needed for eigenvalue design sensitivity. In particular, treatment of the repeated eigenvalue illustrates the need for care in establishing and utilizing the properties of these functionals, since it is shown that repeated eigenvalues are not differentiable.

As shown in Section 3.2, eigenvalue problems for vibration and buckling of elastic systems are best described by variational equations, in the form

$$a_{\boldsymbol{u}}(\boldsymbol{y}, \bar{\boldsymbol{y}}) = \zeta d_{\boldsymbol{u}}(\boldsymbol{y}, \bar{\boldsymbol{y}}), \qquad \forall \bar{\boldsymbol{y}} \in Z, \tag{5.110}$$

where Z is the space of kinematically admissible displacements, and ζ is the eigenvalue associated with eigenfunction $\boldsymbol{y}$. Since (5.110) is homogeneous in eigenfunction $\boldsymbol{y}$, a normalizing condition must be added to uniquely define the eigenfunction. The normalizing condition employed is

$$d_{\boldsymbol{u}}(\boldsymbol{y}, \boldsymbol{y}) = 1. \tag{5.111}$$

The energy bilinear form on the left side of (5.110) is in the same form as the static problem treated in Section 5.1. It therefore shares the same properties discussed in that section. The kinetic bilinear form $d_u(\bullet,\bullet)$ on the right side of (5.110) represents the mass effect that occurs in vibration problems and the geometric effect that appears in buckling problems. In most cases $d_u(\bullet,\bullet)$ is even more regular in its dependence on design $\boldsymbol{u}$ and eigenfunction $\boldsymbol{y}$ than the energy bilinear form $a_u(\bullet,\bullet)$. In the exceptional case of column buckling, the kinetic bilinear form involves derivatives of the eigenfunction and must be treated somewhat more carefully. As proved in Section 2.5.1 in [5], the variation of $d_u(\bullet,\bullet)$ with respect to the design is given by

$$d'_{\delta u}(\boldsymbol{y},\bar{\boldsymbol{y}}) \equiv \frac{d}{d\tau} d_{u+\tau\delta u}(\tilde{\boldsymbol{y}},\bar{\boldsymbol{y}})\bigg|_{\tau=0}, \tag{5.112}$$

where $\tilde{\boldsymbol{y}}$ denotes holding $\boldsymbol{y}$ constant for differentiation with respect to τ. Thus, (5.112) represents the dependence of the kinetic bilinear form on the design, and is applicable to all problems.

Simple Eigenvalues

In the case of a simple eigenvalue (that is, an eigenvalue with only one independent eigenfunction), Section 2.5 in [5] demonstrates that eigenvalue ζ is differentiable. Kato [53] showed that the corresponding eigenfunction $\boldsymbol{y}$ is also differentiable. Thus, the following variations are well defined:

$$\begin{aligned} \zeta' = \zeta'(\boldsymbol{u},\delta\boldsymbol{u}) &\equiv \frac{d}{d\tau}[\zeta(\boldsymbol{u}+\tau\delta\boldsymbol{u})]\bigg|_{\tau=0} = \frac{\partial\zeta}{\partial\boldsymbol{u}}\delta\boldsymbol{u} \\ \boldsymbol{y}' = \boldsymbol{y}'(x;\boldsymbol{u},\delta\boldsymbol{u}) &\equiv \frac{d}{d\tau}[\boldsymbol{y}(x;\boldsymbol{u}+\tau\delta\boldsymbol{u})]\bigg|_{\tau=0} = \frac{\partial\boldsymbol{y}}{\partial\boldsymbol{u}}\delta\boldsymbol{u}. \end{aligned} \tag{5.113}$$

In fact, both eigenvalue and eigenfunction variations are linear in $\delta\boldsymbol{u}$; hence, they are Fréchet derivatives (Appendix A.3) of the eigenvalue and eigenfunction. Proof of these results is far from obvious; details are provided in Section 2.5 in [5].

Given this differentiability result, the variation of both sides of (5.110) can be taken to obtain

$$a_u(\boldsymbol{y}',\bar{\boldsymbol{y}}) + a'_{\delta u}(\boldsymbol{y},\bar{\boldsymbol{y}}) = \zeta' d_u(\boldsymbol{y},\bar{\boldsymbol{y}}) + \zeta d_u(\boldsymbol{y}',\bar{\boldsymbol{y}}) + \zeta d'_{\delta u}(\boldsymbol{y},\bar{\boldsymbol{y}}), \qquad \forall \bar{\boldsymbol{y}} \in Z, \tag{5.114}$$

where $\bar{\boldsymbol{y}}$ is independent of the design. Since (5.114) holds for all $\bar{\boldsymbol{y}} \in Z$, this equation may be evaluated with specific $\bar{\boldsymbol{y}} = \boldsymbol{y}$, using the symmetry of bilinear forms $a_u(\bullet,\bullet)$ and $d_u(\bullet,\bullet)$, to obtain

$$\zeta' d_u(\boldsymbol{y},\boldsymbol{y}) = a'_{\delta u}(\boldsymbol{y},\boldsymbol{y}) - \zeta d'_{\delta u}(\boldsymbol{y},\boldsymbol{y}) + [a_u(\boldsymbol{y},\boldsymbol{y}') - \zeta d_u(\boldsymbol{y},\boldsymbol{y}')] \tag{5.115}$$

Noting that $\boldsymbol{y}' \in Z$, it can be observed that the term in brackets on the right side of (5.115) vanishes from the relation in (5.110). Furthermore, due to the normalizing condition in (5.111), a simplified equation for the eigenvalue variation can be obtained as

$$\zeta' = a'_{\delta u}(\boldsymbol{y},\boldsymbol{y}) - \zeta d'_{\delta u}(\boldsymbol{y},\boldsymbol{y}). \tag{5.116}$$

For a precise proof of this result, see Corollary 2.5.1, in Section 2.5 in [5].

This result, obtained with little effort, is the foundation for a large body of work on structural optimization that has eigenvalue constraints. It is a remarkably simple result, clearly showing that the eigenvalue's directional derivative is indeed linear in $\delta\boldsymbol{u}$, since the variation of bilinear forms on the right side of (5.116) is linear in $\delta\boldsymbol{u}$. It should be emphasized, however, that the validity of this result depends on the existence of eigenvalue variations and eigenfunctions, as defined in (5.113). As will be seen in the following section, formal expression of this analysis for repeated eigenvalues would lead to erroneous results.

Repeated Eigenvalues

Now, consider a situation in which eigenvalue ζ associates with its s linearly independent eigenfunctions, that is,

$$\begin{aligned} a_{\boldsymbol{u}}(\boldsymbol{y}^i, \bar{\boldsymbol{y}}) &= \zeta d_{\boldsymbol{u}}(\boldsymbol{y}^i, \bar{\boldsymbol{y}}), \quad && \forall \bar{\boldsymbol{y}} \in Z \\ d_{\boldsymbol{u}}(\boldsymbol{y}^i, \boldsymbol{y}^i) &= 1, && i = 1, \ldots, s. \end{aligned} \tag{5.117}$$

It is easy to show that any linear combination of $\boldsymbol{y}^i$ eigenfunctions in (5.117) is also an eigenfunction. Therefore, an infinite variety of choices exists as the basis of the eigenspace associated with repeated eigenvalue ζ. One practical requirement on the family of eigenfunctions is that they be orthonormal with respect to bilinear form $d_{\boldsymbol{u}}(\bullet,\bullet)$, that is,

$$d_{\boldsymbol{u}}(\boldsymbol{y}^i, \boldsymbol{y}^j) = \delta_{ij}, \qquad i, j = 1, \ldots, s. \tag{5.118}$$

It is assumed throughout this text that such an orthonormalization of eigenfunctions corresponding to a repeated eigenvalue has been carried out. Nevertheless, an infinite choice of eigenfunctions still exists.

Theorem 2.5.1 (in Section 2.5) in [5] proves that while repeated eigenvalue ζ is a continuous function of the design, eigenfunctions are not. Although the eigenvalue is continuous, it is not Fréchet differentiable, but only directionally differentiable (Appendix A.3). In addition, Theorem 2.5.2 and Corollary 2.5.2 in [5] demonstrate that when eigenvalue ζ is repeated s times at design $\boldsymbol{u}$ for a perturbation in the design of $\boldsymbol{u} + \tau\delta\boldsymbol{u}$, the eigenvalue may branch into s number of eigenvalues, given by

$$\zeta_i(\boldsymbol{u} + \tau\delta\boldsymbol{u}) = \zeta(\boldsymbol{u}) + \tau\zeta_i'(\boldsymbol{u}, \delta\boldsymbol{u}) + o(\tau), \qquad i = 1, \ldots, s, \tag{5.119}$$

where the directional derivatives $\zeta'_i(\boldsymbol{u},\delta\boldsymbol{u})$ are the eigenvalues of the matrix

$$\mathcal{M} = [a'_{\delta\boldsymbol{u}}(\boldsymbol{y}^i, \boldsymbol{y}^j) - \zeta d'_{\delta\boldsymbol{u}}(\boldsymbol{y}^i, \boldsymbol{y}^j)]_{s\times s}. \tag{5.120}$$

The notation $\zeta'_i(\boldsymbol{u},\delta\boldsymbol{u})$ has been selected in order to emphasize the dependence of the directional derivative on $\delta\boldsymbol{u}$. The term $o(\tau)$ in (5.119) is defined as a quantity that approaches zero more rapidly than τ [that is, $\lim_{\tau\to 0} o(\tau)/\tau = 0$]. All characteristics of repeated eigenvalues discussed in Section 4.2.5 hold true in the distributed-parameter model. Moreover, the directional derivatives of twice-repeated eigenvalues are given in (4.156) and (4.157), and are rewritten here as

$$\begin{aligned} \zeta_1'(\boldsymbol{u},\delta\boldsymbol{u}) = \hat{\mathcal{M}}_{11} &= \cos^2\phi(\delta\boldsymbol{u})\mathcal{M}_{11}(\delta\boldsymbol{u}) \\ &\quad + \sin 2\phi(\delta\boldsymbol{u})\mathcal{M}_{12}(\delta\boldsymbol{u}) + \sin^2\phi(\delta\boldsymbol{u})\mathcal{M}_{22}(\delta\boldsymbol{u}) \end{aligned} \tag{5.121}$$

and

$$\begin{aligned} \zeta_2'(\boldsymbol{u},\delta\boldsymbol{u}) = \hat{\mathcal{M}}_{22} &= \sin^2\phi(\delta\boldsymbol{u})\mathcal{M}_{11}(\delta\boldsymbol{u}) \\ &\quad - \sin 2\phi(\delta\boldsymbol{u})\mathcal{M}_{12}(\delta\boldsymbol{u}) + \cos^2\phi(\delta\boldsymbol{u})\mathcal{M}_{22}(\delta\boldsymbol{u}), \end{aligned} \tag{5.122}$$

where ϕ is the rotation parameter, given as

$$\phi = \phi(\delta \boldsymbol{u}) = \frac{1}{2}\arctan\left[\frac{2\mathcal{M}_{12}(\boldsymbol{y}^1, \boldsymbol{y}^2, \delta \boldsymbol{u})}{\mathcal{M}_{11}(\boldsymbol{y}^1, \boldsymbol{y}^1, \delta \boldsymbol{u}) - \mathcal{M}_{22}(\boldsymbol{y}^2, \boldsymbol{y}^2, \delta \boldsymbol{u})}\right], \tag{5.123}$$

and $\mathcal{M}_{ij}(\delta \boldsymbol{u})$ is a component of matrix $\boldsymbol{\mathcal{M}}$ given in (5.120).

5.2.2 Analytical Examples of Eigenvalue Design Sensitivity

To illustrate the preceding results, a design sensitivity analysis of the eigenvalue problems presented in Section 3.2 will now be studied. Numerical considerations that are made possible by the resulting formulas will be taken into account in Section 5.2.3.

Vibration of a String

Consider the string in Fig. 5.11, with variable mass density $\rho(x)$ and tension T. The energy and kinetic bilinear forms of (3.94) and (3.95) are

$$\begin{aligned} a_{\boldsymbol{u}}(y, \bar{y}) &= T \int_0^l y_{,1} \bar{y}_{,1}\, dx \\ d_{\boldsymbol{u}}(y, \bar{y}) &= \int_0^l \rho y \bar{y}\, dx. \end{aligned} \tag{5.124}$$

Variations of these bilinear forms yield

$$\begin{aligned} a'_{\delta \boldsymbol{u}}(y, \bar{y}) &= \delta T \int_0^l y_{,1} \bar{y}_{,1}\, dx \\ d'_{\delta \boldsymbol{u}}(y, \bar{y}) &= \int_0^l \delta\rho\, y \bar{y}\, dx. \end{aligned} \tag{5.125}$$

Given the fact that only simple eigenvalues can occur in Sturm-Liouville problems [67], only the variation of a simple eigenvalue is of interest. The variations of the bilinear forms in (5.125) can be substituted into the eigenvalue sensitivity formula in (5.121) to obtain

$$\zeta' = \left[\int_0^l (y_{,1})^2\, dx\right] \delta T - \zeta \int_0^l y^2\, \delta\rho\, dx. \tag{5.126}$$

It is interesting to note that since the coefficient of variation δT is positive, it is clear that the natural frequency increases with an increase in tension. Similarly, since the coefficient of variation $\delta\rho$ is positive, any increase in density decreases the natural frequency. While both results are obvious on an intuitive level, the plotting of $y(x)^2$ for a uniform string if $l = 1$, $\rho_0 = 2.0$, and $T = 1.0$ (Fig. 5.12) shows that a unit increase of $\rho(x)$ near the string center has a substantially larger effect on the smallest eigenvalue than unit increases elsewhere on the string. Thus, an indication is obtained of the most profitable design change areas.

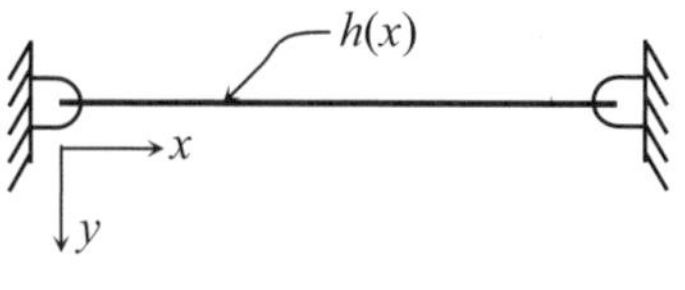

Figure 5.11. Vibrating string with linear mass density $\rho(x)$.

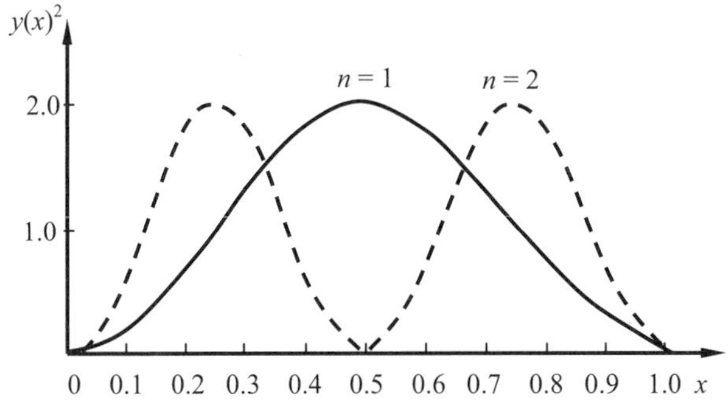

Figure 5.12. First two eigenfunctions of vibrating string.

Vibration of a Beam

For a beam with a variable cross section, Young's modulus, and mass density (as shown in Fig. 3.11), the strain and kinetic energy bilinear forms in (3.99) and (3.100) are rewritten as

$$\begin{aligned} a_{\boldsymbol{u}}(y,\bar{y}) &= \int_0^l E\alpha A^2 y_{,11}\bar{y}_{,11}\,dx \\ d_{\boldsymbol{u}}(y,\bar{y}) &= \rho\int_0^l Ay\bar{y}\,dx. \end{aligned} \tag{5.127}$$

If we let the design function be $\boldsymbol{u} = [E,\ A,\ \rho]^T$, the design variations of these bilinear forms are

$$\begin{aligned} a'_{\delta\boldsymbol{u}}(y,\bar{y}) &= \delta E\int_0^l \alpha A^2 y_{,11}\bar{y}_{,11}\,dx + \int_0^l 2E\alpha A\, y_{,11}\bar{y}_{,11x}\,\delta A\,dx \\ d'_{\delta\boldsymbol{u}}(y,\bar{y}) &= \delta\rho\int_0^l A\,y\bar{y}\,dx + \rho\int_0^l \delta A\,y\bar{y}\,dx. \end{aligned} \tag{5.128}$$

Combining the simple eigenvalue in (5.116) with the expressions in (5.128) yields

$$\begin{aligned} \zeta' = &\left[\int_0^l \alpha A^2(y_{,11})^2\,dx\right]\delta E - \left[\zeta\int_0^l A\,y^2\,dx\right]\delta\rho \\ &+ \int_0^l [2E\alpha A(y_{,11})^2 - \zeta\rho y^2]\delta A\,dx. \end{aligned} \tag{5.129}$$

As in the static response case, the sensitivity formula in (5.129) is valid for other boundary conditions in (3.19) through (3.21). This result clearly shows that increasing Young's modulus increases the natural frequency, and increasing the material density decreases the natural frequency. However, since the coefficient of δA in the integral may have either a positive or negative sign, it is not clear how a change in the cross-sectional area will influence the natural frequency. Consider, for example, a uniform cantilever beam (Fig. 5.13) with the nominal properties of $l = 1$ m, $E = 2 \times 10^5$ MPa, $\alpha = 1/6$, $\rho = 7.87$ Mg/m^3, and $h = 0.005$ m^2. In this example, the smallest eigenvalue is $\zeta = 0.00157$ and the corresponding eigenfunction can be obtained analytically as

$$y(x) = 0.159\left[\cosh k_n x - \cos k_n x - \frac{\cos k_n + \cosh k_n}{\sin k_n + \sinh k_n}(\sinh k_n x - \sin k_n x)\right],$$

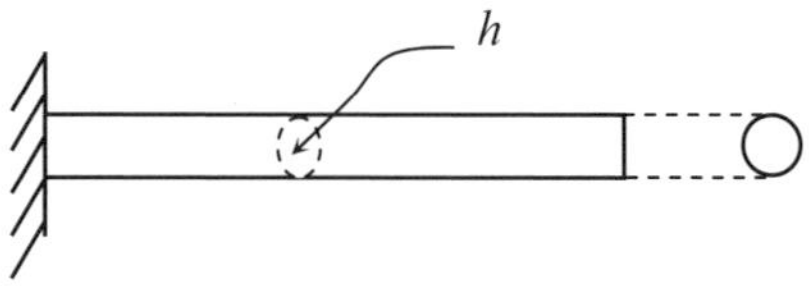

Figure 5.13. Uniform cantilever beam.

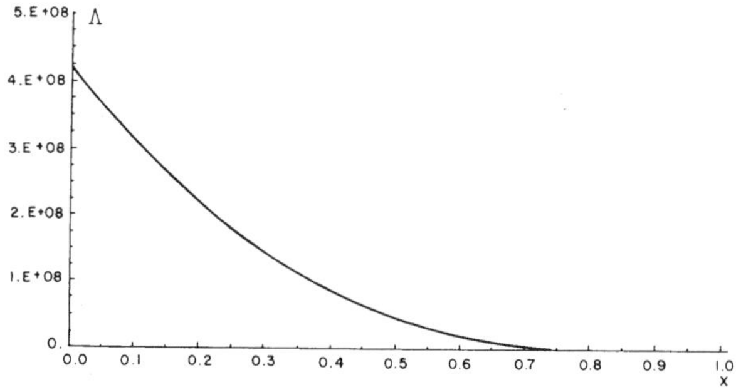

Figure 5.14. Design sensitivity coefficient of δA for cantilever beam.

where $k_1 = 1.875$, $k_2 = 4.694$, …. Evaluating the coefficient of δA in the integral form of (5.129), a curve is obtained in the form shown in Fig. 5.14. The design sensitivity coefficient of Fig. 5.14 shows that a change in the cross-sectional area at the clamped end of the beam has a substantially greater effect on the smallest eigenvalue than a change at the free end.

Buckling of a Column

Let us now consider the problem of column buckling, with the design variables as the distribution of the cross-sectional area along the column, and Young's modulus of the material, that is, $\boldsymbol{u} = [A, E]^T$. The energy and geometric bilinear forms in (3.103) and (3.104) are rewritten as

$$\begin{aligned} a_u(y,\bar{y}) &= \int_0^l E\alpha A^2 y_{,11}\bar{y}_{,11}\, dx \\ d_u(y,\bar{y}) &= \int_0^l y_{,1}\bar{y}_{,1}\, dx. \end{aligned} \tag{5.130}$$

The variations of these bilinear forms become

$$\begin{aligned} a'_{\delta u}(y,\bar{y}) &= \delta E \int_0^l \alpha A^2 y_{,11}\bar{y}_{,11}\, dx + \int_0^l 2E\alpha A\, y_{,11}\bar{y}_{,11}\, \delta A\, dx \\ d'_{\delta u}(y,\bar{y}) &= 0. \end{aligned} \tag{5.131}$$

The variation of the buckling load for a simple eigenvalue is given by (5.116) and combined with the expressions in (5.131) as

$$\zeta' = \left[\int_0^l \alpha A^2 (y_{,11})^2\, dx\right]\delta E + \int_0^l [2E\alpha A (y_{,11})^2]\delta A\, dx. \tag{5.132}$$

Clearly, increasing Young's modulus increases the buckling load, and any increase in the cross-sectional area similarly increases the buckling load. Both results are expected.

Consider the clamped-clamped column in Fig. 3.12, with a uniform cross section and with the following values: $l = 1$ m, $A_0 = 0.005$ m^2, $\alpha = 1/6$, and $E = 2 \times 10^5$ MPa. A plot of the first and second mode shapes and their second derivatives is shown in Fig. 5.15(a). Using these functions, the coefficient of δA may be evaluated in the integral form of (5.132), obtaining the curve shown in Fig. 5.15(b). In order to increase the buckling load in the first mode, material in the vicinity of points A and C, where the sensitivity coefficient of δh for ζ_1 is zero, may be removed and added either at point B or at the ends, where the sensitivity coefficient is at maximum value. This process, however, may decrease the buckling load in the second mode, since its sensitivity coefficient is positive at points A and C and zero at point B. In fact, it has been shown by Olhoff and Rasmussen [62] that when attempting to maximize the fundamental buckling load for a clamped-clamped column, the systematic occurrence of a repeated eigenvalue may be forced, much as shown in the examples in Section 4.3.5. It is therefore in our interest to obtain expressions for directional derivatives of this column for a repeated eigenvalue.

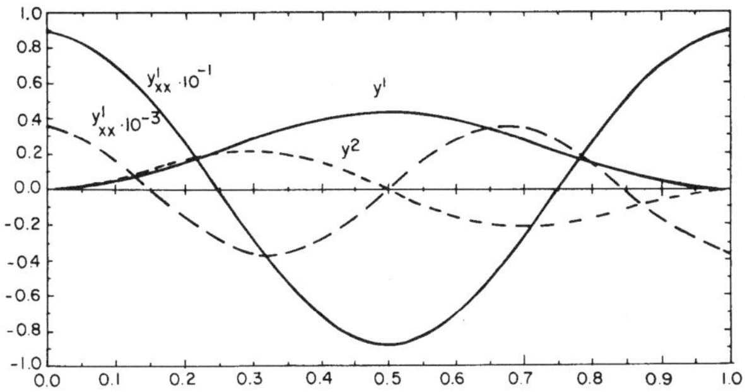

(a)

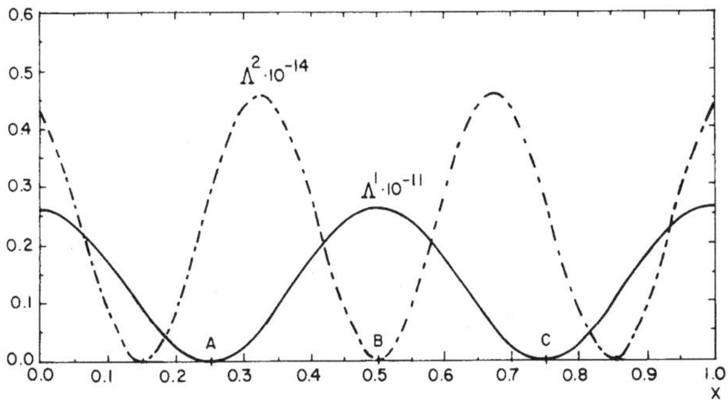

(b)

Figure 5.15. Mode and sensitivities for uniform clamped-clamped column.

If y^1 and y^2 are eigenfunctions corresponding to a repeated eigenvalue with multiplicity $s = 2$, then from (5.120)

$$\mathcal{M}_{ij} = \int_0^1 2E\alpha A\, y^i_{,11} y^j_{,11}\, \delta A\, dx, \tag{5.133}$$

where the effect of variation in Young's modulus has been suppressed. Note that if we limit attention to those designs $A(x)$ that are symmetric near the center of the column, and if $\delta A(x)$ is symmetric near the center, then, as is indicated by Fig. 5.15(a), the second derivatives of the first and second eigenfunctions will be symmetric and antisymmetric, respectively, near the center of the column. Thus, the product $A\delta A\, y^1_{,11}$ is an even function near the center, while $y^2_{,11}$ is an odd function near the center. Therefore, $\mathcal{M}_{12} = 0$. Since this is true for all design variations in the class of designs that are symmetric near the center, the directional derivatives of the repeated eigenvalue for symmetric columns are given by (5.132), with the symmetric and antisymmetric modes being y^1 and y^2, respectively. Since the resulting expression is linear in the design variation, the repeated eigenvalues, which are ordered by symmetric and antisymmetric modes, will be differentiable for symmetric columns. In addition, the derivatives of these repeated eigenvalues may be obtained using (5.132) with symmetric and antisymmetric modes.

However, if asymmetric designs are allowed, then in general $\mathcal{M}_{12}$ will not be zero, and moreover the eigenvalue will only be directionally differentiable, not Fréchet differentiable. In this case, the angle of rotation that is required can be obtained from (5.123) as

$$\phi = \phi(\delta A) = \frac{1}{2}\arctan\left[\frac{2\int_0^1 A y^1_{,11} y^2_{,11}\delta A\, dx}{\int_0^1 A[(y^1_{,11})^2 - (y^2_{,11})^2]\delta A\, dx}\right] \tag{5.134}$$

The directional derivatives of the repeated eigenvalue are then given by (5.121) and (5.122) as

$$\begin{aligned}\zeta_1'(A,\delta A) &= 2E\alpha\int_0^1 A[\cos^2\phi(\delta A)(y^1_{,11})^2 + \sin 2\phi(\delta A) y^1_{,11} y^2_{,11} + \sin^2\phi(\delta A)(y^2_{,11})^2]\delta A\, dx\\ \zeta_2'(A,\delta A) &= 2E\alpha\int_0^1 A[\sin^2\phi(\delta A)(y^1_{,11})^2 - \sin 2\phi(\delta A) y^1_{,11} y^2_{,11} + \cos^2\phi(\delta A)(y^2_{,11})^2]\delta A\, dx.\end{aligned} \tag{5.135}$$

It is clear from the above equation that in general the directional derivatives of repeated eigenvalues are not linear in δA; hence, they are not differentiable.

Vibration of a Membrane

Consider a vibrating membrane with variable mass density $\rho(\mathbf{x})$ and tension T as the design variables. Without repeating the definition of the energy and kinetic bilinear forms from (3.108) and (3.109), we can write the first variation of these bilinear forms, which are evaluated at $\bar{y} = y$, as

$$\begin{aligned}a'_{\delta u}(y,y) &= \delta T \iint_\Omega (y_{,1}^2 + y_{,2}^2)\, d\Omega\\ d'_{\delta u}(y,y) &= \iint_\Omega y^2\, \delta\rho\, d\Omega.\end{aligned} \tag{5.136}$$

For a simple eigenvalue problem, (5.116) and (5.136) will yield

$$\zeta' = \delta T \iint_\Omega (y_{,1}^2 + y_{,2}^2)\, d\Omega - \zeta \iint_\Omega y^2\, \delta\rho\, d\Omega. \tag{5.137}$$

As with the vibrating string problem, it is clear that the natural frequency increases with an increasing amount of tension, and decreases with increasing density.

In the case of a repeated eigenvalue problem [73] with a multiplicity of $s = 2$, if y^1 and y^2 are eigenfunctions corresponding to a repeated eigenvalue ζ, then from (5.120) we obtain

$$\mathcal{M}_{ij} = \delta T \iint_\Omega \nabla y^{i^T} \nabla y^j \, d\Omega - \zeta \iint_\Omega y^i y^j \, \delta\rho \, d\Omega, \qquad i,j = 1,2. \tag{5.138}$$

The required rotational angle is obtained from (5.123) and (5.138) as

$$\phi(\delta\rho) = \frac{1}{2}\arctan\left[\frac{2[\delta T \iint_\Omega \nabla y^{1^T} \nabla y^2 \, d\Omega - \zeta \iint_\Omega y^1 y^2 \, \delta\rho \, d\Omega]}{\delta T \iint_\Omega [(\nabla y^1)^2 - (\nabla y^2)^2] d\Omega - \zeta \iint_\Omega [(y^1)^2 - (y^2)^2] \delta\rho \, d\Omega}\right]. \tag{5.139}$$

The directional derivatives of the repeated eigenvalue are then given by (5.121) and (5.122) as

$$\begin{aligned}
\zeta_1'(\boldsymbol{u},\delta\boldsymbol{u}) &= \delta T \iint_\Omega [(\nabla y^1)^2 \cos^2\phi + (\nabla y^{1^T} \nabla y^2)\sin 2\phi + (\nabla y^2)^2 \sin^2\phi] d\Omega \\
&\quad - \zeta \iint_\Omega [(y^1)^2 \cos^2\phi + (y^1 y^2)\sin 2\phi + (y^2)^2 \sin^2\phi] \delta\rho \, d\Omega \\
\zeta_2'(\boldsymbol{u},\delta\boldsymbol{u}) &= \delta T \iint_\Omega [(\nabla y^1)^2 \sin^2\phi - (\nabla y^{1^T} \nabla y^2)\sin 2\phi + (\nabla y^2)^2 \cos^2\phi] d\Omega \\
&\quad - \zeta \iint_\Omega [(y^1)^2 \sin^2\phi - (y^1 y^2)\sin 2\phi + (y^2)^2 \cos^2\phi] \delta\rho \, d\Omega.
\end{aligned} \tag{5.140}$$

Vibration of a Plate

As a final example, consider the variable thickness vibrating plate in Fig. 3.14. Plate thickness $h(\boldsymbol{x})$, Young's modulus E, and mass density ρ are taken as design variables. Without repeating the definition for energy and kinetic bilinear forms given by (3.113) and (3.114), the first variation of these bilinear forms, evaluated at $\bar{y} = y$, may be written as

$$\begin{aligned}
a'_{\delta\boldsymbol{u}}(y,y) &= \frac{\delta E}{12(1-\nu^2)} \iint_\Omega h^3 \boldsymbol{\kappa}(y)^T \boldsymbol{C} \boldsymbol{\kappa}(y) \, d\Omega \\
&\quad + \frac{E}{4(1-\nu^2)} \iint_\Omega h^2 \boldsymbol{\kappa}(y)^T \boldsymbol{C} \boldsymbol{\kappa}(y) \delta h \, d\Omega \\
d'_{\delta\boldsymbol{u}}(y,y) &= \delta\rho \iint_\Omega h y^2 \, d\Omega + \rho \iint_\Omega y^2 \delta h \, d\Omega.
\end{aligned} \tag{5.141}$$

The derivative of a simple eigenvalue is therefore given by (5.116) as

$$\begin{aligned}
\zeta' &= \frac{\delta E}{12(1-\nu^2)} \iint_\Omega h^3 \boldsymbol{\kappa}(y)^T \boldsymbol{C} \boldsymbol{\kappa}(y) \, d\Omega - \left[\zeta \iint_\Omega h y^2 \, d\Omega\right] \delta\rho \\
&\quad + \frac{E}{4(1-\nu^2)} \iint_\Omega h^2 \boldsymbol{\kappa}(y)^T \boldsymbol{C} \boldsymbol{\kappa}(y) \delta h \, d\Omega - \zeta\rho \iint_\Omega y^2 \delta h \, d\Omega.
\end{aligned} \tag{5.142}$$

It is clear that increasing Young's modulus also increases the natural frequency, and that increasing the density decreases the natural frequency, as one would intuitively expect. The effect of a thickness variation, however, is not obvious, since the coefficient of δh in the integral may be either positive or negative. Numerical examples of the effect of thickness variations will be considered in Section 5.2.4.

In the case of a repeated eigenvalue [74], if y^1 and y^2 are eigenfunctions corresponding to a repeated eigenvalue ζ, then as with the membrane problem, (5.121) and (5.122) provide the directional derivatives of the repeated eigenvalues, where the rotation parameter ϕ is given by (5.123) and where

$$\begin{aligned}\mathcal{M}_{ij} &= \frac{\delta E}{12(1-\nu^2)} \iint_\Omega h^3 \boldsymbol{\kappa}(y^i)^T \boldsymbol{C\kappa}(y^j)\, d\Omega \\ &+ \frac{E}{4(1-\nu^2)} \iint_\Omega h^2 \boldsymbol{\kappa}(y^i)^T \boldsymbol{C\kappa}(y^j)\, \delta h\, d\Omega \\ &- \zeta \left[\delta\rho \iint_\Omega h y^i y^j \, d\Omega + \rho \iint_\Omega y^i y^j \delta h \, d\Omega \right], \qquad i,j = 1,2.\end{aligned} \tag{5.143}$$

5.2.3 Numerical Considerations

The numerical aspects involved in evaluating design sensitivity formulas for a simple eigenvalue in (5.116), or for a repeated eigenvalue in (5.121) and (5.122), follow the same pattern as those considerations presented in Section 5.1.4 on the computational aspects of static design sensitivity. However, with conservative eigenvalue problems, it is an additional advantage that an adjoint variable does not need to be calculated as the solution to a separate problem. This feature allows the design sensitivity to be directly computed once the analysis problem has been solved. It should be added that direct computation is not possible for nonconservative problems [74] and [75].

As in the case of static response presented in Section 5.1.4, the design can be parameterized as shown in Section 12.1 and explicit design derivatives of eigenvalues can be obtained with respect to design parameters. Since the functionals arising in the present formulation are identical to those appearing in Section 5.1.4, the reader is referred to that section for further details on numerical considerations involved in evaluating parameterized design sensitivity.

5.2.4 Numerical Examples of Eigenvalue Design Sensitivity

Plate

To account for the effect of plate thickness on variations in the natural frequency, consider the application in (5.142): a clamped, square plate with a dimension of 1 m and a uniform thickness of h = 0.05 m, with E = 200 GPa , ν = 0.3, and ρ = 7870 kg/m^3. For this design, the eigenvalue is $\zeta = 0.3687 \times 10^6$ (rad/sec^2). If plate thickness is considered as a design variable and if piecewise-constant thickness is assumed, as with the plate example in Section 5.1.5, then (5.142) can be rewritten as

$$\zeta' = \sum_{i=1}^{n} \left[\iint_{\Omega_i} \left\{ \frac{E b_i^2}{4(1-\nu^2)} \boldsymbol{\kappa}(y)^T \boldsymbol{C\kappa}(y) - \zeta \rho y^2 \right\} d\Omega \right] \delta b_i. \tag{5.144}$$

As with the plate example in Section 5.1.5, a nonconforming rectangular plate element with 12 degrees of freedom is used. The graph of the coefficient of δb_i presented in Fig. 5.16 explains how the addition or deletion of material affects the eigenvalue. The maximum value of the coefficient of δb_i is $\Lambda_{max} = 7.949 \times 10^6$ in the middle of the edge elements, while its minimum value occurs at the corner elements, with $\Lambda_{max} = 3.365 \times 10^3$. Thus, in order to increase the eigenvalue most effectively, material should be removed from the vicinity of the four corners and added to the middle of the four edges.

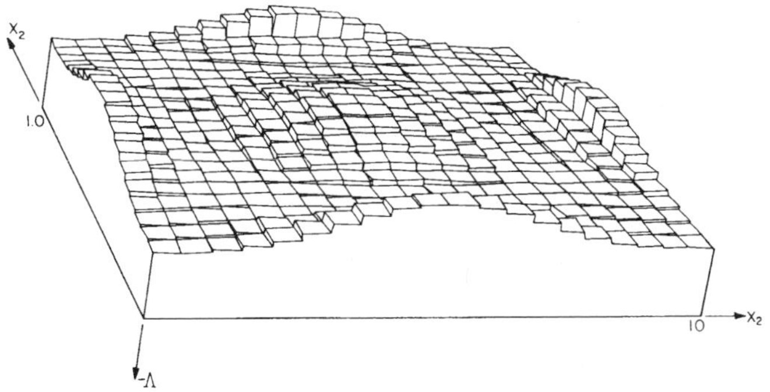

Figure 5.16. Eigenvalue sensitivity Λ for plate.

5.3 Transient Dynamic Response Design Sensitivity

Design sensitivity of a transient dynamic response in a distributed parameter system has received a minimum amount of attention in the literature [76], in contrast to the large amount of attention devoted to design sensitivity analysis and optimization of a dynamic control system. A development by Rousselet [73] provided the beginning of a foundation for design sensitivity analysis of the transient dynamic problem, but work remains to be done. Since the theory of dynamic, distributed-parameter design sensitivity analysis is not as well developed as that of static response and eigenvalues, a more formal treatment of the subject is presented in this section. A variational formulation of the structural dynamics problem, as outlined in Section 2.4, is used here. Under the assumption of design differentiability of the state variable, the adjoint variable method presented in Section 5.1 is extended to the transient dynamic response problem and analytical examples are provided.

5.3.1 Design Sensitivity of Structural Dynamics Performance

For derivational simplicity, it is presumed that the design vector is a scalar function. Let Ω be the structural domain and let $t \in [0, t_T]$ be the time interval. Consider a general integral functional in transient dynamics, in the form

$$\psi = \int_0^{t_T} \iint_\Omega g(z, \nabla z, u)\, d\Omega\, dt. \tag{5.145}$$

Since solution z of the structural equation is design dependent, dependence in such a functional appears both explicitly and, through argument z, implicitly. As in static response and eigenvalue problems, something must be known about the nature of the dependence of the state variable on the design, that is, $z = z(\boldsymbol{x},t;u)$. Using a slightly more restrictive set of hypotheses than those employed in Section 2.4 in [5] for the static response problem, it has been shown by Rousselet [73] that z is Fréchet differentiable with respect to design u. This fact allows us to develop explicit expressions for the sensitivity of functional ψ in (5.145) with respect to design u, much in the fashion that sensitivity results were developed for static problems in Section 5.1.

As with the static problem, design u is perturbed in the direction of δu, and is controlled by parameter τ. Take the variation of (5.145) with respect to τ to obtain

$$\begin{aligned}\psi' &= \frac{d}{d\tau}\left[\int_0^T \iint_\Omega g(z(x,t;u+\tau\delta u),\nabla z(x,t;u+\tau\delta u),u+\tau\delta u)\,d\Omega\,dt\right]\Bigg|_{\tau=0} \\ &= \int_0^T \iint_\Omega (g_{,z}z' + g_{,\nabla z}:\nabla z' + g_{,u}\delta u)\,d\Omega\,dt.\end{aligned} \tag{5.146}$$

The objective here is to rewrite the first two terms on the right side of (5.146) explicitly in terms of the design variation δu.

The governing variational equation for the structural dynamic problem in (2.37) is rewritten here as

$$\begin{aligned}&\int_0^T \left\{ \iint_\Omega [\bar{z}^T \rho(u) z_{,tt} + \bar{z}^T C(u) z_{,t}]\,d\Omega + a_u(z,\bar{z}) \right\} dt \\ &\qquad = \int_0^T \iint_\Omega \bar{z}^T f\,d\Omega\,dt, \qquad \forall \bar{z} \in Z,\end{aligned} \tag{5.147}$$

with initial conditions

$$\begin{aligned} z(x,0;u) &= z^0(x), \\ z_{,t}(x,0;u) &= z_{,t}^0(x), \end{aligned} \qquad x \in \Omega. \tag{5.148}$$

Presuming that $\rho(u)$, $C(u)$, $f(u)$, and $a_u(\bullet,\bullet)$ are differentiable with respect to the design, along with solution z of the dynamic problem, take the variation of both sides of (5.147) to obtain

$$\begin{aligned}&\int_0^T \left\{ \iint_\Omega [\bar{z}^T \rho_{,u} z_{,tt} + \bar{z}^T C_{,u} z_{,t} - \bar{z}^T f_{,u}]\delta u\,d\Omega + a'_{\delta u}(z,\bar{z}) \right\} dt \\ &\qquad + \int_0^T \left\{ \iint_\Omega [\bar{z}^T \rho z'_{,tt} + \bar{z}^T C z'_{,t}]\,d\Omega + a_u(z',\bar{z}) \right\} dt = 0, \qquad \forall \bar{z} \in Z.\end{aligned} \tag{5.149}$$

To take advantage of this equation, terms in the second integral may be integrated by parts to move the time derivatives from z' to $\bar{z}$. To carry out this calculation it is necessary to reverse the order of integration, to carry out the integration by parts with respect to time, and to again change the order of integration, to obtain

$$\begin{aligned}&\int_0^T \left\{ \iint_\Omega [z'^T \rho \bar{z}_{,tt} - z'^T C \bar{z}_{,t}]\,d\Omega + a_u(z',\bar{z}) \right\} dt \\ &\qquad + \iint_\Omega [\bar{z}^T \rho z'_{,t} - \bar{z}_{,t}^T \rho z' - \bar{z}^T C z']\Big|_0^{T} \,d\Omega \\ &\qquad = \int_0^T \left\{ \iint_\Omega [\bar{z}^T f_{,u} - \bar{z}^T \rho_{,u} z_{,tt} - \bar{z}^T C_{,u} z_{,t}]\delta u\,d\Omega - a'_{\delta u}(z,\bar{z}) \right\} dt, \qquad \forall \bar{z} \in Z.\end{aligned} \tag{5.150}$$

Note that as a result of the initial conditions given by (5.148), for which the right side does not depend on u, the variation yields

$$\begin{aligned} z'(x,0;u) &= 0, \\ z'_{,t}(x,0;u) &= 0, \end{aligned} \qquad x \in \Omega. \tag{5.151}$$

which eliminates the initial terms in (5.150) that arose due to integration by parts.

To take advantage of the identity in (5.150), which must hold for all $\bar{z} \in Z$, let us define an adjoint equation by replacing z' with an arbitrary virtual displacement $\bar{\lambda} \in Z$ in (5.150) and (5.146), defining the variational adjoint equation for $\lambda \in Z$ as

$$\int_0^{t_T}\left\{\iint_\Omega[\bar{\boldsymbol{\lambda}}^T\rho\boldsymbol{\lambda}_{,tt}-\bar{\boldsymbol{\lambda}}^TC\boldsymbol{\lambda}_{,t}]d\Omega+a_u(\boldsymbol{\lambda},\bar{\boldsymbol{\lambda}})\right\}dt$$
$$=\int_0^{t_T}\iint_\Omega[g_{,z}\bar{\boldsymbol{\lambda}}+g_{,\nabla z}:\nabla\bar{\boldsymbol{\lambda}}]d\Omega dt,\qquad\forall\bar{\boldsymbol{\lambda}}\in Z,\tag{5.152}$$

where the additional terminal condition on $\boldsymbol{\lambda}$ is defined as

$$\begin{aligned}\boldsymbol{\lambda}(\boldsymbol{x},t_T;u)&=\boldsymbol{0},\\ \boldsymbol{\lambda}_{,t}(\boldsymbol{x},t_T;u)&=\boldsymbol{0},\end{aligned}\qquad\boldsymbol{x}\in\Omega.\tag{5.153}$$

The terminal conditions of (5.153) are introduced to eliminate those terms that have arisen at $t=t_T$ in (5.150) due to the integration by parts.

Since (5.150) must hold for all $\bar{\boldsymbol{z}}\in Z$, this equation may be evaluated at a specific $\bar{\boldsymbol{z}}=\boldsymbol{\lambda}$ using (5.153), to obtain

$$\int_0^{t_T}\left\{\iint_\Omega[\boldsymbol{z}'^T\rho\boldsymbol{\lambda}_{,tt}-\boldsymbol{z}'^TC\boldsymbol{\lambda}_{,t}]d\Omega+a_u(\boldsymbol{z}',\boldsymbol{\lambda})\right\}dt$$
$$=\int_0^{t_T}\left\{\iint_\Omega[\boldsymbol{\lambda}^T\boldsymbol{f}_{,u}-\boldsymbol{\lambda}^T\rho_{,u}\boldsymbol{z}_{,tt}-\boldsymbol{\lambda}^TC_{,u}\boldsymbol{z}_{,t}]\delta u\,d\Omega-a'_{\delta u}(\boldsymbol{z},\boldsymbol{\lambda})\right\}dt.\tag{5.154}$$

Similarly, (5.152) must hold for all $\bar{\boldsymbol{\lambda}}\in Z$, so by evaluating this equation at $\bar{\boldsymbol{\lambda}}=\boldsymbol{z}'$ we obtain

$$\int_0^{t_T}\left\{\iint_\Omega[\boldsymbol{z}'^T\rho\boldsymbol{\lambda}_{,tt}-\boldsymbol{z}'^TC\boldsymbol{\lambda}_{,t}]d\Omega+a_u(\boldsymbol{z}',\boldsymbol{\lambda})\right\}dt$$
$$=\int_0^{t_T}\iint_\Omega[g_{,z}\boldsymbol{z}'+g_{,\nabla z}:\nabla\boldsymbol{z}']d\Omega dt.\tag{5.155}$$

Note that the right side of (5.155) consists of exactly the same terms in (5.146) that are written in terms of δu. Furthermore, the left side of (5.154) and (5.155) are identical, such that

$$\int_0^{t_T}\iint_\Omega[g_{,z}\boldsymbol{z}'+g_{,\nabla z}:\nabla\boldsymbol{z}']d\Omega dt$$
$$=\int_0^{t_T}\left\{\iint_\Omega[\boldsymbol{\lambda}^T\boldsymbol{f}_{,u}+\boldsymbol{\lambda}_{,t}^T\rho_{,u}\boldsymbol{z}_{,t}-\boldsymbol{\lambda}^TC_{,u}\boldsymbol{z}_{,t}]\delta u\,d\Omega-a'_{\delta u}(\boldsymbol{z},\boldsymbol{\lambda})\right\}dt\tag{5.156}$$
$$+\iint_\Omega[\boldsymbol{\lambda}(\boldsymbol{x},0;u)^T\rho_{,u}\boldsymbol{z}_{,t}^0(\boldsymbol{x})]\delta u\,d\Omega,$$

where an integration by parts has been carried out and the initial conditions in (5.148) have been used to reduce the order of differentiation of $\boldsymbol{z}$ with respect to t. Substituting this result into (5.146) yields the explicit expression of ψ' in terms of δu, as

$$\psi'=\int_0^{t_T}\left\{\iint_\Omega[g_{,u}+\boldsymbol{\lambda}^T\boldsymbol{f}_{,u}+\boldsymbol{\lambda}_{,t}^T\rho_{,u}\boldsymbol{z}_{,t}-\boldsymbol{\lambda}^TC_{,u}\boldsymbol{z}_{,t}]\delta u\,d\Omega-a'_{\delta u}(\boldsymbol{z},\boldsymbol{\lambda})\right\}dt$$
$$+\iint_\Omega[\boldsymbol{\lambda}(\boldsymbol{x},0;u)^T\rho_{,u}\boldsymbol{z}_{,t}^0(\boldsymbol{x})]\delta u\,d\Omega.\tag{5.157}$$

Note that the variational adjoint equation in (5.152) and (5.153) is not the same as the variational state equation in (5.147) and (5.148). There are two fundamental differences between these two sets of equations. First, while the state equation includes the initial conditions in (5.148), the adjoint equation contains those terminal conditions in (5.153). Second, the sign of the damping term in (5.147) is different from (5.152). These facts somewhat complicate calculations associated with dynamic design sensitivity analysis, since the adjoint dynamic problem will be different from the original dynamic problem. As will be seen in the examples presented in the following section, however, more similarities exist than at first appear.

5.3.2 Analytical Examples

String

The equation of motion for a vibrating string in a viscous medium is given in the form

$$\begin{aligned} \rho(u)z_{,tt} + C(u)z_{,t} - Tz_{,11} &= f(x,t,u), & 0<t<t_T,\ x\in[0,l] \\ z(0,t;u) &= 0, & 0\le t\le t_T \\ z(l,t;u) &= 0, & 0\le t\le t_T \\ z(x,0;u) &= z^0(x), & t=0 \\ z_{,t}(x,0;u) &= z_{,t}^0(x), & t=0. \end{aligned} \tag{5.158}$$

where $\rho(u)$ is the mass per unit length along the string, $C(u)$ is the damping coefficient per unit length, and T is the string tension. Space Z of all kinematically admissible displacements is $H_0^1(0,l)$, that is, space Z is the set of all functions in Sobolev space $H^1(0,l)$ that vanish at the boundary (interval's end points). The energy bilinear form is given in (3.94) as

$$a_u(z,\bar{z}) = T\int_0^l z_{,1}\bar{z}_{,1}\,dx. \tag{5.159}$$

Consider the mean square displacement as a performance measure, that is,

$$\psi_1 = \frac{1}{t_T}\int_0^{t_T}\int_0^l z^2\,dx\,dt. \tag{5.160}$$

The adjoint equations for this problem, derived from (5.152) and (5.153), are

$$\begin{aligned} &\int_0^{t_T}\left\{\int_0^l[\bar{\lambda}\rho\lambda_{,tt} - \bar{\lambda}C\lambda_{,t}]dx + T\int_0^l \bar{\lambda}_{,1}\lambda_{,1}\,dx\right\}dt \\ &= \frac{1}{t_T}\int_0^{t_T}\int_0^l 2z\bar{\lambda}\,dx\,dt, \qquad \forall\bar{\lambda}\in Z, \end{aligned} \tag{5.161}$$

and

$$\begin{aligned} \lambda(x,t_T) = \lambda_{,t}(x,t_T) &= 0, & 0\le x\le l \\ \lambda(0,t) = \lambda(l,t) &= 0, & 0\le t\le t_T. \end{aligned} \tag{5.162}$$

The energy bilinear form for the string in (5.159) has been employed. Integration by parts in (5.161), using the boundary conditions in (5.162) and the fact that $\bar{\lambda}$ satisfies the same boundary conditions, yields

$$\int_0^{t_T}\int_0^l\left\{\rho\lambda_{,tt} - C\lambda_{,t} - T\lambda_{,11} - \frac{2}{t_T}z\right\}\bar{\lambda}\,dx\,dt = 0, \qquad \forall\bar{\lambda}\in Z. \tag{5.163}$$

Since $\bar{\lambda}$ is arbitrary except for its boundary conditions, its coefficient in (5.163) must be zero, yielding the following differential equation:

$$\rho\lambda_{,tt} - C\lambda_{,t} - T\lambda_{,11} = \frac{2}{t_T}z. \tag{5.164}$$

Note that this differential equation differs in form from (5.158) only by the algebraic sign of the damping term and by the applied load on the right side.

To verify that the adjoint problem in (5.162) and (5.164) can be rewritten in a form closer to that of the physical structure [(5.158)], a backward time, $\tau \equiv t_T - t$, may be defined. With this time variable, $d/dt = -d/d\tau$, and the terminal conditions in (5.162) for the t variable become initial condition for the τ variable. Thus, the backward time initial-boundary-value problem for the new variable $\tilde{\lambda}(x,\tau) = \lambda(x, t_T - t)$ becomes

$$
\begin{aligned}
\rho\tilde{\lambda}_{,\tau\tau} + C\tilde{\lambda}_{,\tau} - T\tilde{\lambda}_{,11} &= \frac{2}{t_T} z(x, t_T - \tau), && 0 < \tau < t_T,\ 0 < x < l \\
\tilde{\lambda}(0,\tau) = \tilde{\lambda}(l,\tau) &= 0, && 0 \le \tau < t_T \\
\tilde{\lambda}(x,0) = \tilde{\lambda}_{,\tau}(x,0) &= 0, && 0 < x < l.
\end{aligned}
\tag{5.165}
$$

Consequently, the *adjoint structure* is physically the same as the original structure, but with the addition of a backward clock and an applied load, $2z(x, t_T - \tau)/t_T$.

If the applied load $f(x,t,u)$ in (5.158) is the string self-weight added to an excitation $F(x,t)$, then $f = g\rho + F(x,t)$, where g is the acceleration of gravity. Equation (5.157) then yields the sensitivity of functional ψ_1 in (5.160) as

$$
\begin{aligned}
\psi_1' = &\int_0^{t_T} \left\{ \int_0^l [(\lambda g + \lambda_{,t} z_{,t})\delta\rho - \lambda z_{,t}\delta C]\,dx - \delta T \int_0^l \lambda_{,1} z_{,1}\,dx \right\} dt \\
&+ \int_0^l \lambda(x,0) z_{,t}^0(x) \delta\rho\, dx.
\end{aligned}
\tag{5.166}
$$

Since $\delta\rho$ and δC depend on x and not on t, the integration order for the first term in (5.166) may be reversed, yielding the explicit relation

$$
\begin{aligned}
\psi_1' = &\int_0^l \left\{ \int_0^{t_T} [(\lambda g + \lambda_{,t} z_{,t})\,dt + \lambda(x,0) z_{,t}^0(x) \right\} \delta\rho\, dx \\
&- \int_0^l \left\{ \int_0^{t_T} \lambda z_{,t}\,dt \right\} \delta C\, dx - \left\{ \int_0^{t_T} \int_0^l \lambda_{,1} z_{,1}\,dx\,dt \right\} \delta T.
\end{aligned}
\tag{5.167}
$$

Note that the sensitivity coefficients of $\delta\rho$ and δC are explicitly a function of x since the time variable has been integrated in calculating the coefficients of $\delta\rho$ and δC. This fortunate circumstance allows the time variable to be eliminated from the design sensitivity formula, which is natural since the design vector $\boldsymbol{u} = [\rho(x), C(x), T]^T$ is only dependent on x.

Beam

The equation for beam motion in a viscous fluid is

$$
\begin{aligned}
m(u) z_{,tt} + C(u) z_{,t} + (EI(u) z_{,11})_{,11} &= f(x,t,u), \qquad 0 < t < t_T,\ x \in [0,l] \\
z(x,0;u) &= z^0(x) \\
z_{,t}(x,0;u) &= z_{,t}^0(x),
\end{aligned}
\tag{5.168}
$$

where $m(u)$ is the mass per unit length of the beam and $C(u)$ is the damping coefficient per unit length. Boundary conditions may be the same as any reasonable set of boundary conditions, as in (3.19) through (3.21). For the clamped-clamped beam, $Z = H_0^2(0,l)$. The energy bilinear form for the beam is given in (3.14) as

$$
a_u(z,\bar{z}) = \int_0^l EI(u) z_{,11} \bar{z}_{,11}\, dx. \tag{5.169}
$$

Consider the dynamics of a clamped-clamped beam, with functional ψ_2 representing the mean value over time of the square of displacement at a given point $\hat{x}$.

$$\begin{aligned}\psi_2 &= \frac{1}{t_T}\int_0^{t_T} z^2(\hat{x},t)\,dt \\ &= \frac{1}{t_T}\int_0^{t_T}\int_0^l \delta(x-\hat{x})z^2(x,t)\,dx\,dt,\end{aligned} \tag{5.170}$$

with cross-sectional area A as the design variable, $m = \rho A$, $I = \alpha A^2$, and $f = gh + F(x,t)$. In this case, the adjoint equation of (5.152) is

$$\begin{aligned}&\int_0^{t_T}\left\{\int_0^l [\rho A\bar{\lambda}\lambda_{,tt} - C\bar{\lambda}\lambda_{,t}]dx + \int_0^l EI\bar{\lambda}_{,11}\lambda_{,11}\,dx\right\}dt \\ &= \frac{2}{t_T}\int_0^{t_T}\int_0^l \delta(x-\hat{x})z\bar{\lambda}\,dx\,dt, \qquad \forall\bar{\lambda}\in Z,\end{aligned} \tag{5.171}$$

with the terminal and boundary conditions for a clamped-clamped beam,

$$\begin{aligned}&\lambda(x,t_T) = \lambda_{,t}(x,t_T) = 0, && 0 \le x \le l \\ &\lambda(0,t) = \lambda_{,1}(0,t) = \lambda(l,t) = \lambda_{,1}(l,t) = 0, && 0 \le t \le t_T.\end{aligned} \tag{5.172}$$

To reduce the variational equation in (5.171) to a differential equation, carry out integration by parts using the boundary conditions in (5.172) to obtain

$$\int_0^{t_T}\int_0^l \bar{\lambda}\left\{\rho A\lambda_{,tt} - C\lambda_{,t} + (EI\lambda_{,11})_{,11} - \frac{2}{t_T}\delta(x-\hat{x})z\right\}dx\,dt = 0, \qquad \forall\bar{\lambda}\in Z. \tag{5.173}$$

Since this equation must hold for all $\bar{\lambda}$ satisfying boundary conditions, its coefficient must be zero, leading to the following differential equation:

$$\rho A\lambda_{,tt} - C\lambda_{,t} + (EI\lambda_{,11})_{,11} = \frac{2}{t_T}\delta(x-\hat{x})z, \tag{5.174}$$

which, except for the sign for damping terms, is nothing other than a beam equation with point load $2z(\hat{x},t)/t_T$ applied at point $\hat{x}$. As with a string in (5.165), a backward time τ could be defined and the equations rewritten to obtain the adjoint equations with a backward time variable, in exactly the same form as the basic structural equation.

The design sensitivity result from (5.157) may thus be directly written as

$$\begin{aligned}\psi_2' = &\int_0^{t_T}\left\{\int_0^l [(\lambda\gamma + \rho\lambda_{,t}z_{,t})\delta A - \lambda z_{,t}\delta C]dx - \int_0^l 2E\alpha A\delta A\lambda_{,11}z_{,11}\,dx\right\}dt \\ &+ \int_0^l \lambda(x,0)\rho z_{,t}^0(x)\delta A\,dx.\end{aligned} \tag{5.175}$$

Reversing the integration order in the first integral of (5.175) yields

$$\begin{aligned}\psi_2' = &\int_0^l\left\{\int_0^{t_T}[(\lambda\gamma + \rho\lambda_{,t}z_{,t} - 2E\alpha A\lambda_{,11}z_{,11})dt + \lambda(x,0)\rho z_{,t}^0(x)\right\}\delta A\,dx \\ &- \int_0^l\left\{\int_0^{t_T}\lambda z_{,t}\,dt\right\}\delta C\,dx,\end{aligned} \tag{5.176}$$

which again provides a design sensitivity coefficient of δA only as a function of x.

Membrane

The equation of motion for a membrane in a viscous fluid is

$$\begin{aligned} \rho(u)z_{,tt} + C(u)z_{,t} - T\nabla^2 z &= f(\boldsymbol{x},t,u), && 0<t<t_T,\ \boldsymbol{x}\in\Omega \\ z(\boldsymbol{x},t;u) &= 0, && 0\le t\le t_T,\ \boldsymbol{x}\in\Gamma \\ z(\boldsymbol{x},0;u) &= z^0(\boldsymbol{x}), && \boldsymbol{x}\in\Omega \\ z_{,t}(\boldsymbol{x},0;u) &= z_{,t}^0(\boldsymbol{x}), && \boldsymbol{x}\in\Omega, \end{aligned} \tag{5.177}$$

where $\rho(u)$ is the mass per unit area of the membrane and $C(u)$ is the damping coefficient per unit area. Space Z of kinematically admissible displacements is $H_0^1(\Omega)$. The energy bilinear form for this problem is given in (3.108) as

$$a_u(z,\bar{z}) = T\iint_\Omega (z_{,1}\bar{z}_{,1} + z_{,2}\bar{z}_{,2})\,d\Omega. \tag{5.178}$$

Consider a vibrating membrane with the mean square displacement as the functional

$$\psi_3 = \frac{1}{t_T}\int_0^{t_T}\iint_\Omega z^2\,d\Omega\,dt. \tag{5.179}$$

In this case, the adjoint problem in (5.152) would be

$$\begin{aligned} &\int_0^{t_T}\left\{\iint_\Omega [\bar{\lambda}\rho\lambda_{,tt} - \bar{\lambda}C\lambda_{,t}]d\Omega + T\iint_\Omega [\bar{\lambda}_{,1}\lambda_{,1} + \bar{\lambda}_{,2}\lambda_{,2}]d\Omega\right\}dt \\ &\qquad = \frac{1}{t_T}\int_0^{t_T}\iint_\Omega 2z\bar{\lambda}\,d\Omega\,dt, \qquad \forall\bar{\lambda}\in Z, \end{aligned} \tag{5.180}$$

with the following terminal condition and boundary conditions:

$$\begin{aligned} \lambda(\boldsymbol{x},t_T) = \lambda_{,t}(\boldsymbol{x},t_T) &= 0, && \boldsymbol{x}\in\Omega \\ \lambda(\boldsymbol{x},t) &= 0, && \boldsymbol{x}\in\Gamma,\ 0\le t\le t_T. \end{aligned} \tag{5.181}$$

To reduce the adjoint variational (5.180) to a differential equation, carry out integration by parts using the boundary conditions provided by (5.181) to obtain

$$\int_0^{t_T}\iint_\Omega \bar{\lambda}\left\{\rho\lambda_{,tt} - C\lambda_{,t} - T\nabla^2\lambda - \frac{2}{t_T}z\right\}d\Omega\,dt = 0, \qquad \forall\bar{\lambda}\in Z. \tag{5.182}$$

Since $\bar{\lambda}$ is arbitrary except for its boundary conditions, its coefficient must be zero, yielding

$$\rho\lambda_{,tt} - C\lambda_{,t} - T\nabla^2\lambda = \frac{2}{t_T}z. \tag{5.183}$$

Except for the sign of the damping term and a different applied load on the right side, (5.183) is the membrane equation. As in the case of a string, a backward time τ could be defined to obtain a backward time initial-boundary-value problem, which is the same membrane equation, but with a different load.

For a given load $f(\boldsymbol{x},t)$, one may obtain a design sensitivity for ψ_3 from (5.157) as

$$\begin{aligned} \psi_3' = &\int_0^{t_T}\iint_\Omega \lambda_{,t}z_{,t}\,\delta\rho\,d\Omega\,dt - \int_0^{t_T}\iint_\Omega \lambda z_{,t}\,\delta C\,d\Omega\,dt \\ &+ \iint_\Omega \lambda(\boldsymbol{x},0)z_{,t}^0\,\delta\rho\,d\Omega. \end{aligned} \tag{5.184}$$

Since $\delta\rho$ is independent of t, reversing the integration order in the first term of (5.184) yields

$$\psi_3' = \iint_\Omega \left\{ \int_0^{t_T} \lambda_{,t} z_{,t}\, dt + \lambda(\boldsymbol{x},0) z_{,t}^0 \right\} \delta\rho\, d\Omega - \int_0^{t_T} \iint_\Omega \lambda z_{,t}\, \delta C\, d\Omega\, dt. \tag{5.185}$$

Plate

The equation of motion for a plate in a viscous fluid is

$$\begin{aligned} & m(u) z_{,tt} + C(u) z_{,t} + [D(\boldsymbol{u})(z_{,11} + \nu z_{,22})]_{,11} + [D(\boldsymbol{u})(z_{,22} + \nu z_{,11})]_{,22} \\ & \quad +2(1-\nu)[D(\boldsymbol{u}) z_{,12}]_{,12} = f(\boldsymbol{x},t,u) \\ & \quad z(\boldsymbol{x},0,u) = z^0(\boldsymbol{x},u), \qquad \boldsymbol{x} \in \Omega \\ & \quad z_{,t}(\boldsymbol{x},0,u) = z_{,t}^0(\boldsymbol{x},u), \qquad \boldsymbol{x} \in \Omega \\ & \quad z(\boldsymbol{x},t,u) = 0, \qquad \boldsymbol{x} \in \Gamma,\ 0 \le t \le t_T \\ & \quad \frac{\partial z}{\partial n}(\boldsymbol{x},t,u) = 0, \qquad \boldsymbol{x} \in \Gamma,\ 0 \le t \le t_T, \end{aligned} \tag{5.186}$$

where $m(u)$ is the mass per unit area of the plate and $C(u)$ is the damping coefficient per unit area. For a clamped plate in (5.186), $Z = H_0^2(\Omega)$. The energy bilinear form for the plate is given in (3.41) as

$$a_u(z,\bar{z}) = \iint_\Omega \frac{Eh^3}{12(1-\nu^2)} \boldsymbol{\kappa}(\bar{z})^T \boldsymbol{C} \boldsymbol{\kappa}(z)\, d\Omega. \tag{5.187}$$

Consider the dynamics of a clamped plate with a damping coefficient of zero, given load $f(\boldsymbol{x},t)$, variable thickness h, and $m = \rho h$. The functional ψ_4 is the work done by the applied load during the plate's motion, that is

$$\psi_4 = \int_0^{t_T} \iint_\Omega f z_{,t}\, d\Omega\, dt. \tag{5.188}$$

Presuming that load function f is differentiable with respect to time, and that $f(\boldsymbol{x},0) = f(\boldsymbol{x},t_T) = 0$, integrate the term on the right side of (5.188) by parts with respect to time, to obtain

$$\psi_4 = -\int_0^{t_T} \iint_\Omega f_{,t} z\, d\Omega\, dt. \tag{5.189}$$

In this case, the adjoint variational problem in (5.152) becomes

$$\begin{aligned} & \int_0^{t_T} \left\{ \iint_\Omega \bar{\lambda} \rho h \lambda_{,tt}\, d\Omega + \iint_\Omega \frac{Eh^3}{12(1-\nu^2)} \boldsymbol{\kappa}(\bar{\lambda})^T \boldsymbol{C} \boldsymbol{\kappa}(\lambda)\, d\Omega \right\} dt \\ & \quad = \int_0^{t_T} \iint_\Omega f_{,t} \bar{\lambda}\, d\Omega\, dt, \qquad \forall \bar{\lambda} \in Z, \end{aligned} \tag{5.190}$$

with the following boundary conditions for the clamped plate and terminal condition:

$$\begin{aligned} & \lambda(\boldsymbol{x},t_T) = \lambda_{,t}(\boldsymbol{x},t_T) = 0, \qquad \boldsymbol{x} \in \Omega \\ & \lambda(\boldsymbol{x},t) = \frac{\partial \lambda}{\partial n}(\boldsymbol{x},t) = 0, \qquad \boldsymbol{x} \in \Gamma,\ 0 \le t \le t_T. \end{aligned} \tag{5.191}$$

By using the definition of the plate differential operator A, and the spatial integration by parts on the left of (5.190), the following equation is obtained

$$\int_0^T \iint_\Omega \bar{\lambda}\{\rho h \lambda_{,tt} + A\lambda + f_{,t}\} d\Omega\, dt = 0, \tag{5.192}$$

which must hold for all arbitrary virtual displacements $\bar{\lambda}$ that satisfy the boundary conditions. Therefore, the differential equation for λ is

$$\rho h \lambda_{,tt} + A\lambda = -f_{,t}, \tag{5.193}$$

which is of the same fundamental form as the basic plate equation without a damping term.

Using the solution λ from the adjoint equations, we can bring the design sensitivity for ψ_4 directly from (5.157) as

$$\psi_3' = \iint_\Omega \left\{ \int_0^T [\rho\lambda_{,t} z_{,t} - \frac{Eh^2}{4(1-\nu^2)} \boldsymbol{\kappa}(z)^T \boldsymbol{C}\boldsymbol{\kappa}(\lambda)] dt + \lambda(\boldsymbol{x},0)\rho z_{,t}^0(\boldsymbol{x}) \right\} \delta h\, d\Omega. \tag{5.194}$$

5.4 Frequency Response Design Sensitivity

The dynamic frequency response of mechanical and structural systems is of interest in design problems that are subjected to harmonically varying external loads caused by a reciprocating power train or by such other rotating machine parts as motors, fans, compressors, and forging hammers [19]. Airplane body and wing structures are also subjected to a harmonic load, transmitted from the propulsion system. In addition, ship vibration resulting from the propeller and from engine excitation can cause noise problems, cracks, fatigue failure of the tail-shaft, and discomfort to the crew. When a machine or structure oscillates in some form of periodic or random motion, that motion generates alternating pressure waves that propagate from the moving surface at the velocity of sound. For example, interior sound pressure in an automobile compartment can occur when the input force transmitted from the road and power train excites the vehicle compartment boundary panels. Motion with frequencies between 20 Hz and 20 kHz stimulates the human hearing mechanism [20].

In order to find the relationship between a variation in the frequency response and a variation in the design parameters, the design derivative of the variational governing equation is taken. Since no mathematical proof is available regarding the existence and uniqueness of sensitivity, presuming that all variables are smooth, a formal approach is taken. Explicit design sensitivity formulas for dynamic frequency response performance measures are then obtained using the adjoint variable and direct differentiation methods [28] and [29]. One major difference compared with previous sections is that the complex conjugate of the adjoint variable is used in the design sensitivity expression. The bilinear and linear forms in the previous sections must be understood as the sesquilinear and semilinear forms in frequency response analysis.

5.4.1 Design Sensitivity Analysis of Frequency Response

A general performance measure representing a variety of structural responses can be defined as

$$\psi = \iint_{\Omega} g(z,\nabla z,\boldsymbol{u})\, d\Omega, \tag{5.195}$$

where the function $g(z,\nabla z,\boldsymbol{u})$ is assumed to be continuously differentiable with respect to its arguments, and $\nabla z = [\partial z_i/\partial x_j]$. In frequency response analysis, displacement z is a complex variable, while design $\boldsymbol{u}$ is a vector of real functions. Let design $\boldsymbol{u}$ be perturbed in the direction of $\delta\boldsymbol{u}$ and be controlled by parameter τ. The variation of the performance measure with respect to the design variable becomes

$$\begin{aligned}\psi' &= \frac{d}{d\tau}\left[\iint_{\Omega} g(z(\boldsymbol{x};\boldsymbol{u}+\tau\delta\boldsymbol{u}),\nabla z(\boldsymbol{x};\boldsymbol{u}+\tau\delta\boldsymbol{u}),\boldsymbol{u}+\tau\delta\boldsymbol{u})\, d\Omega\right]\Bigg|_{\tau=0} \\ &= \iint_{\Omega} (g_{,z}z' + g_{,\nabla z} : \nabla z' + g_{,\boldsymbol{u}}\delta\boldsymbol{u})\, d\Omega.\end{aligned} \tag{5.196}$$

Recall that z' and $\nabla z'$ depend on direction $\delta\boldsymbol{u}$ of the design change. The objective is to obtain an explicit expression of ψ' in terms of $\delta\boldsymbol{u}$, which requires rewriting the first two terms under the integral on the right of (5.196) explicitly in terms of $\delta\boldsymbol{u}$.

Consider a frequency response analysis of the dynamic structure under harmonic excitation, as introduced in Section 2.6.1 of Chapter 2. Let ω be the excitation frequency. The variational equation of the dynamic response problem can be rewritten from (2.55) as

$$-\omega^2 d_{\boldsymbol{u}}(z,\bar{z}) + j\omega c_{\boldsymbol{u}}(z,\bar{z}) + a_{\boldsymbol{u}}(z,\bar{z}) = \ell_{\boldsymbol{u}}(\bar{z}), \qquad \forall \bar{z}^* \in \mathbb{Z}. \tag{5.197}$$

where $\mathbb{Z}$ is the complex space of kinematically admissible virtual displacements. In (5.197), the following sesquilinear forms are used:

$$d_{\boldsymbol{u}}(z,\bar{z}) = \iint_{\Omega} \rho(\boldsymbol{x},\boldsymbol{u}) z^T \bar{z}^*\, d\Omega \tag{5.198}$$

and

$$c_{\boldsymbol{u}}(z,\bar{z}) = \iint_{\Omega} C(\boldsymbol{x},\boldsymbol{u}) z^T \bar{z}^*\, d\Omega, \tag{5.199}$$

and the sesquilinear form $a_{\boldsymbol{u}}(z,\bar{z})$ and semilinear form $\ell_{\boldsymbol{u}}(\bar{z})$ are similar to the energy bilinear and load linear forms presented for the static problem in Section 3.2 of Chapter 3. The difference is that the response variable z is now a complex variable and, $\bar{z}^*$ is the complex conjugate of admissible virtual displacement $\bar{z}$. Variational (5.197) determines the steady-state response of the structure under harmonic excitation.

The design variations of sesquilinear forms $a_{\boldsymbol{u}}(\bullet,\bullet)$ and $d_{\boldsymbol{u}}(\bullet,\bullet)$ and semilinear form $\ell_{\boldsymbol{u}}(\bullet)$ are given in (5.2), (5.112), and (5.3), respectively. Those design variations can be used for the design variations of the sesquilinear forms $a_{\boldsymbol{u}}(z,\bar{z})$ and $d_{\boldsymbol{u}}(z,\bar{z})$, and the semilinear form $\ell_{\boldsymbol{u}}(\bar{z})$. Similarly, the design variation of the damping sesquilinear form can be obtained as

$$c'_{\delta\boldsymbol{u}}(z,\bar{z}) \equiv \frac{d}{d\tau} c_{\boldsymbol{u}+\tau\delta\boldsymbol{u}}(\tilde{z},\bar{z})\Bigg|_{\tau=0}, \tag{5.200}$$

where the dependence of z on the design is fixed in $\tilde{z}$. Thus, by taking a variation of both sides of (5.197) with respect to the design, and by moving the terms explicitly dependent on the design to the right side, we obtain

$$\begin{aligned}&-\omega^2 d_{\boldsymbol{u}}(z',\bar{z}) + j\omega c_{\boldsymbol{u}}(z',\bar{z}) + a_{\boldsymbol{u}}(z',\bar{z}) \\ &= \ell'_{\delta\boldsymbol{u}}(\bar{z}) + \omega^2 d'_{\delta\boldsymbol{u}}(z,\bar{z}) - j\omega c'_{\delta\boldsymbol{u}}(z,\bar{z}) - a'_{\delta\boldsymbol{u}}(z,\bar{z}), \qquad \forall \bar{z}^* \in \mathbb{Z},\end{aligned} \tag{5.201}$$

where $\bar{z}^*$ is independent of the design. Presuming that displacement z is known as the solution to (5.197), (5.201) is a variational equation with the same sesquilinear forms for the displacement variation z'. It can be noted that the stiffness matrices in (5.197) and (5.201) are the same and that the right side of (5.201) can be considered a fictitious load term. If design perturbation $\delta\boldsymbol{u}$ is defined, and if the right side of (5.201) is evaluated with state solution z from (5.197), then (5.201) can be numerically solved to obtain z' using the finite element method. Design sensitivity ψ' in (5.196) can then be evaluated using z' in the direct differentiation method.

Adjoint Variable Method

To obtain an explicit expression for ψ' in terms of $\delta\boldsymbol{u}$, it is necessary to rewrite the first two terms in (5.196) explicitly in terms of $\delta\boldsymbol{u}$. Much in the same way as the static problem, an adjoint equation can be introduced by replacing z' in (5.196) with the complex virtual displacement $\bar{\lambda}$ and by equating it to the variational equation (5.197) with respect to adjoint variable λ as

$$\begin{aligned} &-\omega^2 d_u(\bar{\lambda},\lambda) + j\omega c_u(\bar{\lambda},\lambda) + a_u(\bar{\lambda},\lambda) \\ &\qquad = \iint_\Omega (g_{,z}\bar{\lambda} + g_{,\nabla z} : \nabla\bar{\lambda})\, d\Omega, \qquad \forall\bar{\lambda} \in \mathbb{Z}, \end{aligned} \tag{5.202}$$

where the adjoint response $\lambda^* \in \mathbb{Z}$ is desired. Since (5.202) satisfies for all $\bar{\lambda} \in \mathbb{Z}$, and since $z' \in \mathbb{Z}$, (5.202) may be evaluated at $\bar{\lambda} = z'$ to obtain

$$\begin{aligned} &-\omega^2 d_u(z',\lambda) + j\omega c_u(z',\lambda) + a_u(z',\lambda) \\ &\qquad = \iint_\Omega (g_{,z}z' + g_{,\nabla z} : \nabla z')\, d\Omega. \end{aligned} \tag{5.203}$$

In addition, since sensitivity equation (5.201) satisfies for all $\bar{z}^* \in \mathbb{Z}$, and since $\lambda^* \in \mathbb{Z}$,, (5.201) may be evaluated at $\bar{z}^* = \lambda^*$ to obtain

$$\begin{aligned} &-\omega^2 d_u(z',\lambda) + j\omega c_u(z',\lambda) + a_u(z',\lambda) \\ &\qquad = \ell'_{\delta u}(\lambda) + \omega^2 d'_{\delta u}(z,\lambda) - j\omega c'_{\delta u}(z,\lambda) - a'_{\delta u}(z,\lambda). \end{aligned} \tag{5.204}$$

When (5.203) and (5.204) are compared, their left sides are found to be exactly the same. Thus, from (5.203) and (5.204) we have

$$\begin{aligned} &\iint_\Omega (g_{,z}z' + g_{,\nabla z} : \nabla z')\, d\Omega \\ &\qquad = \ell'_{\delta u}(\lambda) + \omega^2 d'_{\delta u}(z,\lambda) - j\omega c'_{\delta u}(z,\lambda) - a'_{\delta u}(z,\lambda). \end{aligned} \tag{5.205}$$

Therefore, the terms that are implicitly dependent on the design in (5.196) are explicitly expressed in terms of $\delta\boldsymbol{u}$. By substituting the relation in (5.205) into (5.196), ψ' is explicitly represented in terms of $\delta\boldsymbol{u}$, as

$$\begin{aligned} \psi' &= \iint_\Omega (g_{,z}z' + g_{,\nabla z} : \nabla z' + g_{,u}\delta\boldsymbol{u})\, d\Omega \\ &= \iint_\Omega g_{,u}\delta\boldsymbol{u}\, d\Omega + \ell'_{\delta u}(\lambda) + \omega^2 d'_{\delta u}(z,\lambda) - j\omega c'_{\delta u}(z,\lambda) - a'_{\delta u}(z,\lambda). \end{aligned} \tag{5.206}$$

Note that the design sensitivity in (5.206) is in terms of the adjoint response λ^*. Thus, the adjoint variable design sensitivity analysis is extended to a problem where the energy forms are not symmetric. Specific expressions of ψ' for different performance measures and different structural components will be developed in detail in the following analytical examples.

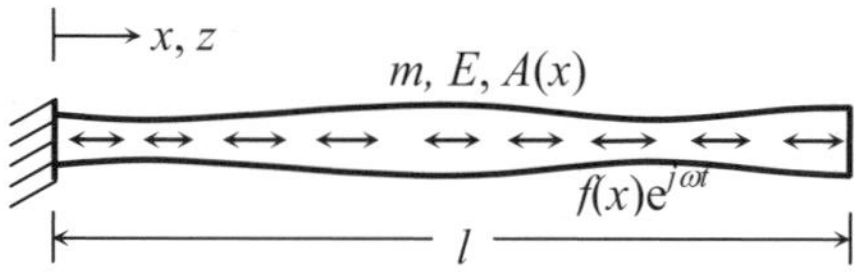

Figure 5.17. Truss component under harmonic load.

5.4.2 Analytical Examples

In order to demonstrate the basic principles in implementing design sensitivity analysis results, the sesquilinear and semilinear forms of various structural design components are derived. The design sensitivity expressions for such dynamic frequency responses as displacement and stress are obtained using the adjoint variable and direct differentiation method.

Truss

Consider a unit-length truss component under an oscillating excitation $F(x,t) = f(x)e^{j\omega t}$, as shown in Fig. 5.17. When the structural damping in (2.55) is used with φ as the structural damping coefficient, then the variational equation for frequency response can be written as

$$\int_0^l -\omega^2 mz\bar{z}^* \, dx + (1 + j\varphi)\int_0^l EAz_{,1}\bar{z}_{,1}^* \, dx = \int_0^l f\bar{z}^* \, dx, \qquad \forall \bar{z} \in \mathbb{Z}, \tag{5.207}$$

where $m(x;\boldsymbol{u}) = \rho A$ is the mass, ρ is the density, A is the cross-sectional area, E is Young's modulus, and φ is the structural damping coefficient. Let the design vector be $\boldsymbol{u} = [E, A]^T$.

Variations in sesquilinear forms on the left of (5.207) can be obtained as

$$d'_{\delta\boldsymbol{u}}(z,\bar{z}) = \int_0^l \rho z\bar{z}^* \delta A \, dx \tag{5.208}$$

and

$$a'_{\delta\boldsymbol{u}}(z,\bar{z}) = \int_0^l E\delta Az_{,1}\bar{z}_{,1}^* \, dx + \int_0^l \delta EAz_{,1}\bar{z}_{,1}^* \, dx. \tag{5.209}$$

In many applications, harmonic excitation $f(x)$ is usually the prescribed magnitude and frequency; thus, it is independent of the design.

Consider a special functional that defines the complex displacement value at an isolated point $\hat{x}$, that is,

$$\psi_1 \equiv z(\hat{x}) = \int_0^l \delta(x - \hat{x})z(x) \, dx, \tag{5.210}$$

where $\delta(x)$ is the Dirac delta measure at zero. The variation of this functional is thus written as

$$\psi_1' = \int_0^l \delta(x - \hat{x})z'(x) \, dx. \tag{5.211}$$

By using (5.202), the adjoint equation of the functional in (5.211) can be written as

$$-\omega^2 d_u(\bar{\lambda},\lambda)+(1+j\varphi)a_u(\bar{\lambda},\lambda)=\int_0^l \delta(x-\hat{x})\bar{\lambda}\,dx,\qquad \forall\bar{\lambda}\in\mathbb{Z},\tag{5.212}$$

and the adjoint response λ^* is desired. Interpreting the Dirac delta measure as a unit load applied at point $\hat{x}$, a physical interpretation of λ^* is immediately obtained as the complex displacement, due to a positive unit excitation at $\hat{x}$. From the design sensitivity expression in (5.206), the following sensitivity of complex displacement can be obtained:

$$\begin{aligned}\psi_1' &= \omega^2 d'_{\delta u}(z,\lambda)-(1+j\varphi)a'_{\delta u}(z,\lambda)\\ &= \int_0^l\left\{\omega^2\rho z\lambda^* -(1+j\varphi)Ez_{,1}\lambda^*_{,1}\right\}\delta A\,dx\\ &-\left\{(1+j\varphi)\int_0^l Az_{,1}\lambda^*_{,1}\,dx\right\}\delta E.\end{aligned}\tag{5.213}$$

The finite element method can be used to compute state variable z in (5.207) and the design sensitivity in (5.213). In the adjoint variable method, the design sensitivity of a complex displacement at a discrete point can be obtained from (5.213) by using (5.208) and (5.209), where the adjoint equation is given in (5.212). Using the direct differentiation method, design sensitivity z' of the complex displacement can be obtained by solving (5.201).

Next, let us consider a complex performance measure that represents an averaged amount of axial stress over a small subinterval of the beam, as

$$\psi_2=\int_0^l Ez_{,1}m_p\,dx,\tag{5.214}$$

where m_p is a characteristic function that is independent of the design, and is only nonzero on the small subinterval (x_a, x_b), defined in (5.31). The first variation of averaged stress ψ_2 is

$$\psi_2'=\int_0^l Ez'_{,1}m_p\,dx+\delta E\int_0^l z_{,1}m_p\,dx.\tag{5.215}$$

In this case, the adjoint equation is

$$-\omega^2 d_u(\bar{\lambda},\lambda)+(1+j\varphi)a_u(\bar{\lambda},\lambda)=\int_0^l E\bar{\lambda}_{,1}m_p\,dx,\qquad \forall\bar{\lambda}\in\mathbb{Z}.\tag{5.216}$$

Using the adjoint variable method, the design sensitivity of complex stress appears in the form

$$\psi_2'=\delta E\int_0^l z_{,1}m_p\,dx+\omega^2 d'_{\delta u}(z,\lambda)-(1+j\varphi)a'_{\delta u}(z,\lambda).\tag{5.217}$$

Beam

Consider a unit-length beam component under harmonic excitation $F(x,t) = f(x)e^{j\omega t}$, as shown in Fig. 5.18. The beam's steady-state response with structural damping is determined by the variational equation of frequency response, as

$$\int_0^l -\omega^2 mz\bar{z}^*\,dx+(1+j\varphi)\int_0^l EIz_{,11}\bar{z}^*_{,11}\,dx=\int_0^l f\bar{z}^*\,dx,\qquad \forall\bar{z}^*\in\mathbb{Z},\tag{5.218}$$

where $m = \rho A$ is the mass per unit length, I is the moment of inertia, and φ is the structural damping coefficient. Since the moment of inertia is dependent of cross-

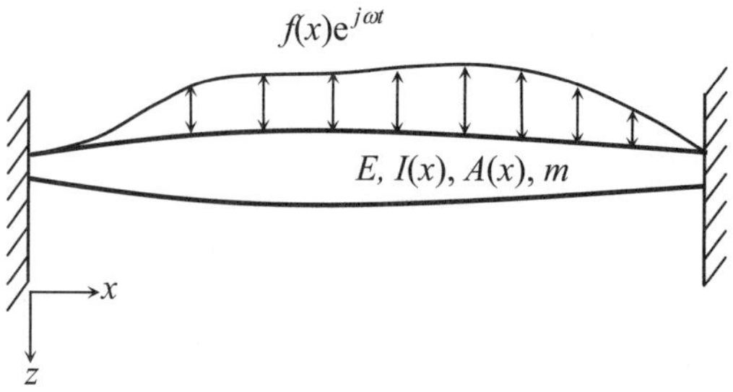

Figure 5.18. Beam component under harmonic excitation.

sectional area A, presume that $I = \alpha A^2$ without any loss of generality, as with the static problem. Similar to the formula for a truss component, here the design vector is $\boldsymbol{u} = [E, A]^T$.

The first variation of the sesquilinear forms on the left side of (5.218) can be taken as

$$d'_{\delta u}(z,\bar{z}) = \int_0^l \rho z\bar{z}^* \delta A\, dx \tag{5.219}$$

and

$$a'_{\delta u}(z,\bar{z}) = \int_0^l 2E\alpha A\delta A z_{,11}\bar{z}^*_{,11}\, dx + \int_0^l \delta E\alpha A^2 z_{,11}\bar{z}^*_{,11}\, dx, \tag{5.220}$$

and once again, the semilinear form is presumed to be independent of the design.

Displacement performance can be defined in exactly the same way as the truss component in (5.210): by interpreting $z(x;\boldsymbol{u})$ as a vertical deflection. The first sensitivity expression in (5.213) is still applicable if variations of the sesquilinear forms from (5.219) and (5.220) are used in place of (5.208) and (5.209), respectively.

Consider a complex performance measure representing an averaged amount of bending stress over a small subinterval of the beam as

$$\psi_3 = \int_0^l E\beta A^{1/2} z_{,11} m_p\, dx, \tag{5.221}$$

where m_p is given in (5.31) and $\beta A^{1/2}$ is the half-depth of the beam. The first variation of the averaged stress performance measure ψ_3 becomes

$$\begin{aligned}\psi'_3 &= \int_0^l E\beta A^{1/2} z'_{,11} m_p\, dx \\ &\quad + \int_0^l E\beta A^{-1/2} z_{,11} m_p \delta A\, dx + \delta E \int_0^l \beta A^{1/2} z_{,11} m_p\, dx.\end{aligned} \tag{5.222}$$

The first integral on the right side of (5.222) is used to define the adjoint equation as

$$-\omega^2 d_u(\bar{\lambda},\lambda) + (1 + j\varphi)a_u(\bar{\lambda},\lambda) = \int_0^l E\beta A^{1/2}\bar{\lambda}_{,11} m_p\, dx, \quad \forall \bar{\lambda} \in \mathbb{Z}. \tag{5.223}$$

Using the adjoint variable method, the design sensitivity of the complex stress takes the form

$$\psi_3' = \int_0^l E\beta A^{-1/2} z_{,11} m_p \delta A\, dx + \delta E \int_0^l \beta A^{1/2} z_{,11} m_p\, dx \\ +\omega^2 d_{\delta u}'(z,\lambda) - (1+j\varphi) a_{\delta u}'(z,\lambda). \tag{5.224}$$

The design sensitivity of (5.224) can be obtained using z from the response analysis in (5.218), as well as by using λ^* from the adjoint equation (5.223). If the finite element method is used, numerical integration is involved in the evaluation of (5.224).

Plate

Consider the plate component in Fig. 5.19, under the oscillating excitation $F(\boldsymbol{x},t) = f(\boldsymbol{x})e^{j\omega t}$. Structural domain Ω is parallel to the x_1-x_2 plane such that the material point is denoted by $\boldsymbol{x} = [x_1, x_2]^T$. A clamped boundary condition is assumed along boundary Γ. The sizing design variable $\boldsymbol{u} = [h(\boldsymbol{x})]$ represents the thickness of the structural component. Because of harmonic excitation, the steady-state response of the plate structure is determined using a variational equation, with structural damping, as

$$\iint_\Omega -\omega^2 m z\bar{z}^*\, d\Omega + (1+j\varphi)\iint_\Omega D(\boldsymbol{u})\boldsymbol{\kappa}(\bar{z}^*)^T \boldsymbol{C}\boldsymbol{\kappa}(z)\, d\Omega = \iint_\Omega f\bar{z}^*\, d\Omega, \quad \forall \bar{z}^* \in \mathbb{Z}, \tag{5.225}$$

where $m = \rho h$ is the mass, and curvature vector $\boldsymbol{\kappa}(z)$, flexural rigidity $D(\boldsymbol{u})$, φ is the structural damping coefficient, and stiffness matrix $\boldsymbol{C}$ are defined in (5.45). The space of kinematically admissible displacements for a clamped boundary is $Z = H_0^2(\Omega)$.

The first variation of sesquilinear forms in (5.225) can be obtained as

$$d_{\delta u}'(z,\bar{z}) = \iint_\Omega \rho z\bar{z}^* \delta h\, d\Omega \tag{5.226}$$

and

$$a_{\delta u}'(z,\bar{z}) = \iint_\Omega \frac{Eh^2\,\delta h}{4(1-\nu^2)} \boldsymbol{\kappa}(\bar{z}^*)^T \boldsymbol{C}\boldsymbol{\kappa}(z)\, d\Omega. \tag{5.227}$$

Again, the applied load is presumed to be independent of the design.

As with the truss components, the adjoint variable method allows the design sensitivity of a complex displacement at a discrete point to be obtained from the first sensitivity expression in (5.213) by using (5.226) and (5.227) with the adjoint equation given in equation (5.212). The only difference is that a two-dimensional domain is now being considered.

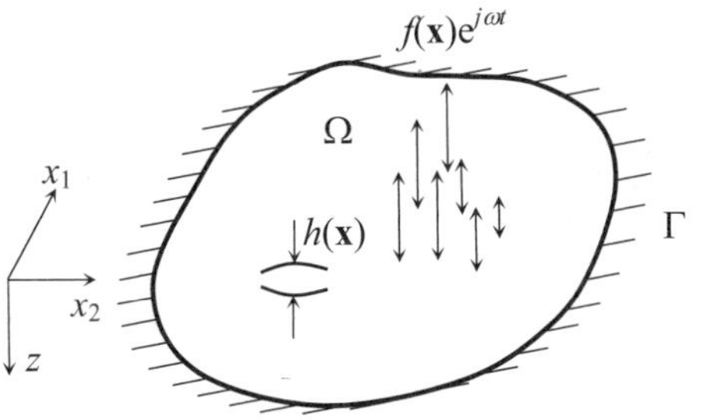

Figure 5.19. Plate structural component under harmonic load.

Consider a general form of the locally averaged stress measure on the structural component, written as

$$\psi_4 = \iint_\Omega g(\boldsymbol{\sigma}(z))\, m_p \, d\Omega, \tag{5.228}$$

where $m_p(\boldsymbol{x})$ is similar to the $m_p(\boldsymbol{x})$ of (5.31), but is extended to a two-dimensional domain such that it is nonzero only in the small fixed subdomain Ω_a. $g(\boldsymbol{\sigma})$ is assumed to be a continuously differentiable function of the stress components. Stress vector $\boldsymbol{\sigma}$ is a function of displacement z, with the following set of relations

$$\boldsymbol{\sigma}(z) = \begin{bmatrix} \sigma_{11} \\ \sigma_{22} \\ \sigma_{12} \end{bmatrix} = -\frac{h}{2}\boldsymbol{C}\begin{bmatrix} z_{,11} \\ z_{,22} \\ z_{,12} + z_{,21} \end{bmatrix} = -\frac{h}{2}\boldsymbol{C}\boldsymbol{\kappa}(z). \tag{5.229}$$

Since the magnitude of stress is at a maximum level either at the top or bottom surface of the plate, the half-thickness $h/2$ is used in (5.229). As a result, the stress vector has an explicitly dependent term on the design.

The first variation of the stress performance measure in (5.228) is

$$\psi_4' = \iint_\Omega g_{,\sigma}\boldsymbol{\sigma}(z') m_p \, d\Omega + \iint_\Omega \frac{1}{2} g_{,\sigma}\boldsymbol{C}\boldsymbol{\kappa}(z) m_p \delta h \, d\Omega. \tag{5.230}$$

The adjoint equation can then be defined by using the first integral in (5.230) as

$$-\omega^2 d_u(\bar{\lambda},\lambda) + (1+j\varphi)a_u(\bar{\lambda},\lambda) = \iint_\Omega g_{,\sigma}\boldsymbol{\sigma}(\bar{\lambda}) m_p \, d\Omega, \quad \forall \bar{\lambda} \in \mathbb{Z}, \tag{5.231}$$

where adjoint response λ^* is desired. Using the adjoint variable method, the design sensitivity of the complex stress can be obtained as

$$\begin{aligned} \psi_3' = & \iint_\Omega \frac{1}{2} g_{,\sigma}\boldsymbol{C}\boldsymbol{\kappa}(z) m_p \delta h \, d\Omega \\ & + \omega^2 d_{\delta\boldsymbol{u}}'(z,\lambda) - (1+j\varphi) a_{\delta\boldsymbol{u}}'(z,\lambda). \end{aligned} \tag{5.232}$$

Numerical Implementation with an Established Finite Element Analysis Code

For a design sensitivity analysis using the adjoint variable method, the adjoint load for each performance measure needs to be computed. To calculate the displacement sensitivity at the specified node, a unit harmonic load is applied to the adjoint structure at the same node in the same direction as the displacement. To compute the adjoint load associated with a stress performance measure, finite element shape functions can be used. The adjoint structural response, that is, the solutions to (5.216), (5.223) and (5.231), can be efficiently obtained by using the restart capability of the established FEA code. Using an original and adjoint response, the design sensitivity information in (5.217), (5.224), and (5.232) can be obtained by carrying out numerical integration.

When design sensitivity analysis is performed using the direct differentiation method, the fictitious load on the right of (5.201) is computed using finite element shape functions and numerical integration methods. The difficulties involved in numerical implementation for either of these two methods are the same. An efficient solution to (5.201) can also be obtained using the restart capability, as with the adjoint variable method. Note that structural response z' itself contains design sensitivity information. To calculate the stress design sensitivity, integration in (5.215), (5.222), and (5.230) can be evaluated numerically using z' and shape functions.

5.4.3 Numerical Examples

Plate Supported by Shock Absorbers

Consider a plate with shock absorbers, as shown in Fig. 5.20 [28]. The plate dimension is 1.02 m × 1.02 m (40 in × 40 in), and it contains a concentrated vertical dynamic harmonic load at the plate center. The first three natural frequencies of this plate are 10.94 Hz, 37.76 Hz, and 47.24 Hz. The load magnitude is 44.5 N (10 lb) and the two cyclic frequencies are $\omega = 10.8$ Hz and $\omega = 10.9$ Hz. Note that these cyclic frequencies are very close to the fundamental frequency $\omega_c = 10.94$ Hz. Due to symmetry, only one quarter of the plate is analyzed. Young's modulus is $E = 206.8$ GPa (3.0×10^7 psi), mass density is $\rho = 20.3$ kg/m^3 (7.34×10^{-4} lb/in^3), and Poisson's ratio is $\nu = 0.3$. The spring and damping coefficients are 656.5 kN/m and 17.5 kN·s/m, respectively. The nominal design has a uniform thickness of 2.54 mm (0.1 in) and the structural damping is $\varphi = 0.04$.

The finite element model consists of 25 square bending elements from COSMIC/NASTRAN QUAD2 [77], 36 nodal points, and 108 degrees of freedom. The bending element QUAD2 uses two sets of overlapping triangular elements, and stresses in the subtriangle are computed at the intersection point of the two diagonals and averaged [78]. Although the element has both bending and membrane capabilities, only bending part is considered. To numerically compute design sensitivity, the original displacements and stresses and the adjoint displacements and strains are required.

To demonstrate the accuracy of the design sensitivity results in the vicinity of resonance, a design sensitivity analysis is performed with excitation frequencies of 10.8 Hz and 10.9 Hz. Thus, the frequency ratios are $r = \omega/\omega_c = 0.987$ and $r = 0.996$. A nine-point Gauss integration is used for the numerical integration of design sensitivity expressions. The design sensitivity results of real, imaginary, and maximum displacements, and the phase angle for the selected nodal points are shown in Table 5.5, where ND and Hz denote the node number and load frequency, respectively. Real displacement z^1, imaginary displacement z^2, maximum displacement z, and phase angle α are denoted by the letters R, I, D, and P, respectively. In addition, $\psi(\boldsymbol{u} - \Delta\boldsymbol{u})$ and $\psi(\boldsymbol{u} + \Delta\boldsymbol{u})$ represent performance measure values at the perturbed designs $\boldsymbol{u} - \Delta\boldsymbol{u}$ and $\boldsymbol{u} + \Delta\boldsymbol{u}$, respectively, where $\Delta\boldsymbol{u}$ is seen as the amount of design perturbation.

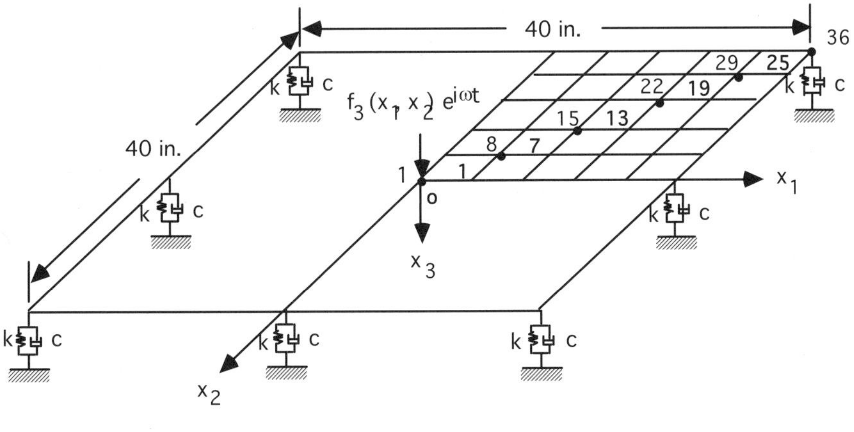

Figure 5.20. Plate supported by shock absorbers.

Table 5.5. Design sensitivity of plate supported by shock absorbers (displacement, 0.1% perturbation).

Hz	ND		$\psi(\boldsymbol{u}-\Delta\boldsymbol{u})$	$\psi(\boldsymbol{u}+\Delta\boldsymbol{u})$	$\Delta\psi$	ψ'	$\psi'/\Delta\psi$ (%)	$\lvert\Delta\psi/\psi\rvert$ (%)
10.8	1	R	0.270E+1	0.275E+1	0.267E–1	0.259E–1	97.0	1.0
		I	–.361E+1	–.330E+1	0.151E+0	0.149E+0	98.3	4.4
		D	0.450E+1	0.430E+1	–.102E+0	–.100E+0	98.4	2.3
		P	0.306E+3	0.309E+3	0.149E+1	0.146E+1	98.1	0.5
10.8	15	R	0.183E+1	0.187E+1	0.187E–1	0.182E–1	97.1	1.0
		I	–.250E+1	–.229E+1	0.105E+0	0.103E+0	98.3	4.4
		D	0.310E+1	0.296E+1	–.716E–1	–.705E–1	98.4	2.4
		P	0.306E+3	0.309E+3	0.149E+1	0.146E+1	98.1	0.5
10.8	29	R	0.495E+0	0.505E+0	0.535E–2	0.523E–2	97.9	1.1
		I	–.695E+0	–.637E+0	0.290E–1	0.285E–1	98.1	4.4
		D	0.853E+0	0.813E+0	–.200E–1	–.196E–1	98.1	2.4
		P	0.305E+3	0.308E+3	0.149E+1	0.146E+1	98.1	0.5
10.9	1	R	0.174E+1	0.204E+1	0.147E+0	0.144E+0	98.1	7.7
		I	–.498E+1	–.467E+1	0.154E+0	0.152E+0	98.6	3.2
		D	0.528E+1	0.510E+1	–.897E–1	–.885E–1	98.7	1.7
		P	0.289E+3	0.293E+3	0.213E+1	0.217E+1	101.8	0.7
10.9	15	R	0.117E+1	0.138E+1	0.102E+0	0.100E+0	98.1	8.0
		I	–.345E+1	–.324E+1	0.106E+0	0.105E+0	98.5	3.2
		D	0.365E+1	0.352E+1	–.632E–1	–.623E–1	98.6	1.8
		P	0.288E+3	0.293E+3	0.213E+1	0.217E+1	101.8	0.7
10.9	29	R	0.310E+0	0.367E+0	0.284E–1	0.279E–1	98.2	8.4
		I	–.957E+0	–.898E+0	0.294E–1	0.289E–1	98.3	3.2
		D	0.100E+1	0.971E+0	–.178E–1	–.175E–1	98.1	1.8
		P	0.288E+3	0.292E+3	0.213E+1	0.217E+1	101.9	0.7

Table 5.6. Design sensitivity of plate Supported by shock absorbers (stress, psi, 0.1% perturbation).

Hz	EL		$\psi(\boldsymbol{u}-\Delta\boldsymbol{u})$	$\psi(\boldsymbol{u}+\Delta\boldsymbol{u})$	$\Delta\psi$	ψ'	$\psi'/\Delta\psi$ (%)	$\lvert\Delta\psi/\psi\rvert$ (%)
10.8	1	R	0.327E+5	0.334E+5	0.326E+3	0.316E+3	97.1	1.0
		I	–.410E+5	–.376E+5	0.168E+4	0.166E+4	98.3	4.3
		S	0.525E+5	0.503E+5	–.108E+4	–.106E+4	98.4	2.1
10.8	13	R	0.177E+5	0.181E+5	0.201E+3	0.195E+3	97.1	1.1
		I	–.245E+5	–.225E+5	0.101E+4	0.996E+3	98.3	4.3
		S	0.303E+5	0.289E+5	–.685E+3	–.674E+3	98.4	2.3
10.8	25	R	0.386E+4	0.395E+4	0.450E+2	0.446E+2	99.0	1.2
		I	–.546E+4	–.501E+4	0.225E+3	0.220E+3	97.8	4.3
		S	0.669E+4	0.638E+4	–.153E+3	–.149E+3	97.5	2.3
10.9	1	R	0.219E+5	0.253E+5	0.168E+4	0.165E+4	98.1	7.1
		I	–.566E+5	–.532E+5	0.170E+4	0.168E+4	98.5	3.1
		S	0.608E+5	0.589E+5	–.900E+3	–.888E+3	98.7	1.5
10.9	13	R	0.112E+5	0.132E+5	0.101E+4	0.998E+3	98.1	8.3
		I	–.339E+5	–.319E+5	0.102E+4	0.100E+4	98.6	3.1
		S	0.357E+5	0.345E+5	–.603E+3	–.595E+3	98.7	1.7
10.9	25	R	0.242E+4	0.287E+4	0.226E+3	0.222E+3	98.3	8.5
		I	–.754E+4	–.709E+4	0.227E+3	0.222E+3	97.9	3.1
		S	0.792E+4	0.765E+4	–.136E+3	–.132E+3	97.4	1.8

The central finite difference between these performance measure values is denoted by $\Delta\psi = [\psi(\boldsymbol{u} + \Delta\boldsymbol{u}) - \psi(\boldsymbol{u} - \Delta\boldsymbol{u})]/2$, and ψ' is the design sensitivity prediction. The ratio between ψ' and $\Delta\psi$ times 100% is used as a measure of accuracy for design sensitivity computations. In other words, a 100% agreement means that the predicted change is exactly the same as the finite difference. When $\Delta\psi$ is too small, this accuracy measure may fail to give the correct information because $\Delta\psi$ may lose significant digits during numerical computation. On the other hand, if $\Delta\psi$ is too large, the finite difference may contain nonlinear terms. To monitor the magnitude of $\Delta\psi$, the ratio $|\Delta\psi/\psi| \times 100$ (%) is given in the table. In order to avoid nonlinear behavior, a 0.1% perturbation of the design is the size used for computational purposes. As shown in Table 5.5, agreement is excellent between the finite difference results and the results predicted by sensitivity analysis.

Now, consider a design sensitivity computation for bending stress. Since NASTRAN provides element stress at the centroid, the pointwise stress performance is measured at the element centroid. To evaluate the design sensitivity equation, an adjoint solution to (5.231) is required. A nine-point Gauss integration is used for numerical integration. For pointwise stress σ_{11}, design sensitivity analysis is carried out with a 0.1% design perturbation; sensitivity results are listed in Table 5.6. Real, imaginary, and maximum stresses are denoted as R, I, and S, respectively, and EL denotes the element number. As shown in Table 5.6, excellent agreement is obtained where $|\Delta\psi/\psi|$ is in the range of 1.0 to 8.5 %.

Vehicle Chassis Frame Structure

For a second example, a vehicle chassis frame model [76] is presented in Fig. 5.21, and a finite element model used for design sensitivity purposes is shown in Fig. 5.22 [28]. The structure is 7.35 m (289.37 in) long and 0.8 m (31.5 in) wide, with two hollow longitudinal frames and five hollow cross members. The model in Fig. 5.21 has suspension coil springs, shock absorbers, linear springs and dampers representing the vehicle tire stiffness and damping effects, and lumped masses of suspension sprung. The engine and body masses are attached to the frame using linear springs [79], since such masses are not welded to the chassis. The chassis is subjected to harmonic loads, Fe1, Fe2, and Fe3, excited by the engine. In addition, harmonic loads Fs1 and Fs2 are applied to the frame to simulate a sinusoidal-shaped road surface. Table 5.7 lists the design specifications for mechanical properties, chassis natural frequencies, and loading conditions.

As illustrated in Fig. 5.22, the finite element model contains two longitudinal and five transverse design components with 68 hollow rectangular beam elements (individually numbered 1 to 68 in bold face), 13 spring elements, 8 damping elements, 65 nodal points (numbers 1 to 65 not in bold face), and 15 scalar points (numbers 66 to 80 not in bold face) including body and engine mount spring attachments with 397 degrees of freedom. The nominal design has a uniform thickness $t = w$ of 5.08 mm (0.2 in), a width b of 10.16 cm (4.0 in), and a height h of 15.24 cm (6.0 in). Young's modulus is $E = 206.8$ GPa (3.0×10^7 psi), Poisson's ratio is $\nu = 0.3$, mass density is $\rho = 20.3$ kg/m^3 (7.34×10^{-4} lb/in^3), and the structural damping coefficient is $\varphi = 0.04$.

To prevent vehicle rigid body motion, x_1- and x_2-displacements at nodes 24 and 49 are constrained and the ends of the spring and damping elements, which represent tires, are fixed on the ground. A COSMIC/NASTRAN Direct Frequency Response Analysis is used for the original and adjoint finite element analysis [77].

Using the design sensitivity analysis method, displacement sensitivity is calculated with a 1% design perturbation on width, height, and thickness. The results for randomly selected nodal points are provided in Table 5.8. The design sensitivity results for the averaged maximum bending stress for each element are given in Table 5.9 for randomly selected finite elements with a 1% design perturbation. As Tables 5.8 and 5.9 show, there is excellent agreement between the results of the proposed method and the finite difference method.

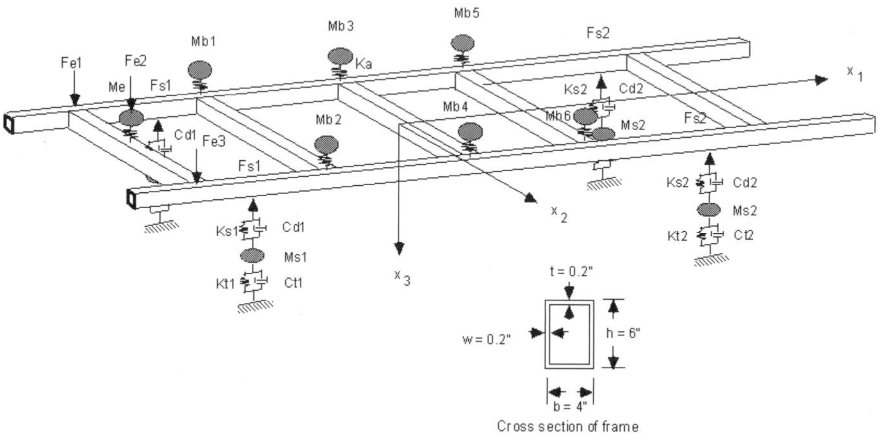

Figure 5.21. Vehicle chassis frame structure.

Table 5.7. Design specification for vehicle chassis frame.

• BODY MASS Mb1=Mb2=250 kg, Mb3=Mb4=340 kg, Mb5=Mb6=227 kg • ENGINE MASS: Me=254 kg • SUSPENSION SPRUNG MASS Ms1=52 kg, Ms2=75 kg • NATURAL FREQUENCIES 2.89 Hz, 3.61 Hz, 3.97Hz • LOADING CONDITION: ω = 2.0 Hz Fe1 , Fe3: 1779 $e^{j\omega t}$ N Fe2: 3558 $e^{j\omega t}$ N Fs1 , Fs2: 445 $e^{j\omega t}$ N	• SPRING COEFFICIENT FRONT COIL SPRING: Ks1=32.3 kN/m REAR COIL SPRING: Ks2=65.7 kN/m FRONT TIRE STIFFNESS: Kt1=306.0 kN/m REAR TIRE STIFFNESS: Kt2=306.0 kN/m ATTACHMENT SPRING: Ka=4991.1 kN/m • DAMPING COEFFICIENT FRONT SHOCK ABSORBER: Cd1=1.7 kN-s/m REAR SHOCK ABSORBER: Cd2=1.7 kN-s/m FRONT TIRE DAMPING: Ct1=0.3 kN-s/m REAR TIRE DAMPING: Ct2=0.3 kN-s/m

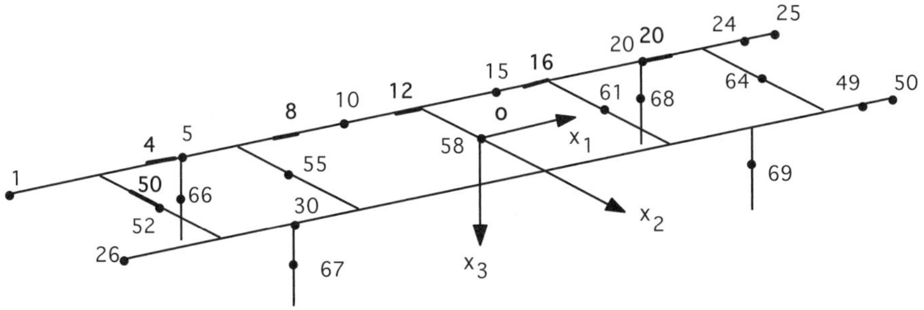

Figure 5.22. Finite element model of vehicle chassis frame structure.

Table 5.8. Design sensitivity of vehicle chassis frame (displacement).

ND		$\psi(\boldsymbol{u}-\Delta\boldsymbol{u})$	$\psi(\boldsymbol{u}+\Delta\boldsymbol{u})$	$\Delta\psi$	ψ'	$\psi'/\Delta\psi(\%)$
5	D	0.599E+1	0.609E+1	0.472E–1	0.472E–1	99.8
	P	0.310E+3	0.313E+3	–.478E+0	–.478E+0	99.9
10	D	0.410E+1	0.417E+1	0.369E–1	0.368E–1	99.6
	P	0.311E+3	0.310E+3	–.467E+0	–.466E+0	99.7
15	D	0.202E+1	0.207E+1	0.228E–1	0.227E–1	99.5
	P	0.304E+3	0.303E+3	–.480E+0	–.479E+0	99.7
20	D	0.501E+0	0.510E+0	0.445E–2	0.443E–2	99.6
	P	0.193E+3	0.194E+3	0.485E+0	0.487E+0	100.3
25	D	0.288E+1	0.291E+1	0.172E–1	0.170E–1	99.1
	P	0.144E+3	0.144E+3	–.273E+0	–.271E+0	99.4
52	D	0.712E+1	0.723E+1	0.519E–1	0.519E–1	100.0
	P	0.315E+3	0.314E+3	–.489E+0	–.489E+0	100.0
55	D	0.528E+1	0.537E+1	0.438E–1	0.436E–1	99.7
	P	0.313E+3	0.312E+3	–.473E+0	–.472E+0	99.8
58	D	0.289E+1	0.295E+1	0.289E–1	0.288E–1	99.6
	P	0.308E+3	0.307E+3	–.471E+0	–.469E+0	99.7
61	D	0.118E+1	0.121E+1	0.167E–1	0.166E–1	99.6
	P	0.295E+3	0.295E+3	–.483E+0	–.482E+0	99.7
64	D	0.184E+1	0.186E+1	0.956E–2	0.947E–2	99.0
	P	0.149E+3	0.149E+3	–.200E+0	–.198E+0	99.1

Table 5.9. Design sensitivity of vehicle chassis frame (stress, psi).

EL		$\psi(\boldsymbol{u}-\Delta\boldsymbol{u})$	$\psi(\boldsymbol{u}+\Delta\boldsymbol{u})$	$\Delta\psi$	ψ'	$\psi'/\Delta\psi(\%)$
4	S	0.334E+4	0.318E+4	–.842E+2	–.841E+2	99.9
8	S	0.304E+4	0.286E+4	–.876E+2	–.874E+2	99.7
12	S	0.183E+4	0.172E+4	–.561E+2	–.560E+2	99.8
16	S	0.893E+3	0.843E+3	–.246E+2	–.246E+2	99.7
20	S	0.240E+3	0.243E+3	0.142E+1	0.136E+1	96.2
50	S	0.714E+3	0.677E+3	–.183E+2	–.201E+2	109.5

5.5 Structural-Acoustic Design Sensitivity Analysis

The objective of structural-acoustic design sensitivity analysis is to predict the variation of acoustic and structural performance measures resulting from the change in design variables of structural components [80]. Such sizing design variables as material properties, panel thickness, and beam cross-sectional area are taken into account in the development of design sensitivity analysis.

A performance measure variation is expressed in terms of the state variable variation, which can be obtained by solving an equation that is derived from the state equation. Alternatively, the sensitivity expression can be reduced using adjoint variables and adjoint equations to a form that does not include the variations of state variables.

5.5.1 Design Sensitivity Analysis of Structural-Acoustic Response

A state variable variation with respect to the design was introduced in Section 5.1. This concept can be extended to those state variables of structural-acoustic problems in which the acoustic pressure and the structural displacement are considered as state variables. If we let current design $\boldsymbol{u}$ be perturbed in the direction of $\delta\boldsymbol{u}$ and controlled by parameter τ, then the variations of displacement pressure in the direction of $\delta\boldsymbol{u}$ can be defined as

$$z' = \frac{d}{d\tau} z(\boldsymbol{x}, \boldsymbol{u} + \tau\delta\boldsymbol{u})\Big|_{\tau=0} \tag{5.233}$$

and

$$p' = \frac{d}{d\tau} p(\boldsymbol{x}, \boldsymbol{u} + \tau\delta\boldsymbol{u})\Big|_{\tau=0}. \tag{5.234}$$

To develop a design sensitivity analysis for structural-acoustic problems, the variational formulation of Section 2.6 is rewritten

$$q_u(z,\bar{z}) + b(p,\bar{p}) - \phi(z,\bar{p}) - \chi(p,\bar{z}) = \ell_u(\bar{z}), \quad \forall\{\bar{z}^*, \bar{p}^*\} \in Q, \tag{5.235}$$

where Q is a complex vector space defined in (2.74). Note that the sesquilinear forms $b(\bullet,\bullet)$, $\phi(\bullet,\bullet)$, and $\chi(\bullet,\bullet)$ do not explicitly depend on the sizing design variables. However, their arguments z and p do depend on the design through the response analysis in (5.235), as shown in (5.233) and (5.234).

Since the kinematically admissible space Q is independent of the design, a variation of (5.235) can be taken to obtain the following design sensitivity equation:

$$\begin{aligned} &q_u(z',\bar{z}) + b(p',\bar{p}) - \phi(z',\bar{p}) - \chi(p',\bar{z}) \\ &\quad = \ell'_{\delta u}(\bar{z}) - q'_{\delta u}(z,\bar{z}), \quad \forall\{\bar{z}^*, \bar{p}^*\} \in Q, \end{aligned} \tag{5.236}$$

where $\ell'_{\delta u}(\bar{z})$ is the same as in (5.3), if a complex variable is used instead of a real function, and where

$$q'_{\delta u}(z,\bar{z}) = \frac{d}{d\tau} q_{u+\delta u}(\tilde{z},\bar{z})\Big|_{\tau=0} \tag{5.237}$$

can be calculated using the terms in (5.201). In fact, the right side of (5.236) is the same as the right side of (5.201). This is true because the acoustic medium does not explicitly depend on the sizing design.

Adjoint Variable Method

The harmonic performance measures of interest when designing a coupled system are expressed as a function of complex variables, which correspond to acoustic pressure and structural displacement. Consider a pressure-related performance in the acoustic cavity

$$\psi_p = \iiint_{\Omega^a} h(p,\nabla p,\boldsymbol{u})\, d\Omega^a, \tag{5.238}$$

where it is assumed that function h is continuously differentiable with respect to its arguments and that the domain of the integral Ω^a is independent of design variation. Taking the variation of ψ_p, we obtain

$$\psi'_p = \iiint_{\Omega^a} (h_{,p} p' + h_{,\nabla p} \nabla p' + h_{,\boldsymbol{u}} \delta \boldsymbol{u})\, d\Omega^a. \tag{5.239}$$

The variation of ψ_p includes the variation of the state variable p. If sensitivity equation (5.236) is solved for z' and p', then (5.239) can also be evaluated by substituting p' to express the variation of ψ_p explicitly in terms of $\delta \boldsymbol{u}$, which is the direct differentiation method.

In the adjoint variable method, an adjoint equation is adapted by using the first two integrands in (5.239). Let the adjoint equation and the adjoint response $\{\lambda^*, \eta^*\}$ that correspond to the state variable $\{z, p\}$ be defined such that $\{\lambda^*, \eta^*\} \in Q$ satisfy the adjoint equation. We can then define the adjoint equation of a structural-acoustic problem as

$$\begin{aligned} &q_{\boldsymbol{u}}(\bar{\lambda},\lambda) + b(\bar{\eta},\eta) - \phi(\bar{\lambda},\eta) - \chi(\bar{\eta},\lambda) \\ &\quad = \iiint_{\Omega^a} (h_{,p}\bar{\eta} + h_{,\nabla p}\nabla \bar{\eta})\, d\Omega, \qquad \forall \{\bar{\lambda},\bar{\eta}\} \in Q, \end{aligned} \tag{5.240}$$

where adjoint solutions $\{\lambda^*, \eta^*\}$ are desired. To utilize the adjoint equation, every $\{\bar{\lambda},\bar{\eta}\} \in Q$ in (5.240) is replaced by $\{z', p'\} \in Q$. We obtain

$$\begin{aligned} &q_{\boldsymbol{u}}(z',\lambda) + b(p',\eta) - \phi(z',\eta) - \chi(p',\lambda) \\ &\quad = \iiint_{\Omega^a} (h_{,p} p' + h_{,\nabla p}\nabla p')\, d\Omega. \end{aligned} \tag{5.241}$$

On the other hand, since design sensitivity equation (5.236) satisfies for all $\{\bar{z}^*,\bar{p}^*\} \in Q$, we can replace $\{\bar{z}^*,\bar{p}^*\}$ with $\{\lambda^*,\eta^*\}$, since the solution $\{\lambda^*,\eta^*\}$ to adjoint equation (5.240) belongs to Q. Thus, we have

$$\begin{aligned} &q_{\boldsymbol{u}}(z',\lambda) + b(p',\eta) - \phi(z',\eta) - \chi(p',\lambda) \\ &\quad = \ell'_{\delta \boldsymbol{u}}(\lambda) - q'_{\delta \boldsymbol{u}}(z,\lambda). \end{aligned} \tag{5.242}$$

Notice that the left sides of (5.241) and (5.242) are identical, yielding the following relation:

$$\iiint_{\Omega^a} (h_{,p} p' + h_{,\nabla p}\nabla p')\, d\Omega^a = \ell'_{\delta \boldsymbol{u}}(\lambda) - q'_{\delta \boldsymbol{u}}(z,\lambda). \tag{5.243}$$

Thus, the first two terms on the right side of (5.239) are expressed as adjoint variables. Finally, the design sensitivity of ψ_p becomes

$$\psi'_p = \ell'_{\delta \boldsymbol{u}}(\lambda) - q'_{\delta \boldsymbol{u}}(z,\lambda) + \iiint_{\Omega^a} h_{,\boldsymbol{u}} \delta \boldsymbol{u}\, d\Omega^a. \tag{5.244}$$

Now, consider a displacement-related performance on the structure

$$\psi_z = \iint_{\Omega^s} g(z,\nabla z,\boldsymbol{u})\, d\Omega^s, \tag{5.245}$$

where function g is presumed to be differentiable with respect to its arguments and the domain of the integral Ω^s is independent of design variation. Taking the variation of ψ_z, we obtain

$$\psi'_z = \iint_{\Omega^s} (g_{,z}z' + g_{,\nabla z} : \nabla z' + g_{,u}\delta \boldsymbol{u})\, d\Omega^s. \tag{5.246}$$

A similar procedure used to obtain ψ'_p in (5.239) can now be used to obtain ψ'_z. Define the adjoint equation of the performance in (5.245) as

$$\begin{aligned} &q_u(\bar{\lambda},\lambda) + b(\bar{\eta},\eta) - \phi(\bar{\lambda},\eta) - \chi(\bar{\eta},\lambda) \\ &\quad = \iint_{\Omega^s} (g_{,z}\bar{\lambda} + g_{,\nabla z} : \nabla\bar{\lambda})\, d\Omega^s, \qquad \forall \{\bar{\lambda},\bar{\eta}\} \in Q. \end{aligned} \tag{5.247}$$

Following the same procedure as with (5.241) and (5.242), the first two terms of (5.246) can be expressed in terms of the response result z and the adjoint result $\boldsymbol{\lambda}^*$, as

$$\iint_{\Omega^s} (g_{,z}z' + g_{,\nabla z} : \nabla z')\, d\Omega^s = \ell'_{\delta u}(\lambda) - q'_{\delta u}(z,\lambda), \tag{5.248}$$

and, accordingly, the sensitivity of performance ψ_z is obtained,

$$\psi'_z = \ell'_{\delta u}(\lambda) - q'_{\delta u}(z,\lambda) + \iint_{\Omega^s} g_{,u}\delta \boldsymbol{u}\, d\Omega^s. \tag{5.249}$$

Note that even if ψ_p and ψ_z are completely different performance measures, the expressions of (5.244) and (5.249) are quite similar. Only the adjoint solutions of (5.240) and (5.247) will be different based on the type of performance measure. Thus, the same adjoint equation can be used for different kinds of design variables. This similarity in the adjoint variable method makes it convenient from the viewpoint of computational implementation. For different performance measures, only the right sides of (5.240) and (5.247) are different, which are known as the adjoint loads. After obtaining the adjoint variables, the remaining numerical integration process is basically the same for all performance measures.

5.5.2 Analytical Example

Design components that appear in the structural-acoustic problem are exactly the same as those that appear in the frequency response problem. For example, the expression of $q'_{\delta u}(z,\bar{z})$ for each design component can be found in Section 5.4.2 as

$$q'_{\delta u}(z,\bar{z}) = -\omega^2 d'_{\delta u}(z,\bar{z}) + j\omega c'_{\delta u}(z,\bar{z}) + a'_{\delta u}(z,\bar{z}). \tag{5.250}$$

The other sesquilinear forms, $b(\bullet,\bullet)$, $\phi(\bullet,\bullet)$, and $\chi(\bullet,\bullet)$, are independent of the sizing design variables.

The design sensitivity formulation developed in previous sections can be illustrated with a simple example in which design sensitivity is derived using the adjoint variable method. Consider an acoustic cavity with a flexible panel, as illustrated in Fig. 5.23. The cavity is surrounded on all sides by rigid walls except for one side, which is closed by a clamped panel of linear elastic material with the structural damping coefficient φ. The panel's uniform thickness h is selected as the design variable, that is, $\boldsymbol{u}(\boldsymbol{x}) = [h]$. Let us consider such performance measures as the acoustic pressure $p(\boldsymbol{x}^a)$ at point $\boldsymbol{x}^a$ in the acoustic cavity, and the x^3-directional displacement $z(\boldsymbol{x}^s)$ at point $\boldsymbol{x}^s$ on the structural panel. A harmonic force $f(\boldsymbol{x},t)$ with frequency ω is applied to the plate. Here, $f(\boldsymbol{x},t)$ is not assumed to be dependent on the design variable $\boldsymbol{u}(\boldsymbol{x})$.

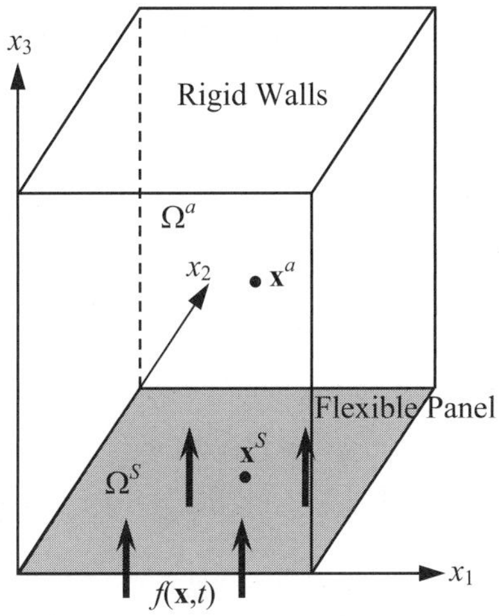

Figure 5.23. Cavity with flexible wall.

The variational equation of harmonic motion of this coupled system is given as (5.235). The pressure performance measure is

$$\psi_1 = p(\boldsymbol{x}^a) = \iiint_{\Omega^a} \delta(\boldsymbol{x}-\boldsymbol{x}^a)p\, d\Omega^a, \tag{5.251}$$

where $\delta(\boldsymbol{x})$ is the Dirac delta measure. Equation (5.251) is a simple form of (5.238), a general form of acoustic performance measure. The variation of the performance measure, corresponding to (5.251), is

$$\psi_1' = p'(\boldsymbol{x}^a) = \iiint_{\Omega^a} \delta(\boldsymbol{x}-\boldsymbol{x}^a)p'\, d\Omega^a. \tag{5.252}$$

The adjoint equation for ψ_1' is formed from (5.240) as

$$\begin{aligned} q_{\boldsymbol{u}}(\bar{\boldsymbol{\lambda}},\boldsymbol{\lambda}) + b(\bar{\eta},\eta) - \phi(\bar{\boldsymbol{\lambda}},\eta) - \chi(\bar{\eta},\boldsymbol{\lambda}) \\ = \iiint_{\Omega^a} \delta(\boldsymbol{x}-\boldsymbol{x}^a)\bar{\eta}\, d\Omega^a, \qquad \forall\{\bar{\boldsymbol{\lambda}},\bar{\eta}\} \in Q. \end{aligned} \tag{5.253}$$

The term on the right side of this equation is referred to as the acoustic adjoint load. The physical meaning of the acoustic adjoint load, which corresponds to the acoustic pressure at a point, is the unit pressure source at the point. From (5.244), the design sensitivity of the acoustic pressure is

$$\psi_1' = -q'_{\delta \boldsymbol{u}}(z,\boldsymbol{\lambda}). \tag{5.254}$$

If the panel in this example is modeled as a flat plate, the variation of sesquilinear form $q_{\boldsymbol{u}}(\bullet,\bullet)$ is given in (5.226) and (5.227). Substituting the variations of these sesquilinear forms into (5.254), the design sensitivity of (5.251) is

$$\psi_1' = \omega^2 \iint_{\Omega^s} \rho z \lambda^* \delta h\, d\Omega^s - (1+j\varphi)\iint_{\Omega^s} \frac{Eh^2\,\delta h}{4(1-\nu^2)} \boldsymbol{\kappa}(\lambda^*)^T \boldsymbol{C}\boldsymbol{\kappa}(z)\, d\Omega^s, \tag{5.255}$$

where φ is the structural damping coefficient.

The other performance measure in this example is the displacement at point $\boldsymbol{x}^s$. Its mathematical expression is

$$\psi_2 = z(\boldsymbol{x}^s) = \iint_{\Omega^s} \delta(\boldsymbol{x}-\boldsymbol{x}^s) z \, d\Omega^s. \tag{5.256}$$

Equation (5.256) is a simple form of (5.245), which is the general form for a structural performance measure. The variation of ψ_2 is

$$\psi_2' = z'(\boldsymbol{x}^s) = \iint_{\Omega^s} \delta(\boldsymbol{x}-\boldsymbol{x}^s) z' \, d\Omega^s. \tag{5.257}$$

Working from (5.247), the corresponding adjoint equation is

$$\begin{aligned} &q_u(\bar{\lambda},\lambda) + b(\bar{\eta},\eta) - \phi(\bar{\lambda},\eta) - \chi(\bar{\eta},\lambda) \\ &\quad = \iint_{\Omega^s} \delta(\boldsymbol{x}-\boldsymbol{x}^s)\bar{\lambda} \, d\Omega^s, \quad \forall \{\bar{\lambda},\bar{\eta}\} \in Q. \end{aligned} \tag{5.258}$$

In (5.258), the term on the right side is the adjoint load for the structural displacement. The physical meaning of the adjoint load, which corresponds to the harmonic displacement at a point, is a unit harmonic force applied at point $\boldsymbol{x}^s$. From (5.249), the design sensitivity is

$$\psi_2' = -q_{\delta u}'(z,\lambda). \tag{5.259}$$

The fact that the primary state equation (5.235) and the adjoint equations (5.253) and (5.258) represent the same structure with different loads provides an efficient method for numerical implementation, since only one finite element model is required to solve the primary and adjoint equations. As was indicated in Section 5.1.3, the design sensitivity expressions in (5.254) and (5.259) also have identical forms, which is convenient in the design sensitivity computation of a coupled system.

5.5.3 Numerical Examples

Numerical Considerations

Structural-acoustic systems can be solved with either the finite element or the boundary element method. In this section, the finite element method is utilized for analysis [77] and [81]. The variational equation of the harmonic motion of a continuum model, (5.235), can be reduced to a set of linear algebraic equations by discretizing the model into finite elements and by introducing shape functions and nodal variables for each element. The acoustic pressure $p(\boldsymbol{x})$ and the structural displacement $z(\boldsymbol{x})$ are approximated using shape functions and nodal variables for each element in the discretized model, as

$$\left.\begin{aligned} z(\boldsymbol{x}) &= \boldsymbol{N}_s(\boldsymbol{x})\boldsymbol{z}^e \\ p(\boldsymbol{x}) &= \boldsymbol{N}_a(\boldsymbol{x})\boldsymbol{p}^e \end{aligned}\right\}, \tag{5.260}$$

where $\boldsymbol{N}_s(\boldsymbol{x})$ and $\boldsymbol{N}_a(\boldsymbol{x})$ are matrices of shape functions for displacement and pressure, respectively, and $\boldsymbol{z}^e$ and $\boldsymbol{p}^e$ are the element nodal variable vectors. Substituting (5.260) into (5.235) and carrying out integration yields the following matrix equation:

$$\begin{bmatrix} -\omega^2\boldsymbol{M}_s + j\omega\boldsymbol{C}_s + \boldsymbol{K}_s & \boldsymbol{K}_{as} \\ -\omega^2\boldsymbol{M}_{as} & -\omega^2\boldsymbol{M}_a + \boldsymbol{K}_a \end{bmatrix} \begin{bmatrix} \boldsymbol{z} \\ \boldsymbol{p} \end{bmatrix} = \begin{bmatrix} \boldsymbol{F} \\ \boldsymbol{0} \end{bmatrix}, \tag{5.261}$$

where $\boldsymbol{M}_s$, $\boldsymbol{C}_s$, and $\boldsymbol{K}_s$ are the mass, damping, and stiffness matrices of the structure, respectively, and $\boldsymbol{F}$ is the loading vector, obtained from the right side of (5.235). Similarly, $\boldsymbol{M}_a$ and $\boldsymbol{K}_a$ are, respectively, the equivalent mass and stiffness matrices of the acoustic medium. The coupling terms between the structure and the acoustic medium are off-diagonal submatrices $\boldsymbol{M}_{as}$ and $\boldsymbol{K}_{as}$, and correspond to the coupling terms in (5.235). As a result, the global matrix in (5.261) is not symmetric due to the off-diagonal coupling submatrices.

In solving (5.261), efficiency cannot be overlooked when considering practical application. Either direct or modal frequency FEA methods can be used to solve the coupled equation. In the direct frequency FEA method, (5.261) is directly solved as a linear algebraic equation with complex variables [81]. Although the method is straightforward in application and provides a very accurate solution, it requires a large amount of computational costs for the repeated analyses required of a large system at several frequencies and with several different loading conditions. Modal frequency FEA is an efficient and a practical method for solving a large size coupled system [34]. In this method, a finite number of modes of the structure and acoustic medium are obtained independently, and a set of selected modes are used to diagonalize the mass and stiffness submatrices, even though the off-diagonal submatrices in (5.261) cannot be diagonalized in this process, since the modes are not orthogonal with respect to the off-diagonal submatrices.

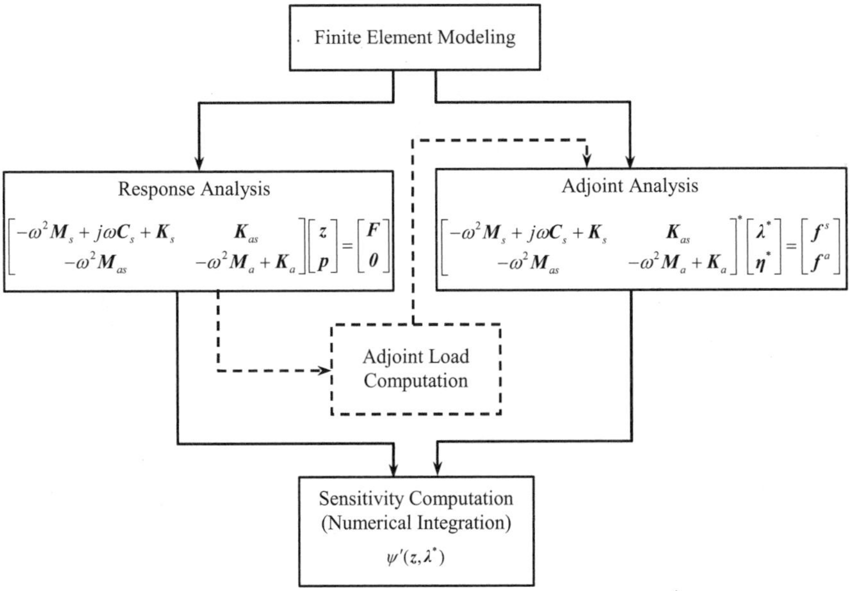

Figure 5.24. Computational procedure of design sensitivity analysis.

Figure 5.24 shows the computational procedure for the adjoint variable method with an established FEA code. A finite element model is constructed by discretizing both the structural components and the acoustic medium. Structural members consist of one- and two-dimensional elements. Triangular and quadrilateral flat elements are used for surface design components, and line segments are used for line design components. The acoustic medium, which is a three-dimensional volume, is modeled using tetrahedrons and hexagons. Identifying the boundary conditions, the interface conditions, and the external load at the nodal points completes the finite element modeling process.

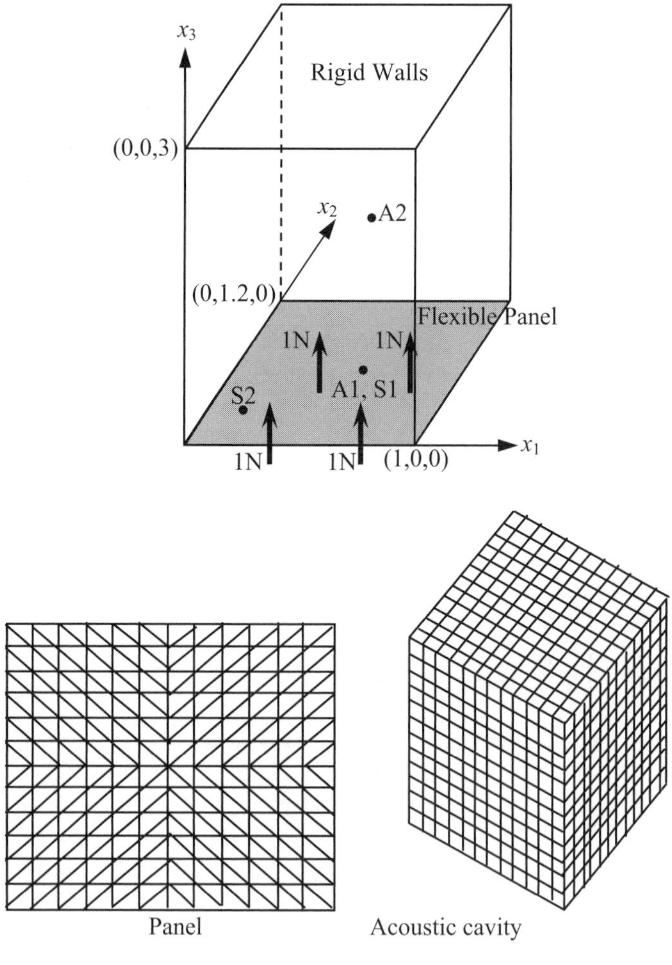

Figure 5.25. Acoustic cavity with flexible wall and finite element models.

An initial finite element analysis of the coupled system gives the primary response z and p at the nodal points. However, as indicated in Section 5.5.2, only a structural response is required in order to perform a design sensitivity analysis.

The adjoint equations are numerically solved using the FEA code with the same finite element model used in the initial analysis. However, in general computation of such adjoint loads as those in (5.253) and (5.258) requires the solutions z and p from the primary analysis, as well as numerical integration. As mentioned before, the solution to the adjoint analysis is the complex conjugate λ^* and η^*. Moreover, this complex conjugate is directly used to evaluate design sensitivity expressions.

The numerical solutions are used to compute design sensitivity, and the integration of the design sensitivity expressions in (5.244) can be evaluated using such numerical integration methods as Gaussian quadrature [46]. The integrands are functions of the state variable, the complex conjugate of the adjoint variable, and the gradients of both variables, as illustrated in (5.255). The function values at Gauss points in each element are required.

Cavity with Flexible Wall

Figure 5.25 depicts the acoustic cavity and panel, previously discussed in Section 5.5.2 [80]. The acoustic medium in the cavity is air, with an equilibrium mass density ρ_0 of 0.1205 Kg/m^3 and an adiabatic bulk modulus β of 139298 N/m^2. The panel is an aluminum plate with a thickness of 0.01 m, a mass density ρ_s of 2700 Kg/m^3, a Young's modulus E of 7.1×10^{10} Pa, a Poisson's ratio ν of 0.334, and a structural damping coefficient φ of 0.06. Harmonic forces of 1.0 N in the x_3-direction are applied at the four points on the plate. The finite element model shown in Fig. 5.25 includes 1728 linear hexagonal elements and 288 triangular shell elements for the panel.

Panel thickness is chosen as the design variable. The design sensitivities of the following performance measures are considered: the acoustic pressure at points A1 (0.5, 0.6, 0.) and A2 (0.5, 0.6, 1.5), and the x_3-direction displacement at points S1 (0.5, 0.6, 0.) and S2 (1/12, 0.2, 0). The ABAQUS code [81] is used for the direct frequency analysis of primary and adjoint problems. Fig. 5.26 provides the primary analysis results.

Design sensitivities are computed at 55 Hz and 60 Hz, which are close to the resonant and the antiresonant frequencies, respectively, as shown in Fig. 5.26. The three-point Gaussian quadrature formula is used for numerical integration over triangular elements. Design sensitivity results are shown in Tables 5.10 and 5.11.

In Tables 5.10 and 5.11, $\psi(\boldsymbol{u} - \Delta\boldsymbol{u})$ and $\psi(\boldsymbol{u} + \Delta\boldsymbol{u})$ are the frequency responses of perturbed designs $\boldsymbol{u} - \Delta\boldsymbol{u}$ and $\boldsymbol{u} + \Delta\boldsymbol{u}$, respectively, where $\delta\boldsymbol{u}$ is the amount of variation in the design. The central finite difference in the design sensitivity is denoted by $\Delta\psi = (\psi(\boldsymbol{u} + \Delta\boldsymbol{u}) - \psi(\boldsymbol{u} - \Delta\boldsymbol{u}))/2$, and ψ' is the predicted design sensitivity. Design perturbation of $\pm 1.0 \times 10^{-4}$ m is used, and predicted design sensitivity values are compared with the central finite difference results. Table 5.10 presents design sensitivity results for acoustic pressure in pascals (Pa), while Table 5.11 shows design sensitivity results for structural displacement in the x_3-direction in meters (m). In Tables 5.10 and 5.11, the real and imaginary parts of complex phasors are denoted by R and I, respectively, and the magnitude is denoted by D, which is the harmonic response amplitude. Table 5.11 shows the design sensitivities of the velocity and acceleration amplitudes V and A, as well as the structural displacement. Excellent agreement is obtained between the design sensitivity predictions ψ' and the finite differences $\Delta\psi$.

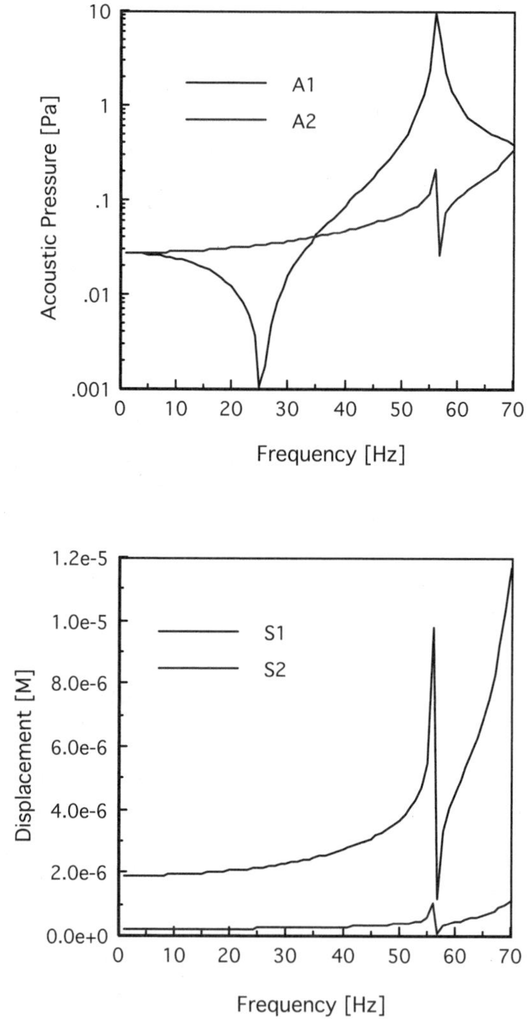

Figure 5.26. Analysis results of acoustic cavity with flexible wall.

Table 5.10. Design sensitivity of cavity with flexible wall (acoustic frequency response, unit Pa).

Frequency	Location		$\psi(\boldsymbol{u}-\Delta\boldsymbol{u})$	$\psi(\boldsymbol{u}+\Delta\boldsymbol{u})$	$\Delta\psi$	ψ'	$\psi'/\Delta\psi(\%)$
		R	–.2513E+1	–.2197E+1	0.1581E+0	0.1582E+0	100.0
55 Hz	A1	I	0.4885E+0	0.3953E+0	–.4661E–1	–.4715E–1	101.2
		D	0.2560E+1	0.2232E+1	–.1640E+4	–.1641E+0	100.1
		R	0.1126E+1	0.1013E+1	–.5652E–1	–.5668E–1	100.3
60 Hz	A1	I	–.1903E+0	–.1630E+0	0.1365E–1	0.1369E–1	100.3
		D	0.1142E+1	0.1026E+1	–.5800E–1	–.5816E–1	100.3
		R	0.1223E+0	0.1070E+0	–.7661E–2	–.7698E–2	100.5
55Hz	A2	I	–.2380E–1	–.1923E–1	0.2283E–2	0.2295E–2	100.5
		D	0.1246E+0	0.1087E+0	–.7950E–2	–.7989E–2	100.5
		R	0.1073E+0	0.9645E–1	–.5413E–2	–.5402E–2	99.8
60 Hz	A2	I	–.1815E–1	–.1554E–1	0.1306E–2	0.1306E–2	100.0
		D	0.1088E+0	0.9769E–1	–.5553E–2	–.5543E–2	99.8

Table 5.11. Design sensitivity of cavity with flexible wall (structural frequency response in x_3-direction, unit m).

Frequency	Location		$\psi(\boldsymbol{u}-\Delta\boldsymbol{u})$	$\psi(\boldsymbol{u}+\Delta\boldsymbol{u})$	$\Delta\psi$	ψ'	$\psi'/\Delta\psi(\%)$
		R	0.5801E–5	0.5064E–5	–.3688E–6	–.3690E–6	100.1
		I	–.1146E–5	–.9239E–6	0.1108E–6	0.1109E–6	100.1
55 Hz	S1	D	0.5913E–5	0.5147E–5	–.3831E–6	–.3832E–6	100.0
		V	0.2044E–2	0.1779E–2	–.1324E–3	–.1324E–3	100.0
		A	0.7062E+0	0.6147E+0	–.4574E–1	–.4577E–1	100.0
		R	0.4695E–5	0.4215E–5	–.2403E–6	–.2410E–6	100.3
		I	–.8130E–6	–.6949E–6	0.5904E–7	0.5923E–7	100.3
60 Hz	S1	D	0.4765E–5	0.4272E–5	–.2468E–6	–.2474E–6	100.3
		V	0.1796E–2	0.1610E–2	–.9302E–4	–.9329E–4	100.3
		A	0.6773E+0	0.6071E+0	–.3517E–1	–.3517E–1	100.3
		R	0.5963E–6	0.5211E–6	–.3755E–7	–.3755E–7	100.0
		I	–.1158E–6	–.9339E–7	0.1120E–7	0.1117E–7	99.7
55 Hz	S2	D	0.6074E–6	0.5295E–6	–.3897E–7	–.3896E–7	100.0
		V	0.2099E–3	0.1830E–3	–.1347E–4	–.1346E–4	100.0
		A	0.7254E–1	0.6323E–1	–.4653E–2	–.4653E–2	100.0
		R	0.4565E–6	0.4117E–6	–.2243E–7	–.2249E–7	100.2
		I	–.7536E–7	–.6471E–7	0.5323E–8	0.5335E–8	100.2
60 Hz	S2	D	0.4627E–6	0.4167E–6	–.2300E–7	–.2305E–7	100.2
		V	0.1744E–3	0.1571E–3	–.8673E–5	–.8689E–5	100.2
		A	0.6576E–1	0.5922E–1	–.3270E–2	–.3276E–2	100.2

Simple Box with Elastic Supports

Figure 5.27 is the model of a simple box vehicle, an example of a built-up structure [29]. The body structure is made of thin aluminum plates of uniform thickness that enclose the acoustic medium (air), and the structure is mounted on a simplified suspension system with springs and dampers. The air has an equilibrium density ρ_0 of 0.1205 Kg/m^3 and an adiabatic bulk modulus β of 139,298 N/m^2. The material properties of the structure are a Poisson's ratio ν of 0.334, a structural damping coefficient φ of 0.06, a mass density ρ_s of 2700 Kg/m^3, and a Young's modulus E of 7.1×10^{10} N/m^2. Body panel thickness is chosen as a design variable, and the current design value is 0.01 m.

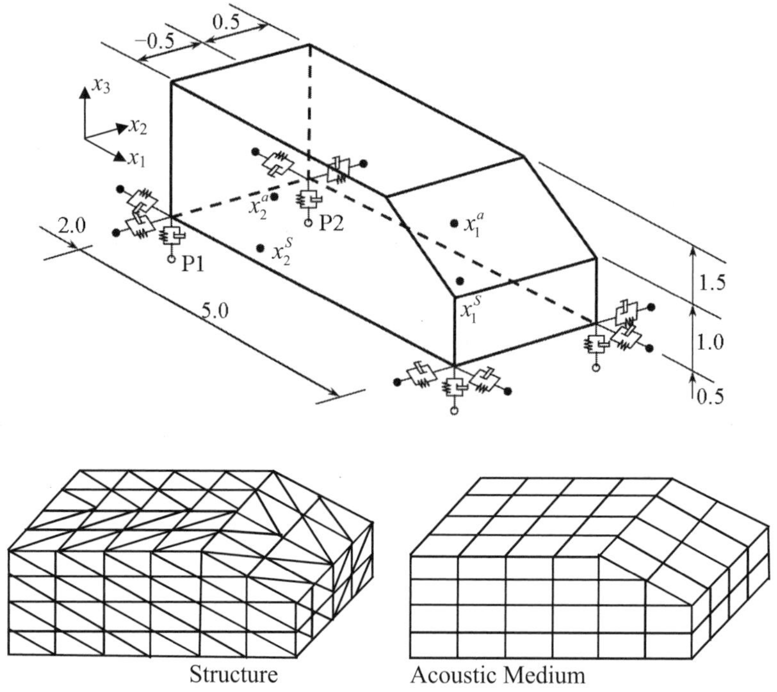

Figure 5.27. Simple box with elastic supports and finite element method.

The finite element model includes 688 hexagonal and 32 tetrahedral acoustic elements, and 928 triangular structural shell elements for the panels. Twelve spring elements and twelve viscous dampers support the structure in three directions at each attachment point. The rear suspension supports P1 and P2 are excited with harmonic displacements in the x_3-direction with amplitudes of 1.0×10^{-4} m, and the front supports are fixed on the ground. The direct frequency response analysis of ABAQUS [81] is used for the analysis of both primary and adjoint problems.

As in the previous example, the predicted design sensitivity results of the harmonic responses at 54 and 62 Hz are compared with the central finite difference results. For this test, the $\pm 1.0 \times 10^{-5}$ m thickness perturbations of the body panels are taken as the design variations. Table 5.12 shows the results for the acoustic pressures in pascals (Pa) at points $\boldsymbol{x}_1^a$ (4.0, 0.25, 1.0), and $\boldsymbol{x}_2^a$ (3.0, –0.25, 1.0). Table 5.13 shows the results for structural displacements, velocities, and accelerations in the x_3-direction at points $\boldsymbol{x}_1^s$ (4.0, 0.25, 0.5) and $\boldsymbol{x}_2^s$ (3.0, –0.25, 0.5), both located on the floor panel. The unit of displacement is meters (m). The same notation used in the previous example is used here. Both tables have good agreement between predicted and finite difference results.

Table 5.12. Design sensitivity of simple box (acoustic frequency response, unit Pa).

Frequency	Location		$\psi(\boldsymbol{u}-\Delta\boldsymbol{u})$	$\psi(\boldsymbol{u}+\Delta\boldsymbol{u})$	$\Delta\psi$	ψ'	$\psi'/\Delta\psi(\%)$
54 Hz	x_1^a	R	0.1002E–1	0.9630E–2	–.1961E–3	–.1978E–3	100.8
		I	0.1294E–1	0.1334E–1	0.1992E–3	0.2010E–3	100.9
		D	0.1637E–1	0.1645E–1	0.4200E–4	0.4254E–4	101.3
62 Hz	x_1^a	R	–.1729E–1	–.1651E–1	0.3874E–3	0.3914E–3	101.0
		I	–.1131E–1	–.1235E–1	–.5202E–3	–.5294E–3	101.8
		D	0.2066E–1	0.2062E–1	–.1900E–4	–.1700E–4	89.4
54Hz	x_2^a	R	–.1724E–2	–.1642E–2	0.4139E–4	0.3985E–4	96.3
		I	–.9126E–2	–.9387E–2	–.1303E–3	–.1310E–3	100.6
		D	0.9288E–2	0.9529E–2	0.1208E–3	0.1218E–3	100.8
62 Hz	x_2^a	R	0.6635E–2	0.5604E–2	–.5154E–3	–.5204E–3	101.0
		I	0.1315E–1	0.1387E–1	0.3621E–3	0.3677E–3	101.6
		D	0.1473E–1	0.1496E–1	0.1170E–3	0.1200E–3	102.6

Table 5.13. Design sensitivity of simple box (structural frequency response in x_3-direction, unit m).

Frequency	Location		$\psi(\boldsymbol{u}-\Delta\boldsymbol{u})$	$\psi(\boldsymbol{u}+\Delta\boldsymbol{u})$	$\Delta\psi$	ψ'	$\psi'/\Delta\psi(\%)$
54 Hz	x_1^s	R	0.1070E–8	0.3797E–9	–.3454E–9	–.3465E–9	100.3
		I	–.3445E–7	–.3509E–7	–.3203E–9	–.3231E–9	100.9
		D	0.3446E–7	0.3509E–7	0.3130E–9	0.3159E–9	100.9
		V	0.1169E–4	0.1191E–4	0.1062E–6	0.1072E–6	100.9
		A	0.3967E–2	0.4039E–2	0.3603E–4	0.3637E–4	100.9
62 Hz	x_1^s	R	–.4879E–7	–.4777E–7	0.5085E–9	0.5083E–9	100.0
		I	–.3730E–7	–.3924E–7	–.9691E–9	–.9823E–9	101.4
		D	0.6141E–7	0.6182E–7	0.2035E–9	0.2118E–9	104.1
		V	0.2392E–4	0.2408E–4	0.7927E–7	0.8251E–7	104.1
		A	0.9320E–2	0.9381E–7	0.3088E–4	0.3214E–4	104.1
54 Hz	x_2^s	R	0.5631E–7	0.5635E–7	0.2412E–10	0.2134E–10	88.5
		I	0.1242E–6	0.1266E–6	0.1195E–8	0.1210E–8	101.3
		D	0.1364E–6	0.1386E–6	0.1100E–8	0.1113E–8	101.2
		V	0.4626E–4	0.4701E–4	0.3732E–6	0.3776E–6	101.2
		A	0.1570E–1	0.1595E–1	0.1266E–3	0.1281E–3	101.2
62 Hz	x_2^s	R	0.2030E–7	0.1734E–7	–.1481E–8	–.1499E–8	101.2
		I	0.9691E–7	0.9793E–7	0.5076E–9	0.5203E–9	102.5
		D	0.9902E–7	0.9945E–7	0.2175E–9	0.2264E–9	104.1
		V	0.3857E–4	0.3874E–4	0.8473E–7	0.8818E–7	104.1
		A	0.1503E–1	0.1509E–1	0.3301E–4	0.3435E–4	104.1

6
Continuum Shape Design Sensitivity Analysis

Chapter 5 treats the design sensitivity analysis of structural components whose shapes are defined by cross-sectional area and thickness variables. In such systems, a function that specifies the structural shape is defined on a fixed physical domain. The design function, known as design variable u, explicitly appears in the variational equation and may explicitly appear in a performance measure in which integration is taken over a fixed domain Ω.

There is an important class of structural design problems needed to determine the shape of a two- or three-dimensional structure (that is, the domain it occupies), subject to such constraints as natural frequency, displacement, and stress. Such problems cannot always be reduced to a formulation that can express the structural shape as a design function, and which appears explicitly in the formulation. Rather, the shape of physical domain Ω must be treated as the design variable. The notion of a material derivative taken from continuum mechanics, and the adjoint variable method in design sensitivity analysis (presented in a similar fashion in Chapter 5) will be used in this chapter to obtain a computable expression for the effect of the shape design on the performance measure. In order to simplify technical complexities as well as to give a clear concept of shape design sensitivity, the variation of the conventional design variable u treated in Chapter 5 will be eliminated.

6.1 Material Derivatives for Shape Design Sensitivity Analysis

To carry out shape design optimization, the design space needs to be defined. For this purpose, a shape design parameterization method, which describes the boundary shape of a structure as a function of design variables, needs to be developed. The design parameterization methods developed in Section 12.2 of Chapter 12 represent changes in the structural shape as a function of the design variable. In this and subsequent sections, the effect of this shape variation on structural performance is developed in a similar way to the size design sensitivity analysis conducted in Section 5.1. The first step in a shape design sensitivity analysis is to develop the relationship between a variation in shape and the resulting variations in functionals, which arise in the shape design problems of Chapter 3. Since the domain shape a structural component occupies is treated as a design variable, it is convenient to think of domain Ω as a continuous medium, and to utilize the material derivative concept from continuum mechanics. In this section, a basic material derivative definition is introduced, and several material derivative formulas that will be used in later sections are derived.

6.1.1 Material Derivative

Domain Change and Design Velocity

Consider a domain Ω in one-, two-, or three-dimensions, as shown schematically in Fig. 6.1. The initial structural geometry Ω is changed to the new geometry Ω_τ by using the design parameterization method in Section 6.1. Here, a scalar parameter τ denotes the amount of shape change in the design variable direction, such that $\tau = 0$ represents the initial geometry Ω. This shape perturbation can be considered a mapping or transformation from Ω to Ω_τ, which is denoted as $\boldsymbol{T}$, as shown in Fig. 6.1. The mapping $\boldsymbol{T}: \boldsymbol{x} \rightarrow \boldsymbol{x}_\tau(\boldsymbol{x})$, $\boldsymbol{x} \in \Omega$ is given by

$$\begin{aligned} \boldsymbol{x}_\tau &\equiv \boldsymbol{T}(\boldsymbol{x},\tau) \\ \Omega_\tau &\equiv \boldsymbol{T}(\Omega,\tau). \end{aligned} \tag{6.1}$$

The process of deforming Ω to Ω_τ by mapping (6.1) may be viewed as a dynamic process of deforming a continuum, with τ playing the role of (design) time. At the initial time $\tau = 0$, the domain is Ω. The trajectory of a point $\mathbf{x} \in \Omega$, beginning at $\tau = 0$, can now be followed. The initial point moves to $\boldsymbol{x}_\tau = \boldsymbol{T}(\boldsymbol{x},\tau)$. By thinking of τ as time, a *design velocity* can be defined as

$$V(\boldsymbol{x}_\tau,\tau) \equiv \frac{d\boldsymbol{x}_\tau}{d\tau} = \frac{d\boldsymbol{T}(\boldsymbol{x},\tau)}{d\tau} = \frac{\partial \boldsymbol{T}(\boldsymbol{x},\tau)}{\partial \tau}, \tag{6.2}$$

where the last equality is due to the fact that the initial point $\boldsymbol{x}$ does not depend on τ. This velocity can also be expressed in terms of the particle position at time τ. If it is assumed that $\boldsymbol{T}^{-1}$ exists, that is, $\boldsymbol{x} = \boldsymbol{T}^{-1}(\boldsymbol{x}_\tau, \tau)$, then the design velocity at $\boldsymbol{x}_\tau = \boldsymbol{T}(\boldsymbol{x},\tau)$ is

$$V(\boldsymbol{x}_\tau,\tau) = \frac{d\boldsymbol{x}_\tau}{d\tau} = \frac{\partial \boldsymbol{T}}{\partial \tau}\left(\boldsymbol{T}^{-1}(\boldsymbol{x}_\tau,\tau),\tau\right). \tag{6.3}$$

The design trajectory of the particle that was at $\boldsymbol{x}$ when $\tau = 0$ is now defined by the initial-value problem

$$\begin{aligned} \dot{\boldsymbol{x}}_\tau &= V(\boldsymbol{x}_\tau,\tau) \\ \boldsymbol{x}_0 &= \boldsymbol{x}, \end{aligned} \tag{6.4}$$

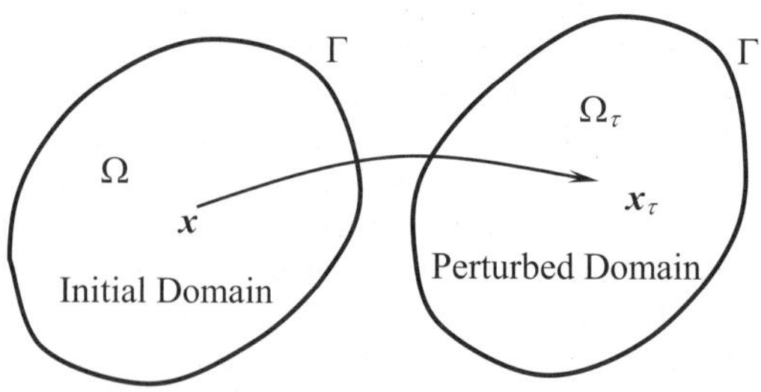

Figure 6.1. Domain perturbation induced by mapping $\boldsymbol{T}$.

where $\dot{\boldsymbol{x}}_\tau = d\boldsymbol{x}_\tau / d\tau$. Thus, if $\boldsymbol{T}$ is given, then design velocity $\boldsymbol{V}$ can be constructed. Conversely, if design velocity field $\boldsymbol{V}(\boldsymbol{x}_\tau, \tau)$ is given, then mapping $\boldsymbol{T}$ can be defined as

$$\boldsymbol{T}(\boldsymbol{x},\tau) = \boldsymbol{x}_\tau(\boldsymbol{x}),$$

where $\boldsymbol{x}_\tau$ is the solution to the initial-value problem in (6.4).

If the transformation $\boldsymbol{T}(\boldsymbol{x},\tau)$ is assumed to be regular enough in the neighborhood of $\tau = 0$, then it can be expanded using the Taylor series around the initial mapping point $\boldsymbol{T}(\boldsymbol{x},0)$ as

$$\boldsymbol{T}(\boldsymbol{x},\tau) = \boldsymbol{T}(\boldsymbol{x},0) + \tau \frac{\partial \boldsymbol{T}}{\partial \tau}(\boldsymbol{x},0) + \cdots = \boldsymbol{x} + \tau \boldsymbol{V}(\boldsymbol{x},0) + \cdots.$$

To develop first-order shape design sensitivity, the first two terms in the Taylor series expansion of $\boldsymbol{T}(\boldsymbol{x},\tau)$ are used. Thus, by ignoring higher-order terms, the following linear mapping relation is obtained:

$$\boldsymbol{T}(\boldsymbol{x},\tau) = \boldsymbol{x} + \tau \boldsymbol{V}(\boldsymbol{x}), \tag{6.5}$$

where $\boldsymbol{V}(\boldsymbol{x}) \equiv \boldsymbol{V}(\boldsymbol{x},0)$. In this text, transformation $\boldsymbol{T}$ from (6.5) will be used, the geometry of which is shown in Fig. 6.2. Equation (6.5) provides the approximation of the transformation by using the linear design velocity field. Variations of domain Ω due to the velocity field $\boldsymbol{V}(\boldsymbol{x})$ are denoted as $\Omega_\tau = \boldsymbol{T}(\Omega,\tau)$, and the boundary of Ω_τ is denoted as Γ_τ. (Note: Henceforth, the term "design velocity" will be referred to simply as "velocity" unless clarity necessitates the longer title.) Even if the linear approximation of mapping defined in (6.5) is enough for design sensitivity analysis, the domain may still be updated during design changes, according to the mapping of (6.1). Note that if design parameterization is defined such that the mapping relation in (6.1) is linear, then (6.5) is exact.

Example 6.1. Consider a beam design component with initial length l, as shown in Fig. 6.3. The initial domain is $\Omega = [0,l]$. In this example, l is the shape design variable. Then a mapping $T(x,\tau)$ can be defined such that the length of the beam is changed to $l + \tau\delta l$. The design sensitivity that will be developed in Sections 6.2.3 and 6.2.5 is based on velocity information at the boundary and domain, respectively. It is clear in this example that the design velocities at the boundaries are $V(0) = 0$ and $V(l) = \delta l$. However, if design velocity

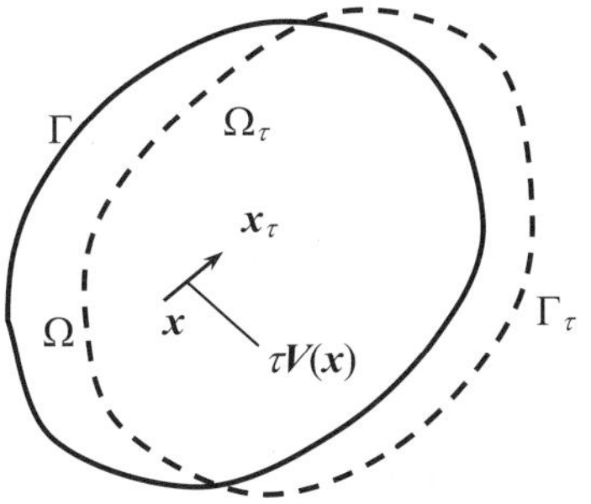

Figure 6.2. Variation of domain using linear mapping.

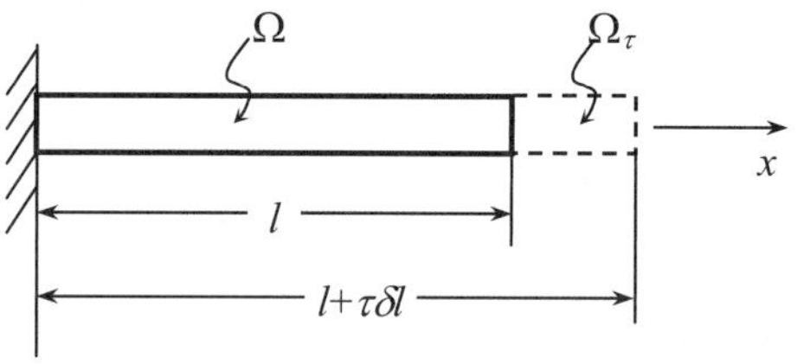

Figure 6.3. Domain perturbation of beam.

information is required for the domain, as with the domain method that will be developed in Section 6.2.5, then there is no unique way to define the design velocity field in the domain. For example, if linear and quadratic velocity fields are assumed, then corresponding mapping relations can be defined, respectively, as

$$T_1(x,\tau) = x + \tau\frac{x}{l}\delta l \tag{a}$$

$$T_2(x,\tau) = x + \tau\frac{x^2}{l^2}\delta l. \tag{b}$$

Note that since two mappings yield the same velocity field at the boundaries, the boundary velocity method provides the same sensitivity results for both (a) and (b) mappings. The domain velocity method also yields the same sensitivity results if a certain regularity requirement is satisfied for the velocity field. However, if the finite element method is used and the initial domain $\Omega = [0,l]$ is subdivided into elements with equal length, then the mapping of (a) provides a perturbed domain Ω_τ with equal-length elements, whereas the mapping of (b) yields an irregular mesh distribution. Thus, it is better to select a mapping that yields a more regular mesh distribution after perturbation, which will be addressed in Section 13.3 of Chapter 13.

Regularity of Perturbed Domain

Let Ω be a C^k-regular open set, that is, its boundary Γ is a compact manifold of C^k in R^n (n = 1, 2, or 3), so that boundary Γ is closed and bounded in R^n, and can be locally (or piecewisely) represented by a C^k function [82]. Let velocity $\boldsymbol{V}(\boldsymbol{x}) \in R^n$ from (6.5) be a vector defined in the neighborhood U of the closure $\bar{\Omega}$, and let $\boldsymbol{V}(\boldsymbol{x})$ and its derivatives up to the order of $k \geq 1$ be continuous. Using these hypotheses, it has been shown [83] that for small τ, $\boldsymbol{T}(\boldsymbol{x},\tau)$ is a homeomorphism (that is, a one-to-one, continuous map with a continuous inverse) from U to $U_\tau = \boldsymbol{T}(U,\tau)$, and that $\boldsymbol{T}(\boldsymbol{x},\tau)$ and its inverse mapping $\boldsymbol{T}^{-1}(\boldsymbol{x}_\tau,\tau)$ are C^k-regular. In conclusion, Ω_τ is also C^k-regular.

Material Derivative of State Variable

Suppose $z_\tau(\boldsymbol{x}_\tau)$ is a smooth, classical solution to the following formal operator equation in deformed domain Ω_τ:

$$\begin{aligned} Az_\tau &= f, \qquad & x \in \Omega_\tau \\ z_\tau &= 0, \qquad & x \in \Gamma_\tau, \end{aligned} \tag{6.6}$$

where A is a differential operator that appears in structural problems. Then the mapping $z_\tau(x_\tau) \equiv z_\tau(x+\tau V(x))$ is defined in Ω, and $z_\tau(x_\tau)$ in Ω_τ depends on τ in two ways. First, it is the solution to the boundary-value problem in Ω_τ. Second, it is evaluated at point x_τ, which moves with τ. If the *pointwise material derivative* exists at $x \in \Omega$, then it is defined as

$$\dot{z} = \dot{z}(x:\Omega,V) \equiv \frac{d}{d\tau} z_\tau(x+\tau V(x))\bigg|_{\tau=0} = \lim_{\tau\to 0}\left[\frac{z_\tau(x+\tau V(x)) - z(x)}{\tau}\right]. \tag{6.7}$$

If z_τ has a regular extension into the neighborhood U_τ of $\bar{\Omega}_\tau$, denoted again as z_τ, then the material derivative of (6.7) can be separated into two contributions as

$$\begin{aligned} \dot{z}(x) &= \lim_{\tau\to 0}\left[\frac{z_\tau(x+\tau V(x)) - z(x)}{\tau}\right] \\ &= \lim_{\tau\to 0}\left[\frac{z_\tau(x) - z(x)}{\tau}\right] + \lim_{\tau\to 0}\left[\frac{z_\tau(x+\tau V(x)) - z_\tau(x)}{\tau}\right] \\ &= z'(x) + \nabla z V(x), \end{aligned} \tag{6.8}$$

where

$$z' = z'(x;\Omega,V) \equiv \lim_{\tau\to 0}\left[\frac{z_\tau(x) - z(x)}{\tau}\right] \tag{6.9}$$

is the partial derivative of z and $\nabla z V = [z_{i,j}]V_j = [\partial z_i/\partial x_j]V_j$ is the convective term. The pointwise material derivative $\dot{z}$ in (6.7) can be defined only for the solution to the classical differential equation (6.6). For the solution to the variational problem, $\dot{z}$ can be defined in the Sobolev norm sense as follows.

If $z_\tau(x_\tau)$ is the solution to the variational equation on the deformed domain Ω_τ, written as

$$a_{\Omega_\tau}(z_\tau, \bar{z}_\tau) = \ell_{\Omega_\tau}(\bar{z}_\tau), \qquad \forall \bar{z}_\tau \in Z_\tau, \tag{6.10}$$

then $z_\tau \in Z_\tau \subset H^m(\Omega_\tau)$, where $Z_\tau \subset H^m(\Omega_\tau)$ is the space of kinematically admissible displacements. When z_τ belongs to the space $H^m(\Omega_\tau)$, the material derivative $\dot{z}$ at Ω is defined as

$$\lim_{\tau\to 0}\left\| \frac{z_\tau(x+\tau V(x)) - z(x)}{\tau} - \dot{z}(x) \right\|_{H^m(\Omega)} = 0. \tag{6.11}$$

Note that for the solution $z_\tau \in H^m(\Omega_\tau)$, the pointwise derivative of (6.7) is meaningless. It was shown by Zolesio [83] that since $T(x,\tau)$ is a C^k homeomorphism, the Sobolev space $H^m(\Omega)$ for $m \le k$ is preserved by $T(x,\tau)$, that is,

$$H^m(\Omega) = \left\{ z_\tau(x+\tau V(x) \middle| \; z_\tau \in H^m(\Omega_\tau) \right\}. \tag{6.12}$$

This observation is used in Section 3.5 in [5] to prove the existence of material derivative $\dot{z}$ for those structural problems presented in Chapter 3.

If $m > k/2$, then according to the Sobolev imbedding theorem (see Appendix A.2), vector space $H^m(\Omega_\tau)$ is a topological subspace of $C^0(\bar{\Omega}_\tau)$, and the pointwise material derivative can be defined. However if $m \leq k/2$, then z_τ is exclusively defined in the sense of "almost everywhere" in Ω_τ, and the pointwise derivative makes no sense.

For the situation in which $z_\tau \in H^m(\Omega_\tau)$, Adams [22] showed that for a C^k-regular open set Ω_τ and for a large enough k, an extension of z_τ exists within the neighborhood U_τ of $\bar{\Omega}_\tau$; hence, the partial derivative z' is defined in the same way as in (6.9). In such a case, the equality in (6.9) must be interpreted in the $H^m(\Omega)$ norm, as in (6.11). The reader who is interested in the exact conditions that are placed on k to produce an extension of z_τ is referred to Adams [22].

One attractive feature of the partial derivative is that, with an assumption of smoothness, the differentiation order between it and the spatial derivative are interchangeable, because both are independent, that is,

$$\begin{aligned}\left(\frac{\partial z}{\partial \boldsymbol{x}}\right)' &= \frac{\partial}{\partial \boldsymbol{x}}(z') \\ (\nabla z)' &= \nabla z'.\end{aligned} \tag{6.13}$$

6.1.2 Basic Material Derivative Formulas

A number of material derivative formulas used throughout the remainder of this text are derived in this section. The reader who is primarily interested in applications may wish to concentrate on results rather than derivations. The most important results obtained have been stated as lemmas.

Material Derivative of Jacobian

Let $\boldsymbol{J}$ be the Jacobian matrix of the mapping $\boldsymbol{T}(\boldsymbol{x},\tau)$, that is,

$$\begin{aligned}\boldsymbol{J} \equiv \frac{\partial \boldsymbol{T}}{\partial \boldsymbol{x}} &= \boldsymbol{I} + \tau\frac{\partial \boldsymbol{V}}{\partial \boldsymbol{x}} \\ &= \boldsymbol{I} + \tau\nabla \boldsymbol{V}(\boldsymbol{x}),\end{aligned} \tag{6.14}$$

where $\boldsymbol{I} = [\delta_{ij}]$ is the identity matrix and $\nabla \boldsymbol{V}(\boldsymbol{x})$ is the Jacobian matrix of $\boldsymbol{V}(\boldsymbol{x})$. Then, from its definition it is trivial to find that $\boldsymbol{J}|_{\tau=0} = \boldsymbol{J}^{-1}|_{\tau=0} = \boldsymbol{I}$. Using the definition in Eq.(6.14), the material derivative of $\boldsymbol{J}$ becomes the Jacobian matrix of the velocity as

$$\left.\frac{d\boldsymbol{J}}{d\tau}\right|_{\tau=0} = \nabla \boldsymbol{V}(\boldsymbol{x}). \tag{6.15}$$

In addition, the material derivative of $\boldsymbol{J}^{-1}$ can be calculated from the relation of $\boldsymbol{J}\boldsymbol{J}^{-1} = \boldsymbol{I}$ as

$$0 = \left.\frac{d}{d\tau}(\boldsymbol{J}\boldsymbol{J}^{-1})\right|_{\tau=0} = \left.\frac{d\boldsymbol{J}}{d\tau}\boldsymbol{J}^{-1}\right|_{\tau=0} + \left.\boldsymbol{J}\frac{d\boldsymbol{J}^{-1}}{d\tau}\right|_{\tau=0}.$$

Since $\boldsymbol{J}|_{\tau=0} = \boldsymbol{J}^{-1}|_{\tau=0} = \boldsymbol{I}$, (6.15) combined with the above equation gives

$$\left.\frac{d\boldsymbol{J}^{-1}}{d\tau}\right|_{\tau=0} = -\nabla \boldsymbol{V}(\boldsymbol{x}). \tag{6.16}$$

Similarly,

$$\left.\frac{d\boldsymbol{J}^{-T}}{d\tau}\right|_{\tau=0} = -\nabla \boldsymbol{V}(\boldsymbol{x})^T, \tag{6.17}$$

where $\boldsymbol{J}^{-T} = (\boldsymbol{J}^{-1})^T = (\boldsymbol{J}^T)^{-1}$.

By denoting $|\boldsymbol{J}|$ as the determinant of $\boldsymbol{J}$, it can be verified by direct calculation that the material derivative of $|\boldsymbol{J}|$ can be obtained as

$$\left.\frac{d}{d\tau}|\boldsymbol{J}|\right|_{\tau=0} = div\boldsymbol{V}(\boldsymbol{x}) = \frac{\partial V_i}{\partial x_i}. \tag{6.18}$$

The material derivative of $|\boldsymbol{J}^{-1}|$ can be calculated by using the relation of $|\boldsymbol{J}\boldsymbol{J}^{-1}| = 1$ and the product rule of differentiation as

$$0 = \left.\frac{d}{d\tau}|\boldsymbol{J}\boldsymbol{J}^{-1}|\right|_{\tau=0} = \left.|\boldsymbol{J}^{-1}|\frac{d}{d\tau}|\boldsymbol{J}|\right|_{\tau=0} + \left.|\boldsymbol{J}|\frac{d}{d\tau}|\boldsymbol{J}^{-1}|\right|_{\tau=0}.$$

Since $|\boldsymbol{J}||_{\tau=0} = |\boldsymbol{J}^{-1}||_{\tau=0} = 1$ at the initial design, (6.18) combined with the above equation gives

$$\left.\frac{d}{d\tau}|\boldsymbol{J}^{-1}|\right|_{\tau=0} = -div\boldsymbol{V}(\boldsymbol{x}). \tag{6.19}$$

These formulas are useful to compute the material derivative of the domain at the perturbed design point. Let $d\Omega$ and $d\Omega_\tau$ be the infinitesimal volume of domains Ω and Ω_τ, respectively. Using the fact that $d\Omega_\tau = |\boldsymbol{J}|d\Omega$, the variation of $d\Omega_\tau$ can be obtained as

$$\left.\frac{d}{d\tau}d\Omega_\tau\right|_{\tau=0} = \left.\frac{d}{d\tau}|\boldsymbol{J}|\right|_{\tau=0} d\Omega = div\boldsymbol{V}\, d\Omega. \tag{6.20}$$

Material Derivative of Surface Area

Let $\boldsymbol{n}$ be the unit normal vector on infinitesimal area $d\Gamma$ of the parallelogram shown in Fig. 6.4, with two edges ($d\boldsymbol{x}$ and $\delta\boldsymbol{x}$) on undeformed surface Γ. Let $\boldsymbol{n}_\tau$ be the unit normal vector on infinitesimal area $d\Gamma_\tau$ of deformed surface Γ_τ, with edges $d\boldsymbol{x}_\tau$ and $\delta\boldsymbol{x}_\tau$. Since Ω and Ω_τ are C^k-regular, $\boldsymbol{n}$ and $\boldsymbol{n}_\tau$ are C^{k-1}-regular. The objective is to create a relation between $d\Gamma$ and $d\Gamma_\tau$ so that the material derivative of $d\Gamma_\tau$ may be obtained, like the material derivative of $d\Omega_\tau$ in (6.20).

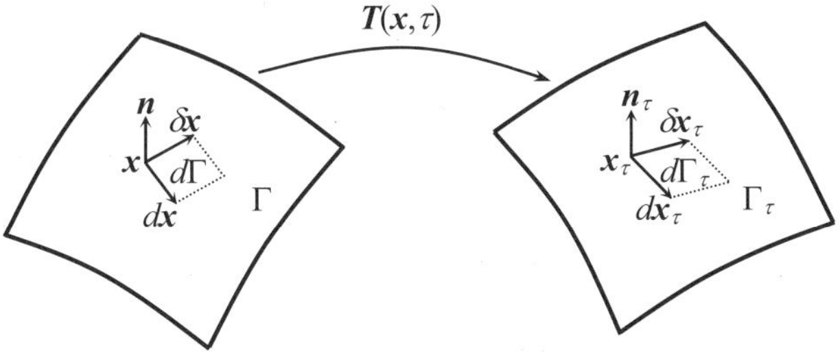

Figure 6.4. Transformation of area.

The edges $d\boldsymbol{x}_\tau$ and $\delta\boldsymbol{x}_\tau$ can be represented by using the Jacobian matrix and the edges on the initial boundary as

$$\begin{aligned} d\boldsymbol{x}_\tau &= \boldsymbol{J}d\boldsymbol{x} \\ \delta\boldsymbol{x}_\tau &= \boldsymbol{J}\delta\boldsymbol{x}. \end{aligned} \tag{6.21}$$

Since $\boldsymbol{J}^{-1}$ exists, the inverse relations can also be obtained as

$$\begin{aligned} d\boldsymbol{x} &= \boldsymbol{J}^{-1}d\boldsymbol{x}_\tau \\ \delta\boldsymbol{x} &= \boldsymbol{J}^{-1}\delta\boldsymbol{x}_\tau. \end{aligned} \tag{6.22}$$

Then, the infinitesimal areas of two boundaries can be denoted by using the vector product as

$$\begin{aligned} \boldsymbol{n}d\Gamma &= d\boldsymbol{x} \times \delta\boldsymbol{x} \\ \boldsymbol{n}_\tau d\Gamma_\tau &= d\boldsymbol{x}_\tau \times \delta\boldsymbol{x}_\tau. \end{aligned} \tag{6.23}$$

The above vector notation can be represented in Cartesian rectangular components as

$$\begin{aligned} n_i d\Gamma &= e_{ijk} dx_j \delta x_k \\ n_{\tau_r} d\Gamma_\tau &= e_{rst} dx_{\tau_s} \delta x_{\tau_t}, \end{aligned} \tag{6.24}$$

where e_{ijk} is a permutation symbol, defined as

$$e_{ijk} = \begin{cases} 0 & \text{when any two indices are equal} \\ +1 & \text{when } i,j,k \text{ are 1, 2, 3 or an even} \\ & \text{permulation of 1, 2, 3} \\ -1 & \text{when } i,j,k \text{ are an odd permutation of 1, 2, 3.} \end{cases} \tag{6.25}$$

From the first equation of (6.24), and by using (6.22),

$$n_i d\Gamma = e_{ijk} \frac{\partial x_j}{\partial x_{\tau_s}} \frac{\partial x_k}{\partial x_{\tau_t}} dx_{\tau_s} \delta x_{\tau_t}. \tag{6.26}$$

Multiplying both sides of (6.26) by $\partial x_i / \partial x_{\tau_r}$ and summing on i,

$$\frac{\partial x_i}{\partial x_{\tau_r}} n_i d\Gamma = e_{ijk} \frac{\partial x_i}{\partial x_{\tau_r}} \frac{\partial x_j}{\partial x_{\tau_s}} \frac{\partial x_k}{\partial x_{\tau_t}} dx_{\tau_s} \delta x_{\tau_t}. \tag{6.27}$$

For any 3 × 3 matrix with elements a_{mn}, the following identity can be proved by direct calculation:

$$e_{rst} \det[a_{mn}] = e_{ijk} a_{ir} a_{js} a_{kt}. \tag{6.28}$$

Since the Jacobian matrix $\boldsymbol{J}$ has $\partial \boldsymbol{x}_\tau / \partial \boldsymbol{x}$ as elements, the following relations hold:

$$\begin{aligned} e_{ijk} |\boldsymbol{J}| &= e_{rst} \frac{\partial x_{\tau_r}}{\partial x_i} \frac{\partial x_{\tau_s}}{\partial x_j} \frac{\partial x_{\tau_t}}{\partial x_k} \\ e_{rst} |\boldsymbol{J}^{-1}| &= e_{ijk} \frac{\partial x_i}{\partial x_{\tau_r}} \frac{\partial x_j}{\partial x_{\tau_s}} \frac{\partial x_k}{\partial x_{\tau_t}}. \end{aligned} \tag{6.29}$$

By substituting the second part of (6.29) into (6.27), and by recalling that $|\boldsymbol{J}^{-1}| = |\boldsymbol{J}|^{-1}$, we can obtain a simplified form as

$$\frac{\partial x_i}{\partial x_{\tau_r}} n_i d\Gamma = |\boldsymbol{J}|^{-1} e_{rst} dx_{\tau_s} \delta x_{\tau_t},$$

which can be rewritten using (6.24) as

$$\boldsymbol{n}_\tau d\Gamma_\tau = |\boldsymbol{J}| \boldsymbol{J}^{-T} \boldsymbol{n} d\Gamma. \tag{6.30}$$

Thus, $\boldsymbol{n}_\tau$ is parallel to $\boldsymbol{J}^{-T}\boldsymbol{n}$. The explicit form of $\boldsymbol{n}_\tau$ can be obtained by normalizing the right side of (6.30) as

$$\boldsymbol{n}_\tau = \frac{\boldsymbol{J}(\boldsymbol{x}_\tau)^{-T} \boldsymbol{n}(\boldsymbol{x})}{\left\| \boldsymbol{J}(\boldsymbol{x}_\tau)^{-T} \boldsymbol{n}(\boldsymbol{x}) \right\|}, \tag{6.31}$$

where $\|\boldsymbol{a}\| = (\boldsymbol{a}^T\boldsymbol{a})^{1/2}$ is the Euclidean norm. By applying (6.31) to (6.30), we finally obtain the desired relation between $d\Gamma$ and $d\Gamma_\tau$ as

$$d\Gamma_\tau = |\boldsymbol{J}| \left\| \boldsymbol{J}(\boldsymbol{x}_\tau)^{-T} \boldsymbol{n}(\boldsymbol{x}) \right\| d\Gamma. \tag{6.32}$$

To calculate the material derivative of $d\Gamma_\tau$, it is necessary to differentiate the right side of (6.32) as

$$\frac{d}{d\tau} \left\| \boldsymbol{J}(\boldsymbol{x}_\tau)^{-T} \boldsymbol{n}(\boldsymbol{x}) \right\| \Big|_{\tau=0} \equiv \frac{d}{d\tau} (\boldsymbol{J}^{-T}\boldsymbol{n}, \boldsymbol{J}^{-T}\boldsymbol{n})^{1/2} \Big|_{\tau=0} = -\boldsymbol{n}^T \nabla \boldsymbol{V} \boldsymbol{n}, \tag{6.33}$$

where $(\boldsymbol{a},\boldsymbol{b}) \equiv \boldsymbol{a}^T\boldsymbol{b}$. The material derivative of the normal vector can also be calculated from the relation in (6.31) as

$$\begin{aligned} \dot{\boldsymbol{n}} \equiv \frac{d\boldsymbol{n}_\tau}{d\tau} \Big|_{\tau=0} &= \frac{1}{\left\| \boldsymbol{J}^{-T}\boldsymbol{n} \right\|^2} \left[\left\| \boldsymbol{J}^{-T}\boldsymbol{n} \right\| \frac{d\boldsymbol{J}^{-T}}{d\tau} \boldsymbol{n} - (\boldsymbol{J}^{-T}\boldsymbol{n}) \frac{d}{d\tau} \left\| \boldsymbol{J}^{-T}\boldsymbol{n} \right\| \right] \Big|_{\tau=0} \\ &= (\boldsymbol{n}^T \nabla \boldsymbol{V} \boldsymbol{n})\boldsymbol{n} - \nabla \boldsymbol{V}^T \boldsymbol{n}. \end{aligned} \tag{6.34}$$

Also, using (6.18) and (6.33), we obtain the following formula:

$$\frac{d}{d\tau} \left(|\boldsymbol{J}| \left\| \boldsymbol{J}^{-T}\boldsymbol{n} \right\| \right) \Big|_{\tau=0} = div\boldsymbol{V} - \boldsymbol{n}^T \nabla \boldsymbol{V} \boldsymbol{n}. \tag{6.35}$$

Material Derivative of Domain Functional

Lemma 6.1. Let ψ_1 be a domain functional, defined as an integral over Ω_τ, namely,

$$\psi_1 = \iint_{\Omega_\tau} f_\tau(\boldsymbol{x}_\tau)\, d\Omega_\tau \tag{6.36}$$

where f_τ is a regular function defined in Ω_τ. If Ω is C^k-regular, then the material derivative of ψ_1 at Ω is

$$\begin{aligned} \psi_1' &= \iint_\Omega [f'(\boldsymbol{x}) + \nabla f(\boldsymbol{x})^T \boldsymbol{V} + f(\boldsymbol{x}) div\boldsymbol{V}]\, d\Omega \\ &= \iint_\Omega [f'(\boldsymbol{x}) + div(f\boldsymbol{V})]\, d\Omega \end{aligned} \tag{6.37}$$

or,

$$\psi_1' = \iint_\Omega f'(\boldsymbol{x})\, d\Omega + \int_\Gamma f(\boldsymbol{x}) V_n \, d\Gamma, \tag{6.38}$$

where $V_n = \boldsymbol{V}^T\boldsymbol{n}$ is the normal component of $\boldsymbol{V}(\boldsymbol{x})$ on boundary Γ.

Proof. Function f can be a scalar, vector, or tensor. By transforming the variables of integration from (6.36) to the initial design domain Ω, we achieve

$$\psi_1 = \iint_{\Omega_\tau} f_\tau(\boldsymbol{x}_\tau)\, d\Omega_\tau = \iint_\Omega f_\tau(\boldsymbol{x} + \tau \boldsymbol{V}(\boldsymbol{x}))\, |\boldsymbol{J}|\, d\Omega.$$

Using (6.8) and (6.18), the material derivative of ψ_1 at Ω is obtained as

$$\begin{aligned} \psi_1' &= \frac{d}{d\tau} \iint_\Omega f_\tau(\boldsymbol{x} + \tau\boldsymbol{V}(\boldsymbol{x}))\, |\boldsymbol{J}|\, d\Omega \Big|_{\tau=0} \\ &= \iint_\Omega [\dot{f}(\boldsymbol{x}) + f(\boldsymbol{x}) div\boldsymbol{V}]\, d\Omega \\ &= \iint_\Omega [f'(\boldsymbol{x}) + \nabla f(\boldsymbol{x})^T \boldsymbol{V}(\boldsymbol{x}) + f(\boldsymbol{x}) div\boldsymbol{V}]\, d\Omega \\ &= \iint_\Omega [f'(\boldsymbol{x}) + div(f(\boldsymbol{x})\boldsymbol{V}(\boldsymbol{x}))]\, d\Omega. \end{aligned} \tag{6.39}$$

If Ω is C^k-regular, then the divergence theorem [84] yields (6.38). ■

Note that (6.37) requires design velocity information within domain Ω, whereas (6.38) only needs this information on boundary Γ. The boundary velocity approach provides a simple and convenient way for computing sensitivity information. It also supplies a physical understanding of design sensitivity information. For engineering applications that use the numerical method, however, an accurate evaluation of the function value on the boundary is critical. If a domain method, such as the finite element method, is used to solve a variational equation, it is well known [85] that the results of a finite element analysis may not be satisfactory at the boundaries for a system with a nonsmooth load or with interface problems. The domain velocity approach in (6.37) can be used effectively with a high enough accuracy rate when the finite element method is used. In this text, the boundary velocity approach is relied on to conveniently explain analytical examples, while the domain velocity approach is used to accurately solve numerical examples.

It is interesting and important to note that only the normal component V_n of the boundary velocity appears in (6.38), which is of importance in accounting for the domain variation effect. In fact, Theorem 3.5.3 (Section 3.5.7) in [5] proves that if a general domain functional ψ has a gradient at Ω, and if Ω is C^{k+1}-regular, then only the normal velocity component V_n on the boundary needs to be considered for derivative calculations. The fundamental concept behind this result is that $\Gamma_\tau(V_s) = \Gamma$ for all τ, where V_s is the component of velocity field $\boldsymbol{V}$ in (6.3), which is tangential to boundary Γ. In other words, tangential component V_s of the velocity field does not deform domain Ω.

Material Derivative of Boundary Functional

Consider a boundary functional, defined as an integral over Γ_τ,

$$\psi_2 = \int_{\Gamma_\tau} g_\tau(\boldsymbol{x}_\tau)\, d\Gamma_\tau, \tag{6.40}$$

where g_τ is a regular function defined on Γ_τ. Using (6.32), $d\Gamma_\tau$ is transformed into $d\Gamma$ as

$$\psi_2 = \int_{\Gamma_\tau} g_\tau(\boldsymbol{x}_\tau)\, d\Gamma_\tau = \int_\Gamma g_\tau(\boldsymbol{x} + \tau \boldsymbol{V}(\boldsymbol{x})) |\boldsymbol{J}| \left\| \boldsymbol{J}^{-T} \boldsymbol{n} \right\| d\Gamma, \tag{6.41}$$

and using the formula from (6.35), the material derivative of ψ_2 would be

$$\begin{aligned} \psi_2' &= \frac{d}{d\tau} \int_\Gamma g_\tau(\boldsymbol{x} + \tau \boldsymbol{V}(\boldsymbol{x})) |\boldsymbol{J}| \left\| \boldsymbol{J}^{-T} \boldsymbol{n} \right\| d\Gamma \Big|_{\tau=0} \\ &= \int_\Gamma [\dot{g}(\boldsymbol{x}) + g(\boldsymbol{x})(div \boldsymbol{V}(\boldsymbol{x}) - \boldsymbol{n}^T \nabla \boldsymbol{V}\, \boldsymbol{n})]\, d\Gamma. \end{aligned} \tag{6.42}$$

Suppose that the mapping $\boldsymbol{V} \rightarrow \dot{g}$ is linear and continuous. Then, (6.42) implies that ψ_2 has a gradient at Ω. Further, according to Theorem 3.5.3 (Section 3.5.7) in [5], if Ω is C^{k+1}-regular, then only $\boldsymbol{V} = V_n \boldsymbol{n}$ need be considered, where V_n is a scalar C^k-regular function. If Ω is C^{k+1}-regular, then both $\boldsymbol{n}$ and $\boldsymbol{V} = V_n \boldsymbol{n}$ are C^k-regular. When $\boldsymbol{V} = V_n \boldsymbol{n}$ on Γ, the Jacobian matrix of $\boldsymbol{V}$ can be written as

$$\nabla \boldsymbol{V} = \boldsymbol{n} \nabla V_n^T + V_n \nabla \boldsymbol{n}. \tag{6.43}$$

Since $\boldsymbol{n}$ is the unit normal, the following relation holds:

$$0 = \tfrac{1}{2} \nabla (\boldsymbol{n}^T \boldsymbol{n}) = \nabla \boldsymbol{n}^T \boldsymbol{n}. \tag{6.44}$$

Hence, from (6.43),

$$\nabla \boldsymbol{V}^T \boldsymbol{n} = \nabla V_n \boldsymbol{n}^T \boldsymbol{n} + V_n \nabla \boldsymbol{n}^T \boldsymbol{n} = \nabla V_n. \tag{6.45}$$

From (6.34) and (6.45), with normal velocity $\boldsymbol{V} = V_n \boldsymbol{n}$,

$$\dot{\boldsymbol{n}} = \frac{d\boldsymbol{n}_\tau}{d\tau}\Big|_{\tau=0} = (\boldsymbol{n}^T \nabla \boldsymbol{V}\, \boldsymbol{n}) \boldsymbol{n} - \nabla \boldsymbol{V}^T \boldsymbol{n} = (\nabla V_n^T \boldsymbol{n}) \boldsymbol{n} - \nabla V_n \tag{6.46}$$

and

$$div \boldsymbol{V} - (\nabla \boldsymbol{V} \boldsymbol{n}, \boldsymbol{n}) = \nabla V_n^T \boldsymbol{n} + V_n div \boldsymbol{n} - \nabla V_n^T \boldsymbol{n} = V_n div \boldsymbol{n}. \tag{6.47}$$

It is now to be shown that

$$div \boldsymbol{n} = \kappa, \tag{6.48}$$

where κ is the curvature of Γ in R^2 and twice the mean curvature of Γ in R^3.

In order to prove (6.48), first consider a two-dimensional domain where Γ can be represented by the local graph of a regular function f, say $x_2 = f(x_1)$. Suppose Ω lies below the graph of f. The normal vector is given by

$$\boldsymbol{n}(x_1, f(x_1)) = (1 + f'^2)^{-1/2} \begin{bmatrix} -f' \\ 1 \end{bmatrix}.$$

A direct calculation produces

$$div \boldsymbol{n}(x_1, f(x_1)) = -(1 + f'^2)^{-3/2} f'', \tag{6.49}$$

which is the curvature of Γ. This verifies (6.48) for R^2.

In the case of a three-dimensional domain, Γ is a regular surface. For any point $\boldsymbol{x} \in \Gamma$, consider the R^3-orthonormal basis $\{\boldsymbol{e}_1, \boldsymbol{e}_2, \boldsymbol{n}\}$, as shown in Fig. 6.5, where $\boldsymbol{e}_1$ and $\boldsymbol{e}_2$ are vectors that are tangential to Γ at $\boldsymbol{x}$, such that [86]

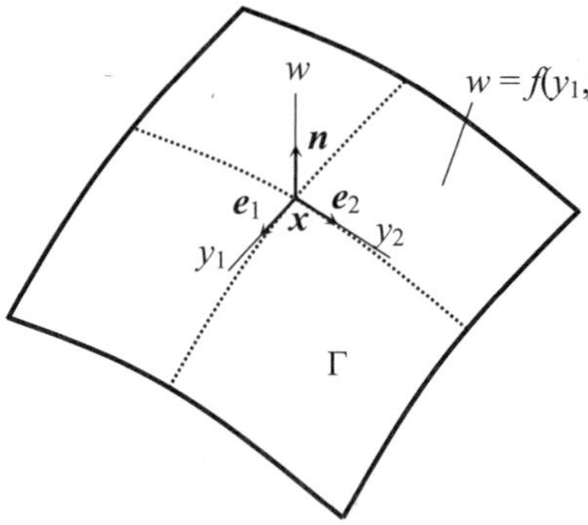

Figure 6.5. Local representation of the boundary Γ.

$$\nabla \boldsymbol{n}\boldsymbol{e}_i = -\kappa_i \boldsymbol{e}_i, \qquad i = 1,2 \text{ (no sum on } i). \tag{6.50}$$

The parameters κ_1 and κ_2 are principal normal curvatures of Γ at $\boldsymbol{x}$, and the vectors $\boldsymbol{e}_1$ and $\boldsymbol{e}_2$ are unit vectors in principal directions. In the neighborhood of $\boldsymbol{x}$, with $\boldsymbol{x}$ taken as the origin, Γ may be represented by the graph of $w = f(y_1, y_2)$ in the (y_1, y_2, w) coordinate system, as shown in Fig. 6.5.

Since the divergence operator is invariant during translation and rotational motion [87], $div\boldsymbol{n}$ can be written in the (y_1, y_2, w) coordinate system, that is,

$$div\boldsymbol{n} = \boldsymbol{e}_i^T \nabla \boldsymbol{n}\boldsymbol{e}_i + \boldsymbol{n}^T \nabla \boldsymbol{n}\boldsymbol{n}.$$

Thus, using (6.44) and (6.50),

$$div\boldsymbol{n} = -(\kappa_1 + \kappa_2), \tag{6.51}$$

which is twice the mean curvature of Γ [86]. This completes the proof of (6.48). ■

If the velocity is normal on Γ, then $\boldsymbol{V} = V_n\boldsymbol{n}$. Consequently, from (6.47),

$$div\boldsymbol{V} - \boldsymbol{n}^T \nabla \boldsymbol{V}\,\boldsymbol{n} = V_n \kappa. \tag{6.52}$$

The choice of $\boldsymbol{n}$, as directed outward from the domain Ω, defines the orientation of boundary Γ. If the orientation of Γ changes, then $\boldsymbol{n}$ is changed to $-\boldsymbol{n}$, and κ must be changed to $-\kappa$.

From (6.42), and using (6.52),

$$\begin{aligned} \psi_2' &= \int_\Gamma [\dot{g}(\boldsymbol{x}) + \kappa g(\boldsymbol{x})V_n]\, d\Gamma \\ &= \int_\Gamma [g'(\boldsymbol{x}) + \nabla g^T \boldsymbol{V} + \kappa g(\boldsymbol{x})V_n]\, d\Gamma \\ &= \int_\Gamma [g'(\boldsymbol{x}) + (\nabla g^T \boldsymbol{n} + \kappa g(\boldsymbol{x}))V_n]\, d\Gamma. \end{aligned}$$

This proves the following lemma:

Lemma 6.2. Suppose that in (6.40) g_τ is a regular function defined on Γ_τ, and that the mapping $V \rightarrow \dot{g}$ is linear and continuous. If Ω is C^{k+1}-regular, then the material derivative of ψ_2 at Ω is

$$\psi_2' = \int_\Gamma [g'(x) + (\nabla g^T \boldsymbol{n} + \kappa g(\boldsymbol{x}))V_n]\, d\Gamma. \tag{6.53}$$

By putting the right sides of (6.42) and (6.53) together, and by using the fact that $\dot{g}(\boldsymbol{x}) = g'(\boldsymbol{x}) + \nabla g^T \boldsymbol{V}$, the following useful relationship for any regular vector field $\boldsymbol{V}$ and regular function g on R^n is obtained:

$$\begin{aligned}\int_\Gamma \nabla g^T \boldsymbol{V}\, d\Gamma = &-\int_\Gamma g(\boldsymbol{x})[div\boldsymbol{V}(\boldsymbol{x}) - (\boldsymbol{n}^T \nabla \boldsymbol{V}\,\boldsymbol{n})]\, d\Gamma \\ &+ \int_\Gamma [\nabla g^T \boldsymbol{n} + \kappa g(\boldsymbol{x})]V_n\, d\Gamma.\end{aligned} \tag{6.54}$$

Finally, consider a special functional that is defined as the integration of a vector function over Γ_τ as

$$\psi_3 = \int_{\Gamma_\tau} \boldsymbol{h}_\tau(\boldsymbol{x}_\tau)^T \boldsymbol{n}_\tau\, d\Gamma_\tau, \tag{6.55}$$

where $\boldsymbol{h}_\tau$ is a regular vector field defined on Γ_τ; hence, $\boldsymbol{h}_\tau^T \boldsymbol{n}_\tau$ is a regular function on Γ_τ. From (6.42), and by using $\boldsymbol{h}_\tau^T \boldsymbol{n}_\tau$ instead of g_τ, we obtain:

$$\psi_3' = \int_\Gamma \left[\dot{\boldsymbol{h}}(\boldsymbol{x})^T \boldsymbol{n}(\boldsymbol{x}) + \boldsymbol{h}(\boldsymbol{x})^T \dot{\boldsymbol{n}}(\boldsymbol{x}) + \boldsymbol{h}(\boldsymbol{x})^T \boldsymbol{n}(\boldsymbol{x})(div\boldsymbol{V}(\boldsymbol{x}) - \boldsymbol{n}^T \nabla \boldsymbol{V}\,\boldsymbol{n})\right] d\Gamma, \tag{6.56}$$

where $\dot{\boldsymbol{n}}$ is given by (6.34). If the mapping $\boldsymbol{V} \rightarrow \dot{\boldsymbol{h}}$ is linear and continuous, then (6.56) implies that ψ_3 has a gradient at Ω. Further, according to Theorem 3.5.3 (Section 3.5.7) in [5], if Ω is C^{k+1}-regular, then only $\boldsymbol{V} = V_n \boldsymbol{n}$ need be considered, with a C^k-regular scalar function V_n. Then, by using (6.46) and (6.52), (6.56) becomes

$$\psi_3' = \int_\Gamma [\dot{\boldsymbol{h}}^T \boldsymbol{n} + \boldsymbol{h}^T((\nabla V_n^T \boldsymbol{n})\boldsymbol{n} - \nabla V_n) + \kappa V_n \boldsymbol{h}^T \boldsymbol{n}]\, d\Gamma. \tag{6.57}$$

Using (6.54) and substituting $\boldsymbol{h}$ for $\boldsymbol{V}$ and V_n for g, we obtain

$$\int_\Gamma \boldsymbol{h}^T \nabla V_n\, d\Gamma = -\int_\Gamma V_n[div\boldsymbol{h} - (\boldsymbol{n}^T \nabla \boldsymbol{h}\,\boldsymbol{n})]\, d\Gamma + \int_\Gamma [\nabla V_n^T \boldsymbol{n} + \kappa V_n](\boldsymbol{h}^T \boldsymbol{n})\, d\Gamma.$$

Hence, (6.57) becomes

$$\begin{aligned}\psi_3' &= \int_\Gamma [\dot{\boldsymbol{h}}^T \boldsymbol{n} + (div\boldsymbol{h} - (\boldsymbol{n}^T \nabla \boldsymbol{h}\,\boldsymbol{n}))V_n]\, d\Gamma \\ &= \int_\Gamma [\boldsymbol{h}'^T \boldsymbol{n} + \boldsymbol{n}^T \nabla \boldsymbol{h} \boldsymbol{V} + (div\boldsymbol{h} - (\boldsymbol{n}^T \nabla \boldsymbol{h}\,\boldsymbol{n}))V_n]\, d\Gamma \\ &= \int_\Gamma [\boldsymbol{h}'^T \boldsymbol{n} + V_n(\boldsymbol{n}^T \nabla \boldsymbol{h}\,\boldsymbol{n}) + (div\boldsymbol{h} - (\boldsymbol{n}^T \nabla \boldsymbol{h}\,\boldsymbol{n}))V_n]\, d\Gamma \\ &= \int_\Gamma [\boldsymbol{h}'^T \boldsymbol{n} + div\boldsymbol{h} V_n]\, d\Gamma.\end{aligned}$$

Thus, the following lemma has been proved.

Lemma 6.3. Suppose that $\boldsymbol{h}_\tau$ in (6.55) is a regular field defined on Γ_τ and that the mapping $\boldsymbol{V} \rightarrow \dot{\boldsymbol{h}}$ is linear and continuous. If Ω is C^{k+1}-regular, then the material derivative of ψ_3 at Ω is

$$\psi_3' = \int_\Gamma [\boldsymbol{h}'(\boldsymbol{x})^T \boldsymbol{n} + div\boldsymbol{h} V_n]\, d\Gamma. \tag{6.58}$$

6.2 Static Response Design Sensitivity Analysis

As was seen in Section 6.1, a static response depends on its domain shape. The existence of material derivative $\dot{z}$, proved in Section 3.5 in [5], and the material derivative formulas derived in Section 6.1 will be used in this section to derive a direct differentiation and adjoint variable method for design sensitivity analysis. As in Chapter 5, an adjoint problem that is closely related to the original structural problem is obtained, and explicit formulas for shape design sensitivity analyses are equally obtained. Finally, numerical methods for parameterizing boundary shape and calculating shape design sensitivity coefficients will also be obtained.

6.2.1 Differentiability of Bilinear Forms and Static Response

Design differentiability of energy bilinear forms and static response are proved in Section 3.5 in [5] for the problems treated in Chapter 3. These differentiability results are used here to develop shape design sensitivity formulas. The reason for this order of presentation is the same as with Chapter 5: because the technical proof of design derivatives does not contribute to a greater understanding of the adjoint variable method. Nevertheless, as noted in Chapter 5, the delicate question concerning the existence of design derivatives should not be ignored.

Within the perturbed domain Ω_τ the variational equations for the problems in Chapter 3 are in the following form:

$$a_{\Omega_\tau}(z_\tau, \bar{z}_\tau) \equiv \iint_{\Omega_\tau} c(z_\tau, \bar{z}_\tau)\, d\Omega_\tau = \iint_{\Omega_\tau} \bar{z}_\tau^T f\, d\Omega_\tau \equiv \ell_{\Omega_\tau}(\bar{z}_\tau), \quad \forall \bar{z}_\tau \in Z_\tau, \tag{6.59}$$

where $Z_\tau \subset H^m(\Omega_\tau)$ is the space of all kinematically admissible displacements, and $c(\bullet,\bullet)$ is a bilinear mapping defined by the integrand of the energy bilinear form. The subscribed Ω_τ denotes that energy bilinear and load linear forms depend on the perturbed domain that is the design variable. Section 3.5 in [5] demonstrates that the energy bilinear and load linear forms are differentiable with respect to the design for the problems in Chapter 3. In general, vector function z is used as the response variable. However, as described in Section 3.1, z has to be interpreted as a scalar function for truss, beam, and plate problems.

A powerful consequence of the proofs in Section 3.5 in [5] is that the solution to (6.59) is differentiable with respect to the design, that is, the material derivative $\dot{z}$ defined in (6.11) exists. Note that material derivative $\dot{z}$ depends on the direction of V (the velocity field). As shown in Section 3.5 in [5], $\dot{z}$ is linear in V and, in fact, is a Fréchet derivative with respect to the design, evaluated in the direction of V. According to Theorem 3.5.3 (Section 3.5.7) in [5], the linearity and continuity of the mapping relation $V \to \dot{z}$ justify the use of normal velocity V_n in the derivation of the material derivative, as (6.38), (6.53) and (6.58) illustrate.

By using the material derivative formula in (6.38) and the fact that the differentiation order can be interchanged between the partial derivative and the spatial derivative, the state variational equation (6.59) can be differentiated with respect to τ as

$$[a_\Omega(z,\bar{z})]' \equiv a'_V(z,\bar{z}) + a_\Omega(\dot{z},\bar{z}) = \ell'_V(\bar{z}), \quad \forall \bar{z} \in Z, \tag{6.60}$$

where

$$\begin{aligned}
[a_\Omega(z,\bar{z})]' &= \iint_\Omega [c(z,\bar{z}') + c(z',\bar{z}) + div\big(c(z,\bar{z})\boldsymbol{V}\big)]\, d\Omega \\
&= \iint_\Omega [c(z,\bar{z}') + c(z',\bar{z})]\, d\Omega + \int_\Gamma c(z,\bar{z})V_n\, d\Gamma \\
&= \iint_\Omega [c(z,\dot{\bar{z}} - \nabla\bar{z}\boldsymbol{V}) + c(\dot{z} - \nabla z\boldsymbol{V},\bar{z})]\, d\Omega + \int_\Gamma c(z,\bar{z})V_n\, d\Gamma
\end{aligned} \tag{6.61}$$

and

$$\begin{aligned}
\ell'_V(\bar{z}) &= \iint_\Omega \{\bar{z}'^T \boldsymbol{f} + div[(\bar{z}^T \boldsymbol{f})\boldsymbol{V}]\}\, d\Omega \\
&= \iint_\Omega \bar{z}'^T \boldsymbol{f}\, d\Omega + \int_\Gamma \bar{z}^T \boldsymbol{f} V_n\, d\Gamma \\
&= \iint_\Omega (\dot{\bar{z}} - \nabla\bar{z}\boldsymbol{V})^T \boldsymbol{f}\, d\Omega + \int_\Gamma \bar{z}^T \boldsymbol{f} V_n\, d\Gamma.
\end{aligned} \tag{6.62}$$

The fact that the partial derivatives of the coefficients (such as cross-sectional area, thickness, etc.) in bilinear mapping $c(\bullet,\bullet)$ are zero is used in derivation of (6.61), while the fact that $\boldsymbol{f}' = 0$ is used in derivation of (6.62). In (6.60), $a'_V(z,\bar{z})$ is defined by suppressing the $a_\Omega(\dot{z},\bar{z})$ term from the expression $[a_\Omega(z,\bar{z})]'$. In other words, $a'_V(z,\bar{z})$ includes explicitly dependent terms on the design, whereas $a_\Omega(\dot{z},\bar{z})$ contains implicitly dependent terms through material derivative $\dot{z}$.

For $\bar{z}_\tau$, let $\bar{z}_\tau(\boldsymbol{x} + \tau\boldsymbol{V}(\boldsymbol{x})) = \bar{z}(\boldsymbol{x})$, that is, select $\bar{z}$ as the constant on the line $\boldsymbol{x}_\tau = \boldsymbol{x} + \tau\boldsymbol{V}(\boldsymbol{x})$. Then, since $H^m(\Omega)$ is preserved by $\boldsymbol{T}(\boldsymbol{x},\tau)$, as shown in (6.12), if $\bar{z}$ is an arbitrary element of $H^m(\Omega)$ that satisfies the kinematic boundary conditions on Γ, then $\bar{z}_\tau$ is equally an arbitrary element of $H^m(\Omega_\tau)$ that satisfies kinematic boundary conditions on Γ_τ. Consequently, and by using (6.8), we can assume

$$\dot{\bar{z}} = \bar{z}' + \nabla\bar{z}\boldsymbol{V} = 0. \tag{6.63}$$

Thus, all terms containing $\dot{\bar{z}}$ in (6.61) and (6.62) can be ignored. However, even if we do not assume $\dot{\bar{z}} = 0$, since $\dot{\bar{z}}$ belongs to space Z, the following relation is valid:

$$a_\Omega(z,\dot{\bar{z}}) = \ell_\Omega(\dot{\bar{z}}), \tag{6.64}$$

Thus, all terms containing $\dot{\bar{z}}$ in (6.61) and (6.62) will be canceled out. In the following derivations, we will not consider the terms that contain $\dot{\bar{z}}$. However, this property of $\bar{z}$ does not mean that its partial derivative vanishes, i.e., $\bar{z}' \neq 0$.

After ignoring the terms containing $\dot{\bar{z}}$, variations in the energy bilinear and the load linear forms can be obtained as

$$\begin{aligned}
a'_V(z,\bar{z}) &= -\iint_\Omega \{c(z,\nabla\bar{z}\boldsymbol{V}) + c(\nabla z\boldsymbol{V},\bar{z}) - div[c(z,\bar{z})\boldsymbol{V}]\}\, d\Omega \\
&= -\iint_\Omega [c(z,\nabla\bar{z}\boldsymbol{V}) + c(\nabla z\boldsymbol{V},\bar{z})]\, d\Omega + \int_\Gamma c(z,\bar{z})V_n\, d\Gamma
\end{aligned} \tag{6.65}$$

and

$$\begin{aligned}
\ell'_V(\bar{z}) &= -\iint_\Omega \{(\nabla\bar{z}\boldsymbol{V})^T \boldsymbol{f} - div[(\bar{z}^T \boldsymbol{f})\boldsymbol{V}]\}\, d\Omega \\
&= -\iint_\Omega (\nabla\bar{z}\boldsymbol{V})^T \boldsymbol{f}\, d\Omega + \int_\Gamma \bar{z}^T \boldsymbol{f} V_n\, d\Gamma.
\end{aligned} \tag{6.66}$$

Then, (6.60) can be rewritten to provide the desired sensitivity equation as

$$
\begin{aligned}
a_{\Omega}(\dot{z},\bar{z}) &= \ell'_V(\bar{z}) - a'_V(z,\bar{z}) \\
&= \iint_{\Omega}[c(z,\nabla\bar{z}V)+c(\nabla zV,\bar{z})-(\nabla\bar{z}V)^T f]\,d\Omega \\
&\quad + \iint_{\Omega} div[-c(z,\bar{z})V+(\bar{z}^T f)V]\,d\Omega \\
&= \iint_{\Omega}[c(z,\nabla\bar{z}V)+c(\nabla zV,\bar{z})-(\nabla\bar{z}V)^T f]\,d\Omega \\
&\quad + \int_{\Gamma}[\bar{z}^T f - c(z,\bar{z})]V_n\,d\Gamma, \qquad \forall\bar{z}\in Z.
\end{aligned}
\tag{6.67}
$$

Notice the similarity between (6.67) and (5.8). As mentioned in Chapter 5, if state variable z [the solution to (6.59)] and velocity field V are known, then (6.67) is a variational equation that solves for $\dot{z} \in H^m(\Omega_\tau)$ with the same energy bilinear form $a_\Omega(\bullet,\bullet)$. Indeed, for such second-order problems as membrane, shaft, and elasticity, kinematic boundary conditions are only imposed on z, so if $z_\tau = 0$ on Γ_τ, then $\dot{z} = 0$ on Γ, and thus, $\dot{z}$ satisfies the kinematic boundary conditions. $\dot{z}$ can also be shown to satisfy kinematic boundary conditions for such higher-order problems as clamped plates. Indeed, for a clamped plate the boundary condition $z = 0$ on Γ also implies $\dot{z} = 0$ on Γ. In addition, the fact that $\partial z/\partial \boldsymbol{n} = 0$ on Γ implies that $\nabla z = (\partial z/\partial \boldsymbol{n})\boldsymbol{n} + (\partial z/\partial \boldsymbol{s})\boldsymbol{s} = 0$ on Γ, which in turn implies that $(\nabla z)^\cdot = 0$ on Γ. To show that $\nabla\dot{z} = 0$ on Γ, the relations in (6.8) and (6.13) can be used to obtain the following useful identity:

$$
\begin{aligned}
(\dot{\nabla z}) &= \nabla(z') + \nabla(\nabla z)V \\
&= \nabla(\dot{z} - \nabla zV) + \nabla(\nabla z)V \\
&= \nabla\dot{z} - \nabla z\nabla V.
\end{aligned}
\tag{6.68}
$$

Since the left side and the second term on the right side both vanish on Γ, we can conclude that $\nabla\dot{z} = 0$ on Γ. Thus, $\partial\dot{z}/\partial\boldsymbol{n} = 0$ on Γ, and $\dot{z}$ satisfies all kinematic boundary conditions.

Note that the right side of (6.67) is linear in $\bar{z}$, while the energy bilinear form on the left side is Z-elliptic. Thus, (6.67) has the unique solution $\dot{z} \in Z$ [16], confirming the previously stated observation that a design derivative exists as the solution to the state equation. As in Chapter 5, (6.67) can be used for the adjoint variable method of design sensitivity analysis.

6.2.2 Adjoint Variable Design Sensitivity Analysis

Consider a general performance measure that may be written in integral form as

$$
\psi = \iint_{\Omega_\tau} g(z_\tau, \nabla z_\tau)\,d\Omega_\tau,
\tag{6.69}
$$

where $z \in H^1(\Omega)$, $\nabla z = [z_{i,j}]$, and function g is continuously differentiable with respect to its arguments. For the case in which $z \in H^2(\Omega)$, the second-order derivatives of z may appear in the integrand of (6.69). This case will be treated separately as specific examples arise. Note that ψ depends on Ω in two ways. First, there is an obvious dependence of ψ on its integral domain. Second, state variable z_τ depends on domain Ω through the variational (6.59).

By using the material derivative formulas from (6.13) and (6.38), the variation of the functional in (6.69) can be obtained as

$$
\begin{aligned}
\psi' &= \iint_{\Omega}[g_{,z}z' + g_{,\nabla z} : \nabla z' + div(gV)]\,d\Omega \\
&= \iint_{\Omega}[g_{,z}z' + g_{,\nabla z} : \nabla z']\,d\Omega + \int_{\Gamma} gV_n\,d\Gamma,
\end{aligned}
\tag{6.70}
$$

where $V_n = \boldsymbol{V}^T\boldsymbol{n}$ is the normal component of the design velocity vector to boundary Γ. Using the relation in (6.8), the partial derivatives in (6.70) can be rewritten in terms of $\dot{\boldsymbol{z}}$ as

$$\begin{aligned}\psi' &= \iint_\Omega [g_{,z}\dot{z} + g_{,\nabla z} : \nabla\dot{z} - g_{,z}(\nabla z\boldsymbol{V}) - g_{,\nabla z} : \nabla(\nabla z\boldsymbol{V}) + div(g\boldsymbol{V})]\,d\Omega \\ &= \iint_\Omega [g_{,z}\dot{z} + g_{,\nabla z} : \nabla\dot{z} - g_{,z}(\nabla z\boldsymbol{V}) - g_{,\nabla z} : \nabla(\nabla z\boldsymbol{V})]\,d\Omega + \int_\Gamma gV_n\,d\Gamma,\end{aligned} \tag{6.71}$$

where ":" is the contraction operator between matrices, such that $\boldsymbol{a} : \boldsymbol{b} = a_{ij}b_{ij}$. Note that $\dot{\boldsymbol{z}}$ and $\nabla\dot{\boldsymbol{z}}$ depend on velocity field $\boldsymbol{V}$. The objective is to obtain an explicit expression for ψ' in terms of $\boldsymbol{V}$, which requires rewriting the first two terms of the first integral on the right side of (6.71) explicitly in terms of $\boldsymbol{V}$, that is, by eliminating $\dot{\boldsymbol{z}}$.

Much in the manner of Chapter 5, an adjoint equation is introduced by replacing $\dot{\boldsymbol{z}} \in Z$ in (6.71) by the virtual displacement $\bar{\boldsymbol{\lambda}} \in Z$ and by equating the sum of terms involving $\bar{\boldsymbol{\lambda}}$ with the energy bilinear form, yielding the adjoint equation for the adjoint variable $\boldsymbol{\lambda}$ as

$$a_\Omega(\boldsymbol{\lambda}, \bar{\boldsymbol{\lambda}}) = \iint_\Omega [g_{,z}\bar{\lambda} + g_{,\nabla z} : \nabla\bar{\lambda}]\,d\Omega, \qquad \forall\bar{\boldsymbol{\lambda}} \in Z. \tag{6.72}$$

Note that the adjoint variational equation (6.72) is the same as (5.11). This fact is advantageous when sizing design and shape design variation are considered simultaneously. As noted in Section 5.2, the Lax-Milgram theorem [16] guarantees that (6.72) has a unique solution $\boldsymbol{\lambda}$, which is called the adjoint variable and is associated with the constraint in (6.69).

To take advantage of the adjoint equation, evaluate (6.72) at $\bar{\boldsymbol{\lambda}} = \dot{\boldsymbol{z}} \in Z$ to obtain the following expression:

$$a_\Omega(\boldsymbol{\lambda}, \dot{\boldsymbol{z}}) = \iint_\Omega [g_{,z}\dot{z} + g_{,\nabla z} : \nabla\dot{z}]\,d\Omega. \tag{6.73}$$

Similarly, the design sensitivity equation (6.67) may be evaluated at $\bar{\boldsymbol{z}} = \boldsymbol{\lambda}$, since both belong to space Z, to obtain

$$a_\Omega(\dot{\boldsymbol{z}}, \boldsymbol{\lambda}) = \ell'_V(\boldsymbol{\lambda}) - a'_V(\boldsymbol{z}, \boldsymbol{\lambda}) \tag{6.74}$$

Recalling that energy bilinear form $a_\Omega(\bullet,\bullet)$ is symmetric in its arguments, we can conclude that the left sides of (6.73) and (6.74) are equal; thus, the right sides of both equations yield the following useful relation:

$$\iint_\Omega [g_{,z}\dot{z} + g_{,\nabla z} : \nabla\dot{z}]\,d\Omega = \ell'_V(\boldsymbol{\lambda}) - a'_V(\boldsymbol{z}, \boldsymbol{\lambda}). \tag{6.75}$$

By substituting (6.75) into (6.71), the expression of ψ' in terms of $\boldsymbol{z}$, $\boldsymbol{\lambda}$, and $\boldsymbol{V}$ is obtained as

$$\begin{aligned}\psi' &= \ell'_V(\boldsymbol{\lambda}) - a'_V(\boldsymbol{z}, \boldsymbol{\lambda}) \\ &\quad - \iint_\Omega [g_{,z}(\nabla z\boldsymbol{V}) + g_{,\nabla z} : \nabla(\nabla z\boldsymbol{V})]\,d\Omega + \int_\Gamma gV_n\,d\Gamma \\ &= \iint_\Omega [c(\boldsymbol{z}, \nabla\boldsymbol{\lambda}\boldsymbol{V}) + c(\nabla\boldsymbol{z}\boldsymbol{V}, \boldsymbol{\lambda}) - \boldsymbol{f}^T(\nabla\boldsymbol{\lambda}\boldsymbol{V})]\,d\Omega \\ &\quad - \iint_\Omega [g_{,z}(\nabla z\boldsymbol{V}) + g_{,\nabla z} : \nabla(\nabla z\boldsymbol{V})]\,d\Omega \\ &\quad + \iint_\Omega div\{[g + \boldsymbol{f}^T\boldsymbol{\lambda} - c(\boldsymbol{z}, \boldsymbol{\lambda})]\boldsymbol{V}\}\,d\Omega \\ &= \iint_\Omega [c(\boldsymbol{z}, \nabla\boldsymbol{\lambda}\boldsymbol{V}) + c(\nabla\boldsymbol{z}\boldsymbol{V}, \boldsymbol{\lambda}) - \boldsymbol{f}^T(\nabla\boldsymbol{\lambda}\boldsymbol{V})]\,d\Omega \\ &\quad - \iint_\Omega [g_{,z}(\nabla z\boldsymbol{V}) + g_{,\nabla z} : \nabla(\nabla z\boldsymbol{V})]\,d\Omega \\ &\quad + \int_\Gamma [g + \boldsymbol{f}^T\boldsymbol{\lambda} - c(\boldsymbol{z}, \boldsymbol{\lambda})]V_n\,d\Gamma.\end{aligned} \tag{6.76}$$

In the following sections, two alternative shape design sensitivity expressions will be developed: the boundary method and the domain method. In the boundary method, the design sensitivity is expressed in terms of boundary integrals by using the variational identities given in Section 3.1 for each structural component and boundary condition. This will be done for each structural component type encountered in the subsequent section. The fact that design sensitivity ψ' can be expressed as a boundary integral provides an advantage in numerical calculation, assuming that accurate boundary information can be obtained. In addition, the boundary method gives excellent physical insight of design sensitivity. On the other hand, in the domain method, the design sensitivity is expressed in terms of domain integrals. The domain method is very much suitable for development of general-purpose computational design sensitivity analysis code, especially for using finite element analysis.

Note that the evaluation of the design sensitivity formula in (6.76) requires solving the variational equation (6.59) for z. Similarly, the variational adjoint equation (6.72) must be solved for adjoint variable λ. Using finite element analysis, solving for λ is efficient if the boundary-value problem for z has already been solved, since all that is required is adapting the solution to the same set of finite element equations with a different right side (the *adjoint load*) .

6.2.3 Boundary Method for Static Design Sensitivity

The adjoint variable method for beam, membrane, torsion, and plate problems as presented in Section 3.1 is now developed to calculate the shape design sensitivity of performance measures using the boundary method. For these problems, scalar displacement function z is used as a state variable. The linear elasticity problem will also be considered in this section, while computational considerations will be taken up in subsequent sections. In the boundary method, the shape design sensitivity is obtained in terms of boundary integrals. The variation with respect to conventional design variable $\boldsymbol{u}$ (the cross-sectional area or thickness considered in Chapter 5) is not discussed in this chapter, and even though there is self-weight in addition to the externally applied load, the total applied load will be expressed as $f(x)$.

Bending of a Beam

Consider the beam bending problem in Section 3.1.2, with a domain $\Omega = (0, l) \subset R^1$ and a moment of inertia $I(x)$. Several structural performance measures are of concern. First, consider the beam weight performance, given as

$$\psi_1 = \int_0^l \gamma A \, dx, \tag{6.77}$$

where γ is the weight density per unit volume, and A is the cross-sectional area. The sensitivity of performance ψ_1 can be obtained by taking the variation of (6.77) and by using the fact that $(\gamma A)' = 0$ as

$$\psi_1' = \gamma A V\Big|_0^l = \gamma A(l)V(l) - \gamma A(0)V(0), \tag{6.78}$$

where $V(0)$ and $V(l)$ are the endpoint perturbations of the beam. These perturbations are determined to be positive if they cause the endpoints to move in a positive x direction. Note that this direct method of calculation provides an explicit form of structural weight variation in terms of shape variation. Consequently, no adjoint problem needs to be defined. For the given boundary velocities $V(0)$ and $V(l)$, ψ_1' can be readily evaluated.

As a second performance measure, consider structural compliance, defined as

$$\psi_2 = \int_0^l fz\, dx. \tag{6.79}$$

Note that the integrand in (6.79) depends on applied load f. However, since $f' = 0$, (6.79) can be treated as the functional form of (6.69). Hence, the adjoint (6.72) can be written as

$$a_\Omega(\lambda, \bar{\lambda}) = \int_0^l f\bar{\lambda}\, dx, \qquad \forall \bar{\lambda} \in Z. \tag{6.80}$$

Since the load functional on the right side is precisely the same as the load functional for the original beam problem in (3.16), adjoint solution λ becomes identical to state solution z, and, in this special case, the sensitivity expression in (6.76) can be used by substituting fz for g, to obtain

$$\psi_2' = \int_0^l [2EI(z_{,1}V)_{,11} z_{,11} - 2f(z_{,1}V)]dx + [2fz - EI(z_{,11})^2]V\Big|_0^l. \tag{6.81}$$

To further simplify it, the variational identity in (3.13) may be used by substituting $(z_{,1}V)$ for $\bar{z}$ to obtain

$$\psi_2' = 2EIz_{,11}(z_{,1}V)_{,1}\Big|_0^l - 2(EIz_{,11})_{,1}(z_{,1}V)\Big|_0^l + [2fz - EI(z_{,11})^2]V\Big|_0^l. \tag{6.82}$$

After applying the boundary conditions $z(0) = z(l) = z_{,1}(0) = z_{,1}(l) = 0$ for a clamped-clamped beam, (6.82) yields

$$\psi_2' = EI(z_{,11})^2 V\Big|_0^l. \tag{6.83}$$

As noted in Section 6.2.2, the design sensitivity result in (6.83) is expressed as boundary values and is explicitly given in terms of the design velocity (perturbation) on the boundary.

Example 6.2. For an example that can be calculated analytically, consider a uniform clamped-clamped beam in which $A = A_0$, with a uniformly distributed load f_0. The displacement under this load is

$$z(x) = \frac{f_0}{24EI}[x^2(l-x)^2].$$

Beam compliance may be calculated from (6.79) as

$$\psi_2 = \frac{f_0^2 l^5}{720EI}.$$

In this special example, beam length l is a design variable. Since the beam has a uniform cross-sectional area A_0, and a uniform load f_0, varying either endpoint $x = 0$ or $x = l$ will have the same effect on compliance. Hence, the variation of compliance with respect to l is

$$\psi_2' = \frac{f_0^2 l^4}{144EI}\delta l,$$

which is an analytical design sensitivity for the compliance performance measure. Alternatively, the design sensitivity of ψ_2 can be calculated by using the formula provided by (6.83), with perturbations $V(l) = \delta l/2$ and $V(0) = -\delta l/2$, as

$$\psi_2' = EI\left[\frac{f_0^2}{12EI}(l^2 - 6lx + 6x^2)\right]^2 V\Bigg|_0^l$$

$$= \frac{f_0^2 l^4}{144EI}\delta l,$$

which is the correct result when compared with the analytical solution.

For those other boundary conditions given by (3.19) through (3.21), the shape design sensitivity formula in (6.82) is valid for compliance performance, since variational equation (3.16) holds for every $\bar{z}$ that satisfies corresponding kinematic boundary conditions. By applying these boundary conditions to the sensitivity expression in (6.82), the following design sensitivity results for the compliance performance measure can be obtained:

1. Simply supported

$$\psi_2' = -2EIz_{,111}(z_{,1}V)\Big|_0^l, \tag{6.84}$$

2. Cantilevered

$$\psi_2' = -EI(z_{,11})^2V\Big|_{x=0} + 2fzV\Big|_{x=l}. \tag{6.85}$$

3. Clamped–simply supported

$$\psi_2' = -EI(z_{,11})^2V\Big|_{x=0} - 2EIz_{,111}(z_{,1}V)\Big|_{x=l}. \tag{6.86}$$

Next, consider a functional that defines the displacement value at an isolated, fixed point $x^* \in (0, l)$, that is,

$$\psi_3 \equiv z(x^*) = \int_0^l \delta(x - x^*) z\, dx. \tag{6.87}$$

When evaluating functional ψ_3, it is important to remember that point x^* does not move with the design change, and that ψ_3 on deformed domain Ω_τ is the displacement value at the same point x^*. Since $m = 2$ and $n = 1$, $m > n/2$ and, according to the Sobolev imbedding theorem [22], $z_\tau \in C^1(\Omega_\tau)$. The functional in (6.87) is thus continuous, and the previously discussed material derivative formulas apply.

Since $\delta(x - x^*)$ is defined in the neighborhood of $[0, l]$ by zero extension, and since x^* is a fixed point, then $\delta'(x - x^*) = 0$. Thus, (6.87) can be treated as the functional form of (6.69), and the adjoint equation from (6.72) would be

$$a_\Omega(\lambda, \bar{\lambda}) = \int_0^l \delta(x - x^*)\bar{\lambda}\, dx, \qquad \forall \bar{\lambda} \in Z. \tag{6.88}$$

As noted in Section 5.2.3, since the right side of this equation defines a bounded linear functional on $H^2(0, l)$, (6.88) has the unique solution $\lambda^{(3)}$, where superscript (i) associates λ with constraint ψ_i. Note that $\lambda^{(3)}$ is the displacement caused by a unit load at x^*, that is, with smoothness assumptions, variational equation (6.88) is equivalent to the following formal differential equation:

$$(EI\lambda_{,11})_{,11} = \delta(x - x^*), \tag{6.89}$$

with λ satisfying the same boundary conditions as the original state variable z. From (6.76), and with $g = \delta(x - x^*)z$, displacement sensitivity can be obtained as

$$\begin{aligned}\psi_3' = &\int_0^l [EIz_{,11}(\lambda_{,1}^{(3)}V)_{,11} + EI(z_{,1}V)_{,11}\lambda_{,11}^{(3)} - f(\lambda_{,1}^{(3)}V) - \delta(x-x^*)(z_{,1}V)]dx \\ &+[f\lambda^{(3)} - EIz_{,11}\lambda_{,11}^{(3)}]V\Big|_0^l.\end{aligned} \tag{6.90}$$

The variational identity in (3.13) may be repeated twice to transform the domain integral to a boundary integral (or to a boundary evaluation in the case of a one-dimensional problem). To do so, first substitute $(\lambda_{,1}^{(3)}V)$ for $\bar{z}$ in (3.13). Next, substitute $\lambda^{(3)}$ for z, $(z_{,1}V)$ for $\bar{z}$, and $\delta(x - x^*)$ for f in (3.13). The displacement sensitivity can now be expressed in terms of boundary values as

$$\begin{aligned}\psi_3' = &[EIz_{,11}(\lambda_{,1}^{(3)}V)_{,1} - (EIz_{,11})_{,1}(\lambda_{,1}^{(3)}V)]\Big|_0^l \\ &+[EI\lambda_{,11}^{(3)}(z_{,1}V)_{,1} - (EI\lambda_{,11}^{(3)})_{,1}(z_{,1}V)]\Big|_0^l \\ &+[f\lambda^{(3)} - EIz_{,11}\lambda_{,11}^{(3)}]V\Big|_0^l.\end{aligned} \tag{6.91}$$

Using the boundary conditions in (3.9) for a clamped beam, (6.91) can be further simplified as

$$\psi_3' = [EIz_{,11}\lambda_{,11}^{(3)}V\Big|_0^l. \tag{6.92}$$

Other boundary conditions can also be applied to simplify (6.91).

Example 6.3. To illustrate, consider the clamped-clamped beam studied earlier in this section. From Example 6.2, the displacement function at arbitrary point x can be written as

$$z(x) = \frac{f_0}{24EI}[x^2(l-x)^2].$$

Since displacement sensitivity information at the center of the beam is desired, let us substitute $x = l/2$ to obtain the value of functional ψ_3 in (6.87):

$$\psi_3 = \frac{f_0 l^4}{384EI}.$$

Since ψ_3 is expressed as a function of design l, analytical shape design sensitivity can be obtained through direct differentiation as

$$\psi_3' = \frac{f_0 l^3}{96EI}\delta l.$$

The objective is to compare the sensitivity result in the above equation with the sensitivity result from (6.91). For the displacement function, the adjoint load from (6.88) is simply a unit point load at the beam's center. The adjoint variable can thus be calculated as

$$\lambda^{(3)} = \frac{1}{48EI}\left[8\left\langle x - \frac{l}{2}\right\rangle^3 - 4x^3 + 3lx^2\right],$$

where

$$\left\langle x-\frac{l}{2}\right\rangle=\begin{cases}0 & \text{for } 0\le x\le \frac{l}{2}\\ x-\frac{l}{2} & \text{for } \frac{l}{2}\le x\le l.\end{cases}$$

In the analytical result, design perturbation is assumed to be δl. To be consistent with this result, let us define the design velocity as $V(0) = -\delta l/2$, and $V(l) = \delta l/2$. Using this adjoint variable, the shape design sensitivity of ψ_3 is obtained from (6.92) as

$$\begin{aligned}\psi_3' &= \frac{f_0}{576EI}\Big[(l^2-6lx+6x^2)(24x-6l)V\big|_{x=0}\\ &\qquad +(l^2-6lx+6x^2)(24x-18l)V\big|_{x=l}\Big]\\ &= \frac{f_0 l^3}{96EI}\delta l.\end{aligned}$$

Note that this is the same as the analytical result.

For simply supported, cantilevered, or clamped–simply supported beams, the sensitivity formula in (6.91) is applicable, in which z and $\lambda^{(3)}$ are solutions to (3.16) and (6.88), respectively, and Z is the appropriate space for all kinematically admissible displacements. Appropriate boundary conditions for z and $\lambda^{(3)}$ can then be applied to (6.91) to obtain useful sensitivity formulas. By applying the boundary conditions from (3.19) through (3.21) to the sensitivity expression in (6.91), the following sensitivity formulas for displacement performance measures can be obtained:

1. Simply supported

$$\psi_3' = -[EI(\lambda^{(3)}_{,111}z_{,1}+z_{,111}\lambda^{(3)}_{,1})]V\big|_0^l, \tag{6.93}$$

2. Cantilevered

$$\psi_3' = -EIz_{,11}\lambda^{(3)}_{,11}V\big|_{x=0} + f\lambda^{(3)}V\big|_{x=l}, \tag{6.94}$$

3. Clamped–simply supported

$$\psi_3' = -EIz_{,11}\lambda^{(3)}_{,11}V\big|_{x=0} - [EI(\lambda^{(3)}_{,111}z_{,1}+z_{,111}\lambda^{(3)}_{,1})]V\big|_{x=l}. \tag{6.95}$$

Shape design sensitivity results from (6.91), or from (6.92) through (6.95) for each boundary condition, are valid for a functional ψ_3 that defines the displacement value at fixed point x^*. It is also possible that a ψ_3 on deformed domain Ω_τ is the displacement at point $x_\tau = x + \tau V(x)$, a situation that will be considered in Section 6.2.5.

Consider another performance measure that is associated with strength constraints, written as

$$\psi_4 = \int_0^l hEz_{,11}m_p\,dx, \tag{6.96}$$

where h is the half-depth of the beam and m_p is a characteristic function defined on a small, open subinterval (x_a, x_b), such that $[x_a, x_b] \subset (0, l)$. The characteristic function m_p is positive and constant on (x_a, x_b), zero outside of (x_a, x_b), and has an integral of one. For the moment, consider the averaged stress on the fixed interval (x_a, x_b); in other words, assume that m_p in (6.96) does not change with τ. It is possible to extend m_p on R^1 by extending it to zero outside (0, l). In this case, m'_p would be equal to zero. In addition, since only the shape design variable is being considered, half-depth h remains constant during the design change process, that is, $h' = 0$.

By using the material derivative formula from (6.38), along with the relation between the partial derivative and the material derivative in (6.13), the design variation of the strength performance measure in (6.96) can be obtained as

$$\begin{aligned} \psi'_4 &= \int_0^l hEz'_{,11}m_p\,dx + hEz_{,11}m_pV\Big|_0^l \\ &= \int_0^l hE[\dot{z}_{,11} - (z_{,1}V)_{,11}]m_p\,dx, \end{aligned} \tag{6.97}$$

where the boundary terms vanish, since $m_p(0) = m_p(l) = 0$. As with the general derivation of adjoint equation (6.72), the adjoint equation may be defined by replacing $\dot{z}$ with $\bar{\lambda}$ in the first term on the right side of the equation, and then equating this adjoint load with the energy bilinear form as

$$a_\Omega(\lambda,\bar{\lambda}) = \int_0^l hE\bar{\lambda}_{,11}m_p\,dx, \qquad \forall\bar{\lambda} \in Z. \tag{6.98}$$

As with adjoint equation (6.72), adjoint equation (6.98) is the same as (5.91), when h is interpreted to be $\beta A^{1/2}$. That is, the adjoint equation is identical for both size and shape design problems. As noted in Section 5.2.3, since the right side of (6.98) is a bounded linear functional on $H^2(0,l)$, it has the unique solution $\lambda^{(4)}$. Taking its smoothness assumptions into account, this variational equation is thus equivalent to the following formal operator equation:

$$(EI\lambda_{,11})_{,11} = (hEm_p)_{,11}, \qquad x \in (0,l), \tag{6.99}$$

with λ satisfying the same boundary conditions as the original structural response z. The differential on the right side of (6.99) is a derivative, in the sense of the distribution theory [17], [22], and [88]. By expanding this derivative, we obtain

$$\begin{aligned} (hEm_p)_{,11} = E\{&h_{,11}m_p + m_p[2h_{,1}(x_a)\delta(x-x_a) \\ &-2h_{,1}(x_b)\delta(x-x_b) + h(x_a)\delta_{,1}(x-x_a) \\ &-h(x_b)\delta_{,1}(x-x_b)]\}. \end{aligned}$$

Thus, the adjoint load consists of a distributed load on the interval (x_a, x_b), point loads at x_a and x_b, and point moments at x_a and x_b. Using the same method employed in Section 3.1.2, the following variational identity can be obtained from (6.99) for the adjoint system:

$$\begin{aligned} &\int_0^l EI\lambda_{,11}\bar{\lambda}_{,11}\,dx - \int_0^l hE\bar{\lambda}_{,11}m_p\,dx \\ &\quad = [EI\lambda_{,11}\bar{\lambda}_{,1} - (EI\lambda_{,11})_{,1}\bar{\lambda}]\Big|_0^l + [(hEm_p)_{,1}\bar{\lambda} - hEm_p\bar{\lambda}_{,1}]\Big|_0^l \\ &\qquad\qquad \forall\bar{\lambda} \in H^2(0,l). \end{aligned} \tag{6.100}$$

Since $\dot{z} \in Z$, (6.98) may be evaluated at $\bar{\lambda} = \dot{z}$ to obtain

$$a_\Omega(\lambda^{(4)},\dot{z}) = \int_0^l hE\dot{z}_{,11}m_p\,dx. \tag{6.101}$$

Similarly, design sensitivity equation (6.67) may be evaluated at $\bar{z} = \lambda^{(4)}$, since both functions are in Z, in order to obtain

$$a_\Omega(\dot{z},\lambda^{(4)}) = \ell'_V(\lambda^{(4)}) - a'_V(z,\lambda^{(4)}). \tag{6.102}$$

Since energy bilinear form $a_\Omega(\bullet,\bullet)$ is symmetric, (6.97), (6.101), and (6.102) can be used to yield

$$\psi'_4 = \ell'_V(\lambda^{(4)}) - a'_V(z,\lambda^{(4)}) - \int_0^l hE(z_{,1}V)_{,11} m_p \, dx,$$

which can be rewritten, using (6.67), as

$$\begin{aligned}\psi'_4 = & \int_0^l [EIz_{,11}(\lambda_{,1}^{(4)}V)_{,11} + EI(z_{,1}V)_{,11}\lambda_{,11}^{(4)} - f(\lambda_{,1}^{(4)}V)]dx \\ & - \int_0^l hE(z_{,1}V)_{,11} m_p \, dx \\ & + [f\lambda^{(4)} - EIz_{,11}\lambda_{,11}^{(4)}]V\Big|_0^l .\end{aligned} \tag{6.103}$$

The variational identities in (3.13) and (6.100) may be used to transform the domain integral into boundary evaluations. After substituting $(\lambda_{,1}^{(4)}V)$ for $\bar{z}$ in (3.13) and substituting $\lambda^{(4)}$ for λ and $(z_{,1}V)$ for $\bar{\lambda}$ in (6.100), the ψ'_4 is expressed in terms of boundary evaluations as

$$\begin{aligned}\psi'_4 = & [EIz_{,11}(\lambda_{,1}^{(4)}V)_{,1} - (EIz_{,11})_{,1}(\lambda_{,1}^{(4)}V)]\Big|_0^l \\ & + [EI\lambda_{,11}^{(4)}(z_{,1}V)_{,1} - (EI\lambda_{,11}^{(4)})_{,1}(z_{,1}V)]\Big|_0^l \\ & + [(hEm_p)_{,1}(z_{,1}V) - hEm_p(z_{,1}V)_{,1}]\Big|_0^l \\ & + [f\lambda^{(4)} - EIz_{,11}\lambda_{,11}^{(4)}]V\Big|_0^l .\end{aligned} \tag{6.104}$$

Since $[x_a, x_b] \subset (0, l)$, and $m_p = 0$ in the neighborhood of $x = 0$ and $x = l$, (6.104) becomes a desired design sensitivity expression, simplified to

$$\begin{aligned}\psi'_4 = & [EIz_{,11}(\lambda_{,1}^{(4)}V)_{,1} - (EIz_{,11})_{,1}(\lambda_{,1}^{(4)}V)]\Big|_0^l \\ & + [EI\lambda_{,11}^{(4)}(z_{,1}V)_{,1} - (EI\lambda_{,11}^{(4)})_{,1}(z_{,1}V)]\Big|_0^l \\ & + [f\lambda^{(4)} - EIz_{,11}\lambda_{,11}^{(4)}]V\Big|_0^l .\end{aligned} \tag{6.105}$$

Using the boundary conditions in (3.9) and the fact that $\lambda^{(4)}$ satisfies the same boundary conditions, (6.105) can be further simplified for a clamped beam to

$$\psi'_4 = EIz_{,11}\lambda_{,11}^{(4)}V\Big|_0^l . \tag{6.106}$$

As before, for simply supported, cantilevered, or clamped–simply supported beams, the shape design sensitivity formula in (6.105) is valid, where z and $\lambda^{(4)}$ are solutions to (3.16) and (6.98), respectively, and Z is the corresponding space for all kinematically admissible displacements. Appropriate boundary conditions can be applied to (6.105) to obtain the following useful shape design sensitivity formulas:

1. Simply supported

$$\psi_4' = -[EI(\lambda_{,111}^{(4)} z_{,1} + z_{,111} \lambda_{,1}^{(4)})]V\Big|_0^l, \tag{6.107}$$

2. Cantilevered

$$\psi_4' = -EIz_{,11}\lambda_{,11}^{(4)}V\Big|_{x=0} + f\lambda^{(4)}V\Big|_{x=l}, \tag{6.108}$$

3. Clamped–simply supported

$$\psi_4' = -EIz_{,11}\lambda_{,11}^{(4)}V\Big|_{x=0} - [EI(\lambda_{,111}^{(4)} z_{,1} + z_{,111}\lambda_{,1}^{(4)})]V\Big|_{x=l}. \tag{6.109}$$

As with displacement functional ψ_3, the design sensitivity formulas found in either (6.105), or (6.106) through (6.109) for each boundary condition, are valid for the functional ψ_4, which defines the averaged stress on a fixed interval (x_a, x_b). The case in which ψ_4 represents the averaged stress on a moving interval (x_a, x_b) within deformed domain Ω_τ will be discussed in Section 6.2.4. A further assumption used to determine the averaged stress functional is that the interval (x_a, x_b) on which stress is averaged is taken such that $[x_a, x_b] \subset (0, l)$. Hence, $x_a \neq 0$ and $x_b \neq l$. Discussion of the situation in which either $x_a = 0$ or $x_b = l$ will take place in Section 6.2.4.

Deflection of a Membrane

The simplest performance measure is the membrane area, defined as

$$\psi_1 = \iint_\Omega d\Omega. \tag{6.110}$$

By using the material derivative formula from (6.38), the variation of ψ_1 can be obtained as

$$\psi_1' = \int_\Gamma V_n \, d\Gamma. \tag{6.111}$$

Note that this direct variational calculation provides the explicit form of area variation in terms of normal velocity V_n on the boundary. Consequently, no adjoint problem needs to be defined for this performance measure.

A second functional, which represents the membrane's strain energy, can be defined in the following form

$$\psi_2 = \frac{T}{2} \iint_\Omega \nabla z^T \nabla z \, d\Omega. \tag{6.112}$$

Since according to the definition proposed by (3.79) strain energy ψ_2 is equal to half of the compliance, we can obtain the following definition

$$\psi_2 = \frac{1}{2} \iint_\Omega fz \, d\Omega. \tag{6.113}$$

Note that the integrand in (6.113) depends on load f. However, since $f' = 0$, (6.113) can be treated as the functional form of (6.69). Hence, the adjoint equation can be defined, using its definition in (6.72), as

$$a_\Omega(\lambda, \bar{\lambda}) = \frac{1}{2} \iint_\Omega f\bar{\lambda} \, d\Omega, \qquad \forall \bar{\lambda} \in Z. \tag{6.114}$$

Because the load functional on the right side of (6.114) is precisely half the load functional of the original membrane problem in (3.80), for this special case $\lambda = z/2$. Using that result, and using the formula in (6.76) such that $z = 0$ on Γ, the sensitivity of ψ_2 can be obtained as

$$\begin{aligned}\psi_2' &= \iint_\Omega [T(\nabla z^T \nabla(\nabla z^T \boldsymbol{V})) - f(\nabla z^T \boldsymbol{V})]\, d\Omega \\ &\quad - \frac{T}{2}\int_\Gamma (\nabla z^T \nabla z) V_n \, d\Gamma.\end{aligned} \tag{6.115}$$

To transform the domain integral in (6.115) into a boundary integral, the variational identity in (3.78) can be used by substituting $(\nabla z^T \boldsymbol{V})$ for $\bar{z}$ in (3.78), to obtain the following sensitivity expression:

$$\psi_2' = T\int_\Gamma \frac{\partial z}{\partial n}(\nabla z^T V)\, d\Gamma - \frac{T}{2}\int_\Gamma (\nabla z^T \nabla z) V_n \, d\Gamma. \tag{6.116}$$

Since $z = 0$ on Γ, the tangential component of ∇z vanishes along the boundary; thus, $\nabla z = (\partial z/\partial n)\boldsymbol{n}$ on Γ, which yields the following simplified result:

$$\psi_2' = \frac{T}{2}\int_\Gamma \left(\frac{\partial z}{\partial n}\right)^2 V_n \, d\Gamma. \tag{6.117}$$

As noted in Section 6.2.2, the design sensitivity expression in (6.117) is given as a boundary integral, and only normal boundary movements V_n appear.

Torsion of an Elastic Shaft

The torsional rigidity provided by (3.83) can be considered a response functional, that is,

$$\psi = 2\iint_\Omega z \, d\Omega. \tag{6.118}$$

As we can determine from (6.72), the adjoint equation is

$$a_\Omega(\lambda, \bar{\lambda}) = 2\iint_\Omega \bar{\lambda}\, d\Omega, \qquad \forall \bar{\lambda} \in Z. \tag{6.119}$$

Thus, in this special case $\lambda = z$. Comparing this equation to the membrane problem, the sensitivity of ψ can be obtained in the following form:

$$\psi' = \int_\Gamma \left(\frac{\partial z}{\partial n}\right)^2 V_n \, d\Gamma. \tag{6.120}$$

Example 6.4. To take an example that can be calculated analytically, consider an elastic shaft with a circular cross section and radius a as its design variable (Fig. 6.6). The Prandtl stress function [89] z for a circular cross section is

$$z = \frac{1}{2}(a^2 - x_1^2 - x_2^2) = \frac{1}{2}(a^2 - r^2).$$

Using the polar coordinates, the torsional rigidity in (6.118) can be calculated analytically as

$$\psi = \int_0^{2\pi}\int_0^a (a^2 - r^2) r \, dr \, d\theta = \frac{\pi a^4}{2}.$$

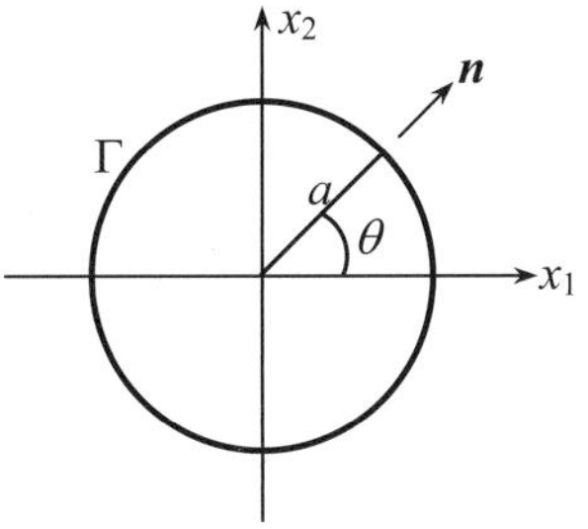

Figure 6.6. Circular cross section of an elastic shaft.

By taking radius a as a design variable, the variation of torsional rigidity with respect to a is obtained by directly differentiating the above analytical solution as

$$\psi' = 2\pi a^3 \delta a,$$

which is the analytical sensitivity of torsional rigidity. Now, let us calculate the sensitivity of torsional rigidity by using the formula provided by (6.120). In the polar coordinate, the unit normal vector to the boundary can be represented by

$$\boldsymbol{n} = [\cos\theta, \sin\theta]^T.$$

In addition, since the radius is a design variable, the design velocity field can be expressed in the polar coordinate as

$$\boldsymbol{V} = [V_1, V_2]^T = [\delta r \cos\theta, \delta r \sin\theta]^T.$$

Since the radial direction is normal to the boundary, the integrand $\partial z/\partial n$ in (6.120) can be equivalently represented as

$$\nabla z^T \boldsymbol{n} = \frac{\partial z}{\partial r} = -r.$$

Hence, from (6.120) the sensitivity of ψ becomes

$$\psi' = \int_0^{2\pi} (-r)^2 \delta r\, r d\theta \bigg|_{r=a} = 2\pi a^3 \delta a,$$

which is the same as the analytical solution.

Bending of a Plate

Consider the plate bending problem as it was presented in Section 3.1.3, with thickness $h(x) \geq h_0 > 0$ and a constant Young's modulus E. A functional definition of the plate weight is

$$\psi_1 = \iint_\Omega \gamma h \, d\Omega, \tag{6.121}$$

where γ is the weight density. In the case of the shape design problem, partial derivative $(\gamma h)' = 0$, and thus, the material derivative of ψ_1 can be calculated by using (6.38) as

$$\psi_1' = \int_\Gamma \gamma h V_n \, d\Gamma. \tag{6.122}$$

Thus, no adjoint variable is necessary, and the explicit design derivative of the weight is obtained. As a result, for a given normal design velocity field, (6.122) can be readily evaluated without recourse to any response analysis or adjoint problem.

Next, consider the compliance functional for the plate, defined as

$$\psi_2 = \iint_\Omega fz \, d\Omega. \tag{6.123}$$

As with (6.113), since $f' = 0$, (6.123) can be treated as the functional form of (6.69). Consequently, by returning to (6.72), the adjoint equation would be

$$a_\Omega(\lambda, \bar{\lambda}) = \iint_\Omega f \bar{\lambda} \, d\Omega, \qquad \forall \bar{\lambda} \in Z. \tag{6.124}$$

For this special case, $\lambda = z$ and the sensitivity expression of ψ_2 can be obtained by substituting fz for g in (6.76) as

$$\begin{aligned} \psi_2' &= \iint_\Omega [2\boldsymbol{\kappa}(z)^T \boldsymbol{C}^b \boldsymbol{\kappa}(\nabla z^T \boldsymbol{V}) - 2f(\nabla z^T \boldsymbol{V})] \, d\Omega \\ &\quad + \int_\Gamma [2fz - \boldsymbol{\kappa}(z)^T \boldsymbol{C}^b \boldsymbol{\kappa}(z)] V_n \, d\Gamma. \end{aligned} \tag{6.125}$$

Equation (6.125) contains the terms of the domain integral, which can be transformed into a boundary integral by employing the variational identity in (3.38). If we substitute $(\nabla z^T \boldsymbol{V})$ for $\bar{z}$ in (3.38), then the following sensitivity expression can be obtained in terms of the boundary integral:

$$\begin{aligned} \psi_2' &= 2\int_\Gamma (\nabla z^T \boldsymbol{V}) Nz \, d\Gamma + 2\int_\Gamma \frac{\partial}{\partial n}(\nabla z^T \boldsymbol{V}) Mz \, d\Gamma \\ &\quad + \int_\Gamma [2fz - \boldsymbol{\kappa}(z)^T \boldsymbol{C}^b \boldsymbol{\kappa}(z)] V_n \, d\Gamma. \end{aligned} \tag{6.126}$$

To further simplify this sensitivity expression, consider the coefficients of the differential operators Mz and Nz in the above equation. If we restrict our attention to the clamped boundary $\Gamma_C \subset \Gamma$, then the first integral on the right side of (6.126) vanishes, since $\nabla z = 0$ on Γ_C. For the second integral,

$$\frac{\partial}{\partial n}(\nabla z^T \boldsymbol{V}) = \sum_{i,j=1}^{2} (V_i z_{,ij} n_j + V_{i,j} z_{,i} n_j).$$

Since $\nabla z = 0$ on Γ_C, the second term on the right side vanishes, and the first term can be expanded to

$$\frac{\partial}{\partial n}(\nabla z^T \boldsymbol{V}) = \sum_{i,j=1}^{2} V_i z_{,ij} n_j = \frac{\partial^2 z}{\partial n^2}(\boldsymbol{V}^T \boldsymbol{n}) + \frac{\partial^2 z}{\partial n \partial s}(\boldsymbol{V}^T \boldsymbol{s}). \tag{6.127}$$

This equation can be verified by expanding the last term in (6.127). Since $\partial z/\partial n = 0$ on Γ_C, $(\partial/\partial s)(\partial z/\partial n) = 0$ on Γ_C, and (6.127) becomes

$$\frac{\partial}{\partial n}(\nabla z^T \boldsymbol{V}) = \frac{\partial^2 z}{\partial n^2}(\boldsymbol{V}^T \boldsymbol{n}). \tag{6.128}$$

Also, since $\partial z/\partial s = 0$, and $\partial^2 z/\partial s^2 = 0$ on Γ_C, the differential operator Mz in (3.36) becomes

$$Mz = D\left[\frac{\partial^2 z}{\partial n^2} + \nu\left(\frac{1}{r}\frac{\partial z}{\partial n} + \frac{\partial^2 z}{\partial s^2}\right)\right] = D\left(\frac{\partial^2 z}{\partial n^2}\right), \qquad \boldsymbol{x} \in \Gamma_C. \tag{6.129}$$

As a result, the sensitivity formula in (6.126) is simplified for the clamped boundary Γ_C as

$$\psi_2' = \int_{\Gamma_C}\left[2D\left(\frac{\partial^2 z}{\partial n^2}\right)^2 - \boldsymbol{\kappa}(z)^T \boldsymbol{C}^b \boldsymbol{\kappa}(z)\right] V_n \, d\Gamma. \tag{6.130}$$

As before, the design sensitivity expression in (6.130) is given as a boundary integral, and only the normal movement V_n appears on the boundary Γ_C.

It was shown by Mikhlin [39] that if the clamped boundary conditions in (3.34) are satisfied, then the following relation could be obtained:

$$\iint_\Omega (z_{,12}^2 - z_{,11}z_{,22})\, d\Omega = 0.$$

Hence, if plate thickness $h(\boldsymbol{x})$ is constant, then the plate variational equation (3.43) is simplified to

$$a_\Omega(z,\bar{z}) \equiv D\iint_\Omega (\nabla^2 z)(\nabla^2 \bar{z})\, d\Omega = \iint_\Omega f\bar{z}\, d\Omega \equiv \ell_\Omega(\bar{z}), \qquad \bar{z} \in Z.$$

Proceeding in exactly the same way as before, instead of (6.130), a more simplified expression can be obtained

$$\begin{aligned}\psi_2' &= \int_{\Gamma_C} D\left[2\left(\frac{\partial^2 z}{\partial n^2}\right)^2 - (\nabla^2 z)^2\right] V_n \, d\Gamma \\ &= D\int_{\Gamma_C}\left(\frac{\partial^2 z}{\partial n^2}\right)^2 V_n \, d\Gamma.\end{aligned} \tag{6.131}$$

Example 6.5. To take an analytical example, consider a clamped circular plate with constant thickness h, radius a, and concentrated load $f = p\delta(\boldsymbol{x})$ at the center of the plate, as illustrated in Fig 6.7 The analytical displacement of the plate [52] is given as

$$z = \frac{p}{16\pi D}\left[a^2 - r^2\left(1 + 2\ln\frac{a}{r}\right)\right],$$

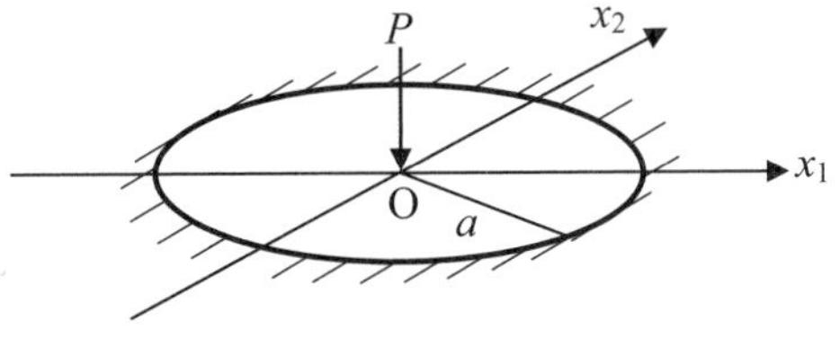

Figure 6.7. Circular plate with concentrated load.

where $r^2 = x_1^2 + x_2^2$. Drawing on (6.123), the compliance functional of the plate is

$$\psi_2 = \iint_\Omega \frac{p^2\delta(\mathbf{x})}{16\pi D}\left[a^2 - r^2\left(1 + 2\ln\frac{a}{r}\right)\right] d\Omega = \frac{p^2 a^2}{16\pi D}.$$

Taking plate radius a to be a design variable, the variation of compliance with respect to a becomes

$$\psi_2' = \frac{p^2 a}{8\pi D}\delta a,$$

which is an analytical sensitivity of the compliance functional. In order to compare this analytical result to the result in (6.131), let us express (6.131) in the polar coordinate, with $V_n = \delta r$ and $\partial^2 z/\partial n^2 = \partial^2 z/\partial r^2$ on the boundary. Then, the compliance sensitivity can be expressed as

$$\begin{aligned}\psi_2' &= D\int_0^{2\pi}\left(\frac{\partial^2 z}{\partial r^2}\right)^2 \delta r\, r\, d\theta\bigg|_{r=a} \\ &= D\left[\frac{p}{16\pi D}\left(-4\ln\frac{a}{r} + 4\right)\right]^2 2\pi r\,\delta r\bigg|_{r=a} \\ &= \frac{p^2 a}{8\pi D}\delta a,\end{aligned}$$

which is the same as the analytical sensitivity.

The compliance sensitivity formula in (6.126) is still valid for other boundary conditions in (3.46) and (3.47), because the variational equation (3.43) is valid for every $\bar{z}$ that satisfies corresponding kinematic boundary conditions. For example, consider the simply supported part of the boundary $\Gamma_S \subset \Gamma$. From the boundary conditions given in (3.46), the fact that $z = 0$ on Γ_S implies that $\partial z/\partial s = 0$ on Γ_S; thus, only the normal component of ∇z exists on the boundary. Hence, $\nabla z = (\partial z/\partial n)\boldsymbol{n}$. By using this condition, the sensitivity formula in (6.126) can be simplified to

$$\psi_2' = \int_{\Gamma_S}\left[2D\left(\frac{\partial z}{\partial n}\right)Nz - \boldsymbol{\kappa}(z)^T \boldsymbol{C}^b \boldsymbol{\kappa}(z)\right]V_n\, d\Gamma. \tag{6.132}$$

From the boundary conditions given by (3.47), i.e., $Mz = Nz = 0$, the sensitivity formula in (6.126) can be simplified for the free edge of the boundary $\Gamma_F \subset \Gamma$ to

$$\psi_2' = \int_{\Gamma_F}\left[2fz - \boldsymbol{\kappa}(z)^T \boldsymbol{C}^b \boldsymbol{\kappa}(z)\right]V_n\, d\Gamma. \tag{6.133}$$

If $\Gamma = \Gamma_C \cup \Gamma_S \cup \Gamma_F$, then the complete shape design sensitivity formula is obtained by adding the terms from (6.130), (6.132), and (6.133).

Next, consider the displacement functional at a discrete point $\boldsymbol{x}^*$, written as

$$\psi_3 = \iint_\Omega \delta(\boldsymbol{x} - \boldsymbol{x}^*) z\, d\Omega, \tag{6.134}$$

where $\boldsymbol{x}^* \in \Omega$ is a fixed point, and $\delta(\mathbf{x})$ is the Dirac delta measure on the plane, acting at the origin. Since $m = 2$ and $n = 2$, $m > n/2$. According to the Sobolev imbedding theorem [22], $z_\tau \in C^0(\Omega_\tau)$. The functional in (6.134) is thus continuous, and the previously presented approach also works.

Since $\delta(\boldsymbol{x}-\boldsymbol{x}^*)$ is defined in the neighborhood of $\bar{\Omega}$ by zero extension, and since $\boldsymbol{x}^*$ is fixed, the partial derivative of $\delta(\boldsymbol{x}-\boldsymbol{x}^*)$ becomes zero. Thus, (6.134) can be treated as the functional form of (6.69). By following the same procedure used to determine (6.72), the adjoint equation is obtained as

$$a_\Omega(\lambda,\bar{\lambda}) = \iint_\Omega \delta(\boldsymbol{x}-\boldsymbol{x}^*)\bar{\lambda}\,d\Omega, \qquad \forall\bar{\lambda}\in Z. \tag{6.135}$$

Equation (6.135) has a unique solution λ, which is the displacement due to the unit load at $\boldsymbol{x}^*$. With the addition of smoothness assumptions, variational equation (6.135) is equivalent to the following operator equation:

$$\begin{aligned}&[D(\lambda_{,11}+\nu\lambda_{,22})]_{,11}+[D(\lambda_{,22}+\nu\lambda_{,11})]_{,22}\\&\qquad+2(1-\nu)[D\lambda_{,12}]_{,12}=\delta(\boldsymbol{x}-\boldsymbol{x}^*), \qquad \boldsymbol{x}\in\Omega,\end{aligned} \tag{6.136}$$

with λ satisfying the same boundary conditions as the original structural response z. Using (6.76) with $g = \delta(\boldsymbol{x}-\boldsymbol{x}^*)z$, the sensitivity of ψ_3 can be expressed as

$$\begin{aligned}\psi_3' = &\iint_\Omega \{\boldsymbol{\kappa}(z)^T \boldsymbol{C}^b\boldsymbol{\kappa}(\nabla\lambda^{(3)T}\boldsymbol{V}) - f(\nabla\lambda^{(3)T}\boldsymbol{V})\\&\qquad+\boldsymbol{\kappa}(\lambda^{(3)})^T\boldsymbol{C}^b\boldsymbol{\kappa}(\nabla z^T\boldsymbol{V}) - \delta(\boldsymbol{x}-\boldsymbol{x}^*)(\nabla z^T\boldsymbol{V})\}\,d\Omega\\&+\int_\Gamma [f\lambda^{(3)} - \boldsymbol{\kappa}(z)^T\boldsymbol{C}^b\boldsymbol{\kappa}(\lambda^{(3)})]V_n\,d\Gamma.\end{aligned} \tag{6.137}$$

As with (6.125), the variational identity in (3.38) may be repeated in order to transform the domain integral in (6.137) to a boundary integral, obtaining

$$\begin{aligned}\psi_3' = &\int_\Gamma \left[(\nabla\lambda^{(3)T}\boldsymbol{V})Nz + \frac{\partial}{\partial n}(\nabla\lambda^{(3)T}\boldsymbol{V})Mz\right]d\Gamma\\&+\int_\Gamma \left[(\nabla z^T\boldsymbol{V})N\lambda^{(3)} + \frac{\partial}{\partial n}(\nabla z^T\boldsymbol{V})M\lambda^{(3)}\right]d\Gamma\\&+\int_\Gamma [f\lambda^{(3)} - \boldsymbol{\kappa}(z)^T\boldsymbol{C}^b\boldsymbol{\kappa}(\lambda^{(3)})]V_n\,d\Gamma.\end{aligned} \tag{6.138}$$

Note that the clamped boundary conditions in (3.34) hold for $\lambda^{(3)}$ as well as for z; thus, the same is true for (6.128) and (6.129). By drawing on these relations, the displacement sensitivity for clamped boundary Γ_C can be expressed as

$$\psi_3' = \int_{\Gamma_C}\left\{2D\left(\frac{\partial^2 z}{\partial n^2}\right)\left(\frac{\partial^2\lambda^{(3)}}{\partial n^2}\right) - \boldsymbol{\kappa}(z)\boldsymbol{C}^b\boldsymbol{\kappa}(\lambda^{(3)})\right\}V_n\,d\Gamma, \tag{6.139}$$

which is valid for variable thickness $h(\boldsymbol{x})$. As with the example of a compliance functional, if plate thickness $h(\boldsymbol{x})$ is constant, then a more simplified expression can be obtained as

$$\psi_3' = D\int_{\Gamma_C}\left(\frac{\partial^2 z}{\partial n^2}\right)\left(\frac{\partial^2\lambda^{(3)}}{\partial n^2}\right)V_n\,d\Gamma. \tag{6.140}$$

Example 6.6. In order to provide an analytical example, consider a clamped circular plate with constant thickness h, radius a, and linearly increasing axisymmetric load $f = (q/a_0)r$, as illustrated in Fig. 6.8, where a_0 represents the present design.

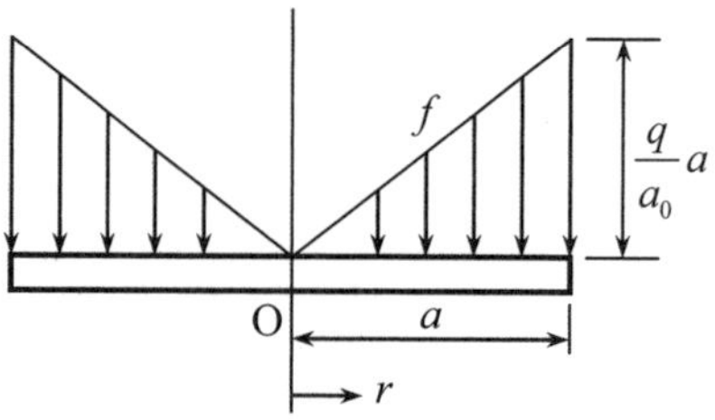

Figure 6.8. Circular plate with axisymmetric load.

The analytical solution to a plate displacement which has a radius of a is given by [52]

$$z_a = \frac{qa}{450Da_0}\left(3a^4 - 5a^2r^2 + 2\frac{r^5}{a}\right).$$

Taking $\boldsymbol{x}^*$ to be the plate center, i.e., $r = 0$, the ψ_3 from (6.134) becomes

$$\psi_3 = \frac{qa^5}{150Da_0}.$$

If plate radius a is a design variable, then the sensitivity of ψ_3 can be obtained by directly differentiating this analytical expression with respect to design a, evaluated at present design a_0 as

$$\psi_3' = \frac{\partial \psi_3}{\partial a}\delta a\bigg|_{a=a_0} = \frac{qa_0^3}{30D}\delta a_0.$$

The objective is to compare the sensitivity result obtained from (6.140) to the above analytical sensitivity result. The sensitivity formula in (6.140) requires an adjoint solution to (6.135), where the adjoint load is a unit point load applied at the plate center. According to classical plate theory [52], the adjoint solution with radius a is

$$\lambda^{(3)} = \frac{1}{16\pi D}\left(a^2 - r^2\left(1 + 2\ln\frac{r}{a}\right)\right).$$

For a circular plate, it is convenient to express (6.140) in the polar coordinate as

$$\begin{aligned}\psi_3' &= D\int_0^{2\pi}\left(\frac{\partial^2 z}{\partial r^2}\right)\left(\frac{\partial^2 \lambda^{(3)}}{\partial r^2}\right)\delta r\, r\, d\theta\bigg|_{r=a_0,a=a_0}\\ &= D\left[\frac{qa}{450Da_0}\left(-10a^2 + 40r^3/a\right)\right]\left[\frac{1}{16\pi D}(-4\ln a/r + 4)\right]\bigg|_{r=a_0,a=a_0}\\ &= \frac{qa_0^3}{30D}\delta a_0,\end{aligned}$$

which is the correct result.

For other boundary conditions in (3.46) and (3.47), the same procedure as with the compliance functional can be used. In the case of a simply supported boundary Γ_S, the sensitivity expression becomes

$$\psi_3' = \int_{\Gamma_S} \left\{ \left(\frac{\partial \lambda^{(3)}}{\partial n} \right) Nz + \left(\frac{\partial z}{\partial n} \right) N\lambda^{(3)} - \boldsymbol{\kappa}(z)^T \boldsymbol{C}^b \boldsymbol{\kappa}(\lambda^{(3)}) \right\} V_n \, d\Gamma \tag{6.141}$$

and, in the case of a free edge Γ_F, the sensitivity of ψ_3 can be written as

$$\psi_3' = \int_{\Gamma_F} [f\lambda^{(3)} - \boldsymbol{\kappa}(z)^T \boldsymbol{C}^b \boldsymbol{\kappa}(\lambda^{(3)})] V_n \, d\Gamma. \tag{6.142}$$

If $\Gamma = \Gamma_C \cup \Gamma_S \cup \Gamma_F$, then the complete shape design sensitivity formula can be obtained by adding the terms from (6.139), (6.141), and (6.142).

As in the case of a beam displacement functional, the sensitivity results in (6.138), or (6.139) through (6.142) for each boundary condition, are valid for displacement at a fixed point $\boldsymbol{x}^*$. For the situation in which ψ_3 is the displacement at a moving point within deformed domain Ω_τ, such that $\boldsymbol{x}^*_\tau = \boldsymbol{x}^* + \tau \boldsymbol{V}(\boldsymbol{x}^*)$, consult Section 6.2.4.

The maximum stress on a thin plate occurs on the plate's surface [23], and is given in the form

$$\boldsymbol{\sigma} = \begin{bmatrix} \sigma_{11} \\ \sigma_{22} \\ \sigma_{12} \end{bmatrix} = -\frac{h}{2} \boldsymbol{C} \boldsymbol{\kappa}(z). \tag{6.143}$$

By using these stress components, The von Mises failure criterion [23] is defined by

$$g(\boldsymbol{\sigma}) = \sqrt{(\sigma_{11} + \sigma_{22})^2 + 3(\sigma_{11} - \sigma_{22})^2 + 12(\sigma_{12})^2} - 4\sigma_p \le 0, \tag{6.144}$$

where σ_p is a given, uniaxial yield stress. This failure criterion is most often used in engineering applications. But, in order to simplify, let us assume that stress σ_{11} in (6.143) is taken as a strength constraint rather than employing the von Mises failure criterion. Once this step is taken, the concept can then be extended to the von Mises failure criterion.

As with a beam, since a pointwise stress constraint is meaningless, the characteristic function approach from (6.96) may be used. That is, a function $m_p(\boldsymbol{x})$ can be defined that is positive and constant on a small, open subset Ω_p, zero outside of Ω_p, and with an integral of one. Then, the averaged value of σ_{11} over this small region can be defined as the following constraint:

$$\begin{aligned} \psi_4 &= \iint_\Omega \sigma_{11} m_p \, d\Omega \\ &= -\iint_\Omega \frac{Eh}{2(1-\nu^2)} (z_{,11} + \nu z_{,22}) m_p \, d\Omega. \end{aligned} \tag{6.145}$$

If the averaged stress on fixed region Ω_p is of concern, then m_p in (6.145) does not change with τ. Thus, $m_p' = 0$.

Now, let us differentiate (6.145) by using the material derivative formula from (6.38), and the property of $h' = 0$ to obtain

$$
\begin{aligned}
\psi_4' &= -\iint_\Omega \frac{Eh}{2(1-\nu^2)}(z'_{,11} + \nu z'_{,22}) m_p \, d\Omega \\
&\quad - \int_\Gamma \frac{Eh}{2(1-\nu^2)}(z_{,11} + \nu z_{,22}) V_n m_p \, d\Gamma \\
&= -\iint_\Omega \frac{Eh}{2(1-\nu^2)}[\dot{z}_{,11} + \nu \dot{z}_{,22} - (\nabla z^T \boldsymbol{V})_{,11} - \nu (\nabla z^T \boldsymbol{V})_{,22}] m_p \, d\Omega.
\end{aligned}
\tag{6.146}
$$

The boundary integral vanishes, since $m_p = 0$ on Γ. As in the general derivation of adjoint equation (6.72), the adjoint load may be defined by replacing $\dot{z}$ with $\bar{\lambda}$ in the first two terms on the right side of (6.146). By equating this adjoint load with the structural bilinear form, the desired adjoint equation can be obtained as

$$
a_\Omega(\lambda, \bar{\lambda}) = -\iint_\Omega \frac{Eh}{2(1-\nu^2)}(\bar{\lambda}_{,11} + \nu \bar{\lambda}_{,22}) m_p \, d\Omega, \qquad \forall \bar{\lambda} \in Z.
\tag{6.147}
$$

Note that adjoint equation (6.147) is the same as adjoint equation (5.64) for the sizing design problem. It can be shown that the linear form of $\bar{\lambda}$ on the right side of (6.147) is bounded in $H^2(\Omega)$. Hence, according to the Lax-Milgram theorem [16], (6.147) has the unique solution $\lambda^{(4)}$.

If smoothness assumptions are added, then variational equation (6.147) is equivalent to the following operator equation:

$$
\begin{aligned}
&[D(\lambda_{,11} + \nu \lambda_{,22})]_{,11} + [D(\lambda_{,22} + \nu \lambda_{,11})]_{,22} + 2(1-\nu)[D\lambda_{,12}]_{,12} \\
&\quad = -\left[\frac{Eh}{2(1-\nu^2)} m_p\right]_{,11} - \nu \left[\frac{Eh}{2(1-\nu^2)} m_p\right]_{,22}, \qquad \boldsymbol{x} \in \Omega,
\end{aligned}
\tag{6.148}
$$

with λ satisfying the same boundary conditions as the original structural response z. As with adjoint equation (6.99) for the beam problem, the derivatives on the right side of (6.148) are interpreted as distributions. Moreover, the distributional derivatives $m_{p,i}$ and $m_{p,ij}$ $(i, j = 1, 2)$ depend on the equation that represents the boundary of Ω_p (The reader is referred to Kecs and Teodorescu [88] for a detailed treatment of the distributional derivative). Following the same procedure set out in Section 3.1.3, a variational identity can be obtained from (6.148) for the adjoint system as

$$
\begin{aligned}
&\iint_\Omega \boldsymbol{\kappa}(\bar{\lambda})^T \boldsymbol{C}^b \boldsymbol{\kappa}(\lambda) \, d\Omega + \iint_\Omega \frac{Eh}{2(1-\nu^2)}(\bar{\lambda}_{,11} + \nu \bar{\lambda}_{,22}) m_p \, d\Omega \\
&= \int_\Gamma \bar{\lambda} N \lambda \, d\Gamma + \int_\Gamma \frac{\partial \bar{\lambda}}{\partial n} M \lambda \, d\Gamma \\
&\quad + \int_\Gamma \left\{ \frac{Eh}{2(1-\nu^2)} m_p \bar{\lambda}_{,1} n_1 - \left[\frac{Eh}{2(1-\nu^2)} m_p\right]_{,1} \bar{\lambda} n_1 \right. \\
&\quad + \left. \int_\Gamma \frac{Eh}{2(1-\nu^2)} m_p \nu \bar{\lambda}_{,2} n_2 - \left[\frac{Eh}{2(1-\nu^2)} m_p\right]_{,2} \nu \bar{\lambda} n_2 \right\} d\Gamma, \qquad \forall \bar{\lambda} \in H^2(\Omega).
\end{aligned}
\tag{6.149}
$$

The sensitivity expression of ψ_4 can now be derived. Since $\dot{z} \in Z$, (6.147) may be evaluated at $\bar{\lambda} = \dot{z}$ to obtain

$$
a_\Omega(\lambda^{(4)}, \dot{z}) = -\iint_\Omega \frac{Eh}{2(1-\nu^2)}(\dot{z}_{,11} + \nu \dot{z}_{,22}) m_p \, d\Omega.
\tag{6.150}
$$

Note that the right side of the equation is exactly the same as the first two terms on the right side of (6.146). Similarly, since both $\bar{z}$ and $\lambda^{(4)}$ belong to Z, design sensitivity (6.67) can be evaluated at $\bar{z} = \lambda^{(4)}$, to yield

$$a_\Omega(\dot{z}, \lambda^{(4)}) = \ell'_V(\lambda^{(4)}) - a'_V(z, \lambda^{(4)}). \tag{6.151}$$

Since the energy bilinear form $a_\Omega(\bullet,\bullet)$ is symmetric, the right sides of both (6.150) and (6.151) are identical. Thus, by drawing on (6.146), the sensitivity expression of ψ_4 is obtained as

$$\psi'_4 = \ell'_V(\lambda^{(4)}) - a'_V(z, \lambda^{(4)}) + \iint_\Omega \frac{Eh}{2(1-\nu^2)}[(\nabla z^T \boldsymbol{V})_{,11} + \nu(\nabla z^T \boldsymbol{V})_{,22}]m_p \, d\Omega,$$

which can be rewritten using (6.67) as

$$\begin{aligned}
\psi'_4 = & \iint_\Omega [\boldsymbol{\kappa}(z)^T \boldsymbol{C}^b \boldsymbol{\kappa}(\nabla \lambda^{(4)T} \boldsymbol{V}) - f \nabla z^T \boldsymbol{V} + \boldsymbol{\kappa}(\nabla z^T \boldsymbol{V})^T \boldsymbol{C}^b \boldsymbol{\kappa}(\lambda^{(4)})] d\Omega \\
& + \iint_\Omega \frac{Eh}{2(1-\nu^2)}[(\nabla z^T \boldsymbol{V})_{,11} + \nu(\nabla z^T \boldsymbol{V})_{,22}]m_p \, d\Omega \\
& + \int_\Gamma [f \lambda^{(4)} - \boldsymbol{\kappa}(z)^T \boldsymbol{C}^b \boldsymbol{\kappa}(\lambda^{(4)})] V_n \, d\Gamma.
\end{aligned} \tag{6.152}$$

As with (6.125), the variational identities in (3.38) and (6.149) can be used to transform the domain integral in (6.152) into the following boundary integral:

$$\begin{aligned}
\psi'_4 = & \int_\Gamma [(\nabla \lambda^{(4)T} \boldsymbol{V}) N z + \frac{\partial}{\partial n}(\nabla \lambda^{(4)T} \boldsymbol{V}) M z] d\Gamma \\
& + \int_\Gamma [(\nabla z^T \boldsymbol{V}) N \lambda^{(4)} + \frac{\partial}{\partial n}(\nabla z^T \boldsymbol{V}) M \lambda^{(4)}] d\Gamma \\
& + \int_\Gamma \left\{ \frac{Eh}{2(1-\nu^2)} m_p (\nabla z^T \boldsymbol{V})_{,1} n_1 - \left[\frac{Eh}{2(1-\nu^2)} m_p \right]_{,1} (\nabla z^T \boldsymbol{V}) n_1 \right. \\
& \left. + \frac{Eh}{2(1-\nu^2)} m_p \nu (\nabla z^T \boldsymbol{V})_{,2} n_2 - \left[\frac{Eh}{2(1-\nu^2)} m_p \right]_{,2} \nu (\nabla z^T \boldsymbol{V}) n_2 \right\} d\Gamma \\
& + \int_\Gamma [f \lambda^{(4)} - \boldsymbol{\kappa}(z)^T \boldsymbol{C}^b \boldsymbol{\kappa}(\lambda^{(4)})] V_n \, d\Gamma.
\end{aligned} \tag{6.153}$$

Since $\bar{\Omega}_p \subset \Omega$, the characteristic function m_p and its derivatives vanish on Γ. Thus, (6.153) can be simplified to express the sensitivity of ψ_4 in terms of the boundary velocity as

$$\begin{aligned}
\psi'_4 = & \int_\Gamma [(\nabla \lambda^{(4)T} \boldsymbol{V}) N z + \frac{\partial}{\partial n}(\nabla \lambda^{(4)T} \boldsymbol{V}) M z] d\Gamma \\
& + \int_\Gamma [(\nabla z^T \boldsymbol{V}) N \lambda^{(4)} + \frac{\partial}{\partial n}(\nabla z^T \boldsymbol{V}) M \lambda^{(4)}] d\Gamma \\
& + \int_\Gamma [f \lambda^{(4)} - \boldsymbol{\kappa}(z)^T \boldsymbol{C}^b \boldsymbol{\kappa}(\lambda^{(4)})] V_n \, d\Gamma.
\end{aligned} \tag{6.154}$$

As was the case for the displacement functional, sensitivity formulas can be obtained for different boundary conditions, as follows.

1. Clamped boundary:

$$\psi'_4 = \int_{\Gamma^C} \left[2\left(\frac{\partial^2 z}{\partial n^2}\right)\left(\frac{\partial^2 \lambda^{(4)}}{\partial n^2}\right) - \boldsymbol{\kappa}(z)^T \boldsymbol{C}^b \boldsymbol{\kappa}(\lambda^{(4)}) \right] V_n \, d\Gamma, \tag{6.155}$$

2. Simply supported boundary:

$$\psi_4' = \int_{\Gamma_S} \left[\frac{\partial \lambda^{(4)}}{\partial n} Nz + \frac{\partial z}{\partial n} N\lambda^{(4)} - \boldsymbol{\kappa}(z)^T \boldsymbol{C}^b \boldsymbol{\kappa}(\lambda^{(4)}) \right] V_n \, d\Gamma, \tag{6.156}$$

3. Free edge boundary:

$$\psi_4' = \int_{\Gamma_F} [f\lambda^{(4)} - \boldsymbol{\kappa}(z)^T \boldsymbol{C}^b \boldsymbol{\kappa}(\lambda^{(4)})] V_n \, d\Gamma. \tag{6.157}$$

For the case in which $\Gamma = \Gamma_C \cup \Gamma_S \cup \Gamma_F$, a complete shape design sensitivity formula can be obtained by adding the terms from (6.155) through (6.157).

As with a beam, the shape design sensitivity results for averaged stress in (6.154) through (6.157) are valid for the fixed region Ω_p. For the case in which ψ_4 is the averaged stress in the moving region $\Omega_{p\tau} = T(\Omega_p, \tau)$, consult Section 6.2.5. It has been assumed that $\bar{\Omega}_p \subset \Omega$, so boundary Γ_p of Ω_p does not meet boundary Γ of Ω. The situation in which Γ_p intersects with Γ will also be considered in Section 6.2.5.

Linear Elasticity Problem

Shape design sensitivity analysis for the linear elasticity problem in Section 3.1.4 is carried out by using the adjoint variable method in this section. For plane stress, or plane strain problems, the formulas derived in Section 3.1.4 remain valid, with the limits of summation running from 1 to 2, and an appropriate modification of the generalized Hooke's law.

Consider the three-dimensional elasticity problem as presented in Section 3.1.4, with a mean stress constraint over fixed volume Ω_p ($\bar{\Omega}_p \subset \Omega$) as

$$\psi = \iiint_{\Omega} g(\boldsymbol{\sigma}(\boldsymbol{z})) m_p \, d\Omega, \tag{6.158}$$

where $\boldsymbol{\sigma}$ denotes the stress vector, Ω_p is an open set, and m_p is a characteristic function that is constant on Ω_p, zero outside of Ω_p, and with an integral of one. It is assumed that function g is continuously differentiable with respect to its arguments. Note that $g(\boldsymbol{\sigma}(\boldsymbol{z}))$ might involve principal stresses, the von Mises failure criterion, or some other material failure criteria. While the integrand in (6.158) could be explicitly written in terms of $\nabla \boldsymbol{z}$ as with the plate problem in Section 6.2.3, it is more effective to continue with the present notation.

In perturbing the boundary, it is assumed that structural boundary $\Gamma = \Gamma^h \cup \Gamma^f \cup \Gamma^s$ is varied along with the design, except the boundary $\partial\Gamma^s$ of the traction surface boundary Γ^s is fixed, so the velocity $\boldsymbol{V}$ at $\partial\Gamma^s$ is zero. For the case in which $\partial\Gamma^s$ is not fixed, variation of the traction term in (3.59) (given as an integral over Γ^s) yields an additional term that was not discussed in Section 6.1.2. For this case, the interested reader is referred to Zolesio [90] for more information. Two kinds of boundary loads may be considered. One is a conservative load that depends on the position but not the boundary shape, while the other is a more general, nonconservative load that depends not only on position but also on boundary shape.

First, consider a conservative load in which traction $\boldsymbol{f}^s$ in (3.59) only depends on the position. By using the material derivative formulas from (6.38) and (6.53) and by using the fact that $\boldsymbol{f}^{b\prime} = \boldsymbol{f}^{s\prime} = 0$, state equation (3.60) is differentiated with respect to the design as

$$\begin{aligned}
&\iiint_{\Omega}[\sigma(z')^T\varepsilon(\bar{z})+\sigma(z)^T\varepsilon(\bar{z}')]d\Omega+\iint_{\Gamma}[\sigma(z)^T\varepsilon(\bar{z})]V_n\,d\Gamma \\
&\quad=\iiint_{\Omega}f^{bT}\bar{z}'\,d\Omega+\iint_{\Gamma^f\cup\Gamma^s}[f^{bT}\bar{z}]V_n\,d\Gamma \\
&\quad+\iint_{\Gamma^s}f^{sT}\bar{z}'\,d\Gamma+\iint_{\Gamma^s}[\nabla(f^{sT}\bar{z})^T n+\kappa(f^{sT}\bar{z})]V_n\,d\Gamma,\qquad\forall\bar{z}\in Z.
\end{aligned}\tag{6.159}$$

After converting z' into $\dot{z}$ using (6.8), and using the property of $\dot{\bar{z}}=0$ in (6.63), the following sensitivity equation is obtained from (6.159):

$$\begin{aligned}
a_{\Omega}(\dot{z},\bar{z})&\equiv\iiint_{\Omega}\sigma(\dot{z})^T\varepsilon(\bar{z})\,d\Omega \\
&=\iiint_{\Omega}[\sigma(z)^T\varepsilon(\nabla\bar{z}^TV)+\sigma(\nabla z^TV)^T\varepsilon(\bar{z})]d\Omega \\
&-\iiint_{\Omega}f^{bT}(\nabla\bar{z}^TV)\,d\Omega-\iint_{\Gamma}[\sigma(z)^T\varepsilon(\bar{z})]V_n\,d\Gamma \\
&+\iint_{\Gamma^f\cup\Gamma^s}[f^{bT}\bar{z}]V_n\,d\Gamma \\
&+\iint_{\Gamma^s}\left\{-f^{sT}(\nabla\bar{z}^TV)+[\nabla(f^{sT}\bar{z})^T n+\kappa(f^{sT}\bar{z})]V_n\right\}d\Gamma,\qquad\forall\bar{z}\in Z.
\end{aligned}\tag{6.160}$$

As in sensitivity equation (6.67), (6.160) is a variational equation for $\dot{z}\in Z$. That is, $\dot{z}\in[H^1(\Omega)]^3$ and satisfies all kinematic boundary conditions.

For the fixed region Ω_p, the characteristic function m_p that appears in the performance measure of (6.158) is independent of the shape design; thus, $m_p'=0$, and $m_p=0$ on the boundary. By using the material derivative formula from (6.38), the sensitivity of ψ can be obtained as

$$\begin{aligned}
\psi'&=\iiint_{\Omega}g_{,\sigma(z)}\sigma(z')m_p\,d\Omega+\iint_{\Gamma}g(\sigma(z))m_pV_n\,d\Gamma \\
&=\iiint_{\Omega}g_{,\sigma(z)}[\sigma(\dot{z})-\sigma(\nabla z^TV)]m_p\,d\Omega.
\end{aligned}\tag{6.161}$$

As with adjoint equation (6.72), in order to define a load functional for the adjoint equation, the material derivative of state variable $\dot{z}\in Z$ may be replaced by virtual displacement $\bar{\lambda}$ in the first term on the right side of (6.161). By equating this load functional with the energy bilinear form, the following adjoint equation can be defined:

$$a_{\Omega}(\lambda,\bar{\lambda})=\iiint_{\Omega}g_{,\sigma(z)}\sigma(\bar{\lambda})m_p\,d\Omega,\qquad\forall\bar{\lambda}\in Z.\tag{6.162}$$

The linear form of $\bar{\lambda}$ on the right side of (6.162) is bounded in $[H^1(\Omega)]^3$. According to the Lax-Milgram theorem [16], (6.162) has a unique solution for displacement field λ.

With its smoothness assumptions, (6.162) is equivalent to the formal operator equation

$$-\sigma_{ij,j}(\lambda)=-(g_{,\sigma_{kl}}C_{klij}m_p)_{,j},\qquad i=1,2,3,\ x\in\Omega,\tag{6.163}$$

with boundary conditions

$$\lambda=0,\qquad x\in\Gamma^h\tag{6.164}$$

$$\sigma n=0,\qquad x\in\Gamma^f\cup\Gamma^s.\tag{6.165}$$

As in adjoint (6.148) for the plate problem, the derivative on the right side of (6.163) is interpreted as a distribution. The distributional derivatives $m_{p,j}$ (j = 1, 2, 3) depend on the equation representing the boundary of Ω_p [88]. By relying on the same method used in Section 3.1.4, a variational identity can be obtained from (6.163) for the adjoint system, that is, by multiplying (6.163) by $\bar{\lambda} \in [H^1(\Omega)]^3$ and integrating by parts, we can obtain

$$\begin{aligned}\iiint_\Omega \sigma_{ij}(\lambda)\bar{\lambda}_{i,j}\, d\Omega - \iint_\Gamma \sigma_{ij}(\lambda) n_j \bar{\lambda}_i\, d\Gamma \\ = \iiint_\Omega g_{,\sigma_{kl}} C_{klij} m_p \bar{\lambda}_{i,j}\, d\Omega \\ - \iint_\Gamma g_{,\sigma_{kl}} C_{klij} m_p n_j \bar{\lambda}_i\, d\Gamma.\end{aligned}$$

Since $\sigma_{ij}(\lambda) = \sigma_{ji}(\lambda)$, and $C_{ijkl} = C_{klji}$, the above equation can be rewritten in the following more convenient variational identity form:

$$\begin{aligned}\iiint_\Omega \sigma_{ij}(\lambda)\varepsilon_{ij}(\bar{\lambda})\, d\Omega - \iiint_\Omega g_{,\sigma_{ij}} \sigma_{ij}(\bar{\lambda}) m_p\, d\Omega \\ = \iint_\Gamma \sigma_{ij}(\lambda) n_j \bar{\lambda}_i\, d\Gamma \\ - \iint_\Gamma g_{,\sigma_{kl}} C_{klij} m_p n_j \bar{\lambda}_i\, d\Gamma, \quad \forall \bar{\lambda} \in [H^1(\Omega)]^3.\end{aligned} \tag{6.166}$$

Note that by imposing boundary conditions in (6.164) and (6.165), and by using the fact that $m_p = 0$ on Γ, the variational equation in (6.162) is obtained.

Since $\dot{z} \in Z$, (6.162) may be evaluated at $\bar{\lambda} = \dot{z}$, to obtain

$$a_\Omega(\lambda, \dot{z}) = \iiint_\Omega g_{,\sigma(z)} \sigma(\dot{z}) m_p\, d\Omega. \tag{6.167}$$

Similarly, since $\bar{z} \in Z$, and $\lambda \in Z$, (6.160) may be evaluated at $\bar{z} = \lambda$, to obtain

$$\begin{aligned}a_\Omega(\dot{z}, \lambda) = \iiint_\Omega [\sigma(z)^T \varepsilon(\nabla\lambda^T V) + \sigma(\nabla z^T V)^T \varepsilon(\lambda)]\, d\Omega \\ - \iiint_\Omega f^{bT}(\nabla\lambda^T V)\, d\Omega - \iint_\Gamma [\sigma(z)^T \varepsilon(\lambda)] V_n\, d\Gamma \\ + \iint_{\Gamma^f \cup \Gamma^s} [f^{bT}\lambda] V_n\, d\Gamma \\ + \iint_{\Gamma^s} \{-f^{sT}(\nabla\lambda^T V) + [\nabla(f^{sT}\lambda)^T n + \kappa(f^{sT}\lambda)] V_n\}\, d\Gamma.\end{aligned} \tag{6.168}$$

According to the Betti's reciprocal theorem [44], the energy bilinear form is symmetric, and can be written as

$$\begin{aligned}a_\Omega(z, \bar{z}) \equiv \iiint_\Omega \sigma(z)^T \varepsilon(\bar{z})\, d\Omega \\ = \iiint_\Omega \sigma(\bar{z})^T \varepsilon(z)\, d\Omega \equiv a_\Omega(\bar{z}, z), \quad \forall \bar{z} \in [H^1(\Omega)]^3.\end{aligned} \tag{6.169}$$

Thus, $a_\Omega(\dot{z}, \lambda) = a_\Omega(\lambda, \dot{z})$, and (6.160), (6.167), and (6.168) yield the sensitivity expression of ψ as

$$\begin{aligned}\psi' = \iiint_\Omega [\sigma(z)^T \varepsilon(\nabla\lambda^T V) + \sigma(\lambda)^T \varepsilon(\nabla z^T V)]\, d\Omega \\ - \iiint_\Omega f^{bT}(\nabla\lambda^T V)\, d\Omega - \iiint_\Omega g_{,\sigma(z)} \sigma(\nabla z^T V) m_p\, d\Omega \\ - \iint_\Gamma [\sigma(z)^T \varepsilon(\lambda)] V_n\, d\Gamma + \iint_{\Gamma^f \cup \Gamma^s} [f^{bT}\lambda] V_n\, d\Gamma \\ + \iint_{\Gamma^s} \{-f^{sT}(\nabla\lambda^T V) + [\nabla(f^{sT}\lambda)^T n + \kappa(f^{sT}\lambda)] V_n\}\, d\Gamma.\end{aligned} \tag{6.170}$$

As before, the variational identities in (3.57) and (6.166) may be used to transform the domain integrals in (6.170) into boundary integrals by substituting $\bar{z}$ in (3.57) and $\bar{\lambda}$ in (6.166) with $(\nabla\lambda^T V)$ and $(\nabla z^T V)$ in (6.170), respectively, obtaining

$$
\begin{aligned}
\psi' = & \iint_{\Gamma} \sigma_{ij}(z) n_j (\nabla \lambda_i^T V) d\Gamma \\
& + \iint_{\Gamma} [\sigma_{ij}(\lambda) n_j (\nabla z_i^T V) - g_{,\sigma_{kl}(z)} C^{klij} m_p n_j (\nabla z_i^T V)] d\Gamma \\
& - \iint_{\Gamma} [\sigma(z)^T \varepsilon(\lambda)] V_n d\Gamma + \iint_{\Gamma^f \cup \Gamma^s} (f^{bT} \lambda) V_n d\Gamma \\
& + \iint_{\Gamma^s} \{ -f^{sT} (\nabla \lambda^T V) + [\nabla (f^{sT} \lambda)^T n + \kappa (f^{sT} \lambda)] V_n \} d\Gamma
\end{aligned}
\tag{6.171}
$$

since $\bar{\Omega}_p \subset \Omega$, $m_p = 0$ on Γ. Using the boundary conditions offered by (3.56) and (6.165), (6.171) becomes

$$
\begin{aligned}
\psi' = & \iint_{\Gamma^h} [\sigma_{ij}(z) n_j (\nabla \lambda_i^T V) + \sigma_{ij}(z) n_j (\nabla z_i^T V)] d\Gamma \\
& - \iint_{\Gamma} [\sigma(z)^T \varepsilon(\lambda)] V_n d\Gamma + \iint_{\Gamma^f \cup \Gamma^s} [f^{bT} \lambda] V_n d\Gamma \\
& + \iint_{\Gamma^s} [\nabla (f^{sT} \lambda)^T n + \kappa (f^{sT} \lambda)] V_n d\Gamma .
\end{aligned}
\tag{6.172}
$$

The fact that $z = \lambda = 0$ on Γ^h implies that $\nabla z_i = (\nabla z_i^T n) n$ and that $\nabla \lambda_i = (\nabla \lambda_i^T n) n$. Hence, (6.172) becomes

$$
\begin{aligned}
\psi' = & \iint_{\Gamma^h} [\sigma_{ij}(z) n_j (\nabla \lambda_i^T V) + \sigma_{ij}(z) n_j (\nabla z_i^T V)] V_n d\Gamma \\
& - \iint_{\Gamma} [\sigma(z)^T \varepsilon(\lambda)] V_n d\Gamma + \iint_{\Gamma^f \cup \Gamma^s} [f^{bT} \lambda] V_n d\Gamma \\
& + \iint_{\Gamma^s} [\nabla (f^{sT} \lambda)^T n + \kappa (f^{sT} \lambda)] V_n d\Gamma ,
\end{aligned}
\tag{6.173}
$$

which is the desired result.

As with the plate problem, the stress sensitivity result for averaged stress in (6.173) is valid for a fixed region Ω_p that satisfies the relation $\bar{\Omega}_p \subset \Omega$. For the case in which ψ is averaged stress on the moving region $\Omega_{p\tau} = T(\Omega_p, \tau)$, or when a part of Γ_p intersects Γ, consult Section 6.2.6.

Next, consider a more general, nonconservative load. A typical example is a pressure load applied in the normal direction to the boundary, defined as

$$
f^s = -p(x) n(x), \qquad x \in \Gamma^s .
\tag{6.174}
$$

In the following derivations, we will consider the case in which $p' = 0$ and $f^{b\prime} = 0$. After substituting f^s into the load linear form in (3.59), the state equation (3.60) is differentiated using the material derivative formulas from (6.38) and (6.58), to obtain

$$
\begin{aligned}
& \iiint_{\Omega} [\sigma(z')^T \varepsilon(\bar{z}) + \sigma(z)^T \varepsilon(\bar{z}')] d\Omega + \iint_{\Gamma} [\sigma(z)^T \varepsilon(\bar{z})] V_n d\Gamma \\
& = \iiint_{\Omega} f^{bT} \bar{z}' d\Omega + \iint_{\Gamma^f \cup \Gamma^s} [f^{bT} \bar{z}] V_n d\Gamma \\
& - \iint_{\Gamma^s} p n^T \bar{z}' d\Gamma - \iint_{\Gamma^s} [div(p \bar{z})] V_n d\Gamma, \qquad \forall \bar{z} \in Z.
\end{aligned}
\tag{6.175}
$$

Since the same performance measure is of interest, the adjoint equation is the same as (6.159) for a nonconservative load. The only difference is the boundary integral terms in (6.175), compared with those in (6.159). Thus, by following a similar procedure as with a conservative load, the sensitivity of ψ can be obtained by

$$\begin{aligned}\psi' = &\iint_{\Gamma_0} [\sigma_{ij}(z)n_j(\nabla\lambda_i^T V) + \sigma_{ij}(z)n_j(\nabla z_i^T V)]V_n\, d\Gamma \\ &- \iint_{\Gamma} [\sigma(z)^T \varepsilon(\lambda)]V_n\, d\Gamma + \iint_{\Gamma^f \cup \Gamma^s} [f^{bT}\lambda]V_n\, d\Gamma \\ &+ \iint_{\Gamma^s} [div(p\lambda)V_n\, d\Gamma,\end{aligned} \tag{6.176}$$

which is the desired result. And as with the conservative loading case, the stress sensitivity result in (6.176) is valid for a fixed region Ω_p that satisfies the relation $\bar{\Omega}_p \subset \Omega$.

6.2.4 Shape Design Sensitivity Analysis of Local Performance Measures

In Section 6.2.3, analytical shape design sensitivity formulas for beams and plates were derived for the functional that defines displacement at an isolated fixed point, $\boldsymbol{x}^* \in \Omega$. For the purpose of shape design sensitivity analysis, it was assumed that this point does not move so that $\delta'(x - x^*) = 0$, that is, that the displacement functional on perturbed domain Ω_τ is evaluated at the same point $\boldsymbol{x}^*$.

In order to numerically determine shape design sensitivity, the finite element method can be employed as a computational tool. A nodal point is the natural choice for evaluating displacement. If the shape (geometry) of domain Ω is perturbed, then the domain finite element grid will also be perturbed, and nodal points will move. Thus, evaluation point $\boldsymbol{x}^*$ moves according to the domain design velocity field, and the additional convective term due to movement of point $\boldsymbol{x}^*$ must be obtained.

The shape design sensitivity of a mean stress functional over fixed, small region Ω_p was considered in Section 6.2.3 for beam, plate, and linear elasticity problems by assuming $m_p' = 0$. Moreover, in order to define the mean stress functional, it was assumed that the functional value on deformed domain Ω_τ is the mean stress value on the same region Ω_p, and that boundary Γ_p of Ω_p does not intersect boundary Γ of Ω. As in the case of displacement, when the finite element method is used, a finite element is a natural choice for Ω_p. Consequently, due to domain perturbation, Ω_p will move as the finite element grid moves, and Γ_p may touch Γ. In this case, the additional convective term due to movement of region Ω_p must be obtained.

Displacement Functional

Consider a scalar displacement functional in either a beam or plate problem, defined by

$$\psi \equiv z(\boldsymbol{x}^*) = \iint_\Omega \delta(\boldsymbol{x} - \boldsymbol{x}^*) z\, d\Omega, \tag{6.177}$$

where point $\boldsymbol{x}^*$ moves to $\boldsymbol{x}^*_\tau = \boldsymbol{x}^* + \tau V(\boldsymbol{x}^*)$. By taking the material derivative of (6.177), the following sensitivity expression is obtained:

$$\psi' = z'(\boldsymbol{x}^*) + \nabla z(\boldsymbol{x}^*)^T V(\boldsymbol{x}^*) = \iint_\Omega \delta(\boldsymbol{x} - \boldsymbol{x}^*) z'\, d\Omega + \nabla z(\boldsymbol{x}^*)^T V(\boldsymbol{x}^*). \tag{6.178}$$

Note that the first term on the right side of (6.178) is the same used in Section 6.2.3 to derive (6.91) for a beam, and (6.138) for a plate. Thus, if point $\boldsymbol{x}^*$ is taken as moving, then the second convective term on the right side of (6.178) needs to be added to (6.91) and (6.138). Consequently, even though the physical domain shape does not change, if point $\boldsymbol{x}^*$ moves, then the contribution from the additional convective term appears.

Example 6.7. To illustrate the use of (6.178), consider a clamped beam with a displacement functional, as studied in Section 6.2.3. Taking beam length l as a design variable, the design velocity on the boundary can be written as $V(0) = 0$ and $V(l) = \delta l$. In the domain it is possible to select $V(x) = x\delta l/l$ $(0 \le x \le l)$ as in Example 6.2, which means that the points on the beam proportionally move to the right. If displacement sensitivity is desired at point $x^* = l/4$, then the above design velocity information can be used to obtain perturbed point $x_\tau^* = (l + \tau\delta l)/4$. In addition, the analytical displacement expression was given in Example 6.2 as $z(x) = (f_0/24EI)[x^2(l - x)^2]$. After substituting x_τ^* into this displacement expression, we obtain a displacement functional at x_τ^* as

$$\psi(\tau) = z_\tau(x_\tau^*) = \frac{f_0}{24EI}[x_\tau^{*2}(l + \tau\delta l - x_\tau^*)^2]. \tag{6.179}$$

Taking the variation of the displacement functional with respect to τ, and evaluating the result at $\tau = 0$, the displacement sensitivity is obtained as

$$\psi' = \frac{3f_0 l^3}{512EI}\delta l.$$

This sensitivity will now be compared with the sensitivity obtained by using the adjoint variable method. The adjoint load from (6.88) is a unit point load at $x^* = l/4$. The adjoint variable is thus obtained as

$$\lambda(x) = \frac{1}{384EI}[64\langle x - l/4\rangle^3 - 54x^3 + 27lx^2].$$

Using these results, combined with the formula in (6.92) for a clamped beam, and the additional term in (6.178), we obtain

$$\psi' = EIz_{,11}\lambda_{,11}V\Big|_0^l + z_{,1}(x^*)V(x^*) = \frac{3f_0 l^3}{512EI}\delta l,$$

which is the same previously obtained result.

From this example it is clear that if different velocity fields are used in the domain, even with the same $V(0) = 0$ and $V(l) = \delta l$, then different sensitivity results will be obtained, since the convective term on the right side of (6.178) depends on velocity $V(x^*)$. However, since the sensitivity of the global functionals such as compliance and eigenvalue only depends on $V(0)$ and $V(l)$, they would yield the same sensitivity results for different domain velocity fields.

Stress Functional

The mean stress functional over a small region Ω_p can be written as

$$\psi = \iiint_\Omega g(\sigma(z))m_p\, d\Omega, \tag{6.180}$$

where m_p is a characteristic function that has the constant value $\bar{m}_p = (\iiint_{\Omega_p} d\Omega)^{-1}$ on Ω_p, and is zero outside Ω_p.

First, consider the case in which Ω_p moves. Since m_p is constant, (6.180)) can be rewritten as

$$\psi = \frac{\iiint_{\Omega_p} g(\sigma(z))\, d\Omega}{\iiint_{\Omega_p} d\Omega}. \tag{6.181}$$

By using the material derivative formula from (6.38), the functional in (6.181) is differentiated with respect to the design as

$$
\begin{aligned}
\psi' = & \left[\left(\iiint_{\Omega_p} g'(\sigma(z)) d\Omega + \iint_{\Gamma_p} g(\sigma(z)) V_n d\Gamma \right) \iiint_{\Omega_p} d\Omega \right. \\
& \left. - \iiint_{\Omega_p} g(\sigma(z)) d\Omega \iint_{\Gamma_p} V_n d\Gamma \right] \bigg/ \left(\iiint_{\Omega_p} d\Omega \right)^2 \\
= & \iiint_{\Omega} g'(\sigma(z)) m_p \, d\Omega + \bar{m}_p \iint_{\Gamma_p} [g(\sigma(z)) - \psi] V_n \, d\Gamma.
\end{aligned}
\tag{6.182}
$$

Note that the first term on the right side of the above equation is the same one used in (6.97), (6.146), and (6.161) for beam, plate, and linear elasticity problems, respectively. If Ω_p is moving, the second term on the right side of the above equation needs to be added to the results in (6.104), (6.153), and (6.171). This convective term is the additional contribution due to the movement of Ω_p. Thus, even without a change in domain shape, the movement of Ω_p will provide a nonzero sensitivity term.

For the case in which a part of Γ_p intersects Γ as shown in Fig. 6.9, $m_p = 0$ cannot be permitted on Γ_p; specifically, it cannot be permitted on $\tilde{\Gamma}_p \equiv \Gamma \cap \Gamma_p$ in (6.104), (6.153), and (6.171). Instead, $m_p = \bar{m}_p$ must be used on $\tilde{\Gamma}_p$, and the distributional derivatives $m_{p,i}$ ($i = 1, 2$) arise on $\tilde{\Gamma}_p \subset \Gamma$. Moreover, even though the kinematic boundary conditions for adjoint response are the same as when $\bar{\Omega}_p \subset \Omega$, the traction boundary conditions will be different on $\tilde{\Gamma}_p$, since $m_p = \bar{m}_p$ and distributional derivatives $m_{p,i}$ ($i = 1, 2$) must be used for the variational identities of the adjoint system given in (6.100), (6.149), and (6.166).

Interpretation of Results

From the shape design sensitivity results derived in this section, it is clear that unlike functionals that define such global measures as compliance and eigenvalues, the shape design sensitivity of local functionals may involve convective design perturbations due to a velocity field in the domain, that is, once the perturbed shape of a domain is given, there is only one way to evaluate global functionals by integrating over the entire perturbed domain. However, perturbations of local functionals may or may not involve design perturbations, depending on the movement of location of the local functional.

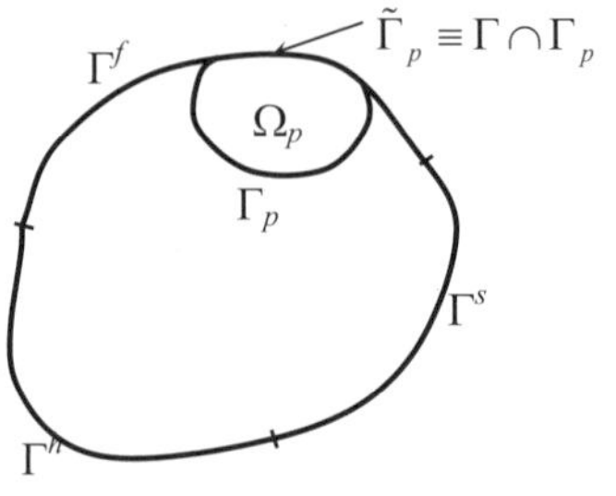

Figure 6.9. Intersection of Γ_p and Γ.

To predict perturbations of local functionals on fixed interior points or regions, the results from Section 6.2.3 can be used. Consequently, the second term on the right side of (6.178) and (6.182) can be ignored. If a perturbation of local functionals on moving interior points or regions is desired, then the domain velocity field must be considered, as in (6.178) and (6.182). In this case, the prediction accounts for the movement of a point or region on which the functional is defined.

6.2.5 Domain Shape Design Sensitivity Method

To numerically calculate the design sensitivity expressions of the boundary method presented in Section 6.2.4, which are given in terms of boundary integrals, stresses, strains, and/or normal derivatives of state and adjoint variables on the boundary must be used. Hence, the accurate evaluation of this boundary information is critical. Thus, when a numerical method such as finite element analysis is used, the accuracy of its results must be verified for those state and adjoint variables that lie on the boundary. It is well known [85] that finite element analysis results may not be satisfactory on the boundary for a system with a nonsmooth load and with interface problems.

Several methods might be devised to overcome this difficulty. The first is using a finite element method that provides accurate results on the boundary. A second choice would be another numerical method, such as the boundary element method [91] and [92]. In the finite element method, the unknown function, e.g, displacement, is approximated by trial functions that do not satisfy the governing equations, but that usually satisfy kinematic boundary conditions. Nodal parameters are then determined by approximate satisfaction of both the differential equations and the nonkinematic boundary conditions, in the sense of a domain integral mean. However, with the boundary element method, approximating functions satisfy the governing equations in the domain, but they do not satisfy the boundary conditions. Nodal parameters are determined by approximate satisfaction of the boundary conditions in the sense of a weighted integral. An important advantage of the boundary element method in shape design sensitivity analysis is that it represents boundary conditions better than the finite element method, and it is usually more accurate in determining stress on the boundary.

A third method uses the domain information in conjunction with finite element analysis, and is the main focus of this section. To develop a domain method, consider the basic material derivative formula in Lemma 6.1. Instead of using (6.38), the result given by (6.37) can be used, which requires velocity information on the domain rather than on the boundary. In this section, the adjoint variable method for beam, plate, and linear elasticity problems as presented in Section 3.1 is now developed to calculate the shape design sensitivity of performance measures using the domain method.

Beam Problem

Consider the beam bending problem in Section 3.1.2, with a domain $\Omega = (0, l) \subset R^1$ and a moment of inertia $I(x)$. The adjoint equation for the domain method is the same as the adjoint equation for the boundary method. Thus, the adjoint equations in Section 6.2.3 can be used for calculating the adjoint response of the beam problem. The difference between the boundary and domain methods is whether those variations $a'_V(z,\bar{z})$ and $\ell'_V(\bar{z})$ are calculated on the boundary or in the domain. For the domain method, these variations are defined as

$$a'_V(z,\bar{z}) = -\int_0^l EI[3z_{,11}\bar{z}_{,11}V_{,1} + (z_{,11}\bar{z}_{,1} + z_{,1}\bar{z}_{,11})V_{,11}]dx \tag{6.183}$$

and

$$\ell'_V(\bar{z}) = \int_0^l (f_{,1}\bar{z}V + f\bar{z}V_{,1})\,dx. \tag{6.184}$$

The sensitivity of a general functional can be calculated by substituting these two definitions into (6.76). Note that the second-order derivative of the design velocity exists in (6.183). Thus, the domain design velocity $V(x)$ must be defined such that $V_{,11}$ is integrable. However, different from the boundary method, the highest derivative of the state variable z is two. Thus, the third-order derivative is not required in the domain method.

When the boundary and domain formulations are compared, an interesting difference appears in the continuity requirement of design velocity between the two formulations. The boundary formulation of the beam component in Section 6.2.3 only requires the first-order derivative of the design velocity, while the domain formulation in (6.183) needs the second-order derivative of the design velocity. In the practical point of view, it is inconvenient to define a design velocity field whose second-order derivative is integrable. If the Timoshenko beam theory in Section 3.1.2 is used, however, it is enough that the first-order derivative of the design velocity is integrable, which is desirable. Thus, it is necessary to develop the shape sensitivity formulation for the Timoshenko beam.

The structural bilinear form for the Timoshenko beam component at the perturbed design can be written from (3.27), as

$$a_{\Omega_\tau}(\mathbf{z}_\tau, \bar{\mathbf{z}}_\tau) = \int_0^{l_\tau} [EI\theta_{,1_\tau}\bar{\theta}_{,1_\tau} + k\mu A(\bar{z}_{,1_\tau} - \bar{\theta}_\tau)(z_{,1_\tau} - \theta_\tau)]\,dx, \tag{6.185}$$

where the state variable $\mathbf{z} = [z, \theta]^T$ and θ is the rotational degree-of-freedom with respect to x_3-coordinate. By differentiating (6.185) and collecting the explicitly dependent terms, the structural fictitious load for the beam component can be defined as

$$a'_V(\mathbf{z}, \bar{\mathbf{z}}) = -\int_0^l [EI\theta_{,1}\bar{\theta}_{,1} + k\mu A(\bar{z}_{,1}z_{,1} - \bar{\theta}\theta)]V_{,1}\,dx. \tag{6.186}$$

Note that the first-order derivative of the design velocity appears in (6.186). Thus, C^0-continuous design velocity is enough to define the structural fictitious load.

The load linear form in (3.28) can be rewritten at the perturbed design as

$$\ell_{\Omega_\tau}(\bar{\mathbf{z}}) = \int_0^{l_\tau} \bar{\mathbf{z}}_\tau^T \boldsymbol{f}\,dx, \tag{6.187}$$

where $\boldsymbol{f} = [f, m]^T$ includes the distributed body force and moment. By assuming that the applied load is independent of design, the variation of the load linear form can be obtained as

$$\ell'_V(\bar{\mathbf{z}}) = \int_0^l [\bar{\mathbf{z}}^T \boldsymbol{f}_{,1}V + \bar{\mathbf{z}}^T \boldsymbol{f}V_{,1}]\,dx. \tag{6.188}$$

Plate Problem

Consider the plate bending problem as it was presented in Section 3.1.3. In the component-fixed local coordinate system, the design velocity field $\boldsymbol{V}(\boldsymbol{x})$ is given in two-dimensional domain Ω. The variations $a'_V(z,\bar{z})$ and $\ell'_V(\bar{z})$ in the domain method can be defined as

$$\begin{aligned} a'_V(z,\bar{z}) = &-\iint_\Omega [\boldsymbol{\kappa}(\bar{z})^T \boldsymbol{C}^b \boldsymbol{\kappa}(\nabla z^T \boldsymbol{V}) + \boldsymbol{\kappa}(\nabla \bar{z}^T \boldsymbol{V})^T \boldsymbol{C}^b \boldsymbol{\kappa}(z)]\,d\Omega \\ &+ \iint_\Omega div\left(\boldsymbol{\kappa}(\bar{z})^T \boldsymbol{C}^b \boldsymbol{\kappa}(z)\boldsymbol{V}\right) d\Omega \end{aligned} \tag{6.189}$$

and

$$\ell'_V(\bar{z}) = \iint_{\Omega} (\bar{z}\nabla f^T V + f\bar{z}divV)\, d\Omega, \tag{6.190}$$

where $\boldsymbol{\kappa}(z)$ is the curvature vector in (3.39), and $\boldsymbol{C}^b$ is the bending stiffness matrix in (3.40). Although the third-order derivatives of state variable z exist in (6.189), they are eventually canceled out each other. Thus, only the second-order derivatives of the state variable are required.

As with beam component, the requirement of the second-order derivative of the design velocity is impractical in engineering applications. Thus, the Mindlin/Reissner plate theory in Section 3.1.3 is used to derive the shape sensitivity formulation. This theory requires the design velocity whose first-order derivative is integrable. The energy bilinear form for the thick plate in (3.55) can be rewritten at the perturbed design, as

$$a_{\Omega_\tau}(z_\tau, \bar{z}_\tau) = \iint_{\Omega_\tau} [\boldsymbol{\kappa}(\bar{z}_\tau)\boldsymbol{C}^b\boldsymbol{\kappa}(z_\tau) + \boldsymbol{\gamma}(\bar{z}_\tau)\boldsymbol{C}^s\boldsymbol{\gamma}(z_\tau)]\, d\Omega, \tag{6.191}$$

where $z = [z, \theta_1, \theta_2]^T$ is the state response. Using the material derivative formulas in Section 6.1, the bilinear form in (6.191) can be differentiated with respect to the design. After collecting explicitly dependent terms, the structural fictitious load can be defined

$$\begin{aligned} a'_V(z, \bar{z}) = &\iint_{\Omega} [\boldsymbol{\kappa}^V(\bar{z})\boldsymbol{C}^b\boldsymbol{\kappa}(z) + \boldsymbol{\kappa}(\bar{z})\boldsymbol{C}^b\boldsymbol{\kappa}^V(z) + \boldsymbol{\kappa}(\bar{z})\boldsymbol{C}^b\boldsymbol{\kappa}(z)divV]\, d\Omega \\ &+ \iint_{\Omega} [\boldsymbol{\gamma}^V(\bar{z})\boldsymbol{C}^s\boldsymbol{\gamma}(z) + \boldsymbol{\gamma}(\bar{z})\boldsymbol{C}^s\boldsymbol{\gamma}^V(z) + \boldsymbol{\gamma}(\bar{z})\boldsymbol{C}^s\boldsymbol{\gamma}(z)divV]\, d\Omega, \end{aligned} \tag{6.192}$$

where

$$\boldsymbol{\kappa}^V(z) = -\frac{1}{2}\left(\frac{\partial\theta_i}{\partial x_k}\frac{\partial V_k}{\partial x_j} + \frac{\partial\theta_j}{\partial x_k}\frac{\partial V_k}{\partial x_i}\right) \tag{6.193}$$

and

$$\boldsymbol{\gamma}^V(z) = -\begin{Bmatrix} \dfrac{\partial z}{\partial x_k}\dfrac{\partial V_k}{\partial x_1} \\ \dfrac{\partial z}{\partial x_k}\dfrac{\partial V_k}{\partial x_2} \end{Bmatrix}. \tag{6.194}$$

The load linear form in (3.55) can be rewritten at the perturbed design as

$$\ell_{\Omega_\tau}(\bar{z}_\tau) = \iint_{\Omega_\tau} \bar{z}_\tau^T \boldsymbol{f}\, d\Omega, \tag{6.195}$$

where $\boldsymbol{f} = [f, m_1, m_2]^T$ includes the distributed body force and moments. By assuming that the applied load is independent of design, the variation of the load linear form can be obtained as

$$\ell'_V(\bar{z}) = \iint_{\Omega} [\bar{z}^T\nabla\boldsymbol{f}V + \bar{z}^T \boldsymbol{f}divV]\, d\Omega. \tag{6.196}$$

Linear Elasticity

Consider the linear elasticity problem in Section 6.2.4. Suppose that the traction loading surface Γ^s is fixed in the design and $\boldsymbol{f}^s$ is a conservative loading. By using the fact that $\boldsymbol{f}^{b\prime} = \boldsymbol{f}^{s\prime} = 0$, and by using the material derivative formulas in (6.37) and (6.53), the state equation (3.60) is differentiated with respect to the design as

$$\begin{aligned}
&\iiint_{\Omega}[\boldsymbol{\sigma}(z')^{T}\boldsymbol{\varepsilon}(\bar{z})+\boldsymbol{\sigma}(z)^{T}\boldsymbol{\varepsilon}(\bar{z}')]\,d\Omega \\
&\quad+\iiint_{\Omega}\nabla[\boldsymbol{\sigma}(z)^{T}\boldsymbol{\varepsilon}(\bar{z})]^{T}\boldsymbol{V}\,d\Omega+\iiint_{\Omega}\boldsymbol{\sigma}(z)^{T}\boldsymbol{\varepsilon}(\bar{z})div\boldsymbol{V}\,d\Omega \\
&=\iiint_{\Omega}\boldsymbol{f}^{bT}\bar{z}'\,d\Omega+\iiint_{\Omega}\nabla(\boldsymbol{f}^{bT}\bar{z})^{T}\boldsymbol{V}\,d\Omega+\iiint_{\Omega}\boldsymbol{f}^{bT}\bar{z}div\boldsymbol{V}\,d\Omega \\
&\quad+\iint_{\Gamma^{s}}\boldsymbol{f}^{sT}\bar{z}'\,d\Gamma+\iint_{\Gamma^{s}}\nabla(\boldsymbol{f}^{sT}\bar{z})^{T}\boldsymbol{V}\,d\Gamma+\iint_{\Gamma^{s}}\kappa\boldsymbol{f}^{sT}\bar{z}V_{n}\,d\Gamma.
\end{aligned} \tag{6.197}$$

After converting the partial derivative z' into $\dot{z}$ by using the formula in (6.8), and after using the property of $\dot{\bar{z}}=0$, the sensitivity equation (6.197) can be rewritten as

$$\begin{aligned}
&\iiint_{\Omega}[\boldsymbol{\sigma}(\dot{z})^{T}\boldsymbol{\varepsilon}(\bar{z})-\boldsymbol{\sigma}(z)^{T}\boldsymbol{\varepsilon}(\nabla\bar{z}^{T}\boldsymbol{V})-\boldsymbol{\sigma}(\nabla z^{T}\boldsymbol{V})^{T}\boldsymbol{\varepsilon}(\bar{z})]\,d\Omega \\
&\quad+\iiint_{\Omega}\nabla[\boldsymbol{\sigma}(z)^{T}\boldsymbol{\varepsilon}(\bar{z})]^{T}\boldsymbol{V}\,d\Omega+\iiint_{\Omega}\boldsymbol{\sigma}(z)^{T}\boldsymbol{\varepsilon}(\bar{z})div\boldsymbol{V}\,d\Omega \\
&=\iiint_{\Omega}\bar{z}^{T}(\nabla\boldsymbol{f}^{bT}\boldsymbol{V})\,d\Omega+\iiint_{\Omega}\boldsymbol{f}^{bT}\bar{z}div\boldsymbol{V}\,d\Omega \\
&\quad-\iint_{\Gamma^{s}}\boldsymbol{f}^{sT}(\nabla\bar{z}^{T}\boldsymbol{V})\,d\Gamma+\iint_{\Gamma^{s}}[\nabla(\boldsymbol{f}^{sT}\bar{z})^{T}\boldsymbol{n}+\kappa(\boldsymbol{f}^{sT}\bar{z})V_{n}]\,d\Gamma.
\end{aligned} \tag{6.198}$$

Direct calculation allows the following to be verified:

$$\begin{aligned}
&\boldsymbol{\sigma}(z)^{T}\boldsymbol{\varepsilon}(\nabla\bar{z}^{T}\boldsymbol{V})+\boldsymbol{\sigma}(\nabla z^{T}\boldsymbol{V})^{T}\boldsymbol{\varepsilon}(\bar{z})-\nabla[\boldsymbol{\sigma}(z)^{T}\boldsymbol{\varepsilon}(\bar{z})]^{T}\boldsymbol{V} \\
&=\boldsymbol{\sigma}(z)^{T}\boldsymbol{\varepsilon}^{V}(\bar{z})+\boldsymbol{\sigma}(\bar{z})^{T}\boldsymbol{\varepsilon}^{V}(z),
\end{aligned} \tag{6.199}$$

where, by using the index notation, the term $\boldsymbol{\varepsilon}^{V}(z)$ can be defined by

$$\varepsilon_{ij}^{V}(z)=\frac{1}{2}\left(\frac{\partial z_{i}}{\partial x_{k}}\frac{\partial V_{k}}{\partial x_{j}}+\frac{\partial z_{j}}{\partial x_{k}}\frac{\partial V_{k}}{\partial x_{i}}\right), \tag{6.200}$$

Note that the second-order derivatives of z in (6.199) cancel each other out. Thus, only the first-order derivatives of z and $\boldsymbol{V}$ appear in the expression of (6.200). This property is important because design sensitivity analysis does not require any higher-order regularity than response analysis. Using these results, design sensitivity (6.198) can be further simplified to

$$\begin{aligned}
a_{\Omega}(\dot{z},\bar{z})&\equiv\iiint_{\Omega}\boldsymbol{\sigma}(\dot{z})^{T}\boldsymbol{\varepsilon}(\bar{z})\,d\Omega \\
&=\iiint_{\Omega}[\boldsymbol{\sigma}(z)^{T}\boldsymbol{\varepsilon}^{V}(\bar{z})+\boldsymbol{\sigma}(\bar{z})^{T}\boldsymbol{\varepsilon}^{V}(z)]\,d\Omega \\
&\quad-\iiint_{\Omega}\boldsymbol{\sigma}(z)^{T}\boldsymbol{\varepsilon}(\bar{z})div\boldsymbol{V}\,d\Omega \\
&\quad+\iiint_{\Omega}\bar{z}^{T}(\nabla\boldsymbol{f}^{T}\boldsymbol{V})\,d\Omega+\iiint_{\Omega}\boldsymbol{f}^{bT}\bar{z}div\boldsymbol{V}\,d\Omega \\
&\quad-\iint_{\Gamma^{s}}\boldsymbol{f}^{sT}(\nabla\bar{z}^{T}\boldsymbol{V})\,d\Gamma+\iint_{\Gamma^{s}}[\nabla(\boldsymbol{f}^{sT}\bar{z})^{T}\boldsymbol{n}+\kappa(\boldsymbol{f}^{sT}\bar{z})]V_{n}\,d\Gamma, \qquad \forall\bar{z}\in Z.
\end{aligned} \tag{6.201}$$

As in (6.160), (6.201) is a variational equation for $\dot{z}\in Z$.

Consider the mean stress functional presented in (6.181) as

$$\psi=\iiint_{\Omega}g(\boldsymbol{\sigma}(z))m_{p}\,d\Omega=\frac{\iiint_{\Omega_{p}}g(\boldsymbol{\sigma}(z))\,d\Omega}{\iiint_{\Omega_{p}}d\Omega}. \tag{6.202}$$

Using the domain material derivative formula in (6.37) the material derivative of (6.202) is obtained as

$$
\begin{aligned}
\psi' &= \left[\iiint_{\Omega} [g' + \nabla g^T V + g \, div V] d\Omega \iiint_{\Omega_p} d\Omega \right. \\
&\quad \left. - \iiint_{\Omega_p} g \, d\Omega \iiint_{\Omega_p} div V \, d\Omega \right] \Big/ \left(\iiint_{\Omega_p} d\Omega \right)^2 \\
&= \iiint_{\Omega} g_{,\sigma(z)} [\sigma(\dot{z}) - \sigma(\nabla z^T V)] m_p \, d\Omega \\
&\quad + \iiint_{\Omega} g_{,\sigma(z)} [\nabla \sigma(z) V] m_p \, d\Omega + \iiint_{\Omega_p} g \, div V m_p \, d\Omega \\
&\quad - \iiint_{\Omega_p} g m_p \, d\Omega \iiint_{\Omega_p} m_p div V \, d\Omega.
\end{aligned}
\tag{6.203}
$$

It can be shown for linear, elastic material that

$$
\sigma(\nabla z^T V) - \nabla \sigma V = C \varepsilon^V(z). \tag{6.204}
$$

Using the above results, (6.203) can be simplified as

$$
\begin{aligned}
\psi' &= \iiint_{\Omega} g_{,\sigma(z)} \sigma(\dot{z}) m_p \, d\Omega \\
&\quad + \iiint_{\Omega} g_{,\sigma(z)} [C \varepsilon^V(z)] m_p \, d\Omega + \\
&\quad \iiint_{\Omega_p} g \, div V m_p \, d\Omega - \iiint_{\Omega_p} g m_p \, d\Omega \iiint_{\Omega_p} m_p div V \, d\Omega.
\end{aligned}
\tag{6.205}
$$

As with the linear elasticity problem in Section 6.2.3, adjoint equation (6.162) can be defined. By using the same method employed in Section 6.2.4, the sensitivity formula is obtained as

$$
\begin{aligned}
\psi' &= \iiint_{\Omega} [\sigma(z)^T \varepsilon^V(\lambda) + \sigma(\lambda)^T \varepsilon^V(z) - \sigma(z)^T \varepsilon(\lambda) div V] d\Omega \\
&\quad + \iiint_{\Omega} \lambda^T [\nabla f^{bT} V] d\Omega + \iiint_{\Omega} \lambda^T f^b div V \, d\Omega \\
&\quad + \iint_{\Gamma^s} \{ -f^{sT} (\nabla \lambda^T V) + [\nabla (f^{sT} \lambda)^T n + \kappa (f^{sT} \lambda)] V_n \} d\Gamma \\
&\quad + \iiint_{\Omega} [g \, div V - g_{,\sigma} C \varepsilon^V(z)] m_p \, d\Omega \\
&\quad - \iiint_{\Omega_p} g m_p \, d\Omega \iiint_{\Omega_p} m_p div V \, d\Omega.
\end{aligned}
\tag{6.206}
$$

There are several advantages and disadvantages to the above discussed domain method. A disadvantage is that a velocity field that satisfies regularity properties, which will be discussed in Section 6.2.7, must be defined in the domain. There is no unique way of defining domain velocity fields for a given normal velocity field V_n on the boundary. In addition, because numerical evaluation of the sensitivity result in the above equation requires domain integration over the entire domain, it is more complicated than the evaluation of (6.182), which only requires integration over the variable boundary.

In addition to numerical accuracy when using finite element method for structural analysis, the domain method has several other advantages. Unlike the boundary method, variational identities are not required to transform domain integrals into boundary integrals. Thus, for a mean stress functional, the special case in which Γ_p intersects Γ, discussed in Section 6.2.4, need not to be separately treated for the domain method. The results presented in this section are valid for both cases. Also, as will be seen shortly, one design sensitivity expression works for all boundary conditions, unlike different design sensitivity expressions that are required for different boundary conditions, as presented in Section 6.1.3. The biggest advantage of the domain method is obtained in built-up structures, which consist of structural components combined in a variety of ways. In applying the domain method, the fact that interface conditions are not required to obtain shape design sensitivity formulas greatly simplifies the derivation process, since contributions from each component are simply added together. As for numerical accuracy, the results of a finite element analysis on interface boundaries are often unsatisfactory for built-up structures, due to abrupt changes in the boundary conditions. Using the domain method and a carefully controlled finite element analysis, stress evaluation at the interface boundaries may be avoided, and thus, accurate sensitivity results can be obtained.

6.2.6 Parameterization of Design Boundary

As with Chapter 5, before proceeding from analytical design sensitivity to numerical implementation, it is helpful to consider the numerical aspects of a computation. It is important to develop an effective boundary parameterization method that can be used in shape design sensitivity analysis. Detailed shape design parameterization methods are presented in Section 12.2 and the associated design velocity computation methods for both boundary and domain are presented in Section 13.3. In this section, a simple idea of using the parameterized boundary for evaluation of shape design sensitivity expressions of the boundary method is presented. To begin, assume that the points on boundary Γ are specified by a position vector, $\boldsymbol{x}(\boldsymbol{\alpha};\boldsymbol{b}) \in R^n$ ($n = 2, 3$), which runs from the origin of the coordinate system to point $\boldsymbol{x}$ on the boundary, as illustrated in Fig 6.10. Here, $\boldsymbol{\alpha} \in R^k$ is a parameter vector that defines the boundary curve ($k = 1$) or boundary surface ($k = 2$) Γ.

Once the design vector $\boldsymbol{b} = [b_1, b_2, ..., b_m]^T$ has been defined, shape design sensitivity formulas can be expressed in terms of the variation of $\boldsymbol{b}$, as $\delta\boldsymbol{b}$. To this end, first denote the perturbed design by

$$\boldsymbol{b}_\tau = \boldsymbol{b} + \tau\delta\boldsymbol{b}, \tag{6.207}$$

where $\boldsymbol{b}$ defines the boundary Γ of Ω, and $\boldsymbol{b}_\tau$ defines the boundary Γ_τ of the perturbed domain Ω_τ. The velocity field in (6.2) can be defined at the boundary as

$$\boldsymbol{V}(\boldsymbol{x}) = \frac{d}{d\tau}[\boldsymbol{x}(\boldsymbol{\alpha};\boldsymbol{b}_\tau)]\bigg|_{\tau=0} = \frac{\partial \boldsymbol{x}}{\partial \boldsymbol{b}}(\boldsymbol{\alpha};\boldsymbol{b})\delta\boldsymbol{b}. \tag{6.208}$$

The shape design sensitivity formula can then be expressed as

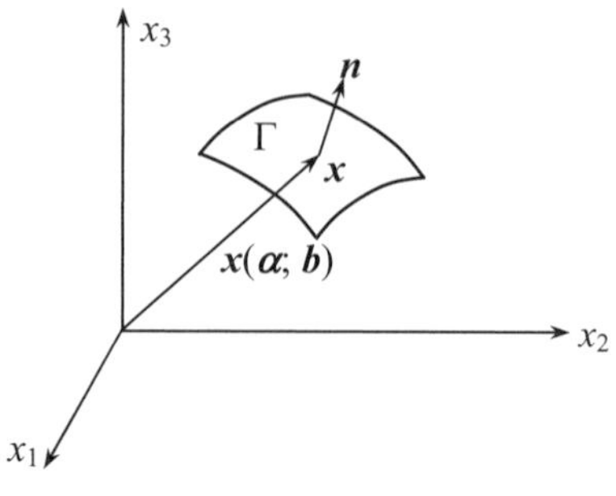

Figure 6.10. Parametric definition of Γ.

$$\psi' = \iint_{\Gamma} G(z,\boldsymbol{\lambda}) V_n \, d\Gamma$$
$$= \left[\iint_{\Gamma} G(z,\boldsymbol{\lambda}) \boldsymbol{n}^T \frac{\partial \boldsymbol{x}}{\partial \boldsymbol{b}}(\boldsymbol{\alpha};\boldsymbol{b}) \, d\Gamma \right] \delta\boldsymbol{b}, \tag{6.209}$$

where variation $\delta\boldsymbol{b}$ can be taken outside the integral, since it is constant. This expression provides the design sensitivity coefficients of ψ associated with variations in the design variables. Hence, once the state and adjoint variables have been determined, all that is required is a numerical calculation of the integral in (6.209).

6.2.7 Regularity of Design Velocity Field

In shape design optimization, a suitable finite element mesh must be constructed in the design iteration for accurate finite element analysis and design sensitivity analysis. In design iteration, the design velocity field used for design sensitivity analysis must be employed to update the finite element mesh. Hou and Cheen [93] pointed out that improper choice of the design velocity field for mesh update may result in a nonoptimal solution. Moreover, an inappropriate design velocity field may even lead to erroneous design sensitivity analysis and inaccurate finite element analysis results.

The design velocity field must meet numerous, stringent theoretical and practical criteria. Theoretically, the design velocity field must have the same regularity (i.e., smoothness), as the weak solution of the governing variational equation [5] and [94] through [97]. Roughly speaking, for truss and two- and three-dimensional elastic solid design components, C^0-continuous design velocity fields with integrable first derivatives are required (i.e., $H^1(\Omega)$), whereas for beam and shell design components, C^1-continuous design velocity fields with integrable second derivatives are required (i.e., $H^2(\Omega)$) [6], [18], and [36]. In addition, the design velocity field must depend linearly on the variation of shape design variables. This requirement stems from the fact that the sensitivity information predicts linear variation of the performance measure with respect to the variation of the shape design variables. In practical terms, the design velocity field must produce a finite element mesh that can support accurate finite element analysis and thus design sensitivity analysis. Moreover, the method for computing the design velocity field should accommodate the use of CAD modelers to define shape design variables, reuse the mathematical rule that determines interior material point movements (domain velocity), and be computationally efficient and applicable to a large class of problems.

In this section, the theoretical and practical requirements of the design velocity field are discussed. Theoretically, the design velocity field must

- Have the same regularity as the displacement field,
- Depend linearly on the variation of shape design variables.

For practical applications, the design velocity computation method must

- Retain the topology of the original finite element mesh,
- Provide finite element boundary nodes that stay on the geometric boundary for all shape changes,
- Use a mathematical rule that guarantees linear dependency of finite element node movements on the variations of shape design variables
- Produce a finite element mesh that is not distorted,
- Be naturally linked to design variables defined on a CAD model,
- Allow the mathematical rule to be reusable,
- Be efficient and general for a large class of applications.

Theoretical Requirements

Regularity Requirement

Design sensitivity requires that the design velocity field be as regular as the weak solution of the governing variational equation, that is, displacement field of the structure. The requirement comes from the mathematical regularity of design velocity field in the design sensitivity expression [5] and [94] through [97]. For truss and two- and three-dimensional elastic solid design components, C^0-continuous design velocity fields with integrable first derivatives are required [5] and [94] through [97]. On the other hand, for beam and shell design components, C^1-continuous design velocity fields with integrable second derivatives are required [6], [18], and [36]. These regularity requirements can be clearly seen from the domain method of design sensitivity expression in (6.183) and (6.189), where the first and second derivatives of the design velocity field appears in the integrands. Naturally, these requirements can be met by using the displacement shape functions to represent design velocity fields.

Linear Dependency Requirement

First-order linear design sensitivity requires that material point movements, that os, design velocity fields, in the structure depend linearly on the variation of shape design variables. This requirement arises from the design sensitivity theory [5] and [95], in which the material derivative of displacement field $\dot{z}$ depends linearly on the design velocity field. Since design sensitivity coefficients provide first-order derivatives of the structural performance measure with respect to design variables and depend linearly on the design velocity field, the design velocity field must depend linearly on variations of shape design variables.

The two-dimensional elastic cantilever beam shown Fig. 6.11 is utilized to explain how the design velocity field depends linearly on the variation of shape design variable. As shown in Fig. 6.11(a), the beam has point load F applied at the tip of the top surface and is modeled as a plane stress problem using a 4 × 3 finite element mesh. The design boundary, that is, the tip edge, is parameterized using a Bezier curve. For a detailed discussion of design parameterization methods, the reader is referred to Section 12.2. Shape design variables are movements of control points $\boldsymbol{p}_1$ and $\boldsymbol{p}_2$, as shown in Fig. 6.11(b).

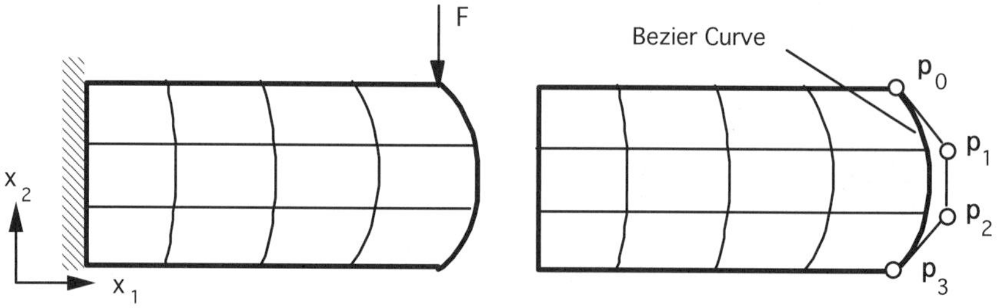

(a) Finite Element Model (b) Design Model

Figure 6.11. Two-dimensional cantilever beam example.

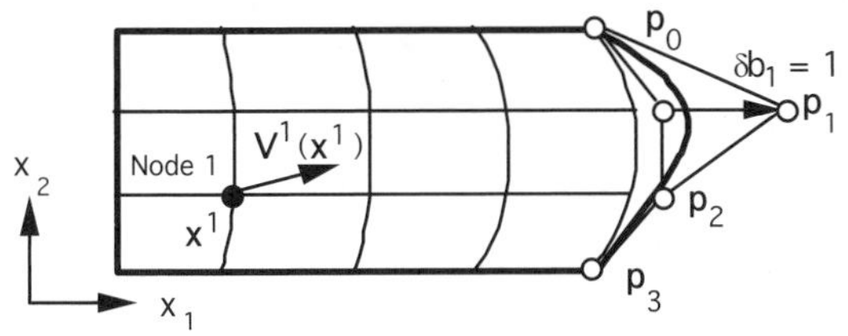

(a) Control Point $\boldsymbol{p}_1$ Moves $\delta b_1 = 1$ in the x_1-Direction

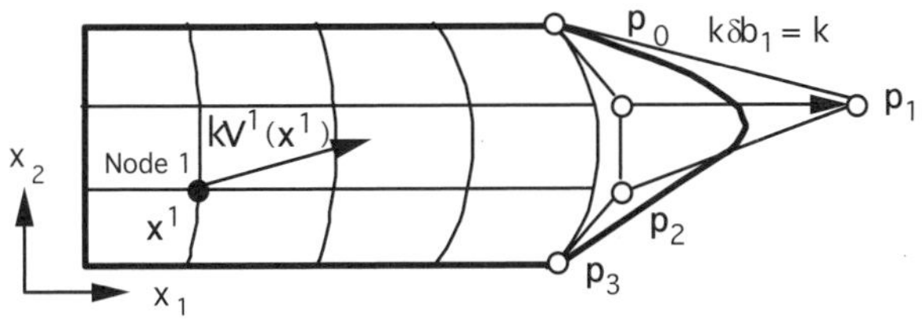

(b) Control Point $\boldsymbol{p}_1$ Moves $k\delta b_1 = k$ in the x_1-Direction

Figure 6.12. Linearity requirement.

Linear dependency requires that if control point $\boldsymbol{p}_1$ moves a distance $\delta b_1 = 1$ in the x_1-direction, producing interior design velocity $\boldsymbol{V}^1(\boldsymbol{x}^1)$ at node 1, as shown in Fig. 6.12(a), then node 1 must move $k\boldsymbol{V}^1(\boldsymbol{x}^1)$ along the same direction when $\boldsymbol{p}_1$ moves $k\delta b_1 = k$ ($k \neq 0$) in the x_1-direction. This linear dependency must be true for all boundary and interior nodes of the structure.

For the discretized finite element model, design sensitivity coefficients predict structural performance measures of the perturbed design with finite element mesh updated by moving nodal points along the direction of the design velocity field using

$$
\begin{aligned}
\boldsymbol{x}_{b+\delta b}^k &= \boldsymbol{x}_b^k + \delta \boldsymbol{x}^k \\
&= \boldsymbol{x}_b^k + \sum_{i=1}^{n} \boldsymbol{V}^i(\boldsymbol{x}^k)\delta b_i,
\end{aligned}
\tag{6.210}
$$

where $\boldsymbol{x}_{b+\delta b}^k$ and $\boldsymbol{x}_b^k$ are the locations of the kth node of the perturbed and the current designs, respectively; $\delta\boldsymbol{x}^k$ is the nodal point movement due to design variations; $\boldsymbol{V}^i$ and δb_i are the design velocity fields associated with the ith design variable and the variation of the ith design variable, respectively; and n is the number of design variables.

It is important to note that design sensitivity information does not predict performance measures of a new design if the finite element mesh is not updated according to the design velocity field, which depends linearly on the variation of shape design variables. For example, consider the design sensitivity coefficient $\partial\psi/\partial b_1$ of the x_2-displacement ψ at node 1, assuming that the design velocity is $\boldsymbol{V}^1(\boldsymbol{x}^1)$, as illustrated in Fig. 6.13. This coefficient predicts x_2-displacement at node 1 moving along $\boldsymbol{V}^1(\boldsymbol{x}^1)$ direction with certain small step size. In case node 1 moves to a new point k, which does not lie on the $\boldsymbol{V}^1(\boldsymbol{x}^1)$ direction of the new design, the design sensitivity $\partial\psi/\partial b_1$ cannot predict x_2-displacement at point k of the new design.

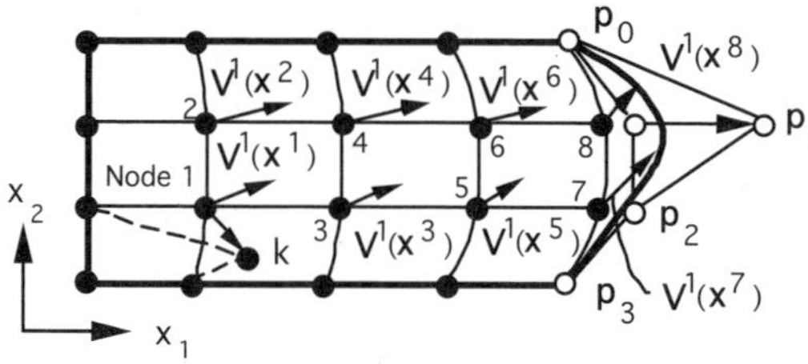

Figure 6.13. Mesh update for the new design.

Practical Requirements

As described in the previous section, the design velocity field is subject to only two theoretical requirements: regularity and linear dependency on the variation of shape design variables. That is, if the analytical solution is used to evaluate the design sensitivity expression, any design velocity field that satisfies the theoretical requirements will yield exact design sensitivity results. However, for many structural problems, the analytical solution cannot be obtained, and the approximated solution is obtained using the finite element method. In this case, different design velocity fields lead to different design sensitivity results. However, the more the finite element model is refined, the less sensitivity results lend to the design velocity fields.

Practically, the design velocity field must produce a quality finite element mesh that is consistent with the structural geometric model to support accurate finite element analysis and design sensitivity analysis in the design optimization process.

Mesh Topology

As discussed earlier, mesh updates must follow the direction of design velocity field and the perturbed design must have the same mesh topology as the current design because materials cannot cross over their relative positions. Of course, the perturbed design can have a different number of finite elements than the original design as long as the material points, and thus the new nodal points, follow the design velocity field. However, changing the number of finite elements during the design process is not practical in defining performance measures, such as nodal displacements and element stresses. Thus, using a mesh generator to define a design velocity field is not desirable if it generates a different mesh topology for the perturbed design.

Nodes on the Geometric Boundary

When the design velocity field is used to update the finite element mesh at the new design, the updated boundary nodes must lie on the geometric boundary so that the finite element and geometric models are consistent. To make sure that the updated boundary nodes lie on the geometric boundary, in addition to the linear dependency requirement, boundary node movements must be restricted. A simple way of restricting boundary nodes stay on the geometric boundary is to ensure that the nodes stay on the same parametric locations of the boundary curve or surface at the original and perturbed designs. For example, suppose that node 8 is located at the parametric location $u = ⅓$ of the Bezier curve of the original design shown in Fig. 6.14. At the perturbed design, node 8 is located at the same parametric location $u = ⅓$ of the perturbed design boundary. By restricting boundary node movements in this way and by using the linear boundary design velocity field, it can be guaranteed that every node always stays on the geometric boundary [98].

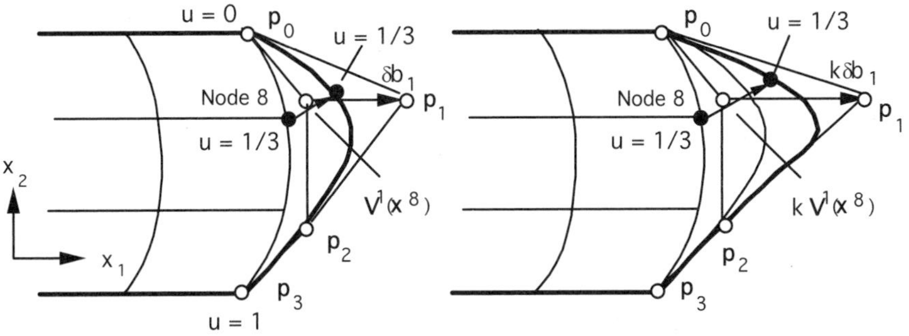

Figure 6.14. Geometric boundary match.

Linearity of the Velocity Field and Quality of the Updated Finite Element Mesh

Once the boundary design velocity field is used to perturb the geometric boundary (hence moving the boundary nodes), the domain design velocity field (hence movements of the interior nodes) must be determined in a way that satisfies the linear dependency requirement and permits a finite element mesh to be generated. Because of the linear dependency requirement, the interior finite element nodes cannot be moved arbitrarily, but must be determined following a specific mathematical rule, which is selected according to the method used for computing the domain velocity field. For the isoparametric mapping method, the rule is an algebraic equation, while for the boundary displacement or fictitious load method, the rule is a finite element matrix equation [95] through [97] and [99] through [101].

It is highly desirable that the design velocity field produces a regular finite element mesh since distorted mesh gives inaccurate finite element analysis results, or fails to provide a solution. In practical applications, however, it is not always convenient to check finite element mesh every design iteration and so finite element error analysis and mesh adaptation must be employed to ensure analysis accuracy [102]. When the finite element mesh is not acceptable, as indicated by an error estimator, the optimization iteration should be stopped. In such cases, a new finite element mesh must be generated, whose topology may differ from that of the previous model, and from then on, new design optimization iteration should be carried out. Computational methods for obtaining domain velocity fields generate more regular (i.e., undistorted) finite element mesh are discussed in Section 13.3.

Link to CAD Modeler

It is widely recognized that a CAD-based shape design optimization method that optimizes a CAD geometric model allows the product developer to take advantage of product modeling, design parameterization, and product data communication among engineering disciplines in supporting a concurrent product design and manufacturing. On the other hand, a finite element–based shape design parameterization method that defines design variables on the finite element model for optimization is problematic. Using the finite element–based design parameterization method, convergence is difficult to achieve due to a large number of design variables being employed, that is, finite element node movements. Moreover, optimizing a finite element model may lead to an optimum design with a nonsmooth design boundary [103]. To promote the connection of shape design optimization to a CAD modeler, the design velocity field computation should naturally be linked to variations of the CAD design variables. This linkage is discussed in Section 13.3.

Reusable Mathematical Rule

Mathematically, the design velocity field is a function of location of the material point $\boldsymbol{x}$, as shown in (6.5). Physically, location of the material point also depends on the shape design variables, which determine the physical domain of the structure. However, in practical applications, the mathematical rule applied to compute the design velocity field may not need physical locations of the material points, but its parametric locations in a decomposed domain, for example, the isoparametric mapping method. In this case, the design velocity field is independent of design variables $\boldsymbol{b}$, which implies that *the design velocity field needs to be computed only once and kept constant during the design optimization iterations*.

General Applicability and Computational Efficiency

In general, there are three types of structural shape design applications in terms of the characteristics of the design boundary [98]. In the first type, shape of a sculptured boundary, such as a fillet or arch dam surface, is to be determined. In the second type of applications, dimensions of predefined shapes, such as the radius of a circular hole, the major and minor axes of an elliptic hole, length of a rectangular membrane, or radius of a rounded corner, are to be found. In the third type, locations or orientations of predefined shapes must be determined relative to the global reference frame, for example, the location of the center of a circular hole, elliptic hole, or slot. The design velocity field computation method must be general enough to support these design applications.

In addition to being generally applicable, the method for computing the design velocity field should also be computationally efficient; although this is of less concern as faster computer hardware and architecture are developed. Furthermore, computational efficiency can be improved by reusing the mathematical rules of design velocity field computation.

6.2.8 Numerical Examples

To illustrate numerical implementation of the shape design sensitivity formulas that have been derived in previous sections, several example problems are now presented. The boundary method is used in the first example, whereas the domain method is used for other examples. The first example clearly demonstrates that accurate evaluation of the original and adjoint responses at the boundary is critical to obtain accurate design sensitivity. On the other hand, the domain method does not suffer from this difficulty.

Fillet

The selection of an optimum fillet shape in a tension bar such that no yielding occurs has long attracted the attention of engineers. The dimensions and variables of the bar and fillet are shown in Fig. 6.15. Since the bar is symmetric, only its upper half has been modeled. Boundary segment Γ_1 is to be varied, with fixed points at A and B. Segment Γ_3 is the central line of the bar and Γ_4 and Γ_2 are uniformly loaded edges.

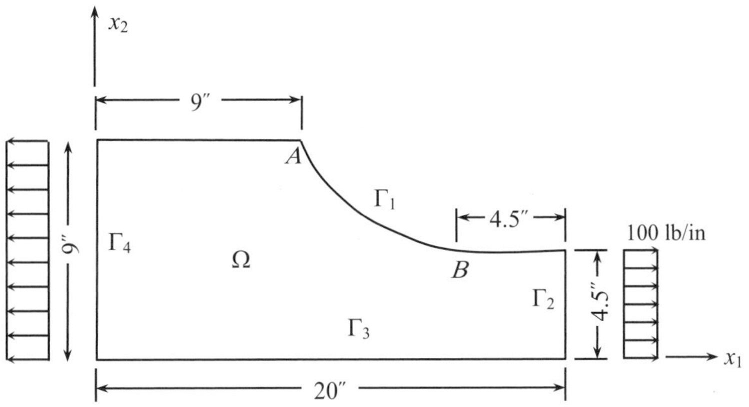

Figure 6.15. Geometric configuration of fillet.

The variational equation of this two-dimensional solid component is

$$\begin{aligned} a_\Omega(z,\bar{z}) &\equiv \iint_\Omega \boldsymbol{\sigma}(z)^T \boldsymbol{\varepsilon}(\bar{z})\, d\Omega \\ &= \int_{\Gamma_2} \bar{z}^T \boldsymbol{f}^s\, d\Gamma \equiv \ell_\Omega(\bar{z}), \qquad \forall \bar{z} \in Z, \end{aligned} \tag{6.211}$$

where the admissible displacement space is defined by

$$Z = \left\{ z \in [H^1(\Omega)]^2 : z_1 = 0, \boldsymbol{x} \in \Gamma_4 \text{ and } z_2 = 0, \boldsymbol{x} \in \Gamma_3 \right\}, \tag{6.212}$$

with no body force acting on the fillet.

Now, consider the von Mises yield stress functional, averaged over a small region Ω_k as

$$\psi_k = \iint_\Omega g m_k\, d\Omega, \tag{6.213}$$

where $g = (\sigma_y - \sigma_a)/\sigma_a$, σ_y is the von Mises yield stress, defined as

$$\sigma_y = \sqrt{\sigma_{11}^2 + \sigma_{22}^2 + 3\sigma_{12}^2 - \sigma_{11}\sigma_{22}}, \tag{6.214}$$

and σ_a is the given allowable stress. In (6.213), m_k is a characteristic function on small region Ω_k. The adjoint equation corresponding to the performance ψ_k is obtained from (6.162) as

$$a_\Omega(\boldsymbol{\lambda},\bar{\boldsymbol{\lambda}}) = \iint_\Omega g_{,\sigma(z)} \boldsymbol{\sigma}(\bar{\boldsymbol{\lambda}}) m_k\, d\Omega, \qquad \forall \bar{\boldsymbol{\lambda}} \in Z. \tag{6.215}$$

Since the design velocity is only defined on Γ_1, the variation of ψ_k is taken from (6.213) as

$$\psi_k' = -\int_{\Gamma_1} \boldsymbol{\sigma}(z)^T \boldsymbol{\varepsilon}(\boldsymbol{\lambda}^{(k)}) V_n\, d\Gamma + \bar{m}_k \int_{\Gamma_k} [g(z) - \psi_k] V_n\, d\Gamma, \tag{6.216}$$

where $\boldsymbol{\lambda}^{(k)}$ is the solution to adjoint equation (6.215), $\bar{m}_k$ is the value of the characteristic function on Ω_k, and Γ_k is the boundary of finite element Ω_k.

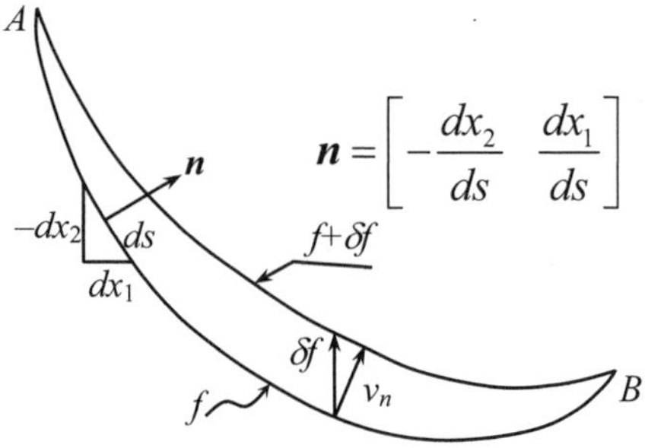

Figure 6.16. Geometry of boundary curve.

Consider the variable boundary Γ_1 of the fillet, as shown in Fig. 6.15, which can be characterized as curve $x_2 = f(x_1)$, with a small, vertical variation $\delta f(x_1)$, as shown in Fig. 6.16. Given the curve's geometry, if only a small vertical change $\delta f(x_1)$ is allowed, the normal movement of the boundary can be written as

$$V_n = \delta f\, n_2 = \delta f \left(\frac{dx_1}{ds} \right), \tag{6.217}$$

where s is the arc length on Γ_1. Thus, the sensitivity formula in (6.216) can be rewritten as

$$\begin{aligned} \psi_k' &= -\int_{\Gamma_1} \boldsymbol{\sigma}(\boldsymbol{z})^T \boldsymbol{\varepsilon}(\boldsymbol{\lambda}^{(k)}) V_n\, ds + \bar{m}_k \int_{\Gamma_k} [g(\boldsymbol{z}) - \psi_k] V_n\, d\Gamma \\ &= -\int_A^B \boldsymbol{\sigma}(\boldsymbol{z})^T \boldsymbol{\varepsilon}(\boldsymbol{\lambda}^{(k)}) \delta f\, dx_1 + \bar{m}_k \int_{\Gamma_k} [g(\boldsymbol{z}) - \psi_k] V_n\, d\Gamma. \end{aligned} \tag{6.218}$$

In (6.218), δf can easily be related to $\delta \boldsymbol{b}$ once curve Γ_1, defined by $x_2 = f(x_1, \boldsymbol{b})$, is parameterized by design variable vector $\boldsymbol{b}$. If the heights of selected boundary points are chosen as design variables, and if the boundary is piecewise linear, then the boundary can be expressed as

$$f(x_1) = \left(\frac{x_1^{i+1} - x_1}{h_i} \right) b_i + \left(\frac{x_1 - x_1^i}{h_i} \right) b_{i+1}, \quad x_1^i \le x_1 \le x_1^{i+1},\ i = 1, \ldots, N, \tag{6.219}$$

where $h_i = x_1^{i+1} - x_1^i$, $f(x_1^i) = b_i$, and N denotes the number of partitions. Then, δf can be obtained by taking variation as

$$\delta f(x_1) = \left(\frac{x_1^{i+1} - x_1}{h_i} \right) \delta b_i + \left(\frac{x_1 - x_1^i}{h_i} \right) \delta b_{i+1}, \quad x_1^i \le x_1 \le x_1^{i+1},\ i = 1, \ldots, N. \tag{6.220}$$

If a cubic spline function is employed to parameterize Γ_1, with $f(x_1^i) = b_i$, the boundary can be expressed as [49]

$$f(x_1) = \frac{M_{i+1}}{6h_i}(x_1 - x_1^i)^3 + \frac{M_i}{6h_i}(x_1^{i+1} - x_1)^3 + \left(\frac{b_{i+1}}{h_i} - \frac{M_{i+1}h_i}{6}\right)(x_1 - x_1^i)$$
$$+ \left(\frac{b_i}{h_i} - \frac{M_i h_i}{6}\right)(x_1^{i+1} - x_1), \qquad x_1^i \le x_1 \le x_1^{i+1},\ i = 1, \ldots, N, \tag{6.221}$$

where $M_i = f''(x_1^i)$ is obtained by solving a system of equations for M_i ($i = 1, 2, \ldots, N+1$) [49]. As a result, the variation of f is

$$\delta f(x_1) = \sum_{j=1}^{N+1} \left\{ \left[\frac{(x_1 - x_1^i)^3}{6h_i} - \frac{h_i}{6}(x_1 - x_1^i) \right] \frac{\partial M_{i+1}}{\partial b_j} + \left[\frac{(x_1^{i+1} - x_1)^3}{6h_i} - \frac{h_i}{6}(x_1^{i+1} - x_1) \right] \frac{\partial M_i}{\partial b_j} + \delta_{i+1,j} \frac{x_1 - x_1^i}{h_i} + \delta_{i,j} \frac{x_1^{i+1} - x_1}{h_i} \right\} \delta b_j,$$
$$x_1^i \le x_1 \le x_1^{i+1},\ i = 1, \ldots, N, \tag{6.222}$$

where δ_{ij} equals one if $i = j$, and is otherwise set at zero. The cubic spline function has two continuous derivatives and a minimum mean curvature property [49]. It also possesses globally controlled properties. Unlike (6.220) for a piecewise-linear function, with a cubic spline function as in (6.222) the perturbation of any design variable $\boldsymbol{b}$ will globally perturb $f(x_1)$.

By using the result from either (6.220) or (6.222), and by expressing the boundary of finite element Ω_k in terms of $\boldsymbol{b}$, (6.218) can be expressed as

$$\psi_k' = \frac{\partial \psi_k}{\partial \boldsymbol{b}} \delta \boldsymbol{b}, \tag{6.223}$$

where $\partial \psi_k / \partial \boldsymbol{b}$ is the desired design sensitivity coefficient for the constraint ψ_k.

For comparative purposes, several different finite elements are used in the numerical calculation of shape design sensitivity: a constant stress triangular (CST), a linear stress triangular (LST), and an eight-node isoparametric (ISP) element. For the ISP element, stresses and strains are evaluated at Gauss points, and boundary stresses and strains are calculated by linearly extrapolating from those Gauss points [104] and [105].

The tension bar is modeled by using triangular and quadrilateral finite elements, as shown in Fig 6.17. Height of the varied boundary Γ_1 is chosen as the design variable, and a piecewise linear boundary parameterization is used in all cases. A cubic spline function is used for the ISP model. For the CST model, 190 elements, 117 nodal points, and 214 degrees of freedom are used. The LST model contains 190 elements, 423 nodal points, and 808 degrees of freedom, while the ISP model contains 111 elements, 384 nodal points, and 716 degrees of freedom. Young's modulus is $E = 30.0 \times 10^6$ psi, Poisson's ratio is $\nu = 0.293$, and the allowable stress is $\sigma_a = 120$ psi. The initial design is chosen as follows:

$$\boldsymbol{b} = [5.55 \quad 5.1 \quad 4.65 \quad 4.2 \quad 3.75 \quad 3.3 \quad 2.85 \quad 2.4 \quad 1.95]^T,$$

which produces a straight boundary for Γ_1, as shown in Fig. 6.17.

In order to compare the accuracy of results obtained with different finite elements, the same small region is used in evaluating the stress functional. As shown in Fig 6.17(b), the small regions selected are located next to variable boundary Γ_1, where high stress occurs. In the ISP model, the characteristic function is applied to each quadrilateral element, while it is applied to the four triangular elements in all other models.

Those numerical results with a 0.1% design change, i.e., $\delta\boldsymbol{b} = \boldsymbol{b} \times 10^{-3}$, are presented in Table 6.1. The abbreviation "ISPS" stands for those isoparametric elements that have a cubic spline function representation for variable boundary Γ_1. As can be seen, the LST model provides good sensitivity results for every region except number 10, whereas the ISP and ISPS models provide good results in every region except number 1. Region 1 has the lowest stress, while region 10 has the highest. When using these results for optimization purposes, those from ISP or ISPS models are preferable to those from the LST model. As expected, results from the CST model are the least accurate, since this model cannot provide accurate stress information on boundary Γ_1.

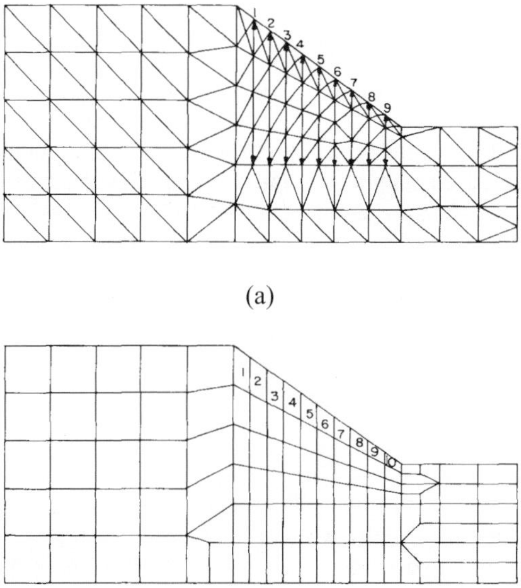

Figure 6.17. Finite element model of fillet: (a) triangular element model, numbers denote node height as a design variable; (b) isoparametric element model, numbers denote the region where the sensitivity is verified.

Table 6.1. Comparison of design sensitivity ($\psi'_k/\Delta\psi_k \times 100$)%.

Region	CST	LST	ISP	ISPS
1	1402.9	108.9	43.3	65.9
2	45.3	99.6	104.6	105.9
3	57.9	99.2	103.2	101.9
4	64.2	99.2	103.4	103.6
5	67.5	99.2	102.8	102.6
6	68.6	99.2	101.8	101.7
7	68.3	99.1	100.0	100.4
8	70.1	99.1	98.4	97.4
9	79.3	98.3	105.2	104.9
10	183.6	87.0	102.8	104.1

6.3 Eigenvalue Shape Design Sensitivity Analysis

The examples presented in Section 3.2 demonstrate that such eigenvalues as natural frequencies of a vibrating structure depend on the structural shape. As in Section 5.3, the objective of this section is to obtain eigenvalue sensitivity with respect to the shape design. As in Chapter 5, no adjoint equations are necessary for conservative systems, and eigenvalue design sensitivity can be expressed directly in terms of eigenvalues, eigenvectors, and variations in bilinear forms. The differentiability of simple eigenvalues and the directional differentiability of repeated eigenvalues are used to obtain explicit formulas for both simple and repeated eigenvalue design sensitivity analysis. Numerical examples of eigenvalue sensitivity computation are also presented.

6.3.1 Differentiability of Bilinear Forms and Eigenvalues

The differentiability of the eigenvalue problems treated in Section 3.2 is proved in Section 3.5 in [5]. The purpose of this section is to summarize important results that are needed for eigenvalue design sensitivity, and to illustrate that repeated eigenvalues are only directionally differentiable.

As was shown in Section 3.2, a vibrating structure's eigenvalue ζ_τ on a deformed domain Ω_τ is determined by a variational equation of the form

$$\begin{aligned} a_{\Omega_\tau}(y_\tau, \bar{y}_\tau) &\equiv \iint_{\Omega_\tau} c(y_\tau, \bar{y}_\tau)\, d\Omega_\tau \\ &= \zeta_\tau \iint_{\Omega_\tau} e(y_\tau, \bar{y}_\tau)\, d\Omega_\tau \equiv \zeta_\tau d_{\Omega_\tau}(y_\tau, \bar{y}_\tau), \quad \forall \bar{y}_\tau \in Z_\tau, \end{aligned} \tag{6.224}$$

where $Z_\tau \subset H^m(\Omega_\tau)$ is the space of kinematically admissible displacements, and $c(\bullet,\bullet)$ and $e(\bullet,\bullet)$ are symmetric bilinear mappings. Since (6.224) is homogeneous in eigenfunction y_τ, a normalizing condition must be used to define a unique eigenfunction. The normalizing condition is

$$d_{\Omega_\tau}(y_\tau, y_\tau) = 1. \tag{6.225}$$

The energy bilinear form on the left side of (6.224) is the same as the bilinear form in those static problems treated in Section 6.2. Therefore, it has the same differentiability properties discussed in that section. The bilinear form $d_\Omega(\bullet,\bullet)$ on the right side of (6.224) represents mass effects in a vibration problem and geometric effects in a buckling problem. Except for the column buckling problem, $d_\Omega(\bullet,\bullet)$ is even more regular than the energy bilinear form in its dependence on the design and on the eigenfunction.

Simple Eigenvalues

It is shown in Section 3.5.5 in [5] that the simple eigenvalue ζ is differentiable with respect to the design. It was shown by Kato [53] that the corresponding eigenfunction $\boldsymbol{y}$ is also differentiable. In fact, material derivatives of both the eigenvalue and the eigenfunction are linear in design velocity $\boldsymbol{V}$; hence, they are Fréchet derivatives. As in the static response problem, linearity and continuity of the mapping $\boldsymbol{V} \to \dot{\boldsymbol{y}}$ only allows (according to Theorem 3.5.3 in Section 3.5.7 in [5]) the use of normal component V_n of velocity field $\boldsymbol{V}$ in order to derive the material derivative, as in (6.38).

By using the material derivative formula from (6.38), and by noting that partial derivatives with respect to τ and $\boldsymbol{x}$ commute with each other, both sides of (6.224) are differentiated with respect to the design, to obtain

$$\begin{aligned}[a_\Omega(y,\bar{y})]' &\equiv a'_V(y,\bar{y}) + a_\Omega(\dot{y},\bar{y}) \\ &= \zeta' d_\Omega(y,\bar{y}) + \zeta[d'_V(y,\bar{y}) + d_\Omega(\dot{y},\bar{y})] \\ &\equiv \zeta' d_\Omega(y,\bar{y}) + \zeta[d_\Omega(y,\bar{y})]', \qquad \forall \bar{y} \in Z,\end{aligned} \tag{6.226}$$

where the variations of each bilinear form can be obtained by using the relation in (6.8) as

$$\begin{aligned}[a_\Omega(y,\bar{y})]' &= \iint_\Omega [c(y',\bar{y}) + c(y,\bar{y}')]\,d\Omega + \int_\Gamma c(y,\bar{y})V_n\,d\Gamma \\ &= \iint_\Omega [c(\dot{y} - \nabla y V,\bar{y}) + c(y,\dot{\bar{y}} - \nabla\bar{y}V)]\,d\Omega + \int_\Gamma c(y,\bar{y})V_n\,d\Gamma\end{aligned} \tag{6.227}$$

and

$$\begin{aligned}[d_\Omega(y,\bar{y})]' &= \iint_\Omega [e(y',\bar{y}) + e(y,\bar{y}')]\,d\Omega + \int_\Gamma e(y,\bar{y})V_n\,d\Gamma \\ &= \iint_\Omega [e(\dot{y} - \nabla y V,\bar{y}) + e(y,\dot{\bar{y}} - \nabla\bar{y}V)]\,d\Omega + \int_\Gamma e(y,\bar{y})V_n\,d\Gamma.\end{aligned} \tag{6.228}$$

As in (6.61), the fact that the partial derivatives of the coefficients (such as cross-sectional area, thickness, etc.) in bilinear mappings $c(\bullet,\bullet)$ and $e(\bullet,\bullet)$ are zero is used in derivation of (6.227) and (6.228). As in the static response problem, perturbation $\bar{y}_\tau$ is selected such that $\bar{y}_\tau(x + \tau V(x)) = \bar{y}(x)$. As explained in the paragraph preceding (6.12), since $H^m(\Omega)$ is preserved by $T(x,\tau)$ if $\bar{y} \in Z$ is arbitrary, then $\bar{y}_\tau$ is an arbitrary element of Z_τ. Also, from (6.8), the following relation can be obtained:

$$\dot{\bar{y}} = \bar{y}' + \nabla\bar{y}V = 0. \tag{6.229}$$

Thus, material derivative $\dot{\bar{y}}$ will be ignored in the following derivations.

Those explicitly dependent terms on the design in (6.226), namely, $a'_V(y,\bar{y})$ and $d'_V(y,\bar{y})$ can be obtained by suppressing the implicit terms from the formulas in (6.227) and (6.228). The following formulas can therefore be obtained:

$$\begin{aligned}a'_V(y,\bar{y}) &= -\iint_\Omega [c(\nabla y V,\bar{y}) + c(y,\nabla\bar{y}V)]\,d\Omega + \int_\Gamma c(y,\bar{y})V_n\,d\Gamma \\ &= -\iint_\Omega [c(\nabla y V,\bar{y}) + c(y,\nabla\bar{y}V) - div\big(c(y,\bar{y})V\big)]\,d\Omega\end{aligned} \tag{6.230}$$

and

$$\begin{aligned}d'_V(y,\bar{y}) &= -\iint_\Omega [e(\nabla y V,\bar{y}) + e(y,\nabla\bar{y}V)]\,d\Omega + \int_\Gamma e(y,\bar{y})V_n\,d\Gamma \\ &= -\iint_\Omega [e(\nabla y V,\bar{y}) + e(y,\nabla\bar{y}V) - div\big(e(y,\bar{y})V\big)]\,d\Omega.\end{aligned} \tag{6.231}$$

Note that the eigenvalue sensitivity equation from (6.226) contains the material derivative of the eigenvalue, as well as that of the eigenfunction. However, the effect of $\dot{y}$ will be eliminated in the following way. Because $\bar{y} \in Z$, (6.226) may be evaluated with $\bar{y} = y$ to obtain

$$\zeta' d_\Omega(y,y) = a'_V(y,y) - \zeta d'_V(y,y) + [a_\Omega(y,\dot{y}) - \zeta d_\Omega(y,\dot{y})]. \tag{6.232}$$

Since $\dot{y} \in Z$, as explained in the paragraph following (6.67), the terms within the brackets are zero. Furthermore, due to the normalizing condition, a simplified expression for eigenvalue sensitivity is obtained as

$$\begin{aligned}\zeta' &= a'_V(y,y) - \zeta d'_V(y,y)\\ &= 2\iint_\Omega [-c(y,\nabla yV) + \zeta e(y,\nabla yV)]\,d\Omega\\ &\quad + \int_\Gamma [c(y,y) - \zeta e(y,y)]V_n\,d\Gamma\\ &= 2\iint_\Omega [-c(y,\nabla yV) + \zeta e(y,\nabla yV)]\,d\Omega\\ &\quad + \iint_\Omega div\big([c(y,y) - \zeta e(y,y)]V\big)\,d\Omega.\end{aligned} \tag{6.233}$$

As with the static response problem, domain integrals in the second term on the right can be transformed into boundary integrals by using the variational identities given in Section 3.2 for each structural component and boundary condition. This will be done below for each class of problem encountered. However, the result of the domain method given in the third term is applicable for all boundary conditions.

Note that the directional derivative of the eigenvalue is linear in V, since the variations of the bilinear forms on the right side of (6.233) are linear in V. As noted in Section 5.3, the validity of this result rests on the existence of eigenvalue and eigenfunction derivatives.

Repeated Eigenvalues

Now consider the situation in which eigenvalue ζ has a multiplicity of $s > 1$ at Ω, that is,

$$\left.\begin{aligned} a_\Omega(y^i,\bar{y}) &= \zeta d_\Omega(y^i,\bar{y}), \quad \forall \bar{y} \in Z\\ d_\Omega(y^i,y^j) &= \delta_{ij}\end{aligned}\right\} \quad i,j = 1,2,\ldots,s. \tag{6.234}$$

It has been shown in Section 3.5 in [5] that repeated eigenvalue ζ is a continuous function of the design, but that corresponding eigenfunctions y^i are not. It will also be shown that when eigenvalue ζ is repeated s times, it is only directionally differentiable, and that the directional derivatives $\zeta_i'(V)$ in the V direction are the eigenvalues of the $s \times s$ matrix $\mathcal{M}$, with the following elements:

$$\begin{aligned}\mathcal{M}_{ij} &= a'_V(y^i,y^j) - \zeta d'_V(y,y)\\ &= \iint_\Omega [-c(\nabla y^i V, y^j) - c(y^i,\nabla y^j V) + \zeta e(y^i,\nabla y^j V) + \zeta e(\nabla y^i V, y^j)]\,d\Omega\\ &\quad + \int_\Gamma [c(y^i,y^j) - \zeta e(y^i,y^j)]V_n\,d\Gamma, \qquad i,j = 1,2,\cdots,s.\end{aligned} \tag{6.235}$$

The notation $\zeta_i'(V)$ is used to emphasize the dependence of the directional derivative on V. As with the simple eigenvalue, the domain integral in (6.235) can be transformed into boundary integrals by using the variational identities given in Section 3.2 for each structural component and boundary condition.

If the d_Ω-orthonormal basis $\{y^i\}_{i=1,\ldots,s}$ of the eigenspace is changed, then matrix $\mathcal{M}$ also changes, but its eigenvalues remain the same. As mentioned in Section 5.3.1, the directional derivatives $\zeta_i'(V)$ are not generally linear in V, although each $\mathcal{M}_{ij}$ is linear in V. Other results on the directional derivatives of repeated eigenvalues from Section 5.3.1 remain valid in this section. When $s = 2$, the directional derivatives of a double eigenvalue are given by

$$\zeta_i'(V) = \frac{1}{2}\left\{(\mathcal{M}_{11} + \mathcal{M}_{22}) \pm [(\mathcal{M}_{11} + \mathcal{M}_{22})^2 - 4(\mathcal{M}_{11}\mathcal{M}_{22} - \mathcal{M}_{12}^2)]^{1/2}\right\}, \tag{6.236}$$

where $i = 1$ corresponds to the minus sign, and $i = 2$ corresponds to the plus sign. Another expression for directional derivatives is

$$\zeta_1'(V) = \cos^2\phi(V)\mathcal{M}_{11} + \sin 2\phi(V)\mathcal{M}_{12} + \sin^2\phi(V)\mathcal{M}_{22} \tag{6.237}$$

$$\zeta_2'(V) = \sin^2\phi(V)\mathcal{M}_{11} - \sin 2\phi(V)\mathcal{M}_{12} + \cos^2\phi(V)\mathcal{M}_{22}, \tag{6.238}$$

where the eigenvector rotation angle ϕ is given as

$$\phi(V) = \frac{1}{2}\arctan\left[\frac{2\mathcal{M}_{12}}{\mathcal{M}_{11} - \mathcal{M}_{22}}\right]. \tag{6.239}$$

6.3.2 Boundary and Domain Methods of Eigenvalue Design Sensitivity

The beam, column, membrane, and plate problems in Section 3.2 are used here as examples for eigenvalue design sensitivity analysis. Both the boundary and domain methods are presented in this section.

Vibration of a Beam

Consider the vibrating beam in Section 3.2, with cross-sectional area $A(x) \geq A_0 \geq 0$, moment of inertia $I(x)$, and Young's modulus E. The eigenvalue sensitivity can be expressed by using the formula in (6.233) as

$$\begin{aligned}\zeta' = {} & 2\int_0^l [-EIy_{,11}(y_{,1}V)_{,11} + \zeta\rho Ay(y_{,1}V)]\,dx \\ & + [EI(y_{,11})^2 - \zeta\rho Ay^2]V\Big|_0^l .\end{aligned} \tag{6.240}$$

To arrive at the boundary representation, the variational identity in (3.67) may be used, identifying $(y_{,1}V)$ in the domain integral term with $\bar{y}$ in (3.67) to obtain

$$\begin{aligned}\zeta' = {} & -2[EIy_{,11}(y_{,1}V)_{,1} - (EIy_{,11})_{,1}(y_{,1}V)]\Big|_0^l \\ & + [EI(y_{,11})^2 - \zeta\rho Ay^2]V\Big|_0^l .\end{aligned} \tag{6.241}$$

By applying the boundary conditions from (3.66) for the clamped-clamped beam, but noting that beam length l is not normalized in this chapter, (6.241) has the following simplified form:

$$\zeta' = -EI(y_{,11})^2 V\Big|_0^l . \tag{6.242}$$

It is interesting to note that since the coefficient of velocity V is negative, the natural frequency decreases as the boundary moves outward, which is physically apparent.

For other boundary conditions in (3.19) through (3.21), the design sensitivity formula in (6.241) is still valid. The following eigenvalue sensitivity expressions can be obtained by applying corresponding boundary conditions:

1. Simply supported:

$$\zeta' = 2EIy_{,111}y_{,1}V\Big|_0^l , \tag{6.243}$$

2. Cantilevered:

$$\zeta' = EI(y_{,11})^2 V\Big|_{x=0} - \zeta\rho A y^2 V\Big|_{x=l}, \tag{6.244}$$

3. Clamped–simply supported:

$$\zeta' = EI(y_{,11})^2 V\Big|_{x=0} + 2EIy_{,111}y_{,1}V\Big|_{x=l}. \tag{6.245}$$

Using the domain method results given in (6.230) through (6.232), the design sensitivity expression is obtained as

$$\zeta' = -\int_0^l [3EIy_{,11}^2 V_{,1} + 2EIy_{,1}y_{,11}V_{,11} + \zeta\rho A y^2 V_{,1}]dx. \tag{6.246}$$

Note that, as in the static response case, using the domain method, only one design sensitivity equation is enough for all boundary conditions, which is very much attractive for numerical implementation. In addition, note that the domain method requires integrable second-order design velocity in the domain because the sensitivity expression in (6.246) includes $V_{,11}$.

Buckling of a Column

Consider the column buckling problem from Section 3.2, with cross-sectional area A, moment of inertia $I(x)$, and Young's modulus E. By substituting variations of the bilinear forms into (6.233), the simple eigenvalue sensitivity (i.e., the buckling load) is obtained as

$$\begin{aligned}\zeta' &= 2\int_0^l [-EIy_{,11}(y_{,1}V)_{,11} + \zeta y_{,1}(y_{,1}V)_{,1}]dx \\ &\quad + [EI(y_{,11})^2 - \zeta(y_{,1})^2]V\Big|_0^l.\end{aligned} \tag{6.247}$$

Paralleling the beam vibration problem, the domain integral is transformed into boundary values as

$$\begin{aligned}\zeta' &= -2[EIy_{,11}(y_{,1}V)_{,1} - (EIy_{,11})_{,1}(y_{,1}V) - \zeta y_{,1}(y_{,1}V)]\Big|_0^l \\ &\quad + [EI(y_{,11})^2 - \zeta(y_{,1})^2]V\Big|_0^l.\end{aligned} \tag{6.248}$$

Using the boundary conditions for a clamped-clamped column in (3.65), (6.248) becomes

$$\zeta' = -EI(y_{,11})^2 V\Big|_0^l. \tag{6.249}$$

As with beam vibration, the coefficient of velocity V is negative. Hence, the buckling load decreases as the boundary moves outward.

As in the case of a vibrating beam problem, if other boundary conditions exist, then the following buckling load sensitivities can be obtained:

1. Simply supported:

$$\zeta' = [2EIy_{,111}y_{,1} + \zeta(y_{,1})^2]V\Big|_0^l, \tag{6.250}$$

2. Cantilevered:

$$\zeta' = EI(y_{,11})^2 V\big|_{x=0} - \zeta(y_{,1})^2 V\big|_{x=l}, \tag{6.251}$$

3. Clamped–simply supported:

$$\zeta' = EI(y_{,11})^2 V\big|_{x=0} + [2EIy_{,111}y_{,1} + \zeta(y_{,1})^2]V\big|_{x=l}. \tag{6.252}$$

For eigenvalue repeated s number of times, and using (6.235), we obtain

$$\begin{aligned}\mathcal{M}_{ij} &= \int_0^l [-EIy^j_{,11}(y^i_{,1}V)_{,11} - EIy^i_{,11}(y^j_{,1}V)_{,11} + \zeta y^j_{,1}(y^i_{,1}V)_{,1} + \zeta y^i_{,1}(y^j_{,1}V)_{,1}]dx \\ &\quad + [EIy^i_{,11}y^j_{,11} - \zeta y^i_{,1}y^j_{,1}]V\big|_0^l, \qquad i,j = 1,2,\ldots,s.\end{aligned} \tag{6.253}$$

Using the variational identity from (3.71) repeated twice in (6.253),

$$\begin{aligned}\mathcal{M}_{ij} &= -[EIy^j_{,11}(y^i_{,1}V)_{,1} - (EIy^j_{,11})_{,1}(y^i_{,1}V) - \zeta y^j_{,1}(y^i_{,1}V)]\big|_0^l \\ &\quad - [EIy^i_{,11}(y^j_{,1}V)_{,1} - (EIy^i_{,11})_{,1}(y^j_{,1}V) - \zeta y^i_{,1}(y^j_{,1}V)]\big|_0^l \\ &\quad + [EIy^i_{,11}y^j_{,11} - \zeta y^i_{,1}y^j_{,1}]V\big|_0^l, \qquad i,j = 1,2,\ldots,s.\end{aligned} \tag{6.254}$$

As with a simple eigenvalue, the result in (6.254) is valid for the boundary conditions given in (3.65) and (3.19) through (3.21). To obtain $\mathcal{M}_{ij}$ for each case, these boundary conditions may apply to (6.254). In the case of a double eigenvalue (s = 2), the directional derivatives of the repeated eigenvalue can be obtained from (6.237) and (6.238), where rotation angle ϕ is given in (6.239).

Using the domain method results given in (6.230) through (6.232), the design sensitivity expression is obtained as

$$\zeta' = -\int_0^l [3EIy^2_{,11}V_{,1} + 2EIy_{,1}y_{,11}V_{,11} - \zeta y^2_{,1}V_{,1}]dx. \tag{6.255}$$

For repeated eigenvalues, similar results can be obtained using (6.236).

Vibration of a Membrane

Consider the vibrating membrane problem in Fig. 3.1, with mass density ρ. By using (6.233) and the fact that y = 0 on Γ for a simple eigenvalue, the following eigenvalue sensitivity expression is obtained:

$$\begin{aligned}\zeta' &= 2\iint_\Omega [-T\nabla y^T \nabla(\nabla y^T V) + \zeta h y(\nabla y^T V)]d\Omega \\ &\quad + T\int_\Gamma (\nabla y^T \nabla y)V_n\, d\Gamma.\end{aligned} \tag{6.256}$$

Applying the variational identity in (3.76) to (6.256) and identifying $(\nabla y^T V)$ in the domain integral of (6.256) with $\bar{y}$ in (3.76), ζ' is expressed in boundary integrals as

$$\zeta' = -2T\int_\Gamma \frac{\partial y}{\partial n}(\nabla y^T V)d\Gamma + T\int_\Gamma (\nabla y^T \nabla y)V_n\, d\Gamma. \tag{6.257}$$

Since y = 0 on Γ, the gradient of y only has the normal component, as $\nabla y = (\partial y/\partial n)\boldsymbol{n}$ on Γ; thus, the above relation can be further simplified as

$$\zeta' = -T \int_\Gamma \left(\frac{\partial y}{\partial n} \right)^2 V_n \, d\Gamma. \tag{6.258}$$

As noted in Section 6.3.1, the eigenvalue design sensitivity in (6.258) is expressed as a boundary integral, and only the normal movement (V_n) of the boundary appears.

It is interesting to note that since the coefficient of (V_n) is negative, the frequency decreases as the boundary moves outward, which is physically apparent. Moreover, moving the boundary outward near a high normal derivative is the most effective way to decrease the fundamental frequency.

For an eigenvalue repeated s number of times, using (6.235) and the fact that $y^i = 0$ on $\Gamma(i = 1, 2, \ldots, s)$ we obtain

$$\begin{aligned} \mathcal{M}_{ij} &= -T \iint_\Omega [\nabla y^{jT} \nabla (\nabla y^{iT} \boldsymbol{V}) + \nabla y^{iT} \nabla (\nabla y^{jT} \boldsymbol{V})] \, d\Omega \\ &\quad + \zeta h \iint_\Omega [y^j (\nabla y^{iT} \boldsymbol{V}) + y^i (\nabla y^{jT} \boldsymbol{V})] \, d\Omega \\ &\quad + T \int_\Gamma (\nabla y^{iT} \nabla y^j) V_n \, d\Gamma, \qquad i, j = 1, 2, \ldots, s. \end{aligned} \tag{6.259}$$

Applying the variational identity from (3.13) twice to (6.259) we obtain

$$\begin{aligned} \mathcal{M}_{ij} &= -T \int_\Gamma \left[\frac{\partial y^j}{\partial n} (\nabla y^{iT} \boldsymbol{V}) + \frac{\partial y^i}{\partial n} (\nabla y^{jT} \boldsymbol{V}) \right] d\Gamma \\ &\quad + T \int_\Gamma (\nabla y^{iT} \nabla y^j) V_n \, d\Gamma, \qquad i, j = 1, 2, \ldots, s. \end{aligned} \tag{6.260}$$

Since $y^i = 0$ on Γ, $\nabla y^i = (\partial y^i / \partial n)\boldsymbol{n}$ on Γ. Thus, the above equation can be simplified to

$$\mathcal{M}_{ij} = -T \int_\Gamma \left(\frac{\partial y^i}{\partial n} \right) \left(\frac{\partial y^j}{\partial n} \right) V_n \, d\Gamma, \quad i, j = 1, 2, \ldots, s. \tag{6.261}$$

Now, consider a double eigenvalue at Ω (i.e., $s = 2$). The directional derivatives of the repeated eigenvalue are given by (6.237) and (6.238) as

$$\begin{aligned} \zeta_1'(\boldsymbol{V}) &= -T \int_\Gamma \left[\cos^2 \phi(\boldsymbol{V}) \left(\frac{\partial y^1}{\partial n} \right)^2 + \sin 2\phi(\boldsymbol{V}) \left(\frac{\partial y^1}{\partial n} \right) \left(\frac{\partial y^2}{\partial n} \right) \right. \\ &\qquad \left. + \sin^2 \phi(\boldsymbol{V}) \left(\frac{\partial y^2}{\partial n} \right)^2 \right] V_n \, d\Gamma \\ \zeta_2'(\boldsymbol{V}) &= -T \int_\Gamma \left[\sin^2 \phi(\boldsymbol{V}) \left(\frac{\partial y^1}{\partial n} \right)^2 - \sin 2\phi(\boldsymbol{V}) \left(\frac{\partial y^1}{\partial n} \right) \left(\frac{\partial y^2}{\partial n} \right) \right. \\ &\qquad \left. + \cos^2 \phi(\boldsymbol{V}) \left(\frac{\partial y^2}{\partial n} \right)^2 \right] V_n \, d\Gamma, \end{aligned} \tag{6.262}$$

where rotation angle ϕ is obtained from (6.239) as

$$\phi(\boldsymbol{V}) = \frac{1}{2} \arctan \left[\frac{\int_\Gamma \frac{\partial y^1}{\partial n} \frac{\partial y^2}{\partial n} V_n d\Gamma}{\int_\Gamma \left(\frac{\partial y^1}{\partial n} \right)^1 V_n d\Gamma - \int_\Gamma \left(\frac{\partial y^2}{\partial n} \right)^2 V_n d\Gamma} \right]. \tag{6.263}$$

It is clear from (6.262) that the directional derivatives of the repeated eigenvalues are not linear in $\boldsymbol{V}$; hence, they are not Fréchet differentiable.

Using the domain method results given in (6.230) through (6.232), the design sensitivity expression is obtained as

$$\begin{aligned}\zeta' = &-T\iint_{\Omega}[2\nabla y^T\nabla(\nabla y^T\boldsymbol{V}) - div(\nabla y^T\nabla y\boldsymbol{V})]\,d\Omega \\ &+\zeta\iint_{\Omega} h[2y(\nabla y^T\boldsymbol{V}) - div(y^2\boldsymbol{V})]\,d\Omega.\end{aligned} \tag{6.264}$$

For repeated eigenvalues, similar results can be obtained using (6.236).

Vibration of a Plate

Consider the vibrating plate from Section 3.2, with thickness h, Young's modulus E, and mass density ρ. Using (6.233), the sensitivity expression of the natural frequency is written as

$$\begin{aligned}\zeta' = 2&\iint_{\Omega}\left[-\boldsymbol{\kappa}(y)^T\boldsymbol{C}^b\boldsymbol{\kappa}(\nabla y^T\boldsymbol{V}) + \zeta\rho h y(\nabla y^T\boldsymbol{V})\right]d\Omega \\ &+\int_{\Gamma}\left[\boldsymbol{\kappa}(y)^T\boldsymbol{C}^b\boldsymbol{\kappa}(y) - \zeta\rho h y^2\right]V_n\,d\Gamma.\end{aligned} \tag{6.265}$$

Using the variational identity in (3.81), and identifying $(\nabla y^T\boldsymbol{V})$ in the domain integral of (6.265) with $\bar{y}$ in (3.81), eigenvalue sensitivity can be expressed by boundary integrals as

$$\begin{aligned}\zeta' = -2&\int_{\Gamma}(\nabla y^T\boldsymbol{V})Ny\,d\Gamma - 2\int_{\Gamma}\frac{\partial}{\partial n}(\nabla y^T\boldsymbol{V})My\,d\Gamma \\ &+\int_{\Gamma}\left[\boldsymbol{\kappa}(y)^T\boldsymbol{C}^b\boldsymbol{\kappa}(y) - \zeta\rho h y^2\right]V_n\,d\Gamma.\end{aligned} \tag{6.266}$$

As with the static response problem, sensitivity formulas resulting from different kinds of boundary conditions can be obtained as follows:

1. Clamped:

$$\zeta' = \int_{\Gamma_C}\left[-2\left(\frac{\partial^2 y}{\partial n^2}\right)^2 + \boldsymbol{\kappa}(y)^T\boldsymbol{C}^b\boldsymbol{\kappa}(y)\right]V_n\,d\Gamma, \tag{6.267}$$

2. Simply supported:

$$\zeta' = \int_{\Gamma_S}\left[-2\left(\frac{\partial y}{\partial n}\right)Ny + \boldsymbol{\kappa}(y)^T\boldsymbol{C}^b\boldsymbol{\kappa}(y)\right]V_n\,d\Gamma, \tag{6.268}$$

3. Free edge:

$$\zeta' = \int_{\Gamma_F}\left[\boldsymbol{\kappa}(y)^T\boldsymbol{C}^b\boldsymbol{\kappa}(y) - \zeta\rho h y^2\right]V_n\,d\Gamma. \tag{6.269}$$

For multiple boundary $\Gamma = \Gamma_C \cup \Gamma_S \cup \Gamma_F$, the complete design sensitivity formula can be obtained by adding the terms from (6.267) through (6.269).

For an s-times repeated eigenvalue ζ on Ω [74], directional derivatives $\zeta_i'(\boldsymbol{V})$ in direction $\boldsymbol{V}$ are the eigenvalues of the $s \times s$ matrix $\mathcal{M}$ with elements

$$\begin{aligned}\mathcal{M}_{ij} &= \iint_{\Omega}\left[-\boldsymbol{\kappa}(y^i)^T \boldsymbol{C}^b \boldsymbol{\kappa}(\nabla y^{jT}\boldsymbol{V}) - \boldsymbol{\kappa}(y^j)^T \boldsymbol{C}^b \boldsymbol{\kappa}(\nabla y^{iT}\boldsymbol{V})\right] d\Omega \\ &+ \iint_{\Omega}\left[\zeta\rho h y^i(\nabla y^{jT}\boldsymbol{V}) + \zeta\rho h y^j(\nabla y^{iT}\boldsymbol{V})\right] d\Omega \\ &+ \int_{\Gamma}\left[\boldsymbol{\kappa}(y^i)^T \boldsymbol{C}^b \boldsymbol{\kappa}(y^j) - \zeta\rho h y^i y^j\right] V_n \, d\Gamma.\end{aligned} \tag{6.270}$$

By using the variational identity in (3.30) twice, the domain integral in (6.270) is converted into boundary integrals as

$$\begin{aligned}\mathcal{M}_{ij} &= -\int_{\Gamma}\left[(\nabla y^{iT}\boldsymbol{V})Ny^j + (\nabla y^{jT}\boldsymbol{V})Ny^i\right] d\Gamma \\ &- \int_{\Gamma}\left[\frac{\partial}{\partial n}(\nabla y^{iT}\boldsymbol{V})My^j + \frac{\partial}{\partial n}(\nabla y^{jT}\boldsymbol{V})My^i\right] d\Gamma \\ &+ \int_{\Gamma}\left[\boldsymbol{\kappa}(y^i)^T \boldsymbol{C}^b \boldsymbol{\kappa}(y^j) - \zeta\rho h y^i y^j\right] V_n \, d\Gamma.\end{aligned} \tag{6.271}$$

As with a simple eigenvalue, the result from (6.271) is valid for the boundary conditions given in (3.26) through (3.28). To obtain $\mathcal{M}_{ij}$ for each condition, these boundary conditions can be applied to (6.271).

Using the domain method results given in (6.230) through (6.232)the design sensitivity expression is obtained as

$$\begin{aligned}\zeta' &= 2\iint_{\Omega}\left[-\boldsymbol{\kappa}(y)^T \boldsymbol{C}^b \boldsymbol{\kappa}(\nabla y^T \boldsymbol{V}) + \zeta\rho h y(\nabla y^T \boldsymbol{V})\right] d\Omega \\ &+ \iint_{\Omega} div\left([\boldsymbol{\kappa}(y)^T \boldsymbol{C}^b \boldsymbol{\kappa}(y) - \zeta\rho h y^2]\boldsymbol{V}\right) d\Omega.\end{aligned} \tag{6.272}$$

For repeated eigenvalues, similar results can be obtained using (6.236).

6.4 Frequency Response Problem

In this section, the domain method of shape design sensitivity of a frequency response problem in Section 2.6 is developed. It is presumed that the excitation frequency is independent of the design variables. The same adjoint equation as (5.240) is obtained for a size design problem using the complex conjugate of the adjoint variable. The regularity requirement of the design velocity field presented in Section 6.2.7 is discussed through numerical examples.

6.4.1 Design Sensitivity of Frequency Response

Consider a general performance measure that defines a variety of dynamic responses, as

$$\psi = \iint_{\Omega} g(z, \nabla z) \, d\Omega. \tag{6.273}$$

In the shape design problem, the structural domain is a design variable. Let that structure be perturbed in the direction of design velocity $\boldsymbol{V}(\boldsymbol{x})$, as explained in Section 6.1.1, and with control parameter τ. The performance measure at this perturbed domain Ω_τ is defined as ψ_τ. The shape variation of the performance measure in (6.273) is the same as differentiating ψ_τ with respect to parameter τ. The first variation of ψ with respect to the shape design variable is obtained as

$$\psi' = \iint_{\Omega}[g_{,z}\dot{z} + g_{,\nabla z} : \nabla\dot{z}] d\Omega + \iint_{\Omega}[-g_{,\nabla z} : (\nabla z \nabla \boldsymbol{V}) + g\,div\boldsymbol{V}] d\Omega. \tag{6.274}$$

From the design velocity field $\boldsymbol{V}(\boldsymbol{x})$ and the result of response analysis $\boldsymbol{z}(\boldsymbol{x})$, the second integral in (6.274) can be calculated through domain integration. The objective is to calculate the first integral in (6.274). As before, two methods can be introduced: the direct differentiation method and the adjoint variable method.

Direct Differentiation Method

The direct differentiation method calculates $\dot{\boldsymbol{z}}$ and $\nabla\dot{\boldsymbol{z}}$ from the shape variation of the state equation. The variational state equation for a frequency response problem (2.56) is rewritten as

$$-\omega^2 d_\Omega(\boldsymbol{z},\bar{\boldsymbol{z}}) + (1+j\varphi)a_\Omega(\boldsymbol{z},\bar{\boldsymbol{z}}) = \ell_\Omega(\bar{\boldsymbol{z}}), \qquad \forall \bar{\boldsymbol{z}}^* \in \mathbb{Z}, \tag{6.275}$$

where subscript Ω denotes the dependency on the domain shape. To simplify the explanation, structural damping in (2.57) is used with damping coefficient φ. For the shape design problem, damping coefficient φ is presumed to be fixed throughout the design change. The shape variation of (6.275) will provide the design sensitivity equation for $\dot{\boldsymbol{z}}$, and the performance sensitivity ψ' in (6.274) can be calculated using $\dot{\boldsymbol{z}}$, which is the direct differentiation method.

In Section 6.2, the material derivatives of $a_\Omega(\boldsymbol{z},\bar{\boldsymbol{z}})$ and $\ell_\Omega(\bar{\boldsymbol{z}})$ were rigorously developed for various structural components. The main difference in our present problem is the use of a complex variable, and thus $a_\Omega(\boldsymbol{z},\bar{\boldsymbol{z}})$ and $\ell_\Omega(\bar{\boldsymbol{z}})$ are sesquilinear and semilinear forms, respectively. The material derivative of the kinetic energy sesquilinear form is derived as

$$\begin{aligned}[d_\Omega(\boldsymbol{z},\bar{\boldsymbol{z}})]' &= \iint_\Omega \rho\dot{\boldsymbol{z}}^T\bar{\boldsymbol{z}}^*\, d\Omega + \iint_\Omega \rho\boldsymbol{z}^T\bar{\boldsymbol{z}}^*\, div\boldsymbol{V}\, d\Omega \\ &\equiv d_\Omega(\dot{\boldsymbol{z}},\bar{\boldsymbol{z}}) + d'_V(\boldsymbol{z},\bar{\boldsymbol{z}}),\end{aligned} \tag{6.276}$$

where $d'_V(\boldsymbol{z},\bar{\boldsymbol{z}})$ can be calculated from known $\boldsymbol{V}(\boldsymbol{x})$ and $\boldsymbol{z}(\boldsymbol{x})$.

Using (6.276) and the results from Section 6.2, the state equation (6.275) is differentiated with respect to the shape design variable to obtain the following design sensitivity equation:

$$\begin{aligned}&-\omega^2 d_\Omega(\dot{\boldsymbol{z}},\bar{\boldsymbol{z}}) + (1+j\varphi)a_\Omega(\dot{\boldsymbol{z}},\bar{\boldsymbol{z}}) \\ &\quad = \ell'_V(\bar{\boldsymbol{z}}) + \omega^2 d'_V(\boldsymbol{z},\bar{\boldsymbol{z}}) - (1+j\varphi)a'_V(\boldsymbol{z},\bar{\boldsymbol{z}}), \qquad \forall \bar{\boldsymbol{z}}^* \in \mathbb{Z}.\end{aligned} \tag{6.277}$$

Note that the left side of (6.277) is the same as that of (6.275), while the right side is different. Thus, the design sensitivity equation is the same as the original state equation with a different harmonic excitation, which is called a fictitious load. Using the same computational method as with response analysis, the design sensitivity equation from (6.277) can be solved for $\dot{\boldsymbol{z}}$, from which ψ' in (6.274) can be obtained.

Adjoint Variable Method

In order to calculate the first integral in (6.274) explicitly in terms of the design change, it is first necessary to define an adjoint equation, as

$$\begin{aligned}&-\omega^2 d_\Omega(\bar{\boldsymbol{\lambda}},\boldsymbol{\lambda}) + (1+j\varphi)a_\Omega(\bar{\boldsymbol{\lambda}},\boldsymbol{\lambda}) \\ &\quad = \iint_\Omega (g_{,\boldsymbol{z}}\bar{\boldsymbol{\lambda}} + g_{,\nabla\boldsymbol{z}} : \nabla\bar{\boldsymbol{\lambda}})\, d\Omega = \ell^a_\Omega(\bar{\boldsymbol{\lambda}}), \qquad \forall \bar{\boldsymbol{\lambda}} \in Z,\end{aligned} \tag{6.278}$$

where Z is the space of kinematically admissible virtual displacements, and a unique solution $\boldsymbol{\lambda}^*$ is desired. To take advantage of the adjoint equation, let us evaluate (6.278) at

$\bar{\lambda} = \dot{z} \in Z$, to obtain the following relation:

$$\begin{aligned}-\omega^2 d_\Omega(\dot{z},\lambda) + (1+j\varphi)a_\Omega(\dot{z},\lambda)\\ = \iint_\Omega (g_{,z}\dot{z} + g_{,\nabla z} : \nabla\dot{z})\, d\Omega.\end{aligned} \tag{6.279}$$

Similarly, evaluate the sensitivity equation (6.277) at $\bar{z}^* = \lambda^*$, to arrive at

$$\begin{aligned}-\omega^2 d_\Omega(\dot{z},\lambda) + (1+j\varphi)a_\Omega(\dot{z},\lambda)\\ = \ell'_V(\lambda) + \omega^2 d'_V(z,\lambda) - (1+j\varphi)a'_V(z,\lambda).\end{aligned} \tag{6.280}$$

Since the left sides of (6.279) and (6.280) are the same, the desired result can be obtained by creating a formula that only uses the right side of each equation as

$$\begin{aligned}\iint_\Omega (g_{,z}\dot{z} + g_{,\nabla z} : \nabla\dot{z})\, d\Omega\\ = \ell'_V(\lambda) + \omega^2 d'_V(z,\lambda) - (1+j\varphi)a'_V(z,\lambda).\end{aligned} \tag{6.281}$$

Note that the left side is now the same as the first integral in (6.274). Therefore, the shape design sensitivity formula for performance ψ is explicitly obtained as

$$\begin{aligned}\psi' &= \iint_\Omega (g_{,z}\dot{z} + g_{,\nabla z} : \nabla\dot{z})\, d\Omega\\ &\quad + \iint_\Omega [-g_{,\nabla z} : (\nabla z \nabla V) + g\,divV]\, d\Omega\\ &= \ell'_V(\lambda) + \omega^2 d'_V(z,\lambda) - (1+j\varphi)a'_V(z,\lambda)\\ &\quad + \iint_\Omega [-g_{,\nabla z} : (\nabla z \nabla V) + g\,divV]\, d\Omega.\end{aligned} \tag{6.282}$$

By using solution z from the state equation and solution λ^* from the adjoint equation, and with given design velocity field $V(\boldsymbol{x})$, (6.282) can be calculated through domain integration, which is the adjoint variable method.

Since the expression of $d'_V(z,\bar{z})$ is simple compared with that of the sesquilinear form $a'_V(z,\bar{z})$, and since other expressions have previously been derived, separate expressions for each structural component are not developed here.

6.4.2 Numerical Examples

In this section, numerical examples are presented to demonstrate the accuracy of shape design sensitivity. A cantilevered beam and a vehicle chassis frame are used to compute shape design sensitivity information using the adjoint variable method. Design velocity fields that satisfy the regularity conditions and domain integration are used to evaluate shape design sensitivity.

Cantilever Beam

Consider a cantilever beam that is 40 in long and subjected to a concentrated harmonic load of $f(x_1) = 10e^{j\omega t}$ at the tip (Fig. 6.18). The cross section has constant width b of 0.4 in and height h of 0.8 in. For numerical tests, finite element models with 2 and 20 elements are considered as coarse and refined models, respectively. Young's modulus is $E = 3.0 \times 10^7$ psi, mass density is $\rho = 7.34 \times 10^{-4}$, Poisson's ratio is $\nu = 0.3$, and the structural damping coefficient is $\varphi = 0.04$.

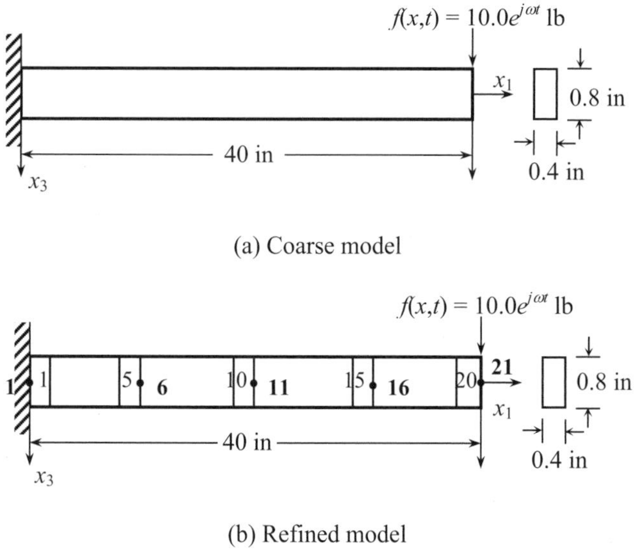

Figure 6.18. Cantilever beam.

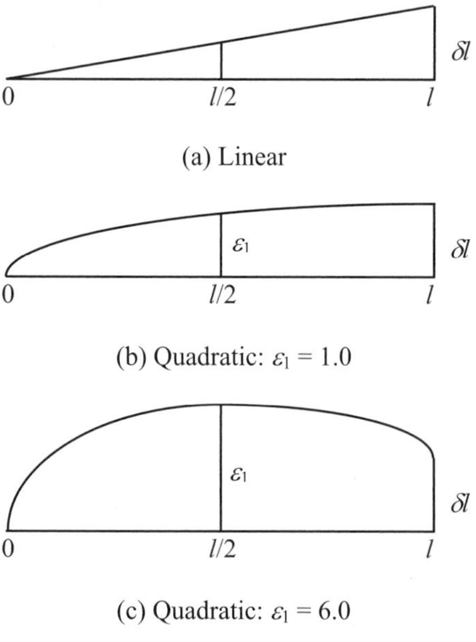

Figure 6.19. Parameterizations of design velocity fields.

Direct frequency response analysis is carried out using a beam element, or BAR, from the COSMIC/NASTRAN finite element code to obtain original and adjoint solutions for the cantilever beam. The first three natural frequencies of the refined model are 16.31, 101.93, and 284.67 Hz. It has been shown that C^1-regular velocity fields are sufficient for shape design sensitivity analysis of truss and plane elastic solid design components, and that C^2-regular velocity fields are sufficient for beam and plate design components [106]. However, as discussed in Section 6.2.7, these regularity conditions can be relaxed. That is, a C^1-regular velocity field with an integrable second derivative can be used for beam and plate design components. Two design velocity fields, the linear and the quadratic, shown in Fig 6.19, are used for this example.

Table 6.2. Shape design sensitivity of beam displacement with excitation frequency of 10 Hz (coarse model, 1% perturbation).

(a) Linear Velocity Field

Hz	ND		ψ^1	ψ^2	$\Delta\psi$	ψ'	$\psi'/\Delta\psi$(%)	$\Delta\psi/\psi$(%)
		R	0.2312E+0	0.2634E+0	0.1609E–1	0.1565E–1	97.2	6.5
		I	–.1688E–1	–.2063E–1	–.1871E–2	–.1800E–2	96.5	10.0
10.0	2	D	0.2318E+0	0.2642E+0	0.1619E–1	0.1574E–1	97.2	6.5
		P	0.3558E+3	0.3555E+3	–.1505E+0	–.1428E+0	94.9	0.0
		R	0.7248E+0	0.8243E+0	0.4976E–1	0.4828E–1	97.0	6.4
		I	–.5233E–1	–.6381E–1	–.5740E–2	–.5510E–2	96.0	9.9
10.0	3	D	0.7267E+0	0.8268E+0	0.5005E–1	0.4856E–1	97.0	6.4
		P	0.3559E+3	0.3556E+3	–.1485E+0	–.1406E+0	94.7	0.0

(b) Quadratic Velocity Field (ε_1 = 1.0)

Hz	ND		ψ^1	ψ^2	$\Delta\psi$	ψ'	$\psi'/\Delta\psi$(%)	$\Delta\psi/\psi$(%)
		R	0.2162E+0	0.2805E+0	0.3214E–1	0.3007E–1	93.6	13.0
		I	–.1592E–1	–.2182E–1	–.2954E–2	–.2642E–2	89.4	15.8
10.0	2	D	0.2168E+0	0.2814E+0	0.3227E–1	0.3018E–1	93.5	13.0
		P	0.3558E+3	0.3556E+3	–.1196E+0	–.8559E–1	71.6	0.0
		R	0.7310E+0	0.8183E+0	0.4368E–1	0.3790E–1	86.8	5.6
		I	–.5321E–1	–.6290E–1	–.4844E–2	–.3961E–2	81.8	8.4
10.0	3	D	0.7329E+0	0.8208E+0	0.4392E–1	0.3809E–1	86.7	5.6
		P	0.3558E+3	0.3556E+3	–.1159E+0	–.8347E–1	72.0	0.0

(c) Quadratic Velocity Field (ε_1 = 6.0)

Hz	ND		ψ^1	ψ^2	$\Delta\psi$	ψ'	$\psi'/\Delta\psi$(%)	$\Delta\psi/\psi$(%)
		R	0.1347E+0	0.4118E+0	0.1386E+0	0.1202E+0	86.7	56.2
		I	–.1085E–1	–.3237E–1	–.1076E–1	–.7903E–2	73.5	57.7
10.0	2	D	0.1351E+0	0.4131E+0	0.1390E+0	0.1205E+0	86.7	56.2
		P	0.3554E+3	0.3555E+3	0.5688E–1	0.2719E+0	478.0	0.0
		R	0.8041E+0	0.8247E+0	0.1029E–1	–.2697E–1	–262.1	1.3
		I	–.6440E–1	–.6404E–1	0.1777E–3	0.5718E–2	3218.1	0.3
10.0	3	D	0.8067E+0	0.8272E+0	0.1024E–1	–.2732E–1	–266.7	1.3
		P	0.3554E+3	0.3556E+3	0.6918E–1	0.2733E+0	395.1	0.0

Numerical results of displacement and stress design sensitivity analysis for the frequency of 10 Hz are given in Tables 6.2 through 6.5. Note that the shape design variable is the beam length l. In the following shape design sensitivity tables, $\Delta\psi = \psi^1 - \psi^2$, where ψ^1 and ψ^2 are the values of displacement or stress of perturbed designs Ω_1 and Ω_2. In this example, $\Omega_1 = [0, l - \Delta l]$ and $\Omega_2 = [0, l + \Delta l]$, where Δl is the amount of beam length perturbation. For a displacement performance, real displacement z^1, imaginary displacement z^2, maximum displacement z, and phase angle α are denoted by the letters R, I, D, and P, respectively. For a stress performance, real stress σ^1, imaginary stress σ^2, and maximum stress σ^{max} are denoted by letters R, I, S, respectively. For the quadratic velocity field, two examples are used, as shown in Fig. 6.19. For the first example, design velocity value ε_1 is set at 1.0 at the center of the beam, while for the second, ε_1 is set at 6.0 at the center.

Table 6.3. Shape design sensitivity of beam stress with excitation frequency of 10 Hz (coarse model, 1% perturbation).

(a) Linear Velocity Field

Hz	EL		ψ^1	ψ^2	$\Delta\psi$	ψ'	$\psi'/\Delta\psi(\%)$	$\Delta\psi/\psi(\%)$
		R	–.1254E+5	–.1371E+5	–.5841E+3	–.5606E+3	96.0	4.4
		I	0.9080E+3	0.1064E+4	0.7813E+2	0.7444E+2	95.3	7.9
10.0	1	S	0.1257E+5	0.1375E+5	0.5883E+3	0.5646E+3	96.0	4.2
		R	–.3854E+4	–.4182E+4	–.1644E+3	–.1535E+3	93.4	4.0
		I	0.2656E+3	0.3088E+3	0.2158E+2	0.2005E+2	92.9	7.5
10.0	2	S	0.3863E+4	0.4194E+4	0.1655E+3	0.1545E+3	93.4	3.9

(b) Quadratic Velocity Field ($\varepsilon_1 = 1.0$)

Hz	EL		ψ^1	ψ^2	$\Delta\psi$	ψ'	$\psi'/\Delta\psi(\%)$	$\Delta\psi/\psi(\%)$
		R	–.1283E+5	–.1342E+5	–.2956E+3	–.1938E+3	65.5	2.2
		I	0.9367E+3	0.1034E+4	0.4856E+2	0.3319E+2	68.4	4.9
10.0	1	S	0.1286E+5	0.1346E+5	0.2984E+3	0.1957E+3	65.6	2.2
		R	–.4088E+4	–.3935E+4	0.7630E+2	0.8638E+2	113.2	1.9
		I	0.2862E+3	0.2860E+3	–.8656E–1	–.2208E+1	2550.9	0.0
10.0	2	S	0.4098E+4	0.3946E+4	–.7611E+2	–.8632E+2	113.4	1.9

(c) Quadratic Velocity Field ($\varepsilon_1 = 6.0$)

Hz	EL		ψ^1	ψ^2	$\Delta\psi$	ψ'	$\psi'/\Delta\psi(\%)$	$\Delta\psi/\psi(\%)$
		R	–.1529E+5	–.1227E+5	0.1510E+4	0.2099E+4	139.0	11.5
		I	0.1228E+4	0.9501E+3	–.1391E+3	–.2246E+3	161.5	14.1
10.0	1	S	0.1534E+5	0.1231E+5	–.1516E+4	–.2110E+4	139.2	12.3
		R	–.5845E+4	–.2610E+4	0.1618E+4	0.1586E+4	98.0	40.3
		I	0.4631E+3	0.1765E+3	–.1433E+3	–.1413E+3	98.6	50.1
10.0	2	S	0.5863E+4	0.2615E+4	–.1624E+4	–.1592E+4	98.0	62.0

In the coarse model (2 elements), design sensitivity results for displacement and stress with a 1% design perturbation of beam length are shown in Tables 6.2 and 6.3, respectively. In both tables, the use of linear velocity yields better agreement between design sensitivity predictions and finite difference results than the use of quadratic velocity fields. However, agreement substantially improves for the refined model (20 elements) with all velocity fields, as can be seen in Tables 6.4 and 6.5 for displacement and stress, respectively. These results confirm that accurate design sensitivity information can be obtained as long as accurate finite element analysis results are used.

Table 6.4. Shape design sensitivity of beam displacement with excitation frequency of 10 Hz (refined model, 1% perturbation).

(a) Linear Velocity Field

Hz	ND		ψ^1	ψ^2	$\Delta\psi$	ψ'	$\psi'/\Delta\psi(\%)$	$\Delta\psi/\psi(\%)$
		R	0.2008E+0	0.2240E+0	0.1161E–1	0.1159E–1	99.8	5.5
		I	–.1272E–1	–.1490E–1	–.1089E–2	–.1086E–2	99.7	7.9
10.0	11	D	0.2012E+0	0.2245E+0	0.1165E–1	0.1163E–1	99.7	5.5
		P	0.3564E+3	0.3562E+3	–.9006E–1	–.8988E–1	99.8	0.0
		R	0.6238E+0	0.6943E+0	0.3521E–1	0.3516E–1	99.8	5.4
		I	–.3879E–1	–.4529E–1	–.3250E–2	–.3241E–2	99.7	7.8
10.0	21	D	0.6250E+0	0.6957E+0	0.3535E–1	0.3529E–1	99.8	5.4
		P	0.3564E+3	0.3563E+3	–.8719E–1	–.8702E–1	99.8	0.0

(b) Quadratic Velocity Field (ε_1 = 1.0)

Hz	ND		ψ^1	ψ^2	$\Delta\psi$	ψ'	$\psi'/\Delta\psi(\%)$	$\Delta\psi/\psi(\%)$
		R	0.1866E+0	0.2398E+0	0.2660E–1	0.2655E–1	99.8	12.6
		I	–.1184E–1	–.1594E–1	–.2052E–2	–.2044E–2	99.6	14.9
10.0	11	D	0.1870E+0	0.2403E+0	0.2668E–1	0.2662E–1	99.8	12.6
		P	0.3564E+3	0.3562E+3	–.8684E–1	–.8649E–1	99.6	0.0
		R	0.6239E+0	0.6942E+0	0.3516E–1	0.3508E–1	99.8	5.3
		I	–.3880E–1	–.4529E–1	–.3244E–2	–.3231E–2	99.6	7.7
10.0	21	D	0.6251E+0	0.6957E+0	0.3530E–1	0.3521E–1	99.8	5.4
		P	0.3564E+3	0.3563E+3	–.8696E–1	–.8661E–1	99.6	0.0

(c) Quadratic Velocity Field (ε_1 = 6.0)

Hz	ND		ψ^1	ψ^2	$\Delta\psi$	ψ'	$\psi'/\Delta\psi(\%)$	$\Delta\psi/\psi(\%)$
		R	0.1073E+0	0.3463E+0	0.1195E+0	0.1200E+0	100.4	56.4
		I	–.6845E–2	–.2290E–1	–.8029E–2	–.8031E–2	100.0	58.4
10.0	11	D	0.1075E+0	0.3471E+0	0.1198E+0	0.1203E+0	100.4	56.4
		P	0.3564E+3	0.3562E+3	–.6696E–1	–.6526E–1	97.5	0.0
		R	0.6244E+0	0.6942E+0	0.3489E–1	0.3457E–1	99.1	5.3
		I	–.3887E–1	–.4529E–1	–.3211E–2	–.3170E–2	98.7	7.7
10.0	21	D	0.6257E+0	0.6957E+0	0.3502E–1	0.3471E–1	99.1	5.3
		P	0.3564E+3	0.3563E+3	–.8548E–1	–.8404E–1	98.3	0.0

Table 6.5. Shape design sensitivity of beam stress with excitation frequency of 10 Hz (refined model, 1% perturbation).

(a) Linear Velocity Field

Hz	EL		ψ^1	ψ^2	$\Delta\psi$	ψ'	$\psi'/\Delta\psi(\%)$	$\Delta\psi/\psi(\%)$
		R	–.7064E+4	–.7511E+4	–.2237E+3	–.2234E+3	99.9	3.1
		I	0.4205E+3	0.4682E+3	0.2385E+2	0.2380E+2	99.8	5.4
10.0	10	S	0.7076E+4	0.7526E+4	0.2248E+3	0.2245E+3	99.9	3.0
		R	–.2442E+3	–.2517E+3	–.3739E+1	–.3336E+1	89.2	1.5
		I	0.1064E+2	0.1111E+2	0.2348E+0	0.2190E+0	93.3	2.2
10.0	20	S	0.2444E+3	0.2519E+3	0.3746E+1	0.3342E+1	89.2	1.5

(b) Quadratic Velocity Field ($\varepsilon_1 = 1.0$)

Hz	EL		ψ^1	ψ^2	$\Delta\psi$	ψ'	$\psi'/\Delta\psi(\%)$	$\Delta\psi/\psi(\%)$
		R	–.7392E+4	–.7165E+4	0.1136E+3	0.1143E+3	100.6	1.6
		I	0.4427E+3	0.4439E+3	0.5874E+0	0.5215E+0	88.8	0.1
10.0	10	S	0.7406E+4	0.7179E+4	–.1133E+3	–.1140E+3	100.6	1.6
		R	–.2649E+3	–.2303E+3	0.1728E+2	0.1626E+2	94.1	7.0
		I	0.1151E+2	0.1014E+2	–.6817E+0	–.6812E+0	99.9	6.3
10.0	20	S	0.2651E+3	0.2305E+3	–.1729E+2	–.1628E+2	94.1	7.5

(c) Quadratic Velocity Field ($\varepsilon_1 = 6.0$)

Hz	EL		ψ^1	ψ^2	$\Delta\psi$	ψ'	$\psi'/\Delta\psi(\%)$	$\Delta\psi/\psi(\%)$
		R	–.9502E+4	–.5082E+4	0.2210E+4	0.2225E+4	100.7	30.4
		I	0.5867E+3	0.3000E+3	–.1434E+3	–.1450E+3	101.1	32.3
10.0	10	S	0.9520E+4	0.5091E+4	–.2214E+4	–.2229E+4	100.7	43.5
		R	–.3927E+3	–.1097E+3	0.1415E+3	0.1388E+3	98.1	57.0
		I	0.1772E+2	0.4785E+1	–.6467E+1	–.6308E+1	97.5	59.6
10.0	20	S	0.3931E+3	0.1098E+3	–.1416E+3	–.1389E+3	98.1	129.0

Table 6.6. Shape design sensitivity of beam displacement with excitation frequency near resonance (refined model, 0.1% perturbation).

(a) Linear Velocity Field

Hz	ND		ψ^1	ψ^2	$\Delta\psi$	ψ'	$\psi'/\Delta\psi(\%)$	$\Delta\psi/\psi(\%)$
		R	0.1246E+1	0.7699E+0	–.2382E+0	–.2387E+0	100.2	23.1
		I	–.2876E+1	–.3257E+1	–.1908E+0	–.1933E+0	101.3	6.2
16.2	11	D	0.3134E+1	0.3347E+1	0.1064E+0	0.1075E+0	101.0	3.3
		P	0.2934E+3	0.2833E+3	–.5067E+1	–.5072E+1	100.1	1.8
		R	0.3706E+1	0.2302E+1	–.7019E+0	–.7031E+0	100.2	22.8
		I	–.8475E+1	–.9599E+1	–.5622E+0	–.5696E+0	101.3	6.2
16.2	21	D	0.9250E+1	0.9872E+1	0.3109E+0	0.3140E+0	101.0	3.2
		P	0.2936E+3	0.2835E+3	–.5066E+1	–.5071E+1	100.1	1.8

(b) Quadratic Velocity Field ($\varepsilon_1 = 1.0$)

Hz	ND		ψ^1	ψ^2	$\Delta\psi$	ψ'	$\psi'/\Delta\psi(\%)$	$\Delta\psi/\psi(\%)$
		R	0.1240E+1	0.7744E+0	–.2326E+0	–.2325E+0	100.0	22.5
		I	–.2861E+1	–.3274E+1	–.2062E+0	–.2085E+0	101.1	6.7
16.2	11	D	0.3118E+1	0.3364E+1	0.1229E+0	0.1239E+0	100.8	3.8
		P	0.2934E+3	0.2833E+3	–.5057E+1	–.5054E+1	99.9	1.8
		R	0.3705E+1	0.2304E+1	–.7005E+0	–.7007E+0	100.0	22.8
		I	–.8476E+1	–.9599E+1	–.5612E+0	–.5678E+0	101.2	6.2
16.2	21	D	0.9251E+1	0.9871E+1	0.3103E+0	0.3130E+0	100.8	3.2
		P	0.2936E+3	0.2835E+3	–.5057E+1	–.5054E+1	99.9	1.8

To test the accuracy of the design sensitivity information near the resonant frequency, a design sensitivity analysis is performed with a 1% design perturbation for the refined model, as shown in Fig. 6.18(b), with a frequency domain of 16.2 Hz where the frequency ratio is r = 16.2/16.3 = 0.993. In order to compare design sensitivity analysis results, only linear and quadratic velocity fields (ε_1 = 1.0 in) are employed. As with the size design problem in Section 5.5, it is found that agreement between displacement and stress design sensitivity predictions and the finite difference results for 1% design perturbation are not good, due to the highly nonlinear behavior of finite difference results. In order to avoid such nonlinear behavior, a 0.1% design perturbation is taken. As shown in Tables 6.6 and 6.7, the agreement becomes excellent for displacement and stress design sensitivity results. Displacement agreements at arbitrarily selected nodes 11 and 21 are also excellent. However, in Table 6.7 agreement for stress at element 20 is not good either for linear or quadratic velocity fields, since $\Delta\psi/\psi'$ is quite small.

Table 6.7. Shape design sensitivity of beam stress with excitation frequencies near resonance (refined model, 0.1% perturbation).

(a) Linear velocity field

Hz	EL		ψ^1	ψ^2	$\Delta\psi$	ψ'	$\psi'/\Delta\psi(\%)$	$\Delta\psi/\psi(\%)$
		R	–.6737E+5	–.4156E+5	0.1291E+5	0.1293E+5	100.2	23.2
		I	0.1548E+6	0.1746E+6	0.9921E+4	0.1005E+5	101.3	6.0
16.2	5	S	0.1688E+6	0.1795E+6	0.5348E+4	0.5402E+4	101.0	3.0
		R	–.3707E+5	–.2327E+5	0.6896E+4	0.6909E+4	100.2	22.3
		I	0.8277E+5	0.9337E+5	0.5300E+4	0.5373E+4	101.4	6.0
16.2	10	S	0.9069E+5	0.9622E+5	0.2768E+4	0.2799E+4	101.1	2.9
		R	–.1278E+5	–.8410E+4	0.2187E+4	0.2190E+4	100.1	20.2
		I	0.2629E+5	0.2964E+5	0.1677E+4	0.1704E+4	101.6	6.0
16.2	15	S	0.2923E+5	0.3081E+5	0.7901E+3	0.8037E+3	101.7	2.6
		R	–.4223E+3	–.3395E+3	0.4140E+2	0.4085E+2	98.7	10.7
		I	0.5044E+3	0.5737E+3	0.3462E+2	0.3172E+2	91.6	6.4
16.2	20	S	0.6579E+3	0.6666E+3	0.4366E+1	0.1962E+1	44.9	0.7

(b) Quadratic Velocity Field (ε_1 = 1.0)

Hz	EL		ψ^1	ψ^2	$\Delta\psi$	ψ'	$\psi'/\Delta\psi(\%)$	$\Delta\psi/\psi(\%)$
		R	–.6748E+5	–.4151E+5	0.1299E+5	0.1300E+5	100.1	23.3
		I	0.1551E+6	0.1743E+6	0.9567E+4	0.9683E+4	101.2	5.8
16.2	5	S	0.1692E+6	0.1792E+6	0.4984E+4	0.5030E+4	100.9	2.8
		R	–.3722E+5	–.2319E+5	0.7017E+4	0.7024E+4	100.1	22.7
		I	0.8318E+5	0.9291E+5	0.4867E+4	0.4929E+4	101.3	5.5
16.2	10	S	0.9113E+5	0.9576E+5	0.2318E+4	0.2342E+4	100.1	2.4
		R	–.1288E+5	–.8357E+4	0.2259E+4	0.2260E+4	100.0	20.9
		I	0.2651E+5	0.2941E+5	0.1452E+4	0.1473E+4	101.4	5.2
16.2	15	S	0.2947E+5	0.3058E+5	0.5536E+3	0.5633E+3	101.7	1.8
		R	–.4316E+3	–.3415E+3	0.4506E+2	0.4350E+2	96.5	11.7
		I	0.5016E+3	0.5534E+3	0.2589E+2	0.2654E+2	102.5	0.9
16.2	20	S	0.6617E+3	0.6502E+3	–.5736E+1	–.3796E+1	66.2	0.9

Table 6.8. Perturbation of shape design variables.

Shape variable	Initial design	Perturbation of shape design variable
α_1	27.56	$\delta\alpha_1 = 0.03628\ \alpha_1 = 1.0$
α_2	55.9	$\delta\alpha_2 = 0.01789\ \alpha_2 = 1.0$
α_3	74.0	$\delta\alpha_3 = 0.01351\ \alpha_3 = 1.0$
α_4	137.0	$\delta\alpha_4 = 0.00730\ \alpha_4 = 1.0$
α_5	183.84	$\delta\alpha_5 = 0.00544\ \alpha_5 = 1.0$
α_6	261.34	$\delta\alpha_6 = 0.00383\ \alpha_6 = 1.0$

$\alpha_1 \sim \alpha_6$, $\delta\alpha_1 \sim \delta\alpha_6$: in.

Vehicle Chassis Frame

The vehicle chassis frame model presented in Chapter 5 will now be used to numerically test shape design sensitivity using Hermite cubic velocity fields. Refer to Figs. 5.21 and 5.22 for structural and finite element models, respectively. The design specification given in Table 6.8 is used for finite element analysis. Two types of shape design variables are considered. The first type includes the location of five cross members and front suspension springs on the longitudinal chassis frames. Six shape design variables (α_1~α_6) of the original chassis frame design are shown in Fig. 6.20(a). The perturbed shape design variables in positive x_1 direction are shown in Fig. 6.20(b).

To define a design velocity field that satisfies the regularity condition described in Section 6.2.7, the Hermite cubic function shown in Fig. 6.21(a) is selected as

$$\phi(x_1) = \gamma_3 x_1^3 + \gamma_2 x_1^2 + \gamma_1 x_1 + \gamma_0. \tag{6.283}$$

With the appropriate boundary and connectivity conditions, the following general expression can be obtained for repeated Hermite cubic functions:

$$\phi_i(x_1) = \begin{cases} \dfrac{2}{(\alpha_i - \alpha_{i-1})^3}\left[(x_1 - \alpha_{i-1})^3 - \tfrac{3}{2}(\alpha_i - \alpha_{i-1})(x_1 - \alpha_{i-1})^2\right](\delta\alpha_{i-1} - \delta\alpha_i) \\ \qquad + \delta\alpha_{i-1}, \quad \alpha_{i-1} \le x_1 \le \alpha_i \\ \dfrac{2}{(\alpha_{i+1} - \alpha_i)^3}\left[(x_1 - \alpha_i)^3 - \tfrac{3}{2}(\alpha_{i+1} - \alpha_i)(x_1 - \alpha_i)^2\right](\delta\alpha_i - \delta\alpha_{i+1}) \\ \qquad + \delta\alpha_i, \quad \alpha_i \le x_1 \le \alpha_{i+1}. \end{cases} \tag{6.284}$$

A design velocity field over the longitudinal chassis frame can then be obtained by summing up the Hermit cubic functions, such as

$$V(x_1) = \sum_{i=1}^{6} \phi_i(x_1). \tag{6.285}$$

A different percentage of perturbation is applied to each shape design variable to obtain the same amount of movement for cross members, as shown in Fig. 6.21(b) and Table 6.8. Therefore, perturbation amounts $\delta\alpha_1$ through $\delta\alpha_6$ have the same value. As shown in Fig. 6.20(b), the ○ marks show the location of the original cross members and front suspension, while the ● marks denote their new, perturbed location.

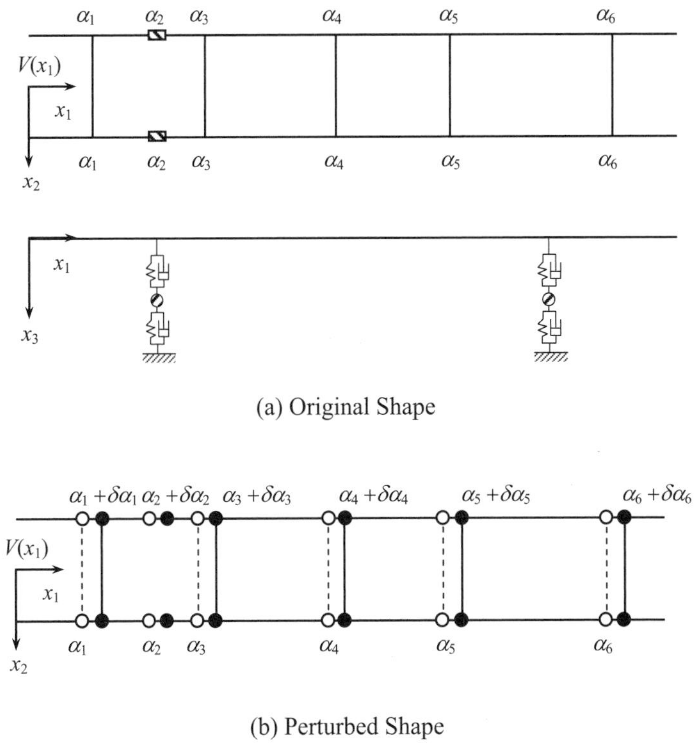

Figure 6.20. Shape design variables for locations of cross members.

For the second type of shape design variable, consider the lengths of five cross members measuring 31.5 inches in the initial design. Since the longitudinal chassis frames are parallel at all times, the lengths of the five cross members are the same. One half of the cross member length is denoted as β. The original shape, a Hermite cubic design velocity field, and the perturbed shape of a cross member, are shown in Figs. 6.22(a), (b), and (c), respectively. Center point C does not move, while endpoints A and B are perturbed to A' and B', respectively, and 5% of the cross member length is taken as design perturbation to be used in a numerical test.

Figure 6.23 shows a perturbed design model where the first set of design variables (α_1~α_6) is perturbed 1.0 in, and the second set of design variables (β) is perturbed 1.58 in. Displacement and stress design sensitivity results are listed in Tables 6.9 and 6.10, respectively. In these tables, ψ^1 denotes the displacement and stress values of the original design and ψ^2 denotes for the perturbed design shown in Fig. 6.23. Excellent sensitivity results are obtained in both displacement and stress performance measures.

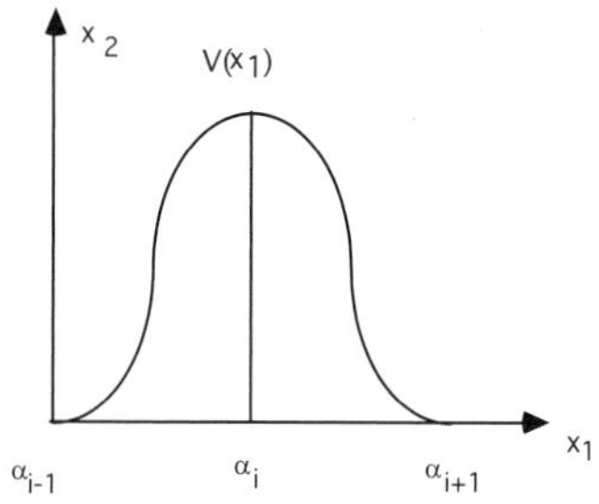

(a) Hermite Cubic Function

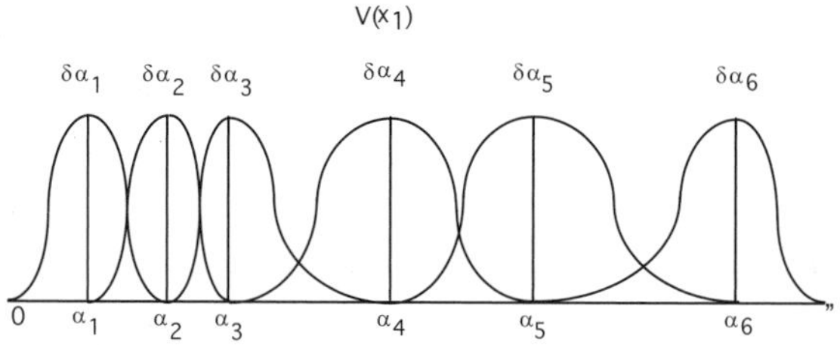

(b) Design Velocity Field

Figure 6.21. Hermite cubic design velocity fields.

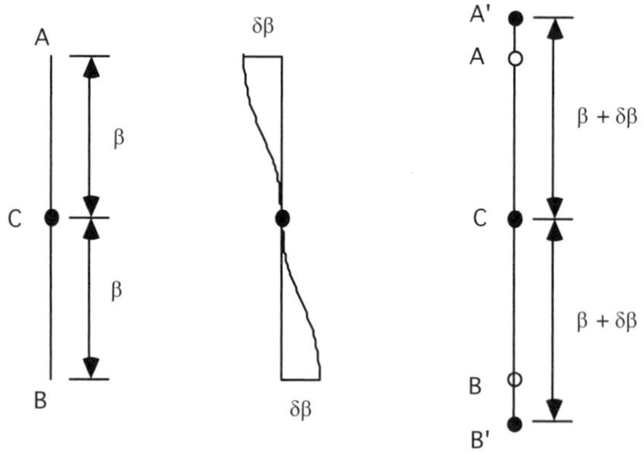

(a) Original Shape (b) Design Velocity Fields (c) Perturbed Shape

Figure 6.22. Shape design variable for length of cross member.

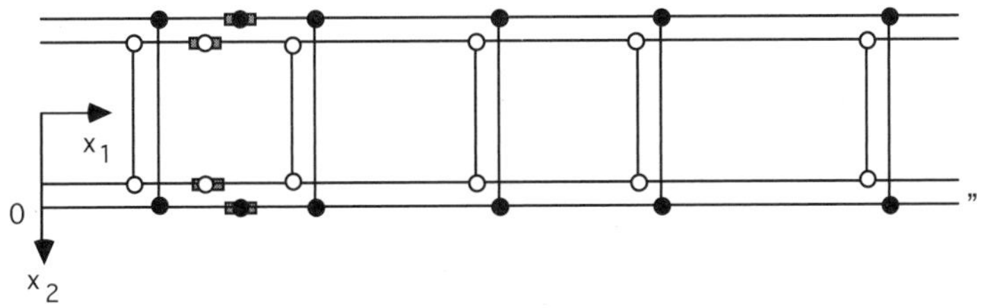

Figure 6.23. Shape of perturbed design.

Table 6.9. Shape design sensitivity of displacement of vehicle chassis frame.

Hz	ND		ψ^1	ψ^2	$\Delta\psi$	ψ'	$\psi'/\Delta\psi(\%)$	$\Delta\psi/\psi(\%)$
		R	0.3896E+1	0.3881E+1	–.7539E–2	–.7453E–2	98.9	0.2
		I	–.4295E+1	–.4429E+1	–.6691E–1	–.6676E–1	99.8	1.5
2.0	5	D	0.5799E+1	0.5888E+1	0.4493E–1	0.4487E–1	99.9	0.8
		P	0.3122E+3	0.3112E+3	–.4918E+0	–.4902E+0	99.7	0.2
		R	0.2531E+1	0.2518E+1	–.6083E–2	–.6020E–2	99.0	0.2
		I	–.3048E+1	–.3145E+1	–.4873E–1	–.4858E–1	99.7	1.6
2.0	10	D	0.3961E+1	0.4029E+1	0.3392E–1	0.3384E–1	99.8	0.8
		P	0.3097E+3	0.3087E+3	–.5092E+0	–.5072E+0	99.6	0.2
		R	0.1060E+1	0.1053E+1	–.3884E–2	–.3808E–2	98.0	0.4
		I	–.1630E+1	–.1688E+1	–.2855E–1	–.2838E–1	99.4	1.7
2.0	15	D	0.1945E+1	0.1989E+1	0.2199E–1	0.2189E–1	99.5	1.1
		P	0.3030E+3	0.3020E+3	–.5422E+0	–.5378E+0	99.2	0.2

Table 6.10. Shape design sensitivity of stress of vehicle chassis frame.

Hz	EL		ψ^1	ψ^2	$\Delta\psi$	ψ'	$\psi'/\Delta\psi(\%)$	$\Delta\psi/\psi(\%)$
		R	–.3127E+4	–.3170E+4	–.2125E+2	–.2193E+2	103.2	0.7
		I	0.6653E+3	0.7355E+3	0.3511E+2	0.3530E+2	100.5	5.0
2.0	4	S	0.3197E+4	0.3254E+4	0.2837E+2	0.2906E+2	102.4	0.9
		R	–.2923E+4	–.2964E+4	–.2068E+2	–.1937E+2	93.7	0.7
		I	0.1239E+3	0.1666E+3	0.2139E+2	0.2078E+2	97.2	14.9
2.0	8	S	0.2925E+4	0.2969E+4	0.2171E+2	0.2036E+2	93.8	0.7
		R	–.1759E+4	–.1787E+4	–.1407E+2	–.1389E+2	98.7	0.8
		I	–.3308E+3	–.3188E+3	0.5987E+1	0.5294E+1	88.4	1.8
2.0	12	S	0.1789E+4	0.1815E+4	0.1276E+2	0.1270E+2	99.5	0.7
		R	–.7851E+3	–.8057E+3	–.1034E+2	–.1023E+2	99.0	1.3
		I	–.3929E+3	–.3915E+3	0.7046E+0	0.5720E+0	81.2	0.2
2.0	16	S	0.8779E+3	0.8958E+3	0.8962E+1	0.8928E+1	99.6	1.0
		R	0.1275E+3	0.1138E+3	–.6856E+1	–.6980E+1	101.8	5.7
		I	–.2091E+3	–.2022E+3	0.3474E+1	0.3375E+1	97.2	1.7
2.0	20	S	0.2449E+3	0.2320E+3	–.6464E+1	–.6439E+1	99.6	2.8
		R	0.6697E+3	0.7239E+3	0.2709E+2	0.2699E+2	99.6	3.9
		I	–.6337E+2	–.7023E+2	–.3429E+1	–.5246E+1	153.0	5.3
2.0	50	S	0.6727E+3	0.7273E+3	0.2730E+2	0.2736E+2	100.3	3.8

6.5 Thermoelastic Problem

A design sensitivity formulation of the thermoelastic problem in Section 2.7 is presented in this section using the domain method. The initially coupled thermoelastic problem is simplified such that the thermal problem is solved first and the temperature field is then considered as an external load for structural analysis purposes. The shape design sensitivity of a stress performance measure is obtained by differentiating the variational forms of thermal and elasticity equations. The adjoint variable method for such a sequential problem cannot be solved in the same way as the coupled problem. A reverse solution procedure is implemented to solve the adjoint problem, in which the structural adjoint problem is first solved to calculate the adjoint load, and the thermal adjoint problem is then solved to calculate the adjoint response. Finally, the sensitivity expression of the stress performance measure is calculated through domain integration.

6.5.1 Design Sensitivity Analysis of Thermal Systems

The variational equation of a thermal problem is presented in Section 2.7.1. Consider a three-dimensional, thermoelastic, isotropic and homogeneous solid, as shown in Fig. 6.24. The variational equation for the temperature field is presented in (2.84), rewritten here as

$$a_\theta(\theta,\bar{\theta}) = \ell_\theta(\bar{\theta}), \quad \forall \bar{\theta} \in \Theta, \tag{6.286}$$

where Θ is the space of kinematically admissible temperature fields, given in (2.82), and where

$$a_\theta(\theta,\bar{\theta}) = \iiint_\Omega k \nabla\theta^T \nabla\bar{\theta}\, d\Omega + \iint_{\Gamma_\theta^2} h\theta\bar{\theta}\, d\Gamma \tag{6.287}$$

and

$$\ell_\theta(\bar{\theta}) = \iiint_\Omega g\bar{\theta}\, d\Omega + \iint_{\Gamma_\theta^1} q\bar{\theta}\, d\Gamma + \iint_{\Gamma_\theta^2} h\theta^\infty \bar{\theta}\, d\Gamma \tag{6.288}$$

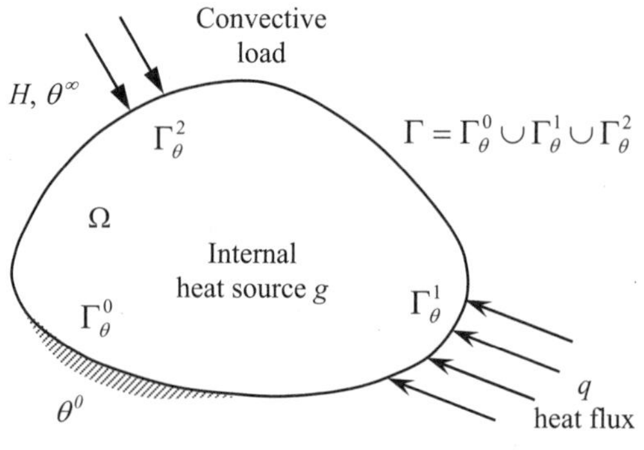

Figure 6.24. Thermoelastic analysis model.

are the energy bilinear and load linear forms, respectively. A subscribed θ is used for $a_\theta(\bullet,\bullet)$ and $\ell_\theta(\bullet)$ to distinguish them from the structural counterparts $a_\Omega(\bullet,\bullet)$ and $\ell_\Omega(\bullet)$. In (6.286) through (6.288), the following notation system is used: $\theta = T - T^0$ is the temperature function, T is the absolute temperature, T^0 is the reference temperature, k is the heat conductivity of the body, h is the convective heat transfer coefficient, g is the internal heat source, q is the heat flux vector, θ^∞ is the ambient temperature, Γ_θ^0 is the boundary where the temperature is prescribed, Γ_θ^1 is the boundary where the heat flux is prescribed, and, finally, Γ_θ^2 is the boundary where the heat convection is prescribed.

For shape design sensitivity analysis, let the domain Ω be perturbed in the direction of design velocity $\boldsymbol{V}(\boldsymbol{x})$ and let τ be a scalar parameter to control perturbation size, such that point $\boldsymbol{x} \in \Omega$ is perturbed to the new point $\boldsymbol{x}_\tau = \boldsymbol{x} + \tau\boldsymbol{V}$. The same material derivative formulas in Section 6.1 are preserved for the temperature field. Similar to the material derivative formula in (6.8), if a material derivative exists in the temperature field, then it is defined as

$$\dot{\theta} = \dot{\theta}(\boldsymbol{x};\Omega,\boldsymbol{V}) \equiv \frac{d}{d\tau}\theta_\tau(\boldsymbol{x}+\tau\boldsymbol{V}(\boldsymbol{x}))\bigg|_{\tau=0} = \lim_{\tau\to 0}\left[\frac{\theta_\tau(\boldsymbol{x}+\tau\boldsymbol{V}) - \theta(\boldsymbol{x})}{\tau}\right]. \tag{6.289}$$

As with the material derivative formulas in Section 6.1, the energy bilinear form (6.287) and the load linear form (6.288) are assumed to be differentiable with respect to the shape design. In the following derivations, it is assumed that the material properties and the boundary conditions remain constant during shape perturbation, that is, $(h\theta^\infty)' = 0$ and $g' = q' = 0$. From the formulas in (6.37) and (6.53), the material derivative of the energy bilinear form is first obtained by differentiating (6.287) at the perturbed design as

$$\begin{aligned}
\frac{d}{d\tau}[a_{\theta_\tau}(\theta_\tau,\bar{\theta}_\tau)]\bigg|_{\tau=0} &= \iiint_\Omega k\nabla\dot{\theta}^T\nabla\bar{\theta}\, d\Omega + \iint_{\Gamma_\theta^2} h\dot{\theta}\bar{\theta}\, d\Gamma \\
&\quad + \iiint_\Omega k[\nabla\theta^T\nabla\bar{\theta}\,div\boldsymbol{V} - (\nabla\theta^T\nabla\boldsymbol{V})\nabla\bar{\theta} - (\nabla\bar{\theta}^T\nabla\boldsymbol{V})\nabla\theta]\, d\Omega \\
&\quad + \iint_{\Gamma_\theta^2} \kappa(h\theta\bar{\theta})V_n\, d\Gamma \\
&\equiv a_\theta(\dot{\theta},\bar{\theta}) + a'_{V_\theta}(\theta,\bar{\theta}).
\end{aligned} \tag{6.290}$$

In the derivation of (6.290), the relation $\frac{d}{d\tau}(\nabla\theta) = \nabla\dot{\theta} - \nabla\theta^T\nabla\boldsymbol{V}$, which is similar to (6.68), is used. As with the structural design sensitivity formulation in Section 6.2, the contribution of $\dot{\bar{\theta}}$ vanishes due to the relation $a_\theta(\theta,\dot{\bar{\theta}}) = \ell_\theta(\dot{\bar{\theta}})$. Equation (6.290) actually defines the term $a'_{V_\theta}(\theta,\bar{\theta})$, which is the explicitly dependent term on the shape design. $a_\theta(\dot{\theta},\bar{\theta})$ represents the implicitly dependent term $(\dot{\theta})$ on the shape design, which must be calculated.

Second, the load linear form in (6.288) is differentiated with respect to the shape design as

$$\begin{aligned}
\frac{d}{d\tau}[\ell_{\theta_\tau}(\bar{\theta}_\tau)]\bigg|_{\tau=0} &= \iiint_\Omega [\nabla g^T\boldsymbol{V}\bar{\theta} + g\bar{\theta}\,div\boldsymbol{V}]\, d\Omega \\
&\quad + \iint_{\Gamma_\theta^1} [\nabla q^T\boldsymbol{V}\bar{\theta} + \kappa q\bar{\theta}V_n]\, d\Gamma \\
&\quad + \iint_{\Gamma_\theta^2} [\nabla(h\theta^\infty)^T\boldsymbol{V}\bar{\theta} + \kappa h\theta^\infty\bar{\theta}V_n]\, d\Gamma \\
&\equiv \ell'_{V_\theta}(\bar{\theta}).
\end{aligned} \tag{6.291}$$

Note that the formula in (6.291) does not depend on the thermal analysis results θ; only the velocity field information is required. In addition, $\ell'_{V_\theta}(\bar{\theta})$ only contains the explicitly dependent terms; no implicitly dependent term exists in (6.291).

By differentiating (6.286) and by using the formulas in (6.290) and (6.291), the design sensitivity equation of the thermal problem can be obtained as

$$a_\theta(\dot{\theta},\bar{\theta}) = \ell'_{V_\theta}(\bar{\theta}) - a'_{V_\theta}(\theta,\bar{\theta}), \quad \forall \bar{\theta} \in \Theta. \tag{6.292}$$

Using the temperature field from the thermal analysis in (6.286) and using design velocity field V, the material derivative of temperature field can be obtained from design sensitivity (6.292).

By comparing (6.286) with (6.292), it is evident that the bilinear forms on the left side are identical with different arguments [θ and $\dot{\theta}$ in (6.286) and (6.292), respectively] while the right sides are different. In finite element analysis, they can be considered as two different loads with the same stiffness matrix. Equation (6.286) is first solved to obtain θ and the decomposed stiffness matrix is reused to obtain $\dot{\theta}$ with (6.292). In this case, the second process is much more efficient than the first.

6.5.2 Design Sensitivity Analysis of Structural Systems

After solving the thermal design sensitivity equation (6.292) for $\dot{\theta}$, the structural design sensitivity equation needs to be solved for $\dot{z}$, which is the material derivative of displacement. The structural part is similar to the forms presented in Section 6.2.7, except for the fact that the material properties are functions of temperature and an additional thermal load exists. The variational equation for the structural part is

$$a_\Omega(\theta; z, \bar{z}) = \ell_\Omega(\theta; \bar{z}), \quad \forall \bar{z} \in Z, \tag{6.293}$$

where Z is the space of kinematically admissible displacements, given in (2.88), and

$$a_\Omega(\theta; z, \bar{z}) = \iiint_\Omega \varepsilon(\bar{z}) : \boldsymbol{C}(\theta) : \varepsilon(z)\, d\Omega \tag{6.294}$$

and

$$\ell_\Omega(\theta; \bar{z}) = \iiint_\Omega [\bar{z}^T \boldsymbol{f}^B + \beta(\theta)\theta div(\bar{z})]\, d\Omega + \iint_{\Gamma^S} \bar{z}^T \boldsymbol{f}^S\, d\Gamma \tag{6.295}$$

are energy bilinear and load linear forms, respectively. In (6.295), $\beta(\theta)$ is the thermal modulus. The notational system that is being employed needs some explanation. The form $a_\Omega(\theta;\bullet,\bullet)$ depends on the temperature θ and is bilinear with respect to its arguments. The form $\ell_\Omega(\theta;\bullet)$ also depends on θ and it is linear with respect to its argument.

By assuming that the forms in (6.294) and (6.295) are differentiable with respect to shape design, the material derivatives of these two forms are derived. Differentiation of (6.294) yields

$$\begin{aligned}
\frac{d}{d\tau}[a_{\Omega_\tau}(\theta_\tau;z_\tau,\bar{z}_\tau)]\bigg|_{\tau=0} &= \iiint_\Omega \varepsilon(\bar{z}):C:\varepsilon(\dot{z})\,d\Omega \\
&+ \iiint_\Omega \varepsilon(\bar{z}):C:\varepsilon^V(z)\,d\Omega \\
&+ \iiint_\Omega \varepsilon^V(\bar{z}):C:\varepsilon(z)\,d\Omega \\
&+ \iiint_\Omega \varepsilon(\bar{z}):C:\varepsilon(z)\,divV\,d\Omega \\
&+ \iiint_\Omega [\varepsilon(\bar{z}):C_{,\theta}:\varepsilon(z)]\dot{\theta}\,d\Omega \\
&\equiv a_\Omega(\theta;\dot{z},\bar{z}) + a'_V(\theta,\dot{\theta};z,\bar{z}),
\end{aligned} \tag{6.296}$$

where the expression of $\varepsilon^V(z)$ is given in (6.200), and where $C_{,\theta} = \partial C/\partial\theta$. Note that the explicitly dependent term $a'_V(\theta,\dot{\theta};z,\bar{z})$ is bilinear with respect to z and $\bar{z}$, and depends on θ and $\dot{\theta}$, which have already been obtained from thermal analysis and its design sensitivity analysis in Section 6.5.1. Actually, (6.296) defines the fictitious load $a'_V(\theta,\dot{\theta};z,\bar{z})$. $a_\Omega(\theta;\dot{z},\bar{z})$ represents the implicitly dependent term ($\dot{z}$) on the shape design, which must be solved.

The material derivative of the load linear form in (6.295) can be derived using a similar procedure, as

$$\begin{aligned}
\frac{d}{d\tau}[\ell_{\Omega_\tau}(\theta;\bar{z}_\tau)]\bigg|_{\tau=0} &= \iiint_\Omega [\bar{z}^T\nabla f^B V + (\beta_{,T}\theta + \beta)\dot{\theta}div\bar{z}]\,d\Omega \\
&+ \iiint_\Omega [\bar{z}^T f^B divV + \beta\dot{\theta}div\bar{z}divV]\,d\Omega \\
&+ \iint_{\Gamma^S} [\bar{z}^T\nabla f^S V + \kappa(\bar{z}^T f^S)V_n]\,d\Gamma \\
&\equiv \ell'_V(\theta,\dot{\theta};\bar{z}).
\end{aligned} \tag{6.297}$$

It is assumed that the body force and surface traction force remain constant during shape perturbation, that is, $f^{B'} = f^{S'} = 0$. Note that the formula in (6.297) does not depend on the structural analysis result z. The velocity field information, thermal analysis results, and thermal design sensitivity analysis results are required.

By using the formulas in (6.296) and (6.297), the design sensitivity equation for the structural problem can be obtained after differentiating the structural variational equation (6.293), as

$$a_\Omega(\theta;\dot{z},\bar{z}) = \ell'_V(\theta,\dot{\theta};\bar{z}) - a'_V(\theta,\dot{\theta};z,\bar{z}), \qquad \forall\bar{z}\in Z. \tag{6.298}$$

Similar to its thermal counterpart, the elasticity equation (6.293) is first solved for displacement z and subsequently the sensitivity equation (6.298) is solved for $\dot{z}$, using the same stiffness matrix. It is important to note that the sensitivity of the displacement field $\dot{z}$ depends on the displacement field z, the temperature field θ, and the sensitivity of the temperature field $\dot{\theta}$.

Consider a stress tensor within a subdomain $\Omega_p \subset \Omega$ as a performance measure. Let m_p be a characteristic function whose value is one within Ω_p and zero otherwise. From the expression of stress in (2.86), the performance measure can defined as

$$\iiint_\Omega \sigma(z,\theta)m_p\,d\Omega = \iiint_\Omega [C(\theta):\varepsilon(z) - \beta(\theta)\theta I]m_p\,d\Omega. \tag{6.299}$$

By taking the material derivative of (6.299), the following formula can be obtained for the stress sensitivity:

$$\begin{aligned}\frac{d}{d\tau}\left[\iiint_{\Omega}\sigma_{\tau}(z_{\tau},\theta_{\tau})m_{p}\,d\Omega\right]\bigg|_{\tau=0} &= \iiint_{\Omega}[C:\varepsilon(\dot{z})]m_{p}\,d\Omega \\ &+ \iiint_{\Omega}[C_{,\theta}:\varepsilon(z)\dot{\theta}-(\beta_{,\theta}\theta+\beta)\dot{\theta}I]m_{p}\,d\Omega \\ &+ \iiint_{\Omega}[C:\varepsilon^{V}(z)]m_{p}\,d\Omega.\end{aligned} \tag{6.300}$$

The first integral on the right side depends on structural design sensitivity results, while the second integral depends on thermal design sensitivity results. The last integral is the explicitly dependent term. Using the direct differentiation method, the design sensitivity of stress in the above equation can be evaluated using the thermal analysis result θ, the sensitivity of the temperature field $\dot{\theta}$, the structural analysis result z, and the sensitivity of the displacement field $\dot{z}$.

6.5.3 Adjoint Variable Method in the Thermoelastic Problem

The design sensitivity formulation in Sections 6.5.1 and 6.5.2 is a sequential process in which the design sensitivity equation of the thermal problem is solved first, and the design sensitivity equation of the structural problem is then solved. The adjoint variable method is complicated in such a sequential process because the adjoint solution requires a process that moves in reverse order. For example, the adjoint problem of the initial-value problem in Section 5.4 becomes the terminal-value problem. In this section, an adjoint variable formulation is presented for the sequential thermoelastic solution process.

The adjoint variable procedure involves removing the implicitly dependent terms from the sensitivity expression of the performance measure. Since the stress sensitivity expression in (6.300) contains two different implicitly dependent terms ($\dot{z}$ and $\dot{\theta}$), let us first take into account the contribution from the displacement sensitivity, which is the first integral on the right side of (6.300). By substituting $\dot{z}$ into the virtual adjoint displacement $\bar{\eta}$ and by equating the first integral on the right side of (6.300) with the bilinear form in (6.294), the adjoint equation of the structural problem is defined as

$$a_{\Omega}(\theta;\eta,\bar{\eta}) = \iiint_{\Omega}[C:\varepsilon(\bar{\eta})]m_{p}\,d\Omega, \quad \forall\bar{\eta}\in Z, \tag{6.301}$$

where the structural adjoint response η is required. The solution procedure in (6.301) is similar to the solution procedure of structural response analysis in (6.293). The only difference is the load on the right side. Thus, (6.301) is equivalent to a structural analysis but with a different loading condition.

The objective is to replace the implicitly dependent term $\dot{z}$ in (6.300) by the structural adjoint response η. In order to do that, let us substitute $\bar{z}$ into η in design sensitivity (6.298), to obtain

$$a_{\Omega}(\theta;\dot{z},\eta) = \ell'_{V}(\theta,\dot{\theta};\eta) - a'_{V}(\theta,\dot{\theta};z,\eta). \tag{6.302}$$

This substitution is valid because both $\bar{z}$ and η belong to the space Z of kinematically admissible displacements. In addition, from the fact that both $\bar{\eta}$ and $\dot{z}$ belong to Z, in the adjoint equation (6.301) $\bar{\eta}$ can be replaced by $\dot{z}$ to yield

$$a_{\Omega}(\theta;\eta,\dot{z}) = \iiint_{\Omega}[C:\varepsilon(\dot{z})]m_{p}\,d\Omega. \tag{6.303}$$

As is clear from the expression in (6.294), $a_{\Omega}(\theta;\,\bullet,\bullet)$ is symmetric with respect to its arguments. Thus, the left sides of (6.302) and (6.303) are the same, which produces the following relation:

$$\iiint_{\Omega}[\boldsymbol{C}:\boldsymbol{\varepsilon}(\dot{\boldsymbol{z}})]m_p\,d\Omega = \ell'_V(\theta,\dot{\theta};\boldsymbol{\eta}) - a'_V(\theta,\dot{\theta};\boldsymbol{z},\boldsymbol{\eta}). \tag{6.304}$$

Equation (6.304) removes the implicitly dependent term $\dot{\boldsymbol{z}}$ from the stress sensitivity expression by replacing it with the structural adjoint response $\boldsymbol{\eta}$. By substituting the relation in (6.304) into (6.300), the stress sensitivity can be rewritten as

$$\begin{aligned}\frac{d}{d\tau}\left[\iiint_{\Omega}\boldsymbol{\sigma}_\tau(\boldsymbol{z}_\tau,\theta_\tau)m_p\,d\Omega\right]\Bigg|_{\tau=0} &= \ell'_V(\theta,\dot{\theta};\boldsymbol{\eta}) - a'_V(\theta,\dot{\theta};\boldsymbol{z},\boldsymbol{\eta})\\ &+ \iiint_{\Omega}[\boldsymbol{C}_{,\theta}:\boldsymbol{\varepsilon}(\boldsymbol{z})\dot{\theta} - (\beta_{,\theta}\theta+\beta)\dot{\theta}\boldsymbol{I}]m_p\,d\Omega\\ &+ \iiint_{\Omega}[\boldsymbol{C}:\boldsymbol{\varepsilon}^V(\boldsymbol{z})]m_p\,d\Omega.\end{aligned} \tag{6.305}$$

The stress sensitivity expression of (6.305) still contains the implicitly dependent term $\dot{\theta}$, which will be removed by defining the thermal adjoint problem. Let us take those terms that contain $\dot{\theta}$ from (6.305) and substitute $\dot{\theta}$ into the virtual adjoint temperature $\bar{\lambda}$. By equating these terms with the energy bilinear form in (6.287), the thermal adjoint problem is obtained as

$$\begin{aligned}a_\theta(\lambda,\bar{\lambda}) &= \ell'_V(\theta,\bar{\lambda};\boldsymbol{\eta}) - a'_V(\theta,\bar{\lambda};\boldsymbol{z},\boldsymbol{\eta})\\ &+ \iiint_{\Omega}[\boldsymbol{C}_{,\theta}:\boldsymbol{\varepsilon}(\boldsymbol{z})\bar{\lambda} - (\beta_{,\theta}\theta+\beta)\bar{\lambda}\boldsymbol{I}]m_p\,d\Omega, \quad \forall\bar{\lambda}\in\Theta,\end{aligned} \tag{6.306}$$

where adjoint response λ is required. Note that the thermal adjoint problem utilizes the result from the structural adjoint problem to define its adjoint load. The solution procedure of (6.306) is similar to the solution procedure of thermal analysis in (6.286). The only difference is the load on the right side. Thus, (6.306) is equivalent to a thermal analysis with a different thermal load.

By following the same procedure as the structural adjoint problem, the virtual temperature $\bar{\theta}$ is replaced by the adjoint temperature λ in the thermal design sensitivity equation (6.292), to obtain

$$a_\theta(\dot{\theta},\lambda) = \ell'_{V_\theta}(\lambda) - a'_{V_\theta}(\theta,\lambda). \tag{6.307}$$

This substitution is valid because both $\bar{\theta}$ and λ belong to the space Θ of kinematically admissible temperatures. In addition, because both $\bar{\lambda}$ and $\dot{\theta}$ belong to Θ, $\bar{\lambda}$ can be replaced by $\dot{\theta}$ in adjoint equation (6.306), to yield

$$\begin{aligned}a_\theta(\lambda,\dot{\theta}) &= \iiint_{\Omega}[\boldsymbol{C}_{,\theta}:\boldsymbol{\varepsilon}(\boldsymbol{z})\dot{\theta} - (\beta_{,\theta}\theta+\beta)\dot{\theta}\boldsymbol{I}]m_p\,d\Omega\\ &+\ell'_V(\theta,\dot{\theta};\boldsymbol{\eta}) - a'_V(\theta,\dot{\theta};\boldsymbol{z},\boldsymbol{\eta}).\end{aligned} \tag{6.308}$$

As is clear from the expression in (6.287), $a_\theta(\bullet,\bullet)$ is symmetric with respect to its arguments. Thus, the left sides of both (6.307) and (6.308) are the same, which produces the following relation:

$$\begin{aligned}&\iiint_{\Omega}[\boldsymbol{C}_{,\theta}:\boldsymbol{\varepsilon}(\boldsymbol{z})\dot{\theta} - (\beta_{,\theta}\theta+\beta)\dot{\theta}\boldsymbol{I}]m_p\,d\Omega + \ell'_V(\theta,\dot{\theta};\boldsymbol{\eta}) - a'_V(\theta,\dot{\theta};\boldsymbol{z},\boldsymbol{\eta})\\ &= \ell'_{V_\theta}(\lambda) - a'_{V_\theta}(\theta,\lambda).\end{aligned} \tag{6.309}$$

Thus, all terms that contain $\dot{\theta}$ in (6.305) are replaced by the known adjoint temperature λ. By using the relation in (6.309), the stress sensitivity of (6.305) is derived as an explicit function of the shape design by

Table 6.11. Design sensitivity procedure for displacement and temperature performance measures.

	Displacement	Temperature
Performance	$\psi_2 = \iiint_\Omega z\,\delta(x-\hat{x})\,d\Omega$	$\psi_3 = \iiint_\Omega \theta\,\delta(x-\hat{x})\,d\Omega$
Sensitivity	$\psi_2' = \iiint_\Omega \dot{z}\,\delta(x-\hat{x})\,d\Omega$	$\psi_3' = \iiint_\Omega \dot{\theta}\,\delta(x-\hat{x})\,d\Omega$
Structural adjoint equation	$a_\Omega(\theta;\eta,\bar{\eta}) = \iiint_\Omega \bar{\eta}\,\delta(x-\hat{x})\,d\Omega, \quad \forall\bar{\eta}\in Z$	Not required
Thermal adjoint equation	$a_\theta(\lambda,\bar{\lambda}) = \ell_V'(\theta,\bar{\lambda};\eta) - a_V'(\theta,\bar{\lambda};z,\eta), \quad \forall\bar{\lambda}\in\Theta$	$a_\theta(\lambda,\bar{\lambda}) = \iiint_\Omega \bar{\lambda}\,\delta(x-\hat{x})\,d\Omega, \quad \forall\bar{\lambda}\in\Theta$
Sensitivity expression	$\psi_2' = \ell_{V_\theta}'(\lambda) - a_{V_\theta}'(\theta,\lambda)$	$\psi_3' = \ell_{V_\theta}'(\lambda) - a_{V_\theta}'(\theta,\lambda)$

$$\frac{d}{d\tau}\left[\iiint_\Omega \sigma_\tau(z_\tau,\theta_\tau)m_p\,d\Omega\right]\bigg|_{\tau=0} = \ell_{V_\theta}'(\lambda) - a_{V_\theta}'(\theta,\lambda) + \iiint_\Omega [C:\varepsilon^V(z)]m_p\,d\Omega. \tag{6.310}$$

Even if the structural adjoint response η does not appear in (6.310), its calculation is required in order to evaluate the structural adjoint load in (6.306).

Different types of performance measures can be treated using a procedure similar to the one described in this section. Table 6.11 summarizes the adjoint variable method procedure for displacement and temperature performance measures. Note that the structural adjoint equation is not required for temperature sensitivity calculations.

6.6 Second-Order Shape Design Sensitivity Analysis

Second-order design sensitivity analysis computes the Hessian information of a performance measure and is useful in many applications, such as a robust (i.e., insensitive) design for improving product quality and a reliability analysis for improving prediction accuracy. To improve product quality through the robust design, product sensitivity needs to be reduced with respect to environmental variations. To minimize this sensitivity, it is necessary to have accurate second-order shape design sensitivity.

A continuum design sensitivity approach with a material derivative is used to derive explicit formulas of the second-order shape design sensitivity for stress and displacement performance measures. Both the direct differentiation and hybrid methods are presented. Efficiency of these methods depends on the number of shape design variables and performance measures. It is interesting to note that the computation of an acceleration field needs little additional effort if the design velocity field is available. Furthermore, by assuming a linear mapping relation in a shape perturbation, the effects of design acceleration are completely eliminated, which is very helpful when developing a practical numerical method for second-order shape design sensitivity analysis. However, further investigation of the regularity conditions of displacement and velocity fields is needed.

6.6.1 Second-Order Material Derivative Formulas

In this section, some preliminary definitions, assumptions, and basic formulations are introduced. Thereafter, all necessary functions, functionals, and shapes are assumed to be smooth enough for their second derivatives to exist.

In the development of shape variation in Section 6.1, the linear mapping relation in (6.5) is used to represent perturbation of the structural domain. This choice of a linear mapping relation is valid since higher-order terms have vanished. In general, however, a quadratic mapping relation can be used in the second-order variation as

$$\boldsymbol{T}(\boldsymbol{x},\tau) = \boldsymbol{x} + \tau \boldsymbol{V}(\boldsymbol{x}) + \frac{\tau^2}{2}\dot{\boldsymbol{V}}(\boldsymbol{x}), \tag{6.311}$$

where

$$\dot{\boldsymbol{V}}(\boldsymbol{x}) = \frac{d}{d\tau}\boldsymbol{V}(\boldsymbol{x}_\tau,\tau)\bigg|_{\tau=0} \tag{6.312}$$

is the material derivative of the design velocity field. With the design velocity field as a function of $\boldsymbol{x}_\tau$, the *design acceleration field* can be derived by taking the material derivative of $\boldsymbol{V}$. From the relation in (6.8) we obtain

$$\boldsymbol{V}'(\boldsymbol{x}) = \dot{\boldsymbol{V}}(\boldsymbol{x}) - \nabla \boldsymbol{V}\boldsymbol{V}(\boldsymbol{x}), \tag{6.313}$$

where the partial derivative $\boldsymbol{V}'(\boldsymbol{x})$ is defined as the design acceleration field [107]. In (6.313), $\nabla\boldsymbol{V} = [V_{i,j}] = [\partial V_i/\partial x_j]$ is the Jacobian matrix of $\boldsymbol{V}$. Here, velocity $\boldsymbol{V}$ is assumed to be smooth enough that the partial derivative $\boldsymbol{V}'(\boldsymbol{x})$ and the material derivative $\dot{\boldsymbol{V}}(\boldsymbol{x})$ exist.

Using the definition of a design acceleration field, the second-order material derivative of state variable $z_\tau(\boldsymbol{x}_\tau)$ at $\boldsymbol{x} \in \Omega$ can be defined as [108] and [109]

$$\begin{aligned}
\ddot{z}_i &= \ddot{z}_i(\boldsymbol{x};\Omega,\boldsymbol{V}) \\
&\equiv \frac{d^2}{d\tau^2} z_{i_\tau}(\boldsymbol{x} + \tau\boldsymbol{V}(\boldsymbol{x}) + \tfrac{\tau^2}{2}\dot{\boldsymbol{V}}(\boldsymbol{x}))\bigg|_{\tau=0} \\
&= (z_i' + \nabla z_i^T\boldsymbol{V})' + \nabla(z_i' + \nabla z_i^T\boldsymbol{V})^T\boldsymbol{V} \\
&= z_i'' + 2(\nabla z_i')^T(\boldsymbol{V}) + \boldsymbol{V}^T\nabla(\nabla z_i)\boldsymbol{V} + (\nabla z_i)^T\boldsymbol{V}' + (\nabla z_i)^T(\nabla\boldsymbol{V})\boldsymbol{V},
\end{aligned} \tag{6.314}$$

where z'' is the second-order partial derivative of z with respect to τ, and $\nabla(\nabla z_i) = [\partial^2 z_i/\partial x_j\partial x_k]$ is the Hessian matrix of z_i. Due to the smoothness assumption, the partial derivative with respect to τ is commutative with the partial derivative with respect to $\boldsymbol{x}$, because they are both derivatives with respect to independent variables.

In order to derive the second-order variation of a functional, we first need to derive the second-order material derivative of the Jacobian matrix of the transformation $\boldsymbol{T}$, and then the second-order material derivative of $\boldsymbol{n}_\tau$, which is the outward unit-normal vector on the boundary Γ_τ of the deformed domain Ω_τ.

From (6.311), the Jacobian matrix of the transformation $\boldsymbol{T}$ can be defined as

$$\begin{aligned}
\boldsymbol{J} &\equiv \left[\frac{\partial T_i}{\partial x_j}\right] = \boldsymbol{I} + \tau\left[\frac{\partial V_i}{\partial x_j}\right] + \frac{\tau^2}{2}\left[\frac{\partial \dot{V}_i}{\partial x_j}\right] \\
&= \boldsymbol{I} + \tau\nabla\boldsymbol{V}(\boldsymbol{x}) + \frac{\tau^2}{2}\nabla\dot{\boldsymbol{V}}(\boldsymbol{x}),
\end{aligned} \tag{6.315}$$

where $\boldsymbol{I}$ is the identity matrix, and $\nabla \boldsymbol{V}(\boldsymbol{x})$ and $\nabla \dot{\boldsymbol{V}}(\boldsymbol{x})$ are Jacobian matrices of $\boldsymbol{V}(\boldsymbol{x})$ and $\dot{\boldsymbol{V}}(\boldsymbol{x})$, respectively. Let $|\boldsymbol{J}|$ be the determinant of $\boldsymbol{J}$. The first-order material derivatives of $\boldsymbol{J}$ and $|\boldsymbol{J}|$ were obtained in (6.15) and (6.18). The second-order material derivatives of $\boldsymbol{J}$ can be calculated as

$$\left.\frac{d^2 \boldsymbol{J}}{d\tau^2}\right|_{\tau=0} = \nabla \dot{\boldsymbol{V}}(\boldsymbol{x}) \tag{6.316}$$

and

$$\left.\frac{d^2 \boldsymbol{J}^{-1}}{d\tau^2}\right|_{\tau=0} = 2\nabla \boldsymbol{V}(\boldsymbol{x})\nabla \boldsymbol{V}(\boldsymbol{x}) - \nabla \dot{\boldsymbol{V}}(\boldsymbol{x}). \tag{6.317}$$

In (6.317), the relation $d^2(\boldsymbol{J}^T\boldsymbol{J}^{-T})/d\tau^2 = \boldsymbol{0}$ is used.

The second-order material derivatives of $|\boldsymbol{J}|$ can be obtained by computing the determinant of (6.315) and by taking the material derivative twice as

$$\left.\frac{d^2 |\boldsymbol{J}|}{d\tau^2}\right|_{\tau=0} = 2M_{ii} + div\dot{\boldsymbol{V}}, \tag{6.318}$$

where M_{ii} is the principal minors of the matrix $\nabla \boldsymbol{V}(\boldsymbol{x})$, and summation rule is used between repeated indices.

Let $\boldsymbol{n}$ be the outward unit-normal vector on boundary Γ of domain Ω, and let $\boldsymbol{n}_\tau$ be the outward unit-normal vector on the boundary Γ_τ of deformed domain Ω_τ. The relation between $\boldsymbol{n}$ and $\boldsymbol{n}_\tau$ is given in (6.31) and the first-order derivative is given in (6.34). To find the second-order derivative of $\boldsymbol{n}_\tau$, consider the second-order derivative of $\left\|\boldsymbol{J}^{-T}(\boldsymbol{x}_\tau)\boldsymbol{n}(\boldsymbol{x})\right\|$, which can be obtained as

$$\begin{aligned}
\left.\frac{d^2}{d\tau^2}\left\|\boldsymbol{J}^{-T}(\boldsymbol{x}_\tau)\boldsymbol{n}(\boldsymbol{x})\right\|\right|_{\tau=0} &\equiv \left.\frac{d^2}{d\tau^2}(\boldsymbol{J}^{-T}\boldsymbol{n},\boldsymbol{J}^{-T}\boldsymbol{n})^{1/2}\right|_{\tau=0} \\
&= (\nabla \boldsymbol{V}^T\boldsymbol{n},\nabla \boldsymbol{V}^T\boldsymbol{n}) + 2(\nabla \boldsymbol{V}\boldsymbol{n},\nabla \boldsymbol{V}^T\boldsymbol{n}) \\
&\quad -(\nabla \boldsymbol{V}\boldsymbol{n},\boldsymbol{n})^2 - (\nabla \dot{\boldsymbol{V}}^T\boldsymbol{n},\boldsymbol{n})^2,
\end{aligned} \tag{6.319}$$

and the second-order material derivative of $\boldsymbol{n}_\tau$ is

$$\begin{aligned}
\ddot{\boldsymbol{n}} \equiv \left.\frac{d^2\boldsymbol{n}_t}{d\tau^2}\right|_{\tau=0} &= \left.\frac{d}{d\tau}\left(\frac{d\boldsymbol{J}(\boldsymbol{x}_\tau)^{-T}}{d\tau}\boldsymbol{n}(\boldsymbol{x}) \Big/ \left\|\boldsymbol{J}^{-T}(\boldsymbol{x}_\tau)\boldsymbol{n}(\boldsymbol{x})\right\|\right)\right|_{\tau=0} \\
&\quad -\left.\frac{d}{d\tau}\left(\boldsymbol{J}^{-T}(\boldsymbol{x}_\tau)\boldsymbol{n}(\boldsymbol{x})\frac{d\left\|\boldsymbol{J}^{-T}(\boldsymbol{x}_\tau)\boldsymbol{n}(\boldsymbol{x})\right\|}{d\tau} \Big/ \left\|\boldsymbol{J}^{-T}(\boldsymbol{x}_\tau)\boldsymbol{n}(\boldsymbol{x})\right\|^2\right)\right|_{t=0} \\
&= 2\nabla \boldsymbol{V}^T\nabla \boldsymbol{V}^T\boldsymbol{n} - \nabla \dot{\boldsymbol{V}}^T\boldsymbol{n} - 2\nabla \boldsymbol{V}^T\boldsymbol{n}(\nabla \boldsymbol{V}\boldsymbol{n},\boldsymbol{n}) - \boldsymbol{n}(\nabla \boldsymbol{V}^T\boldsymbol{n},\nabla \boldsymbol{V}^T\boldsymbol{n}) \\
&\quad + \boldsymbol{n}(\nabla \dot{\boldsymbol{V}}^T\boldsymbol{n},\boldsymbol{n}) - 2\boldsymbol{n}(\nabla \boldsymbol{V}\boldsymbol{n},\nabla \boldsymbol{V}^T\boldsymbol{n}) + 3\boldsymbol{n}(\nabla \boldsymbol{V}\boldsymbol{n},\boldsymbol{n})^2,
\end{aligned} \tag{6.320}$$

where $(\boldsymbol{a},\boldsymbol{b}) \equiv \boldsymbol{a}^T\boldsymbol{b}$. Shape design sensitivity analysis can be carried out using these formulas. However, in the following derivation, we will use special types of the design velocity field with the property of $\dot{\boldsymbol{V}}(\boldsymbol{x}) = 0$ (as explained in Example 6.8). From a practical point of view, these kinds of design velocity fields are very attractive, since the transformation mapping in (6.311) can be simplified to $\boldsymbol{T}(\boldsymbol{x},\tau) = \boldsymbol{x} + \tau\boldsymbol{V}(\boldsymbol{x})$. Thus, we can drop those terms that involve $\dot{\boldsymbol{V}}(\boldsymbol{x})$ in (6.311) and (6.315), as well as from subsequent equations that are derived from these two.

Example 6.8. Linear Mapping. The quadratic mapping in (6.311) is mathematically rigorous but impractical in real applications. Many geometric modelers use parametric representations of the geometry, and those representations are often chosen as design variables, as described in Section 12.2 of Chapter 12. Consider the shape design parameterization of the geometric surface in Fig. 12.27. A three-dimensional geometric surface can be mathematically represented by the $4 \times 4 \times 3$ geometric coefficient matrix $\boldsymbol{B}$, as in (12.20). The geometric coefficients are the positions, tangent vectors, and twist vectors at the four corner points of the geometric surface. Matrix $\boldsymbol{B}$ is rewritten as

$$\boldsymbol{G} = \boldsymbol{B} = \begin{bmatrix} \boldsymbol{p}_{00} & \boldsymbol{p}_{01} & \boldsymbol{p}^{w}_{00} & \boldsymbol{p}^{w}_{01} \\ \boldsymbol{p}_{10} & \boldsymbol{p}_{11} & \boldsymbol{p}^{w}_{10} & \boldsymbol{p}^{w}_{11} \\ \boldsymbol{p}^{u}_{00} & \boldsymbol{p}^{u}_{01} & \boldsymbol{p}^{uw}_{00} & \boldsymbol{p}^{uw}_{01} \\ \boldsymbol{p}^{u}_{10} & \boldsymbol{p}^{u}_{11} & \boldsymbol{p}^{uw}_{10} & \boldsymbol{p}^{uw}_{11} \end{bmatrix}_{4\times4\times3}, \tag{6.321}$$

where $u \in [0,1]$ and $w \in [0,1]$ are the local coordinates on the patch. Each point on the surface has the unique values of u and w. The 48 geometric coefficients of matrix $\boldsymbol{B}$ can be defined as the shape design variable $\boldsymbol{b}$.

By using geometric coefficients as the shape design variables, the isoparametric mapping method can be used to compute the boundary and domain velocity fields. The domain acceleration field can be computed from the domain velocity field with little additional computational costs.

To compute the boundary velocity field, geometric coefficient matrix $\boldsymbol{B}$ should be transformed into the algebraic coefficient matrix $\boldsymbol{A}$ by premultiplying a constant matrix $\boldsymbol{M}$ (see Fig. 12.26). From the algebraic coefficient matrix $\boldsymbol{A}$, the velocity field can be computed as

$$\boldsymbol{V} = \boldsymbol{U}\delta\boldsymbol{A}\boldsymbol{W}^{T}, \tag{6.322}$$

where vectors $\boldsymbol{U} = [1, u, u^2, u^3]^T$ and $\boldsymbol{W} = [1, w, w^2, w^3]^T$ are the locations of the nodes in the parametric direction of the geometric surface. Since an algebraic surface can be used to directly compute the boundary velocity field, the geometric surface must be transformed. For the geometric surface, $\boldsymbol{A} = \boldsymbol{MBM}^T$ and, as a result,

$$\boldsymbol{V} = \boldsymbol{UM}\delta\boldsymbol{BM}^{T}\boldsymbol{W}^{T}. \tag{6.323}$$

Since the coefficient of $\boldsymbol{B}$ is defined as design $\boldsymbol{b}$, the perturbation of design $\delta\boldsymbol{b}$ is the coefficient of matrix $\delta\boldsymbol{B}$. One final note: from (6.323), it is clear that the design velocity field depends on the position of $\boldsymbol{x}$, and on the perturbations of the shape design variables $\delta\boldsymbol{b}$, rather than on design variable $\boldsymbol{b}$. Thus, $\dot{\boldsymbol{V}}(\boldsymbol{x}) = 0$, and the design acceleration field can be computed from

$$\begin{aligned} \dot{\boldsymbol{V}}(\boldsymbol{x}) &= \boldsymbol{V}' + (\nabla\boldsymbol{V})\boldsymbol{V} = 0 \\ &\Rightarrow \boldsymbol{V}' = -(\nabla\boldsymbol{V})\boldsymbol{V}, \end{aligned} \tag{6.324}$$

where $\boldsymbol{V}'$ is the design acceleration field. Once the domain velocity field is obtained, only a small amount of computational effort is required to obtain the domain acceleration field. The relation in (6.324) is used in the development of a second-order design sensitivity analysis.

Using (6.311), (6.313), (6.314), and (6.318), the second-order material derivative of the general domain functional ψ_1, defined as an integral over the domain Ω_τ, can be derived. Let

$$\psi_1 = \iiint_{\Omega_\tau} f_\tau(\boldsymbol{x}_\tau)\, d\Omega_\tau, \tag{6.325}$$

where f_τ is a regular function defined on domain Ω_τ. If the domain Ω_τ is smooth enough, then the second-order variation of ψ_1 at Ω is obtained as

$$\begin{aligned}
\psi_1'' &= \frac{d^2}{d\tau^2} \iiint_{\Omega} f_\tau(\boldsymbol{x}+\tau\boldsymbol{V}(\boldsymbol{x}))\,|\boldsymbol{J}|\,d\Omega\Big|_{\tau=0} \\
&= \iiint_{\Omega} [f'' + 2(\nabla f')^T \boldsymbol{V} + \boldsymbol{V}^T \nabla(\nabla f)\boldsymbol{V} + \nabla f^T \boldsymbol{V}' \\
&\qquad + \nabla f^T (\nabla \boldsymbol{V})\boldsymbol{V} + 2(f' + \nabla f^T \boldsymbol{V}) div\boldsymbol{V} + 2fM_{ii}]\,d\Omega \\
&= \iiint_{\Omega} [f'' + 2(\nabla f')^T \boldsymbol{V} + \boldsymbol{V}^T \nabla(\nabla f)\boldsymbol{V} \\
&\qquad + 2(f' + \nabla f^T \boldsymbol{V}) div\boldsymbol{V} + 2fM_{ii}]\,d\Omega.
\end{aligned} \tag{6.326}$$

Note that the term containing design acceleration is canceled by assuming linear mapping $\boldsymbol{T}(\boldsymbol{x},\tau)$ and by using the relation in (6.324). Thus, virtually no design acceleration information is required in evaluation of the integral in (6.326).

Next, consider a functional defined as an integral over Γ_τ, the boundary of domain Ω_τ, as

$$\psi_2 = \iint_{\Gamma_\tau} g_\tau(\boldsymbol{x}_\tau)\,d\Gamma_\tau = \iint_{\Gamma} g_\tau(\boldsymbol{x}+\tau\boldsymbol{V})\,|\boldsymbol{J}|\,\|\boldsymbol{J}^{-T}\boldsymbol{n}\|\,d\Gamma, \tag{6.327}$$

where the transformation $d\Gamma_\tau = |\boldsymbol{J}(\boldsymbol{x}_\tau)|\,\|\boldsymbol{J}(\boldsymbol{x}_\tau)^{-T}\boldsymbol{n}(\boldsymbol{x})\|\,d\Gamma$ is used. Thus, by using (6.314), (6.315), (6.318), and (6.319), the second-order variation of ψ_2 at Γ is

$$\begin{aligned}
\psi_2'' &= \frac{d^2}{d\tau^2}\Big[\iint_{\Gamma} g_\tau(\boldsymbol{x}+\tau\boldsymbol{V})\,|\boldsymbol{J}|\,\|\boldsymbol{J}^{-T}\boldsymbol{n}\|\,d\Gamma\Big]\Big|_{\tau=0} \\
&= \iint_{\Gamma} [\ddot{g} + 2\dot{g}(div\boldsymbol{V} - (\nabla\boldsymbol{V}\boldsymbol{n},\boldsymbol{n})) \\
&\qquad - 2g(div\boldsymbol{V})(\nabla\boldsymbol{V}\boldsymbol{n},\boldsymbol{n}) + 2gM_{ii} + g(\nabla\boldsymbol{V}^T\boldsymbol{n},\nabla\boldsymbol{V}^T\boldsymbol{n}) \\
&\qquad + 2g(\nabla\boldsymbol{V}\boldsymbol{n},\nabla\boldsymbol{V}^T\boldsymbol{n}) - g(\nabla\boldsymbol{V}\boldsymbol{n},\boldsymbol{n})^2]\,d\Gamma \\
&= \iint_{\Gamma} [g'' + 2(\nabla g')^T\boldsymbol{V} + \boldsymbol{V}^T\nabla(\nabla g)\boldsymbol{V} \\
&\qquad + 2(g' + (\nabla g)^T\boldsymbol{V})(div\boldsymbol{V} - (\nabla\boldsymbol{V}\boldsymbol{n},\boldsymbol{n})) \\
&\qquad - 2g(div\boldsymbol{V})(\nabla\boldsymbol{V}\boldsymbol{n},\boldsymbol{n}) + 2gM_{ii} + g(\nabla\boldsymbol{V}^T\boldsymbol{n},\nabla\boldsymbol{V}^T\boldsymbol{n}) \\
&\qquad + 2g(\nabla\boldsymbol{V}\boldsymbol{n},\nabla\boldsymbol{V}^T\boldsymbol{n}) - g(\nabla\boldsymbol{V}\boldsymbol{n},\boldsymbol{n})^2]\,d\Gamma.
\end{aligned} \tag{6.328}$$

Another functional, also defined as an integral over Γ_τ, is

$$\begin{aligned}
\psi_3 &= \iint_{\Gamma_\tau} \boldsymbol{h}_\tau(\boldsymbol{x}_\tau)^T \boldsymbol{n}_\tau\,d\Gamma_\tau \\
&= \iint_{\Gamma} \boldsymbol{h}_\tau(\boldsymbol{x}+\tau\boldsymbol{V})^T \boldsymbol{n}_\tau\,|\boldsymbol{J}|\,\|\boldsymbol{J}^{-T}\boldsymbol{n}\|\,d\Gamma.
\end{aligned} \tag{6.329}$$

Its second-order variation is derived by using (6.314), (6.315), (6.318), (6.319), and (6.320) as

$$\begin{aligned}\psi_3'' &= \frac{d^2}{d\tau^2}\left[\iint_{\Gamma} \boldsymbol{h}_\tau(\boldsymbol{x}+\tau\boldsymbol{V})^T \boldsymbol{n}_\tau |\boldsymbol{J}| \|\boldsymbol{J}^{-T}\boldsymbol{n}\| d\Gamma\right]_{\tau=0} \\ &= \iint_{\Gamma} [h_i'' n_i + 2(\nabla h_i')^T \boldsymbol{V} n_i + \boldsymbol{V}^T \nabla(\nabla h_i)\boldsymbol{V} n_i \\ &\quad + 2h_i \nabla V^{ji} \nabla V^{kj} n_k + 2h_i n_i M_{jj} \\ &\quad - 2h_i' \nabla V^{ji} n_j - 2(\nabla h_i)^T \boldsymbol{V} \nabla V^{ji} n_j + 2h_i' n_i (div\boldsymbol{V}) \\ &\quad + 2(\nabla h_i)^T \boldsymbol{V} n_i div\boldsymbol{V} - 2h_i \nabla V^{ji} n_j] d\Gamma.\end{aligned} \tag{6.330}$$

In the following sections, (6.325) through (6.330) will be used to derive the second-order variation of the energy bilinear and load linear forms of a three-dimensional elastic solid with respect to shape design variables, and to derive the second-order shape design sensitivity of stress and displacement performance measures.

6.6.2 Direct Differentiation Method

The direct difference method uses the second-order variation of the energy bilinear and load linear forms to obtain second-order sensitivity information.

The variational identity for the three-dimensional linear elastic solid can be written as

$$\begin{aligned}&\iiint_{\Omega} \sigma_{ij}(\boldsymbol{z})\varepsilon_{ij}(\bar{\boldsymbol{z}})\, d\Omega - \iiint_{\Omega} \bar{z}_i f_i^b \, d\Omega \\ &\qquad = \iint_{\Gamma} \sigma_{ij}(\boldsymbol{z}) n_j \bar{z}_i \, d\Gamma, \qquad \forall \bar{\boldsymbol{z}} \in [H^1(\Omega)]^3,\end{aligned} \tag{6.331}$$

where $\boldsymbol{z} = [z_1, z_2, z_3]^T$ is the displacement function, $\boldsymbol{n} = [n_1, n_2, n_3]^T$ is the outward unit normal at the boundary, and $H^1(\Omega)$ is the Sobolev space of order one. If boundary conditions are imposed on boundary Γ, in which Γ^h has a prescribed displacement and Γ^s has a traction load, then

$$\begin{aligned} z_i &= 0, & \boldsymbol{x} &\in \Gamma^h \\ \sigma_{ij} n_j &= f_i^s, & \boldsymbol{x} &\in \Gamma^s.\end{aligned} \tag{6.332}$$

By using these boundary conditions, (6.331) can be rewritten as the variational equation

$$\begin{aligned} a_\Omega(\boldsymbol{z},\bar{\boldsymbol{z}}) &\equiv \iiint_{\Omega} \sigma_{ij}(\boldsymbol{z})\varepsilon_{ij}(\bar{\boldsymbol{z}})\, d\Omega \\ &= \iiint_{\Omega} \bar{z}_i f_i^b \, d\Omega + \iint_{\Gamma^s} \bar{z}_i f_i^s \, d\Gamma \equiv \ell_\Omega(\bar{\boldsymbol{z}}), \qquad \forall \bar{\boldsymbol{z}} \in Z,\end{aligned} \tag{6.333}$$

where Z is the space of kinematically admissible virtual displacements, written as

$$Z = \left\{ \boldsymbol{z} \in [H^1(\Omega)]^3 : \boldsymbol{z} = 0,\ \boldsymbol{x} \in \Gamma^h \right\}. \tag{6.334}$$

To obtain the design sensitivity, the first- and second-order material derivatives of the variational equation (6.333) must be taken. To this end, the first- and second-order material derivatives of general functionals defined by domain and boundary integrals are taken. The first-order derivative of variational equation (6.333) is developed in Section 6.2.6.

For second-order sensitivity information, taking the second-order material derivative of $a_{\Omega_\tau}(\boldsymbol{z}_\tau, \bar{\boldsymbol{z}}_\tau)$ from (6.333) yields

$$\frac{d^2}{d\tau^2}\left[a_{\Omega_\tau}(\boldsymbol{z}_\tau, \bar{\boldsymbol{z}}_\tau)\right]\Big|_{\tau=0} = a_\Omega(\ddot{\boldsymbol{z}}, \bar{\boldsymbol{z}}) + 2a_V'(\dot{\boldsymbol{z}}, \bar{\boldsymbol{z}}) + a_V''(\boldsymbol{z}, \bar{\boldsymbol{z}}), \tag{6.335}$$

where

$$a'_V(\dot{z},\bar{z}) \equiv \iiint_\Omega \{\sigma_{ij}(\bar{z})\varepsilon^V_{ij}(\dot{z}) + \sigma_{ij}(\dot{z})\varepsilon^V_{ij}(\bar{z}) + \sigma_{ij}(\dot{z})\varepsilon_{ij}(\bar{z}) divV\} d\Omega \tag{6.336}$$

$$\begin{aligned} a''_V(z,\bar{z}) \equiv \iiint_\Omega \{ & \sigma_{ij}(\bar{z})\eta^V_{ij}(z) + \sigma_{ij}(z)\eta^V_{ij}(\bar{z}) \\ & -2[\sigma_{ij}(\bar{z})\varepsilon^V_{ij}(z) + \sigma_{ij}(z)\varepsilon^V_{ij}(\bar{z})] divV \\ & +2\varepsilon^V_{ij}(\bar{z})C_{ijkl}\varepsilon^V_{kl}(z) + 2\sigma_{ij}(z)\varepsilon_{ij}(\bar{z})M_{rr}\} d\Omega. \end{aligned} \tag{6.337}$$

In (6.336) and (6.337), $\varepsilon^V_{ij}(\bullet)$ and $\eta^V_{ij}(\bullet)$ are the terms that are explicitly dependent on design velocity. The expression of $\varepsilon^V_{ij}(\bullet)$ has already been introduced in (6.200) when first-order sensitivity was developed. The expression of $\eta^V_{ij}(\bullet)$ is defined as

$$\eta^V_{ij}(z) = \frac{1}{2}[\nabla z \nabla V \nabla V + (\nabla z \nabla V \nabla V)^T]. \tag{6.338}$$

Note that only the first-order derivative of the design velocity field is required in the fictitious load forms in (6.336) and (6.337). This is a very important feature in the application stage, since a C^0-continuous design velocity field that has an integrable first-order derivative is enough to develop a second-order design sensitivity analysis, producing the same regularity requirement as the first-order design sensitivity analysis presented in Section 6.2.7. Such a convenience is made possible from the assumption of a linear mapping relation and from the property $\dot{V}(\boldsymbol{x}) = 0$.

The traction-loaded boundary can be treated by taking a second-order variation of the right side of (6.333), assuming partial derivatives of $\boldsymbol{f}^b$ and $\boldsymbol{f}^s$ are zero (i.e., $\boldsymbol{f}^{b\prime} = \boldsymbol{f}^{b\prime\prime} = \boldsymbol{f}^{s\prime} = \boldsymbol{f}^{s\prime\prime} = 0$), producing

$$\begin{aligned} \frac{d^2}{d\tau^2}\left[\ell_{\Omega_\tau}(\bar{z}_\tau)\right]\Big|_{\tau=0} & \\ = \iiint_\Omega & \bar{z}_i[\boldsymbol{V}^T\nabla(\nabla f_i^b)\boldsymbol{V} + 2(\nabla f_i^b)^T\boldsymbol{V}(div\boldsymbol{V}) + 2f_i^b M_{rr}]d\Omega \\ - \iint_{\Gamma^s} & \bar{z}_i[\boldsymbol{V}^T\nabla(\nabla f_i^s)\boldsymbol{V} + 2(\nabla f_i^s)^T\boldsymbol{V}(div\boldsymbol{V}) \\ & -2(\nabla f_i^s)^T\boldsymbol{V}(\nabla\boldsymbol{V}\boldsymbol{n},\boldsymbol{n}) - 2f_i^s(div\boldsymbol{V})(\nabla\boldsymbol{V}\boldsymbol{n},\boldsymbol{n}) + 2f_i^s M_{rr} \\ & + f_i^s(\nabla\boldsymbol{V}^T\boldsymbol{n},\nabla\boldsymbol{V}^T\boldsymbol{n}) + 2f_i^s(\nabla\boldsymbol{V}\boldsymbol{n},\nabla\boldsymbol{V}^T\boldsymbol{n}) - f_i^s(\nabla\boldsymbol{V}\boldsymbol{n},\boldsymbol{n})^2]d\Gamma \\ \equiv \ell''_V(\bar{z}). & \end{aligned} \tag{6.339}$$

The terms containing $\ddot{\bar{z}}$ and $\dot{\bar{z}}$ in (6.335) and (6.339) have been neglected, since they cancel each other out after the variational identity and the first-order variation of the variational elasticity equation for three-dimensional elastic solids have been used.

From (6.335) and (6.339), the second-order material derivative of the variational equation (6.333) can be written as

$$a_\Omega(\ddot{z},\bar{z}) = \ell''_V(\bar{z}) - 2a'_V(\dot{z},\bar{z}) - a''_V(z,\bar{z}), \qquad \forall \bar{z} \in Z, \tag{6.340}$$

where $\dot{z}$ is the material derivative of the displacement and can be obtained from (6.67). Solving finite element reanalysis with the fictitious load on the right of (6.340), the second-order shape design sensitivity of displacement $\ddot{z}$ can be obtained, yielding the direct differentiation method.

To obtain the second-order variation of a domain stress performance measure, let the general mean stress performance measure be defined over a fixed test volume Ω_p as

$$\psi = \iiint_{\Omega} g(\boldsymbol{\sigma}(z)) m_p \, d\Omega = \frac{\iiint_{\Omega_p} g(\boldsymbol{\sigma}(z)) \, d\Omega}{\iiint_{\Omega_p} d\Omega}, \tag{6.341}$$

where m_p is a characteristic function with a constant value of $m_p = (\iiint_{\Omega_p} d\Omega)^{-1}$ on volume Ω_p, a value of zero outside the volume, and with an integral of one. If the volume Ω_p becomes a point, then m_p becomes the Dirac delta measure and the performance measure ψ in (6.341) is the stress at that point. Note that $g(\boldsymbol{\sigma}(z))$ might be the principal stresses, the von Mises stress, or some other material failure criteria.

The second-order shape design sensitivity of the stress performance measure can be derived by taking the second-order variation of (6.341) as

$$\begin{aligned}
\psi'' = {} & \iiint_{\Omega} g_{,\sigma_{ij}\sigma_{kl}} [\sigma_{ij}(\dot{z})\sigma_{kl}(\dot{z}) - 2\sigma_{ij}(\dot{z}) C_{klst} \varepsilon^V_{st}(z) + C_{ijst}\varepsilon^V_{st}(z) C_{kluv}\varepsilon^V_{uv}(z)] m_p \, d\Omega \\
& + \iiint_{\Omega} g_{,\sigma_{ij}} [\sigma_{ij}(\ddot{z}) - 2C_{ijkl}\varepsilon^V_{kl}(\dot{z}) + C_{ijkl}\eta^V_{kl}(z) + 2\sigma_{ij}(\dot{z}) - 2C_{ijkl}\varepsilon^V_{kl}(z) divV] m_p \, d\Omega \\
& + 2\iiint_{\Omega} gM m_p \, d\Omega - 2\iiint_{\Omega} g divV m_p \, d\Omega \iiint_{\Omega} divV m_p \, d\Omega \\
& - 2\iiint_{\Omega} g m_p \, d\Omega \iiint_{\Omega} M m_p \, d\Omega + 2\iiint_{\Omega} g m_p \, d\Omega \left(\iiint_{\Omega} divV m_p \, d\Omega \right)^2 \\
& - 2\iiint_{\Omega} g_{\sigma_{ij}} [\sigma_{ij}(\dot{z}) - C_{ijkl}\varepsilon^V_{kl}(z)] m_p \, d\Omega \iiint_{\Omega} divV m_p \, d\Omega.
\end{aligned} \tag{6.342}$$

The second-order shape design sensitivity of the stress performance measure can then be obtained by substituting $\ddot{z}$ and $\dot{z}$ into (6.342).

The direct differentiation method is expensive if the design problem has a large number of shape design variables, due to computational costs. For a shape design problem with n shape design variables, this method requires one original finite element analysis and $(3k + k^2)/2$ finite element reanalyses, which include an n number of finite element reanalyses for the first-order shape design sensitivity, and $k(k + 1)/2$ finite element reanalyses (6.340) for the second-order shape design and mixed design sensitivities.

6.6.3 Hybrid Method

As in the discrete method of second-order sensitivity analysis in Section 4.1.5, the hybrid method uses the direct differentiation method to obtain first-order sensitivities and the adjoint variable method to obtain second-order sensitivities.

The displacement performance measure can be defined as

$$\psi_d = \iiint_{\Omega} \boldsymbol{e}^{iT} z \delta(\boldsymbol{x} - \boldsymbol{x}^*) d\Omega, \quad i = 1, 2, \text{or } 3, \tag{6.343}$$

where $\boldsymbol{e}^1 = [1, 0, 0]^T$, $\boldsymbol{e}^2 = [0, 1, 0]^T$, and $\boldsymbol{e}^3 = [0, 0, 1]^T$ are unit vectors. The second-order variation of ψ_d is

$$\psi''_d = \iiint_{\Omega} \boldsymbol{e}^{iT} \ddot{z} \delta(\boldsymbol{x} - \boldsymbol{x}^*) d\Omega. \tag{6.344}$$

The adjoint equation for the displacement performance measure can be defined by replacing $\ddot{z}$ in (6.344) by virtual displacement $\bar{\boldsymbol{\lambda}}$ and by equating the result to the energy bilinear form

$$a_{\Omega}(\boldsymbol{\lambda},\bar{\boldsymbol{\lambda}}) = \iiint_{\Omega} \boldsymbol{e}^{iT}\bar{\boldsymbol{\lambda}}\delta(\boldsymbol{x}-\boldsymbol{x}^*)d\Omega, \quad \forall \bar{\boldsymbol{\lambda}} \in Z. \tag{6.345}$$

The adjoint structural system can be interpreted as the original structure with a positive unit load applied at the point $\boldsymbol{x}^*$.

Let $\bar{\boldsymbol{\lambda}} = \ddot{z} \in Z$ in (6.345), and let $\bar{z} = \boldsymbol{\lambda} \in Z$ in (6.340). Therefore, (6.345) becomes

$$a_{\Omega}(\boldsymbol{\lambda},\ddot{z}) = \iiint_{\Omega} \boldsymbol{e}^{iT}\ddot{z}\delta(\boldsymbol{x}-\boldsymbol{x}^*)d\Omega \tag{6.346}$$

and (6.340) becomes

$$a_{\Omega}(\ddot{z},\boldsymbol{\lambda}) = -2a'_V(\dot{z},\boldsymbol{\lambda}) - a''_V(z,\boldsymbol{\lambda}) + \ell''_V(\boldsymbol{\lambda}). \tag{6.347}$$

Since the energy bilinear form $a_{\Omega}(\bullet,\bullet)$ is symmetric in its arguments, the left sides of (6.346) and (6.347) can be equated, and the second-order sensitivity of the displacement performance measure can be written as

$$\begin{aligned} \psi''_d &\equiv \iiint_{\Omega} \boldsymbol{e}^{iT}\ddot{z}\delta(\boldsymbol{x}-\boldsymbol{x}^*)d\Omega \\ &= -2a'_V(\dot{z},\boldsymbol{\lambda}) - a''_V(z,\boldsymbol{\lambda}) + \ell''_V(\boldsymbol{\lambda}), \end{aligned} \tag{6.348}$$

where $\boldsymbol{\lambda}$ is the adjoint response in (6.345). To evaluate (6.348), $\dot{z}$ needs to be obtained from (6.67) using the direct differentiation method.

Next, let ψ be the stress performance measure defined by (6.341). The adjoint equation for the stress performance measure can be defined by replacing the term that contains $\ddot{z}$ in (6.342) with the virtual displacement $\bar{\boldsymbol{\lambda}}$, and by equating the result to the energy bilinear form

$$a_{\Omega}(\boldsymbol{\lambda},\bar{\boldsymbol{\lambda}}) = \iiint_{\Omega} g_{,\sigma_{ij}}\sigma_{ij}(\bar{\boldsymbol{\lambda}})m_p\, d\Omega, \quad \forall \bar{\boldsymbol{\lambda}} \in Z. \tag{6.349}$$

By setting $\bar{\boldsymbol{\lambda}} = \ddot{z} \in Z$ in (6.345), $\bar{z} = \boldsymbol{\lambda} \in Z$ in (6.340), and by using the symmetric property of the energy bilinear form $a_{\Omega}(\bullet,\bullet)$, the following equation can be obtained

$$\iiint_{\Omega} g_{,\sigma_{ij}}\sigma_{ij}(\ddot{z})m_p\, d\Omega = -2a'_V(\dot{z},\boldsymbol{\lambda}) - a''_V(z,\boldsymbol{\lambda}) + \ell''_V(\boldsymbol{\lambda}), \tag{6.350}$$

where $\boldsymbol{\lambda}$ is the adjoint response in (6.349).

Substituting (6.350) for its corresponding term in (6.342), the second-order shape design sensitivity of the stress performance measure can be written as

$$\begin{aligned} \psi'' &= \iiint_{\Omega} g_{,\sigma_{ij}\sigma_{kl}}[\sigma_{ij}(\dot{z})\sigma_{kl}(\dot{z}) - 2\sigma_{ij}(\dot{z})C_{klst}\varepsilon^V_{st}(z) + C_{ijst}\varepsilon^V_{st}(z)C_{kluv}\varepsilon^V_{uv}(z)]m_p\, d\Omega \\ &+ \iiint_{\Omega} g_{,\sigma_{ij}}[-2C_{ijkl}\varepsilon^V_{kl}(\dot{z}) + C_{ijkl}\eta^V_{kl}(z) + 2\sigma_{ij}(\dot{z}) - 2C_{ijkl}\varepsilon^V_{kl}(z)divV]m_p\, d\Omega \\ &-2a'_V(\dot{z},\boldsymbol{\lambda}) - a''_V(z,\boldsymbol{\lambda}) + \ell''_V(\boldsymbol{\lambda}) \\ &+2\iiint_{\Omega} gMm_p\, d\Omega - 2\iiint_{\Omega} g\,divVm_p\, d\Omega \iiint_{\Omega} divVm_p\, d\Omega \\ &-2\iiint_{\Omega} gm_p\, d\Omega \iiint_{\Omega} Mm_p\, d\Omega + 2\iiint_{\Omega} gm_p\, d\Omega\left(\iiint_{\Omega} divVm_p\, d\Omega\right)^2 \\ &-2\iiint_{\Omega} g_{\sigma_{ij}}[\sigma_{ij}(\dot{z}) - C_{ijkl}\varepsilon^V_{kl}(z)]m_p\, d\Omega \iiint_{\Omega} divVm_p\, d\Omega, \end{aligned} \tag{6.351}$$

where $\dot{z}$ can be obtained from (6.67).

To obtain the second-order shape design sensitivity and mixed second-order design sensitivity using the hybrid method, one original finite element analysis and an k + NC number of finite element reanalyses are required for a design problem that has k number of shape design variables and NC number of performance measures. This includes k

number of finite element reanalyses for the first-order shape design sensitivity, and *NC* number of finite element reanalyses [(6.345) and/or (6.349)] for adjoint structures. Numerical computation using this method is more efficient than direct differentiation if the total number of *NC* performance measures is less than $(k + k^2)/2$.

6.6.4 Numerical Examples

In this section, a numerical method is presented for computing the second-order shape design sensitivity. PATRAN is used to create a geometric model and the ANSYS FEA code is used to analyze structural responses. Numerical results from the examples of a two-dimensional connecting rod and three-dimensional arch dam demonstrate the accuracy and feasibility of the second-order shape design sensitivity analysis method.

The shape design parameterization method explained in Example 6.8 is used to parameterize the three-dimensional geometric model. With the parametric representation of a boundary, an isoparametric mapping method is used to compute the boundary velocity field. The domain velocity field can then be computed using the boundary displacement method (see Section 13.3 of Chapter 13).

Connecting Rod

In order to demonstrate accuracy and feasibility of second-order shape design sensitivity, the two-dimensional connecting rod shown in Fig. 6.25 is chosen as a numerical example. With an in-plane firing load during the engine combustion cycle, this example is treated as a two-dimensional plane stress problem.

A geometric model of the connecting rod is created using PATRAN. There are 24 patches and 62 geometric lines in this geometric model. High stress appears in the shank and neck regions, so the boundary Γ_1 along the neck and shank region is chosen as the shape design boundary.

Six geometric lines are used to represent the design boundary Γ_1: lines 33, 34, 35, 38, 39, and 40. The rest of the boundaries are assumed to be fixed. The x_2-coordinates of grids 22, 23, 24, and 25 are chosen as the independent design variables, so that there are four design variables in this model. The values of each design variable are $b_1 = b_2 = 10.0$ and $b_3 = b_4 = 10.0$.

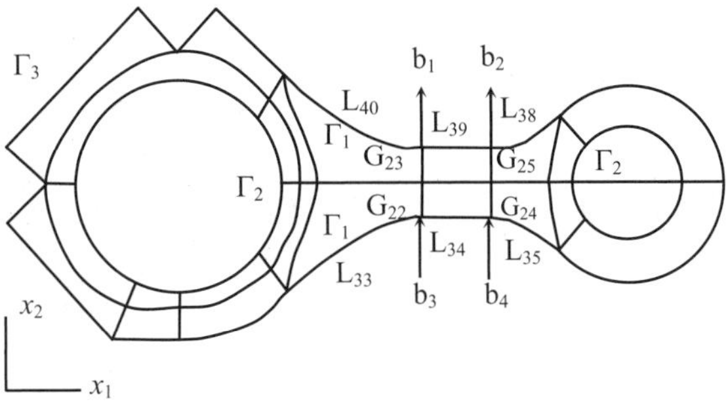

Figure 6.25. Geometric model of connecting rod.

Table 6.12. Firing (compressive) load on the boundaries Γ_2.

θ_1 (deg.)	Force(N)	θ_1 (deg.)	Force(N)	θ_1 (deg.)	Force(N)
−40	0	−10	19587	20	15741
−35	978	−5	20374	25	10335
−30	6868	0	21237	30	6103
−25	11210	5	22243	35	1402
−20	14689	10	20395	40	0
−15	17816	15	17426		

θ_2 (deg.)	Force(N)	θ_2 (deg.)	Force(N)	θ_2 (deg.)	Force(N)
−40	2234	−10	17499	20	15056
−35	6727	−5	17245	25	12622
−30	9808	0	16488	30	9803
−25	12536	5	17365	35	6409
−20	14917	10	17711	40	1055
−15	16308	15	16502		

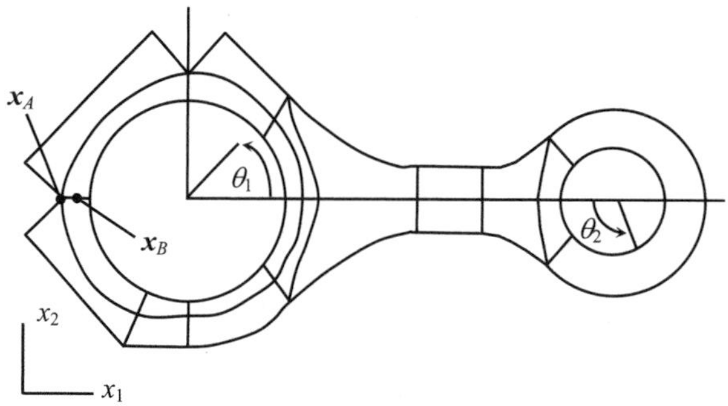

Figure 6.26. Boundary and load conditions.

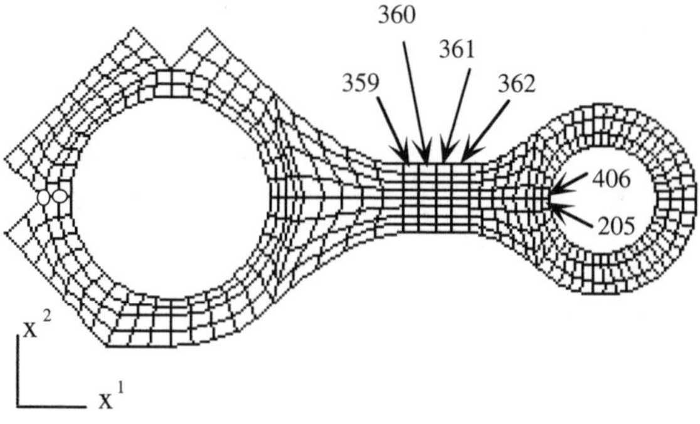

Figure 6.27. Finite element mesh of connecting rod.

In the finite element model, the firing load shown in Table 6.12 is considered. In order to eliminate rigid-body motion, point $\boldsymbol{x}_A$ is fixed in the x_1 and x_2 directions, and point $\boldsymbol{x}_B$ is fixed in the x_2 direction, as shown in Fig. 6.26. Such boundary conditions are reasonable because the firing load is self-equilibrated. The eight-node quadrilateral element STIF82 of ANSYS is used for static analysis. The finite element model of the connecting rod includes 1403 nodes, 406 quadrilateral elements, and 2803 degrees of freedom. Figure 6.27 shows the finite element mesh of the connecting rod.

Four shape design velocity fields are generated corresponding to the four independent design variables. Sixteen shape design acceleration fields are generated from the four shape design velocity fields. The direct differentiation and hybrid methods are used to compute the second-order shape design sensitivity.

As one can see from Fig. 6.27, four finite elements with high von Mises stress and four nodes with large displacement are chosen to verify the second-order shape design sensitivity. Elements 359 and 362 are in the neck region adjacent to the upper boundary Γ_1. Elements 205 and 406 are adjacent to the boundary Γ_2 of the small hole that is connected to a piston pin. The results are shown in Tables 6.13 through 6.15. In each table, design variables are perturbed by 1% for verification, using the finite difference method. The results of second-order shape design sensitivities of the displacement and von Mises stress performance measures with respect to design variables b_1 and b_2 are given in Tables 6.13 and 6.14, respectively. The results of the mixed second-order shape design sensitivity of the displacement and von Mises stress performance measures with respect to b_1 and b_2 are shown in Table 6.15.

Data have been classified as follows in the four tables: the first column is either the element ID or the node ID; the second column is either the stress or the displacement performance measure; the third column is the Gauss point ID; the fourth column are the second-order finite difference results; and the fifth and sixth columns are the second-order design sensitivity results obtained from the direct differentiation and hybrid methods, respectively. The last two columns are the verification results of the second-order shape design sensitivity obtained using two methods. In Tables 6.13 and 6.14,

$$\psi'' = \frac{\partial^2 \psi}{\partial b_i^2}(\delta b_2)^2 \tag{6.352}$$

and

$$\delta^2\psi = \psi(b_i + \delta b_i) - 2\psi(b_i) + \psi(b_i - \delta b_i) \tag{6.353}$$

for the second-order derivative with respect to design variable b_i where δb_i is the perturbation of the design variable b_i, $i = 1, 2$. In Table 6.15,

$$\psi'' = 4\frac{\partial^2 \psi}{\partial b_i \partial b_j}\delta b_i \delta b_j \tag{6.354}$$

and

$$\begin{aligned}\delta^2\psi &= \psi(b_i + \delta b_i, b_j + \delta b_j) - \psi(b_i - \delta b_i, b_j + \delta b_j) \\ &\quad - \psi(b_i + \delta b_i, b_j - \delta b_j) + \psi(b_i - \delta b_i, b_j - \delta b_j)\end{aligned} \tag{6.355}$$

for the mixed second-order derivative with respect to design variables b_i and b_j.

Results in Tables 6.13 through 6.15 indicate that the accuracy of the second-order shape design sensitivity is very good, although stress performance measures are defined for the elements with high stress. These numerical results demonstrate the accuracy and feasibility of the theoretical derivations and their numerical implementation in second-

Table 6.13. Verification of second-order shape design sensitivity of performance measures with respect to design variable b_1[a].

Element or Node	Performance Measure	Gauss Point	$\delta^2\psi(b)$ Finite Diff.	$\psi''(b)$ Direct Method	$\psi''(b)$ Hybrid Method	$\psi''/\delta^2\psi$ (Direct)	$\psi''/\delta^2\psi$ (Hybrid)
205	von Mises	4	–0.1911953E–02	–0.1912507E–02	–0.1912828E–02	100.03	100.04
359	von Mises	1	0.4297248E+01	0.4296987E+01	0.4296987E+01	99.99	99.99
362	von Mises	3	–0.2586931E+00	–0.2585918E+00	–0.2585918E+00	99.96	99.96
406	von Mises	3	–0.2009241E–02	–0.2008747E–02	–0.2008350E–02	99.98	99.96
270	z_x	—	–0.1514805E–04	–0.1514710E–04	–0.1514710E–04	100.01	100.01
279	z_x	—	–0.1514048E–04	–0.1513953E–04	–0.1513953E–04	100.01	100.01
12	z_y	—	0.4075609E–04	0.4075365E–04	0.4075366E–04	100.01	100.01
13	z_y	—	0.4078688E–04	0.4078444E–04	0.4078445E–04	100.01	100.01

[a] $b_1 = 10.0$ and $\delta b_1 = 0.1$

Table 6.14. Verification of second-order shape design sensitivity of performance measures with respect to design variable b_2[a].

Element or Node	Performance Measure	Gauss Point	$\delta^2\psi(b)$ Finite Diff.	$\psi''(b)$ Direct Method	$\psi''(b)$ Hybrid Method	$\psi''/\delta^2\psi$ (Direct)	$\psi''/\delta^2\psi$ (Hybrid)
205	von Mises	4	0.2563881E+00	0.2563777E+00	0.2563776E+00	100.00	100.00
359	von Mises	1	–0.2543052E+00	–0.2542091E+00	–0.2542091E+00	99.96	99.96
362	von Mises	3	0.4395137E+01	0.4394869E+01	0.4394869E+01	99.99	99.99
406	von Mises	3	0.2035245E+00	0.2035262E+00	0.2035265E+00	100.00	100.00
270	z_x	—	–0.1324177E–04	–0.1324094E–04	–0.1324094E–04	100.01	100.01
279	z_x	—	–0.1323826E–04	–0.1323743E–04	–0.1323743E–04	100.01	100.01
12	z_y	—	0.2755482E–04	0.2755317E–04	0.2755317E–04	100.01	100.01
13	z_y	—	0.2757911E–04	0.2757746E–04	0.2757746E–04	100.01	100.01

[a] $b_2 = 10.0$ and $\delta b_2 = 0.1$

Table 6.15. Verification of second-order shape design sensitivity of performance measures with respect to design variable b_1[a] and b_2[b].

Element or Node	Performance Measure	Gauss Point	$\delta^2\psi(b)$ Finite Diff.	$\psi''(b)$ Direct Method	$\psi''(b)$ Hybrid Method	$\psi''/\delta^2\psi$ (Direct)	$\psi''/\delta^2\psi$ (Hybrid)
205	von Mises	4	0.1775493E–01	0.1779886E–01	0.1779885E–01	100.25	100.25
359	von Mises	1	–0.9035997E+00	–0.9033521E+00	–0.9033521E+00	99.97	99.97
362	von Mises	3	–0.8901512E+00	–0.8898907E+00	–0.8898907E+00	99.97	99.97
406	von Mises	3	0.3251864E–02	0.3249104E–02	0.3249142E–02	99.92	99.92
270	z_x	—	–0.1344251E–05	–0.1343677E–05	–0.1343672E–05	99.96	99.96
279	z_x	—	–0.1343977E–05	–0.1343404E–05	–0.1343399E–05	99.96	99.96
12	z_y	—	0.2657888E–05	0.2656586E–05	0.2656553E–05	99.95	99.95
13	z_y	—	0.2660071E–05	0.2658769E–05	0.2658736E–05	99.95	99.95

[a] $b_1 = 10.0$ and $\delta b_1 = 0.1$ [b] $b_2 = 10.0$ and $\delta b_2 = 0.1$

Table 6.16. Second-order shape design sensitivity of stress performance measures.

Element	Criteria	IG	$\partial^2\psi/\partial b_1^2$	$\partial^2\psi/\partial b_1\partial b_2$	$\partial^2\psi/\partial b_1\partial b_3$	$\partial^2\psi/\partial b_1\partial b_4$
359	von Mises	1	0.4296987E+03	–0.2258380E+02	–0.3578100E+01	–0.7854668E+01
360	von Mises	2	0.1741188E+03	0.6666557E+02	–0.3756059E+02	–0.6802056E+00
361	von Mises	3	–0.6541516E+02	0.1028044E+03	–0.4469860E+02	–0.2233664E+01
362	von Mises	3	–0.2585918E+02	–0.2224727E+02	–0.7089901E+01	–0.3412922E+02

order shape design sensitivity. Table 6.16 shows the results of second-order sensitivity analysis for elements 359 through 362. These results illustrate that stress performance measures on elements located near shape design variables (e.g., element 359 is closer to b_1) have higher second-order shape design sensitivity than those stress performance measures defined on other elements.

Arch Dam

A three-dimensional doubly curved arch dam based on the optimum design of Wassermann [110] as shown in Fig. 6.28, can be used to demonstrate the accuracy of derived second-order shape design sensitivities, with the shape changes occurring on loaded boundaries.

The arch dam is subject to water pressure and gravitational force. It is assumed that the dam's foundation is rigid, that its material is homogeneous, and that it behaves elastically. The effects of temperature are ignored. The physical properties include water weight density (10.0 kN/m^3), gravity acceleration (10.0 m/s^2), concrete weight density (25.0 kN/m^3), Young's modulus of concrete (21.0 GPa), and Poisson's ratio of concrete (0.2).

It is assumed that the arch dam is constructed on an idealized valley, and the normal cross section of this valley is shown in Fig. 6.29. Assuming that the structure and loads are symmetric with respect to the crown cross section, only half of the dam's span is analyzed. Two parameterized surfaces, as shown in Fig. 6.30, are used to simulate the water face and air face of the arch dam and are chosen as design boundaries.

The water and air face patches are parameterized as geometric surfaces. The x^2-coordinates of grid points 21, 24, 33, 36, 41, 44, 53, and 56 are chosen as independent design variables, so there are eight design variables in this model. Four of them are defined on the loaded boundary (the water face). The definitions and values of each design variable are given in Table 6.17.

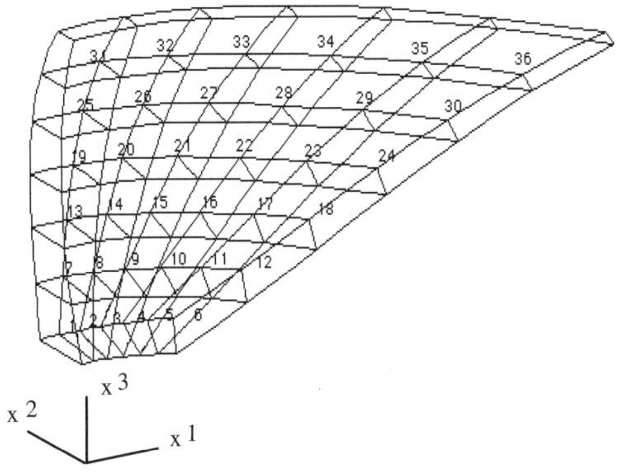

Figure 6.28. Finite element mesh of arch dam.

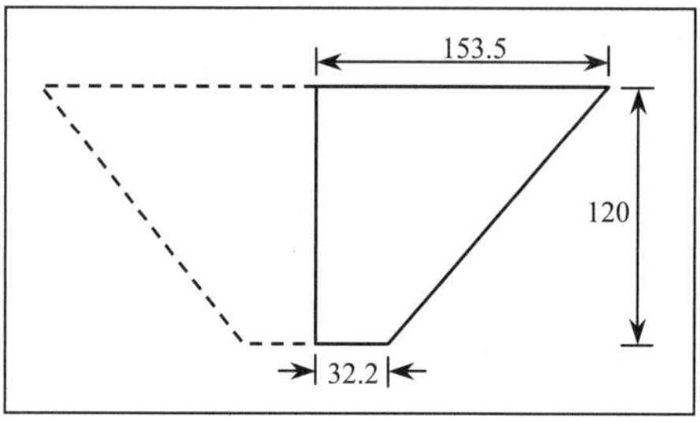

Figure 6.29. Normal cross section of idealized valley.

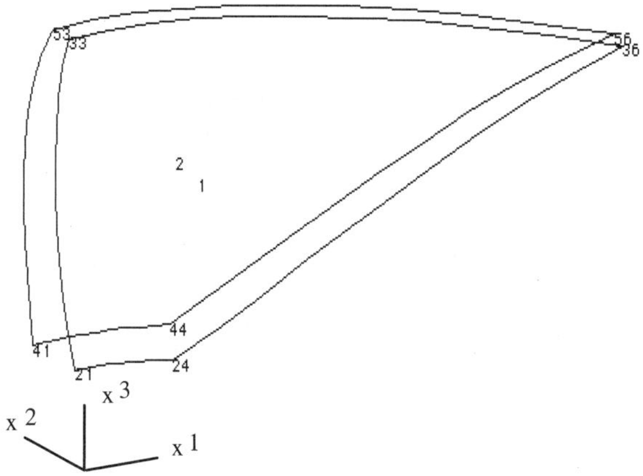

Figure 6.30. Design boundary patches for arch dam.

Table 6.17. Definition of design variables.

Design par number	1	2	3	4	5	6	7	8
Grid number	21	33	36	24	41	53	56	44
Design par value	115	119	10	103	141	129	20	130

The finite element model takes into account the water pressure and the gravity force, which will contribute to second-order shape DSA through (6.339). The water pressure is assumed to be linearly varied, from zero at the top of the dam to 1200 kN/m^2 at the bottom. However, because the ANSYS FEA code is used to analyze the structural response, and because it only allows step pressure to be applied on the loading surface, a step pressure profile is used in the sensitivity analyses. All nodes that connect the dam and foundation are fixed. To maintain symmetry, the nodes on the center crown section are fixed in the x_1-direction. The finite element model of the arch dam includes 315 nodes, 36 20-node isoparametric elements, and 726 active degrees of freedom.

Eight velocity fields are generated that correspond to the eight independent design variables. Sixty-four acceleration fields are then generated from the eight velocity fields. The direct differentiation and hybrid methods are used to compute the second-order shape and mixed second-order design sensitivity.

In the arch dam example, the first principal stresses at Gauss points of finite elements are chosen as stress performance measures. To conduct a numerical test, the first eight elements that have a high tensile and compressible first principal stress are selected. In order to verify the mixed second-order design sensitivity with respect to the shape design variable and material properties, eight nodes with large displacements in the x_2-direction are selected. Second-order sensitivity results are verified using the finite difference method.

Table 6.18 shows the values of selected stress performance measures and the first-order sensitivity of these performance measures with respect to variations in water pressure. The stress performance measures are not sensitive to variations in the water pressure. Tables 6.19 through 6.23 show the verification results of second-order sensitivity. Tables 6.19, 6.20 and 6.22 are for second-order shape design sensitivity, and Table 6.21 is mixed second-order sensitivity with respect to both shape design variable b_1 and distribution of water pressure. Table 6.23 is mixed second-order sensitivity with respect to both shape design variable-b_1 and to Young's modulus.

The first column is the element number in Tables 6.19 through 6.22 and the node number in Table 6.23, and the second column is the Gauss point number in Tables 6.19 through 6.22 and the coordinate direction in Table 6.23. In all five tables, the third column represents the second-order finite difference result, and the fourth and fifth columns are the second-order design sensitivity results from the direct differentiation and hybrid methods, respectively. The last two columns are verification results of the second-order design sensitivity from these two methods.

Results in Tables 6.19 through 6.23 indicate that the accuracy of second-order shape design sensitivity is very good, even though the stress performance measures are defined on elements with high tensile and compressible stresses, and displacement performance measures are defined on nodes with a large deformation. For the design variables defined on the loaded boundary, Tables 6.20 and 6.22 show good results. These numerical results demonstrate the accuracy and feasibility of both the theoretical derivations and the numerical implementations of the second-order shape DSA, while considering shape changes on the loaded boundaries.

Table 6.18. Stress performance measures and sensitivity with respect to distribution of pressure loading (p).

Element Number	Gauss Point	First Principal Stress	1st-Order DSA wrt p
1	1	.21767E+04	.69141D+01
2	1	.21316E+04	.67309D+01
3	1	.20055E+04	.62587D+01
4	1	.17827E+04	.55138D+01
7	14	–.10171E+04	–.92131D+00
8	14	–.97375E+03	–.48038D+00
9	14	–.88323E+03	.85737D+00
10	3	–.75339E+03	–.26008D+00

Table 6.19. Verification of design sensitivity of first principal stress w.r.t. b_1[a].

Element Number	Gauss Point	$\delta^2\psi(b)$ Finite Diff.	$\psi''(b)$ Direct Method	$\psi''(b)$ Hybrid Method	$\psi''/\delta^2\psi$ (Direct)	$\psi''/\delta^2\psi$ (Hybrid)
1	1	0.13090D+02	0.13076D+02	0.13076D+02	100.10	100.10
2	1	0.12198D+02	0.12167D+02	0.12167D+02	100.25	100.25
3	1	0.10751D+02	0.10685D+02	0.10685D+02	100.62	100.62
4	1	0.89328D+01	0.89087D+01	0.89087D+01	100.27	100.27
7	14	0.42968D+01	0.43454D+01	0.43454D+01	98.88	98.88
8	14	0.25103D+01	0.25379D+01	0.25378D+01	98.92	98.92
9	14	−0.10416D+01	−0.10601D+01	−0.10601D+01	98.25	98.25
10	3	0.83281D+00	0.83027D+00	0.83027D+00	100.31	100.31

[a] $b_1 = 115$ and $\delta b_1 = 2.3$.

Table 6.20. Verification of design sensitivity of first principal stress w.r.t b_8[a].

Element No	Gauss Point	$\delta^2\psi(b)$ Finite Diff.	$\psi''(b)$ Direct Method	$\psi''(b)$ Hybrid Method	$\psi''/\delta^2\psi$ (Direct)	$\psi''/\delta^2\psi$ (Hybrid)
1	1	.59819D+01	.59872D+01	.59872D+01	99.91	99.91
2	1	.50765D+01	.50754D+01	.50754D+01	100.02	100.02
3	1	.46340D+01	.45902D+01	.45902D+01	100.96	100.96
4	1	.56897D+01	.56447D+01	.56447D+01	100.80	100.80
7	14	−.16881D+00	−.16548D+00	−.16548D+00	102.02	102.02
8	14	.48064D+01	.47834D+01	.47834D+01	100.48	100.48
9	14	.87771D+01	.87872D+01	.87872D+01	99.89	99.89
10	3	.74884D+01	.75214D+01	.75214D+01	99.56	99.56

[a] $b_8 = 130$ and $\delta b_1 = 2.6$.

Table 6.21. Verification of design sensitivity of first principal stress with respect to design variable b_1[a] and pressure loading p[b].

Element Number	Gauss Point	$\delta^2\psi(b)$ Finite Diff.	$\psi''(b)$ Direct Method	$\psi''(b)$ Hybrid Method	$\psi''/\delta^2\psi$ (Direct)	$\psi''/\delta^2\psi$ (Hybrid)
1	1	.23395D+01	.23392D+01	.23392D+01	99.99	99.99
2	1	.20319D+01	.20316D+01	.20316D+01	99.98	99.98
3	1	.14299D+01	.14295D+01	.14295D+01	99.97	99.97
4	1	.81186D+00	.81152D+00	.81152D+00	99.96	99.96
7	14	−.19913D+01	−.19813D+01	−.19813D+01	99.50	99.50
8	14	−.28154D+01	−.28134D+01	−.28134D+01	99.93	99.93
9	14	−.20867D+01	−.21035D+01	−.21035D+01	100.80	100.80
10	3	−.79824D+00	−.79417D+00	−.79417D+00	99.49	99.49

[a] $b_1 = 115$ and $\delta b_1 = 0.575$; [b] $\delta p = 10$

Table 6.22. Verification of design sensitivity of first principal stress with respect to design variable b_4[a] and pressure loading b_8[b].

Element Number	Gauss Point	$\delta^2\psi(b)$ Finite Diff.	$\psi''(b)$ Direct Method	$\psi''(b)$ Hybrid Method	$\psi''/\delta^2\psi$ (Direct)	$\psi''/\delta^2\psi$ (Hybrid)
1	1	–.87234D–01	–.87230D–01	–.87230D–01	99.99	99.99
2	1	–.86512D–01	–.86507D–01	–.86507D–01	99.99	99.99
3	1	–.82128D–01	–.82127D–01	–.82127D–01	100.00	100.00
4	1	–.68998D–01	–.69006D–01	–.69006D–01	100.01	100.01
7	14	.25730D–02	.25716D–02	.25716D–02	99.95	99.95
8	14	.78610D–02	.78586D–02	.78586D–02	99.97	99.97
9	14	.16764D–01	.16764D–01	.16764D–01	100.00	100.00
10	3	–.61876D–02	–.61912D–02	–.61912D–02	100.06	100.06

[a] $b_4 = 103$ and $\delta b_4 = 0.103$; [b] $b_8 = 130$ and $\delta b_8 = 0.13$.

Table 6.23. Verification of design sensitivity of first principal stress with respect to design variable b_1[a] and Young's modulus E[b].

Node Number	Coord. Direction	$\delta^2\psi(b)$ Finite Diff.	$\psi''(b)$ Direct Method	$\psi''(b)$ Hybrid Method	$\psi''/\delta^2\psi$ (Direct)	$\psi''/\delta^2\psi$ (Hybrid)
87	2	.82990D–05	.82801D–05	.82801D–05	99.77	99.77
88	2	.83416D–05	.83226D–05	.83226D–05	99.77	99.77
89	2	.84393D–05	.84197D–05	.84196D–05	99.77	99.77
90	2	.85935D–05	.85736D–05	.85736D–05	99.77	99.77
300	2	.83275D–05	.83085D–05	.83085D–05	99.77	99.77
301	2	.87985D–05	.87782D–05	.87782D–05	99.77	99.77
302	2	.88751D–05	.88547D–05	.88547D–05	99.77	99.77
303	2	.85244D–05	.85046D–05	.85046D–05	99.77	99.77

[a] $b_1 = 115$ and $\delta b_1 = 0.575$; [b] $\delta E = 1.0E+6$

7
Configuration Design Sensitivity Analysis

A configuration design is applicable to built-up structures with such structural design components as truss, beam, plate, and shell. With these components, the rotation of a component in the global coordinate does not change the integral domain, which in Chapter 6 is considered a shape design. However, the rotation does change the configuration of the built-up structure, thus yields a different structural response. In the built-up structure, the shape change of a component may cause a configuration change in other components. Thus, shape and configuration designs are closely related. In this chapter, the design sensitivity of a structural performance measure with respect to the configuration design is presented. Two configuration design situations might occur. In the first case, straight-line and flat-surface design components simply rotate without changing the component's shape. In this case, the design perturbation is represented first by rotational angles, and then is approximated using the out-of-plane design velocity field from an assumption of small perturbation. In the second case, the design component might change its own configuration, such as in the curvature of a component, in addition to component's rotation. In this case, the transformation between a local and a global coordinate system is identified as a configuration design. In contrast to the first case, the contribution from shape and the configuration designs are strongly coupled.

For the configuration design of built-up structures, changes in structural configuration result in shape and orientation changes to the design components. The first approach to utilize a mathematical programming technique for configuration design optimization was developed by Dorn et al. [111]. Dobbs and Felton [112] dealt with a more general case that included indeterminate structures and multiple loading conditions. Vanderplaats and Moses [113] developed a technique to obtain the minimum weight of a structure subjected to a prescribed set of constraints. They minimized the objective function with respect to geometry variables (coordinate design space) where the member design is updated to maintain optimality with respect to area variables (area design space). Imai and Schmit [114] developed a multiplier method to avoid severe nonlinearities that may occur during configuration design optimization. Recently, Saka and Attili [115] presented configuration optimization of space trusses. In the design problem, they considered displacement, stress, buckling, and minimum size constraints.

The configuration design of a structural component can be characterized by changes in the domain shape and in the component's orientation. A variational approach is then used to incorporate both shape and orientation effects in the same energy equation. A similar approach to the shape design sensitivity analysis method in Chapter 6 is developed in this chapter to account for the effect of directional variation. Variations of energy bilinear and load linear forms, with respect to both shape and orientation design variables, are derived for each structural component. Using the adjoint variable and direct differentiation methods, configuration design sensitivity results are obtained in terms of design velocity fields. Configuration design sensitivity analyses of static and eigenvalue responses, as well as the structural-acoustic responses, are developed.

Configuration and shape design parameterizations are closely related. From the assumption of linear mapping and a small perturbation, shape and configuration parameterizations are uncoupled in Section 7.1. Only straight-line and flat-surface components are considered. The material derivative approach, similar to that employed in Section 6.1, is developed for the configuration problem in Section 7.1. The direct differentiation and adjoint variable methods are developed in Section 7.2 for static and eigenvalue problems. In Section 7.3, the linear transformation relation and the regularity requirement of the configuration design velocity field are discussed. The developed configuration design capability is further applied to the design of the structural-acoustic coupled problem in Section 7.4. A new development of the configuration design sensitivity formulation for the curved beam and surface components is presented in Section 7.5.

7.1 Material Derivatives for Configuration Design Sensitivity Analysis

A configuration design parameterization is represented by the rotation of a component. In the local coordinate system fixed in the components, however, every material point has the same relative position during the design change. It is convenient to introduce a rotational angle to represent the design change. For a line design component, three rotational angles are independent, whereas only two rotational angles are used for a surface design component because the in-plane rotation can be treated as a shape design.

Similar to the shape design sensitivity analysis in Chapter 6, the first step in configuration design sensitivity analysis is to develop a relationship between the variation in the component orientation and the resulting variation in the structural response. The displacement derivative with respect to the orientation will be obtained for the line and surface design components, followed by the performance derivative. For a solid design component, it is unnecessary to carry out the configuration design sensitivity analysis, since any rotational effect can be represented by the shape perturbation in the solid component.

In configuration design sensitivity analysis, the component orientation is treated as a design variable, and two assumptions are used throughout: (1) the design component rotates without any shape change, and (2) only a small design perturbation is considered. Under these assumptions, the design velocity field $\boldsymbol{V}(\boldsymbol{x})$ can be additively decomposed for shape and configuration parts, as

$$\boldsymbol{V}(\boldsymbol{x}) = \boldsymbol{V}_{\Omega}(\boldsymbol{x}) + \boldsymbol{V}_{\Theta}(\boldsymbol{x}). \tag{7.1}$$

In the definition of the above equation, the subscribed $\boldsymbol{V}_{\Theta}$ is used to distinguish it from the shape design velocity $\boldsymbol{V}_{\Omega}$.

If a straight-line or a flat-surface component is considered, $\boldsymbol{V}_{\Omega}(\boldsymbol{x})$ is the projection of $\boldsymbol{V}(\boldsymbol{x})$ onto the component and $\boldsymbol{V}_{\Theta}(\boldsymbol{x})$ is orthogonal to the component. Development of the configuration design sensitivity analysis in Sections 7.1 through 7.4 is limited to straight-line and flat-surface components where the decomposition in (7.1) is valid. Figure 7.1 illustrates the decomposition of $\boldsymbol{V}(\boldsymbol{x})$ for a flat-surface component. If a body-fixed, local coordinate system is established, as shown in Fig. 7.1, $\boldsymbol{V}_{\Theta}(\boldsymbol{x})$ is normal to the surface and parallel to the x_3-direction.

All formulations in Chapter 6 are preserved by interpreting $\boldsymbol{V}_{\Omega}(\boldsymbol{x})$ as the shape design velocity. As with the shape design problem in Chapter 6, it can be supposed that only one parameter τ defines the magnitude of transformation $\boldsymbol{T}_{\Theta}$. In the local coordinate system, the mapping $\boldsymbol{T}_{\Theta} : \boldsymbol{x} \to \boldsymbol{x}_{\tau}(\boldsymbol{x}),\ \boldsymbol{x} \in \Omega$ is then given by

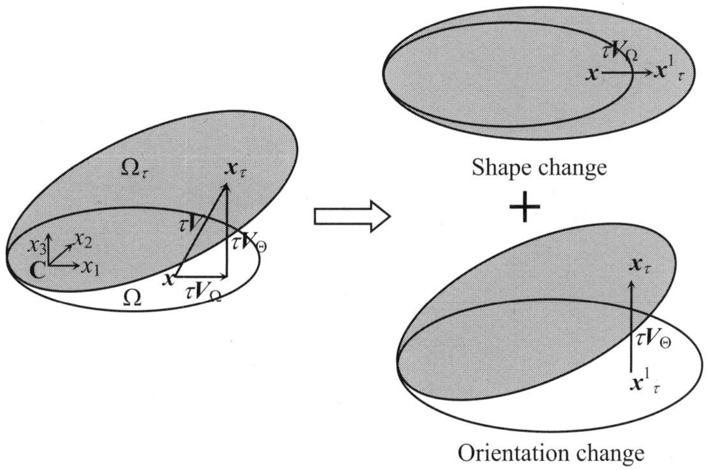

Figure 7.1. Perturbation of shape and orientation in surface design component.

$$
\begin{aligned}
\boldsymbol{x}_\tau &\equiv \boldsymbol{T}_\Theta(\boldsymbol{x},\tau) = \boldsymbol{x} + \tau \boldsymbol{V}_\Theta \\
\Omega_\tau &\equiv \boldsymbol{T}_\Theta(\Omega,\tau),
\end{aligned}
\tag{7.2}
$$

where $\boldsymbol{V}_\Theta(\boldsymbol{x}) = [V_1(\boldsymbol{x}),\ V_2(\boldsymbol{x}),\ V_3(\boldsymbol{x})]^T$ is called the configuration design velocity field. By interpreting τ as time, a configuration design velocity can be defined as a derivative of $\boldsymbol{T}_\Theta(\boldsymbol{x}, \tau)$, that is,

$$
\boldsymbol{V}_\Theta(\boldsymbol{x},\tau) \equiv \frac{d\boldsymbol{x}_\tau}{d\tau} = \frac{d\boldsymbol{T}_\Theta(\boldsymbol{x},\tau)}{d\tau}. \tag{7.3}
$$

From the additive decomposition of a design velocity field and mapping relation in (7.2), the material derivatives of shape and configuration designs can also be additively decomposed, making it is possible to treat them separately. Line and surface components are explained separately in the following subsections.

7.1.1 Line Design Component

Figure 7.2 demonstrates a line design component that represents a truss or a beam. In this case the shape design variable is the change of length l, such that the integration domain changes. The configuration design variable is the component rotation in the three axes, as shown in Fig. 7.2. The original body-fixed coordinate system x_1-x_2-x_3 is first rotated to the $\delta\alpha$ angle with respect to the x_3-axis, resulting in a x_1'-x_2'-x_3' coordinate system. Then, the x_1'-x_2'-x_3' coordinate is rotated to the $\delta\beta$ angle with respect to the x_2'-axis, yielding the x_1''-x_2''-x_3'' coordinate. Finally, rotation $\delta\gamma$ with respect to the x_1''-axis arrives at the perturbed geometry. In the case of small perturbation, the mapping $\boldsymbol{T}_\Theta : \boldsymbol{x} \to \boldsymbol{x}_\tau(\boldsymbol{x}), \boldsymbol{x} \in \Omega$, is given by 7.1.2 and the orientation design velocity is $\boldsymbol{V}_\Theta(\boldsymbol{x}) = [0,\ V_2(\boldsymbol{x}),\ V_3(\boldsymbol{x})]^T$, where $V_1(\boldsymbol{x})$ in the axial direction is zero for a rotational transformation.

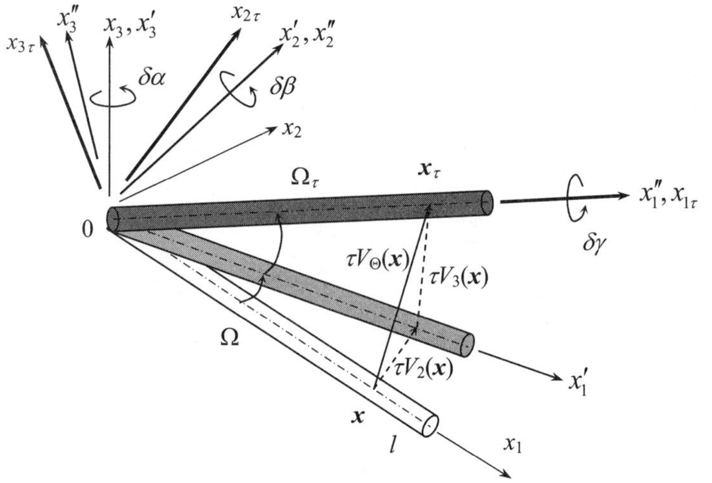

Figure 7.2. Orientation change of a line design component.

Suppose state response $\boldsymbol{z}_\tau(\boldsymbol{x}_\tau) = [z_{1\tau}(\boldsymbol{x}_\tau), z_{2\tau}(\boldsymbol{x}_\tau), z_{3\tau}(\boldsymbol{x}_\tau), \phi_{1\tau}(\boldsymbol{x}_\tau), \phi_{2\tau}(\boldsymbol{x}_\tau), \phi_{3\tau}(\boldsymbol{x}_\tau)]^T$ is a solution to the boundary-value problem where $z_{i\tau}$ and $\phi_{i\tau}$ are the displacements and rotations, respectively, in the perturbed domain Ω_τ. Then, mapping $\boldsymbol{z}_\tau(\boldsymbol{x}_\tau)$ is defined on Ω_τ, and depends on τ in two ways. First, it is the solution to the boundary-value problem in Ω_τ. Second, it is evaluated at point $\boldsymbol{x}_\tau$ that moves with τ. If the pointwise derivative of the displacement with respect to the orientation exists at $\boldsymbol{x} \in \Omega$, then it is defined as

$$\dot{\boldsymbol{z}}_{V_\theta}(\boldsymbol{x}) \equiv \frac{d}{d\tau}\boldsymbol{z}_\tau(\boldsymbol{x} + \tau \boldsymbol{V}_\Theta(\boldsymbol{x}))\bigg|_{\tau=0} = \lim_{\tau\to 0}\left[\frac{\boldsymbol{z}_\tau(\boldsymbol{x} + \tau \boldsymbol{V}_\Theta(\boldsymbol{x})) - \boldsymbol{z}(\boldsymbol{x})}{\tau}\right]. \tag{7.4}$$

Note $\dot{\boldsymbol{z}}_{V_\theta}$ is the material derivative with respect to the configuration design, while $\dot{\boldsymbol{z}}_{V_\Omega}$ is the material derivative with respect to the shape design.

To express the material derivative of $\boldsymbol{z}_\tau(\boldsymbol{x}_\tau)$ in terms of rotational angles, define a regular extension of $\boldsymbol{z}_\tau$ to the initial local coordinate system x_1-x_2-x_3 as

$$\boldsymbol{z}_\tau(\boldsymbol{x}) \equiv \boldsymbol{z}_\tau(\boldsymbol{x}_\tau). \tag{7.5}$$

This regular extension can be viewed as defining $\boldsymbol{z}_\tau(\boldsymbol{x})$ by extending the value of $\boldsymbol{z}_\tau(\boldsymbol{x}_\tau)$ constantly along the direction of rotation. Then, for the line design component, we have the following relationship:

$$\boldsymbol{z}_\tau(\boldsymbol{x}_\tau(\boldsymbol{x})) = \boldsymbol{A}_1(\delta\alpha, \delta\beta, \delta\gamma)\boldsymbol{z}_\tau(\boldsymbol{x}), \tag{7.6}$$

where $\boldsymbol{z}_\tau(\boldsymbol{x}_\tau(\boldsymbol{x}))$ denotes the evaluation of the perturbed solution $\boldsymbol{z}_\tau$ at $\boldsymbol{x}_\tau$ in the original local coordinate system x_1-x_2-x_3, and $\boldsymbol{z}_\tau(\boldsymbol{x})$ is the regular extension of the perturbed solution $\boldsymbol{z}_\tau$ at location $\boldsymbol{x}$. Note that both $\boldsymbol{z}_\tau(\boldsymbol{x}_\tau(\boldsymbol{x}))$ and $\boldsymbol{z}_\tau(\boldsymbol{x})$ are evaluated in the same coordinate system x_1-x_2-x_3. In (7.6), $\boldsymbol{A}_1(\delta\alpha,\delta\beta,\delta\gamma)$ is the rotational transformation matrix, defined by

$$A_1(\delta\alpha,\delta\beta,\delta\gamma)=\begin{bmatrix}\cos\delta\alpha & -\sin\delta\alpha & 0 & 0 & 0 & 0\\ \sin\delta\alpha & \cos\delta\alpha & 0 & 0 & 0 & 0\\ 0 & 0 & 1 & \sin\delta\alpha & 0 & 0\\ 0 & 0 & 0 & \cos\delta\alpha & -\sin\delta\alpha & 0\\ 0 & 0 & 0 & \sin\delta\alpha & \cos\delta\alpha & 0\\ 0 & 0 & 0 & 0 & 0 & 1\end{bmatrix}$$

$$\times\begin{bmatrix}\cos\delta\beta & 0 & \sin\delta\beta & 0 & 0 & 0\\ 0 & 1 & 0 & -\sin\delta\beta & 0 & 0\\ -\sin\delta\beta & 0 & \cos\delta\beta & 0 & 0 & 0\\ 0 & 0 & 0 & \cos\delta\beta & 0 & \sin\delta\beta\\ 0 & 0 & 0 & 0 & 1 & 0\\ 0 & 0 & 0 & -\sin\delta\beta & 0 & \cos\delta\beta\end{bmatrix}$$

$$\times\begin{bmatrix}1 & 0 & 0 & 0 & 0 & 0\\ 0 & \cos\delta\gamma & -\sin\delta\gamma & 0 & 0 & 0\\ 0 & \sin\delta\gamma & \cos\delta\gamma & 0 & 0 & 0\\ 0 & 0 & 0 & 1 & 0 & 0\\ 0 & 0 & 0 & 0 & \cos\delta\gamma & -\sin\delta\gamma\\ 0 & 0 & 0 & 0 & \sin\delta\gamma & \cos\delta\gamma\end{bmatrix}, \tag{7.7}$$

where $\delta\alpha$, $\delta\beta$, and $\delta\gamma$, as shown in Fig. 7.2, are the rotational angles with respect to the x_3, x_2', and x_1'' axes, respectively. For a small design perturbation (i.e., as $\tau \to 0$), τV_2 and τV_3 become arc lengths, and the rotational angles can be approximated by

$$\left.\begin{aligned}\delta\alpha &= \tau V_{2,1}\\ \delta\beta &= -\tau V_{3,1}\\ \delta\gamma &= -\tau V_{2,3} = \tau V_{3,2}\end{aligned}\right\}, \tag{7.8}$$

where $V_{i,j}$ denotes the derivative of V_i with respect to x_j. Thus, the rotational change is represented using V_Θ. A rigorous derivation of (7.8) will be presented in Section 7.3.

If z_τ has a regular extension [see (7.5)] to the neighborhood U_τ of the trajectory $x_\tau \in \Omega_\tau$, then the partial derivative $z'_{V_\theta}(x)$ can be defined as

$$z'_{V_\theta}(x) = \lim_{\tau\to 0}\frac{z_\tau(x) - z(x)}{\tau}. \tag{7.9}$$

Similar to the shape design sensitivity analysis, the differentiation order between the partial derivative and the spatial derivative is interchangeable because they are independent, that is,

$$\left(\frac{\partial z}{\partial x}\right)'_{V_\theta} = \frac{\partial}{\partial x} z'_{V_\theta}. \tag{7.10}$$

The material derivative in (7.4) is expanded into a more convenient form by using (7.6) through (7.9), as

$$\begin{aligned}
\dot{z}_{V_\theta}(x) &= \lim_{\tau\to 0}\left[\frac{z_\tau(x+\tau V_\Theta)-z(x)}{\tau}\right] \\
&= \lim_{\tau\to 0}\left[\frac{z_\tau(x)-z(x)}{\tau}\right]+\lim_{\tau\to 0}\left[\frac{z_\tau(x+\tau V_\Theta)-z_\tau(x)}{\tau}\right] \\
&= z'_{V_\theta}(x)+\lim_{\tau\to 0}\frac{[A_1(\tau V_{2,1},-\tau V_{3,1},-\tau V_{2,3})-A_1(0,-\tau V_{3,1},-V_{2,3})]z_\tau(x)}{\tau} \\
&\quad +\lim_{\tau\to 0}\frac{[A_1(0,-\tau V_{3,1},-\tau V_{2,3})-A_1(0,0,-V_{2,3})]z_\tau(x)}{\tau} \\
&\quad +\lim_{\tau\to 0}\frac{[A_1(0,0,-\tau V_{2,3})-I]z_\tau(x)}{\tau} \\
&= z'_{V_\theta}(x)+\left[\frac{d}{d\tau}A_1(\tau V_{2,1},0,0)\Big|_{\tau=0}+\frac{d}{d\tau}A_1(0,-\tau V_{3,1},0)\Big|_{\tau=0}+\frac{d}{d\tau}A_1(0,0,-\tau V_{2,3})\Big|_{\tau=0}\right]z(x) \\
&= z'_{V_\theta}(x)+V_\theta z(x),
\end{aligned} \tag{7.11}$$

where $\boldsymbol{I}$ is a 6 × 6 identity matrix, and

$$V_\theta = \begin{bmatrix} 0 & -V_{2,1} & -V_{3,1} & 0 & 0 & 0 \\ V_{2,1} & 0 & V_{2,3} & V_{3,1} & 0 & 0 \\ V_{3,1} & -V_{2,3} & 0 & V_{2,1} & 0 & 0 \\ 0 & 0 & 0 & 0 & -V_{2,1} & -V_{3,1} \\ 0 & 0 & 0 & V_{2,1} & 0 & V_{2,3} \\ 0 & 0 & 0 & V_{3,1} & -V_{2,3} & 0 \end{bmatrix}. \tag{7.12}$$

Equation (7.12) is derived using the assumption that the bending curvature and the rotation angle are related, which occurs in the finite element formulation of bending where the interpolation functions for lateral displacements include rotations as nodal parameters [116]. Note that $V_{2,1}$, $V_{3,1}$, and $V_{2,3}$ in (7.12) are constants under the assumption that the line design component remains straight during the design variation.

In the definition of $z'_{V_\theta}(x)$ in (7.9) and of $\dot{z}_{V_\theta}(x)$ in (7.11), subscribed V_θ is used to denote the configuration design variation, in contrast to the partial and material derivatives of shape design variation in Chapter 6 that were denoted as $z'_{V_\Omega}(x)$ and $\dot{z}_{V_\Omega}(x)$. If the shape and configuration designs are considered together, then the following linear superposition can be used:

$$z' = z'_{V_\Omega} + z'_{V_\theta} \tag{7.13}$$

$$\dot{z} = \dot{z}_{V_\Omega} + \dot{z}_{V_\theta} \tag{7.14}$$

Although both (6.8) and (7.11) contain partial derivatives, the convective terms are in different form. The former involves the gradient of the displacement, whereas the latter involves the derivative of the design velocity. Under the assumption that a line design component remains straight after the design change, V_θ becomes a constant matrix and $(z_{,1})^{\cdot}_{V_\theta} = (\dot{z}_{V_\theta})_{,1}$. However, this was not true for the shape design sensitivity analysis, i.e., $(z_{,1})^{\cdot}_{V_\Omega} \neq (\dot{z}_{V_\Omega})_{,1}$.

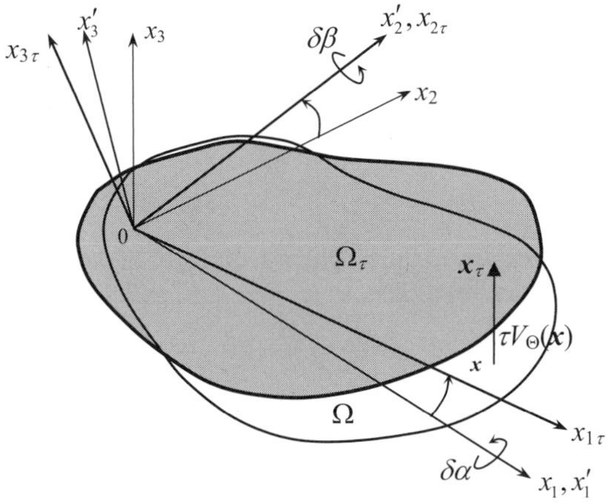

Figure 7.3. Orientation change of a surface design component.

7.1.2 Surface Design Component

Figure 7.3 illustrates a flat-surface design component that rotates in a three-dimensional space. The local coordinate system is fixed on the component with the x_3-axis normal to the surface. The original body-fixed coordinate system x_1-x_2-x_3 is first rotated to the $\delta\alpha$ angle with respect to the x_1-axis, resulting in a x_1'-x_2'-x_3' coordinate system. Then, the x_1'-x_2'-x_3' coordinate is rotated to the $\delta\beta$ angle with respect to the x_2'-axis, yielding the perturbed geometry. In the case of a small perturbation, the mapping $\boldsymbol{T}_\theta : \boldsymbol{x} \rightarrow \boldsymbol{x}_\tau(\boldsymbol{x}), \boldsymbol{x} \in \Omega$ is given by (7.2), and the orientation design velocity field is $\boldsymbol{V}_\Theta(\boldsymbol{x}) = [0, 0, V_3]^T$ in the body-fixed coordinate system. The tangential components $V_1(\boldsymbol{x})$ and $V_2(\boldsymbol{x})$ of the design velocity are zero, because only the normal movement $\tau V_3(\boldsymbol{x})$ changes the orientation of the surface component.

Suppose that state variable $\boldsymbol{z}_\tau(\boldsymbol{x}_\tau) = [z_{1\tau}(\boldsymbol{x}_\tau), z_{2\tau}(\boldsymbol{x}_\tau), z_{3\tau}(\boldsymbol{x}_\tau), \phi_{1\tau}(\boldsymbol{x}_\tau), \phi_{2\tau}(\boldsymbol{x}_\tau), \phi_{3\tau}(\boldsymbol{x}_\tau)]^T$ is a solution to the boundary-value problem in the perturbed domain Ω_τ. The pointwise material derivative of the state variable with respect to the orientation design at $\boldsymbol{x} \in \Omega$ is defined by (7.4). As with the line design component, a regular extension of $\boldsymbol{z}_\tau$ to the initial, local coordinate system x_1-x_2-x_3 is defined by (7.5). According to the definition of a regular extension, the perturbed solution $\boldsymbol{z}_\tau$ at two different locations $\boldsymbol{x}$ and $\boldsymbol{x}_\tau$ can be related by

$$\boldsymbol{z}_\tau(\boldsymbol{x}_\tau) = \boldsymbol{A}_2(\delta\alpha, \delta\beta)\boldsymbol{z}_\tau(\boldsymbol{x}), \tag{7.15}$$

where both $\boldsymbol{z}_\tau(\boldsymbol{x}_\tau(\boldsymbol{x}))$ and $\boldsymbol{z}_\tau(\boldsymbol{x})$ are evaluated in the same x_1-x_2-x_3 local coordinate system, and $\boldsymbol{A}_2(\delta\alpha,\delta\beta)$ is the rotational transformation matrix, defined by

$$\boldsymbol{A}_2(\delta\alpha,\delta\beta) = \begin{bmatrix} 1 & 0 & 0 & 0 & 0 & 0 \\ 0 & \cos\delta\alpha & -\sin\delta\alpha & 0 & 0 & 0 \\ 0 & \sin\delta\alpha & \cos\delta\alpha & 0 & 0 & -\sin\delta\alpha \\ 0 & 0 & 0 & 1 & 0 & 0 \\ 0 & 0 & 0 & 0 & \cos\delta\alpha & -\sin\delta\alpha \\ 0 & 0 & 0 & 0 & \sin\delta\alpha & \cos\delta\alpha \end{bmatrix} \begin{bmatrix} \cos\delta\beta & 0 & \sin\delta\beta & 0 & 0 & 0 \\ 0 & 1 & 0 & 0 & 0 & 0 \\ -\sin\delta\beta & 0 & \cos\delta\beta & 0 & 0 & \sin\delta\beta \\ 0 & 0 & 0 & \cos\delta\beta & 0 & \sin\delta\beta \\ 0 & 0 & 0 & 0 & 1 & 0 \\ 0 & 0 & 0 & -\sin\delta\beta & 0 & \cos\delta\beta \end{bmatrix}, \tag{7.16}$$

where $\delta\alpha$ and $\delta\beta$, as shown in Fig. 7.3, are the rotational angles with respect to the x_1- and x_2'-axes. Again, for a small design perturbation (i.e., as $\tau \to 0$) of the surface design component, those rotational angles can be represented using a design velocity, as

$$\begin{aligned} \delta\alpha &= \tau V_{3,2} \\ \delta\beta &= -\tau V_{3,1}. \end{aligned} \tag{7.17}$$

If $\boldsymbol{z}_\tau$ has a regular extension to the neighborhood U_τ of the trajectory $\boldsymbol{x}_\tau \in \Omega_\tau$, then the partial derivative $\boldsymbol{z}'_{V_\theta}(\boldsymbol{x})$ can be defined, as in (7.9). Using (7.9), and (7.15) through (7.17), (7.4) becomes

$$\begin{aligned} \dot{\boldsymbol{z}}_{V_\theta}(\boldsymbol{x}) &= \lim_{\tau\to 0}\left[\frac{\boldsymbol{z}_\tau(\boldsymbol{x}+\tau\boldsymbol{V}_\Theta)-\boldsymbol{z}(\boldsymbol{x})}{\tau}\right] \\ &= \lim_{\tau\to 0}\left[\frac{\boldsymbol{z}_\tau(\boldsymbol{x})-\boldsymbol{z}(\boldsymbol{x})}{\tau}\right] + \lim_{\tau\to 0}\left[\frac{\boldsymbol{z}_\tau(\boldsymbol{x}+\tau\boldsymbol{V}_\Theta)-\boldsymbol{z}_\tau(\boldsymbol{x})}{\tau}\right] \\ &= \boldsymbol{z}'_{V_\theta}(\boldsymbol{x}) + \lim_{\tau\to 0}\frac{[\boldsymbol{A}_2(\tau V_{3,2},-\tau V_{3,1}) - \boldsymbol{A}_2(0,-\tau V_{3,1}) + \boldsymbol{A}_2(0,-\tau V_{3,1}) - \boldsymbol{I}]\boldsymbol{z}_\tau(\boldsymbol{x})}{\tau} \\ &= \boldsymbol{z}'_{V_\theta}(\boldsymbol{x}) + \left[\frac{d}{d\tau}\boldsymbol{A}(\tau V_{3,2},0)\bigg|_{\tau=0} + \frac{d}{d\tau}\boldsymbol{A}(0,-\tau V_{3,1})\bigg|_{\tau=0}\right]\boldsymbol{z}(\boldsymbol{x}) \\ &= \boldsymbol{z}'_{V_\theta}(\boldsymbol{x}) + \boldsymbol{V}_\theta\boldsymbol{z}(\boldsymbol{x}), \end{aligned} \tag{7.18}$$

where, for the surface design component, $\boldsymbol{V}_\theta$ is given as

$$\boldsymbol{V}_\theta = \begin{bmatrix} 0 & 0 & -V_{3,1} & 0 & 0 & 0 \\ 0 & 0 & -V_{3,2} & 0 & 0 & 0 \\ V_{3,1} & V_{3,2} & 0 & 0 & 0 & -V_{3,1}-V_{3,2} \\ 0 & 0 & 0 & 0 & 0 & -V_{3,1} \\ 0 & 0 & 0 & 0 & 0 & -V_{3,2} \\ 0 & 0 & 0 & V_{3,1} & V_{3,2} & 0 \end{bmatrix}. \tag{7.19}$$

Note that both (7.11) and (7.18) have the same form, with different $\boldsymbol{V}_\theta$ given by (7.12) and (7.19), respectively. For both line and surface design components, $\boldsymbol{V}_\theta$ involves derivatives of the configuration design velocity with respect to the local coordinates on the physical domain.

7.1.3 Material Derivative of a General Functional

In Chapter 6, both boundary and domain methods are used to derive the design sensitivity expression. Although the former has an advantage in providing physical insight of the design sensitivity, the latter is the preferred method for numerical calculations. Since analytical insight of configuration design sensitivity cannot be done readily, only the domain method is used in the following derivations.

If $\boldsymbol{J}_\Theta$ is the Jacobian matrix of the mapping $\boldsymbol{T}_\Theta(\boldsymbol{x},\tau)$, then it is defined as

$$\boldsymbol{J}_\Theta = \left[\frac{\partial \boldsymbol{T}_\Theta}{\partial \boldsymbol{x}}\right] = \boldsymbol{I} + \tau\left[\frac{\partial \boldsymbol{V}_\Theta}{\partial \boldsymbol{x}}\right] = \boldsymbol{I} + \tau\nabla\boldsymbol{V}_\Theta(\boldsymbol{x}), \tag{7.20}$$

where $\boldsymbol{I}$ is the identify matrix and $\nabla\boldsymbol{V}_\Theta(\boldsymbol{x})$ is the Jacobian matrix of $\boldsymbol{V}_\Theta(\boldsymbol{x})$. By using the orientation design velocity defined in (7.3), it can be shown that the determinant of the

Jacobian matrix is independent of the orientation change, i.e., $\frac{d}{d\tau}\left|\boldsymbol{J}_{\Theta}\right|\Big|_{\tau=0}=0$. Physically, this is obvious, since the structural domain remains fixed during the change in orientation.

To derive a material derivative formula, consider a functional defined in domain Ω_τ, as

$$\psi_\tau = \iint_{\Omega_\tau} f_\tau(\boldsymbol{x}_\tau)\,d\Omega_\tau. \tag{7.21}$$

The material derivative of ψ_τ with respect to the direction change is then defined as

$$\psi'_{V_\theta} = \frac{d}{d\tau}\iint_{\Omega_\tau} f_\tau(\boldsymbol{x}_\tau)\,d\Omega_\tau\Big|_{\tau=0}. \tag{7.22}$$

Using the relation in (7.20), the integration domain Ω_τ can be transformed into Ω. With the regular extension defined for the function f_τ as $f_\tau(\boldsymbol{x}_\tau) = f_\tau(\boldsymbol{x})$, the above equation can be rewritten as

$$\psi'_{V_\theta} = \frac{d}{d\tau}\iint_{\Omega} f_\tau(\boldsymbol{x}_\tau)\left|\boldsymbol{J}_\Theta\right| d\Omega\Big|_{\tau=0} = \iint_{\Omega}\frac{d}{d\tau} f_\tau(\boldsymbol{x})\,d\Omega\Big|_{\tau=0} = \iint_{\Omega} f'_{V_\theta}(\boldsymbol{x})\,d\Omega. \tag{7.23}$$

It is interesting to note that only the partial derivative of the integrand appears in the above equation, which is different from the material derivative of the same functional in (6.38).

7.2 Configuration Design Sensitivity Analysis

In this section, the configuration design sensitivity of both the static and eigenvalue responses is formulated using their variational equations. Both the direct differentiation and adjoint variable methods are presented for configuration design sensitivity analysis of static response. No adjoint response is required in eigenvalue design sensitivity analysis.

7.2.1 Variation of the Static Response

The structural variational equation of the boundary-value problem can be written as

$$a_\Omega(z,\bar{z}) = \ell_\Omega(\bar{z}), \qquad \forall \bar{z} \in Z, \tag{7.24}$$

where $a_\Omega(\bullet,\bullet)$ is the energy bilinear form, $\ell_\Omega(\bullet)$ is the load linear form, and Z is the space of kinematically admissible displacements. Before carrying out configuration design sensitivity analysis, the objective is to obtain a relationship between the variation in the configuration design and the resulting variation in the state equation. Since domain Ω, state response z, and virtual displacement $\bar{z}$ are all dependent on the structural configuration, the variational equation at the perturbed design may be written as

$$a_{\Omega_\tau}(z_\tau,\bar{z}_\tau) \equiv \iint_{\Omega_\tau} c(z_\tau,\bar{z}_\tau)\,d\Omega_\tau = \iint_{\Omega_\tau} \bar{z}_\tau^T \boldsymbol{f}\,d\Omega_\tau \equiv \ell_{\Omega_\tau}(\bar{z}_\tau), \qquad \forall \bar{z}_\tau \in Z_\tau, \tag{7.25}$$

where $c(\bullet,\bullet)$ is the bilinear function defined by the integrand of $a_\Omega(\bullet,\bullet)$, and $\boldsymbol{f}$ is the vector of externally applied loads. Without mathematical proof, let us assume that the energy bilinear and load linear forms are differentiable with respect to the configuration design. Using (6.38) and (7.23) and noting that the order of partial derivatives with respect to τ

and $\boldsymbol{x}$ can be interchanged, the first variation of both sides of the above equation can be written as

$$\begin{aligned}[a_\Omega(z,\bar{z})]' &= [a_\Omega(z,\bar{z})]'_{V_\Omega} + [a_\Omega(z,\bar{z})]'_{V_\theta} \\ &\equiv a_\Omega(\dot{z},\bar{z}) + a_\Omega(z,\dot{\bar{z}}) + a'_{V_\Omega}(z,\bar{z}) + a'_{V_\theta}(z,\bar{z})\end{aligned} \tag{7.26}$$

and

$$\begin{aligned}[\ell_\Omega(\bar{z})]' &= [\ell_\Omega(\bar{z})]'_{V_\Omega} + [\ell_\Omega(\bar{z})]'_{V_\theta} \\ &\equiv \ell_\Omega(\dot{\bar{z}}) + \ell'_{V_\Omega}(\bar{z}) + \ell'_{V_\theta}(\bar{z}).\end{aligned} \tag{7.27}$$

Note that both the shape variation and the configuration change contribute to the first variation of the energy bilinear and load linear forms. In (7.26) and (7.27), a prime denotes the first variation of the functional and $\dot{z}$ is the sum of the first variations due to shape variation and configuration change, as in (7.14). Similarly,

$$\dot{\bar{z}} = \dot{\bar{z}}_{V_\Omega} + \dot{\bar{z}}_{V_\theta}. \tag{7.28}$$

The contribution of $\dot{\bar{z}}$ will not be considered in the following derivations due to the same reason as discussed in (6.63).

Using (6.8), (6.38), (7.11), and (7.23), (7.26) and (7.27) become

$$\begin{aligned}[a_\Omega(z,\bar{z})]' &= \iint_\Omega \left\{ c(z,\bar{z}'_{V_\Omega}) + c(z'_{V_\Omega},\bar{z}) + div[c(z,\bar{z})\boldsymbol{V}_\Omega] \right\} d\Omega \\ &\quad + \iint_\Omega [c(z,\bar{z}'_{V_\theta}) + c(z'_{V_\theta},\bar{z})] d\Omega \\ &= \iint_\Omega [c(z,\dot{\bar{z}}) + c(\dot{z},\bar{z})] d\Omega - \iint_\Omega [c(z,\boldsymbol{V}_\theta\bar{z}) + c(\boldsymbol{V}_\theta z,\bar{z})] d\Omega \\ &\quad - \iint_\Omega \left\{ c(z,\nabla\bar{z}\boldsymbol{V}_\Omega) + c(\nabla z\boldsymbol{V}_\Omega,\bar{z}) - div[c(z,\bar{z})\boldsymbol{V}_\Omega] \right\} d\Omega\end{aligned} \tag{7.29}$$

and

$$\begin{aligned}[\ell_\Omega(\bar{z})]' &= \iint_\Omega \left\{ \bar{z}'^T_{V_\Omega} \boldsymbol{f} + \bar{z}^T \boldsymbol{f}'_{V_\Omega} + div[(\bar{z}^T \boldsymbol{f})\boldsymbol{V}_\Omega] \right\} d\Omega \\ &\quad + \iint_\Omega [\bar{z}'^T_{V_\theta} \boldsymbol{f} + \bar{z}^T \boldsymbol{f}'_{V_\theta}] d\Omega \\ &= \iint_\Omega \dot{\bar{z}}^T \boldsymbol{f}\, d\Omega + \iint_\Omega \left\{ \bar{z}^T \boldsymbol{f}'_{V_\Omega} - (\nabla\bar{z}\boldsymbol{V}_\Omega)^T \boldsymbol{f} + div[(\bar{z}^T \boldsymbol{f})\boldsymbol{V}_\Omega] \right\} d\Omega \\ &\quad + \iint_\Omega [\bar{z}^T \boldsymbol{f}'_{V_\theta} - (\boldsymbol{V}_\theta\bar{z})^T \boldsymbol{f}] d\Omega.\end{aligned} \tag{7.30}$$

The differentials $a'_{V_\Omega}(z,\bar{z})$, $a'_{V_\theta}(z,\bar{z})$, $\ell'_{V_\Omega}(\bar{z})$, and $\ell'_{V_\theta}(\bar{z})$ in (7.26) and (7.27) represent the explicit dependence of the energy bilinear and load linear forms on the shape variation and the orientation change, respectively. From (7.26), (7.27), (7.29), and (7.30), these variations can be obtained as

$$a'_{V_\Omega}(z,\bar{z}) = -\iint_\Omega \left\{ c(z,\nabla\bar{z}\boldsymbol{V}_\Omega) + c(\nabla z\boldsymbol{V}_\Omega,\bar{z}) - div[c(z,\bar{z})\boldsymbol{V}_\Omega] \right\} d\Omega \tag{7.31}$$

$$a'_{V_\theta}(z,\bar{z}) = -\iint_\Omega [c(z,\boldsymbol{V}_\theta\bar{z}) + c(\boldsymbol{V}_\theta z,\bar{z})] d\Omega \tag{7.32}$$

$$\ell'_{V_\Omega}(\bar{z}) = \iint_\Omega \left\{ \bar{z}^T \boldsymbol{f}'_{V_\Omega} - (\nabla\bar{z}\boldsymbol{V}_\Omega)^T \boldsymbol{f} + div[(\bar{z}^T \boldsymbol{f})\boldsymbol{V}_\Omega] \right\} d\Omega \tag{7.33}$$

$$\ell'_{V_\theta}(\bar{z}) = \iint_\Omega [\bar{z}^T f'_{V_\theta} - (V_\theta \bar{z})^T f] d\Omega. \tag{7.34}$$

Using the fact that $\dot{\bar{z}} \in Z$ and $a_\Omega(z, \dot{\bar{z}}) = \ell_\Omega(\dot{\bar{z}})$, from (7.26) and (7.27) the first variation of (7.25) becomes

$$a_\Omega(\dot{z}, \bar{z}) = \ell'_{V_\Omega}(\bar{z}) + \ell'_{V_\theta}(\bar{z}) - a'_{V_\Omega}(z, \bar{z}) - a'_{V_\theta}(z, \bar{z}), \qquad \forall \bar{z} \in Z. \tag{7.35}$$

This is the design sensitivity equation for shape and configuration designs. The right side of (7.35) can be readily calculated using the state response z and the design velocity fields V_Ω and V_θ. Thus, (7.35) is similar to variational (7.24) but with a different right side (the fictitious load). As is clear from (7.35), the contributions from shape and configuration designs appear separately, and they are linear with respect to the velocity fields V_Ω and V_θ.

Consider a general performance measure in an integral form as

$$\psi = \iint_{\Omega_\tau} g(z_\tau, \nabla z_\tau) d\Omega_\tau, \tag{7.36}$$

where function g is continuously differentiable with respect to its arguments. The functional ψ includes the first-order derivative of the state response. The situation in which this functional includes a second-order derivative will be addressed in the section on the beam design component. Taking the first variation of ψ and using (6.38) and (7.23),

$$\psi' = \psi'_{V_\Omega} + \psi'_{V_\theta} = \iint_\Omega [g_{,z}(z'_{V_\Omega} + z'_{V_\theta}) + g_{,\nabla z} : (\nabla z'_{V_\Omega} + \nabla z'_{V_\theta}) + div(g V_\Omega)] d\Omega. \tag{7.37}$$

Using (6.8) and (7.11) and the fact that $\dot{z} = \dot{z}_{V_\Omega} + \dot{z}_{V_\theta}$, (7.37) becomes

$$\begin{aligned} \psi' &= \iint_\Omega [g_{,z}\dot{z} + g_{,\nabla z} : \nabla \dot{z}] d\Omega \\ &- \iint_\Omega [g_{,z}(V_\theta z) + g_{,\nabla z} : \nabla(V_\theta z)] d\Omega \\ &- \iint_\Omega [g_{,\nabla z} : (\nabla z \nabla V_\Omega) - g\, div V_\Omega] d\Omega. \end{aligned} \tag{7.38}$$

To obtain ψ' explicitly in terms of the design velocity field, the first integral that includes $\dot{z}$ and $\nabla \dot{z}$ must be written explicitly in terms of the design velocity field.

Adjoint Variable Method

For the adjoint variable method, an adjoint equation is introduced by replacing $\dot{z}$ in (7.38) with a virtual displacement $\bar{\lambda}$ and equating the result to the energy bilinear form, as

$$a_\Omega(\lambda, \bar{\lambda}) = \iint_\Omega [g_{,z}\bar{\lambda} + g_{,\nabla z} : \nabla \bar{\lambda}] d\Omega, \qquad \forall \bar{\lambda} \in Z. \tag{7.39}$$

Since the right side of (7.39) is a bounded linear functional of $\bar{\lambda}$, the adjoint equation has a unique solution λ. Also, since $\dot{z} \in Z$, (7.39) can be evaluated at $\bar{\lambda} = \dot{z}$ to obtain

$$a_\Omega(\lambda, \dot{z}) = \iint_\Omega [g_{,z}\dot{z} + g_{,\nabla z} : \nabla \dot{z}] d\Omega. \tag{7.40}$$

Since $\lambda \in Z$, (7.35) can be evaluated at $\bar{z} = \lambda$, to obtain

$$a_\Omega(\dot{z}, \lambda) = \ell'_{V_\Omega}(\lambda) + \ell'_{V_\theta}(\lambda) - a'_{V_\Omega}(z, \lambda) - a'_{V_\theta}(z, \lambda). \tag{7.41}$$

From (7.38), (7.40), and (7.41), and by using the symmetry of the bilinear form $a_\Omega(\bullet,\bullet)$,

$$\begin{aligned}\psi' = &\ \ell'_{V_\Omega}(\boldsymbol{\lambda}) + \ell'_{V_\theta}(\boldsymbol{\lambda}) - a'_{V_\Omega}(\boldsymbol{z},\boldsymbol{\lambda}) - a'_{V_\theta}(\boldsymbol{z},\boldsymbol{\lambda}) \\ &- \iint_\Omega [g_{,\boldsymbol{z}}(\boldsymbol{V}_\theta \boldsymbol{z}) + g_{,\nabla\boldsymbol{z}} : \nabla(\boldsymbol{V}_\theta \boldsymbol{z})]\, d\Omega \\ &- \iint_\Omega [g_{,\nabla\boldsymbol{z}} : (\nabla\boldsymbol{z}\nabla\boldsymbol{V}_\Omega) - g\, div\boldsymbol{V}_\Omega]\, d\Omega.\end{aligned} \tag{7.42}$$

Once the design velocity fields $\boldsymbol{V}_\Omega$ and $\boldsymbol{V}_\theta$ are defined, with the state response $\boldsymbol{z}$ and adjoint response $\boldsymbol{\lambda}$ obtained from (7.24) and (7.39), the above configuration design sensitivity expression can be evaluated.

Direct Differentiation Method

With the direct differentiation method, (7.35) is the variational equation for the first variation $\dot{\boldsymbol{z}}$. Once the state response $\boldsymbol{z}$ is obtained from (7.24), the right side of (7.35) can be evaluated and (7.35) can be solved for $\dot{\boldsymbol{z}}$. Using the result $\dot{\boldsymbol{z}}$ and the shape and orientation design velocity fields, the configuration design sensitivity expression in (7.38) can be evaluated.

Note that computation of the right side of (7.35) depends on the design velocity fields. Therefore, different types of design parameterization will yield different sets of fictitious loads that are quite different from the adjoint loads, which are associated with performance measures. Further comparisons between the adjoint variable method and the direct differentiation method can be found in References [117] and [118].

7.2.2 Eigenvalue Problems

The variational equation of the eigenvalue problem for vibration and buckling of a structural component can be written as

$$a_\Omega(\boldsymbol{y},\bar{\boldsymbol{y}}) = \zeta d_\Omega(\boldsymbol{y},\bar{\boldsymbol{y}}), \qquad \forall \bar{\boldsymbol{y}} \in Z, \tag{7.43}$$

where ζ is the buckling load for the buckling problem or $\zeta = \omega^2$, with ω being the natural frequency of the vibration problem. In (7.43), Z is the space of kinematically admissible displacements in Ω. Since (7.43) is homogeneous in $\boldsymbol{y}$, a normalizing condition must be used to uniquely define the eigenfunction. The normalizing condition is

$$d_\Omega(\boldsymbol{y},\boldsymbol{y}) = 1. \tag{7.44}$$

Similar to the development of the static problem, the variational equation of the eigenvalue problem on a perturbed domain is

$$\begin{aligned} a_{\Omega_\tau}(\boldsymbol{y}_\tau,\bar{\boldsymbol{y}}_\tau) &\equiv \iint_{\Omega_\tau} c(\boldsymbol{y}_\tau,\bar{\boldsymbol{y}}_\tau)\, d\Omega_\tau \\ &= \zeta_\tau \iint_{\Omega_\tau} e(\boldsymbol{y}_\tau,\bar{\boldsymbol{y}}_\tau)\, d\Omega_\tau \equiv \zeta_\tau d_{\Omega_\tau}(\boldsymbol{y}_\tau,\bar{\boldsymbol{y}}_\tau), \qquad \forall \bar{\boldsymbol{y}}_\tau \in Z_\tau, \end{aligned} \tag{7.45}$$

with the normalizing condition

$$d_{\Omega_\tau}(\boldsymbol{y}_\tau,\boldsymbol{y}_\tau) = 1, \tag{7.46}$$

where $e(\bullet,\bullet)$ is a bilinear function defined by the integrand of $d_\Omega(\bullet,\bullet)$.

The energy bilinear form on the left side of (7.45) is the same as the bilinear form shown in (7.25) for the static problem. The bilinear form $d_\Omega(\bullet,\bullet)$ on the right side of (7.45) represents the mass effect for the vibration problem and the geometric effect for the buckling problem. This form is usually more regular in its dependence on design variables than the energy bilinear form $a_\Omega(\bullet,\bullet)$.

Presuming differentiability of the eigenvalue ζ and the eigenfunction $\boldsymbol{y}$ with respect to the configuration design variable, the first variation of (7.45) is

$$\begin{aligned}[a_\Omega(\boldsymbol{y},\overline{\boldsymbol{y}})]' &\equiv a'_{V_\Omega}(\boldsymbol{y},\overline{\boldsymbol{y}}) + a'_{V_\theta}(\boldsymbol{y},\overline{\boldsymbol{y}}) + a_\Omega(\dot{\boldsymbol{y}},\overline{\boldsymbol{y}}) + a_\Omega(\boldsymbol{y},\dot{\overline{\boldsymbol{y}}})\\ &= [\zeta'_{V_\Omega} + \zeta'_{V_\theta}]d_\Omega(\boldsymbol{y},\overline{\boldsymbol{y}}) + \zeta[d'_{V_\Omega}(\boldsymbol{y},\overline{\boldsymbol{y}}) + d'_{V_\theta}(\boldsymbol{y},\overline{\boldsymbol{y}}) + d_\Omega(\dot{\boldsymbol{y}},\overline{\boldsymbol{y}}) + d_\Omega(\boldsymbol{y},\dot{\overline{\boldsymbol{y}}})]\\ &\equiv \zeta' d_\Omega(\boldsymbol{y},\overline{\boldsymbol{y}}) + \zeta[d_\Omega(\boldsymbol{y},\overline{\boldsymbol{y}})]', \qquad \forall \overline{\boldsymbol{y}} \in Z,\end{aligned} \tag{7.47}$$

where, using (6.8), (6.38), (7.11), and (7.23), the variations of the bilinear forms in (7.47) can be written as

$$\begin{aligned}[a_\Omega(\boldsymbol{y},\overline{\boldsymbol{y}})]' &= \iint_\Omega \{c(\boldsymbol{y}'_{V_\Omega},\overline{\boldsymbol{y}}) + c(\boldsymbol{y},\overline{\boldsymbol{y}}'_{V_\Omega}) + div[c(\boldsymbol{y},\overline{\boldsymbol{y}})\boldsymbol{V}_\Omega]\}\,d\Omega\\ &\quad + \iint_\Omega [c(\boldsymbol{y}'_{V_\theta},\overline{\boldsymbol{y}}) + c(\boldsymbol{y},\overline{\boldsymbol{y}}'_{V_\theta})]\,d\Omega\\ &= \iint_\Omega \{c(\dot{\boldsymbol{y}} - \nabla\boldsymbol{y}\boldsymbol{V}_\Omega,\overline{\boldsymbol{y}}) + c(\boldsymbol{y},\dot{\overline{\boldsymbol{y}}} - \nabla\overline{\boldsymbol{y}}\boldsymbol{V}_\Omega) + div[c(\boldsymbol{y},\overline{\boldsymbol{y}})\boldsymbol{V}_\Omega]\}\,d\Omega\\ &\quad - \iint_\Omega [c(\boldsymbol{V}_\theta\boldsymbol{y},\overline{\boldsymbol{y}}) + c(\boldsymbol{y},\boldsymbol{V}_\theta\overline{\boldsymbol{y}})]\,d\Omega\end{aligned} \tag{7.48}$$

and

$$\begin{aligned}[d_\Omega(\boldsymbol{y},\overline{\boldsymbol{y}})]' &= \iint_\Omega \{e(\boldsymbol{y}'_{V_\Omega},\overline{\boldsymbol{y}}) + e(\boldsymbol{y},\overline{\boldsymbol{y}}'_{V_\Omega}) + div[e(\boldsymbol{y},\overline{\boldsymbol{y}})\boldsymbol{V}_\Omega]\}\,d\Omega\\ &\quad + \iint_\Omega [e(\boldsymbol{y}'_{V_\theta},\overline{\boldsymbol{y}}) + e(\boldsymbol{y},\overline{\boldsymbol{y}}'_{V_\theta})]\,d\Omega\\ &= \iint_\Omega \{e(\dot{\boldsymbol{y}} - \nabla\boldsymbol{y}\boldsymbol{V}_\Omega,\overline{\boldsymbol{y}}) + e(\boldsymbol{y},\dot{\overline{\boldsymbol{y}}} - \nabla\overline{\boldsymbol{y}}\boldsymbol{V}_\Omega) + div[e(\boldsymbol{y},\overline{\boldsymbol{y}})\boldsymbol{V}_\Omega]\}\,d\Omega\\ &\quad - \iint_\Omega [e(\boldsymbol{V}_\theta\boldsymbol{y},\overline{\boldsymbol{y}}) + e(\boldsymbol{y},\boldsymbol{V}_\theta\overline{\boldsymbol{y}})]\,d\Omega.\end{aligned} \tag{7.49}$$

By comparing (7.47) with (7.48) and (7.49) and after using the relation of $a_\Omega(\boldsymbol{y},\dot{\overline{\boldsymbol{y}}}) = \zeta d_\Omega(\boldsymbol{y},\dot{\overline{\boldsymbol{y}}})$, the explicitly dependent terms in (7.47) can be derived as

$$a'_{V_\Omega}(\boldsymbol{y},\overline{\boldsymbol{y}}) = -\iint_\Omega \{c(\nabla\boldsymbol{y}\boldsymbol{V}_\Omega,\overline{\boldsymbol{y}}) + c(\boldsymbol{y},\nabla\overline{\boldsymbol{y}}\boldsymbol{V}_\Omega) - div[c(\boldsymbol{y},\overline{\boldsymbol{y}})\boldsymbol{V}_\Omega]\}\,d\Omega \tag{7.50}$$

$$a'_{V_\theta}(\boldsymbol{y},\overline{\boldsymbol{y}}) = -\iint_\Omega [c(\boldsymbol{V}_\theta\boldsymbol{y},\overline{\boldsymbol{y}}) + c(\boldsymbol{y},\boldsymbol{V}_\theta\overline{\boldsymbol{y}})]\,d\Omega \tag{7.51}$$

$$d'_{V_\Omega}(\boldsymbol{y},\overline{\boldsymbol{y}}) = -\iint_\Omega \{e(\nabla\boldsymbol{y}\boldsymbol{V}_\Omega,\overline{\boldsymbol{y}}) + e(\boldsymbol{y},\nabla\overline{\boldsymbol{y}}\boldsymbol{V}_\Omega) - div[e(\boldsymbol{y},\overline{\boldsymbol{y}})\boldsymbol{V}_\Omega]\}\,d\Omega \tag{7.52}$$

$$d'_{V_\theta}(\boldsymbol{y},\overline{\boldsymbol{y}}) = -\iint_\Omega [e(\boldsymbol{V}_\theta\boldsymbol{y},\overline{\boldsymbol{y}}) + e(\boldsymbol{y},\boldsymbol{V}_\theta\overline{\boldsymbol{y}})]\,d\Omega. \tag{7.53}$$

Both $a'_{V_\Omega}(\boldsymbol{y},\overline{\boldsymbol{y}})$ and $d'_{V_\Omega}(\boldsymbol{y},\overline{\boldsymbol{y}})$ are due to the shape variation, and $a'_{V_\theta}(\boldsymbol{y},\overline{\boldsymbol{y}})$ and $d'_{V_\theta}(\boldsymbol{y},\overline{\boldsymbol{y}})$ are due to the orientation change in the structural system. Note that $\dot{\boldsymbol{y}} = \dot{\boldsymbol{y}}_{V_\Omega} + \dot{\boldsymbol{y}}_{V_\theta}$ and $\dot{\overline{\boldsymbol{y}}} = \dot{\overline{\boldsymbol{y}}}_{V_\Omega} + \dot{\overline{\boldsymbol{y}}}_{V_\theta}$ have been used to obtain (7.47) through (7.49).

In order to obtain a more valuable result for the eigenvalue sensitivity, evaluate (7.47) at $\overline{\boldsymbol{y}} = \boldsymbol{y}$, and use the fact that $a_\Omega(\bullet,\bullet)$ and $d_\Omega(\bullet,\bullet)$ are symmetric, to obtain

$$\begin{aligned}\zeta' d_\Omega(\boldsymbol{y},\boldsymbol{y}) &= [a'_{V_\Omega}(\boldsymbol{y},\boldsymbol{y}) + a'_{V_\theta}(\boldsymbol{y},\boldsymbol{y})] - \zeta[d'_{V_\Omega}(\boldsymbol{y},\boldsymbol{y}) + d'_{V_\theta}(\boldsymbol{y},\boldsymbol{y})]\\ &\quad + [a_\Omega(\boldsymbol{y},\dot{\boldsymbol{y}}) - \zeta d_\Omega(\boldsymbol{y},\dot{\boldsymbol{y}})].\end{aligned} \tag{7.54}$$

Since $\dot{\boldsymbol{y}} \in Z$ [see the paragraph following (6.232)], the term in the last brackets on the right side of (7.54) is zero. Furthermore, due to the normalizing condition in (7.44), a simplified equation may be used, namely,

$$\begin{aligned}\zeta' &= [a'_{V_\Omega}(y,y) + a'_{V_\theta}(y,y)] - \zeta[d'_{V_\Omega}(y,y) + d'_{V_\theta}(y,y)] \\ &= \iint_\Omega \{-2c(y,\nabla y V_\Omega) + 2\zeta e(y,\nabla y V_\Omega) + div[c(y,y)V_\Omega] - \zeta\, div[e(y,y)V_\Omega]\}\, d\Omega \\ &\quad + 2\iint_\Omega [-c(y,V_\theta y) + \zeta e(y,V_\theta y)]\, d\Omega.\end{aligned} \qquad (7.55)$$

Note that the eigenvalue sensitivity formulation in this section is valid for a simple eigenvalue. For repeated eigenvalues, the configuration design sensitivity analysis can be derived in a similar way to the shape design sensitivity analysis discussed in Sections 5.2 and 6.3 in Chapters 5 and 6. respectively.

7.2.3 Analytical Examples

In this section, the general sensitivity expressions presented in Sections 7.2.1 and 7.2.2 are applied to truss, beam, plane elastic solid, and plate design components. For built-up structures, the configuration design sensitivity results are obtained by summing the design sensitivities contributed from each individual design component.

Truss Design Component

A truss design component is the simplest example among structural components. Figure 7.4 shows a simple, two-dimensional truss component whose configuration changes due to the design velocity field $V(x) = [V_1(x), V_2(x)]^T$. Since the design velocity is expressed in the component-fixed local coordinate system, $V_1(x)$ represents the shape design velocity and $V_2(x)$ denotes the configuration design velocity. In this simplified example, the rotational angles in Fig. 7.2. are such that $\delta\beta = \delta\gamma = 0$. The design velocity $V(x)$ is constrained such that the line component remains straight during design perturbation. This constraint provides a design velocity field whose first-order derivative is constant while the second-order derivative vanishes.

The structural bilinear and load linear forms in (3.6) and (3.7) that are used for the truss component are rewritten here as

$$a_\Omega(z,\bar{z}) = \int_0^l EA\bar{z}_{1,1} z_{1,1}\, dx \qquad (7.56)$$

$$\ell_\Omega(\bar{z}) = \int_0^l \bar{z}_1 f_1\, dx, \qquad (7.57)$$

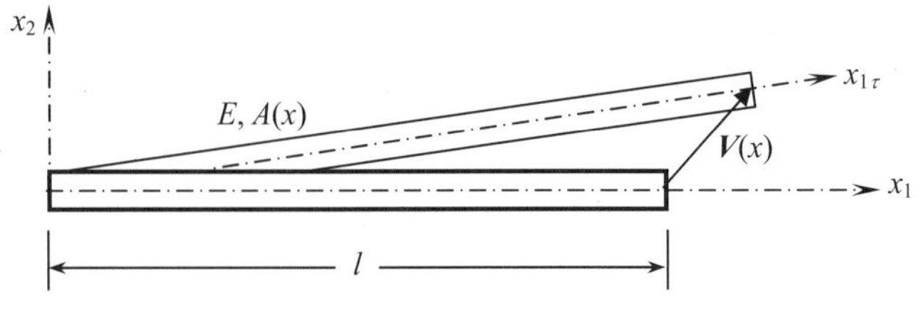

Figure 7.4. Configuration change in the truss component.

where E is Young's modulus, $A(x)$ is the cross-sectional area, and f_1 is the distributed axial force along the x_1-axis.

In this special case, the convective term of the configuration change in (7.11) becomes $\boldsymbol{V}_\theta \boldsymbol{z} = -V_{2,1}z_2$. From the assumption that the line component remains straight, the second-order derivative of the design velocity field vanishes; i.e., $V_{2,11} = 0$. By applying the formulas in (7.31) through (7.34) to the truss component, the explicitly dependent terms are derived as

$$a'_{V_\Omega}(\boldsymbol{z},\bar{\boldsymbol{z}}) = -\int_0^l EAz_{1,1}\bar{z}_{1,1}V_{1,1}\,dx \tag{7.58}$$

$$a'_{V_\theta}(\boldsymbol{z},\bar{\boldsymbol{z}}) = -\int_0^l EA(z_{2,1}\bar{z}_{1,1} + z_{1,1}\bar{z}_{2,1})V_{2,1}\,dx \tag{7.59}$$

$$\ell'_{V_\Omega}(\bar{\boldsymbol{z}}) = \int_0^l (f_{1,1}\bar{z}_1V_1 + f_1\bar{z}_1V_{1,1})\,dx \tag{7.60}$$

$$\ell'_{V_\theta}(\bar{\boldsymbol{z}}) = \int_0^l f_1\bar{z}_2V_{2,1}\,dx. \tag{7.61}$$

In the derivation of (7.60) and (7.61), it is assumed that $f_1' = 0$, which means that the axial force does not vary during the configuration change.

It is interesting to note that the truss component only has the axial displacement z_1, and that the transverse displacement $z_2 = 0$. Thus, the explicitly dependent terms, (7.59) and (7.61), on the configuration design vanish. This is expected because the truss component does not have any rotational degrees-of-freedom. If a built-up structure is composed of beam and truss components, however, then the transverse displacement due to the beam component causes nonzero values to the explicitly dependent terms in (7.59) and (7.61).

Let us calculate the sensitivity of the displacement functional at isolated point $\hat{x}$, which can be represented using the Dirac delta measure as

$$\psi_1 = \int_0^l \delta(x-\hat{x})z_1\,dx. \tag{7.62}$$

The variation of the functional ψ_1 can be simply obtained by

$$\psi_1' = \int_0^l \delta(x-\hat{x})\dot{z}_1\,dx. \tag{7.63}$$

With the adjoint variable method, the integral in (7.63) is used to define the adjoint load when $\dot{z}_1$ is substituted into $\bar{\lambda}_1$, to obtain the adjoint equation as

$$a_\Omega(\boldsymbol{\lambda},\bar{\boldsymbol{\lambda}}) = \int_0^l \delta(x-\hat{x})\bar{\lambda}_1\,dx, \quad \forall \bar{\boldsymbol{\lambda}} \in Z. \tag{7.64}$$

This equation is equivalent to the truss component when a unit force is applied to point $\hat{x}$.

After solving the state response and the adjoint response, the sensitivity of the displacement functional in (7.62) can be calculated using the formula in (7.42), as

$$\begin{aligned}
\psi_1' &= \ell'_{V_\Omega}(\boldsymbol{\lambda}) + \ell'_{V_\theta}(\boldsymbol{\lambda}) - a'_{V_\Omega}(\boldsymbol{z},\boldsymbol{\lambda}) - a'_{V_\theta}(\boldsymbol{z},\boldsymbol{\lambda}) \\
&= \int_0^l (f_{1,1}\lambda_1V_1 + f_1\lambda_1V_{1,1} + f_1\lambda_2V_{2,1})\,dx \\
&\quad + \int_0^l EA[(z_{1,1}\lambda_{1,1})V_{1,1} + (z_{2,1}\lambda_{1,1} + z_{1,1}\lambda_{2,1})V_{2,1}]\,dx.
\end{aligned} \tag{7.65}$$

Note that no explicitly dependent term exists in above design sensitivity expression for the displacement functional.

Next, consider a stress functional defined in a small subdomain $(x_a,x_b) \subset (0,l)$. The stress functional within a subdomain can be defined using a characteristic function m_p in (5.31) whose integral is one in the interval of (x_a,x_b) and zero outside (x_a,x_b), as

$$\psi_2 = \int_0^l EAz_{1,1}m_p\,dx. \tag{7.66}$$

The variation of (7.66) with respect to shape and configuration designs becomes

$$\psi_2' = \int_0^l EA\dot{z}_{1,1}\,dx + \int_0^l EA(z_{2,1}V_{2,1} - z_{1,1}V_{1,1})\,dx, \tag{7.67}$$

where the first integral is used to defined the adjoint equation as

$$a_\Omega(\boldsymbol{\lambda},\bar{\boldsymbol{\lambda}}) = \int_0^l EA\bar{\lambda}_{1,1}\,dx, \quad \forall\bar{\boldsymbol{\lambda}} \in Z. \tag{7.68}$$

After calculating the state response $\boldsymbol{z}$ and the adjoint response $\boldsymbol{\lambda}$, the sensitivity of the stress functional can be obtained as

$$\begin{aligned}\psi_2' &= \ell_{V_\Omega}(\boldsymbol{\lambda}) + \ell_{V_\theta}(\boldsymbol{\lambda}) - a'_{V_\Omega}(\boldsymbol{z},\boldsymbol{\lambda}) - a'_{V_\theta}(\boldsymbol{z},\boldsymbol{\lambda}) \\ &\quad + \int_0^l EA(z_{2,1}V_{2,1} - z_{1,1}V_{1,1})\,dx \\ &= \int_0^l (f_{1,1}\lambda_1V_1 + f_1\lambda_1V_{1,1} + f_1\lambda_2V_{2,1})\,dx \\ &\quad + \int_0^l EA[(z_{1,1}\lambda_{1,1} - z_{1,1})V_{1,1} + (z_{2,1}\lambda_{1,1} + z_{1,1}\lambda_{2,1} + z_{2,1})V_{2,1}]\,dx.\end{aligned} \tag{7.69}$$

The sensitivity expressions of (7.65) and (7.69) can be calculated from the state response $\boldsymbol{z}$, the adjoint response $\boldsymbol{\lambda}$, and the design velocity $\boldsymbol{V}$.

The eigenvalue problem of the truss component uses the same energy bilinear form $a_\Omega(\bullet,\bullet)$ in (7.56) with the argument as eigenfunction $\boldsymbol{y}$. For the mass effect, the bilinear form $d_\Omega(\bullet,\bullet)$ and its variation are defined as

$$d_\Omega(\boldsymbol{y},\bar{\boldsymbol{y}}) = \int_0^l \rho Ay_1\bar{y}_1\,dx \tag{7.70}$$

$$d'_{V_\Omega}(\boldsymbol{y},\bar{\boldsymbol{y}}) = \int_0^l \rho Ay_1\bar{y}_1V_{1,1}\,dx \tag{7.71}$$

$$d'_{V_\theta}(\boldsymbol{y},\bar{\boldsymbol{y}}) = \int_0^l (y_1\bar{y}_2 + y_2\bar{y}_1)V_{2,1}\,dx, \tag{7.72}$$

where ρ is the density and A is the cross-sectional area. Again, the effect of the configuration design appears when a built-up structure is taken into account.

The sensitivity of the eigenvalue in (7.55) can now be evaluated by substituting (7.58), (7.59), (7.71), and (7.72) into (7.55), as

$$\begin{aligned}\zeta' &= [a'_{V_\Omega}(\boldsymbol{y},\boldsymbol{y}) + a'_{V_\theta}(\boldsymbol{y},\boldsymbol{y})] - \zeta[d'_{V_\Omega}(\boldsymbol{y},\boldsymbol{y}) + d'_{V_\theta}(\boldsymbol{y},\boldsymbol{y})] \\ &= -\int_0^l EA[y_{1,1}^2V_{1,1} + 2y_{1,1}y_{2,1}V_{2,1}]\,dx - \zeta\int_0^l \rho A[y_1^2V_{1,1} + 2y_1y_2V_{2,1}]\,dx.\end{aligned} \tag{7.73}$$

The sensitivity expressions of (7.73) can be calculated from the eigenfunction $\boldsymbol{y}$ and the design velocity $\boldsymbol{V}$. The adjoint response is not required in the calculation of the eigenvalue sensitivity.

Beam Design Component

Compared with the truss, the beam component requires more complicated mathematical calculations because the second-order derivatives are involved in the variational equation. Consider the two-dimensional beam design component shown in Fig. 7.5. For the static problem, the energy bilinear and load linear forms of the beam design component are

$$a_{\Omega}(z,\bar{z}) = \int_0^l EIz_{2,11}\bar{z}_{2,11}\,dx \tag{7.74}$$

$$\ell_{\Omega}(\bar{z}) = \int_0^l \bar{z}_2 f_2\,dx, \tag{7.75}$$

where E is Young's modulus, $I(x)$ is the moment of inertia, $f_2(x)$ is the distributed transverse load, and z_2 is the transverse displacement.

Unlike the truss component, the unknown for the beam problem is the transverse displacement z_2. Thus, the convective term of the configuration change in (7.11) becomes $\boldsymbol{V}_{\theta}\boldsymbol{z} = V_{2,1}z_1$ for a two-dimensional beam design component. From the assumption that the line component remains straight, the second-order derivative of the design velocity field vanishes; i.e., $V_{2,11} = 0$. However, the shape design velocity does not have to change linearly, i.e., $V_{1,11} \neq 0$. By applying the formulas in (7.31) through (7.34) to the beam design component, the explicitly dependent terms are derived as

$$a'_{V_{\Omega}}(z,\bar{z}) = -\int_0^l EI[3z_{2,11}\bar{z}_{2,11}V_{1,1} + (z_{2,1}\bar{z}_{2,11} + z_{2,11}\bar{z}_{2,1})V_{1,11}]dx \tag{7.76}$$

$$a'_{V_{\theta}}(z,\bar{z}) = -\int_0^l EI(z_{1,11}\bar{z}_{2,11} + z_{2,11}\bar{z}_{1,11})V_{2,1}\,dx \tag{7.77}$$

$$\ell'_{V_{\Omega}}(\bar{z}) = \int_0^l (f_{2,1}\bar{z}_2V_1 + f_2\bar{z}_2V_{1,1})dx \tag{7.78}$$

$$\ell'_{V_{\theta}}(\bar{z}) = -\int_0^l f_2\bar{z}_1V_{2,1}\,dx. \tag{7.79}$$

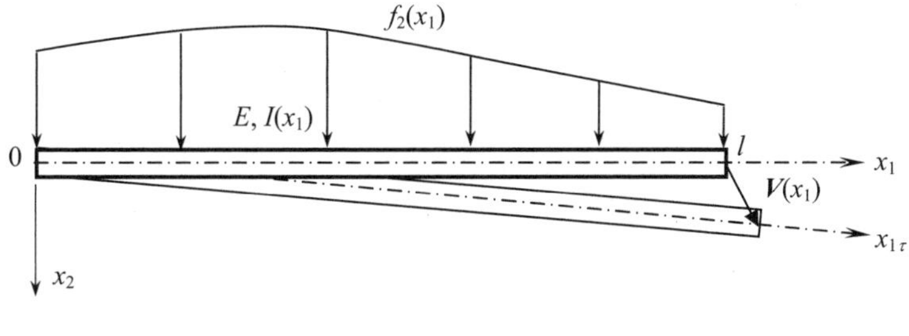

Figure 7.5. Beam design component.

In (7.78) and (7.79), it is assumed that $f_2' = 0$, which means that the transverse force does not vary during the configuration change. Note that the variations in (7.76) through (7.79) linearly depend on the design velocity fields, and are written explicitly in terms of shape and configuration design velocity fields. The axial and transverse displacements are coupled in the explicit term $a'_{V_\theta}(z,\bar{z})$.

The configuration design sensitivity of displacement, stress, and eigenvalue performance measures is now considered for the beam design component. First, a transverse displacement functional can be given in an integral form as

$$\psi_1 = z_2(\hat{x}) = \int_0^l \delta(x-\hat{x}) z_2 \, dx, \tag{7.80}$$

where $\delta(x-\hat{x})$ is the Dirac delta measure and $\hat{x}$ is the location at which displacement z_2 is measured. Taking the first variation of (7.80),

$$\psi_1' = z_2'(\hat{x}) = \int_0^l \delta(x-\hat{x}) \dot{z}_2 \, dx. \tag{7.81}$$

For the displacement performance measure, the adjoint equation from (7.39) is

$$a_\Omega(\boldsymbol{\lambda},\bar{\boldsymbol{\lambda}}) = \int_0^l \delta(x-\hat{x}) \bar{\lambda}_2 \, dx, \qquad \forall \bar{\boldsymbol{\lambda}} \in Z. \tag{7.82}$$

The adjoint load on the right side of (7.82) can be interpreted simply as a unit load applied at point $\hat{x}$ in the x_2-direction. After calculating the adjoint response, the sensitivity expression of the displacement functional becomes

$$\begin{aligned}
\psi_1' &= \ell'_{V_\Omega}(\boldsymbol{\lambda}) + \ell'_{V_\theta}(\boldsymbol{\lambda}) - a'_{V_\Omega}(z,\boldsymbol{\lambda}) - a'_{V_\theta}(z,\boldsymbol{\lambda}) \\
&= \int_0^l [f_{2,1}\lambda_2 V_1 + f_2\lambda_2 V_{1,1} - f_2\lambda_1 V_{2,1}] dx \\
&\quad - \int_0^l EI[3z_{2,11}\lambda_{2,11}V_{1,1} + (z_{2,1}\lambda_{2,11} + z_{2,11}\lambda_{2,1})V_{1,11} + (z_{1,11}\lambda_{2,11} + z_{2,11}\lambda_{1,11})V_{2,1}] dx,
\end{aligned} \tag{7.83}$$

where all terms on the right side are given explicitly in (7.76) through (7.79), with $\boldsymbol{\lambda}$ as the solution to (7.82).

Next, consider a stress performance measure defined over a small interval (x_a, x_b), namely,

$$\psi_2 = \int_0^l Ehz_{2,11} m_p \, dx, \tag{7.84}$$

where h is the half-depth of the beam design component in the x_2-direction, and m_p is the characteristic function that is positive on (x_a, x_b) with a value of $1/(x_b - x_a)$ and zero outside (x_a, x_b). Taking the first variation of (7.84),

$$\psi_2' = \int_0^l Eh\dot{z}_{2,11} m_p \, dx - \int_0^l Eh[2z_{2,11}V_{1,1} + z_{2,1}V_{1,11} + z_{1,11}V_{2,1}] m_p \, dx. \tag{7.85}$$

The adjoint equation of the stress performance measure is defined from (7.39), as

$$a_\Omega(\boldsymbol{\lambda},\bar{\boldsymbol{\lambda}}) = \int_0^l Eh\bar{\lambda}_{2,11} m_p \, dx, \qquad \forall \bar{\boldsymbol{\lambda}} \in Z. \tag{7.86}$$

After calculating the adjoint response $\boldsymbol{\lambda}$, the sensitivity of the stress functional can be evaluated from (7.42), as

$$\psi_2' = \ell_{V_\Omega}'(\lambda) + \ell_{V_\theta}'(\lambda) - a_{V_\Omega}'(z,\lambda) - a_{V_\theta}'(z,\lambda) - \int_0^l Eh[2z_{2,11}V_{1,1} + z_{2,1}V_{1,11} + z_{1,11}V_{2,1}]m_p\, dx. \tag{7.87}$$

Again, $a_{V_\Omega}'(z,\lambda)$, $a_{V_\theta}'(z,\lambda)$, $\ell_{V_\Omega}'(\lambda)$, and $\ell_{V_\theta}'(\lambda)$ are given explicitly in (7.76) through (7.79) and λ is the solution to (7.86). For other kinds of static performance measures, such as compliance or displacement performance at a fixed point, the general design sensitivity expression formulated in Section 7.2.1 can be used.

For the eigenvalue performance measure, the energy bilinear form and the bilinear form due to the mass effect are

$$a_\Omega(\boldsymbol{y},\bar{\boldsymbol{y}}) = \int_0^l EIy_{2,11}\bar{y}_{2,11}\, dx \tag{7.88}$$

$$d_\Omega(\boldsymbol{y},\bar{\boldsymbol{y}}) = \int_0^l \rho A y_2 \bar{y}_2\, dx, \tag{7.89}$$

where ρ is the material density and I is the moment of inertia. If the eigenvalue problem for buckling of the beam is considered, the bilinear form $d_\Omega(\bullet,\bullet)$ should be used to represent the geometric stiffness.

The energy bilinear form in (7.88) is the same as (7.74), except that eigenfunction $\boldsymbol{y}$ is used in place of the displacement z. Therefore, the first variation of the energy bilinear form in (7.88) can be obtained by replacing z with $\boldsymbol{y}$ in (7.76) and (7.77). Using (7.89), the first variation of the bilinear form $d_\Omega(\bullet,\bullet)$ in (7.52) and (7.53) becomes

$$d_{V_\Omega}'(\boldsymbol{y},\bar{\boldsymbol{y}}) = \int_0^l \rho A y_2 \bar{y}_2 V_{1,1}\, dx \tag{7.90}$$

$$d_{V_\theta}'(\boldsymbol{y},\bar{\boldsymbol{y}}) = -\int_0^l \rho A(y_1\bar{y}_2 + \bar{y}_1 y_2)V_{2,1}\, dx. \tag{7.91}$$

The eigenvalue design sensitivity expression given in (7.55) is

$$\psi_3' \equiv \zeta' = a_{V_\Omega}'(\boldsymbol{y},\boldsymbol{y}) + a_{V_\theta}'(\boldsymbol{y},\boldsymbol{y}) - \zeta[d_{V_\Omega}'(\boldsymbol{y},\boldsymbol{y}) + d_{V_\theta}'(\boldsymbol{y},\boldsymbol{y})], \tag{7.92}$$

where all terms on the right side of the above equation are given explicitly in (7.76), (7.77), (7.90), and (7.91).

Plane Elastic Solid

Since the deformation of the two-dimensional solid component is limited to the plane, the configuration design does not affect the state response of the solid component. However, when the two-dimensional solid is used for the membrane effect in the shell structure, the coupled effect of the configuration design appears in the solid component. Consider a two-dimensional, elastic solid component shown in Fig. 7.6.

The energy bilinear and load linear forms for the plane elastic solid component have been developed in (3.75) and (3.59), and are rewritten here,

$$a_\Omega(z,\bar{z}) = \iint_\Omega h\sigma(z)^T \varepsilon(\bar{z})\, d\Omega \tag{7.93}$$

$$\ell_\Omega(\bar{z}) = \iint_\Omega \bar{z}^T \boldsymbol{f}\, d\Omega, \tag{7.94}$$

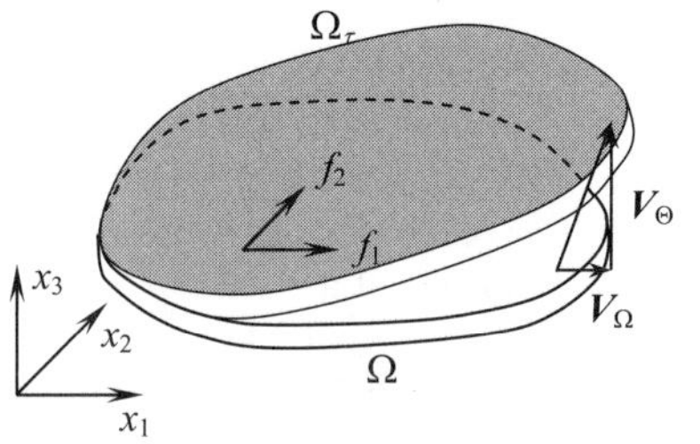

Figure 7.6. Two-dimensional elastic solid component.

where h is the thickness of the solid, $\boldsymbol{f} = [f_1, f_2]^T$ is the body force, $\boldsymbol{z} = [z_1, z_2]$ is the state response, and the stress and strain vectors are defined by

$$\varepsilon(\boldsymbol{z}) = [z_{1,1}, \quad z_{2,2}, \quad z_{1,2} + z_{2,1}]^T \tag{7.95}$$

$$\boldsymbol{\sigma}(\boldsymbol{z}) = [\sigma_{11}, \quad \sigma_{22}, \quad \sigma_{12}]^T . \tag{7.96}$$

In the two-dimensional elastic solid, the constitutive relation between stress and strain is provided by (3.72). Only a body force is considered in (7.94).

In the surface component in Section 7.1.2, the design velocity field $\boldsymbol{V} = [V_1, V_2, V_3]^T$ is split into the shape design velocity field $\boldsymbol{V}_\Omega = [V_1, V_2, 0]^T$ and the configuration design velocity field $\boldsymbol{V}_\Theta = [0, 0, V_3]^T$. The configuration design velocity $\boldsymbol{V}_\Theta$ is constrained such that the component remains flat during design perturbation. This constraint means that the first-order derivative of $\boldsymbol{V}_\Theta$ is constant and its second-order derivative vanishes, i.e., $V_{3,11} = V_{3,12} = V_{3,22} = 0$. Using $\boldsymbol{V}_\Omega$, the shape variations of $a_\Omega(\bullet,\bullet)$ and $\ell_\Omega(\bullet)$ have been developed in Section 6.2. The configuration variations of these forms can be derived from their general expressions in (7.32) and (7.34). In the case of a solid component, the convective part of the configuration design in (7.42) becomes

$$\boldsymbol{V}_\theta \boldsymbol{z} = \begin{bmatrix} -V_{3,1} z_3 \\ -V_{3,2} z_3 \end{bmatrix}. \tag{7.97}$$

As mentioned before, the convective term of the configuration design in (7.97) includes the transverse displacement z_3, which does not appear in the state response $\boldsymbol{z} = [z_1, z_2]^T$. Thus, the configuration design does not affect the state response unless the solid component is coupled with the plate component.

The energy bilinear and load linear forms in (7.93) and (7.94) are differentiated with respect to shape and configuration designs, to obtain the following explicitly dependent terms:

$$a'_{V_\Omega}(\boldsymbol{z},\bar{\boldsymbol{z}}) = -\iint_\Omega h\left\{\boldsymbol{\sigma}(\boldsymbol{z})^T \varepsilon(\nabla\bar{\boldsymbol{z}}\boldsymbol{V}_\Omega) + \boldsymbol{\sigma}(\nabla\boldsymbol{z}\boldsymbol{V}_\Omega)^T \varepsilon(\bar{\boldsymbol{z}}) - div[\boldsymbol{\sigma}(\boldsymbol{z})^T \varepsilon(\bar{\boldsymbol{z}})\boldsymbol{V}_\Omega]\right\} d\Omega \tag{7.98}$$

$$a'_{V_\theta}(\boldsymbol{z},\bar{\boldsymbol{z}}) = -\iint_\Omega h\left\{\boldsymbol{\sigma}(\boldsymbol{z})^T \varepsilon(\boldsymbol{V}_\theta\bar{\boldsymbol{z}}) + \boldsymbol{\sigma}(\boldsymbol{V}_\theta\boldsymbol{z})^T \varepsilon(\bar{\boldsymbol{z}})\right\} d\Omega \tag{7.99}$$

$$\ell'_{V_\Omega}(\bar{z}) = \iint_\Omega [\bar{z}^T \nabla f V_\Omega + \bar{z}^T f div V_\Omega] d\Omega \tag{7.100}$$

$$\ell'_{V_\theta}(\bar{z}) = -\iint_\Omega (V_\theta \bar{z})^T f \, d\Omega, \tag{7.101}$$

where the applied body force is assumed to be independent of the design, such that $f' = 0$. The explicitly dependent terms in (7.98) through (7.101) are linearly dependent on the design velocity field $V(x)$.

Plate Design Component

The configuration design of the plate component is the same as the two-dimensional solid component. However, since the unknown in the plate-bending problem is the transverse displacement z_3, the convective term of the configuration design is different from that of the two-dimensional solid component. Consider the plate design component in Fig. 7.7. The energy bilinear and load linear forms of the plate-bending problem have been provided in (3.41) and (3.42) as

$$a_\Omega(z,\bar{z}) = \iint_\Omega \kappa(\bar{z})^T C^b \kappa(z) d\Omega \tag{7.102}$$

$$\ell_\Omega(\bar{z}) = \iint_\Omega \bar{z}_3 f_3 \, d\Omega, \tag{7.103}$$

where C^b is the bending stiffness matrix provide in (3.40), f_3 is the distributed load on the surface, and $\kappa(z)$ is the curvature vector, defined by

$$\kappa(z) = [z_{3,11}, \quad z_{3,22}, \quad 2z_{3,12}]^T. \tag{7.104}$$

As shown in the above equations, the unknown in the plate bending problem is the transverse displacement z_3. However, the effect of the configuration design is coupled with the membrane displacement z_1 and z_2.

Let the plate component in Fig. 7.7 be perturbed in the direction of design velocity $V(x) = [V_1, V_2, V_3]^T$ in the body-fixed local coordinate system. The in-plane design velocity $V_\Omega(x) = [V_1, V_2, 0]^T$ is the shape design velocity, and the out-of-plane design velocity $V_\Theta(x) = [0, 0, V_3]^T$ is the configuration design velocity. Using (7.18) and (7.19), the convective term of the configuration design becomes a scalar quantity, defined as

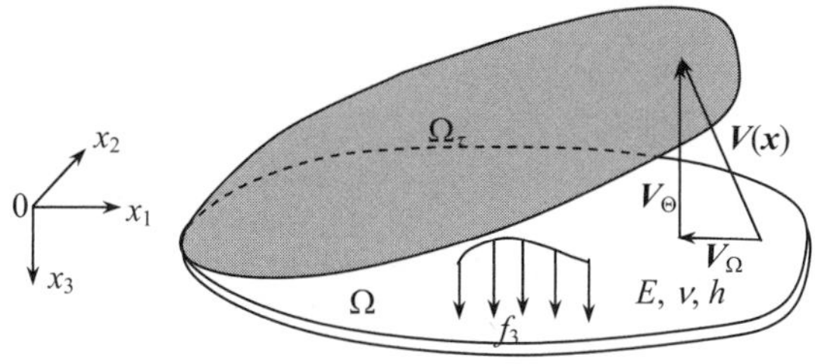

Figure 7.7. Plate design component.

$$\boldsymbol{V}_\theta \boldsymbol{z} = [V_{3,1}z_1 + V_{3,2}z_2]. \tag{7.105}$$

Using the energy bilinear and load linear forms in (7.102) and (7.103), and the formulas in (7.31) through (7.34), the variations of these forms due to shape and configuration designs can be calculated by

$$a'_{V_\Omega}(\boldsymbol{z},\bar{\boldsymbol{z}}) = -\iint_\Omega \left\{ \boldsymbol{\kappa}(\nabla\bar{\boldsymbol{z}}\boldsymbol{V}_\Omega)^T \boldsymbol{C}^b \boldsymbol{\kappa}(\boldsymbol{z}) + \boldsymbol{\kappa}(\bar{\boldsymbol{z}})^T \boldsymbol{C}^b \boldsymbol{\kappa}(\nabla\boldsymbol{z}\boldsymbol{V}_\Omega) - div[\boldsymbol{\kappa}(\bar{\boldsymbol{z}})^T \boldsymbol{C}^b \boldsymbol{\kappa}(\boldsymbol{z})\boldsymbol{V}_\Omega] \right\} d\Omega \tag{7.106}$$

$$a'_{V_\theta}(\boldsymbol{z},\bar{\boldsymbol{z}}) = -\iint_\Omega \left\{ \boldsymbol{\kappa}(\boldsymbol{V}_\theta\bar{\boldsymbol{z}})^T \boldsymbol{C}^b \boldsymbol{\kappa}(\boldsymbol{z}) + \boldsymbol{\kappa}(\bar{\boldsymbol{z}})^T \boldsymbol{C}^b \boldsymbol{\kappa}(\boldsymbol{V}_\theta\boldsymbol{z}) \right\} d\Omega \tag{7.107}$$

$$\ell'_{V_\Omega}(\bar{\boldsymbol{z}}) = \iint_\Omega (\bar{z}_3 \nabla f_3^T \boldsymbol{V}_\Omega + \bar{z}_3 f_3 div \boldsymbol{V}_\Omega) d\Omega \tag{7.108}$$

$$\ell'_{V_\theta}(\bar{\boldsymbol{z}}) = \iint_\Omega (\boldsymbol{V}_\theta\bar{\boldsymbol{z}}) f_3 \, d\Omega. \tag{7.109}$$

Again, variations in (7.106) through (7.109) depend linearly on the shape and the orientation design velocity fields.

Consider the displacement performance measure of the plate component at a point $\hat{\boldsymbol{x}}$,

$$\psi_1 = z_3(\hat{\boldsymbol{x}}) = \iint_\Omega \delta(\boldsymbol{x}-\hat{\boldsymbol{x}}) z_3 \, d\Omega. \tag{7.110}$$

As with the beam design component, the adjoint equation for the displacement performance measure is

$$a_\Omega(\boldsymbol{\lambda},\bar{\boldsymbol{\lambda}}) = \iint_\Omega \delta(\boldsymbol{x}-\hat{\boldsymbol{x}}) \bar{\lambda}_3 \, d\Omega, \qquad \forall \bar{\boldsymbol{\lambda}} \in Z, \tag{7.111}$$

and the same form of the displacement sensitivity can be obtained as

$$\psi'_1 = \ell'_{V_\Omega}(\boldsymbol{\lambda}) + \ell'_{V_\theta}(\boldsymbol{\lambda}) - a'_{V_\Omega}(\boldsymbol{z},\boldsymbol{\lambda}) - a'_{V_\theta}(\boldsymbol{z},\boldsymbol{\lambda}), \tag{7.112}$$

where all terms on the right side are given explicitly in (7.106) through (7.109) and $\boldsymbol{\lambda}$ is the solution to (7.111).

Next, consider a locally averaged stress performance measure, such as the principal stresses or von Mises stress, over a small test region on Ω_p,

$$\psi_2 = \iint_{\Omega_p} g(\boldsymbol{\sigma}(\boldsymbol{z})) m_p \, d\Omega, \tag{7.113}$$

where g is assumed to be continuously differentiable with respect to its argument, $\boldsymbol{\sigma}$ denotes the stress vector, and the characteristic function m_p is defined as

$$m_p = \begin{cases} \dfrac{1}{\iint_{\Omega_p} d\Omega}, & \boldsymbol{x} \in \Omega_p \\ 0 \quad , & \boldsymbol{x} \notin \Omega_p. \end{cases} \tag{7.114}$$

Using (6.38) and (7.23) and the fact that m_p is independent of the orientation change, the first variation of (7.113) is

$$\begin{aligned} \psi'_2 &= \iint_{\Omega_p} g_{,\sigma} \boldsymbol{\sigma}(\dot{\boldsymbol{z}}) m_p \, d\Omega \\ &\quad - \iint_{\Omega_p} g_{,\sigma} [\boldsymbol{\sigma}(\nabla\boldsymbol{z}\boldsymbol{V}_\Omega) + \boldsymbol{\sigma}(\boldsymbol{V}_\theta\boldsymbol{z})] m_p \, d\Omega + \iint_{\Omega_p} \nabla g^T \boldsymbol{V}_\Omega m_p \, d\Omega. \end{aligned} \tag{7.115}$$

The adjoint equation of the performance measure in (7.113) becomes

$$a_{\Omega}(\boldsymbol{\lambda},\bar{\boldsymbol{\lambda}}) = \iint_{\Omega} g_{,\sigma}\boldsymbol{\sigma}(\bar{\boldsymbol{\lambda}})m_p\,d\Omega, \quad \forall\bar{\boldsymbol{\lambda}} \in Z. \tag{7.116}$$

The configuration design sensitivity of the stress performance measure is

$$\begin{aligned}\psi_2' &= \ell'_{V_\Omega}(\boldsymbol{\lambda}) + \ell'_{V_\theta}(\boldsymbol{\lambda}) - a'_{V_\Omega}(\boldsymbol{z},\boldsymbol{\lambda}) - a'_{V_\theta}(\boldsymbol{z},\boldsymbol{\lambda}) \\ &\quad - \iint_{\Omega_p} g_{,\sigma}[\boldsymbol{\sigma}(\nabla\boldsymbol{z}\boldsymbol{V}_\Omega) + \boldsymbol{\sigma}(\boldsymbol{V}_\theta\boldsymbol{z})]m_p\,d\Omega + \iint_{\Omega_p} \nabla g^T \boldsymbol{V}_\Omega m_p\,d\Omega,\end{aligned} \tag{7.117}$$

where $a'_{V_\Omega}(\boldsymbol{z},\boldsymbol{\lambda})$, $a'_{V_\theta}(\boldsymbol{z},\boldsymbol{\lambda})$, $\ell'_{V_\Omega}(\boldsymbol{\lambda})$, and $\ell'_{V_\theta}(\boldsymbol{\lambda})$ are given explicitly in (7.106) through (7.109) and $\boldsymbol{\lambda}$ is the solution to (7.116).

For the eigenvalue performance measure, the energy bilinear form and the bilinear form due to the mass effect are

$$a_{\Omega}(\boldsymbol{y},\bar{\boldsymbol{y}}) = \iint_{\Omega} \boldsymbol{\kappa}(\bar{\boldsymbol{y}})^T \boldsymbol{C}^b \boldsymbol{\kappa}(\boldsymbol{y})\,d\Omega \tag{7.118}$$

$$d_{\Omega}(\boldsymbol{y},\bar{\boldsymbol{y}}) = \iint_{\Omega} \rho h y_3 \bar{y}_3\,d\Omega. \tag{7.119}$$

Again, the first variation of the energy bilinear form in (7.118) can be obtained by replacing $\boldsymbol{z}$ with $\boldsymbol{y}$ in (7.106) and (7.107). Using (7.119), the first variation of bilinear form $d_{\Omega}(\boldsymbol{y},\bar{\boldsymbol{y}})$ in (7.52) and (7.53) becomes

$$d'_{V_\Omega}(\boldsymbol{y},\bar{\boldsymbol{y}}) = \iint_{\Omega} \rho h y_3 \bar{y}_3 \,div \boldsymbol{V}_\Omega\,d\Omega \tag{7.120}$$

$$d'_{V_\theta}(\boldsymbol{y},\bar{\boldsymbol{y}}) = -\iint_{\Omega} \rho h[V_{3,1}(y_1\bar{y}_3 + \bar{y}_1 y_3) + V_{3,2}(y_2\bar{y}_3 + \bar{y}_2 y_3)]\,d\Omega. \tag{7.121}$$

The eigenvalue design sensitivity expression is given in (7.55) as

$$\psi_3' \equiv \zeta' = a'_{V_\Omega}(\boldsymbol{y},\boldsymbol{y}) + a'_{V_\theta}(\boldsymbol{y},\boldsymbol{y}) - \zeta[d'_{V_\Omega}(\boldsymbol{y},\boldsymbol{y}) + d'_{V_\theta}(\boldsymbol{y},\boldsymbol{y})], \tag{7.122}$$

where all terms on the right side are given explicitly in (7.106), (7.107), (7.120), and (7.121).

7.3 Numerical Methods in Configuration Design Sensitivity Analysis

Two questions may arise for configuration design sensitivity analysis in the previous section. First, how can a linear approximation be found between perturbations of the design variables and design velocity fields, in order to obtain first-order sensitivity information? As shown in the previous section, configuration design sensitivity expressions linearly depend on the derivatives of shape and orientation design velocity fields. A linear relationship between perturbations of design variables and design velocity fields is derived in Section 7.3.1 by treating grid point locations of line and surface design components as design variables. The second question is commonly asked in shape design sensitivity analysis: what kind of design velocity fields should be used in the numerical method of design sensitivity analysis for build-up structures? A simple three-bar structure is presented in Section 7.3.2 to provide guidelines for selecting regular velocity fields for configuration design sensitivity analysis.

7.3.1 Linear Approximation between Design Parameterization and Design Velocity Field

Line Design Component

Consider a line design component *cd* in the three-dimensional space, as shown in Fig. 7.8, where X_i, $i = 1, 2, 3$ is the global coordinate system and x_i, $i = 1, 2, 3$ is the local coordinate system. Both X_i and x_i are fixed during design perturbation. Initially, point *c* coincides with the origin of the local reference frame *o*, and the line design component is located on the local x_1-axis. Global coordinates of the grid points $c(b_1, b_2, b_3)$, $d(b_4, b_5, b_6)$ and the line orientation angle γ (or b_7) are treated as independent design variables $\boldsymbol{b} = [b_1, \ldots, b_7]^T$. The domain of a line component is the interval $[0, l]$. Therefore, given a change in length δl, the domain shape design velocity can be defined using a shape function or other smooth functions. As shown in (7.8), the variations in orientation $\delta\alpha$, $\delta\beta$ and $\delta\gamma$ are defined by the derivative of two orthogonal velocity fields. Linear approximations between the variations δl, $\delta\alpha$, $\delta\beta$, and $\delta\gamma$ and the design perturbations δb_i, $i = 1$–7 are derived in this section.

The length of the design component, written in terms of the design variables, is

$$l = \sqrt{(b_4 - b_1)^2 + (b_5 - b_2)^2 + (b_6 - b_3)^2}. \tag{7.123}$$

The Taylor series expansion of length l of the design variables δb_i, $i = 1$–6 is

$$\delta l = \frac{1}{L}[(b_4 - b_1)(\delta b_4 - \delta b_1) + (b_5 - b_2)(\delta b_5 - \delta b_2) + (b_6 - b_3)(\delta b_6 - \delta b_3)] + O(\delta^2 \boldsymbol{b}), \tag{7.124}$$

where $O(\delta^2 \boldsymbol{b})$ denotes higher order terms in the Taylor series expansion. Assuming a small design perturbation, all terms involving products of the small design perturbation can be neglected and the linear approximation can be obtained.

To find the relationship between derivatives of the orientation design velocity field and the design perturbation, the three planes shown in Fig. 7.8 are defined to identify the change in orientation. Plane A is the x_2-x_3 plane in the local coordinate system and is fixed during design perturbation. Plane B contains grid points *c*, *d* and is parallel to direction (0, 0, 1) in the initial local coordinate system. Plane C contains grid points *c*, *d* and is parallel to direction (0, 1, 0) in the local coordinate system. Mathematically, these planes can be described as

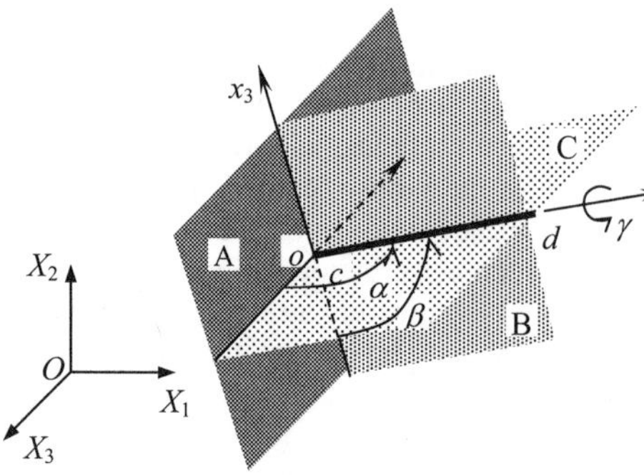

Figure 7.8. A line design component.

$$\text{Plane A}: x_1 = 0 \tag{7.125}$$

$$\text{Plane B}: x_1(x_{2c} - x_{2d}) + x_2(x_{1d} - x_{1c}) - (x_{2c}x_{1d} - x_{1c}x_{2d}) = 0 \tag{7.126}$$

$$\text{Plane C}: x_1(x_{3c} - x_{3d}) + x_3(x_{1d} - x_{1c}) - (x_{3c}x_{1d} - x_{1c}x_{3d}) = 0, \tag{7.127}$$

where,

$$\begin{bmatrix} x_{1c} \\ x_{2c} \\ x_{3c} \end{bmatrix} = \boldsymbol{T} \begin{bmatrix} b_1 - X_{1o} \\ b_2 - X_{2o} \\ b_3 - X_{3o} \end{bmatrix} \tag{7.128}$$

and

$$\begin{bmatrix} x_{1d} \\ x_{2d} \\ x_{3d} \end{bmatrix} = \boldsymbol{T} \begin{bmatrix} b_4 - X_{1o} \\ b_5 - X_{2o} \\ b_6 - X_{3o} \end{bmatrix}. \tag{7.129}$$

In (7.128) and (7.129), $\boldsymbol{T} = [T_{ij}]$ is the direction cosine matrix of the local coordinate system with respect to the global coordinate system, and (X_{1o}, X_{2o}, X_{3o}) is the origin of the local coordinate system in the global coordinate. For the initial design, points c and o coincide and point d is placed on the x_1-axis, so $x_{1c} = x_{2o} = x_{3o} = x_{2d} = x_{3d} = 0$ and $x_{1d} = l$. As the orientation and length of the line design component change, grid points c and d move away from the local coordinate. Thus, planes B and C rotate with respect to the local x_3 and x_2 axes, respectively.

Let α be the angle between two planes A and B, and let β be the angle between planes A and C. Initially, the three planes A, B, and C coincide with planes x_2-x_3, x_1-x_3, and x_1-x_2, respectively, so $\alpha = \beta = 90°$ in the initial design. As the orientation of the line design component changes, planes B and C will depart from the initial local coordinate planes, and angles α and β will be changed. Using (7.125) and (7.126), angle α can be obtained as

$$\alpha = \cos^{-1} \frac{x_{2c} - x_{2d}}{\sqrt{(x_{1d} - x_{1c})^2 + (x_{2d} - x_{2c})^2}}. \tag{7.130}$$

By taking a Taylor expansion of function α with respect to the design variables b_i, $i = 1$–6, the perturbation of angle α is

$$\delta\alpha = \sum_{i=1}^{6} \frac{\partial}{\partial b_i} \left(\cos^{-1} \frac{x_{2c} - x_{2d}}{\sqrt{(x_{1d} - x_{1c})^2 + (x_{2d} - x_{2c})^2}} \right) \Bigg|_{\alpha=\pi/2} \delta b_i + O(\delta^2 \boldsymbol{b}), \tag{7.131}$$

where $O(\delta^2 \boldsymbol{b})$ denotes higher order terms in the Taylor expansion. Assuming a small design perturbation, all higher-order terms in (7.131) can be neglected. Taking the linear approximation and from (7.8), (7.131) can be simplified to

$$\delta\alpha = \sum_{i=1}^{3} \frac{T_{2i}}{L} (\delta b_{i+3} - \delta b_i) = \tau V_{2,1}. \tag{7.132}$$

Similarly, the linear relationship between the angle change $\delta\beta$ and the design perturbation δb is

$$\delta\beta = \sum_{i=1}^{3} \frac{T_{3i}}{L}(\delta b_{i+3} - \delta b_i) = -\tau V_{3,1}, \tag{7.133}$$

and the first variation of the axial rotational angle is

$$\delta\gamma = \delta b_7 = -\tau V_{2,3}. \tag{7.134}$$

The results of (7.132) through (7.134) show a linear relationship between the derivatives of design velocity $V_{2,1}$, $V_{3,1}$, $V_{2,3}$, and the perturbations of design variables. However, for shape variation, (7.124) shows a linear relationship between the shape boundary design velocity field and variations of the design variables. Using a shape function for the design velocity field $V_1(x_1)$ on the domain of the line design component, the design velocity is linear for variations of the design variables. Given these facts and the sensitivity expressions formulated in Section 7.2, the first-order design sensitivity with respect to design variable $\boldsymbol{b}$ is obtained for a line design component.

Surface Design Component

A surface design component in the three-dimensional space is shown in Fig. 7.9. The same notations of the coordinate frames used for the line design component are employed in this section. To simplify the derivation, a planar triangular surface design component is considered. The surface design component *cde* is assumed to stay in plane C during design perturbation, and the domain Ω and the boundary Γ are always coplanar. The locations of grid points *c*, *d*, and *e* in the global coordinate are (b_1, b_2, b_3), (b_4, b_5, b_6), and (b_7, b_8, b_9), respectively, where b_i, i=1–9 are treated as independent design variables. Similar to the line design component, the objective is to find linear relationships between the variations $\delta\Gamma$, $\delta\alpha$, and $\delta\beta$ and the design perturbations δb_i, i = 1–9. Again, $\delta\alpha$ and $\delta\beta$ are defined as the derivatives of design velocity in (7.17).

The locations of the grid points *c*, *d*, and *e* in the local coordinate can be written as

$$\begin{bmatrix} x_{1c} \\ x_{2c} \\ x_{3c} \end{bmatrix} = \boldsymbol{T} \begin{bmatrix} b_1 - X_{1o} \\ b_2 - X_{2o} \\ b_3 - X_{3o} \end{bmatrix} \tag{7.135}$$

$$\begin{bmatrix} x_{1d} \\ x_{2d} \\ x_{3d} \end{bmatrix} = \boldsymbol{T} \begin{bmatrix} b_4 - X_{1o} \\ b_5 - X_{2o} \\ b_6 - X_{3o} \end{bmatrix} \tag{7.136}$$

and

$$\begin{bmatrix} x_{1e} \\ x_{2e} \\ x_{3e} \end{bmatrix} = \boldsymbol{T} \begin{bmatrix} b_7 - X_{1o} \\ b_8 - X_{2o} \\ b_9 - X_{3o} \end{bmatrix}, \tag{7.137}$$

where $\boldsymbol{T} = [T_{ij}]$ is the direction cosine matrix and (X_{1o}, X_{2o}, X_{3o}) is the origin of the local coordinate system in the global coordinate. For the initial design, points *c* and *o* coincide, and $x_{1c} = x_{2c} = x_{3c} = 0$.

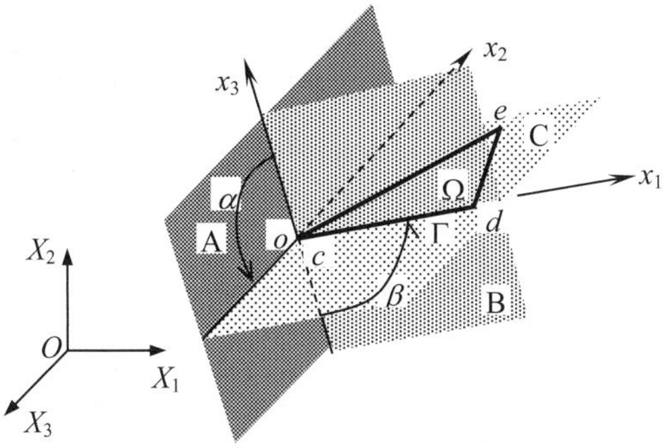

Figure 7.9. A surface design component.

Taking the first variation of (7.135) through (7.137), linear relationships between the variation of the boundary grid points and the design perturbations are obtained as

$$\begin{bmatrix} \delta x_{1c} \\ \delta x_{2c} \end{bmatrix} = \sum_{n=1}^{3} \begin{bmatrix} T_{1n} \\ T_{2n} \end{bmatrix} \delta b_n \tag{7.138}$$

$$\begin{bmatrix} \delta x_{1d} \\ \delta x_{2d} \end{bmatrix} = \sum_{n=1}^{3} \begin{bmatrix} T_{1n} \\ T_{2n} \end{bmatrix} \delta b_{n+3} \tag{7.139}$$

$$\begin{bmatrix} \delta x_{1e} \\ \delta x_{2e} \end{bmatrix} = \sum_{n=1}^{3} \begin{bmatrix} T_{1n} \\ T_{2n} \end{bmatrix} \delta b_{n+6}. \tag{7.140}$$

In (7.138) through (7.140), only variations in the x_1- and x_2-directions are written, because the shape boundary velocity describes perturbations of the domain boundary in x_1-x_2 directions for the surface design component.

To identify the orientation change of the surface, planes A, B, and C shown in Fig. 7.9 are defined as

$$\text{Plane A}: x_1 = 0 \tag{7.141}$$

$$\text{Plane B}: x_2 = 0 \tag{7.142}$$

$$\text{Plane C}: (x_1 - x_{1c})(a_1 a_2 - a_3 a_4) + (x_2 - x_{2c})(a_4 a_5 - a_2 a_6) + (x_3 - x_{3c})(a_6 a_3 - a_5 a_1) = 0. \tag{7.143}$$

In (7.143), $a_1 = x_{2d} - x_{2c}$, $a_2 = x_{3e} - x_{3c}$, $a_3 = x_{2e} - x_{2c}$, $a_4 = x_{3d} - x_{3c}$, $a_5 = x_{1e} - x_{1c}$, and $a_6 = x_{1d} - x_{1c}$. Planes A and B are fixed during the design perturbation. Plane C initially coincides with the plane x_1-x_2 and rotates with respect to the initial local coordinate system as the orientation of the surface design component changes.

The angles α and β shown in Fig. 7.9 are defined between the planes B-C and A-C, respectively. For the initial design, $\alpha = \beta = 90°$. Using (7.142) and (7.143), the angle α is

$$\alpha = \cos^{-1}\frac{a_2a_6 - a_4a_5}{\sqrt{(a_1a_2 - a_3a_4)^2 - (a_4a_5 - a_2a_6)^2 - (a_6a_3 - a_5a_1)^2}}. \tag{7.144}$$

The Taylor expansion of function α with respect to the design variables b_i, $i = 1$–9 is

$$\delta\alpha = \sum_{i=1}^{9}\frac{\partial}{\partial b_i}\left(\cos^{-1}\frac{a_2a_6 - a_4a_5}{\sqrt{(a_1a_2 - a_3a_4)^2 - (a_4a_5 - a_2a_6)^2 - (a_6a_3 - a_5a_1)^2}}\right)\Bigg|_{\alpha=\pi/2}\delta b_i + O(\delta^2\boldsymbol{b}), \tag{7.145}$$

where $O(\delta^2\boldsymbol{b})$ denotes the higher-order terms and can be neglected by assuming a small design perturbation. With several steps of arithmetic simplification and from (7.17), (7.145) becomes

$$\begin{aligned}\delta\alpha &= \sum_{i=1}^{3}\frac{T_{3i}(x_{1e} - x_{1d})}{\sqrt{(x_{1d} - x_{1c})(x_{2e} - x_{2c}) - (x_{1e} - x_{1c})(x_{2d} - x_{2c})}}\delta b_i \\ &+\sum_{i=1}^{3}\frac{T_{3i}(x_{1c} - x_{1e})}{\sqrt{(x_{1d} - x_{1c})(x_{2e} - x_{2c}) - (x_{1e} - x_{1c})(x_{2d} - x_{2c})}}\delta b_{i+3} \\ &+\sum_{i=1}^{3}\frac{T_{3i}(x_{1d} - x_{1c})}{\sqrt{(x_{1d} - x_{1c})(x_{2e} - x_{2c}) - (x_{1e} - x_{1c})(x_{2d} - x_{2c})}}\delta b_{i+6} \\ &= \tau V_{3,2}.\end{aligned} \tag{7.146}$$

Equation (7.146) shows a linear relationship between the variation of the angle $\delta\alpha$ and the design perturbation $\delta\boldsymbol{b}$. Similarly, the linear relationship between the angle change $\delta\beta$ and the design perturbation $\delta\boldsymbol{b}$ is

$$\begin{aligned}\delta\beta &= \sum_{i=1}^{3}\frac{T_{3i}(x_{2d} - x_{2e})}{\sqrt{(x_{1d} - x_{1c})(x_{2e} - x_{2c}) - (x_{1e} - x_{1c})(x_{2d} - x_{2c})}}\delta b_i \\ &+\sum_{i=1}^{3}\frac{T_{3i}(x_{2e} - x_{2c})}{\sqrt{(x_{1d} - x_{1c})(x_{2e} - x_{2c}) - (x_{1e} - x_{1c})(x_{2d} - x_{2c})}}\delta b_{i+3} \\ &+\sum_{i=1}^{3}\frac{T_{3i}(x_{2c} - x_{2d})}{\sqrt{(x_{1d} - x_{1c})(x_{2e} - x_{2c}) - (x_{1e} - x_{1c})(x_{2d} - x_{2c})}}\delta b_{i+6} \\ &= -\tau V_{3,1}.\end{aligned} \tag{7.147}$$

If a quadrilateral surface design component is used, the method discussed in this section can be extended to obtain the linear relationship between the shape boundary velocity field and the design perturbation, except that the design variables b_i, $i = 1$–12 are not independent in this case. A constraint equation should be used in order to ensure that the four corner points remain on a plane during design perturbation.

As with the line design component, the results in (7.146) and (7.147) show a linear relationship between the derivatives of orientation design velocity $V_{3,1}$ and $V_{3,2}$ and perturbations of the design variables. The results in (7.138) through (7.140) show a linear relationship between the shape boundary design velocity field and variations in the design variables. Once the shape boundary velocity is found, the design velocity fields $V_1(x_1, x_2)$ and $V_2(x_1, x_2)$ on the domain of the surface design component can be obtained using the design velocity computation method presented in Section 13.3 of Chapter 13. [96] and [97]. Given these facts and the sensitivity expressions in Section 7.2, the first-order design sensitivity is ensured for the surface design component.

7.3.2 Regularity of Design Velocity Fields

When a built-up structure undergoes configuration design change, the boundary movement of one design component results in the movement of attached design components at their interfaces. During the design change, compatibility must continue at the interfaces so that the built-up structure remains intact. Once the boundary movement is decided, the mesh movement inside the component domain still needs to be decided (i.e., the shape design velocity) using the design velocity computation methods in Section 13.3. As explained in Section 6.2.7, the design sensitivity results of (7.83), (7.87), (7.92), (7.112), (7.117), and (7.122) require that the shape velocity field $V_1(x_1)$ for the line design component and $V_1(x_1,x_2)$ and $V_2(x_1,x_2)$ for the surface design component must be in $H^2(\Omega)$, the second-order Sobolev space, that is, these velocity fields must be C^1-regular with L_2-integrable second derivatives because $V_{1,11}$, $V_{1,12}$, $V_{1,22}$, $V_{2,11}$, $V_{2,12}$, and $V_{2,22}$ appear in these design sensitivity expressions. Therefore, special attention has to be paid to meeting the regularity requirement of domain shape design velocity in order to have an accurate and unified configuration design sensitivity result, as shown in the following example.

Example 7.1. Three-Bar Frame. Consider the three bar frame in Fig. 7.10 with a vertical point load p = 300 lb at node 2. Young's modulus E, moment of inertia I, shear modulus G, and polar moment of inertia J are assumed to be the same for all elements, and are given as 30×10^6 psi, 0.26042×10^{-2} in^4, 0.11539×10^8 psi, and 0.32552×10^{-2} in^4, respectively. The local displacement is $z = [z^1, z^2, z^3]^T$, with $z^i = [z_3^i, \phi_1^i]^T$, where the superscript i denotes the ith design component and the subscript represents the direction in the local coordinate system. The local displacement z belongs to the space Z of kinematically admissible displacement fields that satisfy homogeneous boundary conditions and kinematic interface conditions between components. Mathematically, Z can be written as

$$\begin{aligned} Z = \Big\{ z^i \in H^2(0,l_i) \Big| \; & z_3^1(0) = z_{3,1}^1(0) = \phi_1^1(0) = z_3^3(0) = z_{3,1}^3(0) = \phi_1^3(0) = 0, \\ & \gamma^1 z^1 = \gamma^2 z^2 \text{ at node 2, and } \gamma^2 z^2 = \gamma^3 z^3 \text{ at node 3} \Big\}, \end{aligned} \tag{7.148}$$

where γ^i is interface operator such that $\gamma^i z^i = \gamma^j z^j$ projects displacement fields from design components i and j onto their common boundary nodal points.

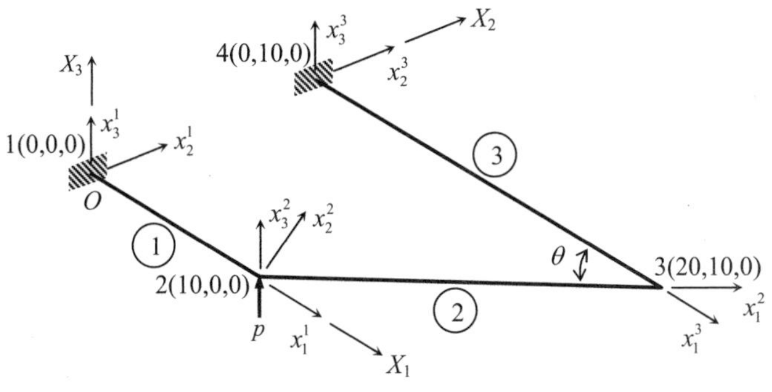

Figure 7.10. Three-bar frame.

As a performance measure, the displacement at a point $\hat{x}$ (node 2) in the x_3-direction is considered as

$$\psi = z_3^1(\hat{x}) = \int_0^{l_1} \delta(x - \hat{x}) z_3^1 \, dx, \tag{7.149}$$

where δ is the Dirac delta measure, and z_3^1 is the local displacement of design component 1. With beam bending and torsion effects only, the energy bilinear and load linear forms of the variational equation can be written as

$$a_\Omega(z, \bar{z}) = \sum_{i=1}^{3} \left[\int_0^{l_i} EI z_{3,11}^i \bar{z}_{3,11}^i \, dx + \int_0^{l_i} GJ \phi_{1,1}^i \bar{\phi}_{1,1}^i \, dx \right] \tag{7.150}$$

$$\ell_\Omega(\bar{z}) = \int_0^{l_1} p \bar{z}_3^1 \delta(x - \hat{x}) \, dx, \tag{7.151}$$

where $\bar{z} = [\bar{z}^1, \bar{z}^2, \bar{z}^3]^T$ and $\bar{z}^i = [\bar{z}_3^i, \bar{\phi}_1^i]^T$ are virtual displacements in the space Z of kinematically admissible displacement fields. Using the length of design component 3 as the design variable b, the configuration design variation is shown in Fig. 7.11. The design perturbation of the first element is zero and $V_3^i = 0$ for all elements (i = 1, 2, 3). In addition, the load is assumed to be independent of the design variation, and the first variation of (7.151) vanishes.

In this bending and torsion problem, the unknowns of the structural problem are $[z_3, \phi_1]^T$. Accordingly, the convective term of the configuration design in (7.11) and (7.12) can be written as

$$(V_\theta z)^i = \begin{bmatrix} V_{2,1}^i \phi_1^i \\ 0 \end{bmatrix}, \tag{7.152}$$

where, in this problem,

$$V_\theta = \begin{bmatrix} 0 & V_{2,1} \\ 0 & 0 \end{bmatrix}.$$

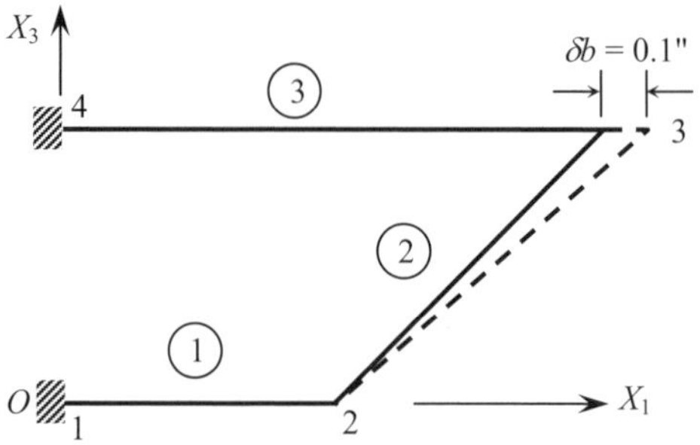

Figure 7.11. Design perturbation of three-bar frame.

Using the above equation and the shape design velocity $V_\Omega = \{V_1\}$, the explicitly dependent terms of the energy bilinear form in (7.150) can be obtained from (7.31) and (7.32) as

$$a'_{V_\Omega}(z,\bar{z}) = -\sum_{i=2}^{3}\int_0^{l_i}[3EIz^i_{3,11}\bar{z}^i_{3,11}V^i_{1,1} + EI(z^i_{3,11}\bar{z}^i_{3,1} + \bar{z}^i_{3,11}z^i_{3,1})V^i_{1,11} + GJ\phi^i_{1,1}\bar{\phi}^i_{1,1}V^i_{1,1}]dx \tag{7.153}$$

$$a'_{V_\theta}(z,\bar{z}) = -\sum_{i=2}^{3}\int_0^{l_i} EI(z^i_{3,11}\bar{\phi}^i_{1,11} + \phi^i_{1,11}\bar{z}^i_{3,11})V^i_{2,1}\,dx. \tag{7.154}$$

Since we are interested in the displacement sensitivity at node 2, the adjoint problem is the same as the state problem with a unit load applied at node 2. Thus, the state response and the adjoint response has the relation of $z = 300\lambda$ (Note that the state problem has a point load of p = 300 lb). Using this relation, the design sensitivity expression can be obtained using (7.83) as

$$\psi' = \frac{1}{300}\sum_{i=2}^{3}\int_0^{l_i}[3EI(z^i_{3,11})^2V^i_{1,1} + 2EIz^i_{3,1}z^i_{3,11}V^i_{1,11} + 2EI\phi^2_{1,11}z^2_{3,11}V^2_{2,1} + GJ(\phi^i_{1,1})^2V^i_{1,1}]dx. \tag{7.155}$$

Note that for the shape variation not only the first derivative $V^i_{1,1}$, but also the second derivative $V^i_{1,11}$ of velocity appears in the design sensitivity expression. Thus, a Dirac delta type of singularity (called the corner term) will occur in the sensitivity computation if the slope of the shape velocity field is discontinuous. The two kinds of shape design velocity fields shown in Fig. 7.12 are used to evaluate the sensitivity expression in (7.155). In Fig. 7.12, design velocity field (a) is represented by a linear function in the domain, whereas

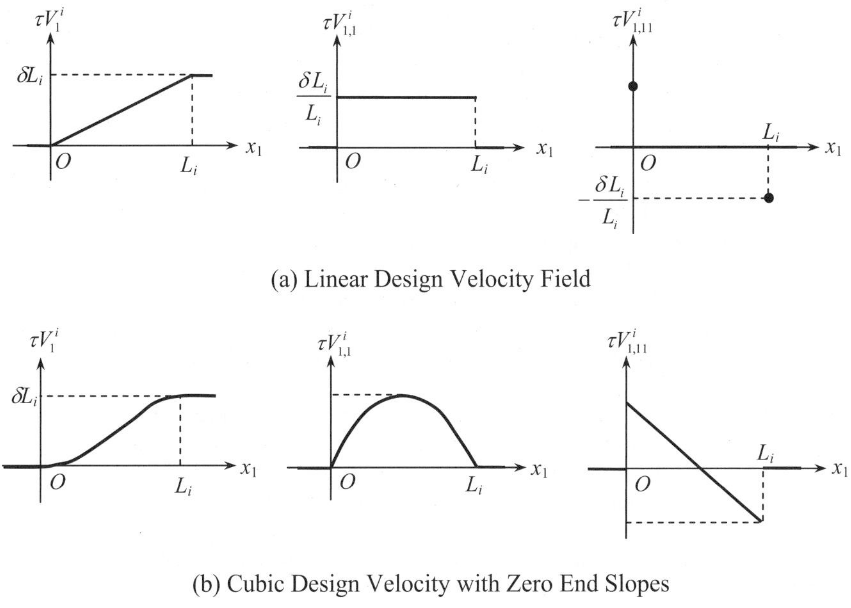

Figure 7.12. Two shape design velocity fields.

Table 7.1. Configuration design sensitivity of displacement at node 2 in z_3 direction of three bar frame.

Design velocity	$\psi(\boldsymbol{b}+\Delta\boldsymbol{b})$	$\psi(\boldsymbol{b}-\Delta\boldsymbol{b})$	$\Delta\psi(\boldsymbol{b})$	$\psi'(\boldsymbol{b})$	$(\psi'/\Delta\psi)\times 100\%$
(a) without (7.156)	0.90357E+0	0.90341E+0	−0.15809E−3	0.92803E−3	−587.02
(a) with (7.156)	0.90357E+0	0.90341E+0	−0.15809E−3	−0.15761E−3	99.70
(b)	0.90357E+0	0.90341E+0	−0.15809E−3	−0.15796E−3	99.92

design velocity (b) is represented by a cubic function with zero end slopes. Note that the second derivative $V^i_{1,11}$ of design velocity (a) is a Dirac delta measure at $x_1 = 0$ and $x_1 = l_i$. Using the Dirac delta measure, the corner terms for the linear design velocity (a) in (7.155) are

$$EI\left[(2z^2_{3,1}z^2_{3,11})\frac{b-10}{l_2^2}\delta b\right]\Bigg|_{x_1^2=0} - EI\left[(2z^2_{3,1}z^2_{3,11})\frac{b-10}{l_2^2}\delta b\right]\Bigg|_{x_1^2=l_2} - EI\left[(2z^3_{3,1}z^3_{3,11})\frac{\delta b}{l_3}\right]\Bigg|_{x_1^3=l_3}. \quad (7.156)$$

Three-point Gauss integration is used to evaluate (7.155). The resulting design sensitivity predictions are compared with the results obtained from the central finite difference, as shown in Table 7.1. In Table 7.1, $\psi(\boldsymbol{b} - \Delta\boldsymbol{b})$ and $\psi(\boldsymbol{b} + \Delta\boldsymbol{b})$ are the values of the performance measure at the backward and forward perturbed designs, respectively; $\Delta\psi(\boldsymbol{b})$ is the central finite difference; and $\psi'(\boldsymbol{b})$ is the predicted change of the performance measure. Because the finite element solution is exact for this problem, the central finite difference should give a reasonable sensitivity, if an appropriate step size is used. The results in Table 7.1 show incorrect design sensitivity for linear design velocity (a) without the corner terms of (7.156). However, if the corner terms are added to the design sensitivity expression of (7.155), accurate design sensitivity is obtained, as shown in Table 7.1. For cubic design velocity (c), the sensitivity expression in (7.155) can be evaluated without the calculation of any corner terms. Again, Table 7.1 shows a good sensitivity prediction for design velocity (b).

To evaluate the corner terms for equations such as (7.156), FEA results at the boundary and/or interfaces must be used. It is well known that the results of finite element analysis at the boundary and/or interfaces may not be accurate for built-up structures. This means that the computation of corner terms for built-up structures may also not be accurate. Moreover, development of a general algorithm that can handle the computation of corner terms is difficult. The result of cubic design velocity (b) shows that the corner terms can be avoided by imposing a design velocity that has zero slopes at the interfaces.

As shown in previous sections, note that configuration design sensitivity expressions for the truss and plane elastic solid (membrane) design components only contain the first-order derivatives of the design velocity fields. Thus, the shape design velocity must be C^0-regular with L_2-integrable first derivatives. This requirement can be easily satisfied by using a linear velocity in the domain. However, for a plate design component, design sensitivity expressions contain second derivatives of the design velocity fields, as does the beam design component. Therefore, the regularity requirement on the shape design velocity for a beam design component discussed in this example should be applied to plate design components.

On the other hand, if the Timoshenko beam or Mindlin/Reissner plate is used, C^0-regular velocity field with L_2-intgrable first derivatives is sufficient without requiring corner terms in configuration design sensitivity analysis. This will be presented in Section 7.5.

7.3.3 Numerical Examples

Helicopter Tail-Boom

Consider the structural configuration design of a helicopter tail-boom shown in Fig. 7.13, where the geometry of the helicopter tail-boom and the maximum in-flight loads are given [48].

The tail-boom structure is modeled as a simple open frame structure without a skin panel. A finite element model of the simplified open tail-boom shown in Fig. 7.14 is created using an ANSYS [119] three-dimensional truss element. There are 28 joints and 108 members, with 72 degrees of freedom. For a wrought aluminum frame, Young's modulus is 10.6×10^6 psi and weight density is 0.1 lb/in^3. The objective of the structural configuration design is to study the effects of a narrower bottom at the tail end of the boom on the performance measures.

To describe the configuration design change, all longeron members are grouped into four design components *ae*, *bf*, *cg*, and *hd*, whereas each diagonal and batten member is treated as a design component. The configuration design variables are the coordinates of the points *e* and *f*. In the perturbed design shown in Fig. 7.15, the coordinates of *e* and *f* are assumed to move by $[0, 0.5, 0.6]^T$ and $[0, -0.5, 0.6]^T$, respectively, and all diagonal and batten members have the same x_1-coordinate as at the initial design. Based on this design perturbation, a linear shape design velocity field is used for each design component, and the derivatives of the orientation design velocity are computed.

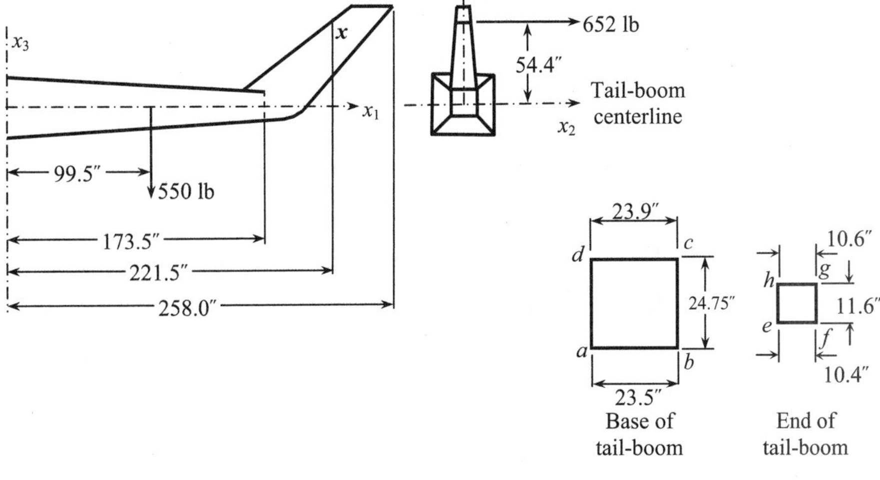

Figure 7.13. Geometry of helicopter tail-boom.

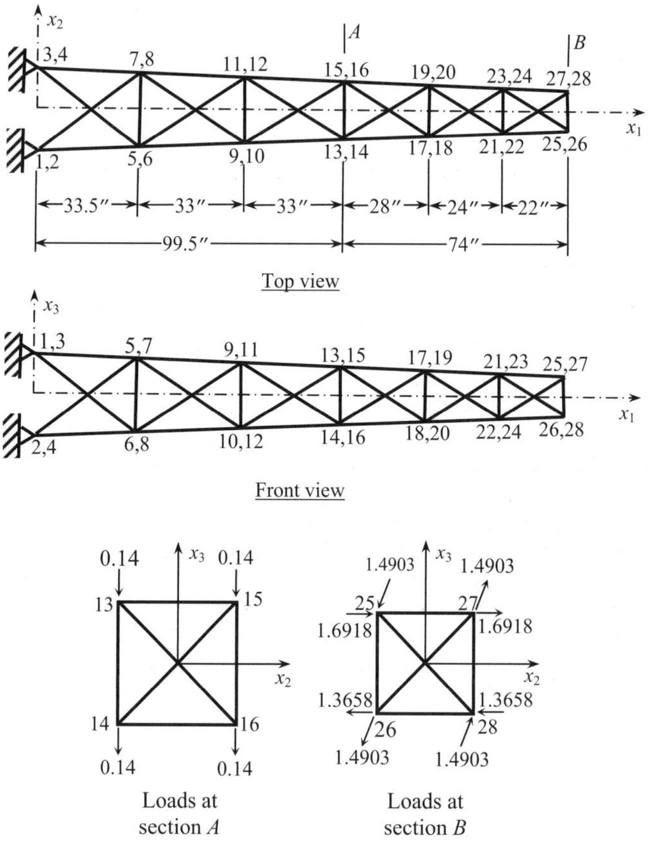

Figure 7.14. Finite element model of the simplified open tail-boom.

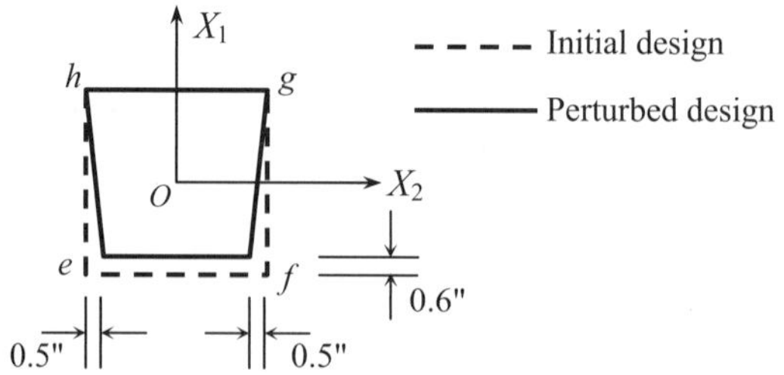

Figure 7.15. Design perturbation at the end of the tail-boom (not drawn to scale).

In Table 7.2, configuration design sensitivity results for the displacements, axial stresses, and eigenvalues are presented. The ratio between the central finite difference and the predicted change is given in the last column, with 100% indicating a complete agreement. Excellent agreement between the predicted design sensitivity results ψ' and the central finite differences $\Delta\psi$ is obtained, as shown in Table 7.2. The configuration design sensitivity expressions shown in previous sections are composed of two parts: shape variation and orientation change. The contribution from shape and orientation effects to configuration design sensitivity results is shown in Table 7.3. In Table 7.3, $\psi'(\boldsymbol{b})$ is the predicted change of the performance measure and is the same as the sixth column in Table 7.2, ψ'_{V_Ω} is the amount contributed from the shape effect, and ψ'_{V_θ} is the contribution from the orientation effect. The percentage of each contribution is computed by $[|\psi'_{V_\Omega}|/(|\psi'_{V_\Omega}|+|\psi'_{V_\theta}|) \times 100]\%$ for the shape effect, and $[|\psi'_{V_\theta}|/(|\psi'_{V_\Omega}|+|\psi'_{V_\theta}|) \times 100]\%$ for the orientation effect. In Table 7.3, it can be seen that the contribution from the orientation effect is more significant than that from the shape effect for the selected displacement and stress performance measures. For eigenvalues, contributions from both shape and orientation effects are significant. The reason for this difference is that the orientation change of design components has the greatest impact on the stiffness of the structure, whereas shape variation influences both stiffness and mass of the system.

Swept Wing

Configuration design sensitivity analysis of the swept wing is considered in this section. The design optimization of the swept wing model shown in Fig. 7.16 has previously been investigated [120]. The wing is made of aluminum with Young's modulus $E = 10.6 \times 10^6$ psi and Poisson's ratio $\nu = 0.3$, and is subjected to a uniform pressure of 0.556 psi acting on the top of the skin panel. The cross-sectional areas are 0.02 in^2 for longitudinal spar caps, and 0.2 in^2 for vertical spar caps. The thickness of the skin panels on the first half of the wing and all shear panels (ribs and spars) is 0.2 in. The thickness of the skin panels on the second half (wing tip) of the wing is 0.1 in. Because of the symmetry of the structure and loading, only half of the wing box is analyzed. The model consists of 60 three-dimensional truss elements and 130 membrane elements. This model has 88 nodal points and 160 degrees-of-freedom.

For a configuration design change, the tip of the swept wing is moved forward as shown in Fig. 7.16. The design velocity fields are defined so that all ribs (across the wing) parallel to the y-axis remain parallel while moving. The orientation of the spars along the wing and the skin panels will then be rotated accordingly, so that each shear panel will remain a plane. Based on the perturbation of nodal points, the movement of the nodal points in the domain direction of each design component is computed. Linear and bilinear shape functions are used to interpolate shape design velocity fields for line and surface design components, respectively.

The displacement at the tip of the wing, averaged axial stress on the spar caps, and averaged von Mises stress on the skin and shear panels are selected as performance measures. For averaged stress performance measures, the general performance measure given in (7.113) can be written explicitly as

$$\psi = \iint_\Omega g(\sigma(z)) m_p \, d\Omega, \tag{7.157}$$

where, for the averaged axial stress performance measure,

$$g(\sigma(z)) = \sigma_{11} = E z_{1,1} \tag{7.158}$$

Table 7.2. Configuration design sensitivity results of helicopter tail-boom.

Node/ Elem.	Dir./ Stress/ Eigval*	$\psi(\boldsymbol{b}-\Delta\boldsymbol{b})$	$\psi(\boldsymbol{b}+\Delta\boldsymbol{b})$	$\Delta\psi(\boldsymbol{b})$	$\psi'(\boldsymbol{b})$	Ratio%
26	x_2	0.43846E+00	0.44749E+00	0.90279E–02	0.90368E–02	100.1
27	x_2	0.56307E+00	0.56733E+00	0.42646E–02	0.42694E–02	100.1
26	x_3	–.88509E–01	–.91167E–01	–.26585E–02	–.26508E–02	99.7
27	x_3	–.19582E+00	–.19565E+00	0.17352E–03	0.18672E–03	107.6
1	σ_{axial}	0.50920E+04	0.50266E+04	–.65373E+02	–.65372E+02	100.0
3	σ_{axial}	–.57162E+04	–.56160E+04	0.10024E+03	0.10025E+03	100.0
6	σ_{axial}	0.30517E+04	0.29443E+04	–.10743E+03	–.10743E+03	100.0
9	σ_{axial}	–.41796E+04	–.40428E+04	0.13681E+03	0.13680E+03	100.0
12	σ_{axial}	0.23750E+04	0.21898E+04	–.18518E+03	–.18518E+03	100.0
13	σ_{axial}	–.21629E+04	–.21596E+04	0.32264E+01	0.32188E+01	99.8
21	σ_{axial}	–.51130E+04	–.49854E+04	0.12755E+03	0.12756E+03	100.0
27	σ_{axial}	–.42783E+04	–.41862E+04	0.92155E+02	0.92120E+02	100.0
1st	ζ	0.19527E+05	0.19764E+05	0.23740E+03	0.23766E+03	100.1
2nd	ζ	0.21974E+05	0.22173E+05	0.19893E+03	0.19934E+03	100.2
3rd	ζ	0.42872E+06	0.42281E+06	–.59063E+04	–.59076E+04	100.0

*Note: ζ denotes the eigenvalue.

Table 7.3. Contributions from shape and orientation effects to configuration design sensitivity results of a helicopter tail-boom.

Node/ Elem.	Dir./ Stress/ Eigval	$\psi'(\boldsymbol{b})$	Shape Effect ψ'_{V_Ω}	(%)	Orientation Effect ψ'_{V_θ}	(%)
26	x_2	0.90368E–02	–.86564E–04	0.9	0.91234E–02	99.1
27	x_2	0.42694E–02	–.16247E–02	21.6	0.58942E–02	78.4
26	x_3	–.26508E–02	–.82414E–04	3.1	–.25684E–02	96.9
27	x_3	0.18672E–03	0.84080E–03	56.2	–.65408E–03	43.8
1	σ_{axial}	–.65372E+02	–.14458E+01	2.2	–.63926E+02	97.8
3	σ_{axial}	0.10025E+03	0.64828E+01	6.5	0.93764E+02	93.5
6	σ_{axial}	–.10743E+03	0.25778E+01	2.3	–.11001E+03	97.7
9	σ_{axial}	0.13680E+03	–.16295E+01	1.2	0.13843E+03	98.8
12	σ_{axial}	–.18518E+03	0.84242E+00	0.5	–.18602E+03	99.5
13	σ_{axial}	0.32188E+01	–.59196E+01	39.3	0.91384E+01	60.7
21	σ_{axial}	0.12756E+03	0.13129E+02	10.3	0.11443E+03	89.7
27	σ_{axial}	0.92120E+02	–.84856E+01	7.8	0.10061E+03	92.2
1st	ζ	0.23766E+03	0.58814E+03	62.7	–.35048E+03	37.3
2nd	ζ	0.19934E+03	0.65374E+03	59.0	–.45440E+03	41.0
3rd	ζ	–.59076E+04	0.10303E+05	38.9	–.16211E+05	61.1

and, for the averaged von Mises stress performance measure,

$$g(\boldsymbol{\sigma}(z)) = \sqrt{\sigma_{11}^2 - \sigma_{11}\sigma_{22} + \sigma_{22}^2 + 3\sigma_{12}^2}. \tag{7.159}$$

The adjoint equation and the configuration design sensitivity expression are given in (7.116) and (7.117). The configuration design sensitivity results of displacement, averaged axial stress, and averaged von Mises stress are presented in Table 7.4. The results in Table 7.4 show accurate predictions for ψ'. Although the contribution of shape and orientation effects is not shown, both effects are significant to the design sensitivity results.

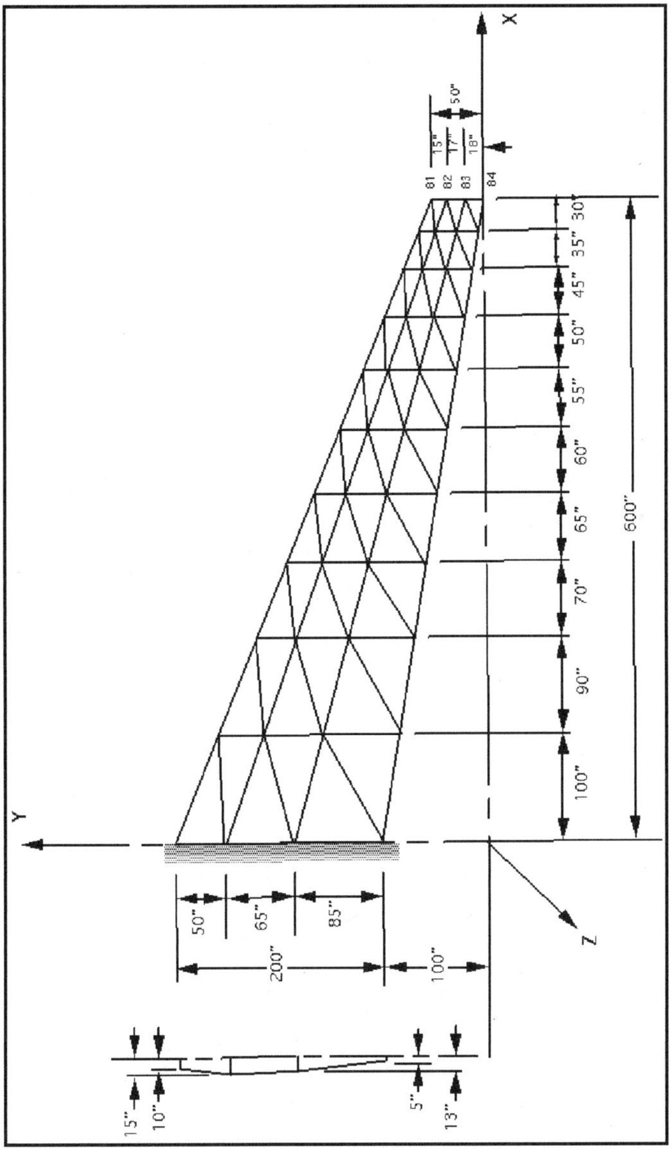

Figure 7.16. Swept wing model.

Table 7.4. Configuration design sensitivity results of swept wing model.

Node/ Elem.	Dir./ Stress	$\psi(\boldsymbol{b}-\Delta\boldsymbol{b})$	$\psi(\boldsymbol{b}+\Delta\boldsymbol{b})$	$\Delta\psi(\boldsymbol{b})$	$\psi'(\boldsymbol{b})$	Ratio%
81	x_3	0.20586E+02	0.20269E+02	–.31699E+00	–.31700E+00	100.0
82	x_3	0.20998E+02	0.20664E+02	–.33409E+00	–.33409E+00	100.0
83	x_3	0.21469E+02	0.21115E+02	–.35364E+00	–.35364E+00	100.0
84	x_3	0.21971E+02	0.21597E+02	–.37451E+00	–.37452E+00	100.0
1	σ_{axial}	–.74291E+04	–.75105E+04	–.81426E+02	–.81426E+02	100.0
2	σ_{axial}	–.14745E+05	–.14643E+05	0.10288E+03	0.10288E+03	100.0
3	$\sigma_{\text{vM}}{}^{*}$	0.85701E+04	0.85801E+04	0.99366E+01	0.99431E+01	100.1
4	σ_{vM}	0.14260E+05	0.14266E+05	0.52348E+01	0.52420E+01	100.1
5	σ_{vM}	0.13482E+05	0.13486E+05	0.37437E+01	0.37431E+01	100.0
6	σ_{vM}	0.14530E+05	0.14424E+05	–.10514E+03	–.10514E+03	100.0
7	σ_{vM}	0.14479E+05	0.14395E+05	–.84104E+02	–.84108E+02	100.0
8	σ_{vM}	0.75624E+04	0.74249E+04	–.13746E+03	–.13747E+03	100.0
9	σ_{vM}	0.38504E+04	0.37776E+04	–.72837E+02	–.72864E+02	100.0
10	σ_{vM}	0.43730E+04	0.42345E+04	–.13849E+03	–.13854E+03	100.0
11	σ_{vM}	0.31716E+04	0.30755E+04	–.96121E+02	–.96311E+02	100.2
63	σ_{vM}	0.67083E+04	0.66969E+04	–.11379E+02	–.11381E+02	100.0
64	σ_{vM}	0.10459E+05	0.10413E+05	–.46390E+02	–.46393E+02	100.0
65	σ_{vM}	0.10149E+05	0.10114E+05	–.35354E+02	–.35360E+02	100.0
66	σ_{vM}	0.88308E+04	0.88073E+04	–.23486E+02	–.23491E+02	100.0
67	σ_{vM}	0.92998E+04	0.92785E+04	–.21310E+02	–.21315E+02	100.0

*Note: σ_{vM} denotes the element averaged von Mises stress.

Vehicle Chassis Frame

Consider the vehicle chassis frame in Fig. 7.17. The chassis frame is 289.37 in long and 31.496 in. wide, with two rectangular longitudinal frames and five cross members. The rectangular cross section is 5 in high and 2 in wide. Young's modulus and the weight density are $E = 30 \times 10^6$ psi and $\rho = 0.283$ lb/in^3, respectively. For static analysis, the loads acting on top of the frame are used to simulate the engine and body weight. The side forces at points *a*, *b*, *c*, and *d* are used to describe the tire side forces. The wheel attachment points *a*, *b*, *c*, and *d* in the chassis frame are assumed to be simply supported boundary conditions. Due to the symmetry of the structure and loading, point *O* is restricted and can only move in the x_3-direction. For the modal (eigenvalue) analysis, no boundary condition is imposed, and there are six rigid body modes. An ANSYS three-dimensional beam element is used to model the chassis frame. There are 48 beam elements and 260 degrees-of-freedom.

The configuration design change of the chassis frame is shown in Fig. 7.18. Four corner points at each end are moved inward by 0.2 and 0.4 in, respectively. Each finite element is treated as a single design component in this example. Based on the design perturbation of nodal points, a cubic shape design velocity with zero slope at both ends is used for each design component. Configuration design sensitivity results for displacement, maximum bending stress, and eigenvalue performance measures are shown in Table 7.5, and excellent agreement can be observed between the predicted sensitivity results and the finite differences. Note that the results in Table 7.6 show that for displacement and stress performance measures the orientation effect contributes more to the sensitivity results for the described configuration design change. For eigenvalue performance measures, both shape and orientation effects are critical, depending on the vibration mode.

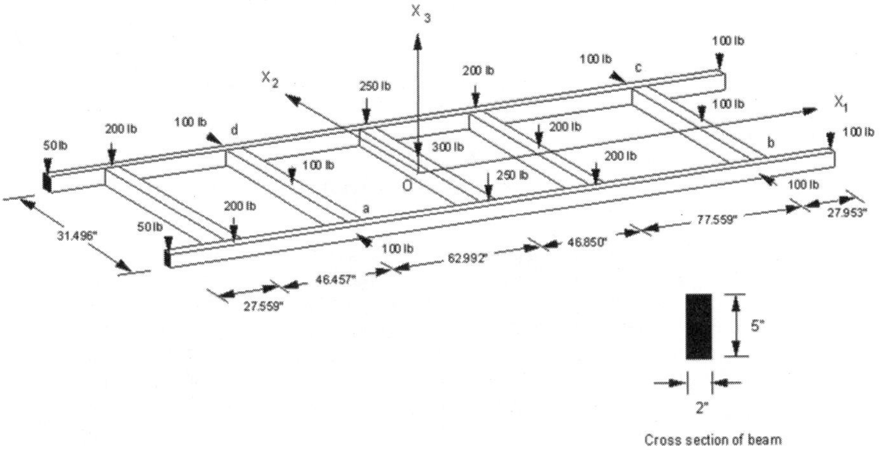

Figure 7.17. Vehicle chassis frame.

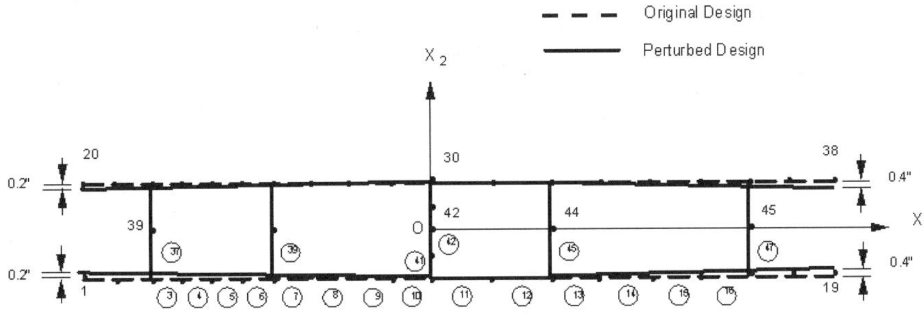

Figure 7.18. Finite element model and perturbed design of vehicle chassis frame.

Platform

The platform given in Fig. 7.19 is assembled by two steel plates and is 100 in long, 60 in wide, and 50 in high. Plate thickness is 0.4 in. The platform is loaded with concentrated forces acting at the middle span and at the end of the horizontal plate. Young's modulus and Poisson's ratio are $E = 30 \times 10^6$ psi and $\nu = 0.3$, respectively. The finite element model shown in Fig. 7.20 contains 240 ANSYS triangular shell elements, 147 nodal points, and 798 degrees of freedom.

The length of the horizontal plate is taken as the design variable. As the design is perturbed, the horizontal plate will only experience shape change. For the inclined plate, both the shape and the orientation of the plate will be changed. A perturbed design of the platform is shown in Fig. 7.21 with $\delta b_1 = 1$ in. Each plate in the platform is treated as a single design component. For the shape effect in configuration design sensitivity computation, a cubic velocity with zero slope at both edges in the x_2 direction is assumed for each design component. The profiles of the shape design velocity fields are shown in Fig. 7.21.

Table 7.5. Configuration design sensitivity results of vehicle chassis frame.

Node/ Elem.	Dir./ Stress / Eigval	$\psi(\boldsymbol{b}-\Delta\boldsymbol{b})$	$\psi(\boldsymbol{b}+\Delta\boldsymbol{b})$	$\Delta\psi(\boldsymbol{b})$	$\psi'(\boldsymbol{b})$	Ratio%
1	x_3	–.17543E–01	–.17550E–01	–.68741E–05	–.68737E–05	100.0
19	x_3	0.20181E–01	0.20179E–01	–.21079E–05	–.21074E–05	100.0
39	x_3	–.64963E–02	–.64906E–02	0.57804E–05	0.57805E–05	100.0
42	x_3	–.42161E–01	–.42146E–01	0.15003E–04	0.15003E–04	100.0
44	x_3	–.49643E–01	–.49622E–01	0.20873E–04	0.20873E–04	100.0
5	σ_z*	0.15461E+04	0.15461E+04	0.13372E–01	0.13373E–01	100.0
6	σ_z	0.19264E+04	0.19264E+04	0.13423E–01	0.13424E–01	100.0
7	σ_z	0.16415E+04	0.16414E+04	–.20966E–01	–.20967E–01	100.0
10	σ_z	–.12674E+04	–.12674E+04	–.96058E–02	–.96097E–02	100.0
11	σ_z	–.92733E+03	–.92736E+03	–.26644E–01	–.26639E–01	100.0
12	σ_z	–.12452E+04	–.12452E+04	–.21106E–01	–.21102E–01	100.0
42	σ_z	–.20257E+03	–.19710E+03	0.54719E+01	0.54718E+01	100.0
45	σ_z	–.99604E+02	–.91400E+02	0.82038E+01	0.82037E+01	100.0
1st	ζ	0.48341E+04	0.48494E+04	0.15305E+02	0.15304E+02	100.0
2nd	ζ	0.76702E+04	0.80034E+04	0.33326E+03	0.33337E+03	100.0
3rd	ζ	0.19359E+05	0.20353E+05	0.99375E+03	0.99946E+03	100.6

*Note: σ_z denotes the element maximum bending stress.

Table 7.6. Contributions from shape and orientation effects to configuration design sensitivity results of vehicle chassis frame.

Node/ Elem.	Dir./ Stress/ Eigval	$\psi'(\mathbf{b})$	Shape Effect ψ'_{V_Ω}	(%)	Orientation Effect ψ'_{V_θ}	(%)
1	x_3	–.68737E–05	0.54287E–18	0.00	–.68737E–05	100.00
19	x_3	–.21074E–05	–.22973E–18	0.00	–.21074E–05	100.00
39	x_3	0.57805E–05	0.11868E–04	66.10	–.60875E–05	33.90
42	x_3	0.15003E–04	0.25902E–06	1.73	0.14744E–04	98.27
44	x_3	0.20873E–04	0.44112E–06	2.11	0.20432E–04	97.89
5	σ_z	0.13373E–01	–.12330E–14	0.00	0.13373E–01	100.00
6	σ_z	0.13424E–01	–.77594E–14	0.00	0.13424E–01	100.00
7	σ_z	–.20967E–01	–.23680E–13	0.00	–.20967E–01	100.00
10	σ_z	–.96096E–02	0.92461E–15	0.00	–.96096E–02	100.00
11	σ_z	–.26639E–01	–.21042E–14	0.00	–.26639E–01	100.00
12	σ_z	–.21102E–01	0.70777E–15	0.00	–.21102E–01	100.00
42	σ_z	0.54719E+01	0.15667E+00	2.86	0.53152E+01	97.14
45	σ_z	0.82038E+01	0.26680E+00	3.25	0.79370E+01	96.75
1st	ζ	0.15304E+02	0.15244E+02	99.61	0.59422E–01	0.39
2nd	ζ	0.33335E+03	0.76070E+02	22.82	0.25728E+03	77.18
3rd	ζ	0.99946E+03	0.86354E+03	86.40	0.13592E+03	13.60

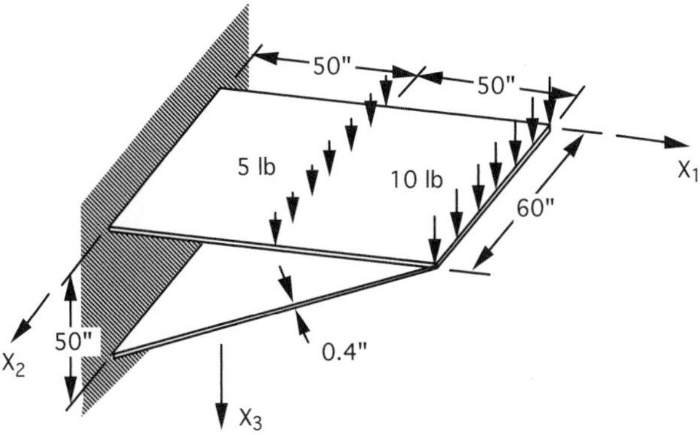

Figure 7.19. Platform.

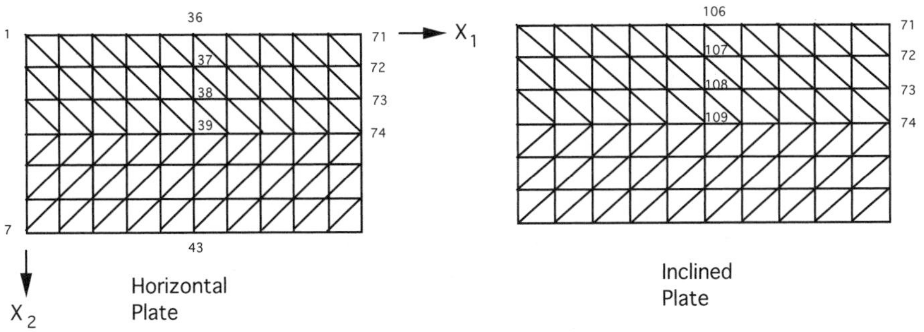

Figure 7.20. Finite element model of platform.

Configuration design sensitivity expressions of the plane elastic solid/plate design component are given in (7.112), (7.117), and (7.122). To evaluate these design expressions, both membrane and bending stresses at Gauss points are required. A stress computation routine is built to calculate stresses at Gauss points using the shape functions presented in [121]. A nine-point Gauss integration is used for the numerical integration of the design sensitivity expressions.

Configuration design sensitivity results of displacement performance measures are evaluated at several points. Table 7.7 shows good sensitivity results using the continuum formulation. The contributions from both shape and orientation effects are shown in Table 7.8. In this example, Table 7.8 shows that the shape effect is significant for those nodal points at the middle span, whereas the orientation effect is dominant for the nodal points at the interface boundary.

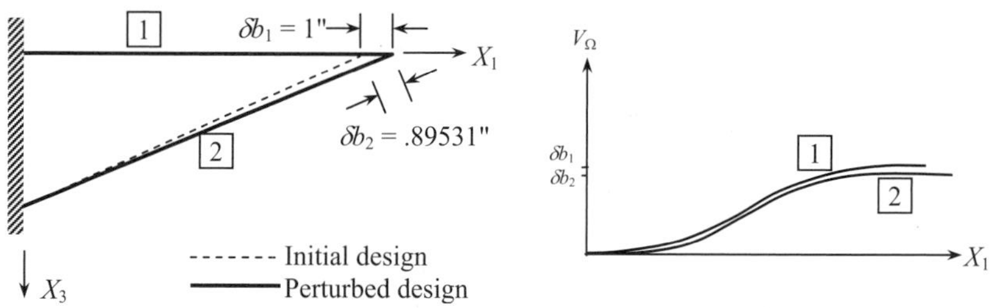

Figure 7.21. Design perturbation of platform and the shape design velocity fields.

Table 7.7. Configuration design sensitivity results of platform.

Node ID	Direct.	$\psi(\mathbf{b}-\Delta\mathbf{b})$	$\psi(\mathbf{b}+\Delta\mathbf{b})$	$\Delta\psi(\mathbf{b})$	$\psi'(\mathbf{b})$	Ratio%
106	x_1	0.37356E–2	0.38775E–2	0.14195E–3	0.14959E–3	105.4
107	x_1	0.35896E–2	0.37297E–2	0.14011E–3	0.14736E–3	105.2
108	x_1	0.35047E–2	0.36436E–2	0.13893E–3	0.14584E–3	105.0
109	x_1	0.34766E–2	0.36152E–2	0.13854E–3	0.14528E–3	104.9
36	x_3	0.26800E–1	0.28360E–1	0.15598E–2	0.15515E–2	99.5
37	x_3	0.24759E–1	0.26265E–1	0.15058E–2	0.14984E–2	99.5
38	x_3	0.23680E–1	0.25154E–1	0.14747E–2	0.14675E–2	99.5
39	x_3	0.23338E–1	0.24803E–1	0.14647E–2	0.14573E–2	99.5
71	x_3	0.10927E–3	0.11512E–3	0.58478E–5	0.58706E–5	100.4
72	x_3	0.10267E–3	0.10829E–3	0.56115E–5	0.56317E–5	100.4
73	x_3	0.10092E–3	0.10646E–3	0.55436E–5	0.55628E–5	100.3
74	x_3	0.10045E–3	0.10598E–3	0.55222E–5	0.55300E–5	100.1
106	x_3	0.74268E–2	0.78645E–2	0.43765E–3	0.45293E–3	103.5
107	x_3	0.71377E–2	0.75658E–2	0.42810E–3	0.44259E–3	103.4
108	x_3	0.69696E–2	0.73919E–2	0.42233E–3	0.43613E–3	103.3
109	x_3	0.69140E–2	0.73344E–2	0.42041E–3	0.43388E–3	103.2

Table 7.8. Contributions from shape and orientation effects to configuration design sensitivity results of platform.

			Shape Effect		Orientation Effect	
Node ID	Direct.	$\psi'(\mathbf{b})$	ψ'_{V_Ω}	(%)	ψ'_{V_θ}	(%)
106	x_1	0.14959E–3	0.21039E–3	77.58	–.60795E–4	22.42
107	x_1	0.14736E–3	0.20581E–3	77.88	–.58447E–4	22.12
108	x_1	0.14584E–3	0.20292E–3	78.05	–.57082E–4	21.95
109	x_1	0.14528E–3	0.20191E–3	78.10	–.56631E–4	21.90
36	x_3	0.15515E–2	0.15503E–2	99.93	0.11147E–5	0.07
37	x_3	0.14984E–2	0.14973E–2	99.92	0.11474E–5	0.08
38	x_3	0.14675E–2	0.14663E–2	99.92	0.11650E–5	0.08
39	x_3	0.14573E–2	0.14562E–2	99.92	0.11706E–5	0.08
71	x_3	0.58706E–5	0.19102E–5	32.54	0.39604E–5	67.46
72	x_3	0.56317E–5	0.19085E–5	33.89	0.37233E–5	66.11
73	x_3	0.55628E–5	0.19033E–5	34.21	0.36595E–5	65.79
74	x_3	0.55300E–5	0.18872E–5	34.13	0.36428E–5	65.87
106	x_3	0.45293E–3	0.42128E–3	93.01	0.31649E–4	6.99
107	x_3	0.44259E–3	0.41212E–3	93.12	0.30469E–4	6.88
108	x_3	0.43613E–3	0.40635E–3	93.17	0.29777E–4	6.83
109	x_3	0.43388E–3	0.40433E–3	93.19	0.29547E–4	6.81

7.4 Structural-Acoustic Problem

In this section, a configuration design sensitivity analysis is developed to predict variations in acoustic and structural performance measures resulting from a variation in the configuration of a coupled system. By definition, the configuration design of a coupled system consists of the geometry of the design components, that is, (1) the shape of the cavity that contains the acoustic medium, and (2) the shape and orientation of the structural components. In general, a variation in the configuration design of a coupled system includes simultaneous variations of the cavity containing the acoustic medium and the design components of a built-up structure.

Although the responses of a coupled system are solutions to a single coupled state equation, the functions representing the responses are defined mathematically in different ways. Acoustic responses, such as the acoustic pressure, are defined on a three-dimensional acoustic domain Ω^a, while structural responses, such as displacement fields, are defined on Ω^s, which is a surface or line depending on the types of design components. Mathematically different ways of describing the variations of the design and responses are required for the acoustic domain and each design component of the structure.

In this section, the variations of harmonic responses of the coupled system resulting from a variation in the configuration are mathematically defined, and the variations of acoustic and structural performance measures are derived. The variations of performance measures are expressed in terms of the variations of the state variables. The variation of the state equation is obtained either to construct an equation for the variation of the state responses, or to derive design sensitivity formulas that are explicit in design variations. The direct differentiation method is briefly discussed, and the adjoint variable method of design sensitivity analysis [5] s derived. The equations used are solved using the finite element analysis (FEA), and a computational method using FEA results is illustrated with example problems. Since the structural-acoustic problem is formulated using the frequency response technique, the state variable becomes a complex variable. In addition, the energy bilinear and load linear forms in the static problem become the sesquilinear and semilinear forms as presented in Section 2.6.

7.4.1 Variations for the Configuration Design

The domains defined in the coupled system are the three-dimensional volume Ω^a of the acoustic medium and its boundary surface Γ^a (= Ω^s), which is also defined as the domain of the surface design components of the built-up structure. The line design components, such as the beam and truss members, also lie on Ω^s, although they do not directly interact with the acoustic medium.

In addition to the two structural components (line and surface) presented in Section 7.1, the variation in the configuration of the acoustic response is required in the structural-acoustic system. In this section, the variation of acoustic response is defined, and the basic formulas for variations of the integral form of functionals are derived, which are required for the variations of performance measures and the state equation.

Variations of Acoustic Response

Consider the structural-acoustic system in Fig. 2.4 with acoustic medium Ω^a surrounded by structural components in Ω^s (= Γ^a). The configuration design variation of the structural components produces shape design variation in the acoustic medium. Thus, only shape design variation is considered for the acoustic medium. Let $p(\boldsymbol{x})$ and $p_\tau(\boldsymbol{x}_\tau)$, which are the complex functions of acoustic pressure, be smooth solutions to the boundary value problem from Section 2.6.2, on Ω^a and Ω_τ^a, respectively. Assuming a regular extension of p_τ in the neighborhood of Ω_τ^a, the variation of $p(\boldsymbol{x})$ is defined as the pointwise material derivative along τ at $\boldsymbol{x} \in \Omega^a$, that is,

$$\begin{aligned}\dot{p} &= \frac{d}{d\tau} p_\tau(\boldsymbol{x} + \tau \boldsymbol{V}_{\Omega^a}(\boldsymbol{x}))\Big|_{\tau=0} \\ &= \lim_{\tau \to 0} \frac{p_\tau(\boldsymbol{x} + \tau \boldsymbol{V}_{\Omega^a}(\boldsymbol{x})) - p(\boldsymbol{x})}{\tau} \\ &= \lim_{\tau \to 0} \frac{p_\tau(\boldsymbol{x} + \tau \boldsymbol{V}_{\Omega^a}(\boldsymbol{x})) - p_\tau(\boldsymbol{x})}{\tau} + \lim_{\tau \to 0} \frac{p_\tau(\boldsymbol{x}) - p(\boldsymbol{x})}{\tau} \\ &= p'(\boldsymbol{x}) + \nabla p^T \boldsymbol{V}_{\Omega^a}(\boldsymbol{x}),\end{aligned} \tag{7.160}$$

where the partial derivative $p'(\boldsymbol{x})$ is defined as

$$p'(\boldsymbol{x}) = \lim_{\tau \to 0} \frac{p_\tau(\boldsymbol{x}) - p(\boldsymbol{x})}{\tau}. \tag{7.161}$$

Assuming that $p(\boldsymbol{x})$ is a smooth function, the partial derivative in (7.161) can be interchanged with the partial derivative with respect to coordinate variable x_i, i= 1, 2, 3, that is,

$$\left(\frac{\partial p}{\partial x_i}\right)' = \frac{\partial}{\partial x_i}(p'). \tag{7.162}$$

The material derivative $\dot{p}$ defined in (7.160) is the variation of $p(\boldsymbol{x})$ with respect to the variations in the shape of the acoustic domain Ω^a and the configuration of the structural domain Ω^s of the coupled system.

Variations of Performance Measure

A harmonic acoustic performance measure can be expressed as an integral over the acoustic domain Ω^a, as

$$\psi_p = \iiint_{\Omega_\tau^a} h_\tau(p_\tau, \nabla p_\tau) \, d\Omega_\tau^a, \tag{7.163}$$

where the function h_τ is continuously differentiable with respect to its arguments. The gradient operator in the acoustic domain of the above equation is $\nabla p = [p_{,1}, p_{,2}, p_{,3}]^T$.

Since the acoustic domain is affected by shape variation, the material derivative of ψ_p in (7.163) can be taken using the formula for the variation of an integral in (6.37), as

$$\psi_p' = \iiint_{\Omega^a} [h_{,p} p' + h_{,\nabla p} \nabla p' + div(h \boldsymbol{V}_{\Omega^a})] \, d\Omega^a. \tag{7.164}$$

Using (7.160) and (7.162), (7.164) becomes

$$\begin{aligned}\psi_p' &= \iiint_{\Omega^a} [h_{,p}(\dot{p} - \nabla p^T \boldsymbol{V}_{\Omega^a}) + h_{,\nabla p} \nabla(\dot{p} - \nabla p^T \boldsymbol{V}_{\Omega^a}) + div(h \boldsymbol{V}_{\Omega^a})] \, d\Omega^a \\ &= \iiint_{\Omega^a} [h_{,p} \dot{p} + h_{,\nabla p} \nabla \dot{p}] \, d\Omega^a \\ &\quad - \iiint_{\Omega^a} [h_{,p} \nabla p^T \boldsymbol{V}_{\Omega^a} + h_{,\nabla p} \nabla(\nabla p^T \boldsymbol{V}_{\Omega^a}) - div(h \boldsymbol{V}_{\Omega^a})] \, d\Omega^a.\end{aligned} \tag{7.165}$$

The first integral in (7.165) is implicit in terms of the variation of the acoustic pressure and its gradients, while the other terms are explicit in terms of the design velocity field V_{Ω^a}, which is the shape design variation.

The acoustic performance measure in (7.163) is defined in the three-dimensional acoustic domain, while the structural performance measure is defined on the two-dimensional structural surface. A harmonic structural performance measure may be given in an integral form as

$$\psi_z = \iint_{\Omega_\tau^s} g_\tau(z_\tau, \nabla z_\tau)\, d\Omega_\tau^s, \tag{7.166}$$

where g is a function continuously differentiable with respect to its arguments. Unlike the acoustic performance measure, the gradient operator in the structural domain of the above equation is $\nabla z = [z_{,1}, z_{,2}]$ in the local coordinate system.

The shape design velocity for the structural component is defined in the local coordinate system by $V_{\Omega^s} = [V_1, V_2, 0]^T$, and the configuration design velocity is given by $V_\Theta = [0, 0, V_3]^T$. Taking the variation of ψ_z in (7.166) by using (6.37),

$$\psi_z' = \iint_{\Omega^s} [g_{,z} z' + g_{,\nabla z} : \nabla z' + div(g V_{\Omega^s})]\, d\Omega^s. \tag{7.167}$$

Using (6.8) and (7.42), (7.167) becomes

$$\begin{aligned} \psi_z' &= \iint_{\Omega^s} [g_{,z}\dot{z} + g_{,\nabla z} : \nabla \dot{z}]\, d\Omega^s \\ &- \iint_{\Omega^s} [g_{,z}(V_\theta z) + g_{,\nabla z} : \nabla (V_\theta z)]\, d\Omega^s \\ &- \iint_{\Omega^s} [g_{,z}(\nabla z V_{\Omega^s}) + g_{,\nabla z} : \nabla(\nabla z V_{\Omega^s}) - div(g V_{\Omega^s})]\, d\Omega^s, \end{aligned} \tag{7.168}$$

where $V_\theta z$ is the convective term of the configuration design in (7.18). The first integral in (7.168) is implicit in terms of the variation of the structural displacement and gradients, while the other terms are explicit in terms of the design velocity field and the state response z.

The expressions in (7.165) and (7.168) include the variations of the acoustic and structural responses that are the solutions to the state (2.73). In order to evaluate the variations ψ_p' and ψ_z', the variations of the acoustic pressure, structural responses, and their gradients may be obtained by solving an equation derived from the state equation. Alternatively, the terms can be converted to a form that is explicit in responses and design velocity fields. The variation of the state equation is required in both cases.

Variation of State Equation

The variational (2.73) of the coupled system can be rewritten with subscripts that identify dependence of each term on the corresponding domain, as

$$q_\Omega(z, \bar{z}) + b_\Omega(p, \bar{p}) - \chi_\Omega(p, \bar{z}) - \phi_\Omega(z, \bar{p}) = \ell_\Omega(\bar{z}), \tag{7.169}$$

which must hold true for all kinematically admissible virtual states $\{\bar{z}^*, \bar{p}^*\} \in Q$ where Q is the complex vector space that satisfies the boundary and interface conditions. Sesqui-linear and semi-linear forms in (7.169) are integrals over the acoustic or structural domain, as defined in (2.73) through (2.79). Note that, as explained in Section 2.6, the complex conjugates of the second arguments (i.e., $\bar{z}$ and $\bar{p}$ in the above state equation) are used in the integration that define the sesquilinear form, whereas the complex conjugate of the argument is used in the integration that defines the semilinear form. This rule is also applicable to the variations of the sesquilinear and semilinear forms in the following derivations. Taking the variation of both sides of (7.169),

$$[q_\Omega(z,\bar{z})]' + [b_\Omega(p,\bar{p})]' - [\chi_\Omega(p,\bar{z})]' - [\phi_\Omega(z,\bar{p})]' = [\ell_\Omega(\bar{z})]', \quad \forall\{\bar{z}^*,\bar{p}^*\} \in Q. \tag{7.170}$$

The variation of each term in (7.170) can be derived using the definitions of the material derivatives in Section 7.1.

First, the variations of the sesquilinear and semilinear forms defined over Ω^s, which is both the structural domain and the interface, can be derived. Because the sesquilinear form $q_\Omega(z,\bar{z})$ has a bilinear function in the integrands, it can be redefined using a bilinear function $e(\bullet,\bullet)$, as

$$q_\Omega(z,\bar{z}) = \iint_{\Omega^s} e(z,\bar{z}^*)\,d\Omega^s. \tag{7.171}$$

The variation of $q_\Omega(z,\bar{z})$ in (7.171) can be obtained using (6.37), the formula for the variation of the integral over Ω^s, as

$$[q_\Omega(z,\bar{z})]' = \iint_{\Omega^s} \left\{ e(z,\bar{z}^{*\prime}) + e(z',\bar{z}^*) + div[e(z,\bar{z}^*)\boldsymbol{V}_{\Omega^s}] \right\} d\Omega^s. \tag{7.172}$$

Using (7.29), the variation of the sesquilinear form in (7.172) can be written as

$$\begin{aligned}
[q_\Omega(z,\bar{z})]' &= \iint_{\Omega^s} e(\dot{z},\bar{z}^*)\,d\Omega^s + \iint_{\Omega^s} e(z,\dot{\bar{z}}^*)\,d\Omega^s \\
&\quad - \iint_{\Omega^s} \left\{ e(z,\nabla\bar{z}^*\boldsymbol{V}_{\Omega^s}) + e(\nabla z\boldsymbol{V}_{\Omega^s},\bar{z}^*) - div[e(z,\bar{z}^*)\boldsymbol{V}_{\Omega^s}] \right\} d\Omega^s \\
&\quad - \iint_{\Omega^s} [e(z,\boldsymbol{V}_\theta\bar{z}^*) + e(\boldsymbol{V}_\theta z,\bar{z}^*)]\,d\Omega^s \\
&\equiv q_\Omega(\dot{z},\bar{z}) + q_\Omega(z,\dot{\bar{z}}) + q_V'(z,\bar{z}).
\end{aligned} \tag{7.173}$$

After collecting those terms that are explicitly dependent on shape and configuration designs, the last term on the right side of the above equation can be defined as

$$\begin{aligned}
q_V'(z,\bar{z}) &\equiv - \iint_{\Omega^s} \left\{ e(z,\nabla\bar{z}^*\boldsymbol{V}_{\Omega^s}) + e(\nabla z\boldsymbol{V}_{\Omega^s},\bar{z}^*) - div[e(z,\bar{z}^*)\boldsymbol{V}_{\Omega^s}] \right\} d\Omega^s \\
&\quad - \iint_{\Omega^s} [e(z,\boldsymbol{V}_\theta\bar{z}^*) + e(\boldsymbol{V}_\theta z,\bar{z}^*)]\,d\Omega^s.
\end{aligned} \tag{7.174}$$

Using a similar procedure, the variations of the interface sesquilinear and the load semilinear forms, which are integrals of bilinear and linear functions over Ω^s, can be derived using (7.160) and (7.172). Results similar to those in (7.173) are obtained, that is,

$$[\chi_\Omega(p,\bar{z})]' \equiv \chi_\Omega(\dot{p},\bar{z}) + \chi_\Omega(p,\dot{\bar{z}}) + \chi_V'(p,\bar{z}) \tag{7.175}$$

$$[\phi_\Omega(z,\bar{p})]' \equiv \phi_\Omega(\dot{z},\bar{p}) + \phi_\Omega(z,\dot{\bar{p}}) + \phi_V'(z,\bar{p}) \tag{7.176}$$

$$[\ell_\Omega(\bar{z})]' \equiv \ell_\Omega(\dot{\bar{z}}) + \ell_V'(\bar{z}). \tag{7.177}$$

Second, the acoustic energy sesquilinear form $b_\Omega(p,\bar{p})$, which is an integral over the acoustic domain Ω^a, can be written using a bilinear function $c(\bullet,\bullet)$, as

$$b_\Omega(p,\bar{p}) = \iiint_{\Omega^a} c(p,\bar{p}^*)\,d\Omega^a. \tag{7.178}$$

Using (6.37), which is the formula for the variation of the integral over Ω^a, the variation of $b_\Omega(p,\bar{p})$ in (7.178) can be obtained, as

$$[b_\Omega(p,\bar{p})]' = \iiint_{\Omega^a} \left\{ c(p,\bar{p}^{*'}) + c(p',\bar{p}^*) + div[c(p,\bar{p}^*)\boldsymbol{V}_{\Omega^a}] \right\} d\Omega^a. \tag{7.179}$$

Using (7.160), which is the variation of the acoustic response, (7.179) is rewritten as

$$\begin{aligned} [b_\Omega(p,\bar{p})]' &= \iiint_{\Omega^a} c(\dot{p},\bar{p}^*)\, d\Omega^a + \iiint_{\Omega^a} c(p,\dot{\bar{p}}^*)\, d\Omega^a \\ &\quad - \iiint_{\Omega^a} \left\{ c(p,\nabla\bar{p}^{*T}\boldsymbol{V}_{\Omega^a}) + c(\nabla p^T\boldsymbol{V}_{\Omega^a},\bar{p}^*) - div[c(p,\bar{p}^*)\boldsymbol{V}_{\Omega^a}] \right\} d\Omega^a \\ &\equiv b_\Omega(\dot{p},\bar{p}) + b_\Omega(p,\dot{\bar{p}}) + b_V'(p,\bar{p}). \end{aligned} \tag{7.180}$$

The variation of $b_\Omega(p,\bar{p})$ with respect to the explicit dependence on the shape design variation is defined, from the third term of (7.180), as

$$b_V'(p,\bar{p}) \equiv - \iiint_{\Omega^a} \left\{ c(p,\nabla\bar{p}^{*T}\boldsymbol{V}_{\Omega^a}) + c(\nabla p^T\boldsymbol{V}_{\Omega^a},\bar{p}^*) - div[c(p,\bar{p}^*)\boldsymbol{V}_{\Omega^a}] \right\} d\Omega^a. \tag{7.181}$$

Note that the explicitly dependent term $b_V'(p,\bar{p})$ only contains the shape design velocity field because the acoustic domain Ω^a does not have the configuration design.

Using (7.173), (7.175) through (7.177), and (7.180), the variation of the state equation (7.168) is written as

$$\begin{aligned} &[q_\Omega(\dot{z},\bar{z}) + b_\Omega(\dot{p},\bar{p}) - \chi_\Omega(\dot{p},\bar{z}) - \phi_\Omega(\dot{z},\bar{p})] \\ &+[q_\Omega(z,\dot{\bar{z}}) + b_\Omega(p,\dot{\bar{p}}) - \chi_\Omega(p,\dot{\bar{z}}) - \phi_\Omega(z,\dot{\bar{p}})] \\ &+[q_V'(z,\bar{z}) + b_V'(p,\bar{p}) - \chi_V'(p,\bar{z}) - \phi_V'(z,\bar{p})] \\ &= \ell_\Omega(\dot{\bar{z}}) + \ell_V'(\bar{z}), \qquad \forall \{\bar{z}^*,\bar{p}^*\} \in Q. \end{aligned} \tag{7.182}$$

Since the space Q of admissible virtual states is preserved during variation [5], i.e., $\{\dot{\bar{z}}^*,\dot{\bar{p}}^*\} \in Q$, $\{\bar{z}^*,\bar{p}^*\}$ in (7.169) can be replaced with $\{\dot{\bar{z}}^*,\dot{\bar{p}}^*\}$ to obtain

$$q_\Omega(z,\dot{\bar{z}}) + b_\Omega(p,\dot{\bar{p}}) - \chi_\Omega(p,\dot{\bar{z}}) - \phi_\Omega(z,\dot{\bar{p}}) = \ell_\Omega(\dot{\bar{z}}). \tag{7.183}$$

Then, by using the relation in (7.183), (7.182) can be reduced to

$$\begin{aligned} &q_\Omega(\dot{z},\bar{z}) + b_\Omega(\dot{p},\bar{p}) - \chi_\Omega(\dot{p},\bar{z}) - \phi_\Omega(\dot{z},\bar{p}) \\ &\quad = \ell_V'(\bar{z}) - [q_V'(z,\bar{z}) + b_V'(p,\bar{p}) - \chi_V'(p,\bar{z}) - \phi_V'(z,\bar{p})], \\ &\qquad\qquad \forall \{\bar{z}^*,\bar{p}^*\} \in Q, \end{aligned} \tag{7.184}$$

which is the variation of (7.169), the state equation of the coupled motion. Equation (7.184) is an equation for $\{\dot{z},\dot{p}\}$, which are the variations of the state variables.

7.4.2 Design Sensitivity Analysis

The variations of the performance measures, or the design sensitivities, derived in Section 7.4.1 include integrals that are explicit in terms of the variations of the state variables and their gradients. The direct differentiation and adjoint variable methods are two ways to treat the variations of state variables in the variations of performance measures.

Direct Differentiation Method

With the direct differentiation method, variations of the performance measures in (7.165) and (7.168) are directly used to compute design sensitivities. Computation of design sensitivity requires the state variables, the variations of the state variables and their gradients, and the design velocity field. The state variables $\{z, p\}$ are obtained from the state equation in (7.169). The variations of the state variables $\{\dot{z}, \dot{p}\}$ can be obtained by solving (7.184), which is the variation of the state equation once the terms on the right side, called the fictitious load, are evaluated using $\{z, p\}$. The variation of the performance measures, or the design sensitivities of (7.165) and (7.168) can then be computed if the design velocity field corresponding to a design variation is given.

Adjoint Variable Method

A set of adjoint variables and an adjoint equation are defined and used to treat the variations of the state variables in design sensitivity expressions. The sesqui-linear form used in the adjoint equation of the structural-acoustic problem is not self-adjoint, as discussed in Section 5.5.

For the acoustic performance measure, the variation of the performance measure in (7.165) is considered. Let the adjoint equation and the adjoint response $\{\lambda^*, \eta^*\}$ that correspond to the state variable $\{z, p\}$ be defined such that $\{\lambda^*, \eta^*\} \in Q$ satisfy the adjoint equation. First, the left side of the relation in (7.169) is taken, and $\{z, p\}$ is replaced by $\{\bar{\lambda}, \bar{\eta}\}$, and $\{\bar{z}^*, \bar{p}^*\}$ is replaced by adjoint variable $\{\lambda^*, \eta^*\}$. Next, the first integral on the right of (7.165) is taken, which includes the variation of the state variable $\dot{p}$ and its gradient $\nabla\dot{p}$ are replaced with the acoustic adjoint virtual field $\bar{\eta}$ and its gradient $\nabla\bar{\eta}$, respectively. Equating the results produces the adjoint equation, as follows:

$$\begin{aligned} &q_\Omega(\bar{\lambda}, \lambda) + b_\Omega(\bar{\eta}, \eta) - \chi_\Omega(\bar{\eta}, \lambda) - \phi_\Omega(\bar{\lambda}, \eta) \\ &\quad = \iiint_{\Omega^a} [h_{,p}\bar{\eta} + h_{,\nabla p}\nabla\bar{\eta}] d\Omega^a, \qquad \forall \{\bar{\lambda}, \bar{\eta}\} \in Q. \end{aligned} \tag{7.185}$$

For the structural performance measure, the variation of the performance measure in (7.168) is considered. The definition of the adjoint equation is similar to (7.185), except for the right side. From the variation of the structural performance measure in (7.168), the first integral on the right of (7.168) is taken that includes the variations of the state response, and $\dot{z}$ and its gradient $\nabla\dot{z}$ are replaced by the adjoint virtual field $\bar{\lambda}$ and its gradient $\nabla\bar{\lambda}$. Equating the results with the left side of (7.185) provides the adjoint equation, as

$$\begin{aligned} &q_\Omega(\bar{\lambda}, \lambda) + b_\Omega(\bar{\eta}, \eta) - \chi_\Omega(\bar{\eta}, \lambda) - \phi_\Omega(\bar{\lambda}, \eta) \\ &\quad = \iint_{\Omega^s} [g_{,z}\bar{\lambda} + g_{,\nabla z} : \nabla\bar{\lambda}] d\Omega^s, \qquad \forall \{\bar{\lambda}, \bar{\eta}\} \in Q. \end{aligned} \tag{7.186}$$

Equations (7.185) and (7.186) are defined as adjoint equations for the acoustic and structural performance measures, respectively. Note that they are the same as (5.240) and (5.247), respectively, the adjoint equations for sizing design sensitivity analysis. The solutions to the adjoint equations are the adjoint responses $\{\lambda^*, \eta^*\}$. As previously mentioned, these equations are identical except for the adjoint loads, which are different depending on the type of performance measures.

Equations (7.165) and (7.168) can be reduced to expressions that are explicit in terms of the variation of the state variables by using the variation of the state (7.184), and the adjoint (7.185) and (7.186). First, by replacing $\{\bar{\lambda}, \bar{\eta}\} \in Q$ in (7.185) and (7.186) with $\{\dot{z}, \dot{p}\} \in Q$, we have

$$\begin{aligned} &q_\Omega(\dot{z}, \lambda) + b_\Omega(\dot{p}, \eta) - \chi_\Omega(\dot{p}, \lambda) - \phi_\Omega(\dot{z}, \eta) \\ &\quad = \iiint_{\Omega^a} (h_{,p}\dot{p} + h_{,\nabla p}\nabla\dot{p}) d\Omega^a \end{aligned} \tag{7.187}$$

and

$$
\begin{aligned}
& q_\Omega(\dot{z},\lambda) + b_\Omega(\dot{p},\eta) - \chi_\Omega(\dot{p},\lambda) - \phi_\Omega(\dot{z},\eta) \\
& \qquad = \iint_{\Omega^s} (g_{,z}\dot{z} + g_{,\nabla z} : \nabla\dot{z})\, d\Omega^s.
\end{aligned} \tag{7.188}
$$

Next, replacing $\{\bar{z}^*, \bar{p}^*\} \in Q$ in (7.184) with the adjoint responses $\{\lambda^*, \eta^*\} \in Q$ gives

$$
\begin{aligned}
& q_\Omega(\dot{z},\lambda) + b_\Omega(\dot{p},\eta) - \chi_\Omega(\dot{p},\lambda) - \phi_\Omega(\dot{z},\eta) \\
& \qquad = \ell'_V(\lambda) - [q'_V(z,\lambda) + b'_V(p,\eta) - \chi'_V(p,\lambda) - \phi'_V(z,\eta)].
\end{aligned} \tag{7.189}
$$

Since the left sides of (7.187), (7.188), and (7.189) are identical, the right sides of (7.187) and (7.188) can be equated with the right side of (7.189). From (7.187) and (7.189),

$$
\begin{aligned}
& \iiint_{\Omega^a} (h_{,p}\dot{p} + h_{,\nabla p}\nabla\dot{p})\, d\Omega^a \\
& \qquad = \ell'_V(\lambda) - [q'_V(z,\lambda) + b'_V(p,\eta) - \chi'_V(p,\lambda) - \phi'_V(z,\eta)]
\end{aligned} \tag{7.190}
$$

and, from (7.188) and (7.189),

$$
\begin{aligned}
& \iint_{\Omega^s} (g_{,z}\dot{z} + g_{,\nabla z} : \nabla\dot{z})\, d\Omega^s \\
& \qquad = \ell'_V(\lambda) - [q'_V(z,\lambda) + b'_V(p,\eta) - \chi'_V(p,\lambda) - \phi'_V(z,\eta)].
\end{aligned} \tag{7.191}
$$

The implicit design sensitivity terms in (7.165) and (7.168), which include $\dot{z}$, $\dot{p}$ and their gradients, can now be replaced using (7.190) and (7.191). Substituting (7.190) and (7.191) into (7.165) and (7.168), respectively, design sensitivity formulas that are explicit in terms of the design velocity field are obtained, as

$$
\begin{aligned}
\psi'_p &= \ell'_V(\lambda) - [q'_V(z,\lambda) + b'_V(p,\eta) - \chi'_V(p,\lambda) - \phi'_V(z,\eta)] \\
& \quad - \iiint_{\Omega^a} [h_{,p}\nabla p^T \boldsymbol{V}_{\Omega^a} + h_{,\nabla p}\nabla(\nabla p^T \boldsymbol{V}_{\Omega^a}) - div(h\boldsymbol{V}_{\Omega^a})]\, d\Omega^a
\end{aligned} \tag{7.192}
$$

and

$$
\begin{aligned}
\psi'_z &= \ell'_V(\lambda) - [q'_V(z,\lambda) + b'_V(p,\eta) - \chi'_V(p,\lambda) - \phi'_V(z,\eta)] \\
& \quad - \iint_{\Omega^s} [g_{,z}(\boldsymbol{V}_\theta z) + g_{,\nabla z} : \nabla(\boldsymbol{V}_\theta z)]\, d\Omega^s \\
& \quad - \iint_{\Omega^s} [g_{,z}(\nabla z \boldsymbol{V}_{\Omega^s}) + g_{,\nabla z} : \nabla(\nabla z \boldsymbol{V}_{\Omega^s}) - div(g\boldsymbol{V}_{\Omega^s})]\, d\Omega^s.
\end{aligned} \tag{7.193}
$$

The configuration design sensitivity formulas for the harmonic performance measures of the coupled structural-acoustic system have been derived. Expressions are explicit in terms of the design velocity. Computation of the design sensitivity using (7.192) and (7.193) requires structural and acoustic responses z and p, which are the solutions to the state equation, and the adjoint responses λ^* and η^*, which are the solutions to the adjoint equations in (7.185) and (7.186). The expressions in the first lines of (7.192) and (7.193) are identical. The same expression can provide an efficient computational procedure. However, depending on the performance measures, different adjoint analysis results are used to evaluate these expressions.

7.4.3 Design Components

The configuration design sensitivity formulas in (7.192) and (7.193) include the explicit variations of the energy sesquilinear, load semilinear and interface sesquilinear forms. In this section, the explicit variations of the sesquilinear and semilinear forms are presented.

The results are then used to compute design sensitivities.

Acoustic Medium

The sesquilinear form of acoustic energy is given in (2.77) as an integral over the acoustic volume. The explicit variation of acoustic energy resulting from the shape change is obtained using (7.181), the definition of the shape variation of a functional, as

$$\begin{aligned} b'_V(p,\overline{p}) &= -\frac{\omega^2}{\beta}\iiint_{\Omega^a} p\overline{p}^* div \boldsymbol{V}_{\Omega^a}\, d\Omega^a \\ &-\frac{1}{\rho_0}\iiint_{\Omega^a}\left\{\nabla(\nabla p^T \boldsymbol{V}_{\Omega^a})^T \nabla\overline{p}^* + \nabla p^T(\nabla\overline{p}^{*T}\boldsymbol{V}_{\Omega^a}) - div[\nabla p^T \nabla \overline{p}^* \boldsymbol{V}_{\Omega^a}]\right\} d\Omega^a, \end{aligned} \tag{7.194}$$

where the shape design velocity is given as a three-dimensional vector $\boldsymbol{V}_{\Omega^a} = [V_1,\ V_2,\ V_3]^T$. As previously mentioned, no variation resulting from the configuration design exists for the functional defined in the acoustic domain.

The built-up structure of a coupled system consists of line and surface design components. The line design components are modeled with truss/beam components, and the flat surface design components are modeled with plane elastic solid/plate design components. The explicit variations of energy sesquilinear and load semilinear forms for these design components are presented below.

Truss/Beam Design Component

The configuration design variations of energy bilinear forms for the static and eigenvalue problems in Section 7.2 are extended to the complex energy sesquilinear and semilinear forms of harmonic vibration. The energy sesquilinear form of (2.76) is written for the truss/beam design components shown in Figs. 7.4 and 7.5 as

$$\begin{aligned} q_\Omega(\boldsymbol{z},\overline{\boldsymbol{z}}) &= -\rho_s\omega^2\int_0^l A(z_1\overline{z}_1^* + z_2\overline{z}_2^*)\,dx \\ &+(1+j\varphi)\int_0^l (EAz_{1,1}\overline{z}_{1,1}^* + EIz_{2,11}\overline{z}_{2,11}^*)\,dx, \end{aligned} \tag{7.195}$$

where ω is the excitation frequency, A is the cross-sectional area, and I is the second moments of inertia. Mass density ρ_s, Young's modulus E, and structural damping coefficient φ are the material properties.

The explicit variation of the energy sesquilinear form in (7.195), with respect to the configuration design variation defined in Section 7.1, can be derived using (7.174), as

$$\begin{aligned} q'_V(\boldsymbol{z},\overline{\boldsymbol{z}}) &= -\rho_s\omega^2\int_0^l A(z_1\overline{z}_1^* + z_2\overline{z}_2^*)V_{1,1}\,dx \\ &-(1+j\varphi)\int_0^l EA[z_{1,1}\overline{z}_{1,1}^*V_{1,1} + (z_{2,1}\overline{z}_{1,1}^* + z_{1,1}\overline{z}_{2,1}^*)V_{2,1}]dx \\ &-(1+j\varphi)\int_0^l EI[3z_{2,11}\overline{z}_{2,11}^*V_{1,1} + (z_{2,1}\overline{z}_{2,11}^* + z_{2,11}\overline{z}_{2,1}^*)V_{1,11} + (z_{1,11}\overline{z}_{2,11}^* + z_{2,11}\overline{z}_{1,11}^*)V_{2,1}]dx, \end{aligned} \tag{7.196}$$

where the shape design velocity $\boldsymbol{V}_\Omega = [V_1,\ 0]^T$ and the configuration design velocity $\boldsymbol{V}_\Theta = [0,\ V_2]^T$, from which the convective part of the configuration design in (7.11) is calculated, by

$$\boldsymbol{V}_\theta \boldsymbol{z} = \begin{bmatrix} -V_{2,1}z_2 \\ V_{2,1}z_1 \end{bmatrix}, \tag{7.197}$$

where $\boldsymbol{z} = [z_1, z_2]^T$ and

$$\boldsymbol{V}_\theta = \begin{bmatrix} 0 & -V_{2,1} \\ V_{2,1} & 0 \end{bmatrix}.$$

From the condition that the line component remains straight during the configuration change, the second-order derivative of the configuration design velocity V_2 disappears in the derivative of (7.196). However, the second-order derivative of the shape design velocity (i.e., $V_{1,11}$) remains, and the regularity problem appears, as discussed in Section 7.3.2.

The external harmonic loads for the truss/beam design component are axial force $f_1(x_1)$ and lateral forces $f_2(x_1)$. The load semilinear form in (2.73) is written as

$$\ell_\Omega(\bar{z}) = \int_0^l \boldsymbol{f}^T \bar{z}^* \, dx, \tag{7.198}$$

where $\boldsymbol{f}(x_1) = [f_1, f_2]^T$ is the complex phasor of the harmonic force.

The explicit variation of (7.198) can be derived using (7.177), as

$$\ell_V'(\bar{z}) = \int_0^l [(f_{1,1}\bar{z}_1^* + f_{2,1}\bar{z}_2^*)V_1 + (f_1\bar{z}_1^* + f_2\bar{z}_2^*)V_{1,1} + f_1\bar{z}_2^* V_{2,1} - f_2\bar{z}_1^* V_{2,1}]dx. \tag{7.199}$$

Plane Elastic Solid/Plate

The energy sesquilinear form for the plane elastic solid/plate (shell) design component shown in Figs. 7.6 and 7.7 is given as

$$\begin{aligned} q_\Omega(z,\bar{z}) = & -\rho_s \omega^2 \iint_{\Omega^s} h z^T \bar{z}^* \, d\Omega^s \\ & +(1+j\varphi) \iint_{\Omega^s} h \boldsymbol{\sigma}(z)^T \boldsymbol{\varepsilon}(\bar{z}^*) \, d\Omega^s \\ & +(1+j\varphi) \iint_{\Omega^s} \boldsymbol{\kappa}(\bar{z}^*)^T \boldsymbol{C}^b \boldsymbol{\kappa}(z) \, d\Omega^s, \end{aligned} \tag{7.200}$$

where h is the thickness of the plate. The flexural rigidity of bending is denoted by $D = Eh^3/12(1 - \nu^2)$, where ν is Poisson's ratio and E is Young's modulus. The plane stress and strain resulting from the in-plane loading are denoted as $\boldsymbol{\sigma} = [\sigma_{11}, \sigma_{22}, \sigma_{12}]^T$ and $\boldsymbol{\varepsilon} = [\varepsilon_{11}, \varepsilon_{22}, 2\varepsilon_{12}]^T$. The curvature vector $\boldsymbol{\kappa} = [z_{3,11}, z_{3,22}, 2z_{3,12}]^T$ and the bending stiffness matrix $\boldsymbol{C}^b$ is defined in (3.40).

The explicit variation of the energy sesquilinear form in (7.200) with respect to configuration design variation is obtained using (7.174), as

$$\begin{aligned} q_V'(z,\bar{z}) = & \rho_s \omega^2 \iint_{\Omega^s} h[z^T \bar{z}^* div\boldsymbol{V}_{\Omega^s} + z^T(\boldsymbol{V}_\theta \bar{z}^*) + (\boldsymbol{V}_\theta z)^T \bar{z}^*] d\Omega^s \\ & -(1+j\varphi) \iint_{\Omega^s} h\{\boldsymbol{\sigma}(z)^T \boldsymbol{\varepsilon}(\nabla \bar{z}^* \boldsymbol{V}_{\Omega^s}) + \boldsymbol{\sigma}(\nabla z \boldsymbol{V}_{\Omega^s})^T \boldsymbol{\varepsilon}(\bar{z}^*) - div[\boldsymbol{\sigma}(z)^T \boldsymbol{\varepsilon}(\bar{z}^*) \boldsymbol{V}_{\Omega^s}]\} d\Omega^s \\ & -(1+j\varphi) \iint_{\Omega^s} h[\boldsymbol{\sigma}(z)^T \boldsymbol{\varepsilon}(\boldsymbol{V}_\theta \bar{z}^*) + \boldsymbol{\sigma}(\boldsymbol{V}_\theta z)^T \boldsymbol{\varepsilon}(\bar{z}^*)] d\Omega^s \\ & -(1+j\varphi) \iint_{\Omega^s} \{\boldsymbol{\kappa}(\nabla \bar{z}^* \boldsymbol{V}_{\Omega^s})^T \boldsymbol{C}^b \boldsymbol{\kappa}(z) + \boldsymbol{\kappa}(\bar{z}^*)^T \boldsymbol{C}^b \boldsymbol{\kappa}(\nabla z \boldsymbol{V}_{\Omega^s}) - div[\boldsymbol{\kappa}(\bar{z}^*)^T \boldsymbol{C}^b \boldsymbol{\kappa}(z) \boldsymbol{V}_{\Omega^s}]\} d\Omega^s \\ & -(1+j\varphi) \iint_{\Omega^s} [\boldsymbol{\kappa}(\boldsymbol{V}_\theta \bar{z}^*)^T \boldsymbol{C}^b \boldsymbol{\kappa}(z) + \boldsymbol{\kappa}(\bar{z}^*)^T \boldsymbol{C}^b \boldsymbol{\kappa}(\boldsymbol{V}_\theta z)] d\Omega^s, \end{aligned} \tag{7.201}$$

where the convective term of the configuration design can be obtained from (7.18) and (7.19), as

$$\boldsymbol{V}_\theta \boldsymbol{z} = \begin{bmatrix} -V_{3,1} z_3 \\ -V_{3,2} z_3 \\ V_{3,1} z_1 + V_{3,2} z_2 \end{bmatrix}, \tag{7.202}$$

where $z = [z_1, z_2, z_3]^T$, and

$$V_\theta = \begin{bmatrix} 0 & 0 & -V_{3,1} \\ 0 & 0 & -V_{3,2} \\ V_{3,1} & V_{3,2} & 0 \end{bmatrix},$$

and the in-plane, shape design velocity is given by $V_{\Omega^s} = [V_1, \ V_2]^T$. Since the surface remains flat during the configuration change, the second-order derivatives of the transverse design velocity V_3 vanish. However, the higher-order derivatives of the shape design velocity do not vanish. Thus, the regularity problem on the boundary appears, as discussed in Section 7.3.2.

The semilinear form of loads shown in Figs. 7.6 and 7.7 is

$$\ell_\Omega(\bar{z}) = \iint_{\Omega^s} \bar{z}^{*T} f \, d\Omega^s, \tag{7.203}$$

where $f(x) = [f_1, \ f_2, \ f_3]^T$ are the complex phasors of the harmonic body force or external load on the structural component. For simplicity, the traction force along the structural boundary is not considered.

With respect to shape and configuration design variation, the explicit variation of the load semilinear form in (7.203) is obtained using (7.177), as

$$\begin{aligned} \ell'_V(\bar{z}) &= \iint_{\Omega^s} [\bar{z}^{*T} \nabla f V_{\Omega^s} + \bar{z}^{*T} f div V_{\Omega^s}] d\Omega^s \\ &+ \iint_{\Omega^s} [(\bar{z}_3^* f_1 - \bar{z}_1^* f_3) V_{3,1} + (\bar{z}_3^* f_2 - \bar{z}_2^* f_3) V_{3,2}] d\Omega^s. \end{aligned} \tag{7.204}$$

In (7.204), it is assumed that $f' = 0$. Since the first integral in (7.204) is the contribution from shape design, the vectors in the first integral are two-dimensional.

Interface

The coupling effect between the acoustic medium and the structural surface in harmonic motion is given by the sesquilinear forms in (2.78) and (2.79), which are the functionals in integral form over the interface Ω^s. Since it is assumed that the surface design component is flat in configuration design sensitivity analysis, the configuration design variation for the interface may be defined as an in-plane shape variation and a rotation of the surface. The explicit variations of interface sesquilinear forms with respect to the configuration design variation are obtained using (7.175) and (7.176), as

$$\chi'_V(p, \bar{z}) = \iint_{\Omega^s} p \bar{z}^{*T} [V_\theta n + n div(V_{\Omega^s})] d\Omega^s \tag{7.205}$$

and

$$\phi'_V(p, \bar{z}) = \omega^2 \iint_{\Omega^s} z^T \bar{p}^* [V_\theta n + n div(V_{\Omega^s})] d\Omega^s, \tag{7.206}$$

where $div(V_{\Omega^s}) = V_{1,1} + V_{2,2}$ for the structural component. If the flat surface structural component is in contact with the acoustic medium, and the local coordinate system fixed in the structure is used, then the normal vector n can be given as

$$n = [0, 0, -1]^T. \tag{7.207}$$

Using (7.207) and $V_\theta z$ in (7.202), (7.205) and (7.206) can be rewritten as

$$\chi'_V(p,\bar{z}) = \iint_{\Omega^s} p[\bar{z}_1^* V_{3,1} + \bar{z}_2^* V_{3,2} - \bar{z}_3^*(V_{1,1} + V_{2,2})]d\Omega^s \tag{7.208}$$

and

$$\phi'_V(p,\bar{z}) = \omega^2 \iint_{\Omega^s} \bar{p}^*[z_1 V_{3,1} + z_2 V_{3,2} - z_3(V_{1,1} + V_{2,2})]d\Omega^s. \tag{7.209}$$

As an alternative, the interface can be treated as the boundary of a volume, Ω^a. In this case, the variations of interface terms can be derived using the shape variation defined in Section 6.2, and, if (7.207) is used, the same results as (7.208) and (7.209) are obtained.

7.4.4 Analytical Example

The formulation of configuration design sensitivity in previous sections of this chapter can be illustrated with an example problem. The design sensitivity formula is derived using the adjoint variable method. The process and basic equations in this example are also applicable to general structural-acoustic systems.

Consider an acoustic cavity with a flexible wall, as shown in Fig. 5.23. As a design variation, the configuration of the system is changed, that is, the shape of the cavity and the shape and orientation of the panel are changed as shown in Fig. 7.22. The variational state equation of harmonic motion of this coupled system is given in (7.169). The panel is modeled using the plane elastic solid/plate design component, and the energy sesquilinear form $q_\Omega(\bullet,\bullet)$ is given in (7.200).

In this example, the acoustic performance measure is the acoustic pressure at $\boldsymbol{x} = \hat{\boldsymbol{x}}$, and can be written as

$$\psi_p = p(\hat{\boldsymbol{x}}) = \iiint_{\Omega^a} \delta(\boldsymbol{x} - \hat{\boldsymbol{x}})p\, d\Omega^a, \qquad \hat{\boldsymbol{x}} \in \Omega^a, \tag{7.210}$$

where $\delta(\bullet)$ is the Dirac delta measure, and Ω^a is the three-dimensional domain containing the acoustic medium. Equation (7.210) is a simple form of (7.163), the general form of an acoustic performance measure. Taking the variation of ψ_p,

$$\psi'_p = \dot{p}(\hat{\boldsymbol{x}}) = \iiint_{\Omega^a} \delta(\boldsymbol{x} - \hat{\boldsymbol{x}})\dot{p}\, d\Omega^a. \tag{7.211}$$

The adjoint equation is formulated using (7.185), the adjoint equation for the general form of the acoustic performance measure, as

$$\begin{aligned} &q_\Omega(\bar{\boldsymbol{\lambda}},\boldsymbol{\lambda}) + b_\Omega(\bar{\eta},\eta) - \chi_\Omega(\bar{\eta},\boldsymbol{\lambda}) - \phi_\Omega(\boldsymbol{\lambda},\eta) \\ &\quad = \iiint_\Omega \delta(\boldsymbol{x} - \hat{\boldsymbol{x}})\bar{\eta}\, d\Omega, \qquad \forall\{\bar{\boldsymbol{\lambda}},\bar{\eta}\} \in Q, \end{aligned} \tag{7.212}$$

where the adjoint response $\{\boldsymbol{\lambda}^*, \eta^*\}$ is calculated. Using the state response $\{\boldsymbol{z}, p\}$ and the adjoint response $\{\boldsymbol{\lambda}^*, \eta^*\}$, (7.192) provides the design sensitivity expression as

$$\psi'_p = \ell'_V(\boldsymbol{\lambda}) - [q'_V(\boldsymbol{z},\boldsymbol{\lambda}) + b'_V(p,\eta) - \chi'_V(p,\boldsymbol{\lambda}) - \phi'_V(p,\eta)], \tag{7.213}$$

where the variation of the load semi-linear form is given in (7.204), and the variations of the sesquilinear forms are given in (7.201), (7.194), (7.208), and (7.209). In (7.213), $\{\boldsymbol{z}, p\}$ is the solution to the variational state equation, (7.169), and $\{\boldsymbol{\lambda}^*, \eta^*\}$ is the solution to the adjoint equation, (7.212).

The other performance measure in this example is the displacement of the structure at point $\hat{\boldsymbol{x}}$. Its mathematical expression is

$$\psi_z = z_3(\hat{\boldsymbol{x}}) = \iint_{\Omega^s} \delta(\boldsymbol{x} - \hat{\boldsymbol{x}})z_3\, d\Omega^s, \tag{7.214}$$

where Ω^s is the two-dimensional domain of the panel. Equation (7.214) is a simple form of (7.166), which is the general form of a structural performance measure. Taking the variation of ψ_z obtains

$$\psi_z' = \dot{z}_3(\hat{\boldsymbol{x}}) = \iint_{\Omega^s} \delta(\boldsymbol{x} - \hat{\boldsymbol{x}})\dot{z}_3 \, d\Omega^s. \tag{7.215}$$

From the general form of adjoint (7.186), the corresponding adjoint equation is

$$\begin{aligned} q_\Omega(\bar{\boldsymbol{\lambda}}, \boldsymbol{\lambda}) + b_\Omega(\bar{\eta}, \eta) - \chi_\Omega(\bar{\eta}, \boldsymbol{\lambda}) - \phi_\Omega(\bar{\boldsymbol{\lambda}}, \eta) \\ = \iint_{\Omega^s} \delta(\boldsymbol{x} - \hat{\boldsymbol{x}})\bar{\lambda}_3 \, d\Omega^s, \quad \forall \{\bar{\boldsymbol{\lambda}}, \bar{\eta}\} \in Q, \end{aligned} \tag{7.216}$$

and from (7.193), the design sensitivity expression is

$$\psi_z' = \ell_V'(\boldsymbol{\lambda}) - [q_V'(\boldsymbol{z}, \boldsymbol{\lambda}) + b_V'(p, \eta) - \chi_V'(p, \boldsymbol{\lambda}) - \phi_V'(p, \eta)], \tag{7.217}$$

where the variations of the semilinear and sesquilinear forms are given by (7.204), (7.201), (7.194), (7.208), and (7.209). In (7.217), $\{\boldsymbol{z}, p\}$ is the solution to the variational state equation, (7.169), and $\{\boldsymbol{\lambda}^*, \eta^*\}$ is the solution to the adjoint equation, (7.216).

As previously indicated, essentially the state and adjoint equations, (7.169), (7.212), and (7.216), have the same form, except for their load terms on the right side. This provides efficiency in numerical implementation with FEA, that is, when a single finite element model is required to solve state and adjoint equations. Also, the design sensitivity formulas in (7.213) and (7.217) are identical, while the adjoint variables are obtained from the different adjoint equations, (7.212) and (7.216).

7.4.5 Numerical Example

For an example of numerical computation of configuration design sensitivity, consider again an acoustic cavity with a flexible wall, shown in Fig. 7.22. As indicated, the shape of the cavity will be changed. The edge *MN* translates to $M_\tau N_\tau$ without rotation and deformation, that is, it remains parallel to the x_2-axis without any change in length. As a result, the shape and orientation of the panel will be changed, along with the boundary of the acoustic volume in contact with the panel. A linear design velocity field $\boldsymbol{V}$ is assumed on the panel Ω^s, and the domain velocity field of the acoustic volume Ω^a is obtained using the boundary displacement method discussed in Section 13.3 of Chapter 13 [122], with $\boldsymbol{V}$ as the boundary velocity field. The distance $\boldsymbol{b}$ of movement of the edge *MN* of the plate is used as a design variable.

ABAQUS 4.9 [81] is used for direct frequency analyses of state and adjoint equations. Tables 7.9 and 7.10 show the results at load frequencies of 55 and 60 Hz. The central finite difference results with ±0.005 m perturbations of the design variable are used to check the accuracy of the predicted design sensitivity. Table 7.9 shows the results for harmonic acoustic performance measures in pascals (Pa) at the points A_1 = (0.5, 0.6, 0.) and A_2 = (0.5, 0.6, 1.5) from Fig. 7.22. Table 7.10 provides data for harmonic displacement performance measures in meters (m) at the points S_1 = (0.5, 0.6, 0.) and S_2 = (1/12, 0.2, 0.). In Tables 7.9 and 7.10, the real and imaginary parts of the complex phasors are denoted by R and I, respectively, and the magnitude is denoted by D, which is the harmonic response amplitude. Table 7.10 shows the design sensitivities of the velocity and acceleration amplitudes V and A, as well as the structural displacement. A good agreement is obtained between the design sensitivity predictions ψ' and the finite differences $\Delta\psi$.

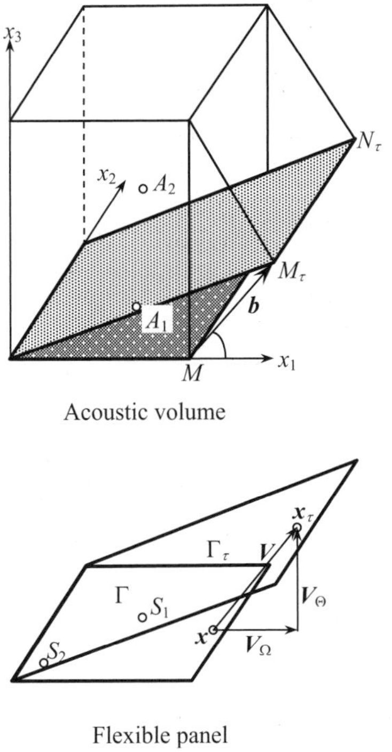

Figure 7.22. Design perturbation and points of measure.

Table 7.9. Configuration design sensitivity of cavity with flexible wall acoustic responses*.

Frequency	Location		$\psi(\boldsymbol{b}-\Delta\boldsymbol{b})$	$\psi(\boldsymbol{b}+\Delta\boldsymbol{b})$	$\Delta\psi$	ψ'	$\psi'/\Delta\psi(\%)$
		R	–.2354E+1	–.2342E+1	0.6060E–2	0.6723E–2	110.9
55 Hz	A_1	I	0.4320E+0	0.4424E+0	–.5231E–2	–.5300E–2	101.3
		D	0.2393E+1	0.2383E+1	–.5000E–2	–.5634E–2	112.7
		R	0.1029E+1	0.1109E+1	0.4006E–1	0.4027E–1	100.5
60 Hz	A_1	I	–.1671E+0	–.1853E+1	–.9064E–2	–.9074E–2	100.1
		D	0.1042E+1	0.1124E+1	0.4100E–1	0.4121E–1	100.5
		R	0.1105E+0	0.1180E+0	–.3727E–2	–.3672E–2	98.5
55Hz	A_2	I	–.2035E–1	–.2238E–1	0.1016E–2	0.1004E–2	98.8
		D	0.1124E+0	0.1201E+0	–.3850E–2	–.3794E–2	98.5
		R	0.9989E–1	0.1034E+0	0.1738E–2	0.1769E–2	101.7
60 Hz	A_2	I	–.1625E–1	–.1730E–1	–.5210E–3	–.5238E–3	100.5
		D	0.1012E+0	0.1048E+0	–.1800E–2	–.1830E–2	101.7

* Unit : Pascal (Pa).

Table 7.10. Configuration design sensitivity of cavity with flexible wall structural responses in x_3-direction*.

Frequency	Location		$\psi(b-\Delta b)$	$\psi(b+\Delta b)$	$\Delta\psi$	ψ'	$\psi'/\Delta\psi(\%)$
		R	0.5293E–5	0.5536E–5	–.1215E–6	–.1204E–7	99.1
		I	–.0989E–5	–.0997E–5	–.3820E–7	0.3807E–7	99.7
55 Hz	S_1	D	0.5385E–5	0.5638E–5	–.1265E–6	–.1254E–7	99.1
		V	0.1861E–2	0.1948E–2	–.4370E–4	–.4332E–4	99.1
		A	0.6431E+0	0.6733E+0	–.1510E–1	–.1497E–1	99.1
		R	0.4350E–5	0.4542E–5	0.9600E–7	0.9519E–7	99.2
		I	–.7244E–6	–.7778E–6	–.2667E–7	–.2638E–7	98.9
60 Hz	S_1	D	0.4410E–5	0.4609E–5	0.9910E–7	0.9825E–7	99.1
		V	0.1663E–2	0.1737E–2	0.3736E–4	0.3704E–4	99.1
		A	0.6268E+0	0.6550E+0	0.1408E–1	0.1396E–1	99.1
		R	0.5454E–6	0.5684E–6	–.1152E–7	–.1146E–7	99.5
		I	–.1001E–6	–.1075E–6	0.3705E–8	0.3656E–8	98.7
55 Hz	S_2	D	0.5545E–6	0.5785E–6	–.1200E–7	–.1194E–7	99.5
		V	0.1916E–3	0.1999E–3	–.4147E–5	–.4125E–5	99.5
		A	0.6622E+1	0.6909E+1	–.1433E–2	–.1426E–2	99.5
		R	0.4248E–6	0.4418E–6	–.8503E–8	–.8461E–8	99.5
		I	–.6745E–7	–.7211E–7	0.2331E–8	0.2314E–8	99.3
60 Hz	S_2	D	0.4301E–6	0.4477E–6	–.8765E–8	–.8721E–8	99.5
		V	0.1622E–3	0.1688E–3	–.3304E–5	–.3288E–5	99.5
		A	0.6113E–1	0.6362E–1	–.1246E–2	–.1240E–2	99.5

* Unit : Meter (m).

7.5 Configuration Design Theory for Curved Structure

In previous sections, variation in the configuration of the coupled system is decomposed into variations in the shape of the acoustic volume, and the shape and orientation of the structural components. The material derivative approach is employed to identify the effects of design variation. The configuration design sensitivity formulations in previous sections are limited to linear geometric perturbation, such that a line component remains straight and a surface component remains flat during the design change.

However, a structural component with curvature could be a more effective design. Thus, a more general theory of the configuration design is required for engineering applications. The purpose of this section is to develop a general configuration design sensitivity formulation that is not limited to linear geometric perturbation. Development of a new design sensitivity analysis method for the configuration design is required to avoid the difficulties caused by a C^0 linear design velocity field. Mathematical models with low-order derivatives can be used in the formulation of a design sensitivity analysis for beam and plate bending. This will require less regularity in design velocity field. Timoshenko beam and Mindlin/Reissener plate theories [123], for example, give second-order differential equations, and the corresponding variational equations only include first-order derivatives of rotation in the integrands.

7.5.1 Geometric Mapping

Although the sensitivity formulation in this section is more general than in previous sections, it is limited to the structures whose geometry can be mapped into a regular parametric domain. Consider a three-dimensional solid structure in domain $\Omega \subset R^3$ that is mapped into the reference domain Ω^r. A material point in domain Ω is denoted by $\boldsymbol{x}$, while its mapped point in the reference domain Ω^r is represented by ξ. It is assumed that a one-to-one mapping relation exists between $\boldsymbol{x}$ and ξ, such that the following notation is valid:

$$\boldsymbol{x} = \boldsymbol{x}(\boldsymbol{b};\xi), \tag{7.218}$$

where $\boldsymbol{x} = [x_1, x_2, x_3]^T$ and $\xi = [\xi_1, \xi_2, \xi_3]^T$. In (7.218), $\boldsymbol{b}$ denotes the design variable that determines the structural shape. Examples of mapping relations for line and surface components are presented in Section 12.2. Using this notation, the Jacobian matrix of the mapping can be represented by

$$\boldsymbol{J} = \left[\frac{\partial x_i}{\partial \xi_j}\right]. \tag{7.219}$$

If a one-to-one mapping relation is preserved in (7.218), the Jacobian matrix in (7.219) is nonsingular, and thus, its inverse $\boldsymbol{J}^{-1}$ exists. All thin structural components can be seen as degenerated forms from the solid component. Thus, the relations in (7.218) and (7.219) are valid for beam and shell components.

Design parameterization of the structural component is related to the perturbation of the material point in (7.218). Let the material point $\boldsymbol{x}(\boldsymbol{b};\ \xi)$ be perturbed to the new position $\boldsymbol{x}_\tau(\boldsymbol{b}+\tau\delta\boldsymbol{b};\ \xi)$. In such a design change, the design velocity field $\boldsymbol{V}(\xi)$ is defined by

$$\boldsymbol{V}(\xi) = \left.\frac{d\boldsymbol{x}_\tau(\boldsymbol{b}+\tau\delta\boldsymbol{b};\xi)}{d\tau}\right|_{\tau=0}. \tag{7.220}$$

Note that the reference coordinate ξ is independent of design perturbation. Since the material point $\boldsymbol{x}(\xi)$ is parameterized using the reference coordinate, the design velocity field $\boldsymbol{V}(\xi)$ is parameterized in the same way in (7.220).

The design velocity field in the three-dimensional space will be applied to structural components using the concept of degeneration. As will be shown with the following structural components, the design velocity field $\boldsymbol{V}(\xi)$ does not have to satisfy the geometric constraint that is required in the previous configuration design velocity field. The original geometry of the beam component, for example, can be perturbed into an arbitrary geometry with curvatures.

Shell Components

A surface geometry is often used to represent the shell component. More specifically, the neutral surface of the shell component is represented by a surface geometry, which uses two parametric coordinates (ξ_1, ξ_2). Thus, a complete three-dimensional coordinate is recovered by taking thickness as the third parametric coordinate. Material point $\boldsymbol{x}(\xi_1,\xi_2,\xi_3)$ of the shell component in Fig. 7.23 can be represented by

$$\boldsymbol{x}(\xi_1,\xi_2,\xi_3) = \boldsymbol{x}^n(\xi_1,\xi_2) + \xi_3\frac{t(\xi_1,\xi_2)}{2}\boldsymbol{n}(\xi_1,\xi_2), \tag{7.221}$$

where $\boldsymbol{x}^n(\xi_1,\xi_2)$ is the coordinate of the neutral surface, $t(\xi_1,\xi_2)$ is the shell thickness, and $\boldsymbol{n}(\xi_1,\xi_2)$ is the unit normal vector, which can be calculated from

$$\boldsymbol{n}(\xi_1,\xi_2) = \frac{\boldsymbol{x}^n_{,\xi_1}\times\boldsymbol{x}^n_{,\xi_2}}{\left\|\boldsymbol{x}^n_{,\xi_1}\times\boldsymbol{x}^n_{,\xi_2}\right\|}. \tag{7.222}$$

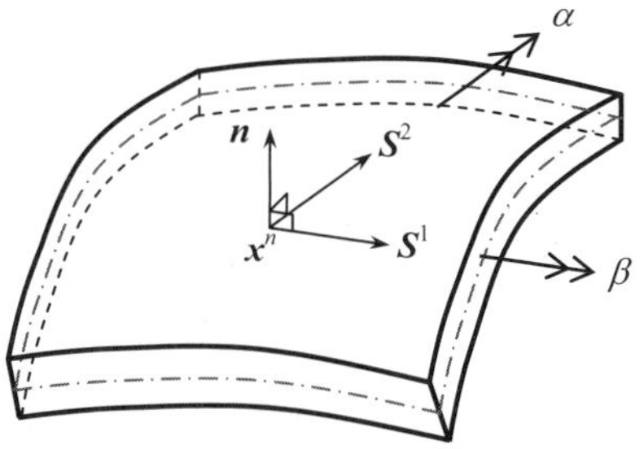

Figure 7.23. A parametric shell component.

The parameterization of $\boldsymbol{x}^n(\xi_1,\xi_2)$ is closely related to shape and configuration design parameterization. Examples of the parametric representation of the neutral surface are presented in Section 12.2.

From the parametric representation of shell component in (7.221), the design velocity field of the beam component can be obtained as

$$\boldsymbol{V}(\xi_1,\xi_2,\xi_3)=\boldsymbol{V}^n(\xi_1,\xi_2)+\xi_3\frac{\delta t}{2}\boldsymbol{n}(\xi_1,\xi_2), \tag{7.223}$$

where $\boldsymbol{V}^n(\xi_1,\xi_2)$ is the design velocity field at the neutral curve, and δt is the perturbation of the thickness. The first term on the right side of (7.223) represents the shape and configuration design variable, while the second term denotes the sizing design variable. The configuration design velocity field $\boldsymbol{V}^n(\xi_1,\xi_2)$ does not contain the limitation it had in Section 7.1. The curvature of the structure can change during the configuration design change.

Example 7.2. Design Velocity Using Geometric Surface. In order to represent the neutral surface $\boldsymbol{x}^n(\xi_1,\xi_2)$, consider the geometric surface given in Fig. 12.27 of Chapter 12. Using the variables (ξ_1,ξ_2), the material point on the neutral surface can be represented by

$$\boldsymbol{x}^n(\xi_1,\xi_2)=\boldsymbol{U}(\xi_1)^T\boldsymbol{M}\boldsymbol{G}(b)\boldsymbol{M}^T\boldsymbol{W}(\xi_2), \tag{7.224}$$

where $\boldsymbol{U}(\xi_1)=[\xi_1^3,\ \xi_1^2,\ \xi_1^1,\ 1]^T$ and $\boldsymbol{W}(\xi_2)=[\xi_2^3,\ \xi_2^2,\ \xi_2^1,\ 1]^T$. The 4×4 matrix $\boldsymbol{G}(b)$ has a three-dimensional vector as its component. The expression of matrix $\boldsymbol{G}(b)$ is presented in 12.20). The parameters $(\xi_1,\ \xi_2)$ are independent of the design, and matrix $\boldsymbol{M}$ is constant. Thus, from (7.220), design velocity at the neutral curve can be obtained as

$$\begin{aligned}\boldsymbol{V}^n(\xi_1,\xi_2)&=\left.\frac{d\boldsymbol{x}^n(b+\tau\delta b)}{d\tau}\right|_{\tau=0}\\&=\boldsymbol{U}(\xi_1)^T\boldsymbol{M}(\frac{\partial\boldsymbol{G}}{\partial b}\delta b)\boldsymbol{M}^T\boldsymbol{W}(\xi_2).\end{aligned} \tag{7.225}$$

For example, when the x-component of $\boldsymbol{p}_{00}$ is chosen as the design variable b, then matrix $\partial \boldsymbol{G}/\partial b$ has all zero components except for the component at (1,1), which has a value of [1, 0, 0]. In that case, the design velocity field $\boldsymbol{V}(\xi_1, \xi_2, \xi_3)$ obtained in (7.223) corresponds to one design variable b.

The advantage of the design velocity computation method in (7.225) is that the design variables can be defined on the geometric model. Design engineers often use a computer-aided design (CAD) tool, which models the structural geometry using a parametric representation. In such a case, the design engineer can choose the structural design variables from the geometric parameter.

7.5.2 Degenerated Shell Formulation

A variational formulation for the shell structure is obtained by degenerating the three-dimensional solid component in the continuum domain. Given the assumption of a constant transverse shear deformation, analytical integration can be performed through the thickness coordinate.

The strain tensor in linear elasticity can be expressed as

$$\varepsilon_{ij}(z) = \frac{1}{2}\left(\frac{\partial z_i}{\partial x_j} + \frac{\partial z_j}{\partial x_i}\right) = sym(z_{i,j}), \tag{7.226}$$

where $sym(\bullet)$ denotes the symmetric part of the tensor, and the subscript after the comma represents the derivative with respect to the spatial coordinate. If the parametric coordinate introduced in the previous section is used to represent the strain tensor in (7.226), then we have

$$\varepsilon_{ij}(z) = \frac{1}{2}\left(\frac{\partial z_i}{\partial \xi_m}\frac{\partial \xi_m}{\partial x_j} + \frac{\partial z_j}{\partial \xi_m}\frac{\partial \xi_m}{\partial x_i}\right) = sym\left(\frac{\partial z_i}{\partial \xi_m} J_{mj}^{-1}\right), \tag{7.227}$$

where the summation rule is used for the repeated indices.

The variational formulation for the structure can be obtained either from the principle of virtual work, or from the principle of minimum potential energy [124]. If we let the structural domain be $\Omega \subset R^3$, and the corresponding parametric domain be $\Omega^r \times R$, $\Omega^r \subset R^2$, then the energy bilinear form for linear elasticity can be obtained as

$$\begin{aligned} a_\Omega(z,\bar{z}) &= \iiint_\Omega \varepsilon_{ij}(\bar{z}) C_{ijkl} \varepsilon_{kl}(z)\, d\Omega \\ &= \iiint_{\Omega^r \times R} sym\left(\frac{\partial \bar{z}_i}{\partial \xi_m} J_{mj}^{-1}\right) C_{ijkl}\, sym\left(\frac{\partial z_j}{\partial \xi_n} J_{ni}^{-1}\right) |\boldsymbol{J}|\, d\xi_1 d\xi_2 d\xi_3, \end{aligned} \tag{7.228}$$

where $\bar{z}$ denotes the displacement variation or the virtual displacement, C_{ijkl} the fourth-order constitutive tensor, Ω^r the (ξ,η) plane, and $|\boldsymbol{J}|$ the determinant of the Jacobian in (7.219).

To further simplify the structural energy form, assume that displacement varies linearly in the thickness direction. This assumption yields a similar result as a Mindlin/Reissner shell formulation described in Section 3.1.3 [125] and [126], in which the flat cross section remains flat during deformation. Accordingly, the displacement is a linear function of ξ_3 and can be represented by the addition of two terms, as

$$z = z^1(\xi_1,\xi_2) + \xi_3 z^2(\xi_1,\xi_2), \tag{7.229}$$

where $z^1(\xi_1,\xi_2)$ represents the displacement of the neutral surface (the membrane deformation) and $\zeta z^2(\xi_1,\xi_2)$ denotes the rotation of the cross section (the bending and shear deformation). In the shear deformable curved shell structure in Fig. 7.23, the second term in (7.229) can be represented by two rotational deformations as

$$z^2(\xi_1,\xi_2) = \frac{t}{2}(\boldsymbol{S}^1\alpha - \boldsymbol{S}^2\beta), \tag{7.230}$$

where

$$\boldsymbol{S}^1 = \frac{\boldsymbol{x}^n_{,\xi_1}}{\left\|\boldsymbol{x}^n_{,\xi_1}\right\|} \tag{7.231}$$

$$\boldsymbol{S}^2 = \boldsymbol{n} \times \boldsymbol{S}^1, \tag{7.232}$$

and α and β are two rotational deformations in the $\boldsymbol{S}^2$ and $\boldsymbol{S}^1$ direction, respectively (see Fig. 7.23).

In addition, since the dimension of the thickness direction is much smaller than the dimensions of two tangential directions, Jacobian $\boldsymbol{J}$ can be presumed to be a function of only ξ_1 and ξ_2 coordinates. Accordingly, given the linear property of the engineering strain, the strain tensor in (7.227) can also be represented by the addition of two terms:

$$\varepsilon_{ij}(z) = \varepsilon^1_{ij}(z) + \xi_3 \varepsilon^2_{ij}(z), \tag{7.233}$$

where

$$\varepsilon^1_{ij}(z) = sym\left(\frac{\partial z^1_i}{\partial \xi_m} J^{-1}_{mj} + z^2_i J^{-1}_{3j}\right) \tag{7.234}$$

$$\varepsilon^2_{ij}(z) = sym\left(\frac{\partial z^2_i}{\partial \xi_m} J^{-1}_{mj}\right) \tag{7.235}$$

are the membrane-shear strain and bending strain, respectively. Unlike the Mindlin/Reissner plate formulation, the membrane and transverse shear strains are coupled in (7.234).

It is clear from the aforementioned assumptions that the energy bilinear form in (7.228) is a quadratic function of the parametric coordinate ξ_3. After analytically integrating (7.228) over the interval $\xi_3 \in [-1, 1]$, the energy bilinear form is simplified to

$$\begin{aligned} a_\Omega(z,\bar{z}) &= 2\iint_{\Omega^r} \varepsilon^1_{ij}(\bar{z}) C_{ijkl} \varepsilon^1_{kl}(z) \left|\boldsymbol{J}\right| d\Omega^r \\ &+ \tfrac{2}{3}\iint_{\Omega^r} \varepsilon^2_{ij}(\bar{z}) C_{ijkl} \varepsilon^2_{kl}(z) \left|\boldsymbol{J}\right| d\Omega^r. \end{aligned} \tag{7.236}$$

Note that the coupled terms of $\varepsilon^1_{ij}(z)$ and $\varepsilon^2_{ij}(z)$ vanish since they are odd functions of ξ_3. The degenerated energy bilinear form in (7.236) is still based on the continuum domain, and domain discretization has not yet been introduced. Since the analytical integration is already carried out over parametric coordinate ξ_3, only domain discretization of the neutral surface is necessary.

If a conservative system were considered, then the applied load would be independent of deformation. If $\boldsymbol{f}^B$ is the body force per unit volume, then the load linear form can be written as

$$\ell_{\Omega}(\bar{z}) = \iiint_{\Omega} \bar{z}^T \boldsymbol{f}^B \, d\Omega. \tag{7.237}$$

After introducing the parametric coordinate and integrating (7.237) along the ξ_3-axis, we obtain

$$\ell_{\Omega}(\bar{z}) = 2\iint_{\Omega^r} \bar{z}^T \boldsymbol{f}^B |\boldsymbol{J}| \, d\Omega^r. \tag{7.238}$$

For simplicity, the traction force is not considered in (7.238).

The structural equilibrium equation can be obtained from (7.236) and (7.238). For a given $\boldsymbol{f}^B$ and Γ_h, the structural variational equation is

$$a_{\Omega}(z, \bar{z}) = \ell_{\Omega}(\bar{z}), \qquad \forall \bar{z} \in Z. \tag{7.239}$$

The continuum form of variational equation (7.239) will be discretized in the following section using the meshfree method.

7.5.3 Material Derivative Formulas

Since the parametric coordinate is independent of design perturbation, the order of differentiation between the material derivative and parametric derivative can be interchanged. However, since the Jacobian matrix in (7.219) relates the physical coordinate to the parametric coordinate, it depends on the design. The material derivative of the Jacobian matrix can be obtained as

$$\left.\frac{d}{d\tau}\boldsymbol{J}_{\tau}\right|_{\tau=0} = \left.\frac{d}{d\tau}\left(\frac{\partial \boldsymbol{x}_{\tau}}{\partial \xi}\right)\right|_{\tau=0} = \frac{\partial \boldsymbol{V}}{\partial \xi}, \tag{7.240}$$

and the material derivative of its inverse can also be obtained, by using the fact of $\boldsymbol{J}\boldsymbol{J}^{-1} = \boldsymbol{I}$, as

$$\left.\frac{d}{d\tau}\boldsymbol{J}_{\tau}^{-1}\right|_{\tau=0} = \left.\frac{d}{d\tau}\left(\frac{\partial \xi}{\partial \boldsymbol{x}_{\tau}}\right)\right|_{\tau=0} = -\boldsymbol{J}^{-1}\frac{\partial \boldsymbol{V}}{\partial \boldsymbol{x}}. \tag{7.241}$$

Finally, the determinant of the Jacobian matrix depends on the design, whose material derivative can be obtained from direct calculation [127] as

$$\left.\frac{d}{d\tau}|\boldsymbol{J}_{\tau}|\right|_{\tau=0} = \frac{\partial V_i}{\partial \xi_i}|\boldsymbol{J}| = div\boldsymbol{V}|\boldsymbol{J}|. \tag{7.242}$$

A performance measure for the shell structure is usually defined on the neutral surface, which can be transformed into the parametric domain. After this transformation, a structural performance measure may be written in integral form as

$$\begin{aligned}\psi &= \iint_{\Omega} g(z,u) \, d\Omega \\ &= \iint_{\Omega^r} g(z,u)|\boldsymbol{J}| \, d\Omega^r.\end{aligned} \tag{7.243}$$

The function g is assumed to be continuously differentiable with respect to its arguments. The functional form of (7.243) represents a variety of structural performance measures. For example, the structural volume can be written with g depending only on u; the averaged stress over a subset of a shell can be written in terms of u and z; and the displacement at a point can be formally written using the Dirac delta measure and z in the integrand.

Since the parametric domain Ω^r is independent of the design perturbation, the integral in (7.243) is interchangeable with the design differentiation. Thus, differentiating the functional with respect to design u yields

$$\begin{aligned} \psi' &= \frac{d}{d\tau}\left[\iint_{\Omega^r} g(z_\tau(\boldsymbol{x}+\tau\boldsymbol{V}),\boldsymbol{b}+\tau\delta\boldsymbol{b})\left|\boldsymbol{J}_\tau\right| d\Omega^r\right]\Bigg|_{\tau=0} \\ &= \iint_{\Omega^r}(g_{,z}\dot{z}+g div\boldsymbol{V}+g_{,b}\delta\boldsymbol{b})\left|\boldsymbol{J}\right| d\Omega^r \\ &= \iint_{\Omega}(g_{,z}\dot{z}+g div\boldsymbol{V}+g_{,b}\delta\boldsymbol{b})\, d\Omega, \end{aligned} \tag{7.244}$$

where the design velocity $\boldsymbol{V}$ corresponds to the design $\boldsymbol{b}$. The chain rule of differentiation, along with the definition in (7.242), has been used to calculate the integrand in (7.244). The objective here is to obtain an explicit expression of ψ' in terms of $\delta\boldsymbol{b}$, which requires rewriting the first term under the integral on the right side of (7.244) explicitly in terms of $\delta\boldsymbol{b}$. Two methods are developed for this purpose, the direct differentiation and the adjoint variable method.

7.5.4 Direct Differentiation Method

The direct differentiation method computes the first integrand on the right side of (7.244) by directly computing $\dot{z}$ from the structural (7.239). Since this equation satisfies for all design ranges, we can differentiate it with respect to the design variable. To that end, the structural bilinear form is differentiated as

$$\frac{d}{d\tau}a_\Omega(z,\bar{z})\bigg|_{\tau=0} \equiv a_\Omega(\dot{z},\bar{z})+a'_V(z,\bar{z}), \tag{7.245}$$

where $a_\Omega(\dot{z},\bar{z})$ is the same as in (7.236) by substituting z into $\dot{z}$, and provides the implicitly dependent terms on the design through $\dot{z}$. $a'_V(z,\bar{z})$ represents the explicitly dependent terms on the design.

The explicit expression of $a'_V(z,\bar{z})$ depends on the shell formulation used in the structural problem, which will be derived as follows. The material derivative for the membrane-shear strain tensor can be obtained from its definition in (7.234) and from the formula in (7.241) as

$$\begin{aligned} \frac{d}{d\tau}(\varepsilon^1_{ij}(z))\bigg|_{\tau=0} &= \frac{d}{d\tau}sym\left(\frac{\partial z^1_i}{\partial\xi_m}J^{-1}_{mi}+z^2_iJ^{-1}_{3j}\right)\Bigg|_{\tau=0} \\ &= sym\left(\frac{\partial\dot{z}^1_i}{\partial x_j}+\dot{z}^2_iJ^{-1}_{3j}\right)-sym\left(\frac{\partial z^1_i}{\partial x_m}\frac{\partial V_m}{\partial x_j}+z^2_iJ^{-1}_{3m}\frac{\partial V_m}{\partial x_j}\right) \\ &\equiv \varepsilon^1_{ij}(\dot{z})+\varepsilon^{V1}_{ij}(z). \end{aligned} \tag{7.246}$$

In (7.246), $\varepsilon^1_{ij}(\dot{z})$ implicitly depends on the design through $\dot{z}$, while $\varepsilon^{V1}_{ij}(z)$ represents the explicitly dependent part that can be computable from both the given analysis result z and the design velocity $\boldsymbol{V}$. Similarly, the material derivative for the bending strain becomes

$$\begin{aligned} \frac{d}{d\tau}(\varepsilon^2_{ij}(z))\bigg|_{\tau=0} &= sym\left(\frac{\partial\dot{z}^2_i}{\partial x_j}\right)-sym\left(\frac{\partial z^1_i}{\partial x_m}\frac{\partial V_m}{\partial x_j}\right) \\ &\equiv \varepsilon^2_{ij}(\dot{z})+\varepsilon^{V2}_{ij}(z). \end{aligned} \tag{7.247}$$

By using (7.242), (7.246), and (7.247), the explicitly dependent term of the structural energy form $a'_V(z,\bar{z})$ can be calculated as

$$\begin{aligned} a'_V(z,\bar{z}) = {} & 2\iint_{\Omega^r} \varepsilon_{ij}^{V1}(\bar{z})C_{ijkl}\varepsilon_{kl}^{1}(z)\left|\boldsymbol{J}\right|d\Omega^r \\ & +2\iint_{\Omega^r} \varepsilon_{ij}^{1}(\bar{z})C_{ijkl}\varepsilon_{kl}^{V1}(z)\left|\boldsymbol{J}\right|d\Omega^r \\ & +2\iint_{\Omega^r} \varepsilon_{ij}^{1}(\bar{z})C_{ijkl}\varepsilon_{kl}^{1}(z)div\boldsymbol{V}\left|\boldsymbol{J}\right|d\Omega^r \\ & +\tfrac{2}{3}\iint_{\Omega^r} \varepsilon_{ij}^{V2}(\bar{z})C_{ijkl}\varepsilon_{kl}^{2}(z)\left|\boldsymbol{J}\right|d\Omega^r \\ & +\tfrac{2}{3}\iint_{\Omega^r} \varepsilon_{ij}^{2}(\bar{z})C_{ijkl}\varepsilon_{kl}^{V2}(z)\left|\boldsymbol{J}\right|d\Omega^r \\ & +\tfrac{2}{3}\iint_{\Omega^r} \varepsilon_{ij}^{2}(\bar{z})C_{ijkl}\varepsilon_{kl}^{2}(z)div\boldsymbol{V}\left|\boldsymbol{J}\right|d\Omega^r . \end{aligned} \tag{7.248}$$

Even though (7.248) looks complicated, every term appears systematically. Note that $a'_V(z,\bar{z})$ is linear in $\delta\boldsymbol{b}$.

The load linear form in (7.238) is also differentiable with respect to the design. More specifically, for design independent $\boldsymbol{f}^B$, the explicitly dependent term of the load linear form is

$$\begin{aligned} \frac{d}{d\tau}\ell_{\Omega}(\bar{z})\bigg|_{\tau=0} &= 2\iint_{\Omega^r} \bar{z}^T\boldsymbol{f}^B div\boldsymbol{V}\left|\boldsymbol{J}\right|d\Omega^r \\ &\equiv \ell'_V(\bar{z}). \end{aligned} \tag{7.249}$$

Since a conservative load is assumed, the variation of the load linear form does not have any implicitly dependent term. As in the case of the energy bilinear form, the variation of the load linear form is linear in $\delta\boldsymbol{b}$. If the concentrated, constant load is applied to the structure, then the variation of the load linear form in (7.249) vanishes.

For the direct differentiation method of design sensitivity analysis, one may take the variation of both sides of (7.239), and use (7.248) and (7.249) to obtain the design sensitivity equation, as

$$a_{\Omega}(\dot{z},\bar{z}) = \ell'_V(\bar{z}) - a'_V(z,\bar{z}), \qquad \forall\bar{z}\in Z. \tag{7.250}$$

Presuming that state variable z is known as the solution to (7.239), (7.250) is a variational equation for the first variation $\dot{z}$ and has the same energy bilinear form. Since (7.250) can be solved directly for $\dot{z}$, it is called the *direct differentiation method* compared with the adjoint variable method, which will be discussed in the next section. Noting that the right side of (7.250) is a linear form in $\bar{z}\in Z$, and that the energy bilinear form on the left side is Z-elliptic, (7.250) has the unique solution $\dot{z}$ [128]. The fact that there is a unique solution agrees with the previously stated observation that a design derivative exists for the solution to the state equation. After solving (7.250) for $\dot{z}$, the sensitivity of ψ can be calculated from (7.244).

7.5.5 Adjoint Variable Method

An adjoint variable method computes the implicitly dependent term, that is, the first integral on the right side of (7.244), by defining an adjoint equation. The adjoint equation is introduced by replacing $\dot{z}$ in (7.244) with a virtual displacement $\bar{\lambda}$, and by equating the terms involving $\bar{\lambda}$ in (7.244) to energy bilinear form $a(\lambda,\bar{\lambda})$, yielding the *adjoint equation* for the *adjoint variable* $\boldsymbol{\lambda}$

$$a_\Omega(\lambda,\bar{\lambda}) = \iint_{\Omega^r} g_{,z}\bar{\lambda}|J|\,d\Omega^r, \qquad \forall\bar{\lambda}\in Z, \tag{7.251}$$

where a solution $\lambda \in Z$ is desired. To take advantage of the adjoint equation, (7.251) may be evaluated at $\bar{\lambda} = \dot{z}$ since $\dot{z} \in Z$, to obtain

$$a_\Omega(\lambda,\dot{z}) = \iint_{\Omega^r} g_{,z}\dot{z}|J|\,d\Omega^r, \tag{7.252}$$

which is the term in (7.244) that is now needed in order to explicitly write it in terms of δu. Similarly, the identity of (7.250) may be evaluated at $\bar{z} = \lambda$, since both are in Z, to obtain

$$a_\Omega(\dot{z},\lambda) = \ell'_V(\lambda) - a'_V(z,\lambda). \tag{7.253}$$

Recalling that energy bilinear form $a_\Omega(\bullet,\bullet)$ is symmetrical in its arguments, the left side of (7.252) and (7.253) are equal, thus yielding the desired result

$$\iint_{\Omega^r} g_{,z}\dot{z}\,d\Omega^r = \ell'_V(\lambda) - a'_V(z,\lambda), \tag{7.254}$$

where the right side is linear in $\delta\boldsymbol{b}$, and can be evaluated once state variable z and adjoint variable λ are determined as solutions to (7.239) and (7.252), respectively. Substituting this result into (7.244), the explicit design sensitivity of ψ is

$$\psi' = \ell'_V(\lambda) - a'_V(z,\lambda) + \iint_{\Omega^r} (g\,div\boldsymbol{V} + g_{,u}\delta u)|\boldsymbol{J}|\,d\Omega^r, \tag{7.255}$$

where the form of the first two terms on the right depends on the specific problem under investigation.

These direct differentiation and adjoint variable methods provide unified design sensitivity analysis for the shell structure with respect to the sizing, shape, and configuration design variables, and using the concept of degeneration from the solid structure. The design velocity field is defined on the solid component level, and is then degenerated by following the same degeneration process for the shell structure.

7.5.6 Numerical Example

During the process of configuration and shape design change, the conventional finite element method often has a mesh distortion problem, that is, the regular mesh shape at the original design becomes distorted through the optimum design process, and the reliability of the analysis results is reduced for the new design. The meshfree methods that were developed recently can be used to relieve the mesh dependence of the analysis result. In the meshfree method, the structural domain is discretized, not by the finite element, but by a set of particles. The state variable is interpolated using the consistency condition between particles, and the domain integration is performed at each particle. In this section, a numerical example using the meshfree method is presented. Even though the meshfree method uses different interpolation and domain integration, the continuum-based design sensitivity theory in the previous sections can still be applied. For a detailed theory and application of the meshfree method, refer to Liu et al. [129] and references therein. For the meshfree discretization of the design sensitivity equation, refer to Kim et al. [130] and [131].

Consider a vehicle roof structure, as shown in Fig. 7.24. Only half of the roof structure is modeled using a single spline surface with a PATRAN geometric modeler [132]. A total of 347 meshfree particles are distributed on the surface, which corresponds to 1735 degrees-of-freedom. A linear elastic material property is assumed with a Young's modulus $E = 26{,}000$ MPa, and a Poisson's ratio $\nu = 0.3$. A constant thickness $t = 2$ mm is used. To evaluate the bending rigidity, three point loads are applied at each pillar location, as described in Fig. 7.24 with $f_1 = 100$ kN, $f_2 = -200$ kN, and $f_3 = 100$ kN.

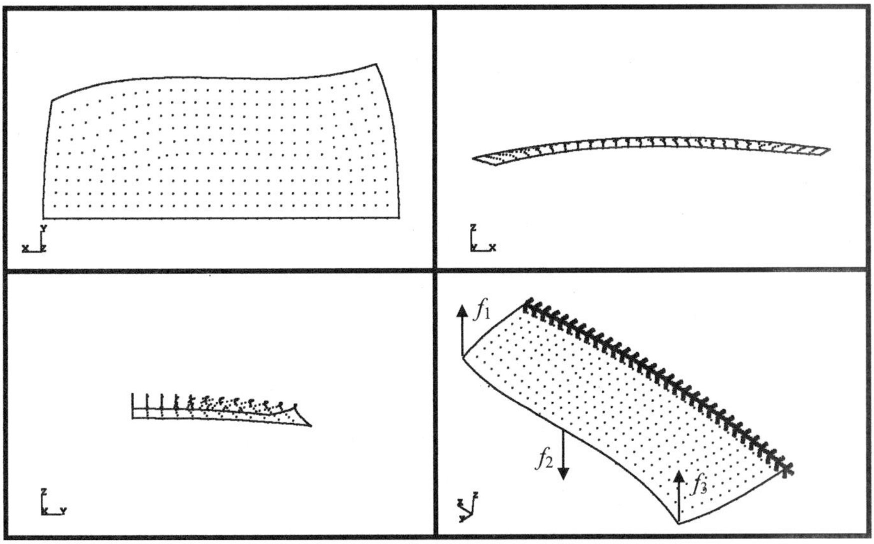

Figure 7.24. Meshfree discretization of a vehicle roof model.

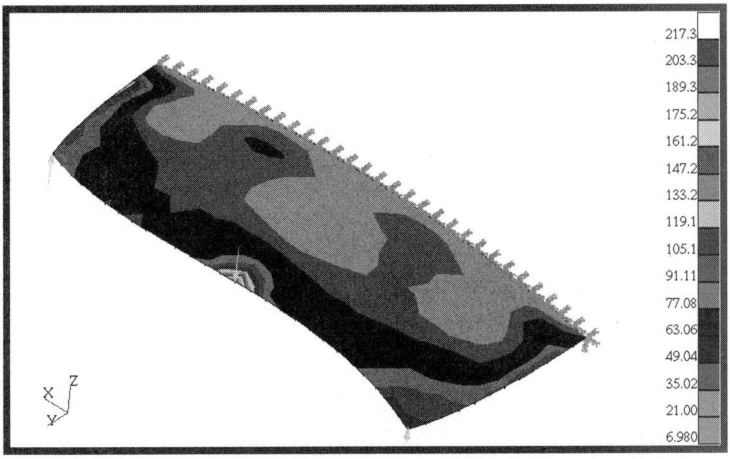

Figure 7.25. Meshfree analysis results (stress plot).

The first step in a meshfree analysis is to compute the nodal area. For a given set of particle distributions, a Voronoi diagram [133] can be used to compute the nodal area, which will act as the integration weight. The meshfree shape function is computed in the parametric domain by imposing the reproducing condition. The domain integration is carried out at each particle point in order to construct the stiffness matrix. After imposing the essential boundary condition, the linear matrix equation is solved by using the LAPACK package [134]. For design sensitivity analysis purposes, the factorized stiffness matrix is retained. Figure 7.25 plots the von Mises stress contour at the shell surface.

All components of the geometric matrix in (7.224) can be treated as shape/configuration design variables. In this specific example, the vertical movement of the tangent vectors is considered as a design variable, such that the curvature of the roof can be changed. Figure 7.26 illustrates the definitions of geometric matrix $\boldsymbol{G}$, while Table 7.11 shows the design variables of the roof structure. A total of eight design variables are chosen for design sensitivity analysis purposes, which include eight shape/configuration designs.

Table 7.11. Design parameterization.

ID	Design Description
1	Vertical movement of $\boldsymbol{p}_{00}^{\xi}$
2	Vertical movement of $\boldsymbol{p}_{10}^{\xi}$
3	Vertical movement of $\boldsymbol{p}_{01}^{\xi}$
4	Vertical movement of $\boldsymbol{p}_{11}^{\xi}$
5	Vertical movement of $\boldsymbol{p}_{00}^{\eta}$
6	Vertical movement of $\boldsymbol{p}_{10}^{\eta}$
7	Vertical movement of $\boldsymbol{p}_{01}^{\eta}$
8	Vertical movement of $\boldsymbol{p}_{11}^{\eta}$

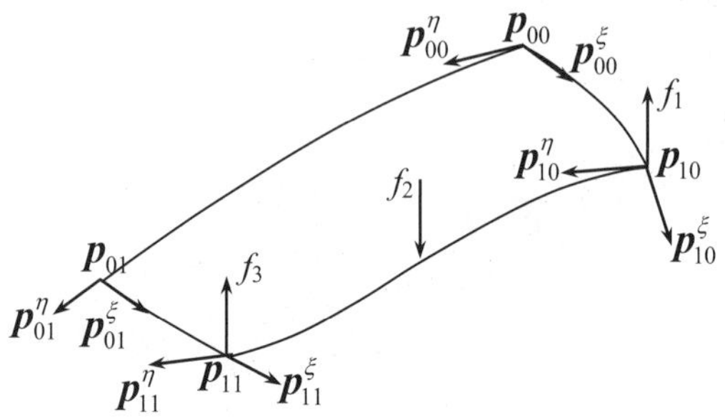

Figure 7.26. Design parameterization of the roof structure.

Table 7.12. Accuracy of sensitivity results

Type	ψ	$\Delta\psi$	ψ'	$(\Delta\psi/\psi')\times 100$
Volume	1.51033E+06	1.95212E–01	1.95180E–01	100.02
σ_{169}	1.34898E+02	9.43339E–03	9.43429E–03	99.99
σ_{182}	2.17291E+02	8.96848E–03	8.96976E–03	99.99
σ_{195}	1.35578E+02	7.27985E–03	7.28063E–03	99.99
σ_{17}	1.21676E+02	1.32524E–03	1.32552E–03	99.98
σ_{31}	1.09005E+02	2.88250E–03	2.88289E–03	99.99
σ_{301}	8.66971E+01	2.66191E–03	2.66156E–03	100.01
σ_{302}	7.34885E+01	1.72660E–03	1.72636E–03	100.01

(a) u_1, $\Delta\tau$ = 1.602E–02

Type	ψ	$\Delta\psi$	ψ'	$(\Delta\psi/\psi')\times 100$
Volume	1.51033E+06	2.08691E–01	2.08665E–01	100.01
σ_{169}	1.34898E+02	6.58233E–03	6.58289E–03	99.99
σ_{182}	2.17291E+02	7.80610E–03	7.80689E–03	99.99
σ_{195}	1.35578E+02	7.50906E–03	7.50947E–03	99.99
σ_{17}	1.21676E+02	2.64271E–04	2.64279E–04	100.00
σ_{31}	1.09005E+02	7.37644E–04	7.37657E–04	100.00
σ_{301}	8.66971E+01	4.77743E–03	4.77783E–03	99.99
σ_{302}	7.34885E+01	1.06057E–03	1.06138E–03	99.92

(b) u_3, $\Delta\tau$ = 1.354E–02

Type	ψ	$\Delta\psi$	ψ'	$(\Delta\psi/\psi')\times 100$
Volume	1.51033E+06	0.00000E+00	0.00000E+00	0.00
σ_{169}	1.34898E+02	–2.12311E–03	–2.12310E–03	100.00
σ_{182}	2.17291E+02	–3.52794E–03	–3.52793E–03	100.00
σ_{195}	1.35578E+02	–2.11374E–03	–2.11373E–03	100.00
σ_{17}	1.21676E+02	–1.37942E–04	–1.37941E–04	100.00
σ_{31}	1.09005E+02	1.94032E–05	1.94029E–05	100.00
σ_{301}	8.66971E+01	–1.72318E–04	–1.72319E–04	100.00
σ_{302}	7.34885E+01	–9.10049E–05	–9.10035E–05	100.00

(c) u_{10}, $\Delta\tau$ = 3.793E–03

Type	ψ	$\Delta\psi$	ψ'	$(\Delta\psi/\psi')\times 100$
Volume	1.51033E+06	0.00000E+00	0.00000E+00	0.00
σ_{169}	1.34898E+02	3.98915E–04	3.98709E–04	100.05
σ_{182}	2.17291E+02	6.15776E–04	6.15426E–04	100.06
σ_{195}	1.35578E+02	3.67303E–04	3.66965E–04	100.09
σ_{17}	1.21676E+02	–4.93757E–05	–4.93850E–05	99.98
σ_{31}	1.09005E+02	–2.60711E–05	–2.60801E–05	99.97
σ_{301}	8.66971E+01	3.88180E–04	3.87581E–04	100.15
σ_{302}	7.34885E+01	1.02153E–03	1.02096E–03	100.06

(d) u_{13}, $\Delta\tau$= 1.853E–05

Since the adjoint variable method is more efficient than the direct differentiation method for many design problems, the former method is used for this example. Eight performance measures are chosen that include the structural volume and the seven von Mises stresses. Since the volume performance measure does not require any adjoint solution, seven adjoint equations are solved by using a factorized stiffness matrix from structural analysis. Again, it is important to stress that the number of adjoint equations is related to the number of performance measures, rather than the number of design variables. The computational costs of design sensitivity analysis include the computation of adjoint load, the solution procedure of adjoint equation (7.251), and the evaluation of the sensitivity in (7.255). A very efficient design sensitivity computation is achieved, which is about 2.5% of the total structural analysis cost. This kind of efficiency can be obtained since the proposed method does not require any stiffness matrix derivative, so that the adjoint equation is solved simply, using a substitution of the factorized matrix.

The accuracy of the design sensitivity results is compared with the finite difference results in Table 7.12. The first column in Table 7.12 denotes the type of performance measure (volume and stresses), while the second column represents their values in the original design. It is well known that the selection of the perturbation size $\Delta\tau$ is a major difficulty in the finite difference method. Given the fact that the solution has an accuracy of at least six numerical digits, perturbation size is chosen such that the performance change $\Delta\psi$ is 10^{-6} times the performance value, as shown in the third column. The fourth column represents the predicted design change using the method proposed by (7.255). This predicted design change is compared with the finite difference $\Delta\psi$ in the last column. Very accurate sensitivity results are obtained, as shown throughout Table 7.12.

Appendix

A.1 Matrix Calculus Notation

In dealing with systems that are described by many variables, it is essential that a precise matrix calculus notation be employed. To explain the notation used in this text, let $\boldsymbol{x}$ be a k vector of real variables (i.e., $k \times 1$ column vector), $\boldsymbol{y}$ be an m vector of real variables, $a(\boldsymbol{x},\boldsymbol{y})$ be a scalar differentiable function of $\boldsymbol{x}$ and $\boldsymbol{y}$, and $\boldsymbol{g}(\boldsymbol{x},\boldsymbol{y}) = [g_1(\boldsymbol{x},\boldsymbol{y}), g_2(\boldsymbol{x},\boldsymbol{y}), \ldots, g_n(\boldsymbol{x},\boldsymbol{y})]^T$ be an n vector of differentiable functions of $\boldsymbol{x}$ and $\boldsymbol{y}$. Using i as the row index and j as the column index, define

$$a_{,\boldsymbol{x}} \equiv \frac{\partial a}{\partial \boldsymbol{x}} = \left[\frac{\partial a}{\partial x_j}\right]_{1\times k} \tag{A.1}$$

$$\boldsymbol{g}_{,\boldsymbol{x}} \equiv \frac{\partial \boldsymbol{g}}{\partial \boldsymbol{x}} = \left[\frac{\partial g_i}{\partial x_j}\right]_{n\times k} \tag{A.2}$$

$$a_{,\boldsymbol{xy}} \equiv \left[\frac{\partial^2 a}{\partial x_i \partial y_j}\right]_{k\times m} = \frac{\partial}{\partial \boldsymbol{y}}\left[\frac{\partial a^T}{\partial \boldsymbol{x}}\right] = \frac{\partial}{\partial \boldsymbol{y}}[a_{,\boldsymbol{x}}^T] = [a_{,\boldsymbol{x}}^T]_{,\boldsymbol{y}}. \tag{A.3}$$

Note that the derivative of a scalar function with respect to a vector variable in (A.1) produces a row vector. This is one of the few vector symbols in the text that is a row vector, rather than the more common column vector. In order to take advantage of this notation, it is important that the correct vector definition of matrix derivatives be used. Note also that the derivative of a vector function with respect to a vector variable in (A.2) produces a matrix. No attempt is made here to define the derivative of a matrix function with respect to a vector variable. Similarly, the second derivative of a scalar function with respect to a vector variable can be defined as in (A.3), but the second derivative of a vector function with respect to a vector variable is not defined.

As an example of the use of this matrix calculus notation, let $\delta\boldsymbol{x}$ and $\delta\boldsymbol{y}$ be small perturbations in $\boldsymbol{x}$ and $\boldsymbol{y}$. The total differential formula of calculus [50] gives

$$\begin{aligned} a(\boldsymbol{x}+\delta\boldsymbol{x}, \boldsymbol{y}+\delta\boldsymbol{y}) - a(\boldsymbol{x},\boldsymbol{y}) &\approx \delta a \\ &\equiv \sum_{j=1}^{k} \frac{\partial a}{\partial x_j}\delta x_j + \sum_{j=1}^{k} \frac{\partial a}{\partial y_j}\delta y_j \\ &= \frac{\partial a}{\partial \boldsymbol{x}}\delta\boldsymbol{x} + \frac{\partial a}{\partial \boldsymbol{y}}\delta\boldsymbol{y} \\ &= a_{,\boldsymbol{x}}\delta\boldsymbol{x} + a_{,\boldsymbol{y}}\delta\boldsymbol{y}. \end{aligned} \tag{A.4}$$

This is just one example of an application of matrix calculus which avoids cumbersome summation notation. Note that both the terms in (A.4) are scalars, since $a_{,\boldsymbol{x}}$ is a row vector and $\delta\boldsymbol{x}$ is a column vector. It is clear that

$$\delta a \neq \delta\boldsymbol{x} a_{,\boldsymbol{x}} + \delta\boldsymbol{y} a_{,\boldsymbol{y}}$$

since the left side is a scalar and the two terms on the right side are $k \times k$ and $m \times m$ matrices, respectively.

Similarly, matrix calculus extensions of ordinary calculus rules can be derived, such as the product rule of differentiation. For example, if $\boldsymbol{A}$ is an $n \times n$ constant matrix, then

$$\begin{aligned} \frac{\partial}{\partial \boldsymbol{x}}(\boldsymbol{A}\boldsymbol{g}(\boldsymbol{x},\boldsymbol{y})) &= \left[\frac{\partial}{\partial x_j}\left(\sum_{l=1}^{n} A_{il} g_l \right) \right] \\ &= \left[\sum_{l=1}^{n} A_{il} \frac{\partial g_l}{\partial x_j} \right] \\ &= \boldsymbol{A}\frac{\partial \boldsymbol{g}}{\partial \boldsymbol{x}} = \boldsymbol{A}\boldsymbol{g}_{,x}. \end{aligned} \tag{A.5}$$

A second example, which provides a result that might not be expected, involves two n-vector functions $\boldsymbol{h}(\boldsymbol{x},\boldsymbol{y})$ and $\boldsymbol{g}(\boldsymbol{x},\boldsymbol{y})$. By careful manipulation,

$$\begin{aligned} \frac{\partial}{\partial \boldsymbol{x}}(\boldsymbol{g}^T\boldsymbol{h}) &= \left[\frac{\partial}{\partial x_j}\left(\sum_{l=1}^{n} g_l h_l \right) \right] \\ &= \left[\sum_{l=1}^{n} \left(\frac{\partial g_l}{\partial x_j} h_l + g_l \frac{\partial h_l}{\partial x_j} \right) \right] \\ &= \left[\sum_{l=1}^{n} h_l \frac{\partial g_l}{\partial x_j} \right] + \left[\sum_{l=1}^{n} g_l \frac{\partial h_l}{\partial x_j} \right] \\ &= \boldsymbol{h}^T \frac{\partial \boldsymbol{g}}{\partial \boldsymbol{x}} + \boldsymbol{g}^T \frac{\partial \boldsymbol{h}}{\partial \boldsymbol{x}} \\ &= \boldsymbol{h}^T \boldsymbol{g}_{,x} + \boldsymbol{g}^T \boldsymbol{h}_{,x}. \end{aligned} \tag{A.6}$$

To see that (A.6) is reasonable, note that $\boldsymbol{g}^T\boldsymbol{h} = \boldsymbol{h}^T\boldsymbol{g}$, and that, in fact, interchanging $\boldsymbol{g}$ and $\boldsymbol{h}$ does not change either side of (A.6). Note also that what might have intuitively appeared to be the appropriate product rule of differentiation is not even defined, much less valid, that is,

$$\frac{\partial}{\partial \boldsymbol{x}}(\boldsymbol{g}^T\boldsymbol{h}) \neq \left(\frac{\partial \boldsymbol{g}}{\partial \boldsymbol{x}} \right)^T \boldsymbol{h} + \boldsymbol{g}^T \frac{\partial \boldsymbol{h}}{\partial \boldsymbol{x}}.$$

In boundary-value problems, derivatives with respect to the independent variable $\boldsymbol{x} \in R^3$ (or R^2) often arise. In these instances, it is convenient to use the gradient notation

$$\nabla a(\boldsymbol{x}) = \left[\frac{\partial a}{\partial x_1} \quad \frac{\partial a}{\partial x_2} \quad \frac{\partial a}{\partial x_3} \right]^T, \tag{A.7}$$

that is,

$$\nabla a = a_{,x}^T. \tag{A.8}$$

Very often in structural mechanics, quadratic forms $\boldsymbol{x}^T\boldsymbol{A}\boldsymbol{x}$ ($\boldsymbol{x} \in R^n$) arise, where $\boldsymbol{A}$ is an $n \times n$ constant matrix, presumed initially not to be symmetric. Using the foregoing definitions,

$$\frac{\partial}{\partial \boldsymbol{x}}(\boldsymbol{x}^T \boldsymbol{A}\boldsymbol{x}) = \left[\frac{\partial}{\partial x_i}\left(\sum_{j,k} x_k A_{kj} x_j \right) \right]$$
$$= \left[\sum_k x_k A_{ki} + \sum_j A_{ij} x_j \right] \qquad \text{(A.9)}$$
$$= \left[\sum_k x_k A_{ki} + \sum_j x_j A_{ji}^T \right]$$
$$= \boldsymbol{x}^T (\boldsymbol{A} + \boldsymbol{A}^T).$$

In particular, if $\boldsymbol{A}$ is symmetric,

$$\frac{\partial}{\partial \boldsymbol{x}}(\boldsymbol{x}^T \boldsymbol{A}\boldsymbol{x}) = 2\boldsymbol{x}^T \boldsymbol{A}. \qquad \text{(A.10)}$$

If a scalar valued function $a(\boldsymbol{x})$ ($\boldsymbol{x} \in R^n$) is twice continuously differentiable, the first-order approximation of (A.4) can be extended to the second-order. Using Taylor's formula [50], we have

$$a(\boldsymbol{x} + \delta\boldsymbol{x}) - a(\boldsymbol{x}) \approx \sum_i \frac{\partial a}{\partial x_i} \delta x_i + \frac{1}{2} \sum_i \sum_j \frac{\partial^2 a(\boldsymbol{x})}{\partial x_i \partial x_j} \delta x_i \delta x_j$$
$$= a_{,\boldsymbol{x}} \delta\boldsymbol{x} + \frac{1}{2} \delta\boldsymbol{x}^T a_{,\boldsymbol{xx}} \delta\boldsymbol{x}. \qquad \text{(A.11)}$$

A.2 Basic Function Spaces

The purpose of this section is to summarize the definitions and properties of function spaces used throughout the text. The mathematical validity of developments presented in the text rest upon fundamental results associated with these spaces, which in many cases are easy to prove. Basic ideas are discussed in this section to assist the engineer in understanding the nature of the spaces and their properties, with references to the literature given for proofs.

A.2.1 R^k; k-Dimensional Euclidean Space

The simplest space encountered in multidimensional analysis is k-dimensional Euclidean space, denoted here as R^k. This is actually a space of column matrices, rather than a function space. The space R^k is quite important in its own right and serves to introduce basic ideas of vector spaces and their properties, prior to the introduction of function spaces. The k-dimensional Euclidean space is defined as

$$R^k = \left\{ \boldsymbol{x} = [x_1 \quad \ldots \quad x_k]^T \mid x_i \text{ real, } i = 1, \ldots, k. \right\} \qquad \text{(A.12)}$$

Note that R^k is simply the collection of all $k \times 1$ matrices (column vectors) whose components are real numbers.

In order to be useful for analyses of finite dimensional structural systems, algebra must be defined on this space to allow for systematic manipulation. As with matrix notation, the addition of two vectors is defined as

$$\boldsymbol{x} + \boldsymbol{y} = [x_1 + y_1 \quad \ldots \quad x_k + y_k]^T, \qquad \text{(A.13)}$$

and multiplication of a vector $\boldsymbol{x}$ by a scalar α is defined as

$$\alpha \boldsymbol{x} = [\alpha x_1 \quad \ldots \quad \alpha x_k]^T. \tag{A.14}$$

These operations have the properties

$$\boldsymbol{x} + \boldsymbol{y} = \boldsymbol{y} + \boldsymbol{x} \tag{A.15}$$

$$(\boldsymbol{x} + \boldsymbol{y}) + \boldsymbol{z} = \boldsymbol{x} + (\boldsymbol{y} + \boldsymbol{z}). \tag{A.16}$$

There is a unique zero vector $\boldsymbol{0} = [0, 0, \ldots, 0]^T$ such that

$$\boldsymbol{0} + \boldsymbol{x} = \boldsymbol{x}, \tag{A.17}$$

and there is also a unique negative vector $-\boldsymbol{x}$ such that

$$\boldsymbol{x} + (-\boldsymbol{x}) = \boldsymbol{0}. \tag{A.18}$$

Additional properties of the operations are

$$\alpha(\boldsymbol{x} + \boldsymbol{y}) = \alpha \boldsymbol{x} + \alpha \boldsymbol{y} \tag{A.19}$$

$$(\alpha + \beta)\boldsymbol{x} = \alpha \boldsymbol{x} + \beta \boldsymbol{x} \tag{A.20}$$

$$\alpha(\beta \boldsymbol{x}) = (\alpha\beta)\boldsymbol{x} \tag{A.21}$$

$$1\boldsymbol{x} = \boldsymbol{x}, \tag{A.22}$$

where $\boldsymbol{x}$ and $\boldsymbol{y}$ are arbitrary vectors in R^k and α and β are arbitrary real constants.

The set of vectors R^k defined in (A.12), with the operations of addition and multiplication by a scalar defined by (A.13) and (A.14) that satisfy (A.15) through (A.22) constitute a *vector space*. As will be seen in Sections A.2.2 through A.2.6, sets of functions that have properties of addition and multiplication by a scalar also obey the properties of (A.15) through (A.22) and define a function space, which is a vector space. The value in such a definition is that functions may be dealt with using an algebra that parallels the arithmetic normally used in the manipulation of column vectors.

Having defined the algebra on the vector space R^k, it is now helpful to define the geometric properties that extend the usual ideas of scalar product and length of a physical vector. The *scalar product* of two vectors in R^k is defined as

$$(\boldsymbol{x}, \boldsymbol{y}) \equiv \boldsymbol{x}^T \boldsymbol{y} \tag{A.23}$$

Much in the same way as the properties of (A.15) through (A.22) for vector addition and multiplication by a scalar, it may be verified that the scalar product defined by (A.23) satisfies the following set of relations:

$$(\boldsymbol{x}, \boldsymbol{y}) = (\boldsymbol{y}, \boldsymbol{x}) \tag{A.24}$$

$$(\boldsymbol{x}, \boldsymbol{y} + \boldsymbol{z}) = (\boldsymbol{x}, \boldsymbol{y}) + (\boldsymbol{x}, \boldsymbol{z}) \tag{A.25}$$

$$(\alpha \boldsymbol{x}, \boldsymbol{y}) = \alpha(\boldsymbol{x}, \boldsymbol{y}) \tag{A.26}$$

$$(\boldsymbol{x},\boldsymbol{x}) \geq 0 \tag{A.27}$$

$$(\boldsymbol{x},\boldsymbol{x}) = 0 \text{ implies } \boldsymbol{x} = 0, \tag{A.28}$$

where $\boldsymbol{x}$, $\boldsymbol{y}$, and $\boldsymbol{z}$ are arbitrary vectors in R^k and α is an arbitrary scalar.

Having defined a scalar product of two vectors, the *norm* of a vector in R^k may be defined as

$$\|\boldsymbol{x}\| \equiv (\boldsymbol{x},\boldsymbol{x})^{1/2}. \tag{A.29}$$

It is not difficult to verify that the norm defined by (A.29) has the following properties:

$$\|\alpha \boldsymbol{x}\| = |\alpha| \|\boldsymbol{x}\| \tag{A.30}$$

$$\|(\boldsymbol{x},\boldsymbol{y})\| \leq \|\boldsymbol{x}\| \|\boldsymbol{y}\| \tag{A.31}$$

$$\|\boldsymbol{x}+\boldsymbol{y}\| \leq \|\boldsymbol{x}\| + \|\boldsymbol{y}\|, \tag{A.32}$$

where $\boldsymbol{x}$ and $\boldsymbol{y}$ are arbitrary vectors and α is an arbitrary scalar. The norm of a vector is the concept of length of a physical vector and allows for an extension of the idea of two vectors $\boldsymbol{x}$ and $\boldsymbol{y}$ being close to one another if the norm of their difference $\|\boldsymbol{x} - \boldsymbol{y}\|$ is small.

It is interesting to note that if the norm is defined by (A.29) in terms of a scalar product, it automatically has the properties of (A.30) through (A.32). There are situations in which a norm can be defined on a vector space without any scalar product. In such a case, an abstract norm is defined as a functional operating on a vector, having the properties of (A.30) through (A.32) and $\|\boldsymbol{x}\| > 0$ for all $\boldsymbol{x} \neq 0$. This last property follows automatically from the definition of (A.29), using the scalar product properties of (A.27) and (A.28). For the case in which no scalar product exists, this latter property must be verified in order to assure properties of the norm.

In addition to allowing for a definition in which two vectors are close, the norm can be used to define *convergence* of a sequence of vectors $\{\boldsymbol{x}^i\}$ (i=1, 2, …) in R^k as follows:

$$\lim_{i\to\infty} \boldsymbol{x}^i = \boldsymbol{x} \quad \text{if and only if} \quad \lim_{i\to\infty} \left\|\boldsymbol{x} - \boldsymbol{x}^i\right\| = 0. \tag{A.33}$$

The concept of convergence in R^k can be shown to be equivalent to convergence of individual components of the vector. This simple property, however, does not carry over to infinite-dimensional vector spaces, such as function spaces that are encountered in the study of boundary-value problems.

A sequence of vectors that cluster near one another as their index i increases is called a Cauchy sequence. More precisely, a sequence $\{\boldsymbol{x}^i\}$ is a *Cauchy sequence* if

$$\lim_{m,n\to\infty} \left\|\boldsymbol{x}^m - \boldsymbol{x}^n\right\| = 0. \tag{A.34}$$

A vector space for which every Cauchy sequence is convergent to a limit in the space is called a *complete vector space.* It is not difficult to show that R^k is a complete vector space under this definition. In fact, any vector space that is complete in the norm defined by a scalar product is called a *Hilbert space*. With this definition, R^k is a Hilbert space.

A *functional* is a mapping from a vector space to a real number. Examples of functionals on R^k include $\|\boldsymbol{x}\|$, and $(\boldsymbol{x}, \boldsymbol{y})$ for a given $\boldsymbol{y}$ in R^k. A functional ℓ is said to be a *linear functional* if

$$\ell(\boldsymbol{x}+\boldsymbol{y}) = \ell(\boldsymbol{x}) + \ell(\boldsymbol{y}) \tag{A.35}$$

$$\ell(\alpha\boldsymbol{x}) = \alpha\ell(\boldsymbol{x}) \tag{A.36}$$

for all $\boldsymbol{x}$ and $\boldsymbol{y}$ in R^k and all scalars α . A linear functional is said to be *bounded*, or *continuous* if a positive constant γ exists, such that

$$|\ell(\boldsymbol{x})| \le \gamma \|\boldsymbol{x}\| \tag{A.37}$$

for all $\boldsymbol{x}$ in R^k.

It is interesting to note that the functional $\|\boldsymbol{x}\|$ is not linear, as is easily verified using the properties of (A.30) through (A.32). It can be verified that the functional $\ell(\boldsymbol{x}) = (\boldsymbol{x}, \boldsymbol{y})$ for a fixed $\boldsymbol{y}$ in R^k is linear, using the properties of a scalar product given in (A.25) and (A.26). Using (A.31), it can also be seen as bounded.

One of the principal reasons that Hilbert spaces are valuable in structural analysis is that any bounded linear functional on a Hilbert space has a very special representation, defined by the *Reisz representation theorem*, that is, any bounded linear functional $\ell(\boldsymbol{x})$ on R^k can be represented as

$$\ell(\boldsymbol{x}) = (\boldsymbol{y}, \boldsymbol{x}) \tag{A.38}$$

for some vector $\boldsymbol{y}$ in R^k. The Reisz representation theorem guarantees the existence of the vector $\boldsymbol{y}$ associated with the bounded linear functional ℓ. While this theorem may not sound like a commonly used idea in mechanics, in fact it is. The concept of generalized force in mechanics follows from the Reisz representation theorem, in which the bounded linear functional ℓ is the virtual work associated with a virtual displacement $\boldsymbol{x}$, and the vector ℓ is defined as the generalized force of the system.

The rather obvious algebra, norm, and convergence properties of the finite-dimensional vector space R^k have been formalized in this section in some detail in order to prepare for the definition of similar properties in function spaces needed in the study of boundary-value problems. The reader unfamiliar with function spaces should recognize the similarity between operations and properties of function spaces, and the more intuitively clear properties of the finite-dimensional vector space R^k.

A.2.2 $C^m(\Omega)$; *m*-Times Continuously Differentiable Functions on Ω

Consider an open set Ω in R^k, with the closure $\bar{\Omega}$ in the norm of R^k. Considerations are limited in this section and in the text to bounded sets, since most structural applications occur on the bounded sets in R^1 through R^3. Bounded sets are sets of points whose distances from the origin are bounded by some finite constant. Restriction to bounded sets has the attractive property that every continuous function on a closed and bounded set in R^k is bounded.

The set of all *m-times continuously differentiable functions* on a set Ω is defined as the *function space*

$$C^m(\Omega) = \left\{ u(\boldsymbol{x}), x \in \Omega \subset R^k \,\middle|\, \frac{\partial^{|j|} u(\boldsymbol{x})}{\partial x_1^{j_1} \cdots \partial x_k^{j_k}} \text{ is continuous for } |\boldsymbol{j}| = 1, 2, \ldots, m \right\}, \tag{A.39}$$

where $\boldsymbol{j}$ is a vector of indices $\boldsymbol{j} = (j_1, \ldots, j_k)$ and $|\boldsymbol{j}| = \Sigma_{i=1}^k j_i$. For simplification in the following, the derivative $\partial^{|j|}u(\boldsymbol{x})/\partial x_1^{j_1}\cdots\partial x_k^{j_k}$ will be denoted simply as $\partial^{|j|}u(\boldsymbol{x})/\partial\boldsymbol{x}^j$. The space of m-times continuously differentiable functions on the closed set $\bar{\Omega}$ is simply defined by replacing Ω in (A.39) by $\bar{\Omega}$. The space $C^m(\Omega)$ is viewed at this point simply as the collection of all possible m-times continuously differentiable functions defined on the set Ω, with no concept of algebra or geometry defined.

To make use of the space of m-times continuously differentiable functions, it is essential to define the algebra on this space. Consider two m-times continuously differentiable functions u and v defined on Ω. The sum of these two functions is defined as

$$(u+v)(\boldsymbol{x}) \equiv u(\boldsymbol{x}) + v(\boldsymbol{x}), \tag{A.40}$$

which must hold for all $\boldsymbol{x} \in \Omega$, that is, the addition of functions is carried out in the natural way of adding their values at points in physical space. Similarly, a scalar α times a function u is defined as

$$(\alpha u)(\boldsymbol{x}) \equiv \alpha u(\boldsymbol{x}) \tag{A.41}$$

for all $\boldsymbol{x} \in \Omega$.

By defining the zero function as

$$0(\boldsymbol{x}) \equiv 0 \tag{A.42}$$

and the negative of a function as

$$(-u)(\boldsymbol{x}) \equiv -u(\boldsymbol{x}) \tag{A.43}$$

it is easy to show that properties of (A.15) through (A.22) follow for addition and multiplication of functions defined in (A.40) and (A.41). Before concluding that $C^m(\Omega)$ is a vector space, however, it must be demonstrated that given two functions u and v in the space and a scalar α, $u + v$ and αu are again in the space, that is, they are m-times continuously differentiable functions. This conclusion follows directly from the following elementary properties of differentiation:

$$\frac{\partial^{|j|}}{\partial\boldsymbol{x}^j}[(u+v)(\boldsymbol{x})] = \frac{\partial^{|j|}}{\partial\boldsymbol{x}^j}[u(\boldsymbol{x})+v(\boldsymbol{x})] = \frac{\partial^{|j|}u(\boldsymbol{x})}{\partial\boldsymbol{x}^j} + \frac{\partial^{|j|}v(\boldsymbol{x})}{\partial\boldsymbol{x}^j} \tag{A.44}$$

$$\frac{\partial^{|j|}(\alpha u)(\boldsymbol{x})}{\partial\boldsymbol{x}^j} = \frac{\partial^{|j|}\alpha u(\boldsymbol{x})}{\partial\boldsymbol{x}^j} = \alpha\frac{\partial^{|j|}u(\boldsymbol{x})}{\partial\boldsymbol{x}^j}. \tag{A.45}$$

Since the sum of two continuous functions and the product of a scalar times a continuous function are continuous, the space $C^m(\Omega)$ is closed under the operations of addition and multiplication by a scalar. It is therefore a vector space. The elements of this space may now be viewed as vectors in the same sense that column matrices are viewed as vectors in R^k. It should not be too surprising that this concept of a vector does not correlate completely with the physical idea of a vector in three-dimensional space as something with magnitude and direction, since for a k larger than 3, these concepts break down even for R^k.

It is possible to directly define a norm on the space $C^m(\Omega)$ as

$$\|u\|_{C^m} \equiv \max_{\substack{\boldsymbol{x}\in\Omega \\ 0\le|j|\le m}} \left|\frac{\partial^{|j|}u(\boldsymbol{x})}{\partial\boldsymbol{x}^j}\right|. \tag{A.46}$$

It can be verified that this is a norm with the properties given in (A.30) through (A.32), and that $\| u \|_{C^m} > 0$ if u $\neq$ 0. In fact, it can be shown that the space $C^m(\Omega)$ is complete in this norm, but that this norm is not generated by any scalar product. Therefore, the space $C^m(\Omega)$ is a complete vector space with a norm, but it is not a Hilbert space. Such spaces are called Banach spaces and have a rather rich mathematical theory. The distinction between Banach and Hilbert spaces, however, will not be required in the analysis presented in this text, since an adequate theory can be developed using Hilbert space properties almost exclusively.

A final space of continuously differentiable functions that is often encountered in applications is the space of functions having all derivatives continuously differentiable, that is,

$$C^\infty(\Omega) \equiv \left\{ u(\boldsymbol{x}), \boldsymbol{x} \in \Omega \middle| u \in C^m(\Omega) \text{ for all } m \right\}. \tag{A.47}$$

It is somewhat remarkable and nontrivial to prove that $C^\infty(\Omega)$ is dense in most of the function spaces dealt with in this text, many of which are composed of functions that have no continuous derivatives. To say that one space is dense in another means that the first space is a subset of the second and that every function in the second can be approximated arbitrarily closely in its own norm by a function in the first space.

A.2.3 $L^2(\Omega)$; The Space of Lebesgue Square Integrable Functions

The concept of the Lebesgue integral is a technical extension of the well-known Riemann integral introduced in basic calculus and used throughout the theory of structural mechanics. The distinction between the definitions of the two integrals is illustrated in Fig. A.l. In defining the Riemann integral of a function, the horizontal axis is partitioned by a grid of points and the sum of the areas of the rectangles shown in Fig. A.1(a) approximates the area beneath the curve defined by the function. It is shown mathematically that for certain classes of regular functions, as the spacing of the grid points approaches zero and hence approaches an infinite number of grid points on the horizontal axis, the sum of the areas converges and is defined as the value of the *Riemann integral.*

In contrast, the Lebesgue integral is defined by placing a grid of points on the vertical axis and drawing a set of horizontal lines that cut the graph of the function being integrated, as shown in Fig. A.1(b). The collection of subintervals on the horizontal axis is associated with a range of values of the function between y_i and y_{i+1}, and a lower bound

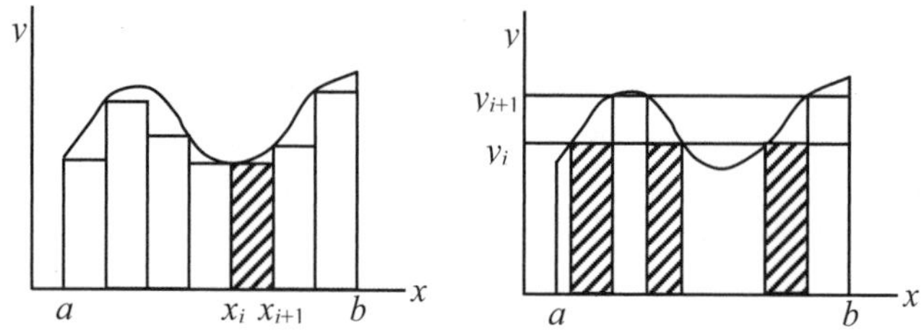

(a) Riemann Integral (b) Lebesgue Integral

Figure A.1. Integral of a function.

on the contribution of the area beneath the curve over these subintervals is calculated as y_i times the sum of the lengths of these intervals. Summing over all grid segments along the vertical axis yields a lower bound on the area beneath the curve defined by the function. A limit is then taken as the spacing of grid points on the vertical axis approaches zero. This limit, if it converges, is called the *Lebesgue integral* of the function and the function is declared to be Lebesgue integrable [35] and [135].

The value of the Lebesgue integral is equal to the value of the Riemann integral if the latter exists for the function under consideration. However, there are pathological functions that do not have Riemann integrals but which do have Lebesgue integrals. Therefore, the Lebesgue integral is an extension of the Riemann integral, with values coinciding for any function that has a Riemann integral. The mathematician exerts considerable effort in defining functions that have a Lebesgue but not a Riemann integral. For the purposes of this text, such studies in pathology are not necessary. The structural engineer should feel quite comfortable that virtually any function he encounters will have a Riemann integral, which must agree with the value of the Lebesgue integral.

Nevertheless, the power of the Lebesgue integral should not be dismissed, since it provides a powerful tool for establishing the mathematical properties of function spaces in which engineers regularly work. Of particular value are properties of the Lebesgue integral in which sequences of functions that are Lebesgue integrable and satisfy certain basic properties have limits that are also Lebesgue integrable. It is shown in the mathematical literature that many sequences of functions that have Riemann integrals either fail to converge or converge to functions for which the Riemann integral is not defined. Thus, if the completeness of function spaces is of concern, then the Lebesgue integral is an essential tool. In particular, using the principle of minimum total potential energy of structural mechanics, Lebesgue integration theory can predict exactly what properties the minimizing function should be expected to have, thereby defining the mathematical properties of solutions to mechanics problems. This is particularly important in structural mechanics, where minimizing sequences are often defined for total potential energy, that is, functions that yield successively lower values of the total potential energy. It is desirable that such minimizing sequences converge and give solutions to the structural problem. Using the theory of Lebesgue integration and associated function spaces, the mathematician has proved that such sequences do converge, and in fact has provided a clear definition of mathematical properties for the solutions.

Lest the engineer dismiss all this as mathematical formality, it is wise to reflect on the fact that limits of minimizing sequences exist in structural analysis and have well-defined mathematical properties. However, if the engineer is seeking to optimize the design of a structure, a minimizing sequence of designs may be obtained, each of which is regular and physically meaningful, and it may be discovered that the limiting function falls outside the class of designs of interest. This dilemma is of very real practical concern if the engineer seeks to use an optimality criterion for discovering the optimum designs. It is well known in the structural optimization literature that certain problems, such as finding the optimum thickness variation for a plate, may lead to a solution that involves an infinite number of infinitesimal ribs, which perhaps approximate a fiber composite structure. Thus, the solution to the plate optimization problem does not exist in the class of smooth thickness distributions. If the engineer writes down the necessary conditions of optimality that would have to hold if there were a smooth solution and attempts to find an optimum design based on these necessary conditions, a surprise is forthcoming since no solution exists.

Without going into a detailed treatment of Lebesgue integration theory, it is still possible to provide an intuitive introduction to technical results that are obtainable with Lebesgue integration theory. For example, the space of *Lebesgue square integrable functions* may be defined as

$$L^2(\Omega) \equiv \left\{ u(\boldsymbol{x}), \boldsymbol{x} \in \Omega \middle| \iint_\Omega u(\boldsymbol{x})^2 \, d\Omega < \infty \right\}, \tag{A.48}$$

where the integral over Ω is the Lebesgue integral, which, as noted above, coincides with the Riemann integral when it exists.

It is possible in this space to define a *scalar product* as the integral of the product of two functions, that is,

$$(u, v)_{L^2(\Omega)} \equiv \iint_\Omega u(\boldsymbol{x}) v(\boldsymbol{x}) \, d\Omega, \tag{A.49}$$

where the integral is in the Lebesgue sense. Using Lebesgue integration theory, it is possible to show that properties given by (A.24) through (A.28) are valid [135]. Therefore, a natural *norm* is defined on this space as

$$\|u\|_{L^2(\Omega)} \equiv (u, u)^{\frac{1}{2}}_{L^2(\Omega)} = \left[\iint_\Omega u(\boldsymbol{x})^2 \, d\Omega \right]^{1/2}, \tag{A.50}$$

which automatically satisfies the properties of (A.30) through (A.32), in particular the important inequality known as the *Schwartz inequality,* given as

$$\left| (u, v)_{L^2(\Omega)} \right| \le \|u\|_{L^2(\Omega)} \|v\|_{L^2(\Omega)} . \tag{A.51}$$

The reader who has studied the Fourier series will recognize these ideas as providing the foundation for the theory of construction of series approximations of functions and their convergence properties.

Using properties of the Lebesgue integral, it is shown that the space $L^2(\Omega)$ is *complete* [135], that is, Cauchy sequences in the L^2 norm converge to square integrable functions. Since the space $L^2(\Omega)$ has a scalar product, it is a *Hilbert space* and has all the desirable properties of Hilbert spaces.

Consider the functional

$$\ell(u) = \iint_\Omega f(\boldsymbol{x}) u(\boldsymbol{x}) \, d\Omega \tag{A.52}$$

defined by a given function f in $L^2(\Omega)$. For any function u in $L^2(\Omega)$, the product of f and u is Lebesgue integrable, and the right side of (A.52) creates a real number. Therefore, $\ell(u)$ is a functional. To see that this is a linear functional, standard properties of integration yield

$$\ell(\alpha u) = \iint_\Omega \alpha f u \, d\Omega = \alpha \iint_\Omega f u \, d\Omega = \alpha \ell(u) \tag{A.53}$$

$$\ell(u + v) = \iint_\Omega f(u + v) \, d\Omega = \iint_\Omega f u \, d\Omega + \iint_\Omega f v \, d\Omega = \ell(u) + \ell(v). \tag{A.54}$$

To see that the functional is bounded, the Schwartz inequality of (A.51) may be applied to obtain

$$|\ell(u)| = \left| (f, u)_{L^2(\Omega)} \right| \le \|f\|_{L^2(\Omega)} \|u\|_{L^2(\Omega)} . \tag{A.55}$$

Thus, the scalar product of an arbitrary function u with a fixed function f in $L^2(\Omega)$ [that is, the right side of (A.52)] defines a bounded linear functional on $L^2(\Omega)$.

Since $L^2(\Omega)$ is a Hilbert space, the Reisz representation theorem guarantees that every bounded linear functional in the space can be represented as the scalar product of u with some function in the space, that is, every linear functional $\ell(u)$ can be written in the form

$$\ell(u) = (g,u)_{L^2(\Omega)} \tag{A.56}$$

for some function g in $L^2(\Omega)$.

A.2.4 $L^\infty(\Omega)$; Space of Essentially Bounded, Lebesgue-Measurable Functions

In Lebesgue integration theory, the *measure* of a set (its length in R^1, area in R^2, or volume in R^3) is defined for very general sets of points. Sets whose measure is zero (for example, sets of discrete points, line segments in R^2 or R^3, and plane segments in R^3), play key rolls in analysis. A function that has a property that holds everywhere in the space except on a *set of measure zero* is said to have that property almost everywhere (abbreviated a.e.). Functions in spaces such as $L^2(\Omega)$ are defined based on properties that are expressed in terms of integral relations. Their values at discrete points do not influence the integrals. Hence, such functions may have irregular properties at discrete points or on sets of measure zero.

As an extension of a collection of integrable functions that are bounded by some finite constant, *essentially bounded functions* are defined as

$$L^\infty(\Omega) = \left\{ u(\boldsymbol{x}), \boldsymbol{x} \in \Omega \,\middle|\, |u(\boldsymbol{x})| \le k < \infty, \text{ a.e. in } \Omega \right\}. \tag{A.57}$$

A *norm* on $L^\infty(\Omega)$ may be defined as

$$\|u\|_{L^\infty(\Omega)} = \inf\left\{ K \,\middle|\, |u(\boldsymbol{x})| \le K \ \text{ a.e. in } \Omega \right\}, \tag{A.58}$$

where the term inf indicates the least upper bound. It is shown in Lebesgue integration theory [135] that this defines a norm on $L^\infty(\Omega)$ and the space is complete in this norm, that is, Cauchy sequences in this norm converge to functions in the space. It is also shown that it is impossible to define a scalar product on this space; hence, the space is not a Hilbert space, even though it is a Banach space.

Note that for a bounded set Ω, $C^m(\bar{\Omega})$ is a subset of $L^\infty(\Omega)$. However, piecewise-continuous functions are also in $L^\infty(\Omega)$. What makes $L^\infty(\Omega)$ so valuable in considering design problems is that minimizing sequences of functions that define mechanical properties, such as the cross-sectional area of a beam or the thickness of a plate, have the property that if they converge in the space $L^\infty(\Omega)$, they remain essentially bounded, which is a physical property that must be preserved. Once such a limiting function is defined, it can be modified only on a set of measure zero to cause it to be finite everywhere.

A.2.5 $H^m(\Omega)$; Sobolev Space of Order m

Because strain energies in structural components are written as integrals of quadratic expressions in the first or second derivatives of displacement fields, and because strain energy must be finite for any physically meaningful displacement field, it is natural to define spaces of functions that can be displacement fields in such a way that strain energy is guaranteed to be finite. Since the derivatives of displacement fields define strain, and strain must be integrable, the regularity of such functions must at least allow for the evaluation of strain energy. These considerations then make it natural to define a *Sobolev space of order m* as

$$H^m(\Omega) \equiv \left\{ u \in L^2(\Omega) \middle| \frac{\partial^{|\boldsymbol{k}|} u}{\partial \boldsymbol{x}^{\boldsymbol{k}}} \in L^2(\Omega),\ |\boldsymbol{k}| \leq m \right\}. \tag{A.59}$$

Such a space may be considered as a space of candidate displacement fields in elasticity for $m = 1$ and for displacement of a beam or plate with $m = 2$.

A *scalar product* may be defined on this Sobolev space as

$$(u, v)_{H^m(\Omega)} \equiv \sum_{|\boldsymbol{k}| \leq m} \iint_\Omega \frac{\partial^{|\boldsymbol{k}|} u}{\partial \boldsymbol{x}^{\boldsymbol{k}}} \frac{\partial^{|\boldsymbol{k}|} v}{\partial \boldsymbol{x}^{\boldsymbol{k}}} d\Omega. \tag{A.60}$$

It is reasonably direct to show that this bilinear functional has the properties of (A.24) through (A.28) and is therefore a scalar product [22]. A *norm* on the Sobolev space can therefore be naturally defined as

$$\|u\|_{H^m(\Omega)} \equiv \left[\sum_{|\boldsymbol{k}| \leq m} \iint_\Omega \left(\frac{\partial^{|\boldsymbol{k}|} u}{\partial \boldsymbol{x}^{\boldsymbol{k}}} \right)^2 d\Omega \right]^{1/2}. \tag{A.61}$$

It is proved in the literature on Sobolev space [22] that an equivalent definition of the Sobolev space can be given in terms of Cauchy sequences of functions in $\left\{ u \in C^m(\Omega) \middle| \|u\|_{H^m(\Omega)} < \infty \right\}$ as follows:

$$\begin{aligned} H^m(\Omega) = \big\{ u \,\big|\, & \text{for some Cauchy sequences } \{\phi^i\} \text{ in } \{u \in C^m(\Omega) \mid \|u\|_{H^m(\Omega)} < \infty\}, \\ & \lim_{i \to \infty} \|\phi^i - u\|_{H^m(\Omega)} = 0 \big\}. \end{aligned} \tag{A.62}$$

Thus,

$$\left\{ u \in C^m(\Omega) \middle| \|u\|_{H^m(\Omega)} < \infty \right\}$$

is *dense* in $H^m(\Omega)$. It is also shown in the literature [22] that $H^m(\Omega)$ is *complete,* hence it is a Hilbert space.

Since convergence of a sequence of functions in the $H^m(\Omega)$ norm involves $L^2(\Omega)$ convergence of derivatives up through order m, it appears reasonable that such convergence should preserve m derivatives of the limit function. As will be seen later, this is indeed the case and provides a natural setting for the study of boundary-value problems using modern variational techniques.

A.2.6 $H_0^m(\Omega)$; Sobolev *m*-Space with Compact Support

A function $u(\boldsymbol{x})$ is said to have *compact support* on Ω if there is a compact set $S \subset \Omega$ such that $u(\boldsymbol{x}) = 0$ for $\boldsymbol{x} \notin S$. Much as in the alternative definition of Sobolev space in (A.62), a new space may be defined as a similar limit of Cauchy sequences of functions that have compact support, that is,

$$\begin{aligned} H_0^m(\Omega) = \big\{ u \in H^m(\Omega) \,\big|\, & \text{for some Cauchy sequences } \{\phi^i\} \text{ of } C^\infty(\Omega) \\ & \text{functions with compact support } \lim_{i \to \infty} \|\phi^i - u\|_{H^m(\Omega)} = 0 \big\}. \end{aligned} \tag{A.63}$$

Since it might be expected that limits of functions in Sobolev space preserve properties of derivatives, as noted above, functions and some of their derivatives that appear in $H_0^m(\Omega)$ should be zero on the boundary of Ω. It will be shown later that this is true.

A.2.7 The Sobolev Imbedding Theorem

Although the proof is not easy, it is shown in the literature [22] that if Ω is a bounded domain in R^n with a smooth boundary and if $2m > n$, then

$$H^{j+m}(\Omega) \subset C^j(\bar{\Omega}). \tag{A.64}$$

Furthermore, identity mapping from $H^{j+m}(\Omega)$ to $C^j(\bar{\Omega})$ is continuous, that is, constants $K_j < \infty$ exist, such that for all u in $H^{j+m}(\Omega)$,

$$\|u\|_{C^j(\bar{\Omega})} \le K_j \|u\|_{H^{j+m}(\Omega)}. \tag{A.65}$$

This theorem gives valuable information concerning properties of functions in Sobolev spaces. In particular, it was noted earlier that functions defined as limits of sequences in the L^2 norm need not have finite values at isolated points. The Sobolev imbedding theorem, however, guarantees that in Sobolev spaces these functions are continuous and in many cases continuously differentiable due to the introduction of L^2-norm convergence of the derivatives of such functions in the Sobolev norm.

As an example, consider the displacement of a string on the interval [0,1] in R^1. To assure finite strain energy, it must be in $H^1(0,1)$. According to the Sobolev imbedding theorem, (A.64) guarantees that

$$H^1(0,1) \subset C^0[0,1] \tag{A.66}$$

and boundary conditions such as $u(0) = u^0$ and $u(1) = u^1$ will be preserved in the convergence of sequences of functions in $H^1(0,1)$.

Similarly, in the case of a beam on the interval [0,1], finiteness of strain energy demands that displacement functions be in $H^2(0,1)$. Thus, following the Sobolev imbedding theorem,

$$H^2(0,1) \subset C^1[0,1]. \tag{A.67}$$

Thus, admissible beam displacements must be continuously differentiable, and boundary conditions of the form $u(0) = u^0$ and $(du/dx)(0) = u'^0$ will be preserved if the limits of sequences of such functions are taken in the H^2 norm.

If $2m > n$ and if $u \in H_0^{j+m}(\Omega)$, a $C^j(\bar{\Omega})$ limit of smooth functions are zero on the boundary Γ of Ω. Thus,

$$\frac{\partial^{|k|} u}{\partial x^k} = 0 \quad k \le j \text{ on } \Gamma. \tag{A.68}$$

For example, if $u \in H_0^2(0,1)$, then since

$$H_0^2(0,1) \subset H^2(0,1) \subset C^1[0,1], \tag{A.69}$$

u must be a $C^1[0,1]$ limit of functions that are zero on the boundary. Hence,

$$u(0) = u(1) = \frac{du(0)}{dx} = \frac{du(1)}{dx} = 0. \tag{A.70}$$

A.2.8 Trace Operator

Projecting a function defined on the interior of a set Ω to its boundary Γ is the process of evaluating the function on the boundary, if the function has a regular extension to that boundary. In general, such a projection is called the *trace* of the function. In particular, for $u \in H^m(\Omega)$, the trace is defined as

$$\gamma u = [\gamma_0(u) \quad \cdots \quad \gamma_{m-1}(u)] \quad \text{on } \Gamma, \tag{A.71}$$

that is, it contains the projection of the function and its first m–1 derivatives to the boundary Γ of Ω, where $\gamma_j(u) = \partial^j u / \partial \boldsymbol{n}^j$ and $\boldsymbol{n}$ is the outward normal to Γ.

The nature of functions projected onto the boundary is somewhat more complicated than has been encountered in spaces of functions on the domain Ω. In particular, it is shown in the literature [22] that γ is a mapping from $H^m(\Omega)$ to a product space (see Section A.2.9) of boundary values of the function, which are *fractional-order Sobolev spaces* on the boundary, that is,

$$\gamma:\ H^m(\Omega) \to \prod_{j=1}^{m-1} H^{m-1-1/2}(\Gamma). \tag{A.72}$$

Due to the technical complexity associated with even defining the fractional-ordered spaces on the boundary, no attempt is made here to describe these spaces (see Adams [22]). This theory, however, makes precise the regularity properties required of functions appearing in boundary conditions of boundary-value problems [16].

Of specific interest is the anticipated result that boundary evaluations of functions appearing in $H_0^m(\Omega)$ are zero. In fact, it is shown that every function in $H_0^m(\Omega)$ is of this kind, that is,

$$H_0^m(\Omega) = \left\{ u \in H^m(\Omega) \middle| \ \gamma u = 0 \right\}. \tag{A.73}$$

Thus, the space $H_0^m(\Omega)$ is exactly the space of candidate solutions of Dirichlet boundary-value problems in which homogeneous boundary conditions are specified for a differential operator equation of order $2m$, to include zero values of the function and its first m–1 derivatives on the boundary. This precisely defines the space of candidate solutions of such a boundary-value problem and provides substantial information on the nature of solutions.

A.2.9 Product Spaces

As a final topic in considering function spaces, it is helpful to define a function space whose elements are groupings of functions of a very different character. For example, consider two function spaces denoted by X and Y. Their *product space* is defined as the collection of all pairs of functions, one from X and one from Y, as

$$X \times Y = \left\{ [u, v] \middle| \ u \in X, v \in Y \right\}. \tag{A.74}$$

A *norm* on this product space can be defined as

$$\| [u, v] \|_{X \times Y} \equiv \| u \|_X + \| v \|_Y . \tag{A.75}$$

As an example of a product space, consider the design of a plate with variable thickness, in which the function h, which defines the thickness in $L^\infty(\Omega)$, and Young's modulus $E \in R^1$ are the design variables. The design space can be defined as the product space of these two spaces consisting of two different types of design variables, as

$$U \equiv L^{\infty}(\Omega) \times R^{1} = \{[h, E] \| h \in L^{\infty}(\Omega),\ E \in R^{1}\}, \tag{A.76}$$

and will have the norm

$$\|[h, E]\|_{U} = \|h\|_{L^{\infty}(\Omega)} + |E|. \tag{A.77}$$

The use of this product space is essential in establishing the regularity of dependence of solutions of boundary-variable problems on design variables.

A.3 Differentials and Derivatives in Normed Space

The purpose of this section is to summarize the definitions of properties of differentials and derivatives of nonlinear mappings or functions, which extend the classical idea of differential and derivative to the calculus of variations and its generalizations. The value of these abstract differentials and derivatives is both practical and theoretical. Practically, the theory allows for first-order approximation or "linearization" of nonlinear functionals that arise in structural design. From a theoretical point of view, differentials and derivatives are used heavily throughout the text to prove existence results and properties of dependence of structural response measures on design variables

A.3.1 Mappings in Normed Spaces

Consider vector spaces X and Y, with norms $\| \bullet \|_X$ and $\| \bullet \|_Y$, respectively. These spaces may be any of the normed spaces discussed in Section A.2. A function $\Phi(\boldsymbol{x})$ that defines a vector in Y, once a vector $\boldsymbol{x}$ in X is specified, may be viewed as a mapping from X into Y, denoted as

$$\Phi : X \to Y. \tag{A.78}$$

$X = R^1$ and $Y = R^1$ is a special case in which Φ is a real-valued function of a single real variable. If, however, $X = L^2(\Omega)$ is a space of the designs and $Y = [H^1(\Omega)]^3$ is the Sobolev space of the displacements of an elastic solid, then Φ may be defined as a mapping from the space X of the designs to the space Y of the solutions to boundary-value problems of elasticity, where $\Phi(\boldsymbol{x})$ is the solution to the boundary value problem for design $\boldsymbol{x}$.

The concept of continuity of a mapping between normed spaces is a direct extension of the concept of continuity of scalar functions of scalar variables. More specifically, the mapping Φ is *continuous* at $\boldsymbol{x}$ if, for every $\varepsilon > 0$, a $\delta > 0$ exists, such that

$$\|\Phi(\boldsymbol{x} + \boldsymbol{\eta}) - \Phi(\boldsymbol{x})\|_{Y} \le \varepsilon \tag{A.79}$$

for all $\boldsymbol{\eta} \in X$, such that

$$\|\boldsymbol{\eta}\|_{X} < \delta. \tag{A.80}$$

If Φ is continuous at every $\boldsymbol{x} \in X$, then it is said to be continuous on X. An algebraic property of the mappings that is of some importance in design sensitivity analysis concerns linearity. A mapping Φ is said to be *homogeneous of degree n* if

$$\Phi(\alpha \boldsymbol{x}) = \alpha^{n} \Phi(\boldsymbol{x}), \tag{A.81}$$

where α is any real number. If (A.81) holds only for $\alpha \geq 0$, then Φ is said to be *positively homogeneous of degree n*. A more important concept is the linearity of a mapping. More specifically, Φ is said to be a linear mapping if

$$\Phi(\alpha \boldsymbol{x} + \beta \boldsymbol{y}) = \alpha\Phi(\boldsymbol{x}) + \beta\Phi(\boldsymbol{y}) \tag{A.82}$$

for all $\boldsymbol{x}$ and $\boldsymbol{y}$ in X, and for all real α and β. Note that a linear mapping is homogeneous of degree one.

A.3.2 Variations and Directional Derivatives

The idea of a derivative or differential of a scalar function of a scalar variable can be profitably extended to general mappings. First, one may define *one-sided Gateaux differentials* as

$$\Phi'_+(\boldsymbol{x},\boldsymbol{\eta}) \equiv \lim_{\substack{\tau \to 0 \\ \tau > 0}} \frac{1}{\tau}[\Phi(\boldsymbol{x} + \tau\boldsymbol{\eta}) - \Phi(\boldsymbol{x})], \tag{A.83}$$

providing that a limit exists on the right side. The term $\Phi'_+(\boldsymbol{x},\boldsymbol{\eta})$ is called the "one-sided Gateaux differential of Φ at point $\boldsymbol{x}$ in the direction $\boldsymbol{\eta}$." This differential exists for large classes of mappings, but it may not possess some of the positive properties usually attributed to derivatives in ordinary calculus. A direct calculation shows that for all $\alpha > 0$,

$$\begin{aligned} \Phi'_+(\boldsymbol{x},\alpha\boldsymbol{\eta}) &= \lim_{\substack{\tau \to 0 \\ \tau > 0}} \frac{1}{\tau}[\Phi(\boldsymbol{x} + \tau\alpha\boldsymbol{\eta}) - \Phi(\boldsymbol{x})] \\ &= \alpha \lim_{\substack{\tau \to 0 \\ \tau > 0}} \frac{1}{\alpha\tau}[\Phi(\boldsymbol{x} + \tau\alpha\boldsymbol{\eta}) - \Phi(\boldsymbol{x})] \\ &= \alpha\Phi'_+(\boldsymbol{x},\boldsymbol{\eta}), \end{aligned} \tag{A.84}$$

which verifies that the one-sided Gateaux differential is positively homogeneous of degree one.

To relate the differential idea to a simple function, consider the real-valued function of a single real variable x,

$$\Phi(x) = |x|. \tag{A.85}$$

A simple check will show that while this function is continuous, it does not have an ordinary derivative at $x = 0$. The one-sided Gateaux differential, however, is defined using (A.83) as

$$\Phi'_+(0,\eta) = \lim_{\substack{\tau \to 0 \\ \tau > 0}} \frac{1}{\tau}[|\tau\eta| - 0] = |\eta|. \tag{A.86}$$

Note that

$$\Phi'_+(0,-\eta) = |-\eta| = |\eta| \neq -\Phi'_+(0,\eta), \tag{A.87}$$

so that the one-sided Gateaux differential is not linear in η and in fact is not homogeneous of degree one. Nevertheless, it predicts the change of the function Φ due to a perturbation η in the independent variable x.

If the limit in (A.83) exists for both $\tau > 0$ and $\tau < 0$, then Φ is said to have a *Gateaux differential* (often called the *differential* or *variation)* at $\boldsymbol{x}$ in the direction $\boldsymbol{\eta}$, given by

$$\Phi'(\boldsymbol{x},\boldsymbol{\eta}) \equiv \lim_{\tau\to 0}\frac{1}{\tau}[\Phi(\boldsymbol{x}+\tau\boldsymbol{\eta})-\Phi(\boldsymbol{x})], \tag{A.88}$$

where the limit may be taken with τ either positive or negative. In this case, the calculations of (A.84) are valid for both positive and negative α, hence the Gateaux differential is homogeneous of degree one.

An example of the Gateaux differential that often arises in structural design sensitivity analysis and in the calculus of variations involves mapping Φ from the space $L^2(\Omega)$ into the real numbers (a functional), defined as

$$\Phi(\boldsymbol{x}) = \iint_\Omega F(\boldsymbol{x})\,d\Omega, \tag{A.89}$$

where the scalar-valued function F is presumed to be continuously differentiable. The Gateaux differential of this functional may be calculated as

$$\begin{aligned}\Phi'(\boldsymbol{x},\boldsymbol{\eta}) &= \lim_{\tau\to 0}\frac{1}{\tau}\iint_\Omega [F(\boldsymbol{x}+\tau\boldsymbol{\eta})-F(\boldsymbol{x})]d\Omega\\ &= \iint_\Omega \lim_{\tau\to 0}\frac{1}{\tau}[F(\boldsymbol{x}+\tau\boldsymbol{\eta})-F(\boldsymbol{x})]d\Omega\\ &= \iint_\Omega \frac{dF}{d\boldsymbol{x}}\boldsymbol{\eta}\,d\Omega,\end{aligned} \tag{A.90}$$

which may be recognized as the *first variation* of the functional Φ in the calculus of variations. Note that in this special case, $\Phi'(\boldsymbol{x},\bullet)$ is a linear mapping from $L^2(\Omega)$ to real numbers.

As will often be the case, the mapping $\Phi'(\boldsymbol{x},\bullet)$ from X to Y may be continuous and linear, in which case it is called the *Gateaux derivative* of Φ at $\boldsymbol{x}$.

A.3.3 Fréchet Differential and Derivative

Let the mapping Φ be given as in (A.78). Then Φ is said to be Fréchet differentiable at $\boldsymbol{x}$ if a continuous linear operator $\Phi'(\boldsymbol{x},\bullet) : X \to Y$ exists, such that

$$\lim_{\|\boldsymbol{\eta}\|_X\to 0}\left[\frac{\|\Phi(\boldsymbol{x}+\tau\boldsymbol{\eta})-\Phi(\boldsymbol{x})-\Phi'(\boldsymbol{x},\boldsymbol{\eta})\|_Y}{\|\boldsymbol{\eta}\|_X}\right]=0 \tag{A.91}$$

holds for any $\boldsymbol{\eta} \in X$. The operator $\Phi'(\boldsymbol{x},\boldsymbol{y})$ in (A.91) is called the Fréchet differential of Φ at $\boldsymbol{x}$. The mapping $\Phi'(\boldsymbol{x},\bullet)$ from X to Y is called the Fréchet derivative of Φ at $\boldsymbol{x}$ and is a continuous linear mapping from X to Y.

It is obvious that if Φ is Fréchet differentiable at $\boldsymbol{x}$, then Φ is Gateaux differentiable at $\boldsymbol{x}$. It is interesting to note that Gateaux and Fréchet derivatives are equivalent for functions defined on R^1, but are not equivalent on higher-dimensional spaces. As an example, consider $X = R^2$ and $Y = R^1$. Define $\Phi : R^2 \to R^1$ as $\Phi(x_1,0) = 0$ and $\Phi(x_1,x_2)=(x_1/x_2)(x_1^2+x_2^2)$, if $x_2 \neq 0$. It is easy to verify that the Gateaux derivative exists at (0,0) and is the zero operator. However, a Fréchet derivative does not exist at (0,0). In fact, Φ is not even continuous at (0,0).

Dieudonne [136] showed that if the Gateaux derivative $\Phi'(\boldsymbol{w},\bullet)$ exists for all $\boldsymbol{w}$ in a neighborhood of $\boldsymbol{x}$ and

$$\lim_{\boldsymbol{w}\to\boldsymbol{x}}\|\Phi'(\boldsymbol{w},\cdot)-\Phi'(\boldsymbol{z},\cdot)\|=0, \tag{A.92}$$

then the Fréchet derivative exists. Note that the norm in (A.92) is for the space of continuous linear mappings [137].

Consider once again the mapping of (A.89) from $L^2(\Omega)$ to real numbers, with the Gateaux differential defined by (A.90). In order to check whether Φ is Fréchet differentiable for the evaluation of (A.91),

$$\Phi(\boldsymbol{x}+\boldsymbol{\eta})-\Phi(\boldsymbol{x})-\Phi'(\boldsymbol{x},\boldsymbol{\eta})=\iint_\Omega\left[F(\boldsymbol{x}+\boldsymbol{\eta})-F(\boldsymbol{x})-\frac{dF}{dx}\boldsymbol{\eta}\right]d\Omega. \tag{A.93}$$

By the remainder form of Taylor's formula,

$$F(\boldsymbol{x}+\boldsymbol{\eta})-F(\boldsymbol{x})=\frac{dF}{dx}\boldsymbol{\eta}+\frac{1}{2}\frac{d^2F}{dx^2}(\bar{x})\boldsymbol{\eta}^2, \tag{A.94}$$

where $\bar{x}=\boldsymbol{x}+\alpha\boldsymbol{\eta}$ and $0<\alpha<1$. If the second derivative of F is bounded by some finite constant K, that is, if

$$\left|\frac{d^2F}{dx^2}\right|<K, \tag{A.95}$$

then from (A.93) through (A.95),

$$\left|\Phi(\boldsymbol{x}+\boldsymbol{\eta})-\Phi(\boldsymbol{x})-\Phi'(\boldsymbol{x},\boldsymbol{\eta})\right|\le\frac{K}{2}\iint_\Omega\boldsymbol{\eta}^2\,d\Omega=\frac{K}{2}\|\boldsymbol{\eta}\|_{L^2}^2. \tag{A.96}$$

Dividing both sides by $\|\boldsymbol{\eta}\|_{L^2}$ and taking the limit as $\|\boldsymbol{\eta}\|_{L^2}$ goes to zero, it is seen that (A.91) is satisfied and that Φ is Fréchet differentiable.

A.3.4 Partial Derivatives and the Chain Rule of Differentiation

Very often in structural design sensitivity analysis, several variables appear in the same expression. Consider a mapping of Φ that depends on a variable from normed space X and a variable from normed space Z, denoted as $\Phi : X \times Z \to Y$. As in ordinary calculus, $\boldsymbol{z}\in Z$ may be held fixed and the Gateaux differential of Φ calculated as a function of $\boldsymbol{x}\in X$. Similarly, $\boldsymbol{x}\in X$ can be held fixed to calculate the Gateaux differential of Φ as a function of $\boldsymbol{z}\in Z$, to obtain

$$\begin{aligned}\Phi'_x(\boldsymbol{x},\boldsymbol{\eta};\boldsymbol{z})&\equiv\lim_{\tau\to0}[\Phi(\boldsymbol{x}+\tau\boldsymbol{\eta};\boldsymbol{z})-\Phi(\boldsymbol{x};\boldsymbol{z})]\\ \Phi'_z(\boldsymbol{x};\boldsymbol{z},\boldsymbol{v})&\equiv\lim_{\tau\to0}[\Phi(\boldsymbol{x};\boldsymbol{z}+\tau\boldsymbol{v})-\Phi(\boldsymbol{x};\boldsymbol{z})],\end{aligned} \tag{A.97}$$

which are called *partial Gateaux differentials* of Φ.

An important result (proved by Dieudonne [136] and Nashed [137]) relates the Gateaux differential of Φ to its partial Gateaux differentials. More specifically, if Φ'_x and Φ'_z in (A.97) exist and are continuous and linear in $\boldsymbol{\eta}$ and $\boldsymbol{v}$, then Φ is Fréchet differentiable on $X \times Z$, and

$$\Phi'(\boldsymbol{x},\boldsymbol{\eta};\boldsymbol{z},\boldsymbol{v})=\Phi'_x(\boldsymbol{x},\boldsymbol{\eta};\boldsymbol{z})+\Phi'_z(\boldsymbol{x};\boldsymbol{z},\boldsymbol{v}). \tag{A.98}$$

This powerful result permits calculations with individual variables and, providing the hypotheses are checked, yields the Gateaux differential of a mapping as the sum of its partial Gateaux differentials.

A related concept extends the classical chain rule of differentiation. Consider a mapping $\Theta : X \rightarrow Z$ and a mapping $\Psi : X \rightarrow Z$, both of which are Fréchet differentiable. Then, the composite mapping $\Phi(\boldsymbol{x}) = \Psi(\Theta(\boldsymbol{x}))$ is Fréchet differentiable and

$$\Phi'(\boldsymbol{x},\boldsymbol{\eta}) = \Psi'(\Theta(\boldsymbol{x}))\Theta'(\boldsymbol{x},\boldsymbol{\eta}). \tag{A.99}$$

This result was proved by Dieudonne [136] and its properties were developed and analyzed by Nashed [137]. The chain rule, however, is not valid for Gateaux derivatives [137]. The concept of chain rule differentiation is used extensively in structural design sensitivity analysis, since structural performance measures are often stated as functionals involving the displacement field, which is itself a function of the design.

References

[1] Bendsoe, M.P., *Optimization of Structural Topology, Shape, and Materials*. 1995, Berlin: Springer-Verlag.

[2] Clark and Fujimoto, *Product Development Performance*. 1991, Boston: Harvard Business School Press.

[3] Choi, K.K., *Simulation-Based Processes for Automotive Design Optimization*. 3rd Edition of Business Briefing: Global Automotive Manufacturing and Technology. 2000.

[4] Zienkiewicz, O.C., *The Finite Element Method*. 1977, New York: McGraw-Hill.

[5] Haug, E.J., K.K. Choi, and V. Komkov, *Design Sensitivity Analysis of Structural Systems*. 1986, London: Academic Press.

[6] Reddy, J.N., *Applied Functional Analysis and Variational Methods in Engineering*. 1986, New York: McGraw-Hill.

[7] Barthelemy, B.M. and R.T. Haftka. Accuracy Analysis of the Semi-analytical Method for Shape Sensitivity Calculation. In *29th AIAA/ASME/ASCE/AHS Structures, Structural Dynamics and Materials Conference*. 1988.

[8] Olhoff, N., J. Rasmussen, and E. Lund, *A Method of Exact Numerical Differentiation for Error Elimination in Finite-Element-Based Semianalytical Shape Sensitivity Analyses*. Mechanics of Structures and Machines, 1993, **21**(1):1–66.

[9] Yang, R.J. and M.E. Botkin, *Accuracy of the Domain Method for the Material Derivative Approach to Shape Design Sensitivities*. In *Sensitivity Analysis in Engineering*. 1987, NASA Conference Publication 2457, pp. 347–353.

[10] Arora, J.S., *Introduction to Optimum Design*. 1989, New York: McGraw-Hill.

[11] Vanderplaats, G.N., *Numerical Optimization Techniques for Engineering Design with Applications*. 1999, Colorado Springs, CO: Vanderplaats Research & Development Inc.

[12] Haftka, R.T. and M.P. Karmat, *Elements of Structural Optimization*. 1985, Netherlands: Nijhoff Publishers.

[13] Luenberger, D.G., *Linear and Nonlinear Programming*. 1984, Massachusetts: Addison-Wesley.

[14] Fletcher, R. and R.M. Reeves, *Function Minimization by Conjugate Gradients*. The Computer Journal, 1964, **7**:149–160.

[15] Fletcher, R. and M.J.D. Powell, *A Rapidly Convergent Descent Method for Minimization*. The Computer Journal, 1963, **6**:163–180.

[16] Aubin, J.-P., *Applied Functional Analysis*. 1979, New York: Wiley.

[17] Fichera, G., *Existence Theorems in Elasticity*. In *Handbuch der Physik*, S. Flügge, ed. 1972, Berlin and New York: Springer-Verlag, pp. 347–389.

[18] Rektorys, K., *Variational Methods in Mathematics, Science, and Engineering*. 1980, Boston: D. Reidel Publishing Co.

[19] Crede, C., *Shock and Vibration Concepts in Engineering Design*. 1965, Englewood Cliffs, NJ: Prentice-Hall.

[20] Fahy, F., *Sound and Structural Vibration, Radiation, Transmission and Response*. 1985, New York: Academic Press.

[21] Haug, E.J. and K.K. Choi, *Methods of Engineering Mathematics*. 1993, Englewood Cliffs, NJ: Prentice-Hall.

[22] Adams, R.A., *Sobolev Space*. 1975, New York: Academic Press.

[23] Fung, Y.C., *Foundations of Solid Mechanics*. 1965, Englewood Cliffs, NJ:

Prentice-Hall.
[24] Washizu, K., *Variational Methods in Elasticity and Plasticity*. 1982, Oxford: Pergamon.
[25] Goldstein, H., *Classical Mechanics*. 1950, Readings, MA: Addison-Wesley.
[26] Lions, J.L. and E. Magenes, *Non-homogeneous Boundary Value Problems and Applications*. Vol. 1. 1972, Berlin and New York: Springer-Verlag.
[27] Treves, F., *Topological Vector Spaces, Distributions and Kernels*. 1967, New York: Academic Press.
[28] Choi, K.K. and J.H. Lee, *Sizing Design Sensitivity Analysis of Dynamic Frequency Response of Vibrating Structures*. ASME Journal of Mechanical Design, 1992, **114**(1):166–173.
[29] Choi, K.K., I. Shim, and S. Wang, *Design Sensitivity Analysis of Structure-Induced Noise and Vibration*. ASME Journal of Vibration and Acoustics, 1997, **119**(2):173–179.
[30] Horvath, J., *Topological Vector Spaces and Distributions*. 1966, London: Addison-Wesley.
[31] Dimarogonas, A.D., *Vibration Engineering*. 1976: West Publishing Co.
[32] Craig, R.R.J., *Structural Dynamics: An Introduction to Computer Methods*. 1981, New York: Wiley.
[33] Dowell, E.H., G.F. Gorman, and D.A. Smith, *Acoustoelasticity: General Theory, Acoustic Natural Modes and Forced Response to Sinusoidal Excitation, Including Comparison with Experiment*. Journal of Sound and Vibration, 1977, **52**(4):519–542.
[34] Flanigan, D.L. and S.G. Borders, *Application of Acoustic Modeling Methods for Vehicle Boom Analysis*. 1988, SAE.
[35] Reed, M. and B. Simon, *Methods of Modern Mathematical Physics*. Vol. 1. 1972, New York: Academic Press.
[36] Strang, G. and G.J. Fix, *An Analysis of the Finite Element Method*. 1973, Englewood Cliffs, NJ: Prentice-Hall.
[37] Ciarlet, P.G., *The Finite Element Method for Elliptic Problems*. 1978, New York: North-Holland.
[38] Mikhlin, S.G., *Mathematical Physics: An Advanced Course*. 1970, Amsterdam: North-Holland.
[39] Mikhlin, S.G., *Variational Methods in Mathematical Physics*. 1964, Oxford: Pergamon.
[40] Pipes, L.A., *Matrix Methods in Engineering*. 1963, Englewood Cliffs, NJ: Prentice-Hall.
[41] Langhaar, H.L., *Energy Methods in Applied Mechanics*. 1962, New York: Wiley.
[42] Gallagher, R.H., *Finite Element Analysis: Fundamentals*. 1975, Englewood Cliffs, NJ: Prentice-Hall.
[43] Hughes, T.J.R., *The Finite Element Method*. 1987, Englewood Cliffs, NJ: Prentice-Hall.
[44] Sokolnikoff, S., *Mathematical Theory of Elasticity*. 1956, New York: McGraw-Hill.
[45] Prezmieniecki, J.S., *Theory of Matrix Structural Analysis*. 1968, New York: McGraw-Hill.
[46] Bathe, K.-J., *Finite Element Procedures in Engineering Analysis*. 1996, Englewood Cliffs, NJ: Prentice-Hall.
[47] Greenwood, D.T., *Principles of Dynamics*. 1965, Englewood Cliffs, NJ: Prentice-Hall.
[48] Haug, E.J. and J.S. Arora, *Applied Optimal Design*. 1979, New York: Wiley.
[49] Atkinson, K.E., *Introduction to Numerical Analysis*. 1978, New York: Wiley.
[50] Goffman, C., *Calculus of Several Variables*. 1965, New York: Haper & Row.

[51] Haftka, R.T., *Second Order Sensitivity Derivatives in Structural Analysis*. AIAA Journal, 1982, **20**:1765–1766.
[52] Roark, R.J. and W.C. Young, *Formulas For Stress and Strain*. 1975, New York: McGraw-Hill.
[53] Kato, T., *Perturbation Theory for Linear Operators*. 1976, Berlin and New York: Springer-Verlag.
[54] Fox, R.L. and M.P. Kapoor, *Rates of Change of Eigenvalue and Eigenvectors*. AIAA Journal, 1968, **6**(12):2426–2429.
[55] Farshad, M., *Variations of Eigenvalues and Eigenvectors in Continuum Mechanics*. AIAA Journal, 1974, **12**(4):560–561.
[56] Reiss, R., *Design Derivatives of Eigenvalues and Eigenvectors for Self-Adjoint Distributed Parameter Systems*. AIAA Journal, 1986, **24**:1169–1172.
[57] Wang, B.P., *Improved Approximate Methods for Computing Eigenvector Derivatives in Structural Dynamics*. AIAA Journal, 1991, **29**(6):1018–1020.
[58] Nelson, R.B., *Simplified Calculation of Eigenvector Derivatives*. AIAA Journal, 1976, **14**:1201–1205.
[59] Wang, S. and K.K. Choi, *Continuum Sizing Design Sensitivity Analysis of Eigenvectors Using Ritz Vectors*. Journal of Aircraft, 1994, **31**:457–459.
[60] Rudisill, C.S. and Y. Chu, *Numerical Methods for Evaluating the Derivatives of Eigenvalues and Eigenvectors*. AIAA Journal, 1975, **13**(6):834–837.
[61] Thompson, J.M.T. and G.W. Hunt, *Dangers of Structural Optimization*. Engineering Optimization, 1974, **2**:99–110.
[62] Olhoff, N. and S.H. Rasmussen, *On Single and Bimodal Optimum Buckling Loads of Clamped Columns*. International Journal of Solids and Structures, 1977, **13**:605–614.
[63] Tadjbakhsh, I. and J.B. Keller, *Strongest Columns and Isoparametric Inequalities for Eigenvalues*. Journal of Applied Mechanics, 1962, **29**:159–164.
[64] Masur, E.F. and Z. Mroz, *Singular Solutions in Structural Optimization Problems*. Variational Methods in Mechanics of Solids, S. Nemat-Nasser ed. 1980, Oxford: Pergamon.
[65] Prager, S. and W. Prager, *A Note on Optimal Design of Columns*. International Journal of Mechanics, 1979, **21**:249–251.
[66] Masur, E.F. and Z. Mroz, *On Non-stationary Optimality Conditions in Structural design*. International Journal of Solids and Structures, 1979, **15**:503–512.
[67] Coddington, E.A. and N. Levinson, *Theory of Ordinary Differential Equations*. 1955, New York: McGraw-Hill.
[68] Taylor, A.E., *Advanced Calculus*. 1955, Boston, MA: Dinn.
[69] Gelfand, M. and S.V. Fomin, *Calculus of Variations*. 1963, Englewood Cliffs, NJ: Prentice-Hall.
[70] Popov, E.P., *Mechanics of Materials*. 2nd ed. 1978, Englewood Cliffs, NJ: Prentice-Hall.
[71] Dems, K. and Z. Mroz, *Variational Approach by Means of Adjoint Systems to Structural Optimization and Sensitivity Analysis*. II Structure Shape Variation. International Journal of Solids and Structures, 1984, **20**(6):527–552.
[72] Mitchell, A.R. and R. Wait, *The Finite Element Method in Partial Differential Equations*. 1977, New York: Wiley.
[73] Rousselet, B., *Quelques Résultats en Optimisation de Domains*. 1982, Nice, France: Université de Nice.
[74] Myslinski, A., *Bimodal Optimal Design of Vibrating Plates Using Theory and Methods of Nondifferentiable Optimization*. Journal of Optimization Theory and Applications, 1985, **46**(2):187–203.
[75] Weisshaar, T.A. and R.H. Plaut, *Structural Optimization under Nonconservative*

Loading, in Optimization of Distributed Parameter Structures. E.J. Haug and J. Cea, ed. 1981, Sijthoff & Noordhoff, The Netherlands: Aiphen aan den Rijn, pp. 843–864.

[76] Haug, E.J., *A Review of Distributed Parameter Structural Optimization Literature*. In *Optimization of Distributed Parameter Structures*, E.J. Haug and J. Cea, ed. 1981, Sijthoff & Nordhoff, The Netherlands: Alphen aan den Rijn, pp. 3–74.

[77] MacNeal-Schwendler, *MSC/NASTRAN User's Manual Vols. I and II*. Vol. 70. 2000, Los Angeles: MacNeal Schwendler Corp.

[78] NASA, *The NASTRAN Theoretical Manual*. 1981, Washington, DC: National Aeronautics and Space Administration.

[79] Kamal, M.M. and J.A.W. Jr., *Modern Automotive Structural Analysis*. 1982, New York: Van Nostrand Reinhold Company.

[80] Choi, K.K. and I. Shim. *Design Sensitivity Analysis of Structural Vibration and Noise*. In *4th AIAA/USAF/NASA/OAI Symposium on Multidisciplinary Analysis and Optimization*. 1992. Cleveland, OH.

[81] Hibbit, Karlsson, and Sorensoen, *ABAQUS Users' Manual*. 1989, Providence, RI: Hibbit, Karlsson & Sorensoen Inc.

[82] Fleming, W.H., *Functionals of Several Variables*. 1965, Readings, MA: Addison-Wesley.

[83] Zolesio, J.P., *Identification de Domains par Deformations*. 1979, Nice, France: Université de Nice.

[84] Malvern, L.E., *Introduction to the Mechanics of a Continuous Medium*. 1969, Englewood Cliffs, NJ: Prentice-Hall.

[85] Babuska, I. and A.K. Aziz, *Survey Lectures on the Mathematical Foundations of the Finite Element Method*. In *The Mathematical Foundations of the Finite Element Methods with Applications to Partial Differential Equations*. A.K. Aziz, ed. 1972, New York: Academic Press, pp. 1–59.

[86] Carme, M.P.D., *Differential Geometry of Curves and Surfaces*. 1976, Englewood Cliffs, NJ: Prentice-Hall.

[87] Kreyszig, E., *Advanced Engineering Mathematics*. 1979, New York: Wiley.

[88] Kecs, W. and P.P. *Teodorescu, Applications of the Theory of Distributions in Mechanics*. 1974, Tunbridge Wells, Kent, England: Abacus Press.

[89] Timoshenko, S.P. and J. N. Goodier, *Theory of Elasticity*. 1951, New York: McGraw-Hill.

[90] Zolesio, J.P., *Gradient des Coüts Governés par des Problémes de Neumann Poses sur des Ouverts Anguleux en Optimisation de Domain*. 1982, University of Montreal: Canada.

[91] Brebbia, C.A. and S. Walker, *Boundary Element Techniques in Engineering*. 1980, Boston, MA: NewnesButterworth.

[92] Banerjee, P.K. and R. Butterfield, *Boundary Element Methods in Engineering Science*. 1981, New York: McGraw-Hill.

[93] Hou, J.W. and J.S. Sheen, *On the Design Velocity Field in the Domain and Boundary Methods for Shape Optimization*. AIAA Paper, 1988, **88-2338**:1032–1040.

[94] Choi, K.K. and H.G. Seong, *A Numerical Method for Shape Design Sensitivity Analysis and Optimization of Built-Up Structures*. In *The Optimum Shape: Automated Structural Design*, J.A. Bennet and M.E. Botkin, ed. 1986, New York: Plenum Press, pp. 329–352.

[95] Choi, K.K., *Shape Design Sensitivity Analysis and Optimal Design of Structural Systems*. In *Computed Aided Optimal Design*, C.A.M. Soares, ed. 1987, Heidelberg: Springer-Verlag, pp. 439–492.

[96] Choi, K.K. and T.M. Yao, 3-D *Shape Modeling and Automatic Regridding in Shape Design Sensitivity Analysis*. In *Sensitivity Analysis in Engineering*. 1987, NASA Conference Publication 2457, pp. 329–345.

[97] Yao, T.M. and K.K. Choi, 3-D *Shape Optimal Design and Automatic Finite Element Regridding*. Internaational Journal for Numerical Methods in Engineering, 1989, **28**:369–384.

[98] Chang, K.H. and K.K. Choi, *A Geometry-Based Parameterization Method for Shape Design of Elastic Solids*. Mechanics of Structures and Machines, 1992, **20**(2):215–252.

[99] Rajan, S.D. and A.D. Belegundu, *Shape Optimization Approach Using Fictitious Loads*. AIAA Journal, 1989, **27**(1):02–107.

[100] Belegundu, A.D. and S.D. Rajan, *A Shape Optimization Approach Based on Natural Design Variables and Shape Functions*. Computer Methods in Applied Mechanics and Engineering, 1988, **66**:89–106.

[101] Zhang, S. and A.D. Belegundu, *A System Approach For Generating Velocity Fields In Shape Optimization*. In *Optimization of Large Scale Structural Systems*, NATO Advanced Study Institute. 1991: Berchtesgaden, Germany.

[102] Chang, K.H. and K.K. Choi, *An Error Analysis and Mesh Adaptation Method for Shape Design of Structural Components*. Computers and Structures, 1992, **44**(6): 1275–1289.

[103] Haftka, R.T. and R.V. Grandhi, *Structural Shape Optimization—A Survey*. Computer Methods in Applied Mechanics and Engineering, 1986, **57**:91–106.

[104] Cook, R.D., *Concepts and Applications of Finite Element Analysis*. 1981, New York: Wiley.

[105] Francavilla, A., C.V. Ramakrishnan, and O.C. Zienkiewicz, *Optimization of Shape to Minimize Stress Concentration*. Journal of Strain Analysis, 1975, **10**(2):63–70.

[106] Pedersen, P. and C.L. Laursen, *Design for Minimum Stress Concentration by Finite Elements and Linear Programming*. Journal of Structural Mechanics, 1982–83, **10**:375–391.

[107] Zolesio, J.P., *The Material Derivative (or Speed) Method for Shape Optimization*. In *Optimization of Distributed Parameter Structures*, E.J. Haug and J. Cea, ed. 1981, Alphen aan den Rijn, The Netherlands: Sijthoff & Noordhoff, pp. 1089–1151.

[108] Delfour, M.C. *Shape Analysis and Optimization in Distributed Parameter System*. In *Proceedings of 5th IFAC Symposium in Control of Distributed Parameter System*. 1989, Perpignan.

[109] Chen, C.-J. and K.K. Choi, *A Continuum Approach for Second-order Shape Design Sensitivity Analysis of 3-D Elastic Solids*. AIAA Journal, 1994, **32**(10):2099–2107.

[110] Wassermann, K., *Three-Dimensional Shape Optimization of Arch Dam with Prescribed Shape Functions*. Journal of Structural Mechanics, 1983, **11**:465–489.

[111] Dorn, W.S., R.E. Gomory, and H.J. Greenberg, *Automatic Design of Optimal Structures*. Journal de Mechanique, 1964, **3**:25–52.

[112] Dobbs, M.W. and L.P. Felton, *Optimization of Truss Geometry*. ASCE Journal of the Structural Division, 1969, **95**:2105–2118.

[113] Vanderplaats, G.N. and F. Moses, *Automated Design of Trusses for Optimum Geometry*. ASCE Journal of Structural Division, 1972, **98**:671–690.

[114] Imai, K. and L.A. Schmit, *Configuration Optimization of Trusses*. ASCE Journal of the Structural Division, 1981, **107**:745–756.

[115] Saka, M.P. and B. Attili, *Shape Optimization of Space Trusses*. In *The Proceedings of the International Conference on the Design and Construction of*

Non-conventional Structures. 1987, London: Civil-Comp Press.
[116] Twu, S.L. and K.K. Choi, *Configuration Design Sensitivity Analysis of Built-Up Structures. Part I: Theory*. International Journal for Numerical Methods in Engineering, 1992, **35**(5):1127–1150.
[117] Dopker, B. and K.K. Choi, *A Study of Solution Algorithms for Shape Design Sensitivity Analysis on a Supermini Computer with an Attached Array Processor*. Engineering with Computers, 1987, **3**:111–119.
[118] Yang, R.J., *A Hybrid Approach for Shape Optimization*. Computers in Engineering, 1988, **2**:107–112.
[119] DeSalvo, G.J. and J.A. Swanson, *ANSYS Engineering Analysis System, User's Manual Vol. I and II*. 1989, Houston, PA: Swanson Analysis Systems Inc.
[120] Schmit, L.A. and H. Miura, *Approximation Concepts for Efficient Structural Synthesis*. 1976, Washington, DC: National Aeronautics and Space Administration.
[121] Sung, S.H., *Automotive Applications of Three-Dimensional Acoustic Finite Element Methods*. 1981, SAE.
[122] Choi, K.K. and K.H. Chang, *A Study of Design Velocity Field Computation for Shape Optimal Design*. Finite Elements in Analysis and Design, 1994, **15**:317–341.
[123] Dym, C.L. and I.H. Shames, *Solid Mechanics*. 1973, New York: McGraw-Hill.
[124] Shames, I.H. and C.L. Dym, *Energy and Finite Element Methods in Structural Mechanics*. 1985, New York: McGraw-Hill.
[125] Reissner, E., *The Effect of Transverse Shear Deformation on the Bending of Elastic Plates*. Journal of Applied Mechanics, 1945, **12**:A69–A77.
[126] Mindlin, R.D., *Influence of Rotary and Shear Deformation on Flexural Motions of Isotropic Plates*. Journal of Applied Mechanics, 1952, **18**:31–38.
[127] Arora, J.S., T.H. Lee, and J.B. Cardoso, *Structural Shape Sensitivity Analysis: Relationship Between Material Derivative and Control Volume Approaches*. AIAA Journal, 1992, **36**:1638–1648.
[128] Choi, K.K. and E.J. Haug, *Shape Design Sensitivity Analysis of Elastic Structures*. Journal of Structural Mechanics, 1983, **11**:231–269.
[129] Liu, W.K., S. Jun, and Y.F. Zhang, *Reproducing Kernel Particle Methods*. International Journal for Numerical Method in Fluids, 1995, **20**:1081–1106.
[130] Kim, N.H., et al., *Meshless Shape Design Sensitivity Analysis and Optimization for Contact Problem with Friction*. Computational Mechanics, 2000, **25**(2–3):157–168.
[131] Kim, N.H., K.K. Choi, and J.S. Chen, *Shape Sensitivity Analysis and Optimization of Elasto-Plasticity with Frictional Contact*. AIAA Journal, 2000, **38**(9):1742–1753.
[132] MacNeal-Schwendler, *MSC/PATRAN User's Manual*. Vol. 70. 2000, Los Angeles: MacNeal Schwendler Corp.
[133] Chen, J.S., et al., *A Stabilized Conforming Nodal Integration for Galerkin Meshfree Methods*. International Journal for Numerical Methods in Engineering, 2001, **50**:435–466.
[134] Anderson, E., et al., *LAPACK User's Guide*. 3rd ed. 1999, Philadelphia: Society for Industrial and Applied Mathematics.
[135] Rudin, W., *Real and Coplex Analysis*. 1974, New York: McGraw-Hill.
[136] Dieudonne, J., *A Treatise on Analysis*. Vol. 1. 1969, New York: Academic Press.
[137] Nashed, M.Z., *Differentiability and Related Properties of Nonlinear Operators: Some Aspects of the Role of Differentials in Nonlinear Functional Analysis*. In *Nonlinear Functional Anaysis and Applications*, L.B. Rall, ed. 1971, New York: Academic Press, pp. 103–109.

Index

Mechanical Engineering Series *(continued from page ii)*

M. Kaviany, **Principles of Convective Heat Transfer, 2nd ed.**

M. Kaviany, **Principles of Heat Transfer in Porous Media, 2nd ed.**

E.N. Kuznetsov, **Underconstrained Structural Systems**

P. Ladevèze, **Nonlinear Computational Structural Mechanics: New Approaches and Non-Incremental Methods of Calculation**

P. Ladevèze and J.-P. Pelle, **Mastering Calculations in Linear and Nonlinear Mechanics**

A. Lawrence, **Modern Inertial Technology: Navigation, Guidance, and Control, 2nd ed.**

R.A. Layton, **Principles of Analytical System Dynamics**

F.F. Ling, W.M. Lai, D.A. Lucca, **Fundamentals of Surface Mechanics: With Applications, 2nd ed.**

C.V. Madhusudana, **Thermal Contact Conductance**

D.P. Miannay, **Fracture Mechanics**

D.P. Miannay, **Time-Dependent Fracture Mechanics**

D.K. Miu, **Mechatronics: Electromechanics and Contromechanics**

D. Post, B. Han, and P. Ifju, **High Sensitivity Moiré:** Experimental Analysis for Mechanics and Materials

F.P. Rimrott, **Introductory Attitude Dynamics**

S.S. Sadhal, P.S. Ayyaswamy, and J.N. Chung, **Transport Phenomena with Drops and Bubbles**

A.A. Shabana, **Theory of Vibration: An Introduction, 2nd ed.**

A.A. Shabana, **Theory of Vibration: Discrete and Continuous Systems, 2nd ed.**